ORGANOSILICON HETEROPOLYMERS AND HETEROCOMPOUNDS

MONOGRAPHS IN INORGANIC CHEMISTRY

Editor: Eugene G. Rochow

Department of Chemistry, Harvard University

I. I. Vol'nov—Peroxides, Superoxides, and Ozonides of Alkaline Earth Metals—1966

M. Tsutsui, M. N. Levy, A. Nakamura, M. Ichikawa, and K. Mori—Introduction to Metal π-Complex Chemistry—1970

S. N. Borisov, M. G. Voronkov, and E. Ya. Lukevits—Organosilicon Heteropolymers and Heterocompounds—1970

ORGANOSILICON HETEROPOLYMERS AND HETEROCOMPOUNDS

S. N. Borisov

Director, Laboratory of Elastomer Synthesis
All-Union Synthetic Rubber Research Institute
Leningrad

M. G. Voronkov

Director, Laboratory of Heteroorganic Compounds
Institute of Organic Synthesis
Academy of Sciences of the Latvian SSR, Riga

E. Ya. Lukevits

Group Leader, Laboratory of Heteroorganic Compounds
Institute of Organic Synthesis
Academy of Sciences of the Latvian SSR, Riga

Translated from Russian by
C. Nigel Turton and Tatiana I. Turton

PLENUM PRESS • NEW YORK • 1970

Sergei Nikolaevich Borisov (1928-1967) was head of the section on heat-resistant elastomers and director of the Laboratory of Elastomer Synthesis, All-Union Synthetic Rubber Research Institute, Leningrad. A graduate student at Leningrad University from 1953 to 1956, in 1958 he defended his Candidate's dissertation on conjugated hydrogenation-halogenation of organosilicon compounds. He was the author of more than 50 articles and reviews on the chemistry and physics of organosilicon, organometallic, and high-molecular compounds and held a number of patents in these fields. Dr. Borisov drowned accidentally on August 6, 1967.

Mikhail Grigor'evich Voronkov was born in 1921. In 1947 he defended his Candidate's dissertation on the action of sulfur on unsaturated compounds. From 1947 to 1954 he was senior scientist of the Organic Chemistry Department, Faculty of Chemistry, Leningrad University. In 1954 he transferred to the Institute of Silicate Chemistry, Academy of Sciences of the USSR, where he worked as senior scientist, and in 1959 he became director of the Inorganic Polymer Laboratory. In 1961 he went to Riga, where he set up the Laboratory of Heteroorganic Compounds, Institute of Organic Synthesis, Academy of Sciences of the Latvian SSR, which he presently directs. In 1961 he was also awarded the degree of Doctor of Chemical Sciences for his work on the heterolytic cleavage of the siloxane bond. With E. Ya. Lukevits Dr. Voronkov is author of *Organic Insertion Reactions of Group IV Elements* (Consultants Bureau, 1966).

Edmund Yanovich Lukevits was born in 1936. In 1966 he defended his Candidate's dissertation on organosilicon derivatives of furan. He currently directs a group of laboratories devoted to the study of heteroorganic compounds in the Institute of Organic Synthesis, Academy of Sciences of the Latvian SSR, Riga.

KREMNEÉLEMENTOORGANICHESKIE SOEDINENIYA
КРЕМНЕЭЛЕМЕНТООРГАНИЧЕСКИЕ СОЕДИНЕНИЯ
С. Н. Борисов, М. Г. Воронков, Э. Я. Лукевиц

Library of Congress Catalog Card Number 68-13393
SBN 306-30379-5

The orginal Russian text, first published by Khimiya Press in Leningrad in 1966, has been corrected by the authors for this edition. The present translation is published under an agreement with Mezhdunarodnaya Kniga, the Soviet book export agency.

A Division of Plenum Publishing Corporation
227 West 17th Street, New York, N. Y. 10011

Distributed in Europe by Heyden & Son Ltd.
Spectrum House, Alderton Crescent, London N.W.4, England

Printed in the United States of America

Foreword

There are numerous criteria for measuring the growth and development of branches of chemistry. This valuable book illustrates a particular aspect of the growth of organosilicon chemistry. The extent of this field has developed so greatly in recent years that it now is desirable to reclassify parts to bring together hitherto fragmented and relatively disparate sections.

This has been accomplished by the presently available large units which have been designated as "organosilicon heterocompounds." Simplified expressions of such classification are structural units of the general type C–Si–heteroelement and heteroelement-C–Si, in which there are attached to the organosilicon moiety elements such as oxygen, nitrogen, metals, etc. This arrangement permits the correlation of extensive material, which will be invaluable to chemists in many areas, both in and out of organosilicon chemistry. Because of the wealth of information, the authors are currently engaged in the preparation of companion volumes arranged on this general principle. The scope is broad, and includes material which will prove highly interesting and useful to those in academic, industrial, and governmental circles. There is not only a wide coverage of the literature generally, but the listings of patent references and of general reviews and books are among the most complete so far presented.

The authors have a special competence for a work of this kind, not only because of their active research contributions but also because of their skill and experience in writing such books or treatises. It is significant to note that some of the outstanding books in organosilicon chemistry have come from eastern European countries. An appropriate illustration is the classic on "Organosilicon Compounds," the three-volume work in English by Bažant, Chvalovsky, and Rathousky. This raises a question on the "Literature Cited" section of the present volume. After the magnificent subsection of

"Monographs, Reviews, and Dissertations," there comes "Articles and Reports of Russian Authors." These comprise about 30% of the total references, the remainder being "Articles and Reports of Foreign Authors." It is, of course, most helpful to have a full coverage of the continuously important and significant studies in this area by Soviet chemists. However, with the current awareness of chemists generally of such contributions, thanks in part to English editions of many Soviet research publications, there appears little to commend the segregation of the common, interdependent interests and studies which are international in scope.

This new series promises to assume classic proportions in the area of organosilicon chemistry. As a corollary of the growth of this segment of chemistry it may indicate to some, if the marked development continues, the possible need for a periodical on organosilicon chemistry. This, as well as one on transition-metal organometallic chemistry, was mentioned several years ago. At this stage, however, one is hardly prepared to state that such specialization would be a happy development.

Ames, Iowa
January 20, 1970

Henry Gilman

Preface to the English Edition

We are pleased that Plenum Press has selected our book as one of the Soviet monographs on the chemistry of heteroorganic compounds being made available to English readers. The list of references at the end of the book shows that about 70% of the material examined was published outside the USSR. We therefore hope that, on the one hand, the English translation will considerably enlarge the number of readers of the book and, on the other, will give foreign scientists a more complete picture of the work of Soviet authors on the chemistry of organosilicon heterocompounds, which takes up approximately 30% of the bibliography. Ultimately, this may further the progress of this rapidly expanding branch of chemistry, which is of interest from both the theoretical and the practical points of view.

We based the arrangement of the material in the book mainly on the extent to which a certain class of compounds was characteristic of the elements of the given group. Therefore, starting with Chapter 3, we went from compounds with Si–O–M bonds to Si–C–M, Si–N–M, and finally, Si–M bonds. In Chapter 2 we deliberately placed compounds of the type Si–M second in the series. In discussing compounds of Group I elements, we placed Si–M compounds first, while compounds with Si–O–M bonds are examined after derivatives containing nitrogen.

In conclusion we would like to thank all the Soviet and foreign scientists who sent us favorable and critical comments on the Russian edition of our monograph. We are particularly grateful to Professor Henry Gilman for his interest in the English edition of this work, and to Professor Eugene Rochow for having kindly agreed to the inclusion of this volume in the series Monographs in Inorganic Chemistry published by Plenum Press.

Riga, July 1969

M. G. V., É. Ya. L.

Preface to the Russian Edition

The marked difference between compounds of silicon and the analogous compounds of its upper neighbor in the periodic system, carbon, their exceptional practical value, and the vast number of theoretical and applied investigations of a wide range of problems in the chemistry and technology of organic and pseudoorganic compounds of this element have led to the appearance of a new and independent field of chemical science – organosilicon chemistry. It is concerned with practically all the nonsilicate compounds of silicon.

In the last two decades the chemistry of organosilicon compounds has developed particularly vigorously. Many countries have experienced the development of high-tonnage industrial production of a wide range of organosilicon monomers, liquids, resins, elastomers, and all kinds of materials based on them. As a result, in recent years a large number of monographs, reviews, and textbooks on general and particular, theoretical and applied problems in organosilicon chemistry have been published.

The establishment and independent development of organosilicon chemistry is a clear example of the general dialectic process of differentiation and integration of contemporary science, which is characteristic of our epoch. However, not all sections of the nonsilicate chemistry of silicon have developed equally. Thus, while in the middle of the present century investigations were concentrated predominantly in the region of polyorganosiloxanes and the monomers required to prepare them, in the sixties there has been a sharp increase in the attention paid to organosilicon heterocompounds containing nitrogen or inorganic elements. As a result a new discipline has developed – the chemistry of organosilicon heterocompounds – a peculiar border zone which closely adjoins and is deeply rooted in the classical divisions of heteroorganic, and organic, chemistry. However, in the literature on the chemistry of organosilicon heterocompounds there has been little work as yet by way of generalizing and systematizing progress.

In the well-known monographs on organosilicon chemistry [4, 78, 83, 85, 110] and also in books and reviews on organic compounds of Group IV elements [100, 105, 112, 117] at best only a few pages are devoted to monomeric organosilicon heterocompounds. The first and still the most complete review of this field is Borisov's article [19]. However, the derivatives of alkali metals and elements of the sulfur subgroup are not examined in this article. This gap has been partly filled by two reviews by Gilman and his co-workers [92, 125] of organosilicon compounds containing an alkali metal atom attached directly to a silicon atom. A recent short article by Wiberg and his co-workers [123] merely generalized the results of several dissertation studies of the chemistry of silyls of some Group II-IV metals.* The individual subjects of the review [19] were subsequently supplemented by new communications describing monomeric siloxy derivatives of metals [60, 101, 114], cyclic silicon compounds containing atoms of other elements [9], and also silylamides of alkali metals and reactions involving them which lead to the formation of heterosilane compounds [121, 122]. These publications, which are brief and limited with respect to the range of the classes of compound examined, constitute the whole of the review literature in the field of organosilicon heterocompounds.

Polymeric organosilicon heterocompounds, particularly polyheterosiloxanes (polymers built up from the alternating groups –Si–O–M–O–) are discussed in a much greater number of reviews [4, 5, 6, 8, 10, 11, 36–39, 43, 73a, 75, 77, 79–81, 92, 99, 102, 108, 110, 120]. However, as a rule, there is a far from exhaustive description of the individual types of heterosiloxane polymers or a brief list of known types without details. The most complete publications in this field are those of K. A. Andrianov [6, 8] and J. Jones [102], but they only cover material published up to 1962. Up to now, no single review has appeared in the literature which classifies work on organosilicon compounds containing elements of the sulfur subgroup or the silicon–oxygen–alkali metal bond.

* Data on the synthesis and properties of compounds with Si–M bonds (M is not an alkali metal) are also given in a review published in August, 1966 [28a].

From this account it is clear that all available material on organosilicon heterocompounds, which is scattered, incomplete, and largely out-of-date, does not give a single clear systematic idea of the diversity and peculiarity of this exceptionally interesting new section of chemical science. This compelled us to attempt to examine systematically the contemporary state of organosilicon heterochemistry as fully and as up-to-date as possible.

The book gives quite a detailed account of preparation methods, physical and chemical properties, analysis methods, and the possibilities of the practical application of monomeric and polymeric organosilicon derivatives of inorganic elements. There are also discussions of theoretical problems connected with the structure and reactivity of heteroorganic and pseudoorganic derivatives of silicon. It is proposed to continue the examination of these problems later in monographs on the chemistry of organosilicon derivatives of the organogens phosphorus and sulfur and also nitrogen.

The book is based on as exhaustive a coverage of the literature data as possible and the authors' own investigations, published mainly up to May 1, 1965. Chapters 2 (apart from section 3.1), 3, 4, 5, 7, and 8 and sections 6 (Ch. 1) and 2 (Ch. 6) were written by S. N. Borisov and M. G. Voronkov. Chapter 1 and sections 3.1 (Ch. 2) and 1 (Ch. 6) were written by É. Ya. Lukevits and M. G. Voronkov.

In writing the monograph the authors were guided by the terminology in the proposed scheme for the nomenclature of organosilicon compounds [62] and the recommendations of the chemical nomenclature section of the Eighth Mendeleev Conference on General and Applied Chemistry.

The authors are grateful to all the many Soviet and foreign scientists who supported the work on this book with their practical help, unquenchable interest, and valuable advice. We are also grateful to those who helped us to prepare the manuscript for publication.

All critical comments and suggestions will be gratefully received and will help us in work on the next books of this series.

S. N. Borisov, M. G. Voronkov, and É. Ya. Lukevits

Leningrad – Riga
March 1966

Contents

CHAPTER 1

ORGANOSILICON COMPOUNDS OF GROUP I ELEMENTS

CHAPTER 2

ORGANOSILICON COMPOUNDS OF GROUP II ELEMENTS

CHAPTER 3

ORGANOSILICON COMPOUNDS OF GROUP III ELEMENTS

Boron

CHAPTER 4

ORGANOSILICON COMPOUNDS OF GROUP IV ELEMENTS

Germanium

Tin

CHAPTER 5

ORGANOSILICON COMPOUNDS OF GROUP V ELEMENTS

LITERATURE CITED

Chapter 1

Organosilicon Compounds of Group I Elements

1. COMPOUNDS CONTAINING AN Si – M BOND*

The simplest compounds containing a bond between silicon and a Group Ia metal are the silicides. The first representative of this class of compound was described as early as 1902 by H. Moissan [964-966]. By fusing excess lithium with finely divided silicon at 500-600°C he obtained lithium silicide and investigated its properties. Depending on the reaction conditions and the ratio of the starting reagents it was possible to obtain the lithium silicides Li_2Si, Li_4Si, and a mixture of them, Li_6Si_2 [71, 564, 884, 942].

Heating silicon with sodium, potassium, rubidium, and cesium [844] at 600-700°C yielded monosilicides of these alkali metals with the general formula **M**Si and also $NaSi_2$ [989]. The properties of the silicides of Group Ia metals have been studied little, but they are apparently very reactive compounds, which are decomposed vigorously on interaction with water with the liberation of hydrogen and silanes [1188].

The reaction of silane SiH_4 with potassium at -78.5°C in 1,2-dimethoxyethane forms silylpotassium [1027], which consists of colorless crystals with a crystal lattice of the NaCl type:

$$2SiH_4 + 2K \longrightarrow 2H_3SiK + H_2 \qquad (1-1)$$

The reaction of silane with sodium in 1,2-dimethoxyethane proceeds analogously. At the same time, if the reaction is carried out in liquid ammonia the silylsodium formed then reacts with the solvent and is converted into sodamide [827].

* In the text of this chapter M denotes a group Ia metal (M = Li, Na, K, Rb, Cs).

The formation of silylpotassium is also observed in the reaction of potassium or its hydride with disilane [1026, 1027, 1218]:

$$H_3SiSiH_3 + KH \longrightarrow H_3SiK + SiH_4 \quad (1\text{-}2)$$

When excess potassium hydride is used, other processes occur which lead to the formation of polysilylpotassium [1026].

Silylpotassium is also a very reactive compound, which is decomposed by water and hydrogen chloride and reacts readily with methyl chloride [1027]:

$$H_3SiK + ClCH_3 \longrightarrow H_3SiCH_3 + KCl \quad (1\text{-}3)$$

It was also to be expected that metal derivatives of organosilanes of the type R_3SiM would also be very reactive compounds. The efforts of many investigators were therefore directed toward the development of a convenient method of preparing such compounds. The hypothesis was put forward that metal derivatives of organosilanes are intermediate products in the synthesis of disilanes from halosilanes and alkali metals [255, 594, 611, 742, 754, 794, 796, 798, 875–879, 1070, 1185, 1192, 1193, 1271, 1694]:

$$R_3SiCl + Na \longrightarrow R_3SiNa + NaCl \quad (1\text{-}4)$$

$$R_3SiCl + R_3SiNa \longrightarrow R_3SiSiR_3 + NaCl \quad (1\text{-}5)$$

Analogous schemes were also proposed to explain the reaction between sodium and organoxysilanes [1007, 1008, 1018]. However, many attempts to obtain stable organosilicon compounds with and Si–M bond were unsuccessful [898, 899, 1025] and it was not possible [559] to reproduce the apparently successful original experiments of Kraus [897].

It must be assumed that the first true organosilane with an Si–M bond was the triphenylsilylpotassium obtained in 1951 by cleavage of (dimethylphenylmethyl)triphenylsilane by metallic potassium [560]:

$$(C_6H_5)_3SiC(CH_3)_2C_6H_5 + 2K \longrightarrow (C_6H_5)_3SiK + C_6H_5(CH_3)_2CK \quad (1\text{-}6)$$

However, this reaction is of no preparative interest since a mixture of organosilicopotassium and organopotassium compounds is obtained. Therefore, the present development of the chemistry of organosilicon heterocompounds of Group Ia elements began with the preparation by Gilman and his co-workers of pure triphenylsilylpotassium [125, 791]:

$$(C_6H_5)_3SiSi(C_6H_5)_3 \xrightarrow{Na|K} 2(C_6H_5)_3SiK \quad (1-7)$$

1.1. Synthesis Methods

1.1.1. Cleavage of Si–H Bond

Depending on the nature of the organic radical attached to the silicon atom, organohydrosilanes behave differently on reacting with alkali metals. Trialkylsilanes, for example, do not react with Na|K alloy even when stirred for 24 hr [556, 754]. On the other hand, triarylsilanes readily form triarylsilylpotassium under these conditions. Thus, a suspension of triphenylsilylpotassium is formed when triphenylsilane is stirred with Na|K alloy (1:5) in diethyl ether even at room temperature for 12 hr. By hydrolysis of the reaction mixture it is possible to isolate 41% of triphenylsilanol [549]. Increasing the reaction time to 48 hr raises the yield of triphenylsilylpotassium to 67% [556]:

$$(C_6H_5)_3SiH \xrightarrow{Na|K} (C_6H_5)_3SiK \quad (1-8)$$

Triphenylsilane reacts slowly with lithium in diethyl ether. Triphenylsilanol and hexaphenyldisilane are the products from hydrolysis of the reaction mixture [742]. When the ether is replaced by tetrahydrofuran, the rate of reaction of alkali metals with triphenylsilane is increased [92]. The silyllithium derivative is also formed by the reaction of tris(triphenylgermyl)silane with lithium in ethylamine [961]. The reaction of silacyclopentadiene with potassium in ether or tetrahydrofuran forms silacyclopentadienylpotassium [552].

In all the reactions of triphenylsilane with lithium and sodium-potassium alloy, a small amount of tetraphenylsilane is obtained in addition to the main metallation products. In the case of the reaction of tribenzylsilane with sodium in decalin, tetrabenzylsilane is the main product [550]. This indicates that either the alkali metal or the R_3SiM formed reacts with the starting organohydrosilane, splitting off an aryl (aralkyl) group. The cleavage power of alkali metals was demonstrated in a study of the reaction of triphenylsilane with sodium in toluene [550] or sodium-potassium alloy with diphenylsilane and phenylsilane [549, 556]. Sodium reacts with triphenylsilane in toluene only in the presence of traces of potassium. The reaction products include benzyltriphenylsilane and a compound of an inorganic nature, which contains silicon. The presence of benzyltriphenylsilane may be explained by metallation of toluene by triphenylsilylsodium (or phenylsodium) and the subsequent reaction of the benzylsodium formed with triphenylsilane:

$$\begin{array}{ccc} (C_6H_5)_3SiH & \xrightarrow{Na} & C_6H_5Na + HSiNa_3 \\ \downarrow{\scriptstyle Na} & & \downarrow{\scriptstyle C_6H_5CH_3} \\ (C_6H_5)_3SiNa & \xrightarrow{C_6H_5CH_3} & C_6H_5CH_2Na \end{array} \quad (1\text{-}9)$$

$$(C_6H_5)_3SiH + C_6H_5CH_2Na \longrightarrow (C_6H_5)_3SiCH_2C_6H_5 + NaH \quad (1\text{-}10)$$

The formation of series of products in the reactions of $C_6H_5SiH_3$ and $(C_6H_5)_2SiH_2$ with Na|K alloy may be explained analogously.

In these reactions tetraphenylsilane is formed by the reaction of phenylsilanes with phenylpotassium, which is obtained as a result of the splitting out of a phenyl group from phenylsilanes:

$$(C_6H_5)_3SiH + C_6H_5K \longrightarrow (C_6H_5)_4Si + KH \quad (1\text{-}11)$$

$$(C_6H_5)_2SiH_2 + 2C_6H_5K \longrightarrow (C_6H_5)_4Si + 2KH \quad (1\text{-}12)$$

Tetraphenylsilane is also formed by the reaction of a triphenylsilylmetal with triphenylsilane. This indicates that part of the tetraphenylsilane observed in reaction (1-8) together with hexaphenyldisilane may also be formed in accordance with another scheme:

$$(C_6H_5)_3SiNa + HSi(C_6H_5)_3 \longrightarrow (C_6H_5)_3SiSi(C_6H_5)_3 + NaH + (C_6H_5)_4Si \text{ and other products} \quad (1\text{-}13)$$

We must also adhere to this scheme in explaining the formation of tetraphenylsilane and hexaphenyldisilane in the reaction of triphenylsilane with triphenylmethylsodium, the initial stage of which is apparently an exchange process:

$$(C_6H_5)_3SiH + NaC(C_6H_5)_3 \rightleftarrows (C_6H_5)_3SiNa + HC(C_6H_5)_3 \quad (1\text{-}14)$$

which is followed by reaction (1-13).

From the above it follows that the reaction of organohydrosilanes with alkali metals is not a convenient method of preparing compounds of the type R_3SiM as this reaction is accompanied by a series of side reactions.

1.1.2. Cleavage of Si – C Bond by Alkali Metals

The cleavage of an Si – C bond in tetraorganosilanes by alkali metals is not very suitable for preparative purposes. In a number of cases, the Si – C bond cannot be cleaved at all by the alkali metal. Thus, for example, tetraalkylsilanes are not cleaved by lithium [92]. It is not possible to cleave the Si – C bond in many trialkylarylsilanes by treating them with Na|K alloy in diethyl ether [754] or lithium in diethyl ether, tetrahydrofuran, or 1,1-dimethyoxyethane [92]. Na|K alloy does not react (in decalin) with alkyltriphenylsilanes or dialkyldiphenylsilanes [550]. 9,9-Dimethyl-9-silafluorene likewise is not cleaved by lithium in tetrahydrofuran [750].

The Si – C bond in trialkyl(arylalkyl)silanes [in trimethylbenzylsilane, trimethyl(diphenylmethyl)silane, and trimethyl(triphenylmethyl)silane] may be cleaved by potassamide in liquid ammonia [864]. In this case an aralkylpotassium and the N-potassium derivative of hexamethyldisilazan are formed instead of a triorganosilylpotassium.

A triorganosilylpotassium is formed by cleavage of a triphenyl(aralkyl)silane with Na | K alloy in diethyl ether [560, 585]:

$$(C_6H_5)_3\,Si{-}C\,(C_6H_5)_3 + 2K \longrightarrow (C_6H_5)_3\,SiK + (C_6H_5)_3\,CK \qquad (1\text{-}15)$$

Triphenylbenzylsilane is cleaved analogously by lithium in tetrahydrofuran [768].

Tetraphenylsilane reacts with Na|K alloy in diethyl ether to form triphenylsilylpotassium [796] and with lithium in tetrahydrofuran to form triphenylsilyllithium [92]:

$$(C_6H_5)_4\,Si + 2Li \longrightarrow (C_6H_5)_3\,SiLi + C_6H_5Li \qquad (1\text{-}16)$$

1,1,2-Triphenyl-1-silacyclobutane reacts exothermally with lithium in tetrahydrofuran. However, it is not possible to demonstrate the formation of an organosilyllithium derivative in this case as it reacts rapidly with the starting 1,1,2-triphenyl-1-silacyclobutane to form a polymer [735].

Cleavage of an Si–C bond with the possible intermediate formation of an Si–M bond is also observed in the reaction of triarylsilanes with alkali metals [92, 549, 550, 556]. However, the hydrolysis products corresponding to diphenylsilylpotassium and phenylsilylpotassium have not been isolated.

1.1.3. Cleavage of Si–Si Bond

As in the case of an Si–C bond, the cleavage of an Si–Si bond depends strongly on the nature of the organic radicals attached to the silicon atoms. All attempts to cleave hexaalkyldisilanes in order to obtain trialkylsilylmetals have been unsuccessful. Thus, for example, Na|K alloy does not cleave hexamethyldisilane in diethyl ether [594, 754]. Hexaethyldisilane is not cleaved by lithium in tetrahydrofuran, 1,1-dimethoxyethane, or triethylamine [92], sodium in liquid ammonia [754], Na|K alloy in diethyl ether [92], tetrahydrofuran, or 1,2-dimethoxyethane, or in the absence of a solvent at 150°C [754]. Hexaethyldisilane is not cleaved either by rubidium or cesium [754].

It was possible to cleave 1,1,1-trialkyl-2,2,2-triaryldisilanes by alkali metals. A mixture of triarylsilylmetals and trialkylsilylmetals was formed. Thus, for example, 1,1,1-triethyl-2,2,2-triphenyldisilane reacts with lithium in tetrahydrofuran. After hydrolysis of the reaction mixture it was possible to isolate triethylsilane (11%), hexaethyldisilane (14%), and triphenylsilane (92%). The Si–Si bond in 1,1,1-triethyl-2,2,2-triphenyldisilane is also cleaved by Na|K alloy [754]. This reagent also cleaves 1,1,1-trimethyl-2,2,2-triphenylsilane, with which lithium does not react [754, 796]:

$$(C_6H_5)_3Si\text{—}Si(CH_3)_3 \xrightarrow{Na\,|\,K} (C_6H_5)_3SiK + (CH_3)_3SiK \qquad (1\text{-}17)$$

There is analogous cleavage of 1,1,2,2-tetramethyl-1,2-diphenyldisilane and 1,2-dimethyl-1,1,2,2-tetraphenyldisilane by lithium in tetrahydrofuran [758] and by Na|K alloy in diethyl ether [754] with the formation of the dimethylphenyl- and methyldiphenylsilyl derivatives of Li and K, respectively.

Lithium in tetrahydrofuran will cleave the Si–Si bond in silacyclanes, in which the ring containing silicon is condensed with benzene rings. For example, bi-9,9'-(9-methyl-9-silafluorenyl) is cleaved in accordance with the scheme 1-18 [750]:

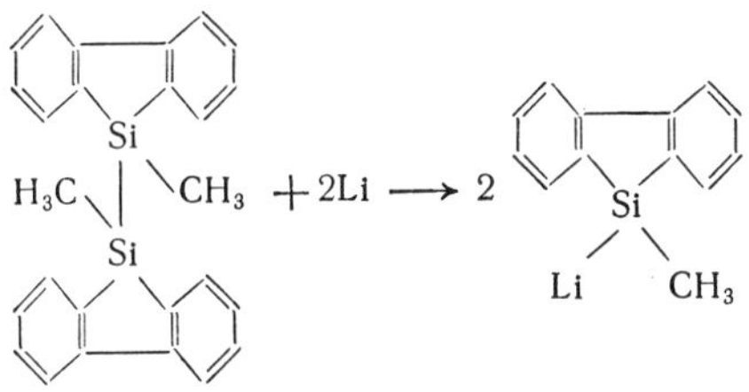

$$\text{(dibenzosilole)Si(CH_3)–Si(CH_3)(dibenzosilole)} + 2Li \longrightarrow 2\ \text{(dibenzosilole)Si(Li)(CH_3)} \tag{1-18}$$

Hexaaryldisilanes may be cleaved in various solvents by any of the five alkali metals:

$$Ar_3Si{-}SiAr_3 + 2M \longrightarrow 2Ar_3SiM \tag{1-19}$$

The most convenient solvent for the preparation of triphenylsilyllithium from hexaphenyldisilane is tetrahydrofuran, in which the reaction proceeds rapidly and in good yield. The solution of triphenylsilyllithium in tetrahydrofuran formed is quite stable (some reaction with the solvent is observed only after prolonged heating) [1278]. Tetrahydropyran and dioxan are more resistant to triphenylsilyllithium; however, the cleavage of hexaphenyldisilane by lithium in these solvents is slower than in tetrahydrofuran [714, 1274]. Hexaphenyldisilane is cleaved readily by lithium in 1,2-dimethoxyethane [581, 1274] and pyridine [1277], but the triphenylsilyllithium formed reacts to a considerable extent with the solvent. While the cleavage of hexaphenyldisilane proceeds quantitatively in pyridine, lithium does not react with hexaphenyldisilane in quinoline, even after prolonged heating [92]. The Si–Si bond in hexaphenyldisilane is not cleaved by lithium in acetal at room temperature [787]. The reaction of lithium with sym-tetraphenyldisilane forms diphenylsilyllithium, in which the silicon atom is attached simultaneously to a hydrogen atom and a lithium atom [780]:

$$(C_6H_5)_2SiH{-}SiH(C_6H_5)_2 + 2Li \longrightarrow 2\,(C_6H_5)_2SiH{-}Li \tag{1-20}$$

A hypothesis on the possibility of the existence of this compound was put forward earlier [1275].

Triphenylsilylsodium may be prepared by the reaction of hexaphenyldisilane with sodium in 1,1-dimethoxyethane [581], but not in diethyl ether, tetralin, xylene, or dioxan [92, 791].

Triphenylsilylpotassium is formed by the reaction of Na|K alloy (1:5) with hexaphenyldisilane in diethyl ether [791, 796, 798], 1, 2-dimethoxyethane [581], petroleum ether, and benzene [796]. If necessary, excess alloy may be removed by amalgamation. Triphenylsilylpotassium can also be prepared by the reaction of hexaphenyldisilane with potassium in dibutyl ether [791]. The cleavage of hexaaryldisilanes by Na|K alloy in diethyl ether is also used to prepare a triarylsilylpotassium containing a substituent in the benzene ring [586, 748, 794].

Lithium [123], sodium [123, 798, 1287], and potassium [123] cleave hexaphenyldisilane in liquid ammonia. In 4 hr at -50°C, lithium cleaves 60% of the initial hexaphenyldisilane and sodium 70%, while the reaction proceeds quantitatively with potassium [123]. The triphenylsilylsodium formed then reacts with ammonia and is converted to hexaphenyldisilazan [798, 1287].

Cleavage of hexaphenyldisilane by rubidium and cesium in diethyl and dibutyl ethers yields triphenylsilylrubidium and triphenylsilylcesium [796]. The yield of the latter does not exceed 50% because of its high reactivity.

Lithium reacts with perphenylpolysilanes in tetrahydrofuran. In the case of acyclic polysilanes this yields a mixture of triphenylsilyllithium and a lithium derivative of the polysilane.

Thus, for example, the reaction of lithium with octaphenyltrisilane in tetrahydrofuran forms triphenylsilylithium and pentaphenyldisilyllithium. Cleavage of decaphenyltetrasilane under these conditions leads to the formation of three products [1275]:

$$(C_6H_5)_3Si{-}Si(C_6H_5)_2{-}Si(C_6H_5)_2{-}Si(C_6H_5)_3 \xrightarrow{Li} \underset{(30.8\%)}{(C_6H_5)_3SiLi} + \underset{(45.2\%)}{(C_6H_5)_3Si{-}Si(C_6H_5)_2{-}Li} + \underset{(17.1\%)}{(C_6H_5)_3Si{-}Si(C_6H_5)_2{-}Si(C_6H_5)_2{-}Li} \quad (1\text{-}21)$$

The formation of dodecamethylcyclohexasilane in the reaction of 2,2-dimethyl-1,1,1,3,3,3-hexaphenyltrisilane with lithium is unexpected [782]. The following scheme (1-22) was proposed to explain this fact:

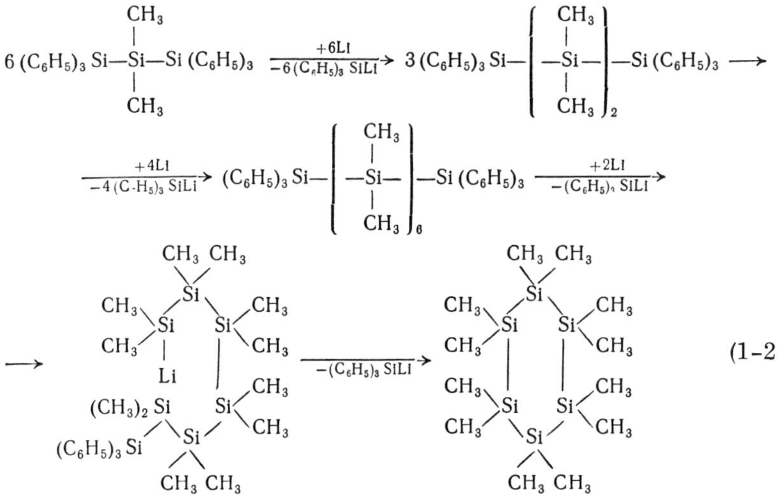

$$6\,(C_6H_5)_3Si{-}\underset{CH_3}{\overset{CH_3}{|}}\!\!\!\!Si{-}Si(C_6H_5)_3 \xrightarrow[-6(C_6H_5)_3SiLi]{+6Li} 3\,(C_6H_5)_3Si{-}\left[{-}\underset{CH_3}{\overset{CH_3}{|}}\!\!\!\!Si{-}\right]_2{-}Si(C_6H_5)_3 \longrightarrow$$

$$\xrightarrow[-4(C_6H_5)_3SiLi]{+4Li} (C_6H_5)_3Si{-}\left[{-}\underset{CH_3}{\overset{CH_3}{|}}\!\!\!\!Si{-}\right]_6{-}Si(C_6H_5)_3 \xrightarrow[-(C_6H_5)_3SiLi]{+2Li}$$

$$\longrightarrow \text{(C}_6\text{H}_5)_3\text{Si}{-}[Si(CH_3)_2]_6{-}\text{Li} \xrightarrow{-(C_6H_5)_3SiLi} \text{cyclo-}[Si(CH_3)_2]_6 \qquad (1\text{-}22)$$

The products of the reaction of octaphenylcyclotetrasilane [91, 752, 772, 854, 1272, 1273], decaphenylcyclopentasilane [91,725, 726, 739, 777, 778], and dodecaphenylcyclohexasilane [91, 782,1272, 1273] with lithium in tetrahydrofuran are the corresponding dilithium derivatives of the linear perphenyl polysilanes, e.g.:

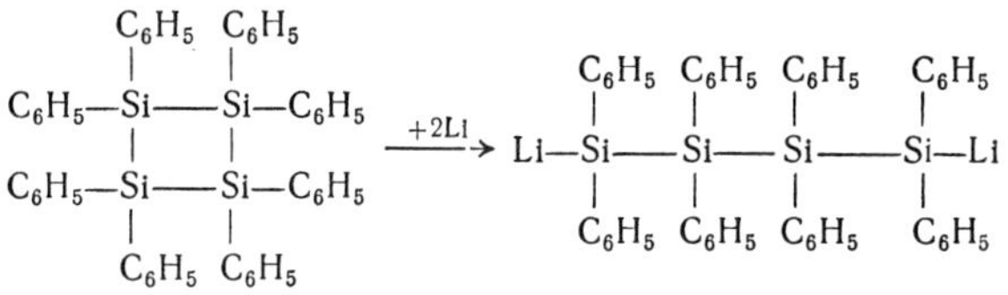

$$\text{cyclo-}[Si(C_6H_5)_2]_4 \xrightarrow{+2Li} Li{-}\left[\underset{C_6H_5}{\overset{C_6H_5}{|}}\!\!\!\!Si{-}\right]_4{-}Li \qquad (1\text{-}23)$$

However, the yields of the 1,4-disubstituted derivatives of the reaction of 1,4-dimethyloctaphenyltetrasilane with trimethyl phosphate, diphenylchlorosilane, methyldiphenylchlorosilane, and dilute hydrochloric acid do not exceed 25-27%. At the same time, compounds with a shorter chain of silicon atoms (for example, sym-tetraphenyldisilane) were detected in the hydrolysis products. The yield of 1,4-dilithiooctaphenyltetrasilane may be raised to 35-40% by reducing its contact time with unreacted octaphenylcyclotetrasilane [752]. Hence it may be concluded that the cleavage of octaphenylcyclotetrasilane by lithium competes with its cleavage by the

1,4-dilithiooctaphenyltetrasilane formed (and possibly also with the cleavage of the latter by lithium).

In contrast to octaphenylcyclotetrasilane, the cleavage of decaphenylcyclopentasilane by lithium in tetrahydrofuran leads to the formation of 1,5-dilithiodecaphenylpentasilane in high yield (the yields of the corresponding 1,5-derivatives in reactions with trimethyl phosphate, tributyl phosphate, and trimethylchlorosilane are 73-83%) [91, 724, 725, 777, 778]:

$$[(C_6H_5)_2Si]_5 \xrightarrow{+2Li} Li{-}\underset{C_6H_5}{\overset{C_6H_5}{Si}}{-}\left[\underset{C_6H_5}{\overset{C_6H_5}{Si}}{-}\right]_3{-}\underset{C_6H_5}{\overset{C_6H_5}{Si}}{-}Li \qquad (1\text{-}24)$$

This reaction is usually complete in 2 hr at room temperature. However, if the reaction time is increased there are various side cleavage processes and the reaction products contain derivatives of disilane and trisilane [778].

Dodecaphenylcyclohexasilane is cleaved by lithium in tetrahydrofuran with much more difficulty than octaphenylcyclotetrasilane and decaphenylcyclopentasilane [91, 782, 1272, 1273]. The reaction forms a mixture of 1,6-dilithiododecaphenylhexasilane and the products of its further cleavage (1,3-dilithiohexaphenyltrisilane, 1,4-dilithiooctaphenyltetrasilane, and 1,2-dilithiotetraphenyldisilane). The main reaction product is 1,3-dilithiohexaphenyltrisilane, whose yield is 55.6% after 14 hr at 25°C [91]:

$$[(C_6H_5)_2Si]_6 \xrightarrow{+2Li} Li{-}\underset{C_6H_5}{\overset{C_6H_5}{Si}}{-}\left[\underset{C_6H_5}{\overset{C_6H_5}{Si}}{-}\right]_4{-}\underset{C_6H_5}{\overset{C_6H_5}{Si}}{-}Li \xrightarrow{+2Li}$$

$$\longrightarrow 2Li{-}\underset{C_6H_5}{\overset{C_6H_5}{Si}}{-}\underset{C_6H_5}{\overset{C_6H_5}{Si}}{-}\underset{C_6H_5}{\overset{C_6H_5}{Si}}{-}Li \longrightarrow Li{-}\underset{C_6H_5}{\overset{C_6H_5}{Si}}{-}\underset{C_6H_5}{\overset{C_6H_5}{Si}}{-}Li + Li{-}\underset{C_6H_5}{\overset{C_6H_5}{Si}}{-}\left[\underset{C_6H_5}{\overset{C_6H_5}{Si}}{-}\right]_2{-}\underset{C_6H_5}{\overset{C_6H_5}{Si}}{-}Li \qquad (1\text{-}25)$$

The Si–Si bond in hexaphenyldisilane may also be cleaved by phenyllithium:

$$(C_6H_5)_3Si{-}Si(C_6H_5)_3 + C_6H_5Li \longrightarrow (C_6H_5)_3SiLi + (C_6H_5)_4Si \qquad (1\text{-}26)$$

The process is very slow in diethyl ether, as is indicated by the very low yield of triphenylsilanol (2.7%) on hydrolysis of the reaction mixture after the latter has been boiled for 120 hr [712]. The reaction rate is increased if the diethyl ether is replaced by tetrahydrofuran or a mixture of it with ether.

In the latter case the yield of triphenylsilyllithium is 67.5% after boiling for 48 hr, but this is still less than in the case of the reaction of hexaphenyldisilane with lithium.

Butyllithium does not react with hexaphenyldisilane in tetrahydrofuran at -50°C [788]. Tolyllithium behaves analogously at 0°C and triphenylmethyllithium at 20°C [712]. At the same time benzyllithium readily cleaves the Si–Si bond in hexaphenyldisilane [779].

The reaction of 2,2'-dilithio-4,4'-dimethyl-N-methyldiphenylamine with sym-tetraphenyldisilane evidently also involves cleavage of the Si–Si bond and the intermediate formation of an Si–Li bond. This is indicated by the fact that the reaction product contains only one silicon atom [799]:

$$H_3C{-}C_6H_3(Li){-}N(CH_3){-}C_6H_3(Li){-}CH_3 + (C_6H_5)_2\underset{H}{Si}{-}\underset{H}{Si}(C_6H_5)_2 \longrightarrow H_3C{-}C_6H_3{<}^{N(CH_3)}_{Si(C_6H_5)_2}{>}C_6H_3{-}CH_3 \qquad (1\text{-}27)$$

The reaction of phenyllithium with perphenylcyclosilanes is more complex since the lithium derivative formed initially reacts both with the starting cyclosilane and with excess phenyllithium. In this case the direction of the reaction depends markedly on the ratio of the reagents [853]. The reaction of phenyllithium with octaphenylcyclotetrasilane in a mixture of tetrahydrofuran and diethyl ether leads to decaphenylcyclopentasilane. With excess phenyllithium the main reaction product is pentaphenyldisilanyllithium [853]. The cleavage of octaphenylcyclotetrasilane by methyllithium forms 1-lithio-4-methyloctaphenyltetrasilane:

$$\begin{array}{c}(C_6H_5)_2\,Si{-}Si\,(C_6H_5)_2\\ |\quad\quad | \\ (C_6H_5)_2\,Si{-}Si\,(C_6H_5)_2\end{array} + CH_3Li \longrightarrow CH_3{-}\overset{C_6H_5}{\underset{C_6H_5}{\overset{|}{\underset{|}{Si}}}}{-}\left[\overset{C_6H_5}{\underset{C_6H_5}{\overset{|}{\underset{|}{Si}}}}{-}\right]_2{-}\overset{C_6H_5}{\underset{C_6H_5}{\overset{|}{\underset{|}{Si}}}}{-}Li \qquad (1\text{-}28)$$

If the reaction time is no longer than 0.5 hr and the amount of methyllithium is two equivalents, the yield of 1-lithio-4-methyloctaphenyltetrasilane is 66%.

Increasing the reaction time and using a large excess of methyllithium promotes further cleavage, leading to the formation of 2-methyl-1,1,2,2-tetraphenyldisilane [91].

Phenyllithium and methyllithium react similarly with decaphenylcyclopentasilane [91]. Dodecaphenylcyclohexasilane is not cleaved by methyllithium [91].

1.1.4. Cleavage of Si–Ge and Si–Sn Bonds by Alkali Metals

Si–Ge and Si–Sn bonds are cleaved by alkali metals analogously to Si–Si bonds (see also Ch. 4, Sections 4 and 8).

Lithium in ethylamine reacts vigorously with triethylsilyltriphenylgermane [898]. The hydrolysis products of the reaction mixture are triethylsilane and triphenylgermane. Treatment of the mixture with ethyl bromide leads to the formation of tetraethylsilane and ethyltriphenylgermane. This indicates the presence of triethylsilyllithium in the reaction mixture:

$$(C_6H_5)_3\,GeSi\,(C_2H_5)_3 + 2Li \longrightarrow (C_6H_5)_3\,GeLi + (C_2H_5)_3\,SiLi \xrightarrow[-2LiBr]{+2C_2H_5Br}$$

$$\longrightarrow (C_6H_5)_3\,GeC_2H_5 + (C_2H_5)_4\,Si \qquad (1\text{-}29)$$

The reaction of triphenylsilyltriphenylgermane with Na|K alloy in diethyl ether with tetrahydrofuran added forms a mixture of triphenylsilylpotassium and triphenylgermylpotassium, which, after treatment with carbon dioxide and hydrolysis gives triphenylsilanol (86%) and triphenylgermanecarboxylic acid (74%) [747].

Cleavage of triphenylsilyltrimethylstannane by sodium in liquid ammonia may be described by the scheme [897]:

$$(C_6H_5)_3SiSn(CH_3)_3 + 2Na \longrightarrow (C_6H_5)_3SiNa + (CH_3)_3SnNa \xrightarrow[-2NaI]{+2CH_3I}$$

$$\longrightarrow (C_6H_5)_3SiCH_3 + (CH_3)_4Sn \qquad (1\text{-}30)$$

None of these reactions are of preparative value.

1.1.5. Cleavage of Si–O–C and Si–O–Si Bonds by Alkali Metals

Trialkylalkoxysilanes do not react with lithium in either diethyl ether or tetrahydrofuran. No changes are observed when they are treated with Na|K alloy (in diethyl ether and tetrahydrofuran) and even cesium (in diethyl ether) [92].

Triarylalkoxysilanes are cleaved by Na|K alloy in ether with the formation of triphenylsilylpotassium in a yield varying from 40 to 70% [556, 589, 791, 796]:

$$(C_6H_5)_3Si{-}O{-}C_2H_5 \xrightarrow{Na\,|\,K} (C_6H_5)_3SiK + C_2H_5OK \qquad (1\text{-}31)$$

Triphenylalkoxysilanes do not react with sodium in boiling xylene [797].

It has been suggested that the reaction of phenoxysilanes with sodium also proceeds through the intermediate formation of compounds containing an Si–Na bond [1007, 1008, 1018].

Alkali metals react with hexaorganodisiloxanes of definite structure with the formation of silanolates and compounds containing an Si–alkali metal bond.

Hexaalkyldisiloxanes are not cleaved by lithium or cesium in tetrahydrofuran or by Na|K alloy in ether. 1,1,1-Trialkyl-3,3,3-triaryldisiloxanes react with lithium in tetrahydrofuran to form triarylsilyllithium derivatives:

$$(CH_3C_6H_4)_3Si{-}O{-}Si(CH_3)_3 + 2Li \longrightarrow (CH_3C_6H_4)_3SiLi + (CH_3)_3SiOLi \qquad (1\text{-}32)$$

1,1,3,3-Tetramethyl-1,3-diphenyldisiloxane is cleaved by an analogous scheme. Hexaphenyldisiloxane is readily cleaved by rubidium, cesium, and also lithium in tetrahydrofuran, methyltetrahydrofuran, and tetrahydropyran [92]:

$$(C_6H_5)_3Si{-}O{-}Si(C_6H_5)_3 + 2Li \longrightarrow (C_6H_5)_3SiLi + (C_6H_5)_3SiOLi \quad (1.33)$$

1.1.6. Cleavage of Si–Halogen Bond

The synthesis of compounds containing an Si–M bond from triorganohalosilanes and alkali metals is only possible if one of the organic groups attached to the silicon atom is aryl. In most of the reactions of this type which have been carried out successfully the halogen atom replaced by the metal is chlorine and in only one case, fluorine. Attempts to obtain R_3SiM from trialkylchloro-, trialkylbromo-, and trialkyliodosilanes were unsuccessful and lead to the formation of hexaalkyldisilanes which are not cleaved by alkali metals. It was not possible to obtain triethylsilyllithium from triethylbromosilane by treatment with lithium in ethylamine [898]; the reaction product was triethyl-N-ethylaminosilane. The product of an analogous reaction with triphenylbromosilane [897] was also a compound containing an Si–N bond (triphenyl-N-ethylaminosilane) and not triphenylsilyllithium since the reaction mixture did not react with bromobenzene [559].

Triphenylsilylpotassium could be obtained in 92% yield in accordance with the scheme [716, 791, 796]:

$$(C_6H_5)_3SiCl \xrightarrow{Na\,|\,K} (C_6H_5)_3SiK + KCl \quad (1\text{-}34)$$

Tri-(p-tolyl)chlorosilane reacts analogously with this alloy [586]. A certain amount of triphenylsilylsodium is apparently formed by boiling with sodium in xylene since treatment of the reaction mixture with trimethylchlorosilane leads to the formation of 1,1,1-trimethyl-2,2,2-triphenyldisilane [788].

Triphenylsilyllithium is obtained in high yield from triphenylchlorosilane and lithium [716, 773, 1204]. The synthesis is carried out by adding a solution of triphenylchlorosilane in tetrahydrofuran to lithium wire in a nitrogen atmosphere. In another method, tetrahydrofuran is added to a suspension of lithium in triphenylchlorosilane. The use of 2-methyltetrahydrofuran and tetrahydropyran reduces the yield of triphenylsilyllithium from 80-90% (in the case of tetrahydrofuran) to 60-65% [716]. The yield of triphenylsilyllithium is 84% when it is synthesized from triphenylfluorosilane in tetrahydrofuran [937].

Triphenylsilylrubidium and triphenylsilylcesium are obtained from triphenylchlorosilane and the corresponding metals in lower yield than triphenylsilyllithium since they react with the solvent with the result that the reaction must be carried out at low temperature (-50°C).

Methyldiphenylsilyllithium [773, 1274], dimethylphenylsilyllithium [716, 773], and 1-lithio-1-phenyl-1-silacyclohexane [773] have been prepared by reactions of type (1-34). By this method it was also possible to prepare diphenylsilyllithium containing both lithium and hydrogen atoms attached to the silicon atom [780].

Most reactions of type (1-34) proceed in several stages, including the intermediate formation of a hexaorganodisilane and its subsequent cleavage by the alkali metal. This is confirmed by the fact that in the investigation of the products from the reaction of triphenylchlorosilane with lithium, before its completion it is possible to isolate up to 64% of hexaphenyldisilane. At the end of the reaction hexaphenyldisilane is absent from the reaction mixture. The hexaphenyldisilane is formed either by condensation of two molecules of triphenylchlorosilane under the action of lithium or by the reaction of the triphenylsilyllithium formed initially with triphenylchlorosilane:

$$(C_6H_5)_3SiCl + Li \rightarrow \begin{cases} \rightarrow (C_6H_5)_3SiLi + LiCl \\ \qquad \downarrow (C_6H_5)_3SiCl \\ \rightarrow (C_6H_5)_3SiSi(C_6H_5)_3 \xrightarrow{Li} (C_6H_5)_3SiLi \end{cases} \qquad (1\text{-}35)$$

The product from the reaction of diphenyldichlorosilane with lithium in tetrahydrofuran is 1,4-dilithiooctaphenyltetrasilane. It is most probably formed as a result of cleavage of octaphenylcyclotetrasilane obtained initially [91].

In the case of triarylchlorosilanes containing such radicals as o-tolyl [716] and o-xenyl [92], the formation of the corresponding hexaaryldisilanes is impossible due to steric hindrance. In this case lithium and cesium derivatives are formed directly by cleavage of the Si–Cl bond by these metals.

In the reaction of lithium, sodium, and potassium derivatives of triarylmethane [585], triarylgermane [747], and triarylstannane [123, 1205] with triorganohalosilanes, hexaorganodisilanes are sometimes formed in addition to the main reaction products. This may be explained by exchange of the halogen for the metal with the intermediate formation of triorganosilyl derivatives of lithium, sodium, or potassium. Thus, for example, hexaphenyldisilane is a product of the reaction of triphenylmethylsodium with triphenylbromosilane. In this case there is probably some halogen–metal exchange between the triphenylbromosilane and triphenylmethylsodium with the formation of triphenylsilylsodium, which then reacts with triphenylbromosilane:

$$(C_6H_5)_3CNa + (C_6H_5)_3SiBr \rightleftarrows (C_6H_5)_3CBr + (C_6H_5)_3SiNa \quad (1-36)$$

$$(C_6H_5)_3SiNa + (C_6H_5)_3SiBr \longrightarrow (C_6H_5)_3Si{-}Si(C_6H_5)_3 + NaBr \quad (1-37)$$

1.2. Physical Properties

Organosilicon compounds containing an Si–M bond are very reactive. They decompose under the action of oxygen and carbon dioxide and also moisture. Therefore, all operations in their preparation are carried out in dry nitrogen with the use of an appropriate solvent. This yields suspensions or solutions of R_3SiM, which are used for subsequent synthesis without isolation of the individual organosilicon metal compounds.

Triphenylsilyllithium decomposes in tetrahydrofuran, 2-methyltetrahydrofuran, tetrahydropyran, dioxan, and 1,2-dimethoxyethane. Solutions of it in tetrahydrofuran gradually decompose because of the reaction with the solvent. The replacement of the aryl radicals in R_3SiLi by methyl radicals reduces the stability of tetrahydrofuran solutions of compounds of this type, which falls in the series [756, 758]

$$(C_6H_5)_3SiLi > CH_3(C_6H_5)_2SiLi > C_6H_5(CH_3)_2SiLi$$

The solutions of R_3SiLi in 1,2-dimethoxyethane are still less stable [581].

Up to now only one attempt to isolate R_3SiM in a pure form is known [123]. When a yellow solution of triphenylsilyllithium in liquid ammonia is evaporated a yellow solid separates and this partly decomposes. It is apparently free triphenylsilyllithium. From a solution of triphenylsilylsodium in liquid ammonia it is also possible to isolate yellow triphenylsilylsodium. Triphenylsilylpotassium forms a yellow-brown suspension in diethyl ether and a red-brown solution in liquid ammonia. After evaporation of the solvent, yellow triphenylsilylpotassium separates. In all cases, during the evaporation of solutions of R_3SiM in liquid ammonia their partial decomposition occurs:

$$2x\,(C_6H_5)_3\,SiM \longrightarrow 2xM + x\,(C_6H_5)_4\,Si + [(C_6H_5)_2\,Si]_x\,. \qquad (1\text{-}38)$$

In the UV spectra of $(C_6H_5)_3SiLi$ there is an intense maximum ($\log \varepsilon \sim 4.0$) at 335 m$\mu$, which indicates considerable delocalization of the electrons in this compound [1220].

1.3. Chemical Properties

1.3.1. Reaction with Inorganic Compounds

Reaction with Elements. Triphenylsilyllithium reacts readily with oxygen in tetrahydrofuran and tetrahydropyran even at temperatures below 0°C. If the reaction is carried out at -25°C in tetrahydrofuran, after hydrolysis of the reaction mixture it is possible to isolate triphenylsilanol (60.3%), hexaphenyldisilane (0.6%), and triphenylsilane (12.8%). With a rise in the reaction temperature the amount of hexaphenyldisilane increases (6.7% at 0°C and 19.3% at 25°C), while the yield of triphenylsilanol decreases (47% at 0°C and 32.6% at 25°C). In the opinion of George and Gilman [714], the first stage in the reaction of triphenylsilyllithium with oxygen is the formation of the lithium salt of triphenylsilyl hydroperoxide, which subsequently decomposes into radicals. The final reaction products are formed by the reaction of these radicals with the solvent or with unreacted triphenylsilyllithium or as a result of their dimerization. It may be surmised that the first stage of the interaction of triphenylsilyllithium with oxygen proceeds through an intermediate four-membered cyclic complex:

$$\begin{array}{l}(C_6H_5)_3\,Si\text{—}Li\\ \qquad\;\;\vdots\quad\;\vdots\\ \qquad\;\; O\text{±}O\end{array} \longrightarrow (C_6H_5)_3\,SiOOLi \longrightarrow (C_6H_5)_3\,SiO\cdot + \cdot OLi \qquad (1\text{-}39)$$

$$(C_6H_5)_3\,SiO\cdot + (C_6H_5)_3\,SiLi \longrightarrow (C_6H_5)_3\,SiOLi + (C_6H_5)_3\,Si \qquad (1\text{-}40)$$

$$(C_6H_5)_3\,Si\cdot + \text{ solvent } \longrightarrow (C_6H_5)_3\,SiH \qquad (1\text{-}41)$$

$$2\,(C_6H_5)_3\,Si\cdot \longrightarrow (C_6H_5)_3\,Si\text{—}Si\,(C_6H_5)_3 \qquad (1\text{-}42)$$

The hydroperoxide salt may also react with triphenylsilyllithium to form lithium triphenylsilanolate:

$$(C_6H_5)_3\,SiOOLi + (C_6H_5)_3\,SiLi \longrightarrow 2\,(C_6H_5)_3\,SiOLi \qquad (1\text{-}43)$$

The reaction of triphenylsilylithium with sulfur, in contrast to the reaction with oxygen, leads to the formation of lithium triphenylsilylmercaptide in high yield [763]:

$$(C_6H_5)_3SiLi + S \longrightarrow (C_6H_5)_3SiSLi \qquad (1\text{-}44)$$

Triphenylsilyllithium is converted quantitatively into hexaphenyldisilane by the action of halogens [756]:

$$2(C_6H_5)_3SiLi + Br_2 \longrightarrow (C_6H_5)_3Si{-}Si(C_6H_5)_3 + 2LiBr \qquad (1\text{-}45)$$

Reaction with Oxides of the Elements. Triphenylsilyllithium, triphenylsilylsodium, and triphenylsilylpotassium react vigorously with water. The hydrolysis product of triphenylsilylpotassium is triphenylsilanol, since the triphenylsilane formed initially then reacts with water in the alkaline medium with the liberation of hydrogen [796]:

$$(C_6H_5)_3SiK + H_2O \longrightarrow (C_6H_5)_3SiH + KOH \qquad (1\text{-}46)$$

$$(C_6H_5)_3SiH + H_2O \xrightarrow{KOH} (C_6H_5)_3SiOH + H_2 \qquad (1\text{-}47)$$

When the hydrolysis is carried out in an acid medium, the reaction is limited to the formation of triphenylsilane [560].

The product of the reaction of triphenylsilyllithium with carbon dioxide is the lithium salt of triphenylsilanecarboxylic acid, which is converted by treatment with dilute mineral acids into free $(C_6H_5)_3SiCOOH$ (93.4% yield) [714]. In the case of triphenylsilylpotassium the yield of this acid is somewhat lower (87%) [560, 584]. Triphenylsilanecarboxylic acid (or, more accurately, its salt) is decarboxylated rapidly:

$$(C_6H_5)_3SiLi + CO_2 \xrightarrow{H_2O} (C_6H_5)_3SiCOOH \xrightarrow{OH^{\ominus}} (C_6H_5)_3SiOH + CO \qquad (1\text{-}48)$$

The reaction of different R_3SiLi with carbon dioxide makes it possible to obtain a whole series of triorganosilanecarboxylic acids such as methyldiphenylsilanecarboxylic (66.5% yield) [783], dimethylphenylsilanecarboxylic (47%) [783], and tri(p-tolyl)silanecarboxylic acids (56%) [586].

The reaction of R_3SiM with halides of the elements will be described in later chapters.

1.3.2. Reaction with Aromatic and Unsaturated Organic Compounds (Metallation and Addition)

In the reaction of R_3SiM with hydrocarbons the latter may undergo metallation. However, this reaction occurs only if the hydrocarbon contains a labile hydrogen atom. Triphenylsilyllithium is considerably less active in metallation than alkyllithiums. Thus, for example, it metallates xanthene, thiaxanthene, 10-ethylphenothiazine, diphenylmethane [767], fluorene [739, 767], and phenylacetylene [767, 792], but does not react with naphthalene, dibenzofuran, dibenzothiophene, 5,5-dioxobenzothiophene [1283], or 5-oxo-10-ethylphenothiazine [763]. The metallation of toluene by triphenylsilyllithium with the formation of benzyllithium proceeds very slowly [779].

Triphenylsilylpotassium reacts rapidly and quantitatively with triarylmethanes [583]:

$$(C_6H_5)_3SiK + R_3CH \longrightarrow (C_6H_5)_3SiH + R_3CK \quad (1\text{-}49)$$

The formation of a small amount of tetraphenylsilane is observed. If the ratio of triphenylsilylpotassium to triarylmethane is increased from 1:1 to 2.5:1, the yield of tetraphenylsilane increases from 3 to 26%. Tetraphenylsilane is evidently formed as a result of the reaction of triphenylsilane with excess triphenylsilylpotassium. This also follows from the fact that a by-product from the reaction of triphenylsilylpotassium with tritolylmethane is tetraphenylsilane, while no tetrasubstituted silanes containing tolyl groups are formed.

The reaction of a triorganosilyllithium with fluorene was used to establish the relation between the reactivity of RR'R"SiM and the structure of the radicals R, R', and R". Triorganosilyllithium compounds and triphenylgermyllithium may be arranged in the following order with respect to the yield of the metallation products of fluorene [739]:

$$\underset{61.3\%}{CH_3(C_6H_5)_2SiLi} > \underset{58.1\%}{(CH_3)_2C_6H_5SiLi} > \underset{49.6\%}{(C_6H_5)_3GeLi} > \underset{42.2\%}{(C_6H_5)_3SiSi(C_6H_5)_2Li} \approx \underset{41.1\%}{(C_6H_5)_3SiLi}$$

Depending on their structure, unsaturated hydrocarbons either do not react with R_3SiM or add them at the double bond or are polymerized. In some cases, metallation also occurs.

Triphenylsilylpotassium adds to alkenes and cycloalkenes. On the other hand, arylalkenes readily add triphenylsilyllithium and triphenylsilylpotassium. Thus, for example, trans-stilbene forms an adduct with triphenylsilylpotassium in 55% yield [790]:

$$C_6H_5CH{=}CHC_6H_5 + (C_6H_5)_3SiK \longrightarrow C_6H_5CH(K){-}CH(Si(C_6H_5)_3)C_6H_5 \qquad (1\text{-}50)$$

The reaction of stilbene with triphenylsilyllithium is more complex:

$$RCH{=}CHR + R_3SiLi \longrightarrow \left[R_3SiCH(R){-}CH(Li)R\right]$$

$$\left[R_3SiCH(R){-}CH(Li)R\right] \begin{cases} \xrightarrow{H_2O} R_3SiCH(R)CH_2R \\ \xrightarrow[(H_2O)]{R_3SiLi} R_3SiCH(R)CH(R)SiR_3 \\ \xrightarrow[(H_2O)]{RCH=CHR} R_3Si(CHR)_3CH_2R \end{cases} \qquad (1\text{-}51)$$

$$(R = C_6H_5)$$

Styrene polymerizes when treated with a tetrahydrofuran solution of triphenylsilyllithium [1211].

In the reaction of 1,1-diphenylethylene and triphenylethylene with triphenysilyllithium and triphenylsilylpotassium the triphenylsilyl group adds to the most hydrogenated carbon atom [1288]:

$$(C_6H_5)_2C{=}CH_2 + (C_6H_5)_3SiK \longrightarrow (C_6H_5)_2CK{-}CH_2Si(C_6H_5)_3 \qquad (1\text{-}52)$$

$$(C_6H_5)_2C{=}CHC_6H_5 + (C_6H_5)_3SiLi \longrightarrow (C_6H_5)_2CLi{-}CH(C_6H_5)Si(C_6H_5)_3 \qquad (1\text{-}53)$$

Tetraphenylethylene and 9,9'-bifluorenylidene do not react with triphenylsilylpotassium, probably due to steric hindrance [1288]. It is remarkable that triphenylsilylpotassium does not add to 1,4-diphenylbutadiene-1,3, which is the vinylog of stilbene. At the same time, butadiene-1,3 and isoprene are polymerized by triphenylsilyllithium [656, 1211].

Diphenylacetylene reacts with triphenylsilylpotassium [792] and triphenylsilyllithium [588] with difficulty. In the latter case, the addition product of two molecules of triphenylsilyllithium is formed in low yield (8%):

$$C_6H_5C{\equiv}CC_6H_5 + 2\,(C_6H_5)_3SiLi \xrightarrow{(H_2O)} (C_6H_5)_3SiCH(C_6H_5){-}CH(C_6H_5)Si(C_6H_5)_3 \quad (1\text{-}54)$$

The reaction of 1-triphenylsilylpropyne with triphenylsilyllithium gives triphenylsilane (51.3%), 1,3-bis(triphenylsilyl)propyne (29.9%), and 1, 2-bis(triphenylsilyl) propene (18.5%). The formation of the first two compounds may be explained by the metallation of 1-triphenylsilylpropyne by triphenylsilyllithium with subsequent condensation; 1,2-bis(triphenylsilyl)propene is formed by the addition of triphenylsilyllithium to the triple bond of 1-triphenylsilylpropyne [731]:

$$(C_6H_5)_3SiC{\equiv}CCH_3 + (C_6H_5)_3SiLi \longrightarrow (C_6H_5)_3SiC{\equiv}CCH_2Li + (C_6H_5)_3SiH$$

$$\downarrow \qquad\qquad\qquad\qquad \downarrow$$

$$(C_6H_5)_3SiC(Li){=}C(CH_3)Si(C_6H_5)_3 \qquad (C_6H_5)_3SiC{\equiv}CCH_2Si(C_6H_5)_3 \quad (1\text{-}55)$$

Metallation occurs in the reactions of triphenylsilyllithium with triphenylsilylpropadiene, 1,3-bis(triphenylsilyl)propyne, 1-triphenylsilylbutyne, and triphenylallylsilane [731]. At the same time, addition is observed in the case of 1-triphenylsilylpropene [732].

Triphenylsilyllithium adds to anthracene in the 9,10-position with the formation of 9-lithio-10-triphenylsilyl-9,10-dihydroanthracene [766]:

$$C_{14}H_{10} + (C_6H_5)_3SiLi \longrightarrow \text{9-(H)(}Si(C_6H_5)_3\text{)-10-(H)(Li)-9,10-dihydroanthracene} \quad (1\text{-}56)$$

The results of experiments on the reaction of triphenylsilyllithium and triphenylsilylpotassium with unsaturated hydrocarbons are summarized in Table 1.

TABLE 1. Products of the Reaction of $(C_6H_5)_3SiM$ with Unsaturated Hydrocarbons*

M	Hydrocarbon	Solvent	Addition product	Yield, %	Yield of other reaction products, %					Literature
					$(C_6H_5)_3SiOH$	$[(C_6H_5)_3Si]_2O$	$(C_6H_5)_3SiH$	$(C_6H_5)_4Si$	$[(C_6H_5)_3Si]_2$	
K	$CH_2{=}CH_2$	Ether	Not formed	—	54	—	—	—	—	92
K	[structure]	Ether	» »	—	66—87	—	—	—	—	92
K	[structure]—CH_3	Ether	» »	—	72	—	—	—	—	92
K	$CH_3(CH_2)_5CH{=}CH_2$	Ether	» »	—	63	—	—	—	—	92
K	$CH_3(CH_2)_5CH{=}CH_2$	Ether + DME	» »	—	25	22	—	24	—	1288
K	$CH_3(CH_2)_9CH{=}CH_2$	Ether	» »	—	78	—	—	—	—	92
K	$CH_3(CH_2)_9CH{=}CH_2$	DME	» »	—	21	20	—	36	—	1288
K	$CH_3(CH_2)_{13}CH{=}CH_2$	Ether	» »	—	86	—	—	—	—	92
K	$CH_3(CH_2)_{13}CH{=}CH_2$	DME	» »	—	—	—	—	30	—	1288
K	$CH_3(CH_2)_{15}CH{=}CH_2$	Ether	» »	—	89	—	—	—	—	92
K	$CH_3(CH_2)_{15}CH{=}CH_2$	DME	» »	—	—	—	—	32	—	1288
Li	$C_6H_5CH{=}CH_2$	THF	» »	—	—	—	35	—	—	92
Li	$C_6H_5CH{=}CH_2$	THF	Polymer	—	—	—	—	—	—	1211
Li	$(C_6H_5)_2C{=}CH_2$	THF	$(C_6H_5)_2CHCH_2Si(C_6H_5)_3$	80	—	—	—	—	—	1288
K	$(C_6H_5)_2C{=}CH_2$	Ether	$(C_6H_5)_2CHCH_2Si(C_6H_5)_3$	42	—	—	—	—	—	1288
Li	$C_6H_5CH{=}CHC_6H_5$	DME (0.1 hr)	$(C_6H_5)_3Si(CHC_6H_5)_2Si(C_6H_5)_3$	2	—	—	—	—	—	588
			$(C_6H_5)_3Si(CHC_6H_5)_3CH_2C_6H_5$	16						
Li	$C_6H_5CH{=}CHC_6H_5$	DME (24 hr)	$(C_6H_5)_3Si(CHC_6H_5)_2Si(C_6H_5)_3$	26	—	—	—	—	—	588
			$(C_6H_5)_3Si(CHC_6H_5)_3CH_2C_6H_5$	24						
K	$C_6H_5CH{=}CHC_6H_5$	Ether	$(C_6H_5)_3Si(CHC_6H_5)CH_2C_6H_5$	55	—	—	—	—	—	790
Li	$(C_6H_5)_2C{=}CHC_6H_5$	THF	$(C_6H_5)_3Si(CHC_6H_5)_2C_6H_5$	62	—	—	—	—	—	1288
Li	$(C_6H_5)_2C{=}C(C_6H_5)_2$	THF	Not formed	—	—	—	56	—	—	1288

K	$(C_6H_5)_2C=C(C_6H_5)_2$	Ether	Not formed	—	74	—	—	—	—	1288
Li	$C_6H_5CH=C(CH_3)C_6H_5$	THF	» »	—	—	—	—	—	45	92
Li	$C_6H_5(CH=CH)_2C_6H_5$	Ether	» »	—	42	—	—	—	12	92
K	$C_6H_5(CH=CH)_2C_6H_5$	Ether	» »	—	—	—	—	—	—	1288
K	9,9'-Bifluorenylidene	Ether	Resin	—	—	—	—	—	—	1288
Li	$C_6H_5C\equiv CC_6H_5$	DME	$(C_6H_5)_3Si(CHC_6H_5)_2Si(C_6H_5)_3$	8	—	—	—	—	—	588
K	$C_6H_5C\equiv CC_6H_5$	Ether	Not formed	—	—	—	—	15	—	792
Li	 [structure]	THF	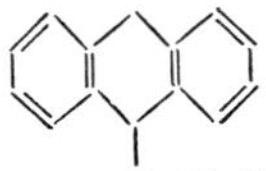 [structure] $Si(C_6H_5)_3$	29.7	—	—	—	—	2.3	766
Li	$(CH_3)_3SiCH=CH_2$	THF	Not formed	—	—	—	—	28	—	92
Li	$(CH_3)_2Si(CH=CH_2)_2$	THF	Polymer	—	—	—	—	—	—	92
Li	$(C_6H_5)_3SiCH_2CH=CH_2$	THF	$(C_6H_5)_3SiCH_2CH=CHSi(C_6H_5)_3$	40	—	—	29	—	—	731
Li	$(C_6H_5)_3SiCH=CHCH_3$	THF	$(C_6H_5)_3SiCH_2CH(CH_3)Si(C_6H_5)_3$	38.4	—	—	—	—	—	732
Li	$[(C_6H_5)_3Si]_3(C=C=CH)$	THF	$[(C_6H_5)_3Si]_2C=C=C[Si(C_6H_5)_3]_2$	2.7 *	—	—	52.6	—	31.4	731
Li	$(C_6H_5)_3SiC\equiv CCH_3$	THF	$(C_6H_5)_3SiCH=C(CH_3)Si(C_6H_5)_3$	18.5			51.3			731
			$(C_6H_5)_3SiC\equiv CCH_2Si(C_6H_5)_3$	29.9						
Li	$(C_6H_5)_3SiC\equiv CCH_3$	THF	$(C_6H_5)_3SiCH=C(CH_3)Si(C_6H_5)_3$	18.0 *	—	—	25.2	—	—	731
			$(C_6H_5)_3SiC\equiv CCH_2Si(C_6H_5)_3$	9.5 *						
			$[(C_6H_5)_3Si]_2C=C=CHSi(C_6H_5)_3$	32.1 *						
Li	$(C_6H_5)_3SiC\equiv CCH_3$	THF + ether	$(C_6H_5)_3SiCH=C(CH_3)Si(C_6H_5)_3$	14.6 *	—	—	54.5	—	15.8	731
			$(C_6H_5)_3SiC\equiv CCH_2Si(C_6H_5)_3$	5.4 *						
			$[(C_6H_5)_3Si]_2C=C=CHSi(C_6H_5)_3$	23.1 *						
Li	$(C_6H_5)_3SiC\equiv CCH_2CH_3$	THF	$(C_6H_5)_3SiC\equiv CCH(CH_3)Si(C_6H_5)_3$	1.8	—	—	—	4.4	—	731
Li	$(C_6H_5)_3SiC\equiv CSi(C_6H_5)_3$	THF	$(C_6H_5)_3SiC\equiv CH$	87	—	—	—	—	93	728
Li	$(C_6H_5)_3SiC\equiv CCH_2Si(C_6H_5)_3$	THF	$[C_6H_5)_3Si]_2C=C=CHSi(C_6H_5)_3$	50.8 *	—	—	88.5	—	2.9	731
			$[(C_6H_5)_3Si]_2C=C=C[Si(C_6H_5)_3]_2$							

* The reaction products obtained after hydrolysis of the reaction mixture are given in the table; the yields of compounds formed after treatment of the reaction mixture with triphenylchlorosilane are marked with an asterisk. In the "Solvent" column ether denotes diethyl ether, THF tetrahydrofuran, and DME 1,2-dimethoxyethane.

1.3.3. Reactions with Organic Halogen Derivatives

The reaction of R_3SiM with organic halogen derivatives proceeds in two directions:

$$R_3SiM + R'X \longrightarrow R_3SiR' + MX \quad (1\text{-}57)$$

$$2R_3SiM + R'X \longrightarrow R_3SiSiR_3 + R'M + MX \quad (1\text{-}58)$$

The latter reaction includes the stage of metal–halogen exchange with the intermediate formation of a triarylhalosilane, which then gives a hexaaryldisilane with R_3SiM. These reactions may proceed through intermediate 4-membered activated complexes [590]. Since silicon is less electronegative than carbon or halogens, there is the possibility of the formation of cyclic complexes of two forms. The first of these leads to the formation of tetraarylsilanes (1-59), while the second leads to metal–halogen exchange (1-60)

$$\begin{array}{ccc} \diagdown\!\!-\text{Si}+\text{M} & & \\ \vdots \quad \vdots & \longrightarrow \; \diagdown\!\!-\text{Si}-\text{C}\!\!-\!\!\diagup \; + \; \text{M}-\text{X} & \\ \diagdown\!\!-\text{C}+\text{X} & & \end{array} \quad (1\text{-}59)$$

$$\begin{array}{ccc} \diagdown\!\!-\text{Si}+\text{M} & & \\ \vdots \quad \vdots & \longrightarrow \; \diagdown\!\!-\text{Si}-\text{X} + \text{M}-\text{C}\!\!-\!\!\diagup & \\ \text{X}+\text{C}\!\!-\!\!\diagup & & \end{array} \quad (1\text{-}60)$$

The direction of the reaction between R_3SiM and an organic halogen derivative depends on both the structure of the hydrocarbon radical in the latter and the nature of the halogen atom. In the case of alkyl and aralkyl halides the capacity of reaction (1-59) increases and that for reaction (1-60) decreases in the series

$$CH_3 < C_2H_5 < n\text{-}C_4H_9 < n\text{-}C_{12}H_{25} < C_6H_5CH_2CH_2 < iso\text{-}C_4H_9$$
$$< tert\text{-}C_4H_9 < C_6H_5CH_2 < (C_6H_5)_3C$$

The reaction according to scheme (1-59) predominates in the case of alkylchlorides; with a change to alkyl bromides and iodides, reaction (1-60) proceeds to an increasing extent.

Reaction (1-60) occurs mainly in the interaction of tert-butyl bromide with triphenylsilyllithium, while in the case of tert-butyl chloride, dehydrochlorination occurs [92]:

$$(C_6H_5)_3SiLi + (CH_3)_3CCl \longrightarrow (C_6H_5)_3SiH + (CH_3)_2C{=}CH_2 + LiCl \quad (1\text{-}61)$$

The ratio of the reaction products forming in accordance with schemes (1-57) and (1-58) also changes, depending on the order of addition of the reagents. In many cases the formation of a hexaaryldisilane in accordance with the scheme (1-58) predominates when the alkyl halides are added to R_3SiM, since in this case there is excess R_3SiM in the reaction mixture. However, when methyl iodide or ethyl bromide or iodide is used, the opposite phenomenon is observed.

In the reaction of triphenylsilyllithium with epichlorohydrin there is competition between cleavage of the epoxide ring and of the C–Cl bond [734]. If the molar ratio of triphenylsilyllithium to epichlorohydrin is 2:1, then both reactions occur and 1,3-bis(triphenylsilyl)propanol-2 is formed after hydrolysis:

$$2\,(C_6H_5)_3SiLi + \underset{\diagdown O \diagup}{CH_2\text{—}CHCH_2Cl} \longrightarrow (C_6H_5)_3SiCH_2\underset{\substack{|\\ OLi}}{CH}CH_2Si(C_6H_5)_3 + LiCl \quad (1\text{-}62)$$

The formation of the products from the reaction of geminal alkyl polyhalides with triphenylsilyllithium may be explained by the occurrence of reaction (1-58) with the subsequent liberation of carbenes, which then react with excess triphenylsilyllithium. Thus, in the reaction of triphenylsilyllithium with methylene chloride, methyltriphenylsilane and chloromethyltriphenylsilane [727, 730] are also formed in addition to hexaphenyldisilane [590]. The chloromethyltriphenylsilane is not detected in the reaction products if the methylene chloride is added to triphenylsilyllithium or if excess triphenylsilyllithium is added to methylene chloride. Instead of chloromethyltriphenylsilane, the reaction products contain bis(triphenylsilyl) methane, which is formed by the reaction of excess $(C_6H_5)_3SiLi$ with the chloromethyltriphenylsilane obtained initially.

With the reverse order of mixing of the reagents a small amount of hexaphenyldisilane is formed. This may be explained by the fact that in this case there are two competing reactions of $(C_6H_5)_3SiLi$ with triphenylchlorosilane and with carbene [730]:

$$\begin{array}{ccccc} (C_6H_5)_3SiLi + CH_2Cl_2 & \longrightarrow & (C_6H_5)_3SiCl + [CH_2ClLi] & \longrightarrow & LiCl + :CH_2 \\ & & \downarrow (C_6H_5)_3SiLi & & \downarrow (C_6H_5)_3SiLi \\ & & (C_6H_5)_3Si\text{—}Si(C_6H_5)_3 & & (C_6H_5)_3SiCH_2Li \end{array} \quad (1\text{-}63)$$

The triphenylsilylmethyllithium formed may react in its turn with triphenylchlorosilane and methylene chloride:

$$(C_6H_5)_3SiCH_2Li \xrightarrow{CH_2Cl_2} (C_6H_5)_3SiCH_2Cl \xrightarrow{(C_6H_5)_3SiLi} (C_6H_5)_3SiCH_2Si(C_6H_5)_3$$
$$(C_6H_5)_3SiCH_2Li \xrightarrow{(C_6H_5)_3SiCl} (C_6H_5)_3SiCH_2Si(C_6H_5)_3$$
$$(C_6H_5)_3SiCH_2Li \xrightarrow{H_2O} (C_6H_5)_3SiCH_3 \quad (1\text{-}64)$$

In the reaction of (dichloromethyl)triphenylsilane with triphenylsilyllithium, hexaphenyldisilane (46.4%) and bis(triphenylsilyl)-methane are formed. The intermediate product of this reaction may be triphenylsilylcarbene:

$$(C_6H_5)_3SiCHCl_2 \xrightarrow{(C_6H_5)_3SiLi} (C_6H_5)_3SiCHClLi + (C_6H_5)_3SiCl$$
$$(C_6H_5)_3SiCHClLi \xrightarrow{-LiCl} [(C_6H_5)_3Si\ddot{C}H] \xrightarrow{(C_6H_5)_3SiLi} (C_6H_5)_3SiCH_2Si(C_6H_5)_3$$
$$(C_6H_5)_3SiCl \xrightarrow{+(C_6H_5)_3SiLi} (C_6H_5)_3Si{-}Si(C_6H_5)_3 \quad (1\text{-}65)$$

The reaction of methylene bromide with triphenylsilyllithium leads to hexaphenyldisilane (42.5–59.4%), methyltriphenylsilane (up to 40%), and bromomethyltriphenylsilane (up to 9.6%). However, in contrast to the reaction with methylene chloride, bis(triphenyl-silyl)methane is not formed in this case. Methylene iodide reacts with triphenylsilyllithium analogously to methylene chloride.

Triphenylsilyllithium reacts with haloforms (trihalomethanes) in the same direction as with dihalomethanes. However, in this case triphenylsilane and (dihalomethyl)triphenylsilane are obtained in addition to methyltriphenylsilane, hexaphenyldisilane, halomethyl-triphenylsilane, and bis(triphenylsilyl)methane [727, 730]. It may be assumed, for example, that with chloroform these compounds are obtained as a result of the intermediate formation of dichloro-carbene and chlorocarbene:

$$(C_6H_5)_3SiLi + CHCl_3 \rightarrow (C_6H_5)_3SiH + [CCl_3Li] \longrightarrow :CCl_2$$
$$(C_6H_5)_3SiLi + CHCl_3 \rightarrow (C_6H_5)_3SiCl + [CHCl_2Li] \longrightarrow :CHCl \quad (1\text{-}66)$$

The subsequent reaction of dichlorocarbene and chlorocarbene formed in accordance with scheme (1-66) with triphenylsilyl-lithium leads to triphenylsilylchlorocarbene and triphenylsilylcarbene. The products from the reaction of the latter with chloroform are chloromethyltriphenylsilane and dichloromethyltriphenylsilane:

$$[(C_6H_5)_3\,Si\ddot{C}Cl] + CHCl_3 \longrightarrow (C_6H_5)_3\,SiCHCl_2 + [:CCl_2] \qquad (1\text{-}67)$$

$$[(C_6H_5)_3\,Si\ddot{C}H] + CHCl_3 \longrightarrow (C_6H_5)_3\,SiCH_2Cl + [:CCl_2] \qquad (1\text{-}68)$$

Of all the compounds in these schemes, methyltriphenylsilane (in the reaction with chloroform) and bis(triphenylsilyl)methane (in the reaction with bromoform) were not detected in the reaction products. When triphenylsilyllithium was added to iodoform (a molar ratio of 3:1), hexaphenyldisilane (60.5%) bis(triphenylsilyl)methane (19.5%), and methyltriphenylsilane (23.7%) were obtained. When this ratio was reduced to 2:1, the last two compounds were not formed. Instead of them, the reaction products contained iodomethyltriphenylsilane (15.6%) and diiodomethyltriphenylsilane (14.5%). With the reverse order of mixing of the reagents it was also possible to isolate a small amount of 1,2-bis(triphenylsilyl)ethane, whose appearance was to be expected in all reactions proceeding through the intermediate formation of triphenylsilylmethyllithium and a halomethyltriphenylsilane [730].

Carbon tetrachloride reacts with triphenylsilyllithium at -60°C to form hexaphenyldisilane (56.6 – 73.5%) and dichloromethyltriphenylsilane (11.2 – 42.2%). The reaction is still more complex with carbon tetrabromide. In this case, a small amount of (triphenylsilyl)bromoacetylene and bis(triphenylsilyl)acetylene is obtained in addition to hexaphenyldisilane and mono- and dibromomethyltriphenylsilanes [730].

Alkylene halides such as 1,2-dichloropropane react with triphenylsilyllithium in accordance with scheme (1-58) with the formation of hexaphenyldisilane. On the other hand, 1,3-dichloropropane reacts with triphenylsilyllithium in accordance with scheme (1-57) and is converted into 1,3-bis(triphenylsilyl)propane [726, 729]. The product from the reaction of 1,3-dichloropropane with 1,5-dilithiodecaphenylpentasilane was found to be the 8-membered cyclic compound 1,1,2,2,3,3,4,4,5,5-decaphenyl-1,2,3,4,5-pentasilacyclooctane [755]. In contrast to the dichloro derivative, 1,3-dibromopropane reacts with $(C_6H_5)_3SiLi$ also in accordance with scheme (1-58) [729]. By reaction with triphenylsilyllithium, 1,2-dichlorobutane is converted into 1,4-bis(triphenylsilyl)butane (62-64% yield) [670, 858]. In the case of 1,4-dibromobutane the main reaction product is hexaphenyldisilane, while the yield of 1,4-bis-(triphenylsilyl) butane does not exceed 3% [1278].

Depending on their structure, the reaction of triphenylsilyllithium with haloalkenes also proceeds in accordance with scheme (1-57) or (1-58). For example, triphenylallylsilane is formed in 56% yield by the reaction of triphenylsilyllithium with allyl chloride [729]. The main product from the reaction of triphenylsilyllithium with 1-bromopropene is hexaphenyldisilane (70%) [731]. However, these reactions of triphenylsilyllithium with haloalkenes are accompanied by many side processes. Thus, in the reaction of 1-chloropropene with triphenylsilyllithium, 1,2-bis(triphenylsilyl)-propane is obtained in addition to hexaphenyldisilane. It would be natural to assume that this compound was formed as a result of the addition of a second molecule of triphenylsilyllithium to 1-(triphenylsilyl)propene, which is the reaction product of triphenylsilyllithium with 1-chloropropene in accordance with scheme (1-57). However, a study of the addition of triphenylsilyllithium to 1-(triphenylsilyl)propene showed that this reaction proceeds much more slowly and even under drastic conditions the yield of 1,2-bis(triphenylsilyl)propane was lower than in the case of the reaction with 1-chloropropene. Therefore, scheme (1-69) was proposed to explain the increased yield of 1,2-bis(triphenylsilyl)propane [732]:

$$(C_6H_5)_3SiLi + CH_3CH{=}CHCl \longrightarrow \underset{(C_6H_5)_3Si\quad\ Li}{CH_3CH{-}CHCl} \xrightarrow{-LiCl}$$

$$\longrightarrow [\underset{CH_3}{(C_6H_5)_3SiCH}{-}\ddot{C}H] \xrightarrow{(C_6H_5)_3SiLi} \underset{CH_3\quad Li}{(C_6H_5)_3SiCH{-}CHSi(C_6H_5)_3} \qquad (1\text{-}69)$$

The reaction of triphenylsilyllithium with 1-bromopropene is much more complex. In this case, in addition to hexaphenyldisilane, the reaction gives a whole series of other products, namely, triphenylsilane, 1,2- bis (triphenylsilyl) propane, 1,3-bis (triphenylsilyl) propene, 1, 3-bis (triphenylsilyl) propyne, 1- (triphenylsilyl) propene, 1- (triphenylsilyl) propyne, and tris (triphenylsilyl) propadiene. The formation of all these substances is explained by the fact that in the reaction of triphenylsilyllithium with compounds which contain several functional groups capable of reacting with it there are many processes which occur in parallel, namely, substitution reactions, addition at a multiple bond, metallation of C–H and C–Br bonds, etc. It is not only the starting reagents which participate in these processes, but also the products of their primary conversions. As intermediate products it is possible to obtain carbenes of various structures, which react rapidly with organosilicon compounds present in the reaction mixture [731].

When excess triphenylsilyllithium is added to 1,1-dichloropropene with subsequent treatment with triphenylchlorosilane there are formed 1-(triphenylsilyl)propyne, 1,3-bis(triphenylsilyl)propyne, tris(triphenylsilyl)propadiene, triphenylsilane, and hexaphenyldisilane. The use of excess 1,1-dichloropropene makes it also possible to isolate 2-chloro-3-(triphenylsilyl)propene [733].

Analogous reaction products are obtained from the interaction of triphenylsilyllithium with 2,3-dichloropropene-1:

$$(C_6H_5)_3SiLi + ClCH_2{-}C(Cl){=}CH_2 \xrightarrow{-LiCl} (C_6H_5)_3SiCH_2{-}C(Cl){=}CH_2 \xrightarrow{+(C_6H_5)_3SiLi}$$

$$\longrightarrow (C_6H_5)_3SiCH(Li){-}C(Cl){=}CH_2 + (C_6H_5)_3SiH \quad (1\text{-}70)$$

$$\downarrow -LiCl$$

$$(C_6H_5)_3SiCH{=}C{=}CH_2 \longrightarrow (C_6H_5)_3SiC{\equiv}C{-}CH_3$$

In the reaction of triphenylsilyllithium with 1,3-dichloropropene, the halogen atom in position 3 reacts first to form 1-chloro-3-(triphenylsilyl)propene, which may contain some 3-chloro-1-(triphenylsilyl)propene. Replacement of the second chlorine atom requires the use of excess triphenylsilyllithium:

$$(C_6H_5)_3SiLi + ClCH_2CH{=}CHCl \xrightarrow{-LiCl} (C_6H_5)_3SiCH_2CH{=}CHCl \quad (1\text{-}71)$$

$$\downarrow (C_6H_5)_3SiLi$$

$$(C_6H_5)_3SiCH_2CH{=}CHSi(C_6H_5)_3$$

The reaction of triphenylsilyllithium with trichloroethylene is more complex [728].

$$(C_6H_5)_3SiLi + Cl_2C{=}CHCl \longrightarrow Cl{-}C(Li){=}CHCl + (C_6H_5)_3SiCl$$

$$\downarrow -LiCl$$

$$ClC{\equiv}CLi + (C_6H_5)_3SiH \xleftarrow{(C_6H_5)_3SiLi} CH{\equiv}CCl$$

$$\downarrow +(C_6H_5)_3SiCl$$

$$(C_6H_5)_3SiC{\equiv}CCl \xrightarrow{(C_6H_5)_3SiLi} (C_6H_5)_3SiC{\equiv}CLi + (C_6H_5)_3SiCl \quad (1\text{-}72)$$

$$(C_6H_5)_3SiC{\equiv}CCl \xrightarrow{(C_6H_5)_3SiLi} (C_6H_5)_3SiC{\equiv}CSi(C_6H_5)_3; \quad (C_6H_5)_3SiC{\equiv}CLi \xrightarrow{(C_6H_5)_3SiCl} (C_6H_5)_3SiC{\equiv}CSi(C_6H_5)_3$$

$$(C_6H_5)_3SiC{\equiv}CLi \xrightarrow{H_2O} (C_6H_5)_3SiC{\equiv}CH; \quad (C_6H_5)_3SiCl \xrightarrow{(C_6H_5)_3SiLi} (C_6H_5)_3SiSi(C_6H_5)_3$$

The action of triphenylsilyllithium on triphenylsilylchloroacetylene actually forms hexaphenyldislane, triphenylsilylacetylene, and bis(triphenylsilyl)acetylene, confirming scheme (1-72).

Perfluorovinyltriethylsilane reacts with triphenylsilyllithium in accordance with scheme (1-57) giving a 67% yield of 1,2-difluoro-1-(triethylsilyl)-2-(triphenylsilyl)ethylene [1157].

The reaction of R_3SiM with aryl halides may also proceed in accordance with a scheme of type (1-57) or (1-58), for example,

$$(C_6H_5)_3SiK + BrC_6H_5 \longrightarrow (C_6H_5)_4Si + KBr \tag{1-73}$$

$$(C_6H_5)_3SiK + BrC_6H_5 \longrightarrow (C_6H_5)_3SiBr + C_6H_5K \tag{1-74}$$

$$(C_6H_5)_3SiBr \xrightarrow{(C_6H_5)_3SiK} (C_6H_5)_3SiSi(C_6H_5)_3 \qquad C_6H_5K \xrightarrow{C_6H_5Br} C_6H_5C_6H_5$$

The formation of phenylpotassium by reaction (1-74) may be demonstrated by the isolation of biphenyl from the reaction mixture and also by the formation of triphenylcarbinol when the process is carried out in the presence of benzophenone [590]. Tetraphenylsilane is formed predominantly in accordance with scheme (1-73) and not due to the reaction of triphenylbromosilane with phenylpotassium obtained by scheme (1-74). This is confirmed by the fact that the yield of tetraphenylsilane is not increased on carrying out the reaction of triphenylsilylpotassium with a mixture of bromobenzene and triphenylchlorosilane.

The amount and nature of the products from the reaction of triphenylsilylpotassium with bromobenzene depends on the order of mixing of the reagents. When bromobenzene is added to a suspension of triphenylsilylpotassium in diethyl ether there are formed 55% of tetraphenylsilane, 22% of hexaphenyldisilane, 5% of triphenylsilanol, 4% of hexaphenyldisiloxane, and 9% of benzene. When the reagents are mixed in reverse order the yield of tetraphenylsilane is 61% and that of hexaphenyldisiloxane 34%. Hexaphenyldisilane is not detected at all in this case, while 8% of biphenyl appears instead of benzene. With the simultaneous addition of triphenylsilylpotassium and bromobenzene to a large amount of diethyl ether the yield of hexaphenyldisilane is low (27%).

In the reaction with aryl halides, triphenylsilylpotassium gives a higher yield of the tetrasubstituted silane than triphenylsilyllithium. Thus, for example, the yields of tetraphenylsilane obtained in the reaction of triphenylsilylpotassium with chlorobenzene, bromobenzene, and iodobenzene are 53, 55, and 63%, respectively, while the yields of tetraphenylsilane when triphenylsilyllithium is used are only 12.2; 18, and 19.5%, respectively. In the case of triphenylsilyllithium a larger amount of hexaphenyldisilane (about 70%) is formed than when triphenylsilylpotassium is used (up to 20%). Fluorobenzene is an exception, which forms 51% of tetraphenylsilane with triphenylsilyllithium and only 12% with triphenylsilylpotassium [796]. The reactions of triphenylsilyllithium and triphenylsilylpotassium with o-bromotoluene proceed in different directions. While 52% of hexaphenyldisilane and only 1.7% of o-tolyltriphenylsilane are formed in the first case, 57% of o-tolyltriphenylsilane containing no trace of hexaphenyldisilane is obtained in the second [796].

The reaction of triphenylsilylpotassium with halogen-substituted heterocyclic compounds makes it possible to synthesize triphenylsilyl derivatives of heterocycles (in higher yield than when triphenylsilyllithium is used [952]).

The investigation of the reactions of aryl halides and certain compounds with different functional groups makes it possible to draw up the following series in which the reactivity toward triphenylsilyllithium falls [784]:

$$(CH_3)_3PO > C_6H_5\underset{\diagdown O \diagup}{CH-CH_2} > C_6H_5COC_6H_5 > C_6H_5CN >$$

$$> C_6H_5Cl > C_8H_{17}F > C_6H_5OCH_3$$

Chlorobenzene is one of the least reactive reagents in this series.

In only rare cases is it possible to introduce more than one triphenylsilyl group by the reaction of triphenylsilyllithium with polyhalides of the aromatic and heterocyclic series. Thus, for example, the action of triphenylsilyllithium on the sulfone of 3,7-dichloro- or 3,7-dibromo-10-ethylphenothiazine yields the sulfone of 3,7-bis(triphenylsilyl)-10-ethylphenothiazine (yields of 17.5 and

19%, respectively) [741]:

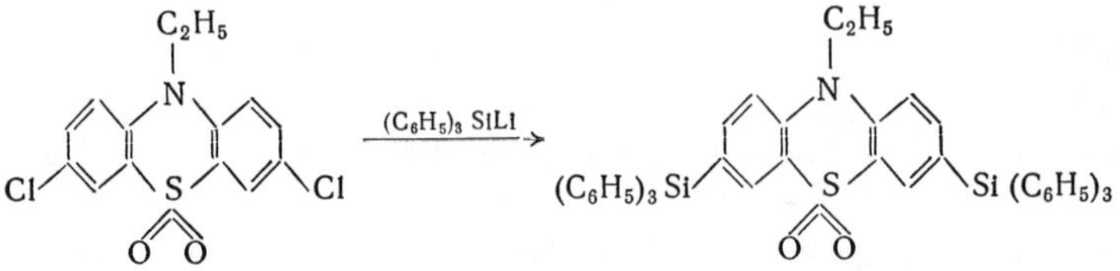

$$\text{Cl–(3,7-dichloro-10-ethylphenothiazine 5,5-dioxide; } C_2H_5\text{ on N, }SO_2\text{)–Cl} \xrightarrow{(C_6H_5)_3SiLi} (C_6H_5)_3Si\text{–(10-ethylphenothiazine 5,5-dioxide)–}Si(C_6H_5)_3 \quad (1\text{-}75)$$

The main product of the reaction of R_3SiM with di- and trihalobenzenes is hexaphenyldisilane [92].

The products of the reaction of triphenylsilyllithium with aromatic dihalides containing different halogen atoms in the ortho positions indicates the intermediate formation of dehydrobenzene:

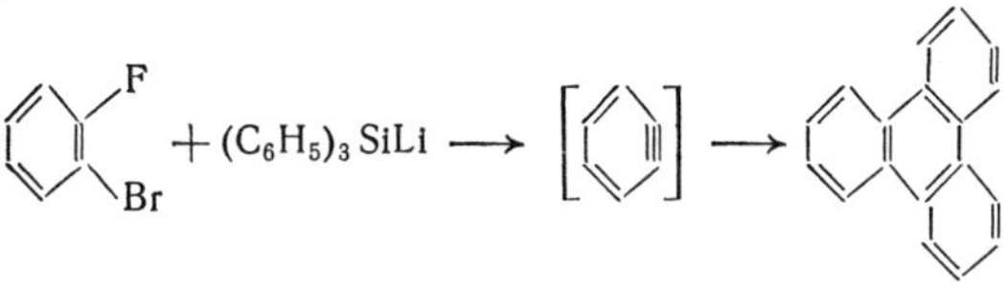

$$o\text{-}FC_6H_4Br + (C_6H_5)_3SiLi \longrightarrow [C_6H_4] \longrightarrow \text{triphenylene} \quad (1\text{-}76)$$

Triphenyl(o-fluorophenyl)silane is also formed in the case of o-chlorofluorobenzene and o-bromofluorobenzene. This indicates the higher reactivity of C – Cl and C – Br bonds in these compounds (in the reaction with triphenylsilyllithium) in comparison with the C – F bond.

The products of the reaction of organosilyl derivatives of alkali metals with organic halides and their yields are given in Tables 2-5.

1.3.4. Reaction with Alcohols, Ethers, and Organic Oxides

There has been little investigation of the reaction of R_3SiM with alcohols. In many cases the product is the corresponding tetraarylsilane. Thus, for example, tetraphenylsilane is formed by the reaction of triphenylsilylpotassium with benzyl alcohol, benzhydrol, triphenylcarbinol, and ditolylcarbinol. This might have been explained by the formation of triphenylsilane in accordance with a scheme of type (1-77)

$$(C_6H_5)_3SiK + C_6H_5CH_2OH \longrightarrow (C_6H_5)_3SiH + C_6H_5CH_2OK \quad (1\text{-}77)$$

which then reacts with triphenylsilylpotassium and is converted into tetraphenylsilane by a scheme of type (1-13). However, the yield of tetraphenylsilane obtained by the action of alcohols on triphenylsilylpotassium is considerably greater than its yield in the direct reaction of triphenylsilylpotassium with triarylsilanes.

Triphenylsilyllithium cleaves ethers:

$$(C_6H_5)_3SiLi + ROR \longrightarrow (C_6H_5)_3SiR + ROLi \qquad (1\text{-}78)$$

Thus, with 1,2-dimethoxyethane methyltriphenylsilane is formed in high yield (84.5%) in only 6 hr at room temperature [1274]. This reaction interferes with the use of 1,2-dimethoxyethane as a solvent in the synthesis of R_3SiM. Triphenylsilylpotassium, which is usually prepared in diethyl ether, though it is not soluble in it, produces no obvious cleavage of this solvent.

Triphenylsilyllithium reacts with methylal on heating to 50°C; the following series of products is formed: methyltriphenylsilane (10.6%); triphenylsilylmethanol (9%); 4-triphenylsilylbutanol (11.5%); tetraphenylsilane (0.9%); and triphenylsilanol (2.9%). Triphenylsilyllithium reacts analogously with dimethylacetal and 2,2,-dimethoxypropane. Diethyl formal and acetal do not react with triphenylsilyllithium even on boiling for 48 hr.

The conversions occurring during the reaction of dimethyl acetal with triphenylsilyllithium may be explained by the scheme (1-79). First of all triphenylsilyllithium removes one methyl group; this yields methyltriphenylsilane, lithium methylate and the corresponding aldehyde. The latter then adds to a second molecule of $(C_6H_5)_3SiLi$ [787]:

$$(C_6H_5)_3SiLi + R\text{—}\underset{OCH_3}{\overset{OCH_3}{C}}\text{—}H \longrightarrow (C_6H_5)_3SiCH_3 + \left[R\text{—}\underset{OCH_3}{\overset{O^-}{C}}\text{—}H \right]^{\ominus} Li^{\oplus} \qquad (1\text{-}79)$$

$$\downarrow$$

$$(C_6H_5)_3Si\underset{R}{C}HOH \xleftarrow{(C_6H_5)_3SiLi} RCHO + LiOCH_3$$

Heating triphenylsilyllithium with anisole at 50°C for 24 hr and subsequent hydrolysis leads to methyltriphenylsilane and

TABLE 2. Products of the Reaction of $(C_6H_5)_3SiM$ with Alkyl, Aralkyl, and Alkenyl Halides and Their Yields†

M	Halide	Reaction products	Yield, %	Yield of $(C_6H_5)_3SiSi(C_6H_5)_3$, %	Literature
Li	CH_3I	$(C_6H_5)_3SiCH_3$	44	29.3	92
Li	CH_3I	»	42.2	22.7	92
Na	CH_3I	»	—	—	897
K	CH_3I	»	33	52	556
Li	C_2H_5Cl	$(C_6H_5)_3SiC_2H_5$	80	0	725
Li	C_2H_5Cl *	»	75.5	0	92
Li	C_2H_5Br	»	45.5	29.3	725
Li	C_2H_5Br *	»	54.5	25.3	92
Li	C_2H_5I	»	28.2	40	725
Li	C_2H_5I *	»	30	36.7	92
Li	$CH_3CH{=}CHCl$ (20°)	$(C_6H_5)_3SiCH_2CH(CH_3)Si(C_6H_5)_3$	26.8	13	92
Li	$CH_3CH{=}CHCl$ (—60°)	»	45.5	13.5	732
Li	$CH_3CH{=}CHCl$ * (0°)	»	36.4	14	92
Li	$CH_3CH{=}CHBr$	$(C_6H_5)_3SiCH{=}CHCH_3$	11.0	68.7	731
		$(C_6H_5)_3SiC{\equiv}CCH_3$	7.1		
		$(C_6H_5)_3SiCH{=}CHCH_2Si(C_6H_5)_3$	1.8		
Li	$CH_3CH{=}CHBr$ (3 : 4)	$(C_6H_5)_3SiCH{=}CHCH_3$	6.7	72.5	731
		$(C_6H_5)_3SiC{\equiv}CCH_3$	15.3		
		$(C_6H_5)_3SiCH_2CH(CH_3)Si(C_6H_5)_3$	0.3 **		
		$[(C_6H_5)_3Si]_2C{=}C{=}CHSi(C_6H_5)_3$	3.7 **		
		$(C_6H_5)_3SiH$	6.5 **		
Li	$CH_3CH{=}CHBr$ (1 : 3)	$(C_6H_5)_3SiCH{=}CHCH_2Si(C_6H_5)_3$	7.2	9.2	731
		$(C_6H_5)_3SiCH_2CH(CH_3)Si(C_6H_5)_3$	6.8		
		$(C_6H_5)_3SiH$	50.2		
Li	$CH_3CH{=}CHBr$ (5 : 8)	$(C_6H_5)_3SiCH{=}CHCH_3$	Formed	Formed	731
		$(C_6H_5)_3SiC{\equiv}CCH_3$	6.6		
		$(C_6H_5)_3SiCH{=}CHCH_2Si(C_6H_5)_3$	1.1 **		
		$(C_6H_5)_3SiCH_2CH(CH_3)Si(C_6H_5)_3$	7.7 **		
		$[(C_6H_5)_3Si]_2C{=}C{=}CHSi(C_6H_5)_3$	6.4 **		

		$(C_6H_5)_3SiH$	Formed **		
Li	$CH_3CH{=}CHBr$ (1 : 3)	$(C_6H_5)_3SiCH{=}CHCH_3$	2.2	Formed	731
		$(C_6H_5)_3SiC{\equiv}CCH_3$	4.3		
		$[(C_6H_5)_3Si]_2C{=}C{=}CHSi(C_6H_5)_3$	3.2 **		
		$(C_6H_5)_3SiH$	36.2 **		
Li	$CH_2{=}CHCH_2Cl$	$(C_6H_5)_3SiCH_2CH{=}CH_2$	56	0	729
Li	OCH_2CHCH_2Cl	$(C_6H_5)_3SiCH_2CH{=}CH_2$	8	0	734
Li	OCH_2CHCH_2Cl (—60°)	$(C_6H_5)_3SiCH_2CH(OH)CH_2Si(C_6H_5)_3$	43.3	0	92
Li	OCH_2CHCH_2Cl (—60°)	$(C_6H_5)_3SiCH_2CH(OH)CH_2Cl$	60.5	—	92
Li	OCH_2CHCH_2Cl	$(C_6H_5)_3SiCH_2CH(OH)CH_2Si(C_6H_5)_3$	22	0	92
Li	OCH_2CHCH_2Br	—	—	68.6	734
Li	OCH_2CHCH_2Br *	$(C_6H_5)_3SiCH_2CH{=}CH_2$	1	72.6	734
Li	$(CH_3)_2CHCl$	$(C_6H_5)_3SiCH(CH_3)_2$	71.2	6.7	92
Li	$(CH_3)_2CHBr$	$(C_6H_5)_3SiCH(CH_3)_2$	5.8	73.5	92
Li	$CH_3CH_2CH{=}CHCl$	$(C_6H_5)_3SiCH_2CH(C_2H_5)Si(C_6H_5)_3$	22.2	0	732
Li	$CH_3CH_2CH{=}CHCl$ (65°)	$(C_6H_5)_3SiCH_2CH(C_2H_5)Si(C_6H_5)_3$	26.0	0	732
Li	$CH_3CH_2CH{=}CHCl$	$(C_6H_5)_3SiCH_2CH(C_2H_5)Si(C_6H_5)_3$	29.6 **	15.6	732
		$(C_6H_5)_3SiC{\equiv}CCH_2CH_3$	7.4		
		$(C_6H_5)_3SiH$	13.0 **		
Li	$n\text{-}C_4H_9Cl$	$(C_6H_5)_3SiC_4H_9\text{-}n$	75	0	123, 729
K	$n\text{-}C_4H_9Cl$	$(C_6H_5)_3C_4H_9\text{-}n$	—	—	123

† In the column "Halide" the molar ratio of halide to $(C_6H_5)_3SiM$ is given in brackets when nonequivalent amounts of reagents were used. In cases where the reaction was not carried out at room temperature, the reaction temperature is given in brackets. The sign * indicates that the halide was added to the R_3SiM in the reaction; in all other cases the R_3SiM was added to the halide. The sign ** denotes the yields of compounds formed after treatment of the reaction mixture with triphenylchlorosilane (Tables 2 and 3) or when a mixture of tetrahydrofuran and ether was used as the solvent (Table 4).

TABLE 2. (Cont'd)

M,	Halide	Reaction products	Yield, %	Yield of $(C_6H_5)_3SiSi(C_6H_5)_3$, %	Literature
Li	n-C_4H_9Br	$(C_6H_5)_3SiC_4H_9$-n	10	60	123
Li	n-C_4H_9Br	$(C_6H_5)_3SiC_4H_9$-n	97	—	756
Li	iso-C_4H_9Cl *	$(C_6H_5)_3SiC_4H_9$-iso	62	3.25	92
Li	sec-C_4H_9Cl *	$(C_6H_5)_3SiC_4H_9$-sec	24.8	13.25	92
Li	tert-C_4H_9Cl *	$(C_6H_5)_3SiH$	56.7	3.5	92
Rb	tert-C_4H_9Cl *	$(C_6H_5)_3SiC_4H_9$-tert	9.5	0	92
		$(C_6H_5)_3SiH$	25.4		
Cs	tert-C_4H_9Cl	$(C_6H_5)_3SiOH$	20	0	92
Li	tert-C_4H_9Br *	$(C_6H_5)_3SiC_4H_9$-tert	1.8	52	92
Li	cyclo-C_5H_9Cl	$(C_6H_5)_3Si—C_5H_9$-cyclo	40	9.3	729
K	cyclo-$C_6H_{11}Br$	cyclo-$C_6H_{11}—C_6H_{11}$-cyclo	2	56	590
		$(C_6H_5)_3SiOSi(C_6H_5)_3$	24		
K	cyclo-$C_6H_{11}Br$	$(C_6H_5)_3SiH$	6	80	590
		$(C_6H_5)_3SiOSi(C_6H_5)_3$	10		
		cyclo-$C_6H_{11}—C_6H_{11}$-cyclo	37		
		Cyclohexene	9		
Li	$C_6H_5CH_2Cl$	$(C_6H_5)_3SiCH_2C_6H_5$	50	7	92
Li	$C_6H_5CH_2Cl$	$(C_6H_5)_3SiCH_2C_6H_5$	39	—	1204
K	$C_6H_5CH_2Cl$	$(C_6H_5)_3SiCH_2C_6H_5$	22	5	796
K	$C_6H_5CH_2Cl$ *	$(C_6H_5)_3SiCH_2C_6H_5$	60	26	590
		$C_6H_5CH_2CH_2C_6H_5$	25		
K	$C_6H_5CH_2Cl$	$(C_6H_5)_3SiCH_2C_6H_5$	40	7	590
		$(C_6H_5)_3SiOSi(C_6H_5)_3$	21		
		$C_6H_5CH_2CH_2C_6H_5$	15		

K	$C_6H_5CH_2CH_2Cl$	$(C_6H_5)_3SiCH_2CH_2C_6H_5$	53	—	92
K	$C_6H_5CH_2CH_2Br$	$(C_6H_5)_3SiCH_2CH_2C_6H_5$	9.3	50.3	92
		$(C_6H_5)_3SiOH$	3		
K	$(C_6H_5)_2CHCH_2Cl$	$(C_6H_5)_3SiCH_2CH(C_6H_5)_2$	47	—	1288
K	$(C_6H_5)_3CCl$	$(C_6H_5)_3SiOSi(C_6H_5)_3$	50	—	585
		$(C_6H_5)_3CC(C_6H_5)_3$	88		
K	$(C_6H_5)_3CCl$	$(C_6H_5)_3SiOSi(C_6H_5)_3$	53	16	585
		$(C_6H_5)_3CC(C_6H_5)_3$	62		
K	$(C_6H_5)_3CCl$ (—30°)	$(C_6H_5)_3SiOSi(C_6H_5)_3$	14	45	585
		$(C_6H_5)_3CC(C_6H_5)_3$	63		
K	n-$C_8H_{17}F$ *	$(C_6H_5)_3SiC_8H_{17}$-n	86.6	—	784
K	n-$C_8H_{17}Br$ *	$(C_6H_5)_3SiC_8H_{17}$-n	26.4	44.4	92
K	n-$C_{10}H_{21}Br$	$(C_6H_5)_3SiC_{10}H_{21}$-n	6.3	62	769
K	n-$C_{12}H_{25}Cl$ *	$(C_6H_5)_3SiC_{12}H_{25}$-n	33	9	769
K	n-$C_{12}H_{25}Cl$	$(C_6H_5)_3SiC_{12}H_{25}$-n	28.7	0	729
K	n-$C_{18}H_{37}Br$	$(C_6H_5)_3SiC_{18}H_{37}$-n	6	59	769
Li	10-(2-Chloroethyl) phenothiazine	10-(2-Triphenylsilylethyl) phenothiazine	32.9	58.6	741
Li	$(CH_3)_3SiCH_2Cl$	$(C_6H_5)_3SiCH_2Si(CH_3)_3$	82	—	730
Li	$(CH_3)_2Si(CH{=}CH_2)CH_2Cl$ *	$(C_6H_5)_3SiCH_2Si(CH{=}CH_2)(CH_3)_2$	63.5	—	730
Li	$(C_6H_5)_3SiCH_2Cl$	$(C_6H_5)_3SiCH_2Si(C_6H_5)_3$	32.4	36.8	730
Li	$(C_6H_5)_3SiCH_2CH_2Cl$ *	$(C_6H_5)_3SiCH_2CH_2Si(C_6H_5)_3$	29.5	74	92
Li	$(C_6H_5)_3SiCH_2CH_2Br$ *	$(C_6H_5)_3SiCH_2CH_2Si(C_6H_5)_3$	2.4	47.4	92
Li	$(C_6H_5)_3SiCH_2CH_2CH(CH_3)Cl$	$(C_6H_5)_3SiCH_2CH_2CH(CH_3)Si(C_6H_5)_3$	67	—	858
Li	$(CH_3)_3SiOCH(CH_3)CH_2CH_2Cl$	$(C_6H_5)_3SiCH_2CH_2CH(OH)CH_3$ after hydrolysis	83	—	858
Li	$(C_6H_5)_3SiC{\equiv}CCl$	$(C_6H_5)_3SiC{\equiv}CSi(C_6H_5)_3$	48.5	22.7	728
		$(C_6H_5)_3SiC{\equiv}CH$	15.8		

TABLE 3. Products of the Reaction of $(C_6H_5)_3SiLi$ with Organic Polyhalides of the Aliphatic Series†

Halide	Products of replacement of the halogen (obtained after hydrolysis of the reaction mixture)	Yield, %	Yield of other reaction products, %			Literature
			$[(C_6H_5)_3Si]_2$	$(C_6H_5)_3SiH$	$(C_6H_5)_3SiCH_3$	
CH_2Cl_2 (1 : 1; —60°)	$(C_6H_5)_3SiCH_2Cl$	3.6	56.4	—	26.0	730
	$(C_6H_5)_3SiCH_2Si(C_6H_5)_3$	5.2				
CH_2Cl_2 (1 : 2; —60°)	$(C_6H_5)_3SiCH_2Cl$	8.5	45.5	—	34.2	730
	$(C_6H_5)_3SiCH_2Si(C_6H_5)_3$	5.4				
CH_2Cl_2 (1 : 2; —60°)	$(C_6H_5)_3SiCH_2Cl$	11.7	40.0	—	15.5	730
	$(C_6H_5)_3SiCH_2Si(C_6H_5)_3$	6.1				
CH_2Cl_2 (2 : 5)	$(C_6H_5)_3SiCH_2Si(C_6H_5)_3$	17.7	43.5	—	21.9	730
CH_2Cl_2 (1 : 2)	$(C_6H_5)_3SiCH_2Si(C_6H_5)_3$	11.7	35.2	—	25.6	730
CH_2Cl_2 (reaction with triphenylsilylpotassium)	$(C_6H_5)_3SiOSi(C_6H_5)_3$	50	—	—	—	
CH_2Cl_2 *	$(C_6H_5)_3SiOSi(C_6H_5)_3$	4	62	—	—	590
CH_2Br_2 (1 : 2; —60°)	$(C_6H_5)_3SiCH_2Br$	6.9	59.4	—	32.4	730
CH_2Br_2 (1 : 2; —60°)	—	—	42.5	—	40.0	730
$CH_2Br_2^*$ (1 : 2,85; —60°)	—	—	57.5	—	Traces	730
CH_2I_2 (1 : 2; —60°)	$(C_6H_5)_3SiCH_2I$	28.3	57.0	—	—	730
	$(C_6H_5)_3SiCH_2Si(C_6H_5)_3$	5.6				
CH_2I_2 (1 : 2; —60°)	$(C_6H_5)_3SiCH_2I$	23.0	50.0	—	16.4	730
CH_2I_2 (1 : 2)		—	65.1			
$CHCl_3$ (1 : 1)	$(C_6H_5)_3SiCH_2Cl$	3.2	51.6	11.3	—	730
	$(C_6H_5)_3SiCH_2Si(C_6H_5)_3$	1.7				
$CHCl_3$ (1 : 2; —60°)	$(C_6H_5)_3SiCH_2Cl$	5.3	53.0	5.3	—	730
	$(C_6H_5)_3SiCHCl_2$	11.7				
	$(C_6H_5)_3SiCH_2Si(C_6H_5)_3$	2.8				
$CHCl_3^*$ (1 : 2; —60°)	$(C_6H_5)_3SiCH_2Cl$	12	60.5	13.0	—	730
	$(C_6H_5)_3SiCHCl_2$					
	$(C_6H_5)_3SiCH_2Si(C_6H_5)_3$	10.3				
$CHBr_3$ (1 : 2; —60°)	$(C_6H_5)_3SiCH_2Br$	20.5	60.5	—	3.3	730
	$(C_6H_5)_3SiCHBr_2$	4.0				
$CHBr_3$ (1 : 2; —60°)	$(C_6H_5)_3SiCH_2Br$	34.0	54.0	0.5	0.9	730
$CHBr_3$ (3 : 8; —60°)	$(C_6H_5)_3SiCH_2Br$	24.5	48.7	—	9.7	730
	$(C_6H_5)_3SiCHBr_2$	3.91				

†See footnote to Table 2.

TABLE 3. (Cont'd)

Halide	Products of replacement of the halogen (obtained after hydrolysis of the reaction mixture)	Yield, %	Yield of other reaction products, %			Literature
			$[(C_6H_5)_3Si]_2$	$(C_6H_5)_3SiH$	$(C_6H_5)_3SiCH_3$	
$CHBr_3^*$ (3 : 8; —60°)	$(C_6H_5)_3SiCH_2Br$	24.0	62.5	1.9	10.3	730
CHI_3 (1 : 3, 45)	$(C_6H_5)_3SiCH_2Si(C_6H_5)_3$	19.7	60.5	—	23.7	730
CHI_3 (1 : 2; —60°)	$(C_6H_5)_3SiCH_2I$	15.6	60.0	86.5	—	730
	$(C_6H_5)_3SiCHI_2$	14.5				
CHI_3^* (3 : 8)	$(C_6H_5)_3SiCH_2Si(C_6H_5)_3$	3.1	64.7	—	11.0	730
	$(C_6H_5)_3SiCH_2CH_2Si(C_6H_5)_3$	2.4				
CCl_4 (1 : 2,6; —60°)	$(C_6H_5)_3SiCHCl_2$	42.2	57.2	Traces	—	730
CCl_4 (1 : 2; —60°)	$(C_6H_5)_3SiCHCl_2$	13.0	56.6	—	—	730
CCl_4^* (1 : 3)	$(C_6H_5)_3SiCHCl_2$	11.2	73.6	—	—	730
CBr_4 (1 : 2; —60°)	$(C_6H_5)_3SiCHBr_2$	1.7	53.1	—	—	730
	$(C_6H_5)_3SiC{\equiv}CSi(C_6H_5)_3$	4.6				
CBr_4^* (1 : 1; —60°)	$(C_6H_5)_3SiC{\equiv}CBr$	2.0	73.9	—	—	730
	$(C_6H_5)_3SiC{\equiv}CSi(C_6H_5)_3$	4.3				
	$(C_6H_5)_3SiCH_2Br$	2.8				
$CH_3CH{=}CCl_2$ (1 : 2; —60°)	$[(C_6H_5)_3Si]_2C{=}C{=}CHSi(C_6H_5)_3$	Traces	70.0	—	Traces [1]	733
$CH_3CH{=}CCl_2^*$ (1 : 2; —60°)	$(C_6H_5)_3SiCH_2CCl{=}CH_2$	0.7	77.6	—	0.5	733
$CH_3CH{=}CCl_2^{**}$ (1 : 3; —60°)	—	—	Traces	23.1	4.0	733
$CH_3CH{=}CCl_2$ (2 : 1)	$(C_6H_5)_3SiCH_2CCl{=}CH_2$	4.2	68.7	3.5	5.7	733
$CH_3CH{=}CCl_2$ (1 : 2)	—	—	57.0	16.9	20.0	733
$CH_3CH{=}CCl_2^{**}$ (1 : 2)	—	—	70.0	18.8	18.5	733
$CH_3CH{=}CCl_2$ (1 : 2,5)	—	—	52.5	16.4	26.3	733
$CH_3CH{=}CCl_2^{**}$ (1 : 3)	$(C_6H_5)_3SiCH{=}C(CH_3)Si(C_6H_5)_3$	5.4	60.2	24.3	41.6	733
	$[(C_6H_5)_3Si]_2C{=}C{=}CHSi(C_6H_5)_3$	5.9				
$CH_2{=}CClCH_2Cl$ (1 : 1)	$(C_6H_5)_3SiCH_2CCl{=}CH_2$	19.5	28.3	10.9	7.8	733
CH_2CClCH_2Cl ** (1 : 2)	$[(C_6H_5)_3Si]_2C{=}C{=}CHSi(C_6H_5)_3$	—	34.7	47.7	20.8	733
$CH_2{=}CClCH_2Cl$ (3 : 8)	$(C_6H_5)_3SiCH{=}C(CH_3)Si(C_6H_5)_3$	4.4	3.3	46.4	9.5	733
	$(C_6H_5)_3SiCH_2C{\equiv}CSi(C_6H_5)_3$	4.5				
	$[(C_6H_5)_3Si]_2C{=}C{=}CHSi(C_6H_5)_3$	18.0				

[1] From here to the end of the table we give the yield of $(C_6H_5)_3SiC{\equiv}CCH_3$.

TABLE 3. (Cont'd)

Halide	Products of replacement of the halogen (obtained after hydrolysis of the reaction mixture)	Yield, %	Yield of other reaction products, %			Literature
			$[(C_6H_5)_3Si]_2$	$(C_6H_5)_3SiH$	$(C_6H_5)_3SiC{\equiv}CCH_3$	
$ClCH{=}CHCH_2Cl$ (1 : 1)	$(C_6H_5)_3SiCH_2CH{=}CHCl$ $(C_6H_5)_3SiCH{=}CHCH_2Cl$	75	—	—	—	733
$ClCH{=}CHCH_2Cl$ (1 : 2)	$(C_6H_5)_3SiCH_2CH{=}CHSi(C_6H_5)_3$	31	12.5	—	—	733
$ClCH{=}CHCH_2Cl$ (1 : 3)	»	50.5	6.7	77.0	—	733
$CH_3CHClCH_2Cl$ (1 : 2)	—	—	93	—	—	725
$CH_3CHClCH_2Cl$ (1 : 3)	—	—	87.5	—	—	92
$CH_2ClCH_2CH_2Cl$	$(C_6H_5)_3SiCH_2CH_2CH_2Si(C_6H_5)_3$	73	—	—	—	734
$CH_2BrCH_2CH_2Br$	—	—	71.5	—	—	734
$ClCH_2CH_2CH_2CH_2Cl$	$(C_6H_5)_3SiCH_2CH_2CH_2CH_2Si(C_6H_5)_3$	64	—	—	—	858
$ClCH_2CH_2CH_2CH_2Cl$	»	62	—	—	—	670
$BrCH_2CH_2CH_2CH_2Br$	»	3	74	—	—	1278
$CH_3CH(CH_2Cl)CH_2CH_2CH_2Cl$	$(C_6H_5)_3SiCH_2CH(CH_3)CH_2CH_2CH_2Si(C_6H_5)_3$	63	—	—	—	858
$(C_6H_5)_3SiCH_2CH{=}CHCl$	$(C_6H_5)_3SiCH_2CH{=}CHSi(C_6H_5)_3$	57.7	—	—	—	733
$(C_6H_5)_3SiCH_2CCl{=}CH_2$	—	—	—	64.1	62,9	733
$(C_6H_5)_3SiCHCl_2$	$(C_6H_5)_3SiCH_2Si(C_6H_5)_3$	34.3	46.4	—	—	730
$Cl_2C{=}CHCl$ (1 : 1)	$(C_6H_5)_3SiC{\equiv}CCl$	5	58.7	6.2	—	728
$Cl_2C{=}CHCl$ (1 : 2)	$(C_6H_5)_3SiC{\equiv}CCl$ $(C_6H_5)_3SiC{\equiv}CSi(C_6H_5)_3$	7.8 11	58.9	10	—	728
$(C_2H_5)_3SiCF{=}CF_2$	$(C_6H_5)_3SiCF{=}CFSi(C_2H_5)_3$	67	—	—	—	1157

TABLE 4. Products of the Reaction of R_3SiM with Aryl Halides†

M	Halide	Products of replacement of the halogen (obtained after hydrolysis of the reaction mixture	Yield, %	Yield of $[(C_6H_5)_3Si]_2$, %	Literature
Li	C_6H_5F	$(C_6H_5)_4Si$	50.9	—	92
Li	C_6H_5F *	»	47.7	—	92
K	C_6H_5F	»	12	3	796
Rb	C_6H_5F *	»	39	—	92
Cs	C_6H_5F	»	25	—	92
Li	C_6H_5Cl	»	11.8	76.5	784
K	C_6H_5Cl	»	53	10	92
K	C_6H_5Cl	$(C_6H_5)_4Si$ $C_6H_5C_6H_5$	52 6	7	590
Cs	C_6H_5Cl	$(C_6H_5)_4Si$	17.4	—	92

†See footnote to Table 2.

TABLE 4. (Cont'd)

M	Halide	Products of replacement of the halogen (obtained after hydrolysis of the reaction mixture	Yield, %	Yield of $[(C_6H_5)_3Si]_2$, %	Literature
Li	C_6H_5Br	$(C_6H_5)_4Si$	17.7	66.3	92
Li	C_6H_5Br *	»	16	75	92
K	C_6H_5Br	»	55	—	796
K	C_6H_5Br *	»	79	—	556
K	C_6H_5Br *	»	45	—	796
K	C_6H_5Br (1 : 2)	$(C_6H_5)_4Si$	70	10	590
		$C_6H_5C_6H_5$	10		
K	C_6H_5Br	$(C_6H_5)_4Si$	61	—	590
		$(C_6H_5)_3SiOSi(C_6H_5)_3$	34	—	
		$C_6H_5C_6H_5$	8	—	
K	C_6H_5Br	$(C_6H_5)_4Si$	55	27	590
K	C_6H_5Br	»	30	—	560
K	C_6H_5Br	»	96	—	791
Li	C_6H_5I	»	19.5—20	68.5—74	92
K	C_6H_5I	»	63	4	796
Li	o-$CH_3C_6H_4Br$	$(C_6H_5)_3SiC_6H_4CH_3$-o	1.7	52	92
K	o-$CH_3C_6H_4Br$	»	57	—	796
K	p-$CH_3C_6H_4I$	$(C_6H_5)_3SiC_6H_4CH_3$-p	95	6	796
Li	o-BrC_6H_4COOH	—	—	71	92
Li	O —Cl	O —$Si(C_6H_5)_3$	19	44.3	741
Li	O —Br	—	—	53.8	741
Li	O —Br **	O —$Si(C_6H_5)_3$	11.8	54	741
K	S —Br	S —$Si(C_6H_5)_3$	57	—	**952**
Li	N(C_2H_5) —Br	N(C_2H_5) —$Si(C_6H_5)_3$	27.5	57.7	741
Li	N(C_2H_5) —Br **	N(C_2H_5) —$Si(C_6H_5)_3$	40.7	44.3	741

TABLE 4. (Cont'd)

M	Halide	Products of replacement of the halogen (obtained after hydrolysis of the reaction mixture)	Yield, %	Yield of $[(C_6H_5)_3Si]_2$, %	Literature
K	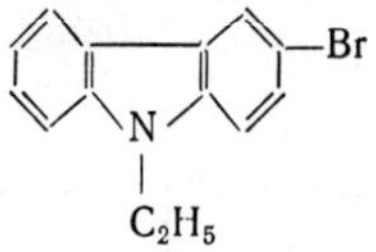 N, —Br; C_2H_5	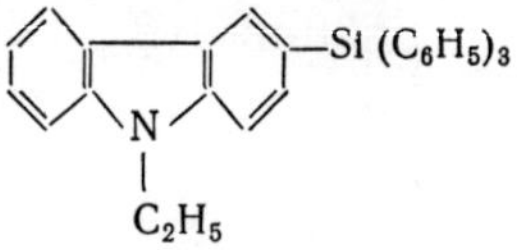 N, —Si $(C_6H_5)_3$; C_2H_5	63	—	952
Li	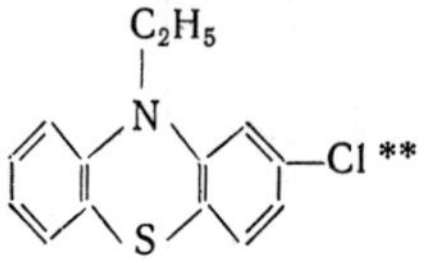 C_2H_5; N; S; —Cl **	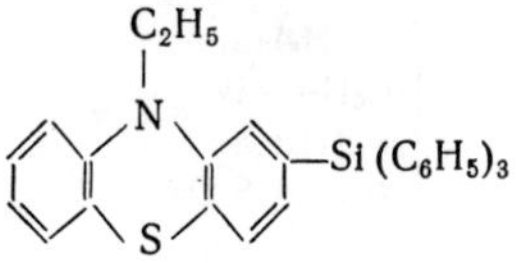 C_2H_5; N; S; —Si $(C_6H_5)_3$	28	57.6	741
Li	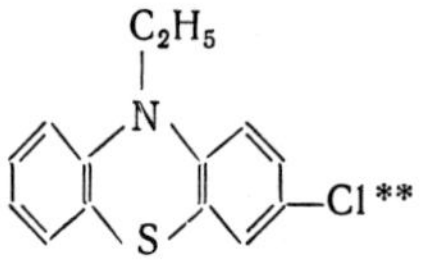 C_2H_5; N; S; —Cl **	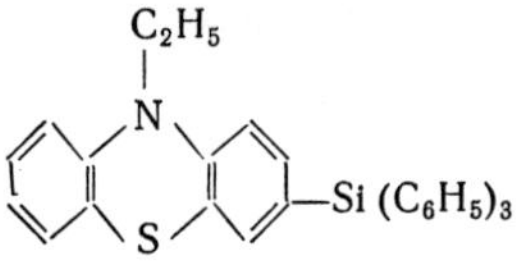 C_2H_5; N; S; —Si $(C_6H_5)_3$	27,8	53.8	741
Li	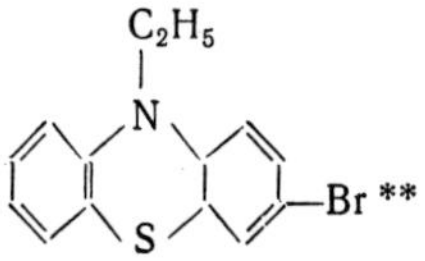 C_2H_5; N; S; —Br **	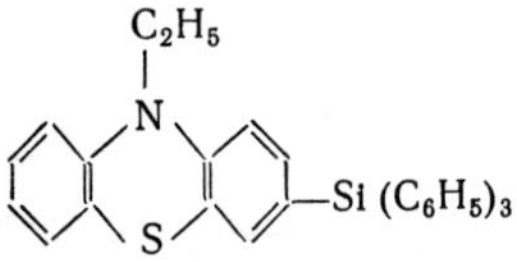 C_2H_5; N; S; —Si $(C_6H_5)_3$	23.7	58	741
K	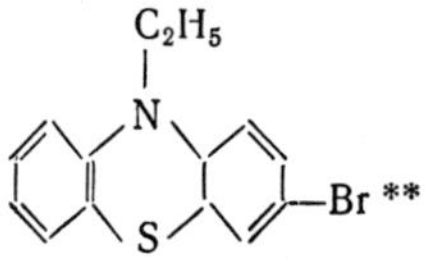 C_2H_5; N; S; —Br **	$(C_6H_5)_3SiOSi(C_6H_5)_3$	20	10	92
Li	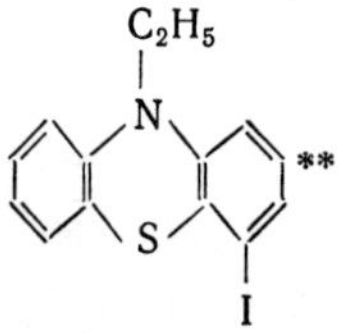 C_2H_5; N; S; **; I	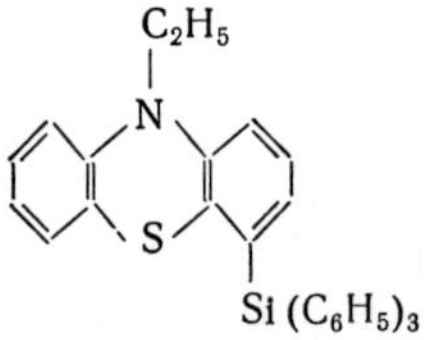 C_2H_5; N; S; Si $(C_6H_5)_3$	14.8	46.3	741
Li	o-C_6H_4ClF * (1 : 2)	$(C_6H_5)_3SiC_6H_4F$-o	28	68.5	92
Li	o-C_6H_4BrF * (1 : 3)	$(C_6H_5)_4Si$	9.5 3,3	60	92

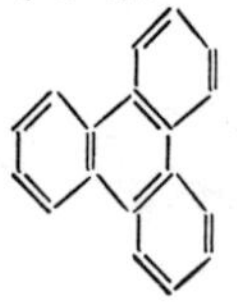

TABLE 4. (Cont'd)

M	Halide	Products of replacement of the halogen (obtained after hydrolysis of the reaction mixture)	Yield, %	Yield of $[(C_6H_5)_3Si]_2$, %	Literature
Li	o-$C_6H_4Cl_2$ (1 : 2)	$(C_6H_5)_4Si$	9.5	79.8	92
Li	o-$C_6H_4Cl_2^*$ (1 : 3)	—	—	69.5	92
Li	o-$C_6H_4Br_2$ (1 : 2)	—	—	71.7	92
Li	o-$C_6H_4Br_2^*$ (1 : 2)	—	—	60.6	92
Li	o-$C_6H_4Br_2^*$ (1 : 3)	—	—	55.9	92
Li	o-C_6H_4IBr * (1 : 2)	$(C_6H_5)_4Si$	3.8	72.5	92
			3.85	—	
Li	o-C_6H_4IBr * (1 : 1)	$(C_6H_5)_4Si$	2.4	68.5	92
Li	m-C_6H_4BrF * (1 : 1)	—	—	51.5	92
Li	m-C_6H_4IBr * (1 : 2)	—	—	84.5	92
Li	m-C_6H_4IBr * (1: 1)	—	—	78	92
Li	p-$C_6H_4Cl_2$ (1 : 3)	—	—	55	92
Li	p-$C_6H_4Br_2$ (1 : 4)	—	—	68	92
Li	p-$C_6H_4I_2$ (1 : 2)	—	—	65.2	92
Li	p-C_6H_4BrF (1 : 2)	$(C_6H_5)_3SiC_6H_4F$-p	13.6	47.2	92
Li	p-C_6H_4BrCl (1 : 3)	—	—	61.5	92
Li	C_2H_5 N Cl S Cl (1 : 2)	C_2H_5 N $(C_6H_5)_3Si$— S —$Si(C_6H_5)_3$	17.5	44.2	741
Li	C_2H_5 N Cl— —Cl SO_2 (1 : 2)	C_2H_5 N $(C_6H_5)_3Si$— —$Si(C_6H_5)_3$ SO_2	19	64.2	741
Li	1,2,3-$C_6H_3Cl_3$ (1 : 3)	—	—	61	92
Li	1,2,4-$C_6H_3Cl_2Br$ * (1 : 3)		—	58	92
Li	1,2,4-$C_6H_3Br_3$ (1 : 3)	p-$[(C_6H_5)_3Si]_2C_6H_4$	0.1	71	92
Li	1,2,4-$C_6H_3Br_3$ (1 : 1)	1, 3, 5-$C_6H_3Br_3$	2.9	65.7	92
Li	1,3,5-$C_6H_3Cl_3$ (1 : 3)	$(C_6H_5)_4Si$	2.2	71	92
Li	1,3,5-$C_6H_3Br_3$ (1 : 3)	»	1.1	65.3	92
Li	1,3,5-$C_6H_3Br_3^*$ (1 : 3)	»	—	66.1	92

TABLE 5. Products of the Reactions of Various Organosilyl Derivatives of Alkali Metals with Organic Halides

Organosilicon compound	Halide	Products of replacement of halogen	Yield, %	Literature
$(CH_3)_2C_6H_5SiLi$	$n\text{-}C_4H_9Br$	$(CH_3)_2C_6H_5SiC_4H_9\text{-}n$	74	756
$CH_3(C_6H_5)_2SiLi$	$n\text{-}C_4H_9Br$	$CH_3(C_6H_5)_2SiC_4H_9\text{-}n$	88	756
	$C_6H_5C_6H_4\text{—}CH_2CH_2Cl$	$C_6H_5C_6H_4\text{—}CH_2CH_2SiCH_3(C_6H_5)_2$	—	1692
	$C_6H_5\text{—}C_6H_4\text{—}CH_2CH_2Cl$	$C_6H_5\text{—}C_6H_4\text{—}CH_2CH_2SiCH_3(C_6H_5)_2$	—	1692
	$C_6H_5\text{—}C_6H_3(C_6H_5)\text{—}CH_2CH_2Cl$	$C_6H_5\text{—}C_6H_3(C_6H_5)\text{—}CH_2CH_2SiCH_3(C_6H_5)_2$	—	1692
$CH_3(C_6H_5)_2SiK$	C_6H_5Br	$(C_6H_5)_3SiCH_3$	—	754
$(p\text{-}C_6H_5\text{—}C_6H_4)_3SiK$	$C_6H_5\text{—}C_6H_4Br\text{-}p$	$(p\text{-}C_6H_5\text{—}C_6H_4)_4Si$	48	556
(cyclic) SiHK	C_6H_5Br	(cyclic) $SiH\text{—}C_6H_5$	—	552
		(cyclic) $Si(C_6H_5)_2$	—	
$Li[Si(C_6H_5)_2]_4Li$	$n\text{-}C_4H_9Cl$	$n\text{-}C_4H_9\text{—}[Si(C_6H_5)_2]_4\text{—}C_4H_9\text{-}n$	—	91
$Li[Si(C_6H_5)_2]_5Li$	$ClCH_2CH_2CH_2Cl$	$(C_6H_5)_2Si[(C_6H_5)_2Si]_4(CH_2)_2CH_2$ (ring closed between Si and CH_2)	—	726, 755

phenol. This indicates that triphenylsilyllithium cleaves the anisole at the CH_3-O bond:

$$(C_6H_5)_3SiLi + C_6H_5OCH_3 \longrightarrow (C_6H_5)_3SiCH_3 + C_6H_5OLi \tag{1-80}$$

Phenetole and phenyl propyl ether are not cleaved by triphenylsilyllithium under the same conditions, even if the reaction time is increased to 72 hr.

The cleavage of ethers may proceed by an S_N2 mechanism. In this case the triphenylsilyl anion attacks the carbon atom of the alkyl group with the elimination of the phenolate anion. Such reactions are very sensitive to a change in steric factors and their rate depends to a considerable extent on the size of the radicals attached to the carbon atom attacked and also the organosilyl anion. With a change from methyl to ethyl and propyl groups in the molecule of the alkyl ether of phenol there is an increase in the steric hindrance to attack on the carbon atom of the alkyl group by the very bulky solvated molecules of triphenylsilyllithium. When triphenylsilyllithium is replaced by a less bulky molecule, namely, dimethylphenylsilyllithium, there is some cleavage of Alk–O bond in phenetole [786].

p-Anisyltriphenylsilane (5.7%) and hexaphenyldisilane (40.7%) are formed by the reaction of triphenylsilyllithium with p-chloroanisole. Cleavage of the ether bond is not observed in this case.

In the case of p-fluoroanisole there is cleavage of both CH_3-O and C–F bonds, but the latter is cleaved to a much lesser extent:

$$(C_6H_5)_3SiLi + F-C_6H_4-OCH_3 \longrightarrow \underset{49.2\%}{(C_6H_5)_3SiCH_3} + \underset{30.5\%}{F-C_6H_4-OLi} + \underset{1.1\%}{(C_6H_5)_3SiC_6H_4OCH_3} \tag{1-81}$$

Methyl ethers of α- and β-naphthols are cleaved more readily than anisole by triphenylsilyllithium. In the reaction with p-dimethoxybenzene, one or two methyl groups may be eliminated, depending on the ratio of the reagents. When a twofold excess of triphenylsilyllithium is used the ratio of hydroquinone to its methyl ether in the reaction products is 1:2.4.

The reaction of triphenylsilyllithium with methyl ether gives after hydrolysis 1-phenylethanol (8%), tetraphenylsilane (42%), triphenylsilane (18.4%), and triphenylbenzylsilane (2%). This shows that triphenylsilyllithium metallates this ether with the formation of a C-lithio derivative which rearranges into an O-lithio derivative. Tetraphenylsilane is then obtained as a result of the reaction of triphenylsilyllithium with triphenylsilane. The presence of a small amount of tribenzylsilane confirms the possibility of cleavage of the ether bond on the side of the larger radical [1274]:

$$(C_6H_5)_3SiLi + C_6H_5CH_2OCH_3 \longrightarrow C_6H_5\underset{\substack{|\\ Li}}{C}HOCH_3 + (C_6H_5)_3SiH \xrightarrow{(C_6H_5)_3SiLi} (C_6H_5)_4Si$$
$$C_6H_5\underset{\substack{|\\ Li}}{C}HOCH_3 \longrightarrow C_6H_5\underset{\substack{|\\ OLi}}{C}HCH_3 \qquad (1\text{-}82)$$

Cleavage of N-(butoxymethyl)piperidine analogously leads to the formation of N-(triphenylsilylmethyl)piperidine but not butyltriphenylsilane. Diphenyl ether is not cleaved by triphenylsilyllithium but is metallated in the o,o'-positions.

Triphenylsilyllithium cleaves cyclic ethers (oxides) of the type $\underbrace{(CH_2)_nO}$ [1282]. The ease of this reaction falls with an increase in n. Thus, ethylene oxide (n = 2) [672, 734] and trimethylene oxide (n=3) [734, 1274] react readily with triphenylsilyllithium:

$$(C_6H_5)_3SiLi + \underbrace{(CH_2)_nO} \xrightarrow{(H_2O)} (C_6H_5)_3Si(CH_2)_nOH \qquad (1\text{-}83)$$

1,2-Cyclohexylene oxide (cyclohexene oxide) also react readily with triphenylsilyllithium and methyldiphenylsilyllithium. The yields of the corresponding triorganosilyl substituted cyclohexanols are 68.6 and 72%, respectively [734]. In the reaction of triphenylsilyllithium with unsymmetrical α-oxides, the C–O bond more remote from the substitutent is broken. This is connected with the fact that the triphenylsilyl group enters the more sterically accessible position [734, 784]:

$$(C_6H_5)_3SiLi + \underset{\diagdown O \diagup}{CH_2—CHC_6H_5} \xrightarrow{(H_2O)} (C_6H_5)_3SiCH_2—\underset{\substack{|\\ OH}}{C}HC_6H_5 \qquad (1\text{-}84)$$

The oxide ring in epichlorohydrin reacts with triphenylsilyllithium more readily than the C–Cl bond. The intermediate product formed by rupture of the C–O bond reacts with a second molecule of $(C_6H_5)_3SiLi$ at the C–Cl bond. Moreover, it may dimerize to a 1,4-dioxane derivative:

$$(C_6H_5)_3SiLi + \underset{\diagdown O \diagup}{CH_2—CHCH_2Cl} \longrightarrow \left[(C_6H_5)_3SiCH_2—\underset{OLi}{\underset{|}{C}}HCH_2Cl\right] \xrightarrow{H_2O} (C_6H_5)_3SiCH_2—\underset{OH}{\underset{|}{C}}HCH_2Cl$$

$$\left[(C_6H_5)_3SiCH_2—\underset{OLi}{\underset{|}{C}}HCH_2Cl\right] \xrightarrow{(C_6H_5)_3SiLi} (C_6H_5)_3SiCH_2\underset{OLi}{\underset{|}{C}}HCH_2Si(C_6H_5)_3 \xrightarrow{H_2O} (C_6H_5)_3SiCH_2\underset{OH}{\underset{|}{C}}HCH_2Si(C_6H_5)_3$$

$$\left[(C_6H_5)_3SiCH_2—\underset{OLi}{\underset{|}{C}}HCH_2Cl\right] \longrightarrow (C_6H_5)_3SiCH_2—HC\overset{O}{\diagup\diagdown}CH_2,\ H_2C\underset{O}{\diagdown\diagup}CH—CH_2Si(C_6H_5)_3 \quad (1\text{-}85)$$

$$(C_6H_5)_3SiCH_2\underset{OH}{\underset{|}{C}}HCH_2Si(C_6H_5)_3 \xrightarrow{\beta\text{-decomposition}} (C_6H_5)_3SiCH_2CH{=}CH_2 + (C_6H_5)_3SiOH$$

Tetrahydrofuran (tetramethylene oxide, (n = 4) is cleaved by R_3SiLi with much more difficulty than ethylene oxide (n = 2) and trimethylene oxide (n=3) [712, 734, 743, 1274]. As a result of this it may be used both as a solvent and for carrying out the above reactions with α- and β-oxides. With a rise in temperature the rate of cleavage of tetrahydrofuran by triphenylsilyllithium increases and the yield of 4-(triphenylsilyl)butanol-1 after heating for 3 hr at 125°C is 71% [1278].

The cleavage of tetrahydropyran (pentamethylene oxide, (n = 5) by triphenylsilyllithium requires a still higher temperature (200-220°C). The products of this reaction were not identified [1274].

1,4-Dioxane reacts slowly with triphenylsilyllithium to form 1,2-bis(triphenylsilyl)ethane. Hence we must assume that in this case there is unsymmetrical cleavage or stepwise cleavage of the two ether groups [1274].

1.3.5. Reaction with Aldehydes and Ketones

Compounds of the type R_3SiM add to the carbonyl group of aliphatic aldehydes. The metal atom is then attached to the oxygen atom and the triorganosilyl group to the carbon atom. The reaction of triphenylsilylpotassium with formaldehyde forms triphenylsilylcarbinol [795, 1284]:

$$(C_6H_5)_3SiK + CH_2O \longrightarrow (C_6H_5)_3SiCH_2OK \quad (1\text{-}86)$$

The additions of dimethylphenylsilyllithium to formaldehyde [783] and of triphenylsilyllithium to acetaldehyde [589, 1279] and propionaldehyde [771] proceed analogously.

The product of the reaction of triphenylsilyllithium or triphenylsilylpotassium with benzaldehyde is triphenylbenzyloxysilane. This may be explained by the rearrangement of the normal adduct which is formed initially. Hexaphenyldisilane, hydrobenzoin, and benzyl alcohol were also isolated [1284]:

$$R_3SiLi + RCHO \longrightarrow [R_3SiCH(OLi)R]$$

$$[R_3SiCH(OLi)R] \xrightarrow{R_3SiLi} R_3SiSiR_3 + RCH\begin{matrix} OLi \\ Li \end{matrix} \xrightarrow{H_2O} RCH_2OH$$

$$[R_3SiCH(OLi)R] \longrightarrow R_3SiOCH(Li)R \xrightarrow{H_2O} R_3SiOCH_2R$$

$$(R{=}C_6H_5) \qquad R_3SiOCH(Li)R \xrightarrow{RCHO} \begin{matrix} R_3SiOCHR \\ | \\ LiOCHR \end{matrix} \xrightarrow{R_3SiLi} R_3SiSiR_3 + \begin{matrix} LiCHR \\ | \\ LiOCHR \end{matrix} \qquad (1\text{-}87)$$

Aliphatic ketones react with R_3SiM analogously to aliphatic aldehydes with the formation of triorganosilylalkylcarbinols. Thus, for example, triphenylsilyllithium forms 2-(triphenylsilyl)propanol-2 in ~50% yield with acetone [579, 759]:

$$(C_6H_5)_3SiLi + CH_3COCH_3 \xrightarrow{(H_2O)} (C_6H_5)_3Si{-}\begin{matrix} CH_3 \\ | \\ C \\ | \\ CH_3 \end{matrix}{-}OH \qquad (1\text{-}88)$$

Dibenzyl ketone does not add triphenylsilyllithium, but is only metallated by it [759].

The main product from the reaction of R_3SiM with benzophenone is a triorganobenzhydryloxysilane [587, 590, 757, 793, 1129]:

$$(C_6H_5)_3SiK + C_6H_5COC_6H_5 \xrightarrow{(H_2O)} (C_6H_5)_3SiOCH(C_6H_5)_2 \qquad (1\text{-}89)$$

This compound may be formed with an anomalous order of addition of R_3SiM to benzophenone or as a result of isomerization of the normal addition products. The second route seems more probable since such an isomerization of triphenylsilyldiphenylcarbinol proceeds instantaneously in the presence of traces of alkali metals [579]. However, when benzophenone is added to triphenylsilylpotassium the first product isolated is $(C_6H_5)_3SiOCK(C_6H_5)_2$ (64% yield) [587, 1129], which may then react with a second molecule of benzophenone to form 2-(triphenylsiloxy)tetraphenylethanol [587]. In addition to this, benzpinacol is isolated and this is obtained from the initially formed metal ketyl of benzophenone $(C_6H_5)_2COK$:

$$(C_6H_5)_3SiK + C_6H_5COC_6H_5 \longrightarrow (C_6H_5)_3SiC(OK)(C_6H_5)_2 \longrightarrow (C_6H_5)_3SiOC(K)(C_6H_5)_2 \xrightarrow[(H_2O)]{(C_6H_5)_2CO} (C_6H_5)_2C(OH)—C(OSi(C_6H_5)_3)(C_6H_5)_2$$

$$(C_6H_5)_3SiK + C_6H_5COC_6H_5 \longrightarrow C_6H_5\dot{C}(OK)C_6H_5 \xrightarrow{(H_2O)} (C_6H_5)_2C(OH)—C(OH)(C_6H_5)_2 \qquad (1\text{-}89a)$$

The products of the reaction of R_3SiM with aldehydes and ketones and their yields are given in Table 6.

1.3.6. Reactions with Derivatives of Carboxylic Acids

Triphenylsilyllithium reacts with acetyl chloride at -50°C with the formation of triphenylsilyl methyl ketone. At the moment of its formation, the latter adds a second molecule of $(C_6H_5)_3SiLi$ and is converted into 1,1-bis(triphenylsilyl)ethanol, which partly isomerizes into triphenyl(1-triphenylsiloxyethyl)silane [1279]:

$$(C_6H_5)_3SiLi + ClCOCH_3 \longrightarrow (C_6H_5)_3SiCOCH_3 \xrightarrow{(C_6H_5)_3SiLi} [(C_6H_5)_3Si]_2C(CH_3)OH \longrightarrow (C_6H_5)_3SiCH(CH_3)OSi(C_6H_5)_3 \qquad (1\text{-}90)$$

Another reaction product is triphenylsilane. If the process is carried out at a temperature between -10 and 20°C, the yield of triphenylsilane increases from 29 to 31%, while the yield of 1,1-bis(triphenylsilyl)ethanol falls from 20.4 to 16.3%. At room temperature the yield of triphenylsilane is increased to 39%. Among the other products of the reaction under these conditions it was possible to detect only triphenyl(1-triphenylsiloxyethyl)silane

TABLE 6. Products of the Reaction of R_3SiM with Aldehydes and Ketones

Aldehyde of ketone	R_3SiM	Products of addition (or subsequent rearrangement)	Yield, %	Other reaction products	Yield, %	Literature
CH_2O	$(C_6H_5)_3SiLi$	$(C_6H_5)_3SiCH_2OH$	12	—	—	795
CH_2O	$(CH_3)_2C_6H_5SiLi$	$(CH_3)_2C_6H_5SiCH_2OH$	45.3	—	—	783
CH_3CHO (—20° C)	$(C_6H_5)_3SiLi$	$(C_6H_5)_3SiCH(OH)CH_3$	39	$(C_6H_5)_3SiH$	38	1279
CH_3CHO	$(C_6H_5)_3SiLi$	»	53	—	—	589
CH_3CH_2CHO	$(C_6H_5)_3SiLi$	$(C_6H_5)_3SiCH(OH)CH_2CH_3$	4.3	$(C_6H_5)_3SiH$	43.6	771
C_6H_5CHO * (—70° C)	$(C_6H_5)_3SiLi$	$(C_6H_5)_3SiOCH_2C_6H_5$	46	—	—	1284
C_6H_5CHO	$(C_6H_5)_3SiLi$	$(C_6H_5)_3SiOCH_2C_6H_5$	4.1	$(C_6H_5)_3SiSi(C_6H_5)_3$	60—62	1284
				$C_6H_5CH(OH)CH(OH)C_6H_5$	49—58	
C_6H_5CHO	$(C_6H_5)_3SiK$	—	—	$(C_6H_5)_4Si$	77	1284
				$(C_6H_5)_3SiOH$	11	
				$C_6H_5CH(OH)CH(OH)C_6H_5$	16	
$(CH_3)_2CO$	$(C_6H_5)_3SiLi$	$(C_6H_5)_3SiCH(OH)(CH_3)_2$	45	—	—	579
$(CH_3)_2CO$	$(C_6H_5)_3SiLi$	»	52	—	—	759
$(CH_3)_2CO$	$CH_3(C_6H_5)_2SiLi$	$CH_3(C_6H_5)_2SiCH(OH)(CH_3)_2$	44			759
$(CH_3)_2CO$	$(CH_3)_2C_6H_5SiLi$	$(CH_3)_2C_6H_5SiCH(OH)(CH_3)_2$	45	—	—	759
C_6H_{10}=O	$(C_6H_5)_3SiLi$	C_6H_{11}—OSi $(C_6H_5)_3$	7	—	—	759
C_6H_{10}=O	$(CH_3)_2C_6H_5SiLi$	C_6H_{11}—OSi $(CH_3)_2C_6H_5$	29	—	—	759

$(n\text{-}C_7H_{15})_2CO$	$(C_6H_5)_3SiLi$	$(C_6H_5)_3SiCH(OH)(C_7H_{15}\text{-}n)_2$	26	—	—	759
$n\text{-}C_{15}H_{31}COC_2H_5$	$(C_6H_5)_3SiLi$	$(C_6H_5)_3SiCH(OH)(C_2H_5)C_{15}H_{31}\text{-}n$	34	—	—	759
$n\text{-}C_{17}H_{35}COCH_3$	$(C_6H_5)_3SiLi$	$(C_6H_5)_3SiCH(OH)(CH_3)C_{17}H_{35}\text{-}n$	33	—	—	759
$n\text{-}C_8H_{17}COC_{10}H_{21}\text{-}n$	$(C_6H_5)_3SiLi$	$(C_6H_5)_3SiCH(OH)(C_8H_{17}\text{-}n)C_{10}H_{21}\text{-}n$	25	—	—	759
$(n\text{-}C_{11}H_{23})_2CO$	$(C_6H_5)_3SiLi$	$(C_6H_5)_3SiCH(OH)(C_{11}H_{23}\text{-}n)_2$	10	—	—	759
$(C_6H_5CH_2)_2CO$	$(C_6H_5)_3SiLi$	—	—	$(C_6H_5)_3SiH$	78	759
$(C_6H_5)_2CO$	$(C_6H_5)_3SiLi$	$(C_6H_5)_3SiOCH(C_6H_5)_2$	50	—	—	757
$(C_6H_5)_2CO$	$CH_3(C_6H_5)_2SiLi$	$CH_3(C_6H_5)_2SiOCH(C_6H_5)_2$	19	$(C_6H_5)_2C(OH)C(OH)(C_6H_5)_2$	Traces	757
$(C_6H_5)_2CO$ *	$(C_6H_5)_3SiK$	$(C_6H_5)_3SiOCH(C_6H_5)_2$	25	$(C_6H_5)_3SiOH$ $[(C_6H_5)_3Si]_2O$ $(C_6H_5)_2CHOH$	13 54 22	793
$(C_6H_5)_2CO$ *	$(C_6H_5)_3SiK$	$(C_6H_5)_3SiOCH(C_6H_5)_2$	21	$(C_6H_5)_3SiH$ $(C_6H_5)_2CHOH$	41 18	790
$(C_6H_5)_2CO$	$(C_6H_5)_3SiK$	$(C_6H_5)_3SiOC(C_6H_5)_2C(OH)(C_6H_5)_2$	48	$(C_6H_5)_3SiOH$	4	587
$(C_6H_5)_2CO$ *	$(C_6H_5)_3SiK$	$(C_6H_5)_3SiOCK(C_6H_5)_2$	64	—	—	587, 1129
$(p\text{-}CH_3C_6H_4)_2CO$	$(C_6H_5)_3SiK$	$(C_6H_5)_3SiOCH(C_6H_4CH_3\text{-}p)_2$	26	—	—	793
$O(C_6H_4)_2C{=}O$	$(C_6H_5)_3SiLi$	$O(C_6H_4)_2C(H)OSi(C_6H_5)_3$	12.7	—	—	92
$S(C_6H_4)_2C{=}O$	$(C_6H_5)SiLi$	$S(C_6H_4)_2C(H)OSi(C_6H_5)_3$	12	$(C_6H_5)_3SiSi(C_6H_5)_3$	34.6	92

* Reaction carried out by adding the carbonyl compound to R_3SiM.

(16.5%). An analogous rearrangement product, namely, triphenyl-(1-triphenylsiloxypropyl)silane, is obtained by the reaction of triphenylsilyllithium with propionyl chloride. In this reaction the acid chloride of phenylacetic acid gives only triphenylsilane [771]. The reaction of triphenylsilylpotassium with benzoyl chloride leads to a small amount of triphenylsilyl phenyl ketone (4%) together with a large amount of hexaphenyldisilane [580].

Like acetyl chloride, acetic anhydride reacts with triphenylsilyllithium in accordance with scheme (1-90) to form 1,1-bis(triphenylsilyl)ethanol (39%), its rearrangement product triphenyl(1-triphenylsiloxyethyl)silane (3%), and also triphenylsilane [589, 771]. Analogous compounds are also formed in the reaction of triphenylsilyllithium with ethyl acetate [771]. The reaction evidently proceeds with the elimination of the ethoxyl group since in the reaction with ethyl benzoate it is possible to isolate triphenylethoxysilane in addition to the other products [784].

N, N-Dimethylbenzamide adds a molecule of triphenylsilyllithium to the carbonyl group. The triphenylsiloxy derivative formed in this way is cleaved by a second molecule of triphenylsilyllithium [711]:

$$C_6H_5CON(CH_3)_2 + (C_6H_5)_3SiLi \longrightarrow C_6H_5\overset{\displaystyle Li}{\underset{\displaystyle OSi(C_6H_5)_3}{C}}\!-\!N(CH_3)_2 \xrightarrow{(C_6H_5)_3SiLi}$$

$$\longrightarrow (C_6H_5)_3SiOLi + (C_6H_5)_3Si\!-\!\overset{\displaystyle C_6H_5}{\underset{\displaystyle Li}{C}}\!-\!N(CH_3)_2 \qquad (1\text{-}91)$$

The reaction of triphenylsilyllithium with N-substituted phthalimides yields a large amount of hexaphenyldisilane and products of the cleavage of the imide ring, namely, o-formyl-N-organobenzamides [92].

Triphenylsilyllithium cleaves the nitrile group from benzonitrile [771, 784]:

$$(C_6H_5)_3SiLi + C_6H_5CN \longrightarrow (C_6H_5)_4Si + LiCN \qquad (1\text{-}92)$$

It was not possible to isolate products of definite composition in its reaction with nitriles of aliphatic carboxylic acids [771].

1.3.7. Reaction with Nitrogen-Containing Organic Compounds

In the previous section we have already examined the reaction of triphenylsilyllithium with the representatives of certain classes of nitrogen-containing organic compounds (amides and nitriles of carboxylic acids). This section is devoted to the reactions of R_3SiM with other classes of nitrogen-containing compounds, namely, amines, ketimines, and azo compounds.

Primary and secondary acyclic and secondary cyclic amines react with triphenylsilyllithium with the formation of N-triphenylsilyl substituted amines [760]:

$$(C_6H_5)_3SiLi + n\text{-}C_4H_9NH_2 \xrightarrow[-LiH]{} (C_6H_5)_3SiNHC_4H_9\text{-}n \qquad (1\text{-}93)$$

$$2(C_6H_5)_3SiLi + HN\langle\bigcirc\rangle NH \xrightarrow[-2LiH]{} (C_6H_5)_3SiN\langle\bigcirc\rangle NSi(C_6H_5)_3 \qquad (1\text{-}94)$$

In the case of acyclic amines the yields are 50-60% and in the case of cyclic amines, 80-90%.

N-Benzhydrylaniline is only metallated by triphenylsilyllithium and does not form a triphenylsilyl substituted amine [1276].

Nitrogen-containing heterocycles (pyridine and acridine) add triphenylsilyllithium to form derivatives of dihydroheterocycles. At the moment of its formation from hexaphenyldisilane and lithium, triphenylsilyllithium adds to pyridine if the latter is used as the solvent:

$$(C_6H_5)_3SiLi + N\langle\bigcirc\rangle \xrightarrow{(H_2O)} HN\langle\bigcirc\rangle\text{—}Si(C_6H_5)_3 \qquad (1\text{-}95)$$

In the case of the reaction of pyridine with triphenylsilyllithium, which has been prepared beforehand in tetrahydrofuran, the yield of the final product is found to be somewhat lower (53%) [1277]. In contrast to this, triphenylsilylpotassium does not give an adduct with pyridine.

TABLE 7. Products of the Reaction of R_3SiM with Nitrogen-Containing Organic Compounds

Nitrogen-containing compound	R_3SiM	Reaction products	Yield, %	Literature
$n\text{-}C_4H_9NH_2$	$(C_6H_5)_3SiLi$	$(C_6H_5)_3SiNHC_4H_9\text{-}n$	56	760
$(n\text{-}C_4H_9)_2NH$	$(C_6H_5)_3SiLi$	$(C_6H_5)_3SiN(C_4H_9\text{-}n)_2$	62	760
$(n\text{-}C_4H_9)_2NH$	$CH_3(C_6H_5)_2SiLi$	$CH_3(C_6H_5)_2SiN(C_4H_9\text{-}n)_2$	41	760
$(n\text{-}C_4H_9)_2NH$	$(CH_3)_2C_6H_5SiLi$	$(CH_3)_2C_6H_5SiN(C_4H_9\text{-}n)_2$	29	760
$(C_6H_5)_2CHNHC_6H_5$	$(C_6H_5)_3SiLi$	$(C_6H_5)_3SiH$	70	123
$(C_6H_5)_3SiCH(C_6H_5)N(CH_3)_2$	$(C_6H_5)_3SiLi$	$(C_6H_5)_3SiSi(C_6H_5)_3$	—	92
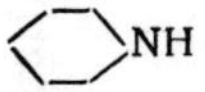 ⬡NH	$(C_6H_5)_3SiLi$	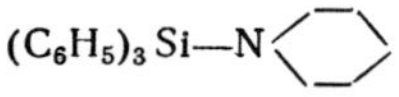 $(C_6H_5)_3Si$—N⬡	86	760
 HN⬡NH	$(C_6H_5)_3SiLi$	 $(C_6H_5)_3Si$—N⬡N—$Si(C_6H_5)_3$	93	760
 O⬡NH	$(C_6H_5)_3SiLi$	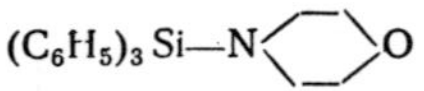 $(C_6H_5)_3Si$—N⬡O	83	760
 ⬡N	$(C_6H_5)_3SiLi$	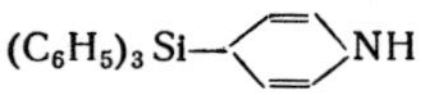 $(C_6H_5)_3Si$—⬡NH	63	1277
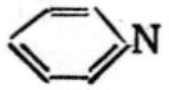 ⬡N	$(C_6H_5)_3SiK$	—	—	1277

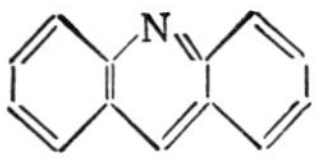 N (acridine ring structure)	$(C_6H_5)_3SiLi$	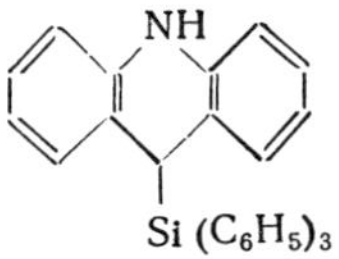 NH (ring structure), $Si(C_6H_5)_3$	27	761
$CH_3CON(CH_3)_2$	$(C_6H_5)_3SiLi$	$(C_6H_5)_3SiCH(CH_3)N(CH_3)_2$	—	92
$C_6H_5CON(CH_3)_2$	$(C_6H_5)_3SiLi$	$(C_6H_5)_3SiCH(C_6H_5)N(CH_3)_2$	—	711
$C_6H_5CON(C_6H_5)_2$	$(C_6H_5)_3SiLi$	$(C_6H_5)_3SiCH(C_6H_5)N(C_6H_5)_2$	—	92
CH_3CN	$(C_6H_5)_3SiLi$	Not identified	—	771
$CH_2{=}CHCN$	$(C_6H_5)_3SiLi$	Polymer	—	771, 1211
C_6H_5CN	$(C_6H_5)_3SiLi$	$(C_6H_5)_4Si$	50	771
$(C_6H_5)_2C{=}NC_6H_5$	$(C_6H_5)_3SiLi$	$(C_6H_5)_3SiN(C_6H_5)CH(C_6H_5)_2$	81.5	1276
$(C_6H_5)_2C{=}NC_6H_5$	$(C_6H_5)_3SiK$	»	81	1276
$C_6H_5N{=}NC_6H_5$	$(C_6H_5)_3SiLi$	$(C_6H_5)_3SiN(C_6H_5)NHC_6H_5$	74	1276
$C_6H_5N{=}NC_6H_5$	$(C_6H_5)_3SiK$	»	67	1276
$C_6H_5N{=}NC_6H_5$	$CH_3(C_6H_5)_2SiLi$	$CH_3(C_6H_5)_2SiN(C_6H_5)NHC_6H_5$	57.9	717
$C_6H_5N{=}N(O)C_6H_5$ * (1 : 2)	$(C_6H_5)_3SiLi$	$(C_6H_5)_3SiN(C_6H_5)NHC_6H_5$	52.9	717
$C_6H_5N{=}N(O)C_6H_5$ * (1 : 1)	$(C_6H_5)_3SiLi$	»	26.8	717
$C_6H_5N{=}N(O)C_6H_5$ * (1 : 2)	$CH_3(C_6H_5)_2SiLi$	$CH_3(C_6H_5)_2SiN(C_6H_5)NHC_6H_5$	53.8	717

* The ratio of azoxybenzene to R_3SiLi is given in brackets.

Acridine adds triphenylsilyllithium analogously to pyridine [761]:

$$(C_6H_5)_3SiLi + \text{acridine} \xrightarrow{(H_2O)} \text{9-}(C_6H_5)_3Si\text{-9,10-dihydroacridine (NH)} \quad (1\text{-}96)$$

Triphenylsilyllithium and triphenylsilylpotassium add to the anil (Schiff's base) of benzophenone with the triphenylsilyl group being attached to the nitrogen atom [1276]:

$$(C_6H_5)_3SiLi + (C_6H_5)_2C{=}NC_6H_5 \xrightarrow{(H_2O)} (C_6H_5)_2CHN(C_6H_5)Si(C_6H_5)_3 \quad (1\text{-}97)$$

The additions of triphenylsilyllithium, triphenylsilylpotassium [1276], and methyldiphenylsilyllithium [717] to azobenzene proceed in high yields:

$$(C_6H_5)_3SiLi + C_6H_5N{=}NC_6H_5 \xrightarrow{(H_2O)} C_6H_5N(Si(C_6H_5)_3){-}NHC_6H_5 \quad (1\text{-}98)$$

In the reaction of triphenylsilyllithium with azoxybenzene there is first reduction of the latter to azobenzene, which adds triphenylsilyllithium in accordance with scheme (1-98) with the formation of 1-triphenylsilyl-1,2-diphenylhydrazine [717].

The products of the reaction of R_3SiM with nitrogen-containing organic compounds and their yields are given in Table 7.

1.3.8. Reaction with Organic Sulfur Derivatives

Triphenylsilyllithium reacts with diphenyl sulfide first with the rupture of one of the C–S bonds with the formation of triphenyl(phenylthio)silane. There is subsequently the rupture of the second C–S bond with the formation of hexaphenyldisilane [1283]:

$$(C_6H_5)_3SiLi + C_6H_5SC_6H_5 \longrightarrow (C_6H_5)_3SiSC_6H_5 + C_6H_5Li \quad (1\text{-}99)$$

$$(C_6H_5)_3SiLi + (C_6H_5)_3SiSC_6H_5 \longrightarrow (C_6H_5)_3SiSi(C_6H_5)_3 + C_6H_5SLi \quad (1\text{-}100)$$

Triphenylsilyllithium cleaves triphenyl(p-tolylthio)silane analogously and the reaction products are hexaphenyldisilane and p-thiocresol [789, 1281]. In contrast to this, only tetraphenylsilane and triphenylsilanol were isolated in the reaction of triphenylsilylpotassium with diphenyl sulfide [1283].

In the reaction with thioanisole, triphenylsilyllithium breaks both C–S bonds in it. As a result there are formed 8.3% of triphenylmethylsilane (rupture of the Alk–S bond), 6.2% of hexaphenyldisilane (rupture of the Ar–S bond with subsequent reaction in accordance with scheme 1-100), and 13.7% of tetraphenylsilane [786].

The product of the reaction of triphenylsilylpotassium with diphenyl sulfone is tetraphenylsilane (36% yield) [1283]. This indicates metallation with the intermediate formation of triphenylsilane, which gives tetraphenylsilane with triphenylsilylpotassium. To avoid these side reactions it is necessary to use triphenylsilyllithium instead of triphenylsilylpotassium for the reaction with diphenyl sulfone. In this case there is the stepwise cleavage of the C–S bonds by triphenylsilyllithium as in the reaction with diphenyl sulfide:

$$\begin{array}{c} (C_6H_5)_4Si + C_6H_5SO_2Li \\ \uparrow (C_6H_5)_3SiLi \\ (C_6H_5)_3SiLi + C_6H_5SO_2C_6H_5 \longrightarrow (C_6H_5)_3SiSO_2C_6H_5 + C_6H_5Li \\ \downarrow + CO_2,\ H_2O \\ (C_6H_5)_3SiOH + C_6H_5SO_2H + C_6H_5COOH \end{array} \tag{1-101}$$

When excess triphenylsilyllithium is used there is a more complex reaction, one of whose products is m-bis(triphenylsilyl)-benzene.

Diphenyl sulfoxide reacts with triphenylsilylpotassium to form tetraphenylsilane, hexaphenyldisilane, and hexaphenyldisiloxane [792]. Hexaphenyldisilane is also obtained by the reaction of triphenylsilyllithium with benzenesulfonyl chloride [1283].

Dialkyl sulfates alkylate R_3SiM to form tetrasubstituted silanes. Thus, for example, the reaction of dimethyl sulfate with 9-lithio-9-methyl-9-silafluorene gives a 54% yield of 9, 9-dimethyl-

9-silafluorene [750]:

$$\text{(9-methyl-9-lithio-9-silafluorene)} \xrightarrow{(CH_3)_2SO_4} \text{(9,9-dimethyl-9-silafluorene)} \quad (1\text{-}102)$$

1.3.9. Reaction with Organosilicon Compounds

Reactions with Compounds Containing an Si – H Bond. The reaction of triphenylsilylpotassium with triarylsilanes forms the corresponding triarylphenylsilane. Compounds of this type are by-products of many reactions of triphenylsilylpotassium with organosilicon compounds in which a hydrogen atom at a silicon can be replaced by potassium. Investigations of the reaction of triphenylsilylpotassium with triarylsilanes showed that there is gradual cleavage of aryl groups from the $(C_6H_5)_3SiK$ molecule right up to the formation of a silicon-containing inorganic compound:

$$Ar_3SiH + (C_6H_5)_3SiK \longrightarrow Ar_3SiC_6H_5 + (C_6H_5)_2SiHK \quad (1\text{-}103a)$$

$$Ar_3SiH + (C_6H_5)_2SiHK \longrightarrow Ar_3SiC_6H_5 + C_6H_5SiH_2K \quad (1\text{-}103b)$$

$$Ar_3SiH + C_6H_5SiH_2K \longrightarrow Ar_3SiC_6H_5 + SiH_3K \quad (1\text{-}103c)$$

This scheme does not exactly reflect the complexity of the process occurring since completion of the reaction requires 3 moles of triarylsilane per mole of triphenylsilylpotassium. In actual fact, more than 1 mole of tetraphenylsilane is formed with a ratio of 1:1. Moreover, in the reaction of a triarylsilylpotassium with triarylsilanes containing different aryl radicals, replacement of the hydrogen atom attached to the silicon by potassium occurs:

$$(C_6H_5)_3SiK + (CH_3C_6H_4)_3SiH \rightleftarrows (C_6H_5)_3SiH + (CH_3C_6H_4)_3SiK \quad (1\text{-}104)$$

$$\downarrow \qquad\qquad\qquad\qquad \downarrow$$

$$(C_6H_5)_4Si;\ CH_3C_6H_4Si(C_6H_5)_3;\ (CH_3C_6H_4)_4Si \qquad (C_6H_5)_4Si;\ (CH_3C_6H_4)_3SiC_6H_5;\ (CH_3C_6H_4)_4Si$$

Triphenylsilylsodium reacts with triphenylsilane analogously to triphenylsilylpotassium, forming tetraphenylsilane, but in a somewhat lower yield. Triphenylsilyllithium differs sharply from triphenylsilylsodium and triphenylsilylpotassium in this respect since in its reaction with triphenylsilane it gives hexaphenyldisilane and only a very small amount of tetraphenylsilane [582]. Hexaphenyldisilane is also formed by the reaction of triphenylsilyllithium with pentaphenyldisilane [1272].

Reactions with Compounds Containing an Si–Si Bond. Triphenylsilyllithium cleaves the Si–Si bond in pentaphenyldisilane [1272, 1275] and phentaphenylchlorodisilane [1275] with the formation of hexaphenyldisilane. The reaction of dimethylphenylsilyllithium with hexaphenyldisilane forms 1,1,2,2-tetramethyl-1,2-diphenyldisilane in 56% yield [764]. The reaction proceeds in two stages. First of all there is cleavage of the Si–Si bond in hexaphenyldisilane:

$$(CH_3)_2C_6H_5SiLi + (C_6H_5)_3SiSi(C_6H_5)_3 \longrightarrow (CH_3)_2C_6H_5SiSi(C_6H_5)_3 + (C_6H_5)_3SiLi \qquad (1\text{-}105)$$

Then dimethylphenylsilyllithium cleaves the Si–Si bond in 1,1-dimethyl-1,2,2,2-tetraphenyldisilane:

$$(CH_3)_2C_6H_5SiLi + (CH_3)_2C_6H_5SiSi(C_6H_5)_3 \longrightarrow$$
$$\longrightarrow (CH_3)_2C_6H_5SiSiC_6H_5(CH_3)_2 + (C_6H_5)_3SiLi \qquad (1\text{-}106)$$

Dimethylphenylsilyllithium cleaves the Si–Si bond in 1,2-dimethyl-1,1,2,2-tetraphenyldisilane. Hydrolysis of the reaction mixture gave methyldiphenylsilane in 77.5% yield. Methyldiphenylsilyllithium also cleaves hexaphenyldisilane, but the yield of 1,2-dimethyl-1,1,2,2-tetraphenyldisilane is only 26% [764]. Thus, the reactivity of triorganosilyllithium derivatives in the cleavage of the Si–Si bond falls in the series

$$(CH_3)_2C_6H_5SiLi > CH_3(C_6H_5)_2SiLi > (C_6H_5)_3SiLi$$

Triphenylsilyllithium [853], methyldiphenylsilyllithium [1273], and dimethylphenylsilyllithium [781] cleave the Si–Si bond in cyclopolysilanes. In the reaction of triphenylsilyllithium with octaphenylcyclotetrasilane there is a whole series of conversions with the formation of new silyllithium derivatives. The latter also cleave the Si–Si bond in the starting octaphenylcyclotetrasilane and in the linear polysilanes formed, which, in their turn, are also cleaved by triphenylsilyllithium. If only 50% excess of triphenylsilyllithium is used, it is possible to isolate from the reaction mixture nonaphenyltetrasilane and dodecaphenylcyclohexasilane. With an eightfold excess of triphenylsilyllithium there is also cleavage of the ring of dodecaphenylcyclohexasilane, which is converted into hexaphenyldisilane and pentaphenyldisilane:

$$[(C_6H_5)_2Si]_6 \xrightarrow[(H_3O^{\oplus})]{(C_6H_5)_3SiLi} (C_6H_5)_3Si-\left[-\overset{C_6H_5}{\underset{C_6H_5}{\overset{|}{\underset{|}{Si}}}}-\right]_2-\overset{C_6H_5}{\underset{C_6H_5}{\overset{|}{\underset{|}{Si}}}}-H + (C_6H_5)_3Si\overset{C_6H_5}{\underset{C_6H_5}{\overset{|}{\underset{|}{Si}}}}H + (C_6H_5)_3SiSi(C_6H_5)_3 \quad (1\text{-}107)$$

Triphenylsilyllithium cleaves decaphenylcyclopentasilane to give a 79% yield of pentaphenyldisilane (after hydrolysis) [91]. Under analogous conditions methyldiphenylsilyllithium [1273] and dimethylphenylsilyllithium [781] form 1-methyl-1,1,2,2-tetraphenyldisilane and 1,1-dimethyl-1,2,2-triphenyldisilane, respectively:

$$[(C_6H_5)_2Si]_5 - \begin{cases} \xrightarrow[(H_3O^{\oplus})]{(C_6H_5)_3SiLi} (C_6H_5)_3SiSiH(C_6H_5)_2 \\ \xrightarrow[(H_3O^{\oplus})]{CH_3(C_6H_5)_2SiLi} (C_6H_5)_2CH_3SiSiH(C_6H_5)_2 \\ \xrightarrow[(H_3O^{\oplus})]{(CH_3)_2C_6H_5SiLi} (CH_3)_2C_6H_5SiSiH(C_6H_5)_2 \end{cases} \quad (1\text{-}108)$$

Reactions with Compounds Containing an Si – O Bond. In the reaction of triphenylsilylpotassium with triphenylsilanol, as with triphenylmethylcarbinol, tetraphenylsilane is formed (86%) [792]. Triphenylsilylpotassium cleaves the Si–O bond in triphenylethoxysilane. The yield of the hexaphenyldisilane obtained is 71% [797]:

$$(C_6H_5)_3SiK + (C_6H_5)_3SiOC_2H_5 \longrightarrow (C_6H_5)_3SiSi(C_6H_5)_3 + C_2H_5OK \quad (1\text{-}109)$$

Hexaphenyldisilane is a product of the reaction of triphenylsilyllithium with tetraethoxysilane or benzhydryloxydiphenylsilane [1275]. In the latter case, pentaphenyldisilane and benzhydrol are also isolated.

Triphenylsilyllithium reacts with hexaphenylcyclotrisiloxane [92]:

$$\text{hexaphenylcyclotrisiloxane } [(C_6H_5)_2SiO]_3 + (C_6H_5)_3SiLi \longrightarrow (C_6H_5)_3Si\left[-\underset{C_6H_5}{\overset{C_6H_5}{|}}{SiO}-\right]_2 \underset{C_6H_5}{\overset{C_6H_5}{|}}{Si}-O-Li \longrightarrow$$

$$\xrightarrow[(H_3O^{\oplus})]{(C_6H_5)_3SiLi} (C_6H_5)_3SiSi(C_6H_5)_3 + H\left[-\underset{C_6H_5}{\overset{C_6H_5}{|}}{SiO}-\right]_3 H \qquad (1\text{-}110)$$

<u>Reaction with Halosilanes.</u> Compounds of the type R_3SiM react with halosilanes to form derivatives containing an Si–Si bond:

$$R_3SiM + X{-}Si\lessdot \longrightarrow R_3Si{-}Si\lessdot + MX \quad (X = \text{halogen}) \qquad (1\text{-}111)$$

However, depending on the structure of the halosilane and the order of mixing of the reagents, together with this reaction there may also be different side processes, leading to the formation of organodisilanes of different structure.

Trialkylchlorosilanes react readily with R_3SiM with the formation of 1,1,1-trialkyl-2,2,2-triaryldisilanes. Thus, in the reaction of trimethylchlorosilane with triphenylsilyllithium [581, 758, 773, 937], triphenylsilylsodium [581], and triphenylsilylpotassium [549, 560, 581] in tetrahydrofuran or in 1,2-dimethoxyethane the yield of 1,1,1-trimethyl-2,2,2-triphenyldisilane is 70-85% if the R_3SiM is added to trimethylchlorosilane:

$$(C_6H_5)_3SiM + ClSi(CH_3)_3 \longrightarrow (C_6H_5)_3SiSi(CH_3)_3 + MCl \qquad (1\text{-}112)$$

With the reverse order of addition, the yields of 1,1,1-trimethyl-2,2,2-triphenyldisilane are lower, namely, 45-75% in the reaction with $(C_6H_5)_3SiK$ [549, 560, 791, 795], 48% in the reaction with $(C_6H_5)_3SiRb$, and 26% in the reaction with $(C_6H_5)_3SiCs$ [795].

In the case of the addition of chlorosilanes to R_3SiM, the hexaorganodisilane formed may be cleaved by excess R_3SiM. This was demonstrated by an investigation of the products from the reaction of dimethylphenylsilyllithium or methyldiphenylsilyllithium with triphenylchlorosilane [757]:

$$CH_3(C_6H_5)_2SiLi + ClSi(C_6H_5)_3 \longrightarrow CH_3(C_6H_5)_2SiSi(C_6H_5)_3 \xrightarrow{CH_3(C_6H_5)_2SiLi} CH_3(C_6H_5)_2SiSi(C_6H_5)_2CH_3 + (C_6H_5)_3SiLi \quad (1\text{-}113)$$

The formation of symmetrical hexaorganodisilanes may be explained by metallation of the Si−Cl bond, but then it should occur rapidly when the reaction mixture contains free R_3SiM and to a very small extent when there is excess triorganochlorosilane.

A study of the competing reactions of triphenylsilyllithium with a mixture of trialkyl- and alkylarylchlorosilanes shows that the yields of the unsymmetrical hexaorganodisilane exceed the yields of hexaphenyldisilane, increasing in the following series of chlorosilanes [788]:

$$CH_3(C_6H_5)_2SiCl < C_6H_5(CH_3)_2SiCl < (C_2H_5)_3SiCl < (CH_3)_3SiCl$$

Triphenylsilyllithium is more reactive toward triphenylchlorosilane than phenyl-, butyl, and benzyllithiums. Thus, the reaction of a mixture of triphenylsilyllithium and butyllithium with triphenylchlorosilane in tetrahydrofuran forms 61% of hexaphenyldisilane and 21.5% of butyltriphenylsilane; in a mixed solvent (tetrahydrofuran and diethyl ether in a ratio of 2:1) their yields are 77.2 and 13.9%, respectively.

In the reaction of triphenylsilyllithium [782, 1275] or triphenylsilylpotassium [795, 798] with diorganodihalosilanes there is the stepwise replacement of halogen atoms by triphenylsilyl groups:

$$(C_6H_5)_3SiK + (C_6H_5)_2SiCl_2 \xrightarrow[-KCl]{} (C_6H_5)_3SiSiCl(C_6H_5)_2 \quad (1\text{-}114)$$

$$2(C_6H_5)_3SiK + (C_6H_5)_2SiCl_2 \xrightarrow[-2KCl]{} (C_6H_5)_3SiSi(C_6H_5)_2Si(C_6H_5)_3 \quad (1\text{-}114a)$$

In the reaction with triphenylsilylpotassium, trichlorosilane gives only a small amount of tris(triphenylsilyl)silane [1275]. Phenyltrichlorosilane forms 34% of 1,1,1,2-tetraphenyl-2,2-dichlorodisilane with triphenylsilylpotassium:

$$(C_6H_5)_3SiK + C_6H_5SiCl_3 \xrightarrow[-KCl]{} (C_6H_5)_3SiSiCl_2C_6H_5 \quad (1\text{-}115)$$

The product of the reaction of triphenylsilylpotassium [794] with carbon tetrachloride is 1,1,1-triphenyl-2,2,2-trichlorodisilane (27%). At the same time, the reaction of carbon tetrachloride with triphenylsilyllithium leads to the formation of only a large amount of hexaphenyldisilane [1275].

In Si-halogen substituted disilanes [725, 726, 772, 781, 854, 1272, 1273, 1275], the halogen atoms are also replaced by triarylsilyl groups by the action of R_3SiM. However, there is often cleavage of the Si–Si bond in the starting disilane. Thus, the action of triphenylsilyllithium on pentaphenylchlorodisilane forms hexaphenyldisilane in addition to octaphenyltrisilane. The action of triphenylsilyllithium on 1,1,2,2-tetraphenyl-1,2-dichlorosilane will give heptaphenylchlorotrisilane [1272]:

$$(C_6H_5)_3SiLi + Cl{-}\underset{C_6H_5}{\overset{C_6H_5}{Si}}{-}\underset{C_6H_5}{\overset{C_6H_5}{Si}}{-}Cl \xrightarrow[-LiCl]{} C_6H_5\left[-\underset{C_6H_5}{\overset{C_6H_5}{Si}}-\right]_3 Cl \quad (1\text{-}116)$$

In the case of methyldiphenylsilyllithium, 1,4-dimethyloctaphenyltetrasilane is formed [772, 854, 1273]:

$$2CH_3(C_6H_5)_2SiLi + Cl{-}\underset{C_6H_5}{\overset{C_6H_5}{Si}}{-}\underset{C_6H_5}{\overset{C_6H_5}{Si}}{-}Cl \xrightarrow[-2LiCl]{} CH_3\left[-\underset{C_6H_5}{\overset{C_6H_5}{Si}}-\right]_4 CH_3 \quad (1\text{-}117)$$

The use of lithium derivatives of polysilanes in these reactions has made it possible to synthesize organopolysilanes containing chains of nine and ten silicon atoms:

$$Li\left[-\underset{C_6H_5}{\overset{C_6H_5}{Si}}-\right]_5 Li + 2(CH_3)_2C_6H_5SiSiCl(C_6H_5)_2 \xrightarrow[-2LiCl]{}$$

$$\longrightarrow (CH_3)_2C_6H_5Si{-}\left[-\underset{C_6H_5}{\overset{C_6H_5}{Si}}-\right]_7{-}SiC_6H_5(CH_3)_2 \quad (1\text{-}118)$$

$$2CH_3\left[-\underset{C_6H_5}{\overset{C_6H_5}{Si}}-\right]_4 Li + Br{-}\underset{C_6H_5}{\overset{C_6H_5}{Si}}{-}\underset{C_6H_5}{\overset{C_6H_5}{Si}}{-}Br \xrightarrow[-2LiBr]{} CH_3\left[-\underset{C_6H_5}{\overset{C_6H_5}{Si}}-\right]_{10} CH_3 \quad (1\text{-}119)$$

TABLE 8. Products of the Reaction of Organosilyl Derivatives of Alkali Meatls with Halosilanes†

Organosilyl derivative of metal	Halosilane	Reaction products	Yield %	Literature
$(C_6H_5)_3SiLi$	$SiCl_4$	$(C_6H_5)_3SiSi(C_6H_5)_3$	72.5	1275
		$(C_6H_5)_3SiH$	7.5	
$(C_6H_5)_3SiK$	$SiCl_4$	$(C_6H_5)_3SiSiCl_3$	27	794
$(C_6H_5)_3SiLi$	$HSiCl_3$	$[(C_6H_5)_3Si]_3SiH$	4.4	1275
		$(C_6H_5)_3SiSi(C_6H_5)_3$	20.6	
		$(C_6H_5)_3SiH$	29.5	
$(C_6H_5)_3SiK$	$C_6H_5SiCl_3$	$(C_6H_5)_3SiSiCl_2C_6H_5$	34	794
		$(C_6H_5)_3SiSi(C_6H_5)_3$	16	
$(C_6H_5)_3SiLi$ (2 : 5)	$(CH_3)_2SiCl_2$	$(C_6H_5)_3SiSi(CH_3)_2Si(C_6H_5)_3$	9.6	782
		$(CH_3)_2Si[(CH_3)_2Si]_4Si(CH_3)_2$ (cyclic)	25.9	
$(C_6H_5)_3SiLi$ (2 : 1)	$(CH_3)_2SiCl_2$	$(C_6H_5)_3SiSi(CH_3)_2Si(C_6H_5)_3$	47.7	782
$Li[(C_6H_5)_2Si]_5Li$	$(CH_3)_2SiCl_2$	$(C_6H_5)_2Si[(C_6H_5)_2Si]_4Si(CH_3)_2$ (cyclic)	85	725
$Li[(C_6H_5)_2Si]_5Li$	$(CH_3)_2SiCl_2$	$(C_6H_5)_2Si[(C_6H_5)_2Si]_4Si(CH_3)_2$ (cyclic)	37	778
$(C_6H_5)_3SiLi$	$(C_6H_5)_2SiCl_2$	$(C_6H_5)_3SiSiH(C_6H_5)_2$ (after reduction with $LiAlH_4$)	24.5	1275
		$(C_6H_5)_3SiSi(C_6H_5)_2Si(C_6H_5)_3$	9.6	
		$(C_6H_5)_3SiSi(C_6H_5)_3$	45	
$(C_6H_5)_3SiK$ (2 : 1)	$(C_6H_5)_2SiCl_2$	$(C_6H_5)_3SiSi(C_6H_5)_2Si(C_6H_5)_3$	29	798
$(C_6H_5)_3SiK$ (1 : 1)	$(C_6H_5)_2SiCl_2$	$(C_6H_5)_3SiSiCl(C_6H_5)_2$	50	748, 798
$CH_3(C_6H_5)_2SiLi$	$(C_6H_5)_2SiCl_2$	$CH_3(C_6H_5)_2SiSi(C_6H_5)_2SiCH_3(C_6H_5)_2$	—	725
$CH_3[(C_6H_4)_2Si]_4Li$	$(C_6H_5)_2SiCl_2$	$CH_3[(C_6H_5)_2Si]_9CH_3$	—	725
$CH_3(C_6H_5)_2SiSiLi(C_6H_5)_2$	$(C_6H_5)_2SiCl_2$	$CH_3[(C_6H_5)_2Si]_5CH_3$	—	778

$Li[(C_6H_5)_2Si]_2Li$ (1 : 1)	$(C_6H_5)_2SiCl_2$	$(C_6H_5)_2Si[(C_6H_5)_2Si]_2Si(C_6H_5)_2$ (cyclic)	13.1	778
		$(C_6H_5)_2Si[(C_6H_5)_2Si]_3Si(C_6H_5)_2$ (cyclic)	24.5	
		$(C_6H_5)_2Si[(C_6H_5)_2Si]_4Si(C_6H_5)_2$ (cyclic)	3	
$Li[(C_6H_5)_2Si]_2Li$ (1 : 3)	$(C_6H_5)_2SiCl_2$	$H[(C_6H_5)_2Si]_7H$	82	778
$(C_6H_5)_3SiLi$	$(CH_3)_3SiCl$	$(C_6H_5)_3SiSi(CH_3)_3$	72—84	581, 712, 716, 758, 773, 937†
$(C_6H_5)_3SiNa$	$(CH_3)_3SiCl$	$(C_6H_5)_3SiSi(CH_3)_3$	68	581
$(C_6H_5)_3SiK$	$(CH_3)_3SiCl$	$(C_6H_5)_3SiSi(CH_3)_3$	45—77	549, 560, 581, 716, 791, 796
$(C_6H_5)_3SiRb$	$(CH_3)_3SiCl$	$(C_6H_5)_3SiSi(CH_3)_3$	48	716, 796
$(C_6H_5)_3SiCs$	$(CH_3)_3SiCl$		26	716, 796
$(C_6H_5)_3SiLi+C_6H_5Li$	$(CH_3)_3SiCl$	$(C_6H_5)_3SiSi(CH_3)_3$	49.2	788
		$C_6H_5Si(CH_3)_3$	20	
$CH_3(C_6H_5)_2SiLi$	$(CH_3)_3SiCl$	$CH_3(C_6H_5)_2SiSi(CH_3)_3$	74	758
$(CH_3)_2C_6H_5SiLi$	$(CH_3)_3SiCl$	$(CH_3)_2C_6H_5SiSi(CH_3)_3$	47	758
$Li[(C_6H_5)_2Si]_4Li$	$(CH_3)_3SiCl$	$(CH_3)_3Si[(C_6H_5)_2Si]_4Si(CH_3)_3$	—	778
$Li[(C_6H_5)_2Si]_5Li$	$(CH_3)_3SiCl$	$(CH_3)_3Si[(C_6H_5)_2Si]_5Si(CH_3)_3$	83	778
$(C_6H_5)_3SiLi$	$(CH_3)_3SiCl+(C_6H_5)_3SiCl$	$(C_6H_5)_3SiSi(CH_3)_3$	45.3	788
		$(C_6H_5)_3SiSi(C_6H_5)_3$	27.2	
$(C_6H_5)_3SiLi$	$(CH_3)_3SiCl+(C_6H_5)_3SiBr$	$(C_6H_5)_3SiSi(CH_3)_3$	48.5	788
		$(C_6H_5)_3SiSi(C_6H_5)_3$	26.1	
$(C_6H_5)_3SiLi$	$(CH_3)_3SiCl+(C_6H_5)_3SiOC_2H_5$	$(C_6H_5)_3SiSi(CH_3)_3$	60.3	788
$(C_6H_5)_3SiLi$	$(C_2H_5)_3SiCl+(C_6H_5)_3SiCl$	$(C_6H_5)_3SiSi(C_2H_5)_3$	44.8	788
		$(C_6H_5)_3SiSi(C_6H_5)_3$	26.1	

† For cases where the reagents were used in nonequivalent amounts, their ratio is given in brackets in the column "Organosilyl derivative."

TABLE 8. (Cont'd)

Organosilyl derivative of metal	Halosilane	Reaction products	Yield, %	Literature
$(C_6H_5)_3SiK$	$(C_2H_5)_3SiCl$	$(C_6H_5)_3SiSi(C_2H_5)_3$	37—87	556, 560
$(C_6H_5)_3SiK$	$(C_2H_5)_3SiI$	$(C_6H_5)_3SiSi(C_2H_5)_3$	80	754
$(C_6H_5)_3SiK$	$(CH_3)_2C_6H_5SiCl$	$(C_6H_5)_3SiSiC_6H_5(CH_3)_2$	68.4	754
$Li[(C_6H_5)_2Si]_5Li$	$(CH_3)_2C_6H_5SiCl$	$(CH_3)_2C_6H_5Si[(C_6H_5)_2Si]_5SiC_6H_5(CH_3)_2$	—	725
$(C_6H_5)_3SiLi$	$(CH_3)_2C_6H_5SiCl + (C_6H_5)_3SiCl$	$(C_6H_6)_3SiSiC_6H_5(CH_3)_2$	40.5	788
		$(C_6H_5)_3SiSi(C_6H_5)_3$	30	
$(C_6H_5)_3SiLi$	$CH_3(C_6H_5)_2SiCl + (C_6H_5)_3SiCl$	$(C_6H_5)_3SiSiCH_3(C_6H_5)_2$	35.9	788
		$(C_6H_5)_3SiSi(C_6H_5)_3$	34.4	
$CH_3[(C_6H_5)_2Si]_4Li$	$CH_3(C_6H_5)_2SiCl$	$CH_3[(C_6H_5)_2Si]_5CH_3$	—	725
$Li[(C_6H_5)_2Si]_4Li$	$CH_3(C_6H_5)_2SiCl$	$CH_3[(C_6H_5)_2Si]_6CH_3$	25	725, 752
$(C_6H_5)_3SiLi$	$(C_6H_5)_2SiHCl$	$(C_6H_5)_3SiSiH(C_6H_5)_2$	42	1272
$Li[(C_6H_5)_2Si]_4Li$	$(C_6H_5)_2SiHCl$	$H[(C_6H_5)_2Si]_6H$	18—40	722, 1272
$Li[(C_6H_5)_2Si]_5Li$	$(C_6H_5)_2SiHCl$	$H[(C_6H_5)_2Si]_7H$	—	91, 725
$(C_6H_5)_3SiLi$	$(C_6H_5)_3SiCl$	$(C_6H_5)_3SiSi(C_6H_5)_3$	89: 92.5	715, 716
$(C_6H_5)_3SiK$	$(C_6H_5)_3SiCl$	$(C_6H_5)_3SiSi(C_6H_5)_3$	69	585, 590
$(CH_3)_2C_6H_5SiLi$	$(C_6H_5)_3SiCl$	$(CH_3)_2C_6H_5SiSiC_6H_5(CH_3)_2$	32	716, 758
		$(C_6H_5)_3SiSi(C_6H_5)_3$	2.3	
$(CH_3)_2C_6H_5SiLi$	$(C_6H_5)_3SiCl$ *	$(CH_3)_2C_6H_5SiSiC_6H_5(CH_3)_2$	58—71	758
		$(C_6H_5)_3SiSi(C_6H_5)_3$	35—38	
$CH_3(C_6H_5)_2SiLi$	$(C_6H_5)_3SiCl$	$CH_3(C_6H_5)_2SiSi(C_6H_5)_3$	36.2: 52	758, 773
$CH_3(C_6H_5)_2SiLi$	$(C_6H_5)_3SiCl$ *	$CH_3(C_6H_5)_2SiSi(C_6H_5)_3$	—	758
		$CH_3(C_6H_5)_2SiSiCH_3(C_6H_5)_2$	—	
		$(C_6H_5)_3SiSi(C_6H_5)_3$	19	
$(C_6H_5)_3SiLi$	$(C_6H_5)_3SiCl$	$(C_6H_5)_3SiSi(C_6H_5)_3$	65	758
$(CH_3)_2C_6H_5SiLi$		$(CH_3)_2C_6H_5SiSiC_6H_5(CH_3)_2$	61	

$(C_6H_5)_3SiLi + C_6H_5Li$	$(C_6H_5)_3SiCl$	$(C_6H_5)_3SiSi(C_6H_5)_3$ $(C_6H_5)_4Si$	73.8 12	788
$(C_6H_5)_3SiLi + C_6H_5Li$	$(C_6H_5)_3SiCl$ **	$(C_6H_5)_3SiSi(C_6H_5)_3$ $(C_6H_5)_4Si$	81.7 7.7	788
$(C_6H_5)_3SiLi + n\text{-}C_4H_9Li$	$(C_6H_5)_3SiCl$	$(C_6H_5)_3SiSi(C_6H_5)_3$ $(C_6H_5)_3SiC_4H_9\text{-}n$	61 21.5	788
$(C_6H_5)_3SiLi + n\text{-}C_4H_9Li$	$(C_6H_5)_3SiCl$ **	$(C_6H_5)_3SiSi(C_6H_5)_3$ $(C_6H_5)_3SiC_4H_9\text{-}n$	77.2 13.9	788
$(C_6H_5)_3SiLi + C_6H_5CH_2Li$	$(C_6H_5)_3SiCl$	$(C_6H_5)_3SiSi(C_6H_5)_3$; $(C_6H_5)_3SiCH_2C_6H_5$	54: 23	788
$(C_6H_5)_2SiHLi$	$(C_6H_5)_3SiCl$	$(C_6H_5)_3SiSiH(C_6H_5)_2$	11.3	780
$(p\text{-}CH_3C_6H_4)_3SiK$	$(C_6H_5)_3SiCl$	$(p\text{-}CH_3C_6H_4)_3SiSi(C_6H_5)_3$	26	794
$C_6H_{11}Si(C_6H_5)Li$	$(C_6H_5)_3SiCl$	$C_6H_{11}Si(C_6H_5)Si(C_6H_5)_3$	—	92
$Li[(C_6H_5)_2Si]_5Li$	$(C_6H_5)_3SiCl$	$C_6H_5[(C_6H_5)_2Si]_7C_6H_5$	—	91, 725
$(C_6H_5)_3SiK$	$p\text{-}CH_3C_6H_4(C_6H_5)_2SiCl$	$(C_6H_5)_3SiSi(C_6H_5)_2C_6H_4CH_3\text{-}p$	77	794
$(C_6H_5)_3SiK$	$(p\text{-}CH_3C_6H_4)_2(C_6H_5)SiCl$	$C_6H_5(p\text{-}CH_3C_6H_4)_2SiSi(C_6H_5)_3$	72	794
$p\text{-}CH_3C_6H_4(C_6H_5)_2SiK$	$(p\text{-}CH_3C_6H_4)_2(C_6H_5)SiCl$	$C_6H_5(p\text{-}CH_3C_6H_4)_2SiSi(C_6H_5)_2C_6H_4CH_3\text{-}p$	52	794
$(C_6H_5)_3SiLi$	$C_6H_{11}Si(C_6H_5)Cl$	$C_6H_{11}Si(C_6H_5)Si(C_6H_3)_3$	—	92
$Li[(C_6H_5)_2Si]_5Li$	$(CH_3)_2C_6H_5SiSiCl(C_6H_5)_2$	$(CH_3)_2C_6H_5Si[(C_6H_5)_2Si]_7SiC_6H_5(CH_3)_2$	26	725, 781
$(C_6H_5)_3SiLi$	$(C_6H_5)_3SiSiCl(C_6H_5)_2$	$(C_6H_5)_3SiSi(C_6H_5)_2Si(C_6H_5)_3$ $(C_6H_5)_3SiSi(C_6H_5)_3$	25.7 43.3	1275
$(C_6H_5)_3SiLi$	$(C_6H_5)_2SiClSiCl(C_6H_5)_2$	$(C_6H_5)_3Si[(C_6H_5)_2Si]_2Cl$	—	1272
$CH_3(C_6H_5)_2SiLi$	$(C_6H_5)_2SiClSiCl(C_6H_5)_2$	$CH_3(C_6H_5)_2Si[(C_6H_5)_2Si]_3CH_3$	—	772, 854
$CH_3(C_6H_5)_2SiLi$	$(C_6H_5)_2SiBrSiBr(C_6H_5)_2$	»	17	1273
$CH_3(C_6H_5)_2Si(C_6H_5)_2SiLi$	$(C_6H_5)_2SiBrSiBr(C_6H_5)_2$	$CH_3[(C_6H_5)_2Si]_6CH_3$	—	1273
$CH_3[(C_6H_5)_2Si]_4Li$	$(C_6H_5)_2SiBrSiBr(C_6H_5)_2$	$CH_3[(C_6H_5)_2Si]_{10}CH_3$	—	725

*The reaction was carried out by addition of the halosilane to R_3SiM.

**The reaction was carried out with the mixed solvent tetrahydrofuran–diethyl ether.

The reaction of α,ω-dilithium derivatives of organopolysilanes with dialkyldichlorosilanes also gives cyclic organopolysilanes [778]:

$$\mathrm{Li}\left[\begin{array}{c} C_6H_5 \\ | \\ -\mathrm{Si}- \\ | \\ C_6H_5 \end{array}\right]_5 \mathrm{Li} + (CH_3)_2SiCl_2 \longrightarrow \begin{array}{ccc} & (CH_3)_2 & \\ & \mathrm{Si} & \\ (C_6H_5)_2\mathrm{Si} & & \mathrm{Si}(C_6H_5)_2 \\ | & & | \\ (C_6H_5)_2\mathrm{Si} & & \mathrm{Si}(C_6H_5)_2 \\ & \mathrm{Si} & \\ & (C_6H_5)_2 & \end{array} \quad (1\text{-}120)$$

Thus, cleavage reactions of organocyclopolysilanes by lithium with subsequent reaction of the dilithium derivatives formed with organochlorosilanes offers a convenient route for the synthesis of very diverse linear or cyclic organopolysilanes with a set number of Si–Si links.

The products of the reaction of R_3SiM with Si-halogen substituted organosilicon compounds and their yields are given in Table 8.

1.3.10. Reaction with Heteroorganic Compounds

Trialkyl phosphates vigorously alkylate triphenylsilyllithium, converting it in high yields (83-97%) into the corresponding alkyltriphenylsilanes [713, 725, 744, 772, 854, 1273]. An investigation of the competing reactions of many organic compounds containing different functional groups (ethers, halogen derivatives, nitriles, and trialkyl phosphates) with triphenylsilyllithium showed that the reaction of the latter with trimethyl phosphate proceeds at the highest rate [784]. In the reaction of triphenylsilyllithium with tributyl phosphate, the lithium atom is replaced by a butyl group:

$$(C_6H_5)_3SiLi + (C_4H_9O)_3PO \longrightarrow (C_6H_5)_3SiC_4H_9 + (C_4H_9O)_2P(O)OLi \quad (1\text{-}121)$$

One butyl group of the tributyl phosphate participates preferentially in the reaction. A second butyl group is removed with much more difficulty, while the third apparently shows no alkylating action at all [744].

In contrast to trialkyl phosphates, triphenyl phosphate and tritolyl phosphate do not arylate triphenylsilyllithium (the reaction products are hexaphenyldisilane and phenol) [743].

By the reaction of triphenylsilyllithium with chloroesters of phosphoric acid it is possible to obtain the corresponding organosilicon phosphorus compounds [977]:

$$(C_6H_5)_3SiLi + ClP(O)(OCH_2C_6H_5)_2 \xrightarrow{-LiCl} (C_6H_5)_3SiP(O)(OCH_2C_6H_5)_2 \quad (1\text{-}122)$$

Triphenylsilyllithium cleaves the Ge–Ge bond in hexaphenyldigermane [92]:

$$(C_6H_5)_3GeGe(C_6H_5)_3 + (C_6H_5)_3SiLi \longrightarrow (C_6H_5)_3GeLi + (C_6H_5)_3SiGe(C_6H_5)_3 \quad (1\text{-}123)$$

Data on the reaction of R_3SiM with other heteroorganic compounds are given in subsequent chapters.

1.4. Analysis

The presence of R_3SiM is readily observed qualitatively by means of a Gilman color test [776]. To 0.5 ml of the solution investigated is added an equal amount of a 1% benzene solution of Michler's ketone. The mixture is hydrolyzed with 1 ml of water by shaking and 1-2 drops of a 0.2% solution of iodine in acetic acid is added. An intense bright green color appears immediately. It is also possible to use benzophenone for a color test: when R_3SiM is added to it, a characteristic greenish blue color appears [92].

R_3SiM is determined quantitatively by acidimetric titration [758, 773]. However, if the R_3SiM reacts with the solvent with the formation of alcoholates, this method leads to high results. Therefore the following method is more suitable [581]. The total amount of alkali in the test sample is first determined separately. The content of alcoholates is determined by titration after decomposition of the R_3SiM with benzyl chloride. The R_3SiM content is calculated from the difference between the first and second titration. Instead of benzyl chloride it is possible to use allyl chloride [725] or bromide [740]. A method has also been proposed based on the decomposition of R_3SiM with butyl bromide followed by determination of the inorganic bromide by the Volhard method [756].

2. COMPOUNDS CONTAINING THE $Si-(C)_n-M$ GROUP

2.1. Synthesis Methods

Organolithium compounds containing silicon in which there is the grouping $Si-(C)_n-Li$ are very important intermediates in many organolithium syntheses of very diverse organic compounds of silicon. They are obtained by methods analogous to those used for the synthesis of organolithium compounds. This refers primarily to the replacement of a halogen atom (X) in haloorganosilanes by a lithium atom:

$$\gt Si-(C)_n-X + 2Li \longrightarrow \gt Si-(C)_n-Li + LiX \qquad (1\text{-}124)$$

Both haloalkyl- and haloarylsilanes are used as organosilicon components for this reaction. The reaction is carried out in ethyl ether [555, 642, 652, 705, 770, 808, 1028, 1153, 1412], pentane [749, 751, 829, 1175, 1176, 1311, 1312, 1547], 2-methylpentane [534, 643], or benzene [751].

The reactivity of triorgano(haloalkyl)silanes in their reaction with lithium depends on the nature of the radicals attached to the silicon atoms. Thus, the yield of trimethylsilylmethyllithium from the reaction of trimethyl(chloromethyl)silane with lithium at room temperature in 2-methylpentane is 60% [643] and in boiling pentane it is 90% [1175]. However, it is not possible to obtain lithium derivatives at all by the action of lithium on methyldiphenyl(chloromethyl)silane [751] or dimethylphenyl(chloromethyl)silane [705] in pentane, ethyl ether, or tetrahydrofuran. These compounds react with lithium only when the reaction temperature is raised to 80°C, but react readily with magnesium in ethyl ether. At the same time, methyldiethoxy(chloromethyl)silane and dimethylethoxy(chloromethyl)silane react smoothly with lithium in pentane (~50% yield) [705]. When treated with lithium in pentane, dimethyl(trimethylsilylmethyl)chloromethylsilane forms a lithium derivative in ~94% yield [1175]. When one methyl group in the silyl radical of this compound is replaced by a phenyl, the yield of the lithium derivative is reduced to 50%. The reaction then proceeds in ether, but not in pentane [705]. Methyl-bis[(dimethylphenylsilyl)methyl](chloromethyl)-silane likewise does not react with lithium in pentane [705].

The reactions of bromoalkylsilanes with lithium occur at lower temperatures (from -25 to -15°C) than in the case of the chloro derivatives. Thus, trimethylsilylmethyllithium [652] and 4-(triphenylsilyl)butyllithium [1278] were obtained in 60-70% yield from the corresponding bromo derivatives.

The use of bis(chloromethyl)organosilanes for the reaction with lithium makes it possible to obtain dilithium derivatives [1153]:

$$(CH_3)_2Si\begin{matrix}\diagup CH_2Cl \\ \diagdown CH_2Cl\end{matrix} + 4Li \longrightarrow (CH_3)_2Si\begin{matrix}\diagup CH_2Li \\ \diagdown CH_2Li\end{matrix} + 2LiCl \qquad (1\text{-}125)$$

Trialkyl(halophenyl)silanes react with lithium with more difficulty than trialkyl(halomethyl)silanes. The yields of the trialkylsilylphenyllithium are 60-80% from bromo derivatives [555, 770, 953] and 30-50% from the corresponding chloro derivatives [554, 555, 1028]. The yields of o-, m-, and p-lithio substituted trimethylbenzylsilanes reach 75-85% [661, 662].

The second general method of synthesizing organosilicon compounds containing the grouping $Si-(C)_n-Li$ is metallation of the corresponding organic derivatives of silicon with an organolithium. Two cases are possible here:

1) replacement of the halogen atom by lithium:

$$\geqslant Si-(C)_n-X + LiR \longrightarrow \geqslant Si-(C)_n-Li + RX \qquad (1\text{-}126)$$

2) replacement of a hydrogen atom by lithium:

$$\geqslant Si-(C)_n-H + LiR \longrightarrow \geqslant Si-(C)_n-Li + RH \qquad (1\text{-}127)$$

The reaction of haloorganosilanes with butyllithium has been used rarely in the chemistry of organosilicon compounds for the synthesis of organolithium derivatives [765, 1219, 1698]. This method has shown no advantages over reaction (1-24). On the contrary, the yield of organolithiosilicon compounds in this case is still lower (~35%) than when lithium is used as the metallating agent (45-50%).

A hypothesis on the intermediate formation of compounds containing the $Si-(C)_n-Li$ group was put forward to explain the mechanism of the formation of trimethylbutylsilane, trimethylamylsilane, 1,2-bis(trimethylsilyl)ethylene, and 1,2-bis(trimethylsilyl)-hexane by the action of butyllithium on trimethyl(chloromethyl)silane. It is assumed that the first stage in this reaction is metallation of the C–H bond of the methylene group by butyllithium with subsequent elimination of LiCl and the formation of trimethylsilylcarbene [282]:

$$(CH_3)_3SiCH_2Cl \xrightarrow[-C_4H_{10}]{C_4H_9Li} (CH_3)_3Si\underset{\displaystyle Li}{\underset{|}{C}}HCl \xrightarrow{LiCl} (CH_3)_3Si\ddot{C}H \qquad (1\text{-}128)$$

The replacement of hydrogen attached to a carbon atom by lithium by the action of butyllithium or phenyllithium in accordance with scheme (1-127) is the main method of synthesizing silicon-containing organolithium derivatives of heterocyclic compounds. The reaction has been used successfully in the furan [355, 547, 548] and thiophene series [392, 547, 548, 737, 774].

Heating with butyllithium produces metallation of triorganocyclopentadienylsilanes [803, 1657] ethynylsilanes [728], fluorenylsilanes [664, 737], benzhydrylsilanes [1288], and carboranylsilanes* [1004]. By more prolonged treatment it was possible to metallate both phenyl groups of trimethyl(p-phenoxyphenyl)silane [785].

Carbonation and hydrolysis of the reaction product of triphenylbenzylsilane and butyllithium forms phenylacetic acid and triphenylsilanol. This may be explained by decomposition of the acid obtained from α-(triphenylsilyl)benzyllithium present in the reaction mixture [753].

The Si–C bond in 9,9'-spyrobi(fluorenyl) is cleaved by phenyllithium in ether at room temperature with the formation of a lithium derivative, which gives after hydrolysis 87% of 9-(2-xenyl)-9-phenyl-9-silafluorene (51% in THF):

Si + C_6H_5Li ⟶ Si, C_6H_5, Li $\xrightarrow{(H_2O)}$ Si, C_6H_5 (1-129)

The third method of synthesizing organolithium compounds containing lithium is the addition of an organolithium to alkenylsi-

* See Ch. 3, Section 2.1.3.

lanes:

$$R_3SiCH{=}CH_2 + R'Li \longrightarrow R_3Si\underset{\displaystyle Li}{\underset{|}{C}}HCH_2R' \quad (1\text{-}130)$$

Whether or not this reaction occurs is determined to a considerable extent by the nature of the radicals R and R'. The addition of butyllithium or phenyllithium to trimethylvinylsilane in ether occurs at room temperature [622]. Polymers are formed in addition to monomeric reaction products, whose yield does not exceed 10%. When treated with ethyllithium, trimethylvinylsilane and dimethylphenylvinylsilane are polymerized completely [468, 1506]. Polymers predominate in the products of the reaction of hexyllithium or naphthyllithium with triphenylvinylsilane [622]. In contrast to this, phenyllithium adds readily to triphenylvinylsilane; after 6 hr at room temperature, 84% of the addition product is formed in accordance with scheme (1-130) [437, 621, 622]. Butyllithium adds to triphenylvinylsilane analogously but with much more difficulty. The yield of the product from a reaction lasting for 44 hr is 67% [621, 622]. Tolyllithium, p-dimethylaminophenyllithium, propyllithium, and methyllithium react with triphenylvinylsilane even more slowly [622].

The reaction of an organolithium with triethylperfluorovinylsilane yields 1-triethylsilyl-1,2-difluoroalkenes [1157, 1158]. Their formation may be explained by direct nucleophilic substitution. However, there is also the possibility that the organolithium first adds to the double bond of the perfluorovinyl group with subsequent elimination of lithium fluoride:

$$(C_2H_5)_3SiCF{=}CF_2 + RLi \longrightarrow \left[(C_2H_5)_3Si\underset{\displaystyle Li}{\underset{|}{C}}FCF_2R\right] \xrightarrow{-LiF} (C_2H_5)_3SiCF{=}CFR \quad (1\text{-}131)$$

It is not possible to add butyllithium [622] or phenyllithium [437] to triorganoallylsilanes.

An alkyllithium reacts vigorously with trialkyl(vinylethynyl)-silanes [437]:

$$R_3SiC{\equiv}C{-}CH{=}CH_2 + R'Li \xrightarrow{(H_2O)} R_3SiC{\equiv}C{-}CH_2CH_2R' + R_3SiCH{=}C{=}CHCH_2R' \quad (1\text{-}132)$$

The ratio of the isomers obtained depends on the structure of the radical R'. Thus, for example, the product of the addition of butyllithium to trimethyl(vinylethynyl)silane is a mixture containing ~55% of a compound with an acetylene structure and ~45% of the isomeric allene. At the same time, the product from the addition of methyllithium contains almost 90% of the acetylene compound [422]. Adducts have also been isolated to which have been assigned the structures of products from the addition of the initially formed lithium derivatives to a second molecule of the starting trialkyl(vinylethynyl)silane [431]. The addition of lithium dialkylamides and piperidide to trialkyl(vinylethynyl)silanes proceeds analogously with the formation of a mixture of different amounts of acetylene and allene amines [423].

The addition of an alkyllithium to 1-trimethylsilylhexadiyne-1,3 occurs in the 3,4-position with attachment of the alkyl radical to the carbon atom in position 4 and the formation of an enyne silicohydrocarbon [493]:

$$(CH_3)_3SiC{\equiv}C{-}C{\equiv}C{-}C_2H_5 + RLi \longrightarrow (CH_3)_3SiC{\equiv}C{-}\underset{\displaystyle Li}{\underset{|}{C}}{=}\underset{\displaystyle R}{\underset{|}{C}}{-}C_2H_5 \qquad (1\text{-}133)$$

The reaction of organosilicon compounds of mercury with lithium may also be regarded as a general method of synthesizing organolithium compounds containing silicon [705-707]:

$$\left[\gt Si{-}(C)_n\right]_2{-}Hg + 2\,Li \longrightarrow 2\ \gt Si{-}(C)_n{-}Li + Hg \qquad (1\text{-}134)$$

However, this method requires a series of preliminary conversions of the starting haloorganosilane into an organomagnesium and then an organomercury derivative. Therefore it is only advantageous to use it in cases where the direct replacement of a halogen atom by lithium in accordance with scheme (1-24) does not occur. Thus, by means of organomercury compounds it is possible to obtain dimethylphenylsilylmethyllithium [705, 707] and methyldiphenylsilylmethyllithium [706]. The reaction is carried out by shaking the organomercury compound containing silicon with lithium and it is followed through the amount of mercury liberated. For example, after bis(dimethylphenylsilylmethyl)mercury has been shaken with a 1.5-fold excess of lithium for 40 hr, 50% of the mercury is liberated. The yield of mercury reaches 86% after 110 hr [705].

In addition to the four general methods given above, specific methods have been developed for synthesizing compounds containing the grouping $Si-(C)_n-M$. They include the metallation of various organosilicon compounds with alkali metals and their organic derivatives. Thus, when tridodecyl(cyclopentadienyl)silane is stirred vigorously with a suspension of sodium in tetrahydrofuran, hydrogen in the cyclopentadiene ring is replaced by sodium [934]. At the same time, heating tetraphenylsilane, methyltriphenylsilane, dimethyldiphenylsilane, and trimethylphenylsilane with sodium in boiling decalin or heptadecene leads to neither metallation nor cleavage of the Si–C bond [550].

Lithium adds readily to trialkylsilyl-substituted naphthalenes:

$$C_{10}H_7SiR_3 + 2Li \longrightarrow \text{1-}R_3Si\text{-1,4-dilithio-1,4-dihydronaphthalene (Li, H; Li, H)} \qquad (1\text{-}135)$$

The lithium then enters the ring without the Si substituent.

The presence of a trialkylsilyl group in an aromatic nucleus produces a manyfold increase in the rate of addition in comparison with unsubstituted naphthalene [477].

Dimethylphenylmethylpotassium metallates the methyl group of tolylsilanes. This is indicated by the formation of tri(o-ethylphenyl)silane after treatment of the product from the reaction of dimethylphenylmethylpotassium with tri(o-tolyl)silane with methyl iodide or the formation of tris(o-carboxymethylphenyl)silane after treatment of the product with carbon dioxide [558]:

$$(o\text{-}CH_3C_6H_4)_3SiH + C_6H_5(CH_3)_2CK \longrightarrow C_6H_5CH(CH_3)_2 + [(o\text{-}KCH_2C_6H_4)_3SiH]$$

$$[(o\text{-}KCH_2C_6H_4)_3SiH] - \begin{cases} \xrightarrow{CH_3I} (o\text{-}C_2H_5C_6H_4)_3SiH \\ \xrightarrow{CO_2} (o\text{-}KOOCCH_2C_6H_4)_3SiH \end{cases} \qquad (1\text{-}136)$$

Dimethylphenylmethylpotassium also metallates triphenyl[(dimethyl)phenylmethyl]silane. The potassium atom then enters the p-position of the phenyl group attached to the methyl [560]:

$$(C_6H_5)_3SiC(CH_3)_2C_6H_5 + C_6H_5(CH_3)_2CK \longrightarrow$$

$$\longrightarrow (C_6H_5)_3SiC(CH_3)_2C_6H_4K + C_6H_5(CH_3)_2CH \qquad (1\text{-}137)$$

Organic derivatives of alkali metals containing silicon are formed in some reactions of triphenylsilyllithium or triphenylsilylpotassium with unsaturated hydrocarbons, compounds with condensed aromatic nuclei, and diaryl ketones (see Ch. 1, Sections 1.3.2 and 1.3.5).

In many reactions of triorganochlorosilanes with dilithium and disodium derivatives of various organic compounds, together with the corresponding bis(triorganosilyl) derivatives it is also possible to isolate substitution products containing only one triorganosilyl group. Hence, we must assume that the reaction mixture may contain a lithium or sodium organosilicon compound, which is converted on hydrolysis into a mono(triorganosilyl) derivative. Such monosubstituted derivatives are formed by the reaction of organochlorosilanes with dilithium derivatives of biphenyl [436, 749] or anthracene [435]. Their existence is assumed to explain the mechanism of the interaction of the sodium derivative of naphthalene with trimethylchlorosilane [1254] and the formation of 3,6-bis-(trimethylsilyl)cyclohexadiene-1,4 in the reaction of benzene with lithium and trimethylchlorosilane in tetrahydrofuran [1253]. The reaction of triphenylchlorosilane with 1,4-dilithiobutane forms 1,4-bis(triphenylsilyl)butane (4.4%), triphenylbutylsilane (5.4%), 1,1-diphenylsilacyclopentane (38.7%), and tetraphenylsilane (48%) [1278]. The presence of the last three compounds in the reaction products may only be explained by the intermediate formation of 4-(triphenylsilyl)butyllithium, which is cyclized with the elimination of phenyllithium, and gives triphenylbutylsilane on hydrolysis:

$$(C_6H_5)_3SiCl + Li(CH_2)_4Li \longrightarrow (C_6H_5)_3SiCH_2CH_2CH_2CH_2Si(C_6H_5)_3$$

$$(C_6H_5)_3SiCl + Li(CH_2)_4Li \longrightarrow (C_6H_5)_3Si(CH_2)_4Li \longrightarrow (C_6H_5)_2Si\left\langle\begin{matrix}CH_2CH_2\\ |\\ CH_2CH_2\end{matrix}\right. + C_6H_5Li \tag{1-138}$$

$$(C_6H_5)_3Si(CH_2)_4Li \xrightarrow{H_2O} (C_6H_5)_3SiC_4H_9 \qquad C_6H_5Li \xrightarrow{(C_6H_5)_3SiCl} (C_6H_5)_4Si$$

The main product from the reaction of triphenylchlorosilane with 1,5-dilithiopentane is 1,5-bis(triphenylsilyl)pentane (75%), while the cyclic compound is formed in only an insignificant amount (~1%) [1278].

The intermediate formation of triphenylsilylmethyllithium is also assumed in explaining the mechanism of the reaction of triphenylsilyllithium with halogen derivatives of methane [727].

2.2. Physical Properties

The physical properties of only one individual organolithiosilicon compound, namely, trimethylsilylmethyllithium, isolated from solution have been studied at the present time. This compound is a white crystalline substance with m.p. 112°C, which sublimes at 100°C and 10^{-5} mm. Molten trimethylsilylmethyllithium is stable on heating to 130°C but with a further rise in temperature it turns yellow and decomposes slowly with the formation of tetramethylsilane and an involatile residue. Its stability at 112-150°C is unusual since alkyllithium compounds decompose even at the melting point.

According to the first determinations, the molecular weight of trimethylsilylmethyllithium, in contrast to alkyllithiums, corresponds to the monomeric state [643]. However, subsequent ebullioscopic [534] and cryoscopic [829] investigations showed that trimethylsilylmethyllithium in benzene at 6 and 80°C and also in 2-methylpentane at 60°C is associated (mean degree of association ~4).

The UV spectra of a solution of trimethylsilylmethyllithium in isooctane show no absorption maxima above 200 mμ. This shows that there is no appreciable delocalization of the electrons in trimethylsilylmethyllithium [534].

The NMR spectrum of a benzene solution of trimethylsilylmethyllithium contains two maxima with chemical shifts of 7.06 (7.07) and 9.25 (9.23) ppm relative to the benzene protons with a relative intensity of 1 : 4.5 [534, 643]. The NMR spectrum of Li^7 of trimethylsilylmethyllithium in hexane has a signal with a chemical shift of -1.74 ppm (relative to an aqueous solution of LiBr as an external standard). It is similar in magnitude to the signal from butyllithium (-1.76 ppm) [829].

2.3. Chemical Properties

Compounds containing the $Si-(C)_n-M$ group take part in all the reactions which are characteristic of organic derivatives of alkali metals. They are therefore used for the synthesis of organosilicon compounds containing either different functional groups or two or several silicon atoms separated from each other by carbon atoms and also to prepare various silicon heteroorganic compounds.

The reaction of silicon-containing organolithium and organopotassium compounds with carbon dioxide yields organosilicon carboxylic acids:

$$\geq Si-(C)_n-M + CO_2 \xrightarrow[-MCl]{+HCl} \geq Si-(C)_n-COOH \tag{1-139}$$

In this way it is possible to obtain 4-(triphenylsilyl)valeric acid (56%) [1278], m-(trimethylsilyl)benzoic acid (48%) [554, 1028], p-(trimethylsilyl)benzoic acid (66%) [555], o-, m-, and p-(trimethylsilylmethyl)benzoic acids (75-85%) [652], p-(tribenzylsilyl)benzoic acid (36.5%) [765], 5-(trimethylsilyl)-2-furancarboxylic acid (62%), 5-(trimethylsilyl)-2-thiophenecarboxylic acid (62%) [547], and 5-(triphenylsilyl)-2-thiophenecarboxylic acid (45%) [737]. The silicon-containing carboxylic acids formed by the reaction of carbon dioxide with 9-lithio-9-trimethylsilylfluorene [737], α-(triphenylsilyl)benzyllithium [753], and diphenyl(triphenylsiloxy)methylpotassium [587, 1129] could not be isolated because of their instability.

Organolithium compounds containing silicon add to the carbonyl group of aldehydes [477, 1129] and ketones [477, 751] to form after hydrolysis the corresponding organosilicon carbinols:

$$\geq Si-(C)_n-Li + O=C\lessdot \xrightarrow[-LiOH]{(H_2O)} \geq Si-(C)_n-\overset{|}{\underset{|}{C}}-OH \tag{1-140}$$

Additions of organolithiosilicon compounds to quinoline [774], quinoxaline [548], and ethylene oxide (with opening of the epoxide ring) [313] have also been described.

Lithium and potassium organosilicon compounds are used most widely for introducing an organic radical into various organosilicon compounds *:

$$\geq Si-(C)_n-M + X-Si\lessdot \xrightarrow{-MX} \geq Si-(C)_n-Si\lessdot \tag{1-141}$$

$$(M = Li, K; \quad X = Cl, Br, I)$$

The organic halides used for this reaction include methyl iodide [558, 664, 1698], benzyl chloride [1129, 1288] and allyl bromide [808]. It is interesting to note that the reaction of triphenylsilylethynyllithium with chloro- or bromoethynyltriphenylsilane

* The reactions with hetero-organic compounds are described in other chapters.

forms bis(triphenylsilyl)acetylene and not the expected diacetylene derivatives [728]:

$$(C_6H_5)_3SiC{\equiv}CLi + ClC{\equiv}CSi(C_6H_5)_3 \longrightarrow (C_6H_5)_3SiC{\equiv}CSi(C_6H_5)_3 \qquad (1\text{-}142)$$

The reaction of trimethylsilylmethyllithium with trimethyl-(chloromethyl)silane also forms tetramethylsilane and bis(trimethylsilyl)methane in addition to 1,2-bis(trimethylsilyl)ethane [643].

The reaction of trimethylsilylmethyllithium with silicon tetrachloride makes it possible to obtain a whole series of (trimethylsilylmethyl)silanes containing from one to four trimethylsilylmethyl groups attached to the central silicon atom [99, 1176, 1311, 1312, 1546, 1547]:

$$n\,(CH_3)_3SiCH_2Li + SiCl_4 \xrightarrow[-nLiCl]{} [(CH_3)_3SiCH_2]_n SiCl_{4-n} \qquad (1\text{-}143)$$
$$n = 1-4$$

In this connection it should be pointed out that the reaction of di-tert-butyldichlorosilane with tert-butyllithium does not form tri-tert-butylchlorosilane even with prolonged heating to 160°C [1213]. In addition, the conditions for the formation of tetrakis(trimethyltrisilylmethyl)silane are more drastic (12 hr at 150°C) than in the preparation of tetrabutylsilane from butyllithium and silicon tetrachloride. This indicates that as regards its steric effect, the trimethylsilylmethyl group is intermediate between normal and tertiary butyl.

The reaction of p-(trimethylsilyl)phenyllithium [953, 1182] and p-(tribenzylsilyl)phenyllithium [765, 1182] with silicon tetrachloride also makes it possible to replace up to four chlorine atoms by p-(trimethylsilyl)phenyl or p-(tribenzylsilyl)phenyl groups. The yields of tetrasubstituted silanes in this case are 77 and 51%, respectively. Tetrakis(triorganosilylphenyl)germanes and tetrakis-(triorganosilylphenyl)stannanes may be synthesized analogously [953].

The products of the reactions of compounds containing the $Si-(C)_n-M$ group with halosilanes and their yields are given in Table 9.

Triorganosilyl(cyclopentadienyl)lithiums are formed as intermediate products in the synthesis of 1,1'-bis(triorganosilyl)ferrocenes (see Ch. 8).

TABLE 9. Products of the Reaction of $R_3Si-(C)_n-M$ with Halosilanes

$R_3Si-(C)_n-M$	Halosilane	Reaction product	Yield, %	Literature
$(CH_3)_3SiCH_2Li$	$SiCl_4$	$[(CH_3)_3SiCH_2]_2SiCl_2$	40	1176
		$[(CH_3)_3SiCH_2]_3SiCl$	50	1176
$(CH_3)_3SiCH_2Li$	$SiCl_4$	$[(CH_3)_3SiCH_2]_3SiCl$	32	1311, 1312, 1547
$(CH_3)_3SiCH_2Li$	$SiCl_4$	$[(CH_3)_3SiCH_2]_4Si$	33	1176, 1311, 1546
$(CH_3)_3SiCH_2Li$	$HSiCl_3$	$[(CH_3)_3SiCH_2]_3SiH$	71	652
$(CH_3)_3SiCH_2Li$	CH_3SiCl_3	$(CH_3)_3SiCH_2SiCl_2CH_3$	25	1176
		$[(CH_3)_3SiCH_2]_2SiClCH_3$	20	1176
$(CH_3)_3SiCH_2Li$	$(CH_3)_2SiCl_2$	$[(CH_3)_3SiCH_2]_2Si(CH_3)_2$	86	1175
$(CH_3)_3SiCH_2Li$	$(CH_3)_2SiClCH_2Cl$	$(CH_3)_3SiCH_2Si(CH_3)_2CH_2Cl$	82	1175
$(CH_3)_2C_6H_5SiCH_2Li$	$(CH_3)_2C_6H_5SiCl$	$[(CH_3)_2C_6H_5Si]_2CH_2$	88	926, 927
$(CH_3)_2C_6H_5SiCH_2Li$	$(CH_3)_2SiClCH_2Cl$	$(CH_3)_2C_6H_5SiCH_2Si(CH_3)_2CH_2Cl$	—	707
$(CH_3)_2C_6H_5SiCH_2Li$	$(CH_3)_2SiCl_2$	$[(CH_3)_2C_6H_5SiCH_2]_2Si(CH_3)_2$	80	705
$(CH_3)_2C_2H_5OSiCH_2Li$	$CH_3SiCl_2CH_2Cl$	$[(CH_3)_2C_2H_5OSiCH_2]_2Si(CH_3)CH_2Cl$	—	705
$CH_3(C_6H_5)_2SiCH_2Li$	$CH_3(C_6H_5)SiClCH_2Cl$	$(C_6H_5)_2CH_3SiCH_2SiCH_3(C_6H_5)CH_2Cl$	57	706
$CH_3(C_2H_5O)_2SiCH_2Li$	$(CH_3)_2SiClCH_2Cl$	$CH_3(C_2H_5O)_2SiCH_2Si(CH_3)_2CH_2Cl$	—	705
$CH_3(C_6H_5CH_2)_2SiCH_2Li$	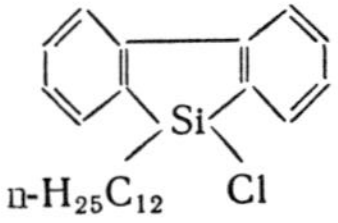 [structure] Si; n-$H_{25}C_{12}$, Cl	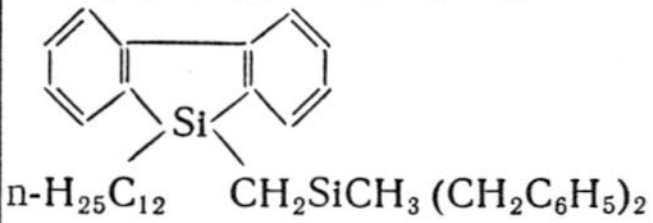 [structure] Si; n-$H_{25}C_{12}$, $CH_2SiCH_3(CH_2C_6H_5)_2$	80	749
$(CH_3)_2Si(CH_2Li)_2$	$(CH_3)_3SiCl$	$[(CH_3)_3SiCH_2]_2Si(CH_3)_2$	36.5	1153
$(CH_3)_3SiCH_2Si(CH_3)_2CH_2Li$	$(CH_3)_2SiCl_2$	$(CH_3)_3Si[CH_2Si(CH_3)_2]_3CH_2Si(CH_3)_3$	85—90	1175
$(CH_3)_3SiCH_2Si(CH_3)_2CH_2Li$	$(CH_3)_2SiClCH_2Cl$	$(CH_3)_3Si[CH_2Si(CH_3)_2]_2CH_2Cl$	85—90	1175
$(CH_3)_2C_6H_5SiCH_2Si(CH_3)_2CH_2Li$	$(CH_3)_2SiClCH_2Cl$	$(CH_3)_2C_6H_5Si[CH_2Si(CH_3)_2]_2CH_2Cl$	—	705

$(CH_3)_3Si[CH_2Si(CH_3)_2]_2CH_2Li$	$(CH_3)_3SiCl$	$(CH_3)_3Si[CH_2Si(CH_3)_2]_3CH_3$	—	1175
$(CH_3)_3Si[CH_2Si(CH_3)_2]_2CH_2Cl$	$(CH_3)_2SiClCH_2Cl$	$(CH_3)_3Si[CH_2Si(CH_3)_2]_3CH_2Cl$	—	1175
$(CH_3)_3Si[CH_2Si(CH_3)_2]_3CH_2Li$	$(CH_3)_3SiCl$	$(CH_3)_3Si[CH_2Si(CH_3)_2]_4CH_3$	—	1175
$(CH_3)_3SiC_6H_4Li$	$SiCl_4$	$[(CH_3)_3SiC_6H_4]_4Si$	77	953
$(CH_3)_3SiC_6H_4Li$	$(C_6H_5)_2SiCl_2$	$[(CH_3)_3SiC_6H_4]_2Si(C_6H_5)_2$	86	953
$(CH_3)_3SiC_6H_4Li$	$(C_6H_5)_3SiCl$	$(CH_3)_3SiC_6H_4Si(C_6H_5)_3$	83	953
$(C_6H_5)_3SiC_6H_4Li$	$(C_6H_5)_2SiCl_2$	$[(C_6H_5)_3SiC_6H_4]_2Si(C_6H_5)_2$	—	1182
$(C_6H_5CH_2)_3SiC_6H_4Li$	$SiCl_4$	$[(C_6H_5CH_2)_3SiC_6H_4]_4Si$	51	765, 1182
$(C_6H_5CH_2)_3SiC_6H_4Li$	$(C_6H_5)_2SiCl_2$	$[(C_6H_5CH_2)_3SiC_6H_4]_2Si(C_6H_5)_2$	43.4	765
$(C_6H_5)_3SiC(CH_3)_2C_6H_4K$	$(C_6H_5)_3SiCl$	$(C_6H_5)_3SiC(CH_3)_2C_6H_5Si(C_6H_5)_3$	—	560
$(C_6H_5)_3SiOCK(C_6H_5)_2$	$(CH_3)_3SiCl$	$(C_6H_5)_3SiOC(C_6H_5)_2Si(CH_3)_3$	25	1129
Li, $Si(CH_3)_3$ (naphthalene ring) Li	$(CH_3)_3SiCl$	$(CH_3)_3Si$, $Si(CH_3)_3$ (naphthalene ring) $(CH_3)_3Si$	—	477
Li, $Si(C_2H_5)_3$ (naphthalene ring) Li	$(C_2H_5)_3SiCl$	$Si(C_2H_5)_3$ (naphthalene ring) $(C_2H_5)_3Si$	—	477
(fluorene ring) $(CH_3)_3Si$, Li	$(CH_3)_3SiCl$	(fluorene ring) $(CH_3)_3Si$, $Si(CH_3)_3$	17	664
$(CH_3)_3Si$—(furan ring, O)—Li	$(CH_3)_3SiCl$	$(CH_3)_3Si$—(furan ring, O)—$Si(CH_3)_3$	31.1	355
$(CH_3)_3Si$—(thiophene ring, S)—Li	$(CH_3)_3SiCl$	$(CH_3)_3Si$—(thiophene ring, S)—$Si(CH_3)_3$	65	355

3. COMPOUNDS CONTAINING THE Si – N – M GROUP

The organosilicon compounds containing the Si–N–M group which have been studied most at the present time are N-lithio, N-sodio, and N-potassio derivatives of linear organosilanes (mainly disilazans), which may also be regarded as organosilyl derivatives of alkali metal amides. However, since most of the preparation methods and reactions are general, in this section we will also examine synthesis methods and properties of N-metal substituted aminosilanes, hydrazinosilanes, and cyclic organosilazans.

3.1. Synthesis Methods

All the many methods of preparing organosilicon compounds containing the Si–N–M group are preparative variants of a general synthesis method which is based on the replacement of the hydrogen atom in the Si–N–H group by an alkali metal. This also refers to reactions in which the starting organosilicon compounds are chlorosilanes, which are converted into aminosilanes immediately before metallation. We will therefore examine below only methods of metallating the N–H bond in nitrogen-containing organosilicon compounds (aminosilanes, silazans, and hydrazinosilanes).

Metallation of the Si–N–H group may be achieved by the direct action of an alkali metal on trialkylaminosilanes [516], hexaalkyldisilazans [386, 387, 603, 805, 1039, 1223], and hexaalkylcyclotrisilazans [555]:

$$2\,{\gt}\mathrm{Si{-}N{-}H} + 2\mathrm{M} \longrightarrow 2\,{\gt}\mathrm{Si{-}N{-}M} + \mathrm{H_2} \qquad (1\text{-}144)$$

The reagents used for this purpose include potassium in liquid ammonia [603, 1233], sodium and potassium in dioxane in the presence of styrene [516, 603, 805, 1039], and sodium in dioxane [576] or in a mixture of naphthalene and tetrahydrofuran [603, 1223]. However, reactions of type (1-144), like the reaction of hexaalkyldisilazans with sodium hydride [603, 1353], have not found wide application.

Hexaorganodisilazans are readily metallated by alkali metal amides:

$$\left(\geq Si\right)_2 N{-}H + MNH_2 \longrightarrow \left(\geq Si\right)_2 N{-}M + NH_3 \qquad (1\text{-}145)$$

The reaction is carried out in benzene or ligroin. The yields of the N-sodio- or N-potassiohexaorganodisilazans formed in this way are close to quantitative. Alkali metal amides are used to metallate hexaalkyldisilazans [122, 603, 1014, 1223, 1231, 1232, 1242], alkylalkoxydisilazans [1228], hexaalkoxydisilazans [1037, 1125, 1224, 1228], bis(triorganosilyl)tetraorganodisilazans [1227], and also N,N'-bis(trimethylsilyl)urea [1017]. It may be assumed that one of the stages in the formation of a N-potassiohexaorganodisilazan by the reaction of potassamide with triethylsilane [898] or trimethylbenzylsilane [831] is also reaction (1-145). The action of sodamide on trimethyl(methylamino)silane forms sodium trimethylsilylamide and methylamine [1067].

An investigation of the reaction of hexamethyldisilazan labeled by the isotope N^{15} with sodamide showed that the ammonia liberated did not contain N^{15} [122, 1232], i.e., the Si–N bond is not touched in this reaction:

$$[(CH_3)_3Si]_2N^*H + NaNH_2 \longrightarrow [(CH_3)_3Si]_2N^*Na + NH_3 \qquad (1\text{-}146)$$

The first attempts to prepare bis(trialkylsilyl)amides of alkali metals by the reaction of sodamide directly with triorganochlorosilanes (without their intermediate conversion to hexaalkyldisilazans) were unsuccessful [1242, 1246]. However, it was observed subsequently that the process may be initiated by the addition of a 50% suspension of sodamide in benzene to 5-6 drops of trimethylchlorosilane. When the reaction begins, the rest of the trimethylchlorosilane is added with stirring. If the reaction is stopped by cooling the reaction mixture to room temperature, subsequent boiling of excess sodamide with trimethylchlorosilane does not lead to their interaction. The reaction proceeds with the intermediate formation of hexamethyldisilazan, which is complete after heating for only half an hour. Further boiling of the reaction mixture for 24 hr leads to the formation of sodium bis(trimethylsilyl)amide in 78% yield [122, 1246]:

$$8NaNH_2 + 6\,(CH_3)_3SiCl \longrightarrow 5NaCl + 4NH_3 + NH_4Cl + 3\,[(CH_3)_3Si]_2NNa \qquad (1\text{-}147)$$

The most common method of synthesizing compounds containing the Si–N–M group is metallation of the Si–N–H bond with an organolithium:

$$\rangle Si{-}\underset{|}{N}{-}H + RLi \xrightarrow{-RH} \rangle Si{-}\underset{|}{N}{-}Li \quad (1\text{-}148)$$

For this purpose it is possible to use methyllithium in ether [1065], propyllithium in petroleum ether [603, 1223], butyllithium in pentane [520, 916, 1033], hexane [528, 691, 693, 695, 1065], petroleum ether [927, 1224], or xylene [1386], and phenyllithium in ethyl ether [603, 928, 978, 1222, 1223, 1235, 1241, 1244, 1270] or a mixture of it with benzene [1242].

The reaction (1-148) is used to prepare N-lithio derivatives of the most diverse classes of nitrogen–containing organosilicon compounds: triorganoaminosilanes [520, 1222], triorgano(alkylamino)silanes [1065, 1067], diorganobis(organoamino)silanes [697, 926-928, 1396], N,N'-bis(triorganosilyl)ethylenediamines [916, 1033], hexaorganodisilazans [693, 982, 1241, 1452], hexaalkoxydisilazans [1224], 1,3-bis(alkylamino)tetraorganodisilazans [926, 927], octaorganotrisilazans [693, 696], hexaorganocyclotrisilazans, octaorganocyclotetrasilazans [691, 695], bis(triorganosilyl)ureas [694, 1017], N-triorganosilyl-N'-organohydrazines [1234, 1235, 1241, 1244], N-triorganosilyl-N',N'-diorganohydrazines [1244], and N,N'-bis(triorganosilyl)hydrazines [528, 1234, 1244].

Thus, the reaction of methyllithium with trimethyl(methylamino)silane in ethyl ether gives a 93% yield of the N-lithio derivative [1065]:

$$(CH_3)_3SiNHCH_3 + CH_3Li \longrightarrow (CH_3)_3Si\underset{\substack{|\\ Li}}{N}CH_3 + CH_4 \quad (1\text{-}149)$$

In the reaction of butyllithium or phenyllithium with diorganobis(organoamino)silanes there occurs the replacement of hydrogen atoms by lithium in both NH groups [697, 926-928, 1386]:

$$R_2Si\langle{}^{NHR'}_{NHR'} + 2R''Li \longrightarrow R_2Si\langle{}^{N(Li){-}R'}_{N(Li){-}R'} + 2R''H \quad (1\text{-}150)$$

The substituents R and R' may be both alkyl and aryl radicals.

The reaction of a N,N'-bis(trialkylsilyl)ethylenediamine with butyllithium yields both mono- and dilithium derivatives [916, 1033]:

$$\begin{matrix}(CH_3)_3SiNH \\ | \\ (CH_2)_2 \\ | \\ (CH_3)_3SiNH\end{matrix} \xrightarrow[-C_4H_{10}]{+C_4H_9Li} \begin{matrix}(CH_3)_3SiN{-}Li \\ | \\ (CH_2)_2 \\ | \\ (CH_3)_3SiN{-}H\end{matrix} + \begin{matrix}(CH_3)_3SiN{-}Li \\ | \\ (CH_2)_2 \\ | \\ (CH_3)_3SiN{-}Li\end{matrix} \quad (1\text{-}151)$$

By changing the ratio of the reagents, it is possible to introduce from one to three lithium atoms into the molecule of a hexaorganocyclotrisilazan [691, 695]:

$$(R_2SiNH)_3 \xrightarrow{+nC_4H_9Li} \text{cyclo-}[R_2Si{-}N(Li){-}R_2Si{-}NH{-}R_2Si{-}NH]\,;\ \text{cyclo-}[R_2Si{-}N(Li){-}R_2Si{-}N(Li){-}R_2Si{-}NH]\,;\ \text{cyclo-}[R_2Si{-}N(Li)]_3 \quad (1\text{-}152)$$

In the reaction of phenyllithium with a N-trialkylsilyl-N'-phenylhydrazine, lithium replaces the hydrogen atom at the nitrogen attached to the phenyl radical [1235], and not the trialkylsilyl group, as was proposed previously [1241]:

$$R_3SiNHNHC_6H_5 + C_6H_5Li \xrightarrow{-C_6H_6} R_3SiNHN(Li)C_6H_5 \quad (1\text{-}153)$$

The reaction of a N,N'-bis(trialkylsilyl)hydrazine with butyllithium or phenyllithium, depending on the amount of the latter, leads to the formation of both mono- [1244] and dilithium derivatives [528]:

$$R_3SiNHNHSiR_3 + R'Li \xrightarrow{-R'H} R_3SiN(Li)NHSiR_3 + R_3SiN(Li)N(Li)SiR_3 \quad (1\text{-}154)$$

The action of methyl iodide on the dilithium derivative of N,N'-bis(trimethylsilyl)hydrazine forms a mixture of two isomers, namely, N,N'-bis(trimethylsilyl)-N,N'-dimethylhydrazine and N,N'-

bis(trimethylsilyl)-N,N'-dimethylhydrazine [528]. It has not been yet been established whether rearrangement occurs in the reaction with methyl iodide or in the metallation by butyllithium.

N-Lithium derivatives of hexaalkoxydisilazans are obtained conveniently by means of an exchange reaction with hexamethyldisilazanyllithium, which occurs when a mixture of the reagents is heated in petroleum ether [1224]:

$$[(RO)_3Si]_2NH + [(CH_3)_3Si]_2NLi \longrightarrow [(RO)_3Si]_2NLi + [(CH_3)_3Si]_2NH \qquad (1\text{-}155)$$

N-Hexamethyldisilazanyllithium is also formed by reduction of hexamethyldisilazan with lithium aluminum hydride [1015]:

$$4[(CH_3)_3Si]_2NH + LiAlH_4 \longrightarrow [(CH_3)_3Si]_2NLi + \{[(CH_3)_3Si]_2N\}_3Al + 4H_2 \qquad (1\text{-}156)$$

To conclude this section we should mention two reactions which proceed with the intermediate formation of the Si–N–N–Li group, but which do not belong to metallation processes of the N–H bond. These reactions are the addition of triphenylsilyllithium and triphenylsilylpotassium to azobenzene [1276] and azoxybenzene [717]:

$$\left.\begin{array}{r} C_6H_5N{=}NC_6H_5 \\ C_6H_5N{=}N(O)C_6H_5 \end{array}\right] \xrightarrow{(C_6H_5)_3SiM} \begin{array}{c} C_6H_5{-}N{-}N{-}C_6H_5 \\ \;\;|\quad\;\; | \\ (C_6H_5)_3Si\;\; M \end{array} \qquad (1\text{-}157)$$

3.2. Physical Properties

With the exception of the trimethylsilyl(triethylsilyl)amide, the bis(triorganosilyl)amides of alkali metals are solid crystalline compounds, which sublime in vacuum, without decomposition in most cases. Sodium bis(dialkylalkylalkoxylsilyl)amides are monomers which are readily soluble in petroleum ether [1231]. Sodium bis(trialkylsilyl)amides, bis(alkyldialkoxysilyl)amides, and bis(trialkoxysilyl)amides are dimers which are difficultly soluble in petroleum ether [1231, 1232]. Sodium bis(triorganosilyl)amides are readily soluble in benzene, toluene, and xylene. Thus, for example, 100 g of lithium bis(trimethylsilyl)amide dissolves in 70 g of xylene at 30°C [1242]. Lithium and solium derivatives of hexaorganodisilazans are dimers in solution. They do not conduct a current [122, 1242] and consequently they cannot have the structure

TABLE 10. Physical Constants of Compounds Containing the Si–N–M Group

Compound	M.p., °C	B.p., °C	Literature
$(CH_3)_3SiNLiCH_3$	~100 *	—	1065
$[(CH_3)_3Si]_2NLi$	70—71	110—115 (1)	1452
	71—72	80—84 (0.01)	520
	70—72	115 (1)	1241, 1242
$[(CH_3)_3Si]_2NNa$	165—167	170 (2)	1242
	170—171	202—204 (1—2)	1246
	172	202 (1—2)	122
	183	204 (2)	121
$[(CH_3)_3Si][(C_2H_5)_3Si]NLi$	—	162—164 (2)	1242
		164 (2)	1222
$[(CH_3)_3SiSi(CH_3)_2]_2NNa$	58—62	—	1227
$[(CH_3)_2(C_2H_5O)Si]_2NNa$	97	—	1231
$[CH_3(CH_2{=}CH)(\text{iso-}C_3H_7O)Si]_2NNa$	70	—	1228, 1231
$[CH_3(CH_3O)_2Si]_2NNa$	171—172	—	1228, 1231
$[CH_2{=}CH(CH_3O)_2Si]_2NNa$	184	—	1231, 1452
$[(CH_3O)_3Si]_2NLi$	145 decomp.	—	1224
$[(CH_3O)_3Si]_2NNa$	182—184	—	1224, 1228
$[(CH_3O)_3Si]_2NK$	202—203	—	1125, 1224
	202	—	1037
$[(C_2H_5O)_3Si]_2NLi$	from −20 to −12	—	1224
$[(C_2H_5O)_3Si]_2NNa$	104	—	1224
$[(\text{iso-}C_3H_7O)_3Si]_2NNa$	204—206	—	1224, 1228
$(CH_3)_3SiNHSi(CH_3)_2NLiSi(CH_3)_3$	—	125—128 (0.2)	693
$[(CH_3)_3SiNLi]_2Si(CH_3)_2$	60—70 (decomp.)	172 (2)	692, 693
$(C_6H_5)_2SiNHSi(C_6H_5)_2NHSi(C_6H_5)_2NLi$ (ring: Si…N linked)	205—206 (decomp.)		695
$\{(CH_3)_3Si[-NSi(CH_3)_2-(Si(CH_3)_3)]_2\}_2NLi$	—	140—140.5 (1.5)	692, 693
$[(CH_3)_3Si]_2NLi(C_2H_5)_2O$	95—110 (decomp.)	—	1241, 1242
$[(CH_3)_3Si]_2NLi \cdot 0.75\ OCH_2CH_2OCH_2CH_2$ (ring)	200	—	1242
$[(CH_3)_3Si]_2NNa \cdot (C_2H_5)_2O$	72—74	—	1242
$[(CH_3)_3Si]_2NNa \cdot OCH_2CH_2CH_2CH_2$ (ring)	70—85 (decomp.)	—	1242
$[(CH_3)_3Si]_2NNa \cdot 0{,}5\ OCH_2CH_2OCH_2CH_2$ (ring)	Carbonized	—	1242
$[(CH_3)_3Si]_2NNa \cdot N{=}CHCH{=}CHCH{=}CH$ (ring)	150—160 (decomp.)	—	1242
$[(CH_3)_3Si]_2NK \cdot 2OCH_2CH_2OCH_2CH_2$ (ring)	Carbonized	—	1242
$[(C_6H_5)_3Si]_2SiLi \cdot (C_2H_5)_2O$	380 (decomp.)	—	1242

* The substance softens and gradually sublimes.

of complexes of the type:

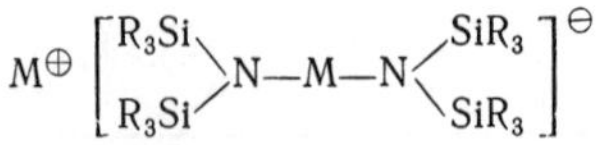

$$M^{\oplus}\left[\begin{array}{l}R_3Si\\R_3Si\end{array}\!\!>N-M-N<\!\!\begin{array}{l}SiR_3\\SiR_3\end{array}\right]^{\ominus}$$

On the basis of the NMR spectra of Li^7, which consists of only one signal, these derivatives are assigned the structure of a dimeric cyclic complex [121]:

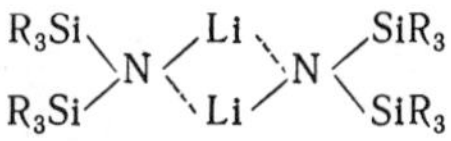

$$\begin{array}{l}R_3Si\\R_3Si\end{array}\!\!>N<\!\!\begin{array}{c}Li\\Li\end{array}\!\!>N<\!\!\begin{array}{l}SiR_3\\SiR_3\end{array}$$

Bis(triorganosilyl)amides of alkali metals form complexes with dioxane, ethyl ether, tetrahydrofuran, and pyridine [121, 1242]. When heated in vacuum, the etherates lose ethyl ether, while the dioxanates are carbonized.

The physical constants of compounds containing the Si–N–M group are given in Table 10.

3.3. Chemical Properties

3.3.1. Reaction with Elements

Bis(triorganosilyl)amides of alkali metals are stable in air. However, lithium derivatives gradually become colored under the action of atmospheric oxygen, even in solution, so that it is better to work with them in an inert atmosphere [691]. When heated in a stream of oxygen or treated with concentrated nitric acid, lithium derivatives are oxidized explosively [121, 1242]. N-Lithio-N-phenyl-N'-trialkylsilylhydrazines are converted into trialkylsilylphenyldiazenes by the action of oxygen [1234]:

$$2R_3SiNHN(Li)C_6H_5 + O_2 \xrightarrow{-2LiOH} 2R_3SiN{=}NC_6H_5 \qquad (1\text{-}158)$$

Trialkylsilylaryldiazenes are also formed by the reaction of N,N'-dilithio-N-phenyl-N'-trialkylsilylhydrazines with bromine and iodine [1234]. The synthesis is carried out in an ether solution at -30°C, using a 20% excess of the lithium derivative. The yield of the trialkylphenyldiazene is 90% when R = n-C_3H_7, 75% when R = CH_3, and only 33% in the case of the triethylsilyl derivative. Raising the reaction temperature and also the use of the equivalent amount of the halogen leads to rupture of the C–N and Si–N bonds and to the liberation of nitrogen; the latter is also observed in the reaction of bromine with monolithio derivatives of N-trialkylsilyl-N-phenylhydrazines and N,N'-bis(trialkylsilyl)hydrazines:

$$2R_3SiN(Li)HNC_6H_5 + Br_2 \longrightarrow N_2 + 2LiBr + 2R_3SiNHC_6H_5 \quad (1\text{-}159)$$

$$2R_3SiN(Li)HNSiR_3 + Br_2 \longrightarrow N_2 + 2LiBr + 2\,(R_3Si)_2NH \quad (1\text{-}160)$$

The reaction of sodium bis(trialkylsilyl)amides with chlorine and iodine in ether at -50°C forms the corresponding N-haloxaalkyldisilazans [1223, 1264]:

$$(R_3Si)_2NM + X_2 \xrightarrow[-MX]{} (R_3Si)_2NX \quad (1\text{-}161)$$

An analogous reaction with sulfur leads to a mixture of polysulfan bis(trialkylsilyl)diamides, from which it is possible to isolate derivatives of disulfan and trisulfan, and also derivatives of tetrasulfan, which decompose during distillation [1059, 1066]. With a large excess of hexaalkyldisilazanylsodium the reaction proceeds according to the scheme:

$$[(CH_3)_3Si]_2NNa + S \longrightarrow [(CH_3)_3Si]_2N\text{—}S\text{—}Na \quad (1\text{-}162)$$

3.3.2. Reaction with Oxides and Sulfides of Elements

Bis(triorganosilyl)amides of alkali metals are decomposed slowly by atmospheric moisture. In solution they are hydrolyzed rapidly to hexaorganodisilazans, which are then converted more slowly into the corresponding disiloxanes [1242]:

$$(R_3Si)_2NM + H_2O \longrightarrow (R_3Si)_2NH + MOH \xrightarrow{H_2O} (R_3Si)_2O + NH_3 \quad (1\text{-}163)$$

Carbon monoxide breaks the Si–N bond in the molecule of a hexaorganodisilazanylsodium with the formation of a hexaorganodisiloxane and sodium cyanide [122, 1239]:

$$(R_3Si)_2NNa + CO \longrightarrow NaCN + (R_3Si)_2O \quad (1\text{-}164)$$

TABLE 11. Products of the Reaction of Compounds Containing the Si–N–M Group with the Halides of Some Elements

Compound with Si–N–M group	Halide	Reaction products	Literature
$[(CH_3)_3Si]_2NLi$	$SiCl_4$	$Cl_3SiN[Si(CH_3)_3]_2$	1243
$[(CH_3)_3Si]_2NNa$	SiF_4	$F_3SiN[Si(CH_3)_3]_2$ $F_2Si\{N[Si(CH_3)_3]_2\}_2$	1223, 1229
$[(CH_3)_3Si]_2NK$	SiF_4	$F_3SiN[Si(CH_3)_3]_2$	1451
$[(CH_3)_3Si]_2NK$	$SiCl_4$	$Cl_3SiN[Si(CH_3)_3]_2$	1451
$[(CH_3)_3Si]_2NK$	$SiBr_4$	$Br_3SiN[Si(CH_3)_3]_2$	1451
$[(CH_3)_3Si]_2NK$	SiI_4	$I_3SiN[Si(CH_3)_3]_2$	1451
$[(CH_3O)_3Si]_2NNa$	$SiCl_4$	$Cl_3SiN[Si(OCH_3)_3]_2$	1224, 1228
$[(C_2H_5O)_3Si]_2NNa$	$SiCl_4$	$Cl_3SiN[Si(OC_2H_5)_3]_2$	1224
$[(iso\text{-}C_3H_7O)_3Si]_2NNa$	$SiCl_4$	$Cl_3SiN[Si(OC_3H_7\text{-}iso)_3]_2$	1224
$[(CH_3)_2CH_2{=}CHSi]_2NNa$	$SiCl_4$	$Cl_3SiN[SiCH{=}CH_2(CH_3)_2]_2$	1247
$(C_2H_5)_3SiNHLi$	SiF_4	$F_3SiNHSi(C_2H_5)_3$	843, 1229
$(CH_3)_2Si(NLiCH_3)_2$	$SiCl_4$	NCH₃ NCH₃ / \ / \ $(CH_3)_2$ Si Si Si $(CH_3)_2$ \ / \ / NCH₃ NCH₃	693, 928
$(CH_3)_2Si(NLiC_6H_4CF_3\text{-}m)_2$	$SiCl_4$	$NC_6H_4CF_3$-m / \ $(CH_3)_2$ Si $SiCl_2$ \ / $NC_6H_4CF_3$-m	697
$(CH_3)_2Si(NLiC_6H_5)_2$	$SiCl_4$	NC_6H_5 / \ $(CH_3)_2$ Si $SiCl_2$ \ / NC_6H_5	1386
$CH_3N[(CH_3)_2SiNLiCH_3]_2$	$SiCl_4$	/Si $(CH_3)_2$—N (CH_3)\ CH_3N $SiCl_2$ \Si $(CH_3)_2$—N (CH_3)/	926, 927
$[(CH_3)_3SiNLiCH_2]_2$	$SiCl_4$	/CH_2CH_2\ $(CH_3)_3$ Si—N N—Si $(CH_3)_3$ \ / Si / \ $(CH_3)_3$ Si—N N—Si $(CH_3)_3$ \CH_2CH_2/	916, 1033

TABLE 11. (Cont'd)

Compound with Si–N–M group	Halide	Reaction products	Literature
$[(CH_3)_3Si]_2NNa$	$Cl_3SiSiCl_3$	$Cl_3SiSiCl_2N[Si(CH_3)_3]_2$ $[(CH_3)_3Si]_2NSiCl_2SiCl_2N[Si(CH_3)_3]_2$	1227
$[(CH_3)_3Si]_2NNa$	PCl_3	$[(CH_3)_3SiNPCl]_x$	1223
$[(CH_3)_3Si]_2NNa$	$POCl_3$	$[(CH_3)_3Si]_2NP(O)Cl—N[Si(CH_3)_3]_2$	122
$[(CH_3)_3Si]_2NNa$	SCl_2	$[(CH_3)_3Si]_2N—S—N[Si(CH_3)_3]_2$ $(CH_3)_3SiN{=}S{=}NSi(CH_3)_3$	1236
$[(CH_3)_3Si]_2NNa$	S_2Cl_2	$[(CH_3)_3Si]_2N—S—S—N[Si(CH_3)_3]_2$	1236

In the reaction of sodium bis(trimethylsilyl)amide with carbon dioxide a whole series of conversions occurs. As a result there are formed bis(trimethylsilyl)carbodiimide, hexamethyldisiloxane, tris(trimethylsilyl)amine, and sodium cyanate, cyanamide, and carbonate [1016, 1223, 1239]. The reactions occurring may be represented in the following way. Carbon dioxide simultaneously attacks both nucleophilic and electrophilic centers of the sodium bis(trimethylsilyl)amide molecule with subsequent elimination of sodium trimethylsilanolate with the formation of trimethylisocyanatosilane:

$$O{=}C\cdots N(Si(CH_3)_3)—Si(CH_3)_3 \;(O\cdots Na) \longrightarrow O(Na)\cdots Si(CH_3)_3,\; O{=}C{=}N—Si(CH_3)_3 \longrightarrow (CH_3)_3SiNCO + (CH_3)_3SiONa \quad (1\text{-}165)$$

Sodium trimethylsilanolate and trimethylisocyanatosilane may react together and also with the starting carbon dioxide and sodium bis(trimethylsilyl)amide. In the first case, hexamethyldisiloxane and sodium cyanate are obtained:

$$[(CH_3)_3Si—N{=}C{=}O \cdots Na—O—Si(CH_3)_3] \longrightarrow (CH_3)_3SiOSi(CH_3)_3 + NaOCN \quad (1\text{-}166)$$

The bulk of the hexamethyldisiloxane is formed as a result of the decomposition of the intermediate sodium trimethylsilylcarbonate:

$$2(CH_3)_3SiONa + 2CO_2 \longrightarrow [2(CH_3)_3SiOCOONa] \longrightarrow [(CH_3)_3Si]_2O + Na_2CO_3 + CO_2 \quad (1\text{-}167)$$

Bis(trimethylsilyl)carbodiimide is the product of a reaction between sodium bis(trimethylsilyl)amide and trimethylisocyanatosilane [1245].

Sulfur dioxide reacts with sodium bis(trimethylsilyl)amide analogously to carbon dioxide to form bis(trimethylsilyl)thiodiimide (but in a lower yield than the corresponding carbodiimide), hexamethyldisiloxane, and sodium sulfite [1233, 1236]:

$$2\,[(CH_3)_3\,Si]_2\,NNa + 2SO_2 \longrightarrow (CH_3)_3\,SiN{=}S{=}NSi\,(CH_3)_3 + [(CH_3)_3\,Si]_2\,O + Na_2SO_3 \quad (1\text{-}168)$$

The reaction of sodium bis(trimethylsilyl)amide with carbon disulfide forms hexamethyldisilthiane (86%), tris(trimethylsilyl)-amine (2%), and bis(trimethylsilyl)carbodiimide (9%) and also sodium sulfide and thiocyanate. An increase in the amount of sodium bis(trimethylsilyl)amide to 2 moles per mole of carbon disulfide raises the yield of bis(trimethylsilyl)carbodiimide to 17%. The yield of hexamethyldisilthiane is then reduced to 70% [1223, 1239]. The conversions which occur in the reaction with carbon disulfide may be explained by schemes analogous to those examined in the discussion of the reaction with carbon dioxide.

3.3.3. Reaction with Halides and Oxyhalides of Elements

Bis(trialkylsilyl)amides of alkali metals react vigorously with halides of metals and metalloids. In most cases the reaction proceeds in accordance with the scheme (1-169)

$$m\,(R_3Si)_2\,NM + M'X_n \longrightarrow [(R_3Si)_2\,N]_m\,M'X_{n-m} + mMX \quad (1\text{-}169)$$

More details on the reaction of bis(trialkylsilyl)amides of alkali metals with halides of Cu, Be, Zn, Cd, B, Sn, Ti, Cr, Mn, Fe, Ni, and Co will be given in the corresponding chapters and sections. The reactions with silicon, sulfur, and phosphorus halides are characterized in Table 11.

In the reaction of a bis(trimethylsilyl)amide with phosgene [1016, 1017] there is both replacement of the chlorine in the C – Cl bond and also addition at the C = O bond with subsequent elimination of sodium trimethylsilanolate and trimethylchlorosilane, which form $[(CH_3)_3Si]_2O$:

$$[(CH_3)_3Si]_2NNa + COCl_2 \xrightarrow[-NaCl]{} \{[(CH_3)_3Si]_2NCOCl\} \xrightarrow{[(CH_3)_3Si]_2NNa}$$

$$\longrightarrow \begin{matrix} & O\cdots Na & \\ & |\!+ \quad + & \\ Cl+C\cdots N-Si(CH_3)_3 & & \\ \vdots \quad \vdots\| \quad | & & \\ (CH_3)_3Si+N \quad Si(CH_3)_3 & & \\ | & & \\ Si(CH_3)_3 & & \end{matrix} \xrightarrow[-(CH_3)_3SiCl]{}$$

$$\longrightarrow \begin{matrix} Na \\ | \\ O\cdots Si(CH_3)_3 \\ + \quad + \\ C\cdots N-Si(CH_3)_3 \\ \| \\ N \\ | \\ Si(CH_3)_3 \end{matrix} \xrightarrow[-(CH_3)_3SiONa]{} (CH_3)_3SiN{=}C{=}NSi(CH_3)_3 \qquad (1\text{-}170)$$

There is also the possibility that an intermediate product of this reaction is trimethylisocyanatosilane, which, as was shown by special investigations [122, 1245], is also converted into bis(trimethylsilyl)carbodiimide and hexamethyldisiloxane by reaction with sodium bis(trimethylsilyl)amide.

3.3.4. Reaction with Organic Compounds

Among the reactions of compounds containing the Si–N–M group with different classes of organic compounds, the most detailed study has been made of their reaction with halohydrocarbons and carbonyl-containing substances. The reaction with halogen derivatives of hydrocarbons proceeds according to the general scheme

$$\gt Si-\underset{|}{N}-M + X-C\lt \xrightarrow{-MX} \gt Si-\underset{|}{N}-C\lt \qquad (1\text{-}171)$$

This describes the reaction of sodium and potassium bis(trimethylsilyl)amides with n-alkly halides and benzyl chloride [1039], trimethyl(N-lithio-N-ethylamino)silane with ethyl bromide [516], N-lithio-N-phenyl-N'-(trimethylsilyl)hydrazine with ethyl iodide [1235, 1244], tertiary butyl chloride, and benzyl chloride [1235], and N-lithio-N'-phenyl-N,N'-bis(trimethylsilyl)urea with bromobenzene [694]. However, in the reaction of potassium bis(trimethyl)silylamide with 2-bromobutane, tertiary butyl bromide, and chlorocyclohexane, instead of reaction (1-171) there is elimination of hydrogen halide and the formation of the corresponding unsaturated hydrocarbon [1039]. The reaction of sodium bis(trimethylsilyl)amide with trifluoroiodomethane gives a 74% yield of bis(trimethylsilyl)carbodiimide [1239]:

$$2\,[(CH_3)_3Si]_2NNa + ICF_3 \longrightarrow NaI + NaF + 2\,(CH_3)_3SiF + [(CH_3)_3SiN{=}]_2C \qquad (1\text{-}172)$$

Bis(trimethylsilyl)carbodiimide [bis(trimethylsilylimino)-methane] is also obtained by the reaction of sodium bis(trimethylsilyl)amide with cyanogen bromide [562]. Methylene chloride, chloroform, and carbon tetrachloride react vigorously with sodium bis(trimethylsilyl)amide. However, the reaction does not proceed according to scheme (1-171) or (1-172) and its products have not been identified [1239]. The reaction of methyl iodide with the dilithium derivative of N,N'-bis(trimethylsilyl)hydrazine forms both the normal product of reaction (1-171) and also N,N'-dimethyl-N, N'-bis(trimethylsilyl)hydrazine, which is isomeric with it [528].

Sodium bis(trimethylsilyl)amide reacts with benzophenone with the formation of N-(trimethylsilyl)diphenylcarbimide (64%) and sodium trimethylsilanolate [907]. The mechanism of this reaction includes two stages, which are general for a reaction of bis-(trialkylsilyl)amides of alkali metals with organic compounds containing a carbonyl group, which is incapable of enolization (and also with CO and CO_2). The first stage is the addition of the sodium bis-(trialkylsilyl)amide to the carbonyl group. The sodium attacks the oxygen atom, while the bis(trimethylsilyl)amino group attacks the carbon atom of the carbonyl group:

$$\underset{\overset{+|\quad|}{O\cdots Na}}{(C_6H_5)_2C{\cdots\cdots}N[Si(CH_3)_3]_2} \longrightarrow \underset{\overset{|}{ONa}}{(C_6H_5)_2C{-}N[Si(CH_3)_3]_2} \qquad (1\text{-}173)$$

In the second stage there is elimination of sodium trimethylsilanolate and the formation of a C = N bond:

$$\underset{\overset{+\quad+}{O\cdots Si(CH_3)_3}\atop{|\atop Na}}{(C_6H_5)_2C{\cdots\cdots}N{-}Si(CH_3)_3} \longrightarrow (C_6H_5)_2C{=}NSi(CH_3)_3 + NaOSi(CH_3)_3 \qquad (1\text{-}174)$$

Reactions (1-173) and (1-174) proceed rapidly, as is confirmed by the fact the NMR spectra of the reaction mixture immediately after mixing of the reagents show signals of protons of CH_3-Si groups in both reaction products [907]. Reactions with benzaldehyde and p-quinone proceed analogously [907]. It may also be assumed that in the reaction with acid chlorides [908, 1014] there is a similar addition to the carbonyl group with the subsequent elimination of sodium trimethylsilanolate or trimethylchlorosilane, which then react again with the intermediate product formed. However, in this case there is a greater probability of the preliminary replacement of the chlorine atom in the C–Cl bond and a subsequent intramolecular rearrangement:

$$\begin{array}{l} \text{Cl}\cdots\text{Na} \\ \;+\quad\; + \\ \text{R—C}\cdots\text{N—Si}(CH_3)_3 \\ \;\;\| \qquad | \\ \;\;\text{O} \quad \text{Si}(CH_3)_3 \end{array} \xrightarrow[-\text{NaCl}]{} \begin{array}{l} \text{R—C}{=\!=}\text{N—Si}(CH_3)_3 \\ \quad \|\!+ \;\; + \\ \quad \text{O}\cdots\text{Si}(CH_3)_3 \end{array} \longrightarrow \begin{array}{l} \text{R—C=N—Si}(CH_3)_3 \\ \quad | \\ \quad \text{OSi}(CH_3)_3 \end{array} \qquad (1\text{-}175)$$

The interconversion of the last two compounds may be a new example of amidoimide tautomerism; the energetically more favorable imide form exists at room temperature [908].

O-Trimethylsiloxy-N-trimethylsilylbenzaldimene is one of the products from the reaction between sodium bis(trimethylsilyl)-amide and trimethylsilyl benzoate (34% yield). However, in this case the reaction does not proceed solely by schemes analogous to (1-173) and (1-174) as a second reaction product is tris(trimethylsilyl)amine, which is formed in 22% yield [908]. It may be assumed that the decomposition of the products from the addition of sodium bis(trimethylsilyl)amide to trimethylsilyl benzoate may occur both in accordance with scheme (1-174) with elimination of sodium trimethylsilanolate and in accordance with scheme (1-176)

$$\begin{array}{l} \qquad\quad \text{Na} \\ \qquad\quad | \\ \qquad\quad \text{O} \;\; \text{Si}(CH_3)_3 \\ \qquad\quad | \quad\; | \\ C_6H_5\text{—C}\!+\!\text{N—Si}(CH_3)_3 \\ \qquad\quad \| \quad \vdots \\ \qquad\quad \text{O}\!+\!\text{Si}(CH_3)_3 \end{array} \longrightarrow \begin{array}{l} C_6H_5\text{—C—O—Na} + \text{N}[\text{Si}(CH_3)_3]_3 \\ \qquad\quad \| \\ \qquad\quad \text{O} \end{array} \qquad (1\text{-}176)$$

There is also the possibility that tris(trimethylsilyl)amine is obtained as a result of direct cleavage of the Si–O bond in trimethylsilyl benzoate without preliminary addition of sodium bis(trimethylsilyl)amide to the carbonyl group. This would also be in accord with the results of the reaction of sodium bis(trimethylsilyl)-amide with methyl benzoate, in which there is observed elimination of both sodium trimethylsilanolate and sodium methylate with subsequent rearrangement in accordance with scheme (1-175) [908]

$$\begin{array}{c} O \\ \| \\ C_6H_5-C+O-CH_3 \\ \vdots \quad \vdots \\ (CH_3)_3Si-N+Na \\ | \\ Si(CH_3)_3 \end{array} \longrightarrow \begin{array}{c} O \quad Si(CH_3)_3 \\ \| \quad | \\ C_6H_5C-N-Si(CH_3)_3 \end{array} + NaOCH_3 \qquad (1\text{-}177)$$

In the reaction of the N-trimethylsilylamide of benzoic acid with sodium bis(trimethylsilyl)amide there is the usual addition to the carbonyl group with subsequent elimination of sodium trimethylsilanolate and the formation of N,N'-bis(trimethylsilyl)benzamidine (24% yield). Simultaneously with this there is metallation of the amide since after treatment of the reaction mixture with trimethylchlorosilane it is possible to isolate 34% of O-trimethylsiloxy-N-trimethylsilylbenzaldimene [908].

Sodium bis(trialkylsilyl)amides forms sodium enolates with carbonyl compounds which are capable of enolization. Thus, for example, acetone reacts with sodium bis(trimethylsilyl)amide to give the enolate, which is converted into trimethyl(isopropenoxy)-silane by the action of trimethylchlorosilane [906, 907]:

$$CH_3COCH_3 \rightleftarrows \underset{\displaystyle OH}{CH_3C=CH_2} \xrightarrow[-HN[Si(CH_3)_3]_2]{+NaN[Si(CH_3)_3]_2} \underset{\displaystyle ONa}{CH_3C=CH_2} \xrightarrow[-NaCl]{+(CH_3)_3SiCl}$$

$$\longrightarrow \underset{\displaystyle OSi(CH_3)_3}{CH_3C=CH_2} \qquad (1\text{-}178)$$

Sodium bis(trimethylsilyl)amide may also be used as the base in many Claisen and Stobbe condensations [904, 906] and also in the Wittig reaction. In the presence of sodium bis(trimethylsilyl)-amide, benzophenone may be converted into 1,1-diphenylethylene in 92% yield [904]:

$$(C_6H_5)_2C=O + (C_6H_5)_3P + CH_3Br + NaN[Si(CH_3)_3]_2 \longrightarrow$$

$$\longrightarrow (C_6H_5)_2C=CH_2 + (C_6H_5)_3PO + NaBr + HN[Si(CH_3)_3]_2 \qquad (1\text{-}179)$$

Acetonitrile is metallated by sodium bis(trimethylsilyl)amide with the formation of a trisodio derivative, from which it is possible to obtain a silylated ketenimine [904]:

$$CH_3C{\equiv}N \xrightarrow[-3HN[Si(CH_3)_3]_2]{+3NaN[Si(CH_3)_3]_2} Na_2C{=}C{=}NNa \xrightarrow[-3NaCl]{+3(CH_3)_3SiCl}$$
$$\longrightarrow [(CH_3)_3Si]_2C{=}C{=}N{-}Si(CH_3)_3 \quad (1\text{-}180)$$

Malononitrile reacts analogously [121]:

$$CH_2(CN)_2 + 2NaN[Si(CH_3)_3]_2 \xrightarrow[-2HN[Si(CH_3)_3]_2,\ -2NaCl]{+2(CH_3)_3SiCl} (CH_3)_3SiN{=}C{=}C{=}C{=}NSi(CH_3)_3 \quad (1\text{-}181)$$

Sodium bis(trimethylsilyl)amide reacts analogously with dithiocyanogen in ether even at 0°C. However, instead of the expected bis(trimethylsilyl)aminothiocyanogen, this gives bis[bis(trimethylsilyl)amino]disulfide, bis(trimethylsilylimino)methane, and sodium thiocyanate [1068]:

$$4[(CH_3)_3Si]_2NNa + 3(SCN)_2 \longrightarrow [(CH_3)_3Si]_2N{-}S{-}S{-}N[Si(CH_3)_3]_2 +$$
$$+ 2(CH_3)_3SiN{=}C{=}NSi(CH_3)_3 + 4NaSCN \quad (1\text{-}182)$$

The reaction of sodium bis(trimethylsilyl)amide with organic disulfides in benzene proceeds quantitatively according to the scheme [1059, 1061]:

$$(R'_3Si)_2NNa + RSSR \longrightarrow RSN(SiR'_3)_2 + RSNa \quad (1\text{-}183)$$

Aryl halosulfans react analogously [1059].

The rate of the reaction depends on the structure of the radicals R and R'; it increases with a change in R in the following series:

$$n\text{-}C_4H_9 < C_2H_5 < C_6H_5$$

When $R' = CH_3$ is replaced by an isopropoxy group, the reaction rate falls [1061].

TABLE 12. Products of the Reaction of Compounds Containing the Si–N–M Group with Organohalosilanes

Compound with the Si–N–M group	Organohalosilane	Reaction products	Yield, %	Literature
$(C_2H_5)_3SiNHLi$	$(CH_3)_3SiCl$	$(C_2H_5)_3SiNHSi(CH_3)_3$	81	1222, 1243
$(CH_3)_3SiNHNa$	$(CH_3)_3SiCl$	$(CH_3)_3SiNHSi(CH_3)_3$	—	1067
$(CH_3)_3SiNLiCH_3$	$(CH_3)_2SiCl_2$	$(CH_3)_2Si[N(CH_3)Si(CH_3)_3]_2$	42	1065, 1067
$(CH_3)_2Si(NLiCH_3)_2$	$(CH_3)_2SiCl_2$	$(CH_3)_2SiN(CH_3)Si(CH_3)_2NCH_3$ (ring)	68	926, 928
$(CH_3)_2Si(NLiCH_3)_2$	$CH_3C_6H_5SiCl_2$	$CH_3C_6H_5SiN(CH_3)Si(CH_3)_2NCH_3$ (ring)	49	926, 928
$(CH_3)_2Si(NLiCH_3)_2$	$(C_6H_5)_2SiCl_2$	$(C_6H_5)_2SiN(CH_3)Si(CH_3)_2NCH_3$ (ring)	59	926, 928
$(CH_3)_2Si(NLiCH_3)_2$	$(CH_3)_2SiClSiCl(CH_3)_2$	$(CH_3)_2SiN(CH_3)Si(CH_3)_2Si(CH_3)_2NCH_3$ (ring)	—	122
$(CH_3)_2Si(NLiCH_3)_2$	$(CH_3)_2SiBrCH_2SiBr(CH_3)_2$	$(CH_3)_2SiN(CH_3)Si(CH_3)_2CH_2Si(CH_3)_2NCH_3$ (ring)	32	927
$(CH_3)_2Si(NLiC_2H_5)_2$	$(CH_3)_2SiCl_2$	$(CH_3)_2SiN(C_2H_5)Si(CH_3)_2N(C_2H_5)$ (ring)	41	697, 1386
$(CH_3)_2Si(NLiC_4H_9\text{-n})_2$	$(CH_3)_2SiCl_2$	$(CH_3)_2SiN(C_4H_9)Si(CH_3)_2N(C_4H_9\text{-n})$ (ring)	83	697, 1386
$(CH_3)_2Si(NLiC_4H_9\text{-tert})_2$	$(CH_3)_2SiCl_2$	$(CH_3)_2SiN(C_4H_9\text{-tert})Si(CH_3)_2N(C_4H_9\text{-tert})$ (ring)	73	1386
$(CH_3)_2Si(NLiC_4H_9\text{-tert})_2$	$(CH_3)_2SiCl_2$	$(CH_3)_2SiN(C_4H_9\text{-tert})Si(CH_3)_2N(C_4H_9\text{-tert})$	75.5	697
$(CH_3)_2Si(NLiC_6H_{11})_2$	$(CH_3)_2SiCl_2$	$(CH_3)_2SiN(C_6H_{11})Si(CH_3)_2N(C_6H_{11})$ (ring)	73.8	1386
$(CH_3)_2Si(NLiC_6H_{11})_2$	$(CH_3)_2SiCl_2$	$(CH_3)_2SiN(C_6H_{11})Si(CH_3)_2N(C_6H_{11})$ (ring)	78.9	697
$(CH_3)_2Si(NLiC_6H_5)_2$	$(CH_3)_2SiCl_2$	$(CH_3)_2SiN(C_6H_5)Si(CH_3)_2N(C_6H_5)$ (ring)	49.5	697

$(CH_3)_2Si(NLiC_6H_5)_2$	$(CH_3)_2SiCl_2$	$(CH_3)_2SiN(C_6H_5)Si(CH_3)_2N(C_6H_5)$	100	1386
$(CH_3)_2Si(NLiC_6H_5)_2$	CH_3SiCl_3	$CH_3SiClN(C_6H_5)Si(CH_3)_2N(C_6H_5)$	85.5	1386
$(CH_3)_2Si(NLiC_6H_4CF_3\text{-}m)_2$	CH_3SiCl_3	$CH_3SiClN(C_6H_4CF_3\text{-}m)Si(CH_3)_2N(C_6H_4CF_3\text{-}m)$	85.5	697
$(C_6H_5)_2Si(NLiC_2H_5)_2$	$(C_6H_5)_2SiCl_2$	$(C_6H_5)_2SiN(C_2H_5)Si(C_6H_5)_2N(C_2H_5)$	63.8	697
$(C_6H_5)_2Si(NLiC_6H_5)_2$	$(CH_3)_2SiCl_2$	$(CH_3)_2SiN(C_6H_5)Si(C_6H_5)_2N(C_6H_5)$	54	697, 1386
$(C_6H_5)_2Si(NLiC_6H_5)_2$	$(C_6H_5)_2SiCl_2$	$(C_6H_5)_2SiN(C_6H_5)Si(C_6H_5)_2N(C_6H_5)$	62.8	1386
$(CH_3)_3SiNLi(CH_2)_2NHSi(CH_3)_3$	$(CH_3)_3SiCl$	$[(CH_3)_3Si]_2NCH_2CH_2NHSi(CH_3)_3$	81	916
$[(CH_3)_3SiNLiCH_2]_2$	$(CH_3)_2SiCl_2$	$(CH_3)_2SiN[Si(CH_3)_3]CH_2CH_2NSi(CH_3)_3$	81.5	916, 1033
$[(CH_3)_3SiNLiCH_2]_2$	CH_3SiCl_3	$(CH_3)_3SiNCH_2CH_2[(CH_3)_3Si]NSiCl(CH_3)$	47	916, 1033
$[(CH_3)_3Si]_2NLi$	$(CH_3)_3SiCl$	$[(CH_3)_3Si]_3N$	85	520, 1222 1241—1243
$[(CH_3)_3Si]_2NLi$	$(n\text{-}C_3H_7)_3SiCl$	$[(CH_3)_3Si]_2NSi(C_3H_7\text{-}n)_3$		1222, 1243
$[(CH_3)_3Si]_2NLi$	$(C_6H_5)_3SiCl$	$[(CH_3)_3Si]_2NSi(C_6H_5)_3$	21	520
$[(CH_3)_3Si]_2NLi$	$(CH_3)_2SiCl_2$	$[(CH_3)_3Si]_2NSiCl(CH_3)_2$	72	520, 1222, 1243
$[(CH_3)_3Si]_2NLi$	$(CH_3)_2SiCl_2$	$\{[(CH_3)_3Si]_2N\}_2Si(CH_3)_2$	35	1243
$[(CH_3)_3Si]_2NLi$	$(C_2H_5)_2SiCl_2$	$[(CH_3)_3Si]_2NSiCl(C_2H_5)_2$	21	520
$[(CH_3)_3Si]_2NLi$	$CH_3(C_2H_5)SiCl_2$	$[(CH_3)_3Si]_2NSiCl(CH_3)C_2H_5$	84	693
$[(CH_3)_3Si]_2NLi$	$(C_6H_5)_2SiCl_2$	$[(CH_3)_3Si]_2NSiCl(C_6H_5)_2$	25	520, 1222, 1243
$[(CH_3)_3Si]_2NLi$	CH_3SiCl_3	$[(CH_3)_3Si]_2NSiCl_2CH_3$	75	978
$[(CH_3)_3Si]_2NNa$	$(CH_3)_3SiCl$	$[(CH_3)_3Si]_3N$	78	122, 805, 1242, 1246
$[(CH_3)_3Si]_2NNa$	$(CH_3)_3SiSiCl(CH_3)_2$	$[(CH_3)_3Si]_2NSi(CH_3)_2Si(CH_3)_3$	90	1227
$[(CH_3)_3Si]_2NNa$	CH_3SiCl_3	$[(CH_3)_3Si]_2NSiCl_2CH_3$	62	1230
$[(CH_3)_3Si]_2NNa$	$n\text{-}C_3H_7SiCl_3$	$[(CH_3)_3Si]_2NSiCl_2C_3H_7\text{-}n$	75.5	1230

TABLE 12. (Cont'd)

Compound with the Si−N−M group	Organohalosilane	Reaction products	Yield, %	Literature
$[(CH_3)_3Si]_2NNa$	$CH_2{=}CHSiCl_3$	$[(CH_3)_3Si]_2NSiCl_2CH{=}CH_2$	57	1230
$[(CH_3)_3Si]_2NNa$	$C_6H_{11}SiCl_3$	$[(CH_3)_3Si]_2NSiCl_2C_6H_{11}$	72	1230
$[(CH_3)_3Si]_2NNa$	$C_6H_5SiCl_3$	$[(CH_3)_3Si]_2NSiCl_2C_6H_5$	60.2	1230
$[(CH_3)_3Si]_2NK$	$(CH_3)_3SiCl$	$[(CH_3)_3Si]_3N$	82	1242, 1451
$[(CH_3)_3Si]_2NK$	$CH_2(CH_2)_2SiClCH_3$	$CH_2(CH_2)_2Si(CH_3)N[Si(CH_3)_3]_2$	41	386, 387
$[(CH_3)_3Si]_2NK$	$(CH_3)_2SiCl_2$	$\{[(CH_3)_3Si]_2N\}_2Si(CH_3)_2$	40	1451
$[(CH_3)_3Si]_2NK$	$C_6H_5SiCl_3$	$[(CH_3)_3Si]_2NSiCl_2C_6H_5$	62	1451
$(CH_3)_3SiNLiSi(C_2H_5)_3$	$(n\text{-}C_3H_7)_3SiCl$	$[(CH_3)_3Si][(C_2H_5)_3Si][(C_3H_7\text{-}n)_3Si]N$	11	1243
$[(CH_3)_2C_2H_5OSi]_2NNa$	$(CH_3)_2SiCl_2$	$[(CH_3)_2C_2H_5OSi]_2NSiCl(CH_3)_2$	52	1231
$[CH_3(CH_2{=}CH)\ iso\text{-}C_3H_7OSi]_2NNa$	$CH_3(CH_2{=}CH)SiCl_2$	$[CH_3(CH_2{=}CH)\ iso\text{-}C_3H_7OSi]_2NSi(CH{=}CH_2)ClCH_3$	45.5	1228, 1231
$[CH_3(CH_3O)_2Si]_2NNa$	CH_3SiCl_3	$[CH_3(CH_3O)_2Si]_2NSiCl_2CH_3$	41.5	1228, 1231
$[CH_2{=}CH(CH_3O)_2Si]_2NNa$	$CH_2{=}CHSiCl_3$	$[CH_2{=}CH(CH_3O)_2Si]_2NSiCl_2CH{=}CH_2$	64	1231
$[(CH_3O)_3Si]_2NNa$	$CH_2{=}CHSiCl_3$	$[(CH_3O)_3Si]_2NSiCl_2CH{=}CH_2$	23	1247
$[(CH_3O)_3Si]_2NNa$	$n\text{-}C_3H_7SiCl_3$	$[(CH_3O)_3Si]_2NSiCl_2C_3H_7\text{-}n$	17	1247
$[(iso\text{-}C_3H_7O)_3Si]_2NNa$	CH_3SiCl_3	$[(iso\text{-}C_3H_7O)_3Si]_2NSiCl_2CH_3$	57	1247
$[(iso\text{-}C_3H_7O)_3Si]_2NNa$	$CH_2{=}CHSiCl_3$	$[(iso\text{-}C_3H_7O)_3Si]_2NSiCl_2CH{=}CH_2$	42	1247
$CH_3N[Si(CH_3)_2NLi(CH_3)]_2$	$(CH_3)_2SiCl_2$	$(CH_3)_2SiN(CH_3)Si(CH_3)_2N(CH_3)Si(CH_3)_2NCH_3$	74	926, 927
$CH_3N[Si(CH_3)_2NLi(CH_3)]_2$	$CH_3(C_6H_5)SiCl_2$	$CH_3(C_6H_5)SiN(CH_3)Si(CH_3)_2N(CH_3)Si(CH_3)_2NCH_3$	56	927

$CH_3N[Si(CH_3)_2NLi(CH_3)]_2$	$(C_6H_5)_2SiCl_2$	$(C_6H_5)_2SiN(CH_3)Si(CH_3)_2N(CH_3)Si(CH_3)_2NCH_3$	59	927
$(CH_3)_3SiNSi(CH_3)_2NSi(CH_3)_3$ Li Li	$(CH_3)_2SiCl_2$	$(CH_3)_3SiNSi(CH_3)_2N[Si(CH_3)_3]Si(CH_3)_2$	81	692, 693
$(CH_3)_3SiNSi(CH_3)_2NSi(CH_3)_3$ Li Li	$(C_2H_5O)_2SiCl_2$	$(CH_3)_3SiNSi(CH_3)_2NSi(CH_3)_3Si(OC_2H_5)_2$	47.8	693
$(CH_3)_3SiNSi(CH_3)_2NSi(CH_3)_3$ Li Li	$[(CH_3)_3Si]_2NSiCl(CH_3)_2$	$(CH_3)_3Si[NSi(CH_3)Si(CH_3)_2]_2NLiSi(CH_3)_3$	—	693, 696
		$(CH_3)_3SiNSi(CH_3)_2NSi(CH_3)_3Si(CH_3)_2$	—	
$(CH_3)_3SiNSi(CH_3)(C_2H_5)NSi(CH_3)_3$ Li Li	$CH_3C_2H_5SiCl_2$	$(CH_3)_3SiNSiCH_3C_2H_5NSi(CH_3)_3SiCH_3C_2H_5$	65	693
$(CH_3)_2Si[NHSi(CH_3)_2]_2NLi$	$(CH_3)_3SiCl$	$(CH_3)_2Si[NHSi(CH_3)_2]_2NSi(CH_3)_3$	38	695
$(CH_3)_2SiNHSi(CH_3)_2NLiSi(CH_3)_2NLi$	$(CH_3)_3SiCl$	$(CH_3)_2SiNHSi(CH_3)_2N[Si(CH_3)_3]Si(CH_3)_2NSi(CH_3)_3$	83.9	691, 695
$[(CH_3)_2SiNLi]_3$	$(CH_3)_3SiCl$	$[(CH_3)_2SiNSi(CH_3)_3]_3$	—	576, 691, 695
$[(C_2H_5)_2SiNLi]_3$	$(CH_3)_3SiCl$	$[(C_2H_5)_2SiNSi(CH_3)_3]_3$	—	691
$[(C_6H_5)_2SiNLi]_3$	$(CH_3)_3SiCl$	$[(C_6H_5)_2SiNSi(CH_3)_3]_3$	—	691
$[(C_6H_5)_2SiNLi]_3$	$(CH_3)_2SiCl_2$	$[(C_6H_5)_2SiNSiCl(CH_3)_2]_3$	—	691
$(C_6H_5)_2Si[NHSi(C_6H_5)_2]_2NLi$	$(CH_3)_3SiCl$	$(C_6H_5)_2Si[NHSi(C_6H_5)_2]_2NSi(CH_3)_3$	70.1	695
$(C_2H_5)_2Si[N(Li)Si(C_2H_5)_2]_2NH$	$(CH_3)_3SiCl$	$(C_2H_5)_2Si[(CH_3)_3SiNSi(C_2H_5)_2]_2NH$	—	691
$(CH_3)_2Si[NHSi(CH_3)_2]_3NLi$	$(CH_3)_3SiCl$	$(CH_3)_2Si[NHSi(CH_3)_2]_3NSi(CH_3)_3$	12.9	697

TABLE 12. (Cont'd)

Compound with the Si–N–M group	Organohalosilane	Reaction products	Yield, %	Literature
$[(CH_3)_2SiNHSi(CH_3)_2NLi]_2$	$(CH_3)_3SiCl$	$[(CH_3)_2SiNHSi(CH_3)_2NSi(CH_3)_3]_2$	68.4	695
$(CH_3)_2Si[NLiSi(CH_3)_2]_3NH$	$(CH_3)_3SiCl$	$(CH_3)_2Si[(CH_3)_3SiNSi(CH_3)_2]_3NH$	55	691, 695
$[(CH_3)_2SiNLi]_4$	$(CH_3)_3SiCl$	$(CH_3)_3SiNSi(CH_3)_2N[Si(CH_3)_3]Si(CH_3)_2$ and other polysilazans	18	695
$(CH_3)_3SiNHCONLiSi(CH_3)_3$	$(CH_3)_3SiCl$	$(CH_3)_3SiNCNSi(CH_3)_3$	80	1017
		$[(CH_3)_3Si]_2O$	—	
$C_6H_5NLiNHSi(CH_3)_3$	$(CH_3)_3SiCl$	$C_6H_5[(CH_3)_3Si]NNHSi(CH_3)_3$	75	1235
$(CH_3)_2NNLiSi(CH_3)_3$	$(CH_3)_3SiCl$	$(CH_3)_2NN[Si(CH_3)_3]_2$	56	1244
$C_2H_5C_6H_5NNLiSi(CH_3)_3$	$(CH_3)_3SiCl$	$C_2H_5(C_6H_5)NN[Si(CH_3)_3]_2$	62	1244
$(C_6H_5)_2NNLiSi(CH_3)_3$	$(CH_3)_3SiCl$	$(C_6H_5)_2NN[Si(CH_3)_3]_2$	67	1244
$(CH_3)_3SiNHNLiSi(CH_3)_3$	$(CH_3)_3SiCl$	$(CH_3)_3SiNHN[Si(CH_3)_3]_2$	56	1244
$(CH_3)_3SiNHNLiSi(CH_3)_3$	$(C_2H_5)_3SiCl$	$(CH_3)_3SiNHNSi(CH_3)_3[Si(C_2H_5)_3]$	70	1244
$(CH_3)_3SiNHNLiSi(C_2H_5)_3$	$(C_2H_5)_3SiCl$	$(CH_3)_3SiNHN[Si(C_2H_5)_3]_2$	45	1244
$(C_2H_5)_3SiNHNLiSi(C_2H_5)_3$	$(CH_3)_3SiCl$	$(C_2H_5)_3SiNHNSi(CH_3)_3Si(C_2H_5)_3$	65	1244
$(CH_3)_3SiNLiNLiSi(CH_3)_3$	$(CH_3)_3SiCl$	$(CH_3)_3SiNHN[Si(CH_3)_3]_2$	77	1234
		$(CH_3)_3SiN[Si(CH_3)_3]_2$	—	

3.3.5. Reaction with Organosilicon Compounds

Compounds containing the Si–N–M group react with halosilanes with the formation of a new Si–N bond:

$$\geq Si{-}\underset{|}{N}{-}M + X{-}Si\leq \xrightarrow[-MX]{} \geq Si{-}\underset{|}{N}{-}Si\leq \qquad (1\text{-}184)$$

In accordance with this scheme, N-lithio derivatives of triorganoaminosilanes form hexaorganodisilazans [1067, 1222, 1243] and derivatives of triorgano(alkylamino)silanes give N-alkylhexaorganodisilazans or N,N'-dialkyloctaorganotrisilazans (if the reaction is carried out with diorganodichlorosilanes) [1065]. Dilithium derivatives of diorgano(N-alkylamino)silanes reacts with diorganodichlorosilanes and organotrichlorosilanes to form derivatives of cyclodisilazan [697, 926, 928, 1386]:

$$R_2Si(NLiR)_2 + Cl_2Si\langle \xrightarrow[-2LiCl]{} \begin{array}{c} R_2Si{-}N{-}R \\ |\quad\;\; | \\ R{-}N{-}Si \end{array} \qquad (1\text{-}185)$$

Under these conditions, silicon tetrachloride may form both monocyclic and spirocyclic compounds [928]. Carrying out the reaction with halogen derivatives of a disilane [122] and disilylmethane [927] leads to the formation of heterocycles:

$$R_2Si(N(R){-}Li)_2 + Cl{-}SiR_2{-}SiR_2{-}Cl \xrightarrow[-2LiCl]{} \begin{array}{c} R_2Si{-}SiR_2 \\ |\qquad\quad | \\ R{-}N \qquad N{-}R \\ \diagdown\;\diagup \\ SiR_2 \end{array} \qquad (1\text{-}186)$$

$$R_2Si(N(R){-}Li)_2 + Br{-}Si(CH_3)_2{-}CH_2{-}Si(CH_3)_2{-}Br \xrightarrow[-2LiBr]{} \begin{array}{c} CH_2 \\ \diagup\;\diagdown \\ (H_3C)_2Si \qquad Si(CH_3)_2 \\ |\qquad\qquad | \\ R{-}N \qquad N{-}R \\ \diagdown\;\diagup \\ SiR_2 \end{array} \qquad (1\text{-}187)$$

The reaction of mono- and dilithio derivatives of N,N'-bis-(trimethylsilyl)ethylenediamine with trimethylchlorosilane proceeds in accordance with scheme (1-184). Dimethyldichlorosilane and methyltrichlorosilane react with the dilithio derivative of N,N'-bis-

(trimethylsilyl)ethylenediamine to form 5-membered heterocycles [916, 1033]:

$$\begin{array}{l}(CH_3)_3SiNLi \\ \quad | \\ \quad (CH_2)_2 \\ \quad | \\ (CH_3)_3SiNLi\end{array} + Cl_2Si\langle \xrightarrow[-2LiCl]{} (CH_3)_3Si{-}N\!\!\begin{array}{c}CH_2{-}CH_2 \\ \diagup \qquad \diagdown \\ \\ \diagdown \qquad \diagup \\ Si \\ \diagup \ \diagdown\end{array}\!\!N{-}Si(CH_3)_3 \qquad (1\text{-}188)$$

A compound with the following structure was also isolated in the reaction with methyltrichlorosilane [1033]:

$$\begin{array}{c}Si(CH_3)_3 \\ | \\ H_2C{-}N \\ | \\ H_2C{-}N \\ | \\ Si(CH_3)_3\end{array}\!\!\!\begin{array}{c}CH_3 \\ | \\ Si{-}NHCH_2CH_2NH{-}Si \\ \end{array}\!\!\!\begin{array}{c}CH_3 \\ | \end{array}\!\!\begin{array}{c}Si(CH_3)_3 \\ | \\ N{-}CH_2 \\ | \\ N{-}CH_2 \\ | \\ Si(CH_3)_3\end{array}$$

The products from the reaction of bis(triorganosilyl)amides of alkali metals with triorganohalosilanes are tris(silyl)substituted amines [386, 387, 520, 693, 805, 978, 1222-1224, 1227, 1228, 1230, 1231, 1241-1243, 1247, 1451]. As a rule, diorganodihalosilanes yield monosubstituted products [520, 1222, 1228, 1231, 1243] and only in the case of dimethyldichlorosilane has it been possible to synthesize a bis-substituted compound [1242, 1451]. On this basis it is possible to replace one of the terminal chlorine atoms in α,ω-dichloropolydimethylsiloxanes by a bis(trimethylsilyl)amino group [903, 905], using the difference in the reactivities of the terminal functions of the oligomers obtained for subsequent conversions. As yet no examples of the substitution of a bis(triorganosilyl)amino group for more than one chlorine atom in $RSiCl_3$ have been described [978, 1228, 1230, 1231, 1247, 1451]. The reactivity of $(R_3Si)_2NM$ toward trimethylchlorosilane depends on the nature of the metal M and increases from lithium to sodium and potassium [1242]. Mono-di-, and trilithio derivatives of linear [692, 693, 696] and cyclic [555, 691, 695] trisilazans react analogously. The reaction of N-lithium derivatives of octamethylcyclotetrasilazan with trimethylchlorosilane is carried out in an autoclave at 150-160°C [691, 692]. In this way it is possible to introduce up to three trimethylsilyl groups into a cyclosilazan molecule. In an attempt to prepare 2,4, 6,8-tetrakis(dimethylsilyl)octamethylcyclotetrasilazan, the formation of bis- and tris(trimethylsilyl) derivatives of linear polysilazans and N,N'-bis(trimethylsilyl)tetramethylcyclodisilazan was observed [697]. The latter compound is also formed in the pyrolysis of various N-lithio derivatives of organosilazans [692, 696].

The reaction of N-lithio derivatives of N-organo-N'-trialkylsilylhydrazines [1235], N-trialkylsilyl-N',N'-diorganohydrazines [1244], and N,N'-bis(triorganosilyl)hydrazines with triorganochlorosilanes also proceeds according to scheme (1-184).

The products from the reaction of compounds containing the group $\overset{Si}{\diagup}\!\!>N{-}M$ with organohalosilanes and their yields are given in Table 12.

In contrast to triorganoalkoxysilanes, triorganophenoxysilanes react with bis(triorganosilyl)amides of alkali metals in boiling petroleum ether or benzene analogously to triorganohalosilanes [121, 122, 1233]:

$$\begin{matrix} R'_3Si + O{-}C_6H_5 \\ (R_3Si)_2N + Na \end{matrix} \longrightarrow (R_3Si)_2NSiR'_3 + NaOC_6H_5 \qquad (1\text{-}189)$$

In benzene solution, triorganoisocyanatosilanes react exothermally with sodium bis(trimethylsilyl)amide to form bis(trimethylsilyl)carbodiimide, hexamethyldisiloxane, and sodium cyanate [122, 1016, 1245]. Triorganoisothiocyanatosilanes (in contrast to the oxygen analogs) react with sodium bis(triorganosilyl)amides only on heating to boiling. The reaction products are bis(trimethylsilyl)carbodiimide, hexamethyldisilthiane, tris(trimethylsilyl)-amine, and sodium thiocyanate. The last two compounds may be formed by a scheme analogous to (1-184) if the isothiocyanate group behaves as a pseudo halogen [1245]:

$$(CH_3)_3SiNCS + NaN[Si(CH_3)_3]_2 \longrightarrow [(CH_3)_3Si]_3N + NaSCN \qquad (1\text{-}190)$$

Reactions with other heteroorganic compounds are described in subsequent chapters.

4. COMPOUNDS CONTAINING THE Si – O – M GROUP (SILANOLATES OF ALKALI METALS)

4.1. Synthesis Methods

4.1.1. Reaction of Silanols with Alkali Metals and Their Hydoxides

Because of the $d_\pi - p_\pi$ interaction of the p-electrons of the oxygen atom with the free 3d-orbitals of the silicon atom, organosilanols have acid properties, which are appreciably stronger than their organic analogs, namely, alcohols. They all react with alkali metals:

$$\geq Si-O-H + 2M \longrightarrow \geq Si-O-M + H_2 \quad (1\text{-}191)$$

On treatment with lithium in petroleum ether or ethyl ether, trimethylsilanol smoothly forms lithium trimethylsilanolate [1206]. The reaction of trimethylsilanol with sodium proceeds vigorously and requires cooling [1206]. It is normally carried out in benzene [1174] or xylene [1177]. Trimethylsilanolates of potassium [1206], rubidium, and cesium [1094] are obtained analogously. Triethylsilanol reacts with sodium much more readily than triethylcarbinol-[917, 920, 1177, 1206]. With an increase in the length of the carbon chain of the alkyl group in trialkylsilanols, the rate of their reaction with sodium falls and heating is required to achieve the reaction [1000, 1207]. However, ethylisobutylsilanol, which is an analog of secondary and not tertiary alcohols, like an overwhelming majority of known silanols, reacts with sodium and potassium even at room temperature [500]. Triphenylsilanol reacts readily with sodium in diethyl ether [388, 389, 631, 736, 1206], benzene [483], toluene [1070], and a mixture of xylene and ethyl ether [72]. The reaction proceeds much more rapidly than in the case of triphenylcarbinol [1070] and the yield of sodium triphenylsilanolate is close to theoretical. Sodium dimethylphenylsilanolate [157] and the disodio derivative of diphenylsilanediol [1206] may be obtained analogously. The sodio derivatives of dimethylsilanediol, methylphenylsilanediol, diphenylsilanediol, phenylsilanetriol, and 1,3-dihydroxytetramethyldisiloxane are obtained by the reaction of these silanols with sodium in liquid ammonia [1126]. The reaction of dimethylchloromethylsilanol with sodium was carried out in liquid ammonia [1269] and this formed the trimethylsilanolate instead of sodium dimethylchloromethylsilanolate. This is the result of a secondary process, namely, reduction by sodium hydride formed in the reaction in accordance with the scheme

$$(CH_3)_2\underset{\substack{|\\ CH_2Cl}}{Si}{-}ONa + NaH \longrightarrow (CH_3)_3SiONa + NaCl \qquad (1\text{-}192)$$

The disodio derivative of 1,3-dihydroxytetraethyldisloxane is formed by reaction of the latter with sodium in xylene in 130°C (up to 86% of the hydroxyl groups react in 1.5 hr) [169]. An analogous reaction made it possible to obtain the disodio derivative of 1,3-dihydroxytetramethyldisiloxane in 40% yield [348]. On heating in boiling toluene, sodium also reacts with free hydroxyl groups of polyphenylsiloxane resins [169]. Sodium tris(trimethylsiloxy)silanolate, which is soluble in benzene, was also obtained in accordance with scheme (1-191) [159].

Organosilanols dissolve in aqueous solutions of NaOH and KOH to form solutions of the corresponding sodium and potassium organosilanolates [709, 873, 874, 882, 939, 1031, 1032, 1171, 1200, 1663]:

$$\rangle Si{-}OH + KOH \longrightarrow \rangle Si{-}OK + H_2O \qquad (1\text{-}193)$$

In some cases it is possible to isolate individual triorganosilanolates from the solutions [709, 1171, 1177, 1663]. A saturated solution of LiOH does not react in accordance with the scheme (1-193) and when trimethylsilanol is shaken with solid LiOH, only dehydration occurs [1206].

4.1.2. Reaction of Halosilanes, Alkoxysilanes, and Acyloxysilanes with Alkali Metal Hydroxides

In many reactions of organohalosilanes [390, 873, 880, 889, 893-895, 950, 951, 1009, 1030, 1315, 1557] and organoalkoxysilanes [701, 871, 917, 919, 922, 1171, 1308] with alkali metal hydroxides, which are carried out in order to prepare (after neutralization) organosilanols, the intermediate products are silanolates of these metals:

$$\geq Si-X + HOM \xrightarrow[-HX]{} \geq Si-O-M \qquad (1\text{-}194)$$

$$(X = \text{halogen}, RO, RCOO)$$

Reactions of this type probably occur in the formation of solutions of sodium silanolates when NaOH is heated with rice husks; in all probability, the latter contain silicon attached to polysaccharides through an Si–O–C group [1163, 1164].

In a number of cases it is possible to obtain silanolates [244, 1527], including those of the type $(RO)_3SiOM$ and R_3SiOM, in a crystalline state by the reaction of alkoxysilanes with anhydrous NaOH or KOH. The reaction of an ether solution of diphenyldimethoxysilane with a 50% aqueous solution of NaOH at 5°C gives a 96% yield of $(C_6H_5)_2Si(ONa)_2$ [896]. A study of the reaction of methylphenyl-α-naphthylacetoxysilane with KOH in xylene showed that the reaction proceeds stereospecifically to a considerable extent with 85% inversion of the configuration [1171].

4.1.3. Preparation of Alkali Metal Silanolates from Siloxanes

Organosiloxanes are cleaved by alkali metal hydroxides with the formation of alkali metal organosilanolates [139, 152, 157, 166, 175, 348, 417, 430, 452, 470, 708, 710, 738, 850, 851, 868, 881, 918, 921, 957, 1023, 1171, 1200, 1380, 1454, 1527, 1554, 1556, 1557]:

$$\geq Si-O-Si \leq + 2MOH \rightleftarrows 2 \geq Si-O-M + H_2O \qquad (1\text{-}195)$$

Both hexaorganodisiloxanes and organopolysiloxanes will take part in reaction (1-195). In a number of cases it is possible to isolate crystalline organosilanolates of alkali metals from the reaction mixture [166, 348, 708, 710, 851, 1171, 1527]. To obtain these compounds successfully by this method it is necessary to completely remove the water liberated during the reaction; it not only retards the process, but also forms hydrates with the organosilanolates obtained. Solvents such as methyl, ethyl, and isopropyl alcohols are used so that the reagents are all in one phase. However, this in its turn complicates the preparation of pure organosilanolates as alcoholates are formed in parallel. Therefore reaction (1-195) is inconvenient for the synthesis of crystalline organosilanolates of alkali metals. Nonetheless, it is irreplaceable for the industrial production of solutions of sodium alkylsiliconates $[RSi(O)ONa]_n$, which are used widely for waterproofing various materials [27, 110, 166, 417, 457, 868, 1527].

The cleavage of 1,3-dimethyl-1,3-diphenyl-1,3-di-α-naphthyldisiloxane by potassium hydroxide in xylene proceeds with retention of the configuration [1171]:

$$(C_6H_5)(CH_3)(\alpha\text{-}C_{10}H_7)Si{-}O{-}Si(CH_3)(\alpha\text{-}C_{10}H_7)(C_6H_5) \xrightarrow{KOH} (C_6H_5)(CH_3)(\alpha\text{-}C_{10}H_7)Si{-}OK + KO{-}Si(CH_3)(\alpha\text{-}C_{10}H_7)(C_6H_5) \quad (1\text{-}196)$$

The formation of organosilanolates of alkali metals is postulated to explain the mechanism of many rearrangements and polymerizations of organosiloxanes, which are catalyzed by alkalis [110, 275, 305, 341-344, 349, 350, 384a, 421, 456, 593, 653, 723, 819, 848, 900, 909-912, 1057, 1298, 1299, 1305, 1316, 1358, 1359, 1396, 1490, 1523, 1531, 1535, 1539-1542, 1551, 1566]; the examination of these reactions is beyond the scope of this monograph.

Organosilanolates of alkali metals are also obtained by the reaction of organosiloxanes with sodium or potassium oxides [851, 1306, 1539] and also with metallic potassium in tetrahydrofuran [835].

When organosiloxanes are treated with sodium or potassium amide in liquid ammonia and also in toluene, cleavage of the siloxane bond occurs with the formation of sodium and potassium organosilanolates [851, 864, 903, 905, 1535]:

$$2\,[(CH_3)_3Si]_2O + 2NaNH_2 \longrightarrow 2\,(CH_3)_3SiONa + [(CH_3)_3Si]_2NH + NH_3 \quad (1\text{-}197)$$

The Si–O–Si group in organosilanes is cleaved by an organolithium with the formation of lithium organosilanolates [736, 830, 864, 903, 905, 1042, 1143, 1160]:

$$R_3SiOSiR_3 + R'Li \longrightarrow R_3SiR' + R_3SiOLi \quad (1\text{-}198)$$

Hexamethyldisiloxane is cleaved quantitatively by methyllithium in the mixed solvent tetrahydrofuran–ethyl ether.

Hexaphenyldisiloxane is cleaved by phenyllithium and p-tolyllithium in a mixture of ethyl ether and xylene [736]. The reaction is catalyzed by copper. Methyllithium does not react with hexaphenyldisiloxane [736].

Heating phenyllithium with 1,1,1-trimethyl-3,3,3-triphenyldisiloxane in boiling ether for 3 hr forms tetraphenylsilane (40%), lithium triphenylsilanolate (25%), and lithium trimethylsilanolate. The reaction of 1,1,1-trimethyl-3,3,3-triphenyldisiloxane with p-tolyllithium (18 hr) gave triphenyl(p-tolyl)silane (58.5%), trimethyl-(p-tolyl)silane (21%), and triphenylsilanol (33%) [736]. This shows that in unsymmetrically substituted hexaorganodisiloxanes there is cleavage of both Si–O bonds, but preferentially at the silicon with the lowest electron density.

Methyllithium also cleaves dimethylpolysiloxanes [1042]:

$$[(CH_3)_2SiO]_n + nCH_3Li \longrightarrow n\ (CH_3)_3SiOLi \qquad (1\text{-}199)$$

4.1.4. Other Reactions

Compounds containing the Si–O–M group form some tetraorganosilanes on alkaline cleavage [258, 546, 612, 663, 738, 872, 1184]:

$$R_3SiR' + KOH \longrightarrow R_3SiOK + R'H \qquad (1\text{-}200)$$

Thus, heating triphenylbenzylsilane with 20% KOH for 16.5 hr yielded toluene (33%) and a white precipitate of potassium triphenylsilanolate, which gave hexaphenyldisiloxane on hydrolysis. Fluorene (100%) and hexaphenyldisiloxene (64%) were obtained analogously from triphenyl-9-fluorenylsilane.

Potassium and sodium derivatives of methylsilanetriol and dimethylsilanediol were obtained by the reaction of methylsilane and dimethylsilane with NaOH and KOH [1189]. Lithium triphenylsilanolate is formed by oxidation of triphenylsilyllithium by oxygen [714].

In a discussion of the mechanism of the reaction of diorganodichlorosilanes with sodium, the hypothesis was put forward that the blue color obtained in this case is due to the formation of the radicals R_2SiONa (from R_2Si; and atmospheric oxygen or a layer of oxide on the sodium surface), which then dimerize to $(R_2SiONa)_2$ [611].

Sodium trimethylsilanolate is also formed by the reaction of sodium with trimethyl(2-chloroisopropoxy)silane [891]:

$$(CH_3)_3SiOCH(CH_3)CH_2Cl + 2Na \xrightarrow[-NaCl]{} (CH_3)_3SiONa + CH_3CH{=}CH_2 \qquad (1\text{-}201)$$

4.1.5. Preparation of Complex Silanolates of Alkali Metals

From a solution of an equimolar mixture of lithium and sodium trimethylsilanolates in carbon tetrachloride there crystallizes a double complex silanolate [1096, 1110]:

$$Na\{Li[OSi(CH_3)_3]_2\}$$

Dissolving potassium trimethylsilanolate in solutions of sodium and lithium trimethylsilanolates also yields double silanolates, which are crystalline substances with high melting points.

Trimethylsilanolates of alkali metals form complex tetrakis(trimethylsiloxy)alanates, tetrakis(trimethylsiloxy)gallates, and tetrakis(trimethylsiloxy)ferrates by reaction with tris(trimethylsiloxy)aluminum, tris(trimethylsiloxy)gallium, and tris(trimethylsiloxy)iron, respectively, and this will be discussed in detail below.

4.2. Physical Properties

Trimethylsilanolates of alkali metals are colorless crystalline substances, which melt with decomposition. Their decomposition points rise from lithium trimethylsilanolate to cesium trimethylsilanolate, but the sodium derivative has a higher decomposition point than the potassium derivative [1094]. Lithium and sodium trimethylsilanolates sublime in vacuum (1 mm) when heated below the decomposition point. Analogous derivatives of potassium, rubidium, and cesium do not sublime when heated in vacuum, but decompose with the elimination of hexamethyldisiloxane.

Lithium and sodium trimethylsilanolates are soluble in ethyl ether, methylene chloride, dimethyl sulfoxide, petroleum ether, benzene, cyclohexane, and carbon tetrachloride [851, 1094, 1206]. In this respect they considerably surpass the corresponding alcoholates. Potassium trimethylsilanolate is difficultly soluble in these solvents, while the derivatives of rubidium and cesium do not dissolve in them at all [1094]. Potassium trimethylsilanolate is soluble to a slight extent in hexamethylsiloxane [1206]. Sodium triphenylsilanolate dissolves in hot toluene and xylene and crystallizes out when the solution is cooled [1070].

In boiling carbon tetrachloride lithium trimethylsilanolate exists as a hexamer and in cold benzene and cyclohexene, heptamers and octomers predominate. Under analogous conditions, sodium trimethylsilanolate is even more associated and forms a decamer in cold benzene. Both of these silanolates are dimeric in boiling ethyl ether [1094]. Complex etherates are formed in this case, evidently due to the donor properties of the ether oxygen. In contrast to this, sodium methylsiliconate, obtained by alkaline cleavage of the hydrolyzate of methyl trichlorosilane, exists to more than 80% in the form of the monomer in molten Glauber's salt at high dilution. The corresponding potassium derivative is completely monomeric under these conditions. At 14% concentration of the sodium and potassium methylsiliconates in molten Glauber's salt about 45% of the sodium and 75% of the potassium compounds are monomeric [969].

According to the data of Voronkov et al. [26, 259], sodium polyorganosiliconates are white powders, which are soluble in water and correspond to the general formula $HO[RSi(O)ONa]_nH$

TABLE 13. Physical Constants of Compounds Containing the Si–O–M Group

Compound	m.p., °C	Decomp. p., °C	Sub. p., °C	Literature
$(CH_3)_3SiOLi$	—	120	115 (1)	1094
	—	—	~180 (1)	1206
$(CH_3)_3SiONa$	—	147—150	145 (1)	1094, 1177, 1539
	—	147—151	—	709
	—	—	130—140 (?)	851
$(CH_3)_3SiOK$	—	131—135	—	1206
	—	135	170 (1)	1094
$(CH_3)_3SiORb$	—	140—150	—	1094
$(CH_3)_3SiOCs$	—	200	—	1094
$(CH_3)_2C_6H_5SiONa$	87—94	—	—	851, 1306
$\{[(CH_3)_3SiO]_2Li\}Na$	232—235	—	—	1096, 1110
$\{[(CH_3)_3SiO]_2Li\}K$	258—260	—	—	1096, 1110
$\{[(CH_3)_3SiO]_2Na\}K$	235—237	—	—	1110

TABLE 14. Vibration Frequencies in IR Absorption Spectra of Alkali Metal Trimethylsilanolates of Type $(CH_3)_3SiOM$ [1094]

Assignment	M; frequency, cm^{-1}				
	Li	Na	K	Rb	Cs
δCH_3Si	1255	1250	1250	1233	1235
δCH_3Si	1250	1247	. . .	. . .	1228
$\nu_{as}SiO(M)$	948	975	990	996	1000
$\nu_{as}SiO(M)$	966	(941)	980	980	(986)
ρ_1CH_3Si	831	825	835	829	820
ρ_2CH_3Si	742	740	733	732	738

where n on an average equals 12 ($R = CH_3$) or 9($R = C_2H_5$). In aqueous solutions these compounds exist largely as monomeric or dimeric molecules.

The melting, decomposition, and sublimation points of triorganosilanolates of alkali metals and their complexes are given in Table 13.

In the IR absorption spectra of trimethylsilanolates of alkali metals [1094, 1206] there is an increase in the frequencies of the antisymmetric valence vibrations of the Si–O (M) bond with a change from Li to Cs. This may be connected with an increase in the d_π-p_π interaction of the p-electrons of the oxygen with the free 3d-orbitals of the silicon atom as a result of the increase in the polarity of the M–O bond. The relatively small changes in the spectrum of $(CH_3)_3SiOM$ with a change from K to Rb and Cs may be due to the fact that even the potassium derivative is an almost completely ionic structure. The vibration frequencies of trimethylsilanolates of alkali metals [1094] are given in Table 14.

In the IR absorption spectra of $[(CH_3)_2SiOM]_2O$ there is a fall in the frequencies ν_{as} and ν_sSiOSi with a change of M from H through Li

TABLE 15. Vibration Frequencies in IR Absorption Spectra of $[(CH_3)_3SiOM]_2O$ [348]

Assignment	M; frequency, cm^{-1}			
	H	Li	Na	K
$\nu_{as}SiOSi$	1049—1037	1055—1022	1010—1000	998 VS
$\nu_{as}SiO(M)$	905 VS	980 VS	963 Sh	941 S
$\nu_{as}SiO(M)$	878 S	939 Sh	940 Sh	925 M
ν_sSiOSi	555 M	565—537 ?	551 W	539 W

and Na to K, since together with an increase in the dynamic coefficient of the Si–O(M) bond there is a decrease in the coefficient of the Si–O(Si) bond [348]. There is also a decrease in the SiOSi angle (by 8-10°). Replacement of hydrogen atoms in the hydroxyl groups of 1,3-dihydroxytetramethyldisiloxane by lithium, sodium, and potassium evidently produces a redistribution of the $d_\pi - p_\pi$ interactions in the $Si-O^\ominus$ and Si–O(Si) bonds. An increase in the electron density at the oxygen atom strengthens this interaction in the $Si-O^\ominus$ bond and this in its turn reduces the order of the Si–O(Si) bond due to a reduction in the effective positive charge of the d-orbitals of silicon [348]. The frequencies of the vibrations of the Si–O–Si and Si–O–M bonds in $[(CH_3)_2SiOM]_2O$ are given in Table 15.

A study of the NMR spectra of alkali metal trimethylsilanolates shows that the signal from the methyl group protons in $(CH_3)_3SiONa$ is at higher fields than in the case of $(CH_3)_4Si$, despite the presence in the silanolate molecule of the oxygen atom, which shifts the signal toward lower fields [1094]. This can only be explained by a strong $d_\pi - p_\pi$ interaction of the p-electrons of the oxygen atom with the 3d-orbitals of silicon in the sodium trimethylsilanolate molecule. An even greater shift toward higher fields is observed in the NMR spectra of the double silanolates $K\{M[OSi(CH_3)_3]_2\}$ [1110].

The chemical shifts of protons of the methyl groups in alkali metal trimethylsilanolates and their complexes and also the spin-spin interaction constants are given in Table 16.

TABLE 16. Chemical Shifts and Spin-Spin Interaction Constants in NMR Spectra of Alkali Metal Trimethylsilanolates and Their Complexes†

Compound	δ*	$J(H^1-C^{13})$	$J(H^1-C-Si^{29})$	Literature
$K\{Li[OSi(CH_3)_3]_2\}$	+3.2	115.5	6.35	1110
$K\{Na[OSi(CH_3)_3]_2\}$	+3.0	114.5	6.40	1110
$(CH_3)_3SiONa$	+1.0	115.0	6.23	1094
$(CH_3)_4Si$	+0.0	118.0	6.78	1094
$(CH_3)_3SiOLi$	—1.25	116.0	6.45	1094
$Na\{Li[OSi(CH_3)_3]_2\}$	—2.75	116.0	6.45	1110
$(CH_3)_3SiOH$	—6.0	119.0	7.05	1094

† The spectra were plotted for solutions in carbon tetrachloride at 60 MHz with $(CH_3)_4Si$ as the internal standard.

Results of crystallographic and x-ray structural investigations of alkali metal triorganosilanolates are given in [851, 1206, 1539].

4.3. Chemical Properties

4.3.1. Hydrolysis

Alkali metal organosilanolates are hydrolyzed very readily by water:

$$\gt Si-O-M + H_2O \rightleftarrows \gt Si-O-H + MOH \qquad (1\text{-}202)$$

They are decomposed by atmospheric moisture and their solutions in water have a strongly alkaline reaction. The hydrolysis of organosilanolates is used to prepare free silanols [73, 83, 101, 110, 280, 388, 851, 864, 873, 880, 889, 895, 896, 1009, 1019, 1030, 1206, 1529]. A solution of the alkali metal organosilanolate is usually obtained for this purpose by dissolving an organohalosilane or organoalkoxysilane in alkali and without being isolated from solution, it is titrated with dilute acetic or hydrochloric acid.

Alkali metal organosilanolates are decomposed even by atmospheric carbon dioxide [110, 388, 868, 896, 917]. Hydrolysis in the presence of strong acids is accompanied by subsequent condensation of the organosilanol formed to a siloxane [1177]. Dry hydrogen chloride and glacial acetic acid convert potassium diphenylsilanediolate into hexaphenylcyclotrisiloxane [1200].

4.3.2. Reaction with Halosilanes*

Alkali metal triorganosilanolates react with halosilanes to form the grouping Si–O–Si:

$$nR_3SiOM + Cl_nSiR'_{4-n} \xrightarrow{-nMCl} [R_3SiO]_n SiR'_{4-n} \qquad (1\text{-}203)$$

This procedure has been used to prepare organic derivatives of disiloxanes [485, 486, 488, 736, 851, 864, 1041, 1171, 1554, 1657, 1663], linear trisiloxanes [482, 485, 488, 851, 864, 1143, 1174], branched tetrasiloxanes [482, 484, 485, 488, 851, 864, 1143, 1174,

* Reactions with halides of the elements and their organic derivatives are described in subsequent chapters.

TABLE 17. Products of the Reaction of Alkali Metal Organosilanolates with Halosilanes

Organosilanolate	Halosilane	Reaction products	Yield, %	Literature
$(CH_3)_3SiOLi$	$(CH_3)_3SiCl$	$(CH_3)_3SiOSi(CH_3)_3$	—	1041
	$(CH_3)_2SiCl_2$	$(CH_3)_2Si[OSi(CH_3)_3]_2$	91	1143
	$CH_3(CH_2{=}CH)SiCl_2$	$CH_3(CH_2{=}CH)Si[OSi(CH_3)_3]_2$	80	1143
	$CH_2{=}CHCH_2SiCl_3$	$CH_2{=}CHCH_2Si[OSi(CH_3)_3]_3$	82	1143
$(CH_3)_3SiONa$	$(C_6H_5)_3SiCl$	$(C_6H_5)_3SiOSi(CH_3)_3$	86	851
	$(C_2H_5)_2SiCl_2$	$(C_2H_5)_2Si[OSi(CH_3)_3]_2$	44	1174
	CH_3SiCl_3	$CH_3Si[OSi(CH_3)_3]_3$	—	851, 1554
	$C_2H_5SiCl_3$	$C_2H_5Si[OSi(CH_3)_3]_3$	52	851, 1174
	$C_6H_5SiCl_3$	$C_6H_5Si[OSi(CH_3)_3]_3$	—	851, 1554
	$SiCl_4$	$Si[OSi(CH_3)_3]_4$	38	851, 1174, 1554
$(CH_3)_3SiOK$	$CH_3(C_6H_5)Si[OSiCH_3(C_6H_5)Cl]_2$	$CH_3(C_6H_5)Si[OSiCH_3(C_6H_5)OSi(CH_3)_3]_2$	58	851
$(CH_3)_2C_6H_5SiONa$	$(CH_3)_3SiCl$	$(CH_3)_3SiOSiC_6H_5(CH_3)_2$	—	851
	$(C_6H_5)_2SiCl_2$	$(C_6H_5)_2Si[OSi(CH_3)_2C_6H_5]_2$	78	851
	$C_6H_5SiCl_3$	$C_6H_5Si[OSi(CH_3)_2C_6H_5]_3$	35	851
$CH_3(C_6H_5)_2SiONa$	$(CH_3)_3SiCl$	$(CH_3)_3SiOSi(C_6H_5)_2CH_3$	83	851, 1554
	$(C_6H_5)_3SiCl$	$(C_6H_5)_3SiOSi(C_6H_5)_2CH_3$	79	851, 1554
$(C_6H_5)_3SiONa$	$(CH_3)_3SiCl$	$(CH_3)_3SiOSi(C_6H_5)_3$	98	851, 1554
	$(CH_3)_3SiCl$	$(CH_3)_3SiOSi(C_6H_5)_3$	67	736

$(CH_3)_2C_6H_5SiCl$	$(CH_3)_2C_6H_5SiOSi(C_6H_5)_3$	70	486
$(C_2H_5)_2C_6H_5SiCl$	$(C_2H_5)_2C_6H_5SiOSi(C_6H_5)_3$	66	486, 488
$CH_3(C_6H_5)_2SiCl$	$CH_3(C_6H_5)_2SiOSi(C_6H_5)_3$	74	488
$(C_6H_5)_3SiCl$	$(C_6H_5)_3SiOSi(C_6H_5)_3$	74.5	736
$(p\text{-}CH_3C_6H_4)_3SiCl$	$(p\text{-}CH_3C_6H_4)_3SiOSi(C_6H_5)_3$	83	736
$p\text{-}CH_3OC_6H_4(CH_3)_2SiCl$	$p\text{-}CH_3OC_6H_4(CH_3)_2SiOSi(C_6H_5)_3$	68	486
$p\text{-}C_6H_5OC_6H_4(CH_3)_2SiCl$	$p\text{-}C_6H_5OC_6H_4(CH_3)_2SiOSi(C_6H_5)_3$	59	486
$[p\text{-}(CH_3)_2NC_6H_4]_3SiCl$	No reaction occurs	—	736
$[(C_6H_5)_3SiO]_2SiHCl$	$[(C_6H_4)_3SiO]_3SiH$	—	482
CH_3SiHCl_2	$CH_3SiHClOSi(C_6H_5)_3$	64.6	485
CH_3SiHCl_2	$CH_3SiH[OSi(C_6H_5)_3]_2$	42.5	485
$(CH_3)_2SiCl_2$	$(CH_3)_2SiClOSi(C_6H_5)_3$	68	486, 488
$(CH_3)_2SiCl_2$	$(CH_3)_2Si[OSi(C_6H_5)_3]_2$	49.8	485, 488
$C_2H_5SiHCl_2$	$C_2H_5SiHClOSi(C_6H_5)_3$	58.4	485
$CH_3(C_6H_5)SiCl_2$	$CH_3(C_6H_5)SiClOSi(C_6H_5)_3$	65—70	486, 488
$(C_6H_5)_3SiOSiHCl_2$	$[(C_6H_5)_3SiO]_2SiHCl$	66	482
$HSiCl_3$	$(C_6H_5)_3SiOSiHCl_2$	87.5	482, 488
$HSiCl_3$	$[(C_6H_5)_3SiO]_3SiH$	54	482
CH_3SiCl_3	$(C_6H_5)_3SiOSiCl_2CH_3$	71	485, 486, 488
CH_3SiCl_3	$[(C_6H_5)_3SiO]_3SiCH_3$	34.5	485, 488
$C_2H_5SiCl_3$	$(C_6H_5)_3SiOSiCl_2C_2H_5$	69.5	485, 488
$C_2H_5SiCl_3$	$[(C_6H_5)_3SiO]_2SiClC_2H_5$	60	488
$C_2H_5SiCl_3$	$[C_6H_5)_3SiO]_3SiC_2H_5$	26	485, 488
$C_6H_5SiCl_3$	$(C_6H_5)_3SiOSiCl_2C_6H_5$	57	485, 488
$(C_6H_5)_3SiOSiCl_3$	$[(C_6H_5)_3SiO]_2SiCl_2$	53	484
$SiCl_4$	$(C_6H_5)_3SiOSiCl_3$	72	484, 488
$SiCl_4$	$[(C_6H_5)_3SiO]_3SiCl$	29.5	484
$SiCl_4$	$[(C_6H_5)_3SiO]_4Si$	—	1289
SiF_4	$[(C_6H_5)_3SiO]_3SiF$	77.4	484

TABLE 17. (Cont'd)

Organosilanolate	Halosilane	Reaction products	Yield, %	Literature
$(o\text{-}CH_3C_6H_4)_3SiOLi$	$(CH_3)_3SiCl$	$(o\text{-}CH_3C_6H_4)_3SiOSi(CH_3)_3$	74.3	736
$(o\text{-}CH_3C_6H_4)_3SiONa$	$(C_6H_5)_3SiCl$	$(o\text{-}CH_3C_6H_4)_3SiOSi(C_6H_5)_3$	76.7	736
	$(p\text{-}CH_3C_6H_4)_3SiCl$	$(o\text{-}CH_3C_6H_4)_3SiOSi(C_6H_4CH_3\text{-}n)_3$	70	736
$(m\text{-}CH_3C_6H_4)_3SiONa$	$(CH_3)_3SiCl$	$(m\text{-}CH_3C_6H_4)_3SiOSi(CH_3)_3$	56.5	736
	$(C_6H_5)_3SiCl$	$(m\text{-}CH_3C_6H_4)_3SiOSi(C_6H_5)_3$	77	736
	$(n\text{-}CH_3C_6H_4)_3SiCl$	$(m\text{-}CH_3C_6H_4)_3SiOSi(C_6H_4CH_3\text{-}n)_3$	62	736
$(p\text{-}CH_3C_6H_4)_3SiONa$	$(CH_3)_3SiCl$	$(p\text{-}CH_3C_6H_4)_3SiOSi(CH_3)_3$	73	736
	$(C_6H_5)_3SiCl$	$(p\text{-}CH_3C_6H_4)_3SiOSi(C_6H_5)_3$	88	736
$[p\text{-}(CH_3)_2NC_6H_4]_3SiONa$	$(C_6H_5)_3SiCl$	No reaction occurs	—	736
$CH_3(C_6H_5)(\alpha\text{-}C_{10}H_7)SiOK$	$CH_3(C_6H_5)(\alpha\text{-}C_{10}H_7)SiCl$	$[CH_3(C_6H_5)(\alpha\text{-}C_{10}H_7)Si]_2O$	—	1171, 1663
$(\alpha\text{-}C_{10}H_7)_3SiONa$	$(CH_3)_3SiCl$	No reaction occurs	—	736
	$(C_6H_5)_3SiCl$	The same	—	736
	$(p\text{-}CH_3C_6H_4)_3SiCl$		—	736
$(C_6H_5CH_2)_3SiONa$	$SiCl_4$	$[(C_6H_5CH_2)_3SiO]_4Si$	75.4	485, 488
$C_6H_5Si(OH)_2ONa$	$(CH_3)_2SiCl_2$	Polymer	98.9	417
	$CH_3(CH_2{=}CH)SiCl_2$	The same	78.3	417
	$CH_3C_6H_5SiCl_2$	" "	96.4	417
	$(C_2H_5)_2SiCl_2$	" "	89.3	417

	$C_2H_5SiCl_3$ $C_6H_5SiCl_3$ $CH_2{=}CHSiCl_3$	" " " " " "	55,6 73,4 58,6	417 417 417
$C_2H_5Si(OH)_2ONa$	$CH_3C_6H_5SiCl_2$	Polymer	69,8	417
$[RSi(OM)O]_n$	R_3SiCl	$[RSi(OSiR_3)O]_n$	—	1527
$(CH_3)_2C_6H_5SiNHSi(OLi)(CH_3)_2$	$(CH_3)_3SiCl$	$(CH_3)_2C_6H_5SiNHSi(CH_3)_2OSi(CH_3)_3$	—	904, 905
$> Si(ONa)_2$	$> SiCl_2$	Polymer	—	335, 1496
$(CH_3)_2Si(ONa)_2$	$(CH_3)_3SiCl + CH_3(C_6H_5)SiCl_2$	Polymer	—	454
$CH_3(C_6H_5)Si(ONa)_2$	$(CH_3)_3SiCl + (CH_3)_2SiCl_2$	Polymer	—	454
$(C_6H_5)_2Si(OK)_2$	$(C_6H_5)_2SiCl_2$ $(C_6H_5)_2Si(OOCCH_3)_2$	Polymer The same	— —	1200 974
$NaO[(CH_3)_2SiO]_3Na$	$(CH_3)_3SiCl$	$(CH_3)_3SiO[(CH_3)_2SiO]_3Si(CH_3)_3$	—	1308, 1554
$NaO[C_2H_5(C_6H_5)SiO]_n Na$	$SiCl_4$	Polymer	—	1308
$CH_3Si(ONa)_3 \cdot nH_2O$	$CH_3(C_6H_5)SiCl_2$	$C_{44}H_{54}O_9Si_8$	—	470
$C_2H_5Si(ONa)_3 \cdot nH_2O$	$CH_3(C_6H_5)SiCl_2$	$C_{46}H_{58}O_9Si_8$	—	470
$C_6H_5Si(ONa)_3 \cdot nH_2O$	$CH_3(C_6H_5)SiCl_2$	$C_{27}H_{29}Cl_3O_3Si_4$	—	470

1554], and pentasiloxanes [483, 488, 851, 1174, 1289, 1535]. The reaction is carried out in ethyl and petroleum ethers, benzene, toluene, and also in mixed solvents (ethyl ether–benzene or toluene). The reactions of sodium triphenylsilanolate with dioganodichlorosilanes, organotrichlorosilanes, and silicon tetrachloride may be carried out in such a way that from one to four chlorine atoms are replaced by triphenylsiloxy groups. The reaction of sodium tripphenylsilanolate with silicon tetrafluoride yields tris- (triphenylsiloxy) fluorosilane and not the tetrasubstituted derivative [484]. With trimethylchlorosilane and triphenylchlorosilane, sodium triphenylsilane forms 1, 1, 1-trimethyl-3, 3, 3-triphenyldisiloxane (67%) and hexaphenyldisiloxane (74.5%), respectively. Under the same conditions, sodium tri (α-naphthyl) silanolate did not react with either trimethylchlorosilane or triphenylchlorosilane [736].

The reaction of alkali metal derivatives of diorganosilanediols and organosilanetriols with diorganodichlorosilanes, organotrichlorosilanes, and silicon tetrachloride leads to the formation of polyorganosiloxanes [335, 417, 974, 1200, 1308, 1496, 1527].

The products of the reaction of alkali metal organosilanolates with halosilanes and their yields are given in Table 17.

4.3.3. Reaction with Organic Compounds

Sodium triphenylsilanolate and tribenzylsilanolate react with chloroform, bromoform [482, 488], carbon tetrachloride, and carbon tris- and tetrakis(trioganosiloxy)methanes:

$$3R_3SiONa + HCX_3 \xrightarrow[-3NaX]{} (R_3SiO)_3CH \qquad (1\text{-}204)$$

$$4R_3SiONa + CX_4 \xrightarrow[-4NaX]{} (R_3SiO)_4C \qquad (1\text{-}205)$$

Reactions with halohydrocarbons proceed with more difficulty than with the analogous silicon halides. Thus, in the reaction of carbon tetrachloride with sodium tribenzylsilanolate in boiling benzene for 8 hr the yield of tetrakis(tribenzylsiloxy)methane is only 42%, while in the analogous reaction with silicon tetrachloride the yield of tetrakis(tribenzylsiloxy)silane reaches 75% after heating for only 4 hr [483].

By treatment with acetyl chloride, sodium and potassium triorganosilanolates are converted into triorganoacetoxysilanolates [1171]:

$$R_3SiONa + ClCOCH_3 \xrightarrow[-NaCl]{} R_3SiOCOCH_3 \quad (1\text{-}206)$$

In the case of optically active potassium methylphenyl(α-naphthyl)silanolate, the reaction (1-206) proceeds with inversion of the configuration [1171].

The reaction of potassium triorganosilanolates with dimethyl sulfate leads to the formation of triorganomethoxysilanes [1171]. With carbon disulfide there is formed a yellow solid with an odor reminiscent of the odor of mercaptans (possibly $R_3SiOCSSK$) [1206].

4.4. Application

Alkali metal alkylsiliconates give water-repellent properties to various materials and this effect is used widely in practice.

Soviet industry produces sodium alkylsiliconates of three grades: GKZh-10, GKZh-11, and GKZh-12. The liquid GKZh-10 consists of a 30% aqueous alcohol solution of sodium ethylsiliconate and GKZh-11 is a 30% aqueous alcohol solution of sodium methylsiliconate. Sodium methylsiliconate is produced abroad under the names of Contraquin I and III (East Germany), SN-20 (Czechoslovakia), Silikonat N/M, Bayer S, and BS-10 (West Germany), SC-50, Silirain (USA), Drisil-29 (England), Rhodorsil, Siliconate 50 K (France), and potassium methylsiliconate is produced under the name of Contraquin IV (East Germany).

Atmospheric carbon dioxide decomposes solutions of sodium alkylsiliconates according to the scheme

$$2HO[RSi(O)ONa]_n H + H_2CO_3 \longrightarrow 2HO[RSi(O)OH]_n H + Na_2CO_3 \quad (1\text{-}207)$$

In the water-proofing process, the alkylsiloxanols formed in this way react with hydroxyl groups of the surface of the material and also participate in polycondensation. As a result of these conversions, on the surface of a material treated with sodium alkylsiliconates there is formed an insoluble water-repellent film of $RSiO_{1.5}$, in which the silicon atoms are attached through oxygen atoms to the surface as well as to each other. In the water-proofing of structural materials, sodium alkylsiliconates are also attached to the surface as a result of exchange reactions with salts present in the material, for example:

$$HO[RSi(O)ONa]_nH + 0.5CaCO_3 \longrightarrow HO[RSi(O)OCa_{0.5}]_n + 0.5Na_2CO_3 \quad (1\text{-}208)$$

5. COMPOUNDS CONTAINING THE Si – S – M GROUP

5.1. Preparation Methods

The simplest method of preparing triorganosilanethiolates of alkali metals is based on the reaction of triorganosilanethiols with alkali metals [561]:

$$2R_3SiSH + 2M \longrightarrow 2R_3SiSM + H_2 \quad (1\text{-}209)$$

Alkylsiliconates of alkali metals are used for water-proofing wool and cotton fabrics [267, 304, 397, 868, 1344, 1393, 1545], fabrics made from glass fiber [30], leather materials for shoes [365, 366], constructional materials [26, 128, 259, 267, 352, 495, 868, 885, 923, 1023, 1024, 1585, 1606, 1634], and a number of other materials and articles [268, 1270, 1392, 1412, 1609, 1629] and also as additives in pigments [1455], aminoplastics [1388], antistick agents [1602], and preserving agents [1605]. An exhaustive bibliography of the use of alkylsiliconates of alkali metals in construction is given in the book of Voronkov and Shorokhov [26] and of their use in the textile and light industries in the book of Orlov et al. [51a].

Alkylsiliconates of alkali metals catalyze the rearrangement and polymerization of siloxanes [110, 237, 238, 305, 340, 452, 633, 849, 866, 959, 960, 1337, 1340, 1343, 1346, 1353, 1362, 1363, 1375, 1426, 1435, 1539, 1648, 1654, 1670]. It is believed that the true catalysts in the alkali polymerization of organosiloxanes are also organosilanolates of alkali metals, formed by cleavage of the siloxane bond by anhydrous alkalis. Sodium trialkylsilanolates catalyze the formation of tetrakis(trialkylsiloxy)titanes from trialkylsilanols and tetraalkoxytitanes [141, 277]. Potassium alkylsiliconates form water-soluble complexes with ascorbic, penicillic, and desoxyribonucleic acids [1719] and this may be of definite practical interest.

When triphenylsiloxanethiol is heated with sodium or potassium in benzene for 4 hr, the corresponding triphenylsilanethiolate is formed in 87-88% yield. Lithium does not react with triphenylsilanethiol under analogous conditions. Therefore lithium triphenylsilanethiolate was obtained by the reaction of triphenylsilanethiol with phenyllithium at room temperature for 1 hr (81% yield) [561]:

$$(C_6H_5)_3SiSH + C_6H_5Li \longrightarrow (C_6H_5)_3SiSLi + C_6H_6 \quad (1\text{-}210)$$

Lithium triorganosilanethiolates may also be obtained by the reaction of a triarylsilyllithium with elementary sulfur [106, 763]:

$$R_3SiLi + S \longrightarrow R_3SiSLi \quad (1\text{-}211)$$

5.2. Physical and Chemical Properties

Triphenylsilanethiolates of alkali metals are soluble in benzene and toluene. Sodium triphenylsilanethiolate melts with decomposition when heated to 200-230°C.

Sodium and lithium triphenylsilanethiolates react with alkyl halides in tetrahydrofuran according to the scheme [561, 763]

$$(C_6H_5)_3SiSM + XR \xrightarrow[-MX]{} (C_6H_5)_3SiSR \quad (1\text{-}212)$$

Their reactions with benzoyl chloride [763] and triphenylchlorosilane [561] proceed analogously. The products of these reactions and their yields are given in Table 18.

The reaction of sodium triphenylsilanethiolate with iodine in benzene gives a quantitative yield of bis(triphenylsilyl)dithiane, which may be reconverted into sodium triphenylsilanethiolate by

TABLE 18. Products of the Reaction of $(C_6H_5)_3SiSM$ with Organic and Organosilicon Halogen Derivatives

M	Halogen derivative	Reaction product	Yield, %	Literature
Li	CH_3I	$(C_6H_5)_3SiSCH_3$	76	763
Na	CH_3I	$(C_6H_5)_3SiSCH_3$	82	561
Na	C_2H_5Br	$(C_6H_5)_3SiSC_2H_5$	81	561
Na	n-C_3H_7Br	$(C_6H_5)_3SiSC_3H_7$-n	77	561
Na	iso-C_3H_7Br	$(C_6H_5)_3SiSC_3H_7$-iso	52	561
Na	$CH_2{=}CHCH_2Br$	$(C_6H_5)_3SiSCH_2CH{=}CH_2$	79	561
Li	$C_6H_5CH_2Cl$	$(C_6H_5)_3SiSCH_2C_6H_5$	45	763
Na	$(C_6H_5)_3CCl$	$(C_6H_5)_3SiSC(C_6H_5)_3$	70	561
Li	C_6H_5COCl	$(C_6H_5)_3SiSCOC_6H_5$	36	763
Na	$(C_6H_5)_3SiCl$	$(C_6H_5)_3SiSSi(C_6H_5)_3$	74	561

boiling with sodium in toluene [561]:

$$2\,(C_6H_5)_3\,SiSNa \underset{2Na}{\overset{I_2}{\rightleftarrows}} (C_6H_5)_3\,SiSSSi\,(C_6H_5)_3 \qquad (1\text{-}213)$$

6. ORGANOSILICON DERIVATIVES OF COPPER AND SILVER

The possibility of preparing organosilicon derivatives of copper and silver has been studied quite inadequately as yet.

Up to now it has not been possible to prepare individual compounds containing the Si–O–Cu group from organosilanolates of alkali metals and copper halides. The only organosilicon compound formed by heating cuprous chloride with sodium diphenylsilanediolate is octaphenylcyclotetrasiloxane [847].

There is a report, however, of the possibility of preparing mixed siliconates of cupric copper by the reaction of $[RSi(OH)_2ONa]_{1.5}$ with copper acetate in aqueous alcohol [397, 1497]. Hydrophobic films of these siliconates on fabrics have a high stability (see Table 97, p. 494).

According to the data of Zhdanov et al. [289], heating equivalent amounts of $CuCl_2$ and the dipotassium salt of α,ω-bis(β-carboxyethyl)polydimethylsiloxane in absolute ethanol gives a 71% yield of a viscous dark green polymer containing the grouping $\{-CuOCoCH_2CH_2[-Si(CH_3)_2O]_mSi(CH_3)_2CH_2CH_2COO\}_n$ where n = 74. We give more details on the synthesis of silicon hetero-organic polymers containing the grouping $-\underset{\substack{\| \\ O}}{C}-O-M-O-\underset{\substack{\| \\ O}}{C}-$ below.

The copper salt of an (organosilylbutyl)dithiocarbamic acid has been described and is proposed as an insectofungicide [1639].

The intermediate formation of an organosilicon derivative of silver is assumed in the postulated mechanism of the reaction of 1,1,3,3-tetramethyl-1,3-disilacyclobutane with silver nitrate [968]:

$$(CH_3)_2Si\begin{matrix} \diagup CH_2 \diagdown \\ \diagdown CH_2 \diagup \end{matrix}Si(CH_3)_2 \xrightarrow{AgNO_3} CH_3-\underset{\substack{| \\ NO_3}}{\overset{\substack{CH_3 \\ |}}{Si}}-CH_2-\underset{\substack{| \\ CH_3}}{\overset{\substack{CH_3 \\ |}}{Si}}-CH_2-Ag \xrightarrow[-Ag]{}$$

$$\longrightarrow CH_3-\underset{NO_3}{\overset{CH_3}{\underset{|}{\overset{|}{Si}}}}-CH_2-\underset{CH_3}{\overset{CH_3}{\underset{|}{\overset{|}{Si}}}}-CH_2CH_2-\underset{CH_3}{\overset{CH_3}{\underset{|}{\overset{|}{Si}}}}-CH_2-\underset{NO_3}{\overset{CH_3}{\underset{|}{\overset{|}{Si}}}}-CH_3 \qquad (1\text{-}214)$$

The reaction of sodium bis(trimethylsilyl)amide with cuprous chloride in pyridine or copper polyiodide in tetrahydrofuran proceeds according to the scheme (1-171) and gives a colorless solid, which sublimes at 180°C (0.2 mm). This is as yet the only individual organosilicon derivative of copper and corresponds in composition to $CuN[Si(CH_3)_3]_2$ and is apparently dimeric [604, 610]; in contact with atmospheric moisture it acquires a green-blue color; hydrolysis is accompanied by disproportionation, metallic copper is liberated, and hydrated copper oxide is formed. The compound binds large amounts of tetrahydrofuran, which separates when the complex is heated to 100°C (1 mm).

The reaction between $[-Si(CH_3)_2NHCH_2CH_2NH-]_n$ and cuprous chloride in xylene involves rearrangement of the polymer with the liberation of ethylenediamine and the formation of small amounts of a polymer containing 1.9% copper. It is assumed that a coordination compound is formed first, in which each copper atom forms a bridge between two polymer chains and then ethylenediamine, which binds a $CuCl_2$ molecule in a complex, is split out [1934]. The polysilazan obtained may coordinate copper with the formation of a copper-containing organosilicon polymer [962]. The copper may be removed from this polymer with ethylenediamine without disrupting the silazan structure.

Triphenylsilyllithium is known to react vigorously with halides of many elements in tetrahydrofuran at room temperature. However, in most cases the main reaction product is hexaphenyldisilane together with other compounds containing a triphenylsilyl group, while the halide of the element used is reduced to the free element. In particular, the reaction of cuprous chloride yields, after hydrolysis of the reaction mixture, 26% of hexaphenyldisilane and 29% of triphenylsilane [715]. In the case of silver chloride, the yield of hexaphenyldisilane reaches 43%.

There are data on use of copper salts of carboxylic acids and other copper compounds as additives for increasing the thermal stability of vulcanizates of siloxane elastomers [1652, 1687] and stabilizing siloxane lubricating fluids [532] and also as components in flame-proofing compositions based on siloxanes [1696] and hardeners for siloxane resins [1440a].

Chapter 2

Organosilicon Compounds of Group II Elements

1. COMPOUNDS CONTAINING THE Si – O – M GROUP

1.1. Preparation Methods

1.1.1. Reaction of Halosilanes, Alkoxysilanes, and Acyloxysilanes with Compounds of Group II Elements

Prolonged boiling of a mixture of silicon tetrachloride with magnesium, calcium, zinc, or mercury oxide in a polar solvent (an organic nitrile or nitro compound) in the presence of mercurous chloride [1555] forms oligomers with the composition $(Cl_3Si(OSiCl_2)_xCl$ (where x = 1-4), which boil over in the range of 190 to 262°C at 15 mm, and solid compounds of the type $ClMOSiCl_3$ (where M = Mg, Ca, Zn, Hg). From the formal point of view the latter would not be called organosilicon heterocompounds since they contain no carbon, but these compounds are very good models for heterosiloxanes, in which the atoms replacing silicon are Group II elements.

The combination of calcium hydroxide with silicon tetrachloride or a tetraalkoxysilane [328] produces an exothermic reaction, which is accompanied by the formation of crystalline polymeric substances that are similar to calcium hydrosilicate, okenite ($CaO \cdot 2SiO_2 \cdot 2H_2O$). The exothermal effect falls in the series SiCl > $SiOCH_3$ > $SiOC_2H_5$. Gel-like products, corresponding in composition to calcium hydrosilicates, are also obtained by alkaline cohydrolysis of tetraethoxysilane with calcium chloride [380].

When tetraethoxysilane is heated with solutions of magnesium, calcium, or zinc acetate in glycerol [296, 297, 333, 481] ethyl acetate is liberated and there are formed thick transparent masses, which are soluble in alcohol, insoluble in organic solvents, and decomposed by water. A molar ratio of tetraethoxysilane to calcium diglyceroacetate of 3:1–4:1 gives amorphous calcium glycerosilicates with a molecular weight of 960–1040. With a ratio of 2:1 the reaction product is calcium glycerodisilicate $CaSi_2(C_3H_6O_3)_6$ and with equimolar amounts it is possible to isolate [481] a crystalline glyceromonosilicate $CaSi(C_3H_6O_3)_4$. After hydrolysis and drying, these products are converted into fine light powders, which are also hydrosilicates of the type $xMO \cdot ySiO_2 \cdot zH_2O$. On the example of calcium derivatives it was established that the hydrolysis product does not contain free CaO, while the ratio of SiO_2:CaO always exceeds the ratio in the starting material.*

Starting from these data, a general method was developed for preparing hydrosilicates in nonaqueous media which was based on the combination of sodium glycerosilicate with sulfates or acetates of metals and subsequent hydrolysis with hot water. In particular, the following hydrosilicates were prepared in this way: $CaO \cdot SiO_2 \cdot 3H_2O$ (N_g 1.528; N_p 1.53), $2CaO \cdot SiO_2 \cdot 3.1\ H_2O$ (N_g 1.553; N_p 1.552), $3.2\ ZnO \cdot SiO_2 \cdot 3H_2O$ (N 1.538), and $1.3\ MgO \cdot SiO_2 \cdot H_2O$.

The reaction of tetraethoxysilane with magnesium, calcium, barium, zinc, cadmium, and mercury acetates in aqueous alcohol yields products containing the Si–O–M group and ethoxy groups at silicon atoms (up to 35 wt.%), which are not eliminated by prolonged heating of the substance at 120°C [32, 272, 322, 335, 481].

The ratio of Si to M in these "organosilicon silicates" is determined by the character of the metal, the molar ratio of the starting components, and also the temperature conditions of the reaction.

Regardless of the reagent ratio and the duration of the reaction, tetraethoxysilane and zinc acetate at room temperature yield a substance with the ratio SiO_2 : ZnO = 1 : 1, for which the following structure is believed to be possible [335]:

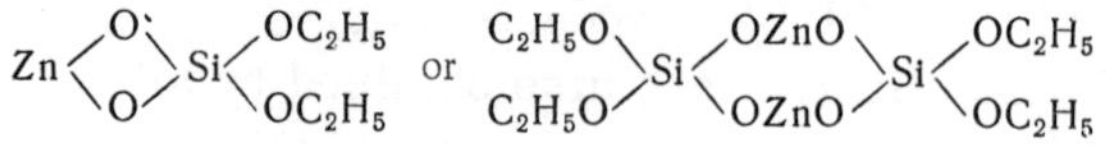

$$Zn\langle{}^{O}_{O}\rangle Si \langle{}^{OC_2H_5}_{OC_2H_5} \quad \text{or} \quad {}^{C_2H_5O}_{C_2H_5O}\rangle Si \langle{}^{OZnO}_{OZnO}\rangle Si \langle{}^{OC_2H_5}_{OC_2H_5}$$

* In the hydrolysis of a mixture of calcium glycerate with tetraethoxysilane without preliminary heating, only silicic acid precipitates [333].

It is interesting that in the reaction of ethyl silicate with zinc acetate, regardless of the conditions, the ethoxy group content of the polymer lies in the range of 25-35%, while in the case of the acetates of other Group II elements, the process conditions have a strong effect on the content of the organic part of the reaction product.

When mixtures of $(RO)_4Si$ with zinc butyrate are heated at ~160°C, the reaction products in the case of $R = CH_3$ and C_2H_5 are viscous liquids with molecular weights from ~1000 and 1800, respectively, containing 27-29% ZnO [1425]. With a change to $R = C_4H_9$, a solid substance is obtained, which has m.p. 156°C and contains ~44% ZnO. Analogous products are obtained by using for the reaction with tetraalkoxysilanes zinc crotonate and benzoate. If the oily substance with a molecular weight of 1400, which is obtained at first as a result of boiling tetramethoxysilane with zinc butyrate and contains 18.6% SiO_2 and 30% ZnO, is then heated with tetramethoxysilane at 120°C, a soluble solid polymer is formed, which has a molecular weight of ~5000 and contains 16.8% SiO_2 and 38.8% ZnO.

The reaction of equimolar amounts of ethyl- or phenyltriethoxysilane with zinc acetate at 100°C yields crystalline zinc "organoethoxysilicates" $[ZnOSiR(OC_2H_5)O]_n$, where n = 5 when $R = CH_3$ and n = 6 when $R = C_6H_5$, which are soluble in benzene and acetone [32, 272, 322, 481]. As a rule, dialkyldiethoxy- and trialkylethoxysilanes do not react with metal acetates under the given conditions. The exception is the reaction of dimethyldiethoxysilane with cadmium acetate, which yields a doughy resinous mass, containing methyl groups, silicon, and cadmium in a ratio of 2 : 1 : 1, corresponding to the linear polymer $[-(CH_3)_2SiO\,CdO-]_n$.

Hydrolysis of dimethyldiethoxysilane with aqueous solutions of sodium beryllate or zincate [331] only with a starting ratio of Si : M = 4 leads to the formation of crystalline products of definite composition $3(CH_3)_2Si(OH)ONa \cdot Na_2BeO_2$ and $3(CH_3)_2Si(OH)ONa \cdot Na_2ZnO_2 \cdot 10H_2O$. Regardless of the starting ratio of Si : M (from 2 : 1 to 1 : 3), the hydrolysis of trimethylethoxysilane also yields crystalline compounds with the composition $3(CH_3)_3SiONa \cdot Na_2BeO_2 \cdot 22H_2O$ and $6(CH_3)_3SiONa \cdot Na_2ZnO_2 \cdot 30H_2O$. These compounds form hexagonal tablets, which have a sharp melting point and true water of crystallization. The structure of these compounds has not been established as yet.

A study of the reaction of tetraacetoxysilane with organomagnesium and cadmium compounds showed that Si–O bonds are cleaved by the action of RMgX. Subsequent hydrolysis leads mainly to the formation of tetraalkylsilanes and tertiary alcohols [507]. In the case of organocadium compounds, hydrolysis yields tertiary alcohols and silicic acid, indicating a reaction involving rupture of the C–O bonds:

$$\geq SiOCOR \xrightarrow{R'CdX} \geq Si{-}O{-}CdX + R'COR \xrightarrow[H_2O]{R'CdX} \geq Si{-}OH + RR'_2COH \qquad (2\text{-}1)$$

Patents [1259, 1390, 1699] contain reports of the formation of organosilicon compounds containing mercury by the reaction of compounds of the type RHgX (where R is an organic radical such as methoxyethyl and X is hydroxyl or a radical from a mineral or acetic acid) with silicon chloride and tetraethoxysilane and also with silicates and silicic acid.

1.1.2. Reactions of Silanols with Group II Elements and Their Compounds

A method of preparing metallosiloxanes with high thermal stability was published in 1948. It is based on the reaction of organosilicon compounds containing silanol groups with finely dispersed metals [1461]. The reaction proceeds at 150-250°C with the liberation of hydrogen and normally does not require the use of any solvents. The author reported the possibility of preparing beryllium, magnesium, calcium, strontium, and zinc derivatives in this way. Nonetheless, without any particular details, only the reaction of metallic magnesium with sym-tetraethyldisiloxanediol was described [5, 169]. As a rule, 63-69% of the silanol groups react in 1.5 hr at 250°C:

$$3HOSiR_2OSiR_2OH + 2Mg \longrightarrow HO\,(SiR_2OSiR_2OMgO)_2\,SiR_2OSiR_2OH + 2H_2 \qquad (2\text{-}2)$$

No reports have appeared up to now on the properties of the magnesiosiloxane obtained in this way. Similarly, there has been no detailed description of the processes mentioned in [1461] for the preparation of heterosiloxanes containing atoms of elements of Group II and other groups by the reaction of silanols and silane(siloxane)diols on metal oxides, hydroxides, and salts. However, the possibility of reactions of type (2-2) should not be forgotten if we take into account the fact that the thermal stability of coatings based on siloxane resins is improved when they are pigmented with powdered zinc or cadmium [800].

According to the data of Hornbaker and Conrad [847], diphenylsilanediol reacts with diethylmagnesium or diethylzinc to form soluble polymers with a metallosiloxane character. However, the white powdery products obtained did not have sufficient stability. A polymer which contained magnesium and corresponded in composition to the formula $(C_6H_5)_2SiO \cdot (MgO)_{1.8}$, decomposed on heating with the formation of hexaphenylcyclotrisiloxane and magnesium oxide. The decomposition of a zincasiloxane, which was close in composition to $(C_6H_5)_2SiO \cdot (ZnO)_{1.35}$, proceeded slowly and was accompanied by the formation of ZnO and linear hexaphenyltrisiloxanediol-1,5.

Triphenylsilanol does not react with $HgCl_2$ or with C_6H_5HgCl in the presence of triethylamine [721].

1.1.3. Reactions of Alkali Metal Silanolates with Metal Halides

It was not possible to isolate any definite compounds from the products of the reaction of alkali metal trimethylsilanolates with magnesium, calcium, strontium, and barium halides [1096].

Mercuric chloride ($HgCl_2$) reacts with sodium trimethylsilanolate in dioxane with the formation of a salt-like product, which is soluble in ether, but extremely sensitive to moisture [1206]:

$$2\,(CH_3)_3SiONa + HgCl_2 \longrightarrow 2NaCl + [(CH_3)_3SiO]_2Hg \xrightarrow{+H_2O}$$

$$\longrightarrow 2\,(CH_3)_3SiOH + HgO \qquad (2\text{-}3)$$

It was not possible to characterize the bis(trimethylsiloxy)-mercurane formed since it was not sufficiently stable. In the reaction of $HgCl_2$ with sodium triphenylsilanolate in a ratio of 1 : 2 in ether there is instantaneous precipitation of sodium chloride and apparently the formation of bis(triphenylsiloxy)mercurane [721]. However, even when the solvent is removed from the filtrate in vacuum below 20°C secondary reactions begin. As a result of this, a 34% yield (on $HgCl_2$) is obtained of (triphenylsiloxy)phenylmercurane (m.p. 140-141°C) and also a sticky solid, which melts in the range of 135-250°C. With a ratio of 1 : 1, a product of similar character is formed together with C_6H_5HgCl, which indicates an analogous process:

$$
nR_3SiONa \xrightarrow{HgCl_2} \begin{cases} \xrightarrow{n=2} [(R_3SiO)_2Hg] \longrightarrow R_3SiOHgR + R_3SiO(R_2SiO)_x HgR \\ \xrightarrow{n=1} [R_3SiOHgCl] \longrightarrow R_3SiO(R_2SiO)_x HgCl + RHgCl \end{cases} \quad (R = C_6H_5) \tag{2-3a}
$$

(Triphenylsiloxy)phenylmercurane is as yet the only individual compound containing the Si–O–Hg group (the band at 885 cm^{-1} in the IR spectrum belongs to its vibrations), which is not hydrolyzed by water. It is cleaved by the action of HCl with the formation of triphenylsilanol and C_6H_5HgCl. This compound is stable at room temperature, but above the melting point it decomposes with the formation of $(C_6H_5)_2Hg$ and $(C_6H_5)_3SiO[(C_6H_5)_2SiO]_xHgC_6H_5$. (Triphenylsiloxy)phenylmercurane may also be obtained in accordance with the following scheme:

$$
(C_6H_5)_3SiOK + C_6H_5HgCl \longrightarrow (C_6H_5)_3SiOHgC_6H_5 + KCl \tag{2-3b}
$$

From the hydrolysis products of the reaction mixture it was also possible to isolate triphenylsilanol and its 1 : 1 adduct with (triphenylsiloxy)phenylmercurane (m.p. 114-117°C). In an examination of the conversions of products from the reaction of sodium triphenylsilanolates with $HgCl_2$, the authors of [721] adhered to a migration mechanism, believing that there was an analogy between this process and "mercuridesilation" (cleavage of Si–Ar bonds in trimethylarylsilanes by mercury acetate in the presence of glacial acetic acid) [553]. For example, this may be represented as follows:

$$
\begin{array}{rl} (C_6H_5)_3SiOHg \,\vdots\, & OSi(C_6H_5)_3 \\ C_6H_5 \,\vdots\, & Si(C_6H_5)_2OHgOSi(C_6H_5)_3 \end{array}
$$

One would expect the formation of polymeric metallosiloxanes by the reaction of sodium diphenylsilanediolate in accordance with scheme (2-3) [847, 1200]. However, available experimental data indicate only the intermediate formation of oligomers with Si–O–M bonds. The character of the products from their decomposition, which contain organosiloxane and inorganic components, depend on the metal chloride which reacted with sodium diphenylsilanediolate.

In the case of $MgCl_2$ and $HgCl_2$, octaphenylcyclotetrasiloxane is obtained; with $CaCl_2$, hexaphenylcyclotrisiloxane is formed quantitatively; in the reaction with $ZnCl_2$, hexaphenyltrisiloxanediol-1,5 is formed. The second product from the decomposition of magnesium- and zinc-containing polymers is the corresponding oxide (as in the reaction of diphenylsilanediol with diethylmetals). Nonetheless, there were no signs of the presence of mercuric oxide in the products from the reaction with $HgCl_2$; a white substance was obtained, which did not contain chlorine, was insoluble in hydrochloric acid, and contained 93.2% of mercury.

Aqueous alcohol solutions of the monosodium salts of organosilanetriols form precipitates with almost all the water-soluble salts of Ca, Sr, Ba, Zn, Cd, and Hg [347]. However, it was possible to isolate only $C_2H_5Si(OH)_2OSi(OH)_2OBaOSi(OH)_2C_2H_5$.

1.1.4. Cleavage of Siloxanes by Organomagnesium Compounds*

In 1911 Kipping and Hackford [881] proposed the use of the reaction of siloxanes with excess of an organomagnesium compound as a method of synthesizing silanols. A classical example is provided by the preparation of triphenylsilanol:

$$(C_6H_5SiO_{1,5})_n + C_6H_5MgX \xrightarrow{\sim 200^\circ C} (C_6H_5)_3SiOMgX \xrightarrow{(H_2O)} \longrightarrow (C_6H_5)_3SiOH \quad (2\text{-}4)$$

That triorganosiloxy derivatives of magnesium actually exist is well illustrated by the work of Sauer [1053], who was able to isolate solid white trimethylsiloxymagnesium iodide from the products of cleaving dimethylpolysiloxanes with methylmagnesium iodide. The same compound was obtained in confirmatory synthesis by the action of methylmagnesium iodide on trimethylsilanol. However, trimethylsiloxymagnesium iodide is practically the only example of a triorganosiloxy derivative of magnesium which has been characterized (in this case by analysis for iodine). Up to now it has generally been the case of the intermediate formation of these compounds.

Organomagnesium compounds do not react with hexaorganodisiloxanes [881] or hexachlorodisiloxane [1127] with cleavage of the Si–O bond. However, in the reaction of phenylmagnesium

* See also the monograph [83] and the review [101].

bromide with hexabromodisiloxane there is cleavage:

$$(Br_3Si)_2O + 7C_6H_5MgBr \longrightarrow (C_6H_5)_4Si + (C_6H_5)_3SiOMgBr + 6MgBr_2 \quad (2-5)$$

In a study of the reaction of organomagnesium compounds with sym-diphenyldisiloxane, data were obtained which indicated two types of interaction [830]. In the first case there is "normal" cleavage:

$$(C_6H_5SiH_2)_2O + RMgX \longrightarrow C_6H_5SiH_2R + C_6H_5SiH_2OMgX \quad (2-6)$$

The formation of phenylsilane and a sym-diphenyldiorganodisiloxane is explained by the following scheme:

$$(C_6H_5SiH_2)_2O + RMgX \longrightarrow C_6H_5SiH_2OSiH(C_6H_5)R + HMgX \longrightarrow$$
$$\longrightarrow C_6H_5SiH_3 + C_6H_5SiHROMgX \longrightarrow (C_6H_5SiHR)_2O \quad (2-7)$$

Eaborn [83] believes that in the latter case cleavage may proceed through a cyclic complex:

$$\begin{array}{c} R \searrow \\ H \longrightarrow SiO \ldots . SiH_2R \\ R' \nearrow \quad | \qquad | \\ XMg \ldots H \end{array}$$

However, the four-center scheme for the cleavage of siloxanes encounters serious objections (see below).

Shostakovskii and his co-workers [310, 498] report that Iotsich's reagent reacts only with linear dimethylsiloxanes with terminal OH groups, while cyclic dimethylsiloxanes are inert. Thus, after treatment with $CH_2=CHC\equiv CHgBr$, more than 90% of octamethylcyclotetrasiloxane is recovered. At the same time, the siloxanediol $HO[(CH_3)_2SiO]_4H$, which has the same number of silicon atoms in the chain, finally is almost half converted into sym-di(vinylacetylenyl)tetramethyldisiloxane; this indicates the intermediate formation of compounds of the type $R'(CH_3)_2SiOMgBr$. To explain their formation there has been put forward the fantastic hypothesis that the siloxanediol decomposes with the formation of the highly reactive species $R_2Si=O$ with a double bond between the silicon and oxygen atoms, to which is added the organomagnesium compound, in analogy with ketones.

There are data on the reaction of organomagnesium compounds with siloxane, its oxidation products, and also powdered quartz and pyrogenic silicon dioxide [869].

1.2. Application

Only the method mentioned already for converting silicon dioxide into diorganosiloxanes through intermediate compounds of the type $R_2Si(OMgX)_2$ is of serious practical interest [869]. However, there is no indication as yet that this has been achieved on a large scale. Reactions of the following type may be of definite preparative value [310, 498]:

$$R'_nSiX_{4-n} + (4-n)\,R''C{\equiv}CSiR_2OMgX \longrightarrow R'_nSi\,[OSiR_2\,(C{\equiv}CR'')]_{4-n} \qquad (2\text{-}8)$$

(R = H or alkyl ; X = halogen; R′ = H, $CH_2{=}CH{-}$, $C_6H_5{-}$, $(CH_3)_3Si{-}$, etc.)

Zincosiloxanes, which are formed in the heterocondensation of alkoxysilanes with zinc acetate, have a high covering power [32]. Silicon-containing derivatives of beryllium and zinc, which are obtained by hydrolysis of alkoxysilanes with solutions of beryllates and zincates, have been proposed as impregnating agents for reducing the adhesion of polymeric materials to cotton and glass fabrics [1425]. The products from the reaction of organomercury compounds with silicon derivatives [1295, 1390, 1699] are recommended as pharmaceutical agents, disinfectants, insecticides, and dips for grain.

A considerable number of communications, largely patents, are concerned with the use of organosilicon compounds of Group II elements, predominantly zinc derivatives – salts of carboxylic and naphthenic acids [614, 845, 1303, 1328, 1330, 1351, 1352, 1372, 1394, 1398, 1400, 1401, 1410, 1424, 1459, 1575, 1599, 1607, 1619, 1689, 1696, 1705, 1710, 1713], diorganodithiocarbamates [1334, 1403, 1557, 1707], mercaptides [1403], complexes of salts with caprolactam [1445], triethylamine [1354], and polyethylenepolyamine [1515], alkenylsiloxanes [1443], peroxides, carbonates, and oxides [1581] as accelerators for the hardening of siloxane resins [614, 845, 1303, 1328, 1330, 1334, 1398, 1400, 1401, 1403, 1410, 1443, 1445, 1459, 1515, 1575, 1607, 1619, 1705, 1710], and impregnating coatings [1354]. They are also used as catalysts for "cold vulcanization" of polysiloxanediols [1372, 1689], stabilizers of siloxane fluids [1394] and resin solutions [1599], components of flame-proofing compositions [1696], additives which improve the properties of water-proofing compositions [397, 1351, 1352, 1424, 1497] and vulcanizates of siloxane rubbers (particularly for improving the compression set) [1581, 1587, 1707, 1713].

The nature of the action of these compounds in all these cases has not been established. However, we cannot exclude the possibility of their participating in the formation of complexes or the introduction of atoms of Group II elements into the siloxane structure.

1.3. Analysis

A gravimetric method (in the form of $Zn_2P_2O_7$) and complexometric titration with Trilon B with Chrome Dark Blue as the indicator have been proposed for the determination of zinc in "organic silicates" [322, 331]. It has been recommended that other Group II elements are determined as the oxides with the preliminary use of wet combustion (treatment of the substance with concentrated sulfuric and nitric acids) [331, 847].

2. COMPOUNDS CONTAINING THE Si – M GROUP

2.1. Derivatives of Magnesium

Numerous attempts to prepare silicon analogs of Grignard reagents have generally been unsuccessful. At the same time, many facts have been accumulated indicating the existence of unstable compounds containing an Si–Mg bond.

Even in 1913 it was shown [940, 941] that the action of methylmagnesium bromide on hexachlorodisilane forms a mixture of methylchlorosilanes. The products from the reaction of hexachlodisilane and phenylmagnesium bomide were diphenyldichlorosilane [762, 1131] and hexaphenyldisilane [1127], while in the case of octachlorotrisilane, tetraphenylsilane and the same hexaphenyldisilane were obtained [1127].

The following mechanism has been proposed [762] for the cleavage of hexachlorodisilane by phenylmagnesium bromide, which is accompanied by the formation of a resin:

$$Cl_3SiSiCl_3 \xrightarrow[-MgBrCl]{C_6H_5MgBr} Cl_3SiSiCl_2C_6H_5 \xrightarrow{C_6H_5MgBr}$$

$$\longrightarrow (C_6H_5)_2SiCl_2 + BrMgSiCl_3 \xrightarrow[-MgBrCl]{} [SiCl_2]_n \qquad (2\text{-}9)$$

In its turn, hexaphenyldisilane is completely inert toward metallic magnesium under the conditions which are usually used to prepare Grignard reagents [581].

The proposal that unstable H_3SiMgI was formed in the reaction of H_3SiI with magnesium in diisoamyl ether [107, 677, 678] was not sufficiently well-founded and was not confirmed in the case of H_3SiBr [524].

Triethyliodosilane does not undergo any changes when treated with magnesium, but catalyzes the reaction of the latter with iodine [255, 659]. Triphenylchlorosilane does not react with magnesium in ether, but in tetrahydrofuran it forms mainly hexaphenyldisilane [716, 796]. In the case of trimethylchlorosilane, the reaction with magnesium in tetrahydrofuran is accompanied by opening of the tetrahydrofuran ring, but the replacement of only the methyl radical in $(CH_3)_3SiCl$ by a phenyl radical is sufficient to direct the process toward coupling [523, 1187]. Therefore, the following scheme is regarded as quite probable:

$$R_3SiX + Mg \longrightarrow R_3SiMgX \xrightarrow{R_3SiX} R_3SiSiR_3 + MgX_2 \qquad (2\text{-}10)$$

The most effective confirmation of the existence of triphenylsilylmagnesium halides is provided by the data of Selin and West [1133, 1134].

Thus, the reaction of triphenylchlorosilane with two equivalents of phenyl-, cyclohexyl-, or 2-methylcyclohexylmagnesium bromide in tetrahydrofuran gives a 60-70% yield of hexaphenyldisillane. No hexamethyldisilane is obtained from trimethylchlorosilane under the same conditions, but a mixture of $(C_6H_5)_3SiCl$ and $(CH_3)_3SiCl$ forms triphenylsilyltrimethylsilane. In the reaction of cyclohexylmagnesium bromide with triphenylchlorosilane, there is formed for every two molecules of the latter, approximately one molecule each of cyclohexene and cyclohexane. All this is interpreted in the following way:

$$R_3SiCl + C_6H_{11}\text{—}MgBr \longrightarrow [\text{cyclic transition state: } C_6H_{10}(H \cdots Cl\text{—}SiR_3)(Mg\text{—}Br)] \longrightarrow R_3SiMgBr + C_6H_{10} + HCl$$

$$R_3SiMgBr + R_3SiCl \longrightarrow R_3SiSiR_3 + MgBrCl$$

$$HCl + C_6H_{11}\text{—}MgBr \longrightarrow C_6H_{12} + MgBrCl \qquad (2\text{-}11)$$

According to the data of Schwartz and Konrad [1130], the action of hydrochloric acid on magnesium silicide does not lead to the simple formation of silane and $MgCl_2$. The first product of the reaction contains up to 20% of magnesium, bound to silicon. It is believed that this indicates the intermediate formation of $(HOMg)_2SiH_2$, which is then hydrolyzed with the liberation of silane and magnesium hydroxide and the formation of polymeric silicon oxyhydrides.

2.2. Derivatives of Calcium, Strontium, and Barium

Suspensions of the alkaline earth metals Ca, Sr, and Ba do not react with hexaphenyldisilane in tetrahydrofuran. At the same time, in liquid ammonia there is cleavage of the Si–Si bond with the formation of triphenylsilyl derivatives of the metals [123]:

$$M + (C_6H_5)_3SiSi(C_6H_5)_3 \longrightarrow [(C_6H_5)_3Si]_2M^* \qquad (2\text{-}12)$$

The rate of cleavage and the resistance of these compounds to solvolysis increase, as in the case of alkali metals, with an increase in the atomic weight of the metal, i.e., in the series $Ca < Sr < Ba$. The rate of formation and resistance to solvolysis of silyl derivatives of alkaline earth metals are substantially lower than for the silyl compounds of alkali metals. $Ca[Si(C_6H_5)_3]_2$ and $Sr[Si(C_6H_5)_3]_2$ are formed so slowly and undergo ammonolysis so rapidly that after the reaction of the corresponding metal with hexaphenyldisilane for many hours, the bulk of the latter (80-90%) re-

---

* It would be logical to name these compounds bis(triorganosilyl)calcane, strontane, barane, etc. These terms are not yet found in the literature.

mains unchanged, while $M[Si(C_6H_5)_3]_2$ can be obtained only in the form of the solvolysis products. On the other hand, $Ba[Si(C_6H_5)_3]_2$ is resistant to the action of ammonia and if a rise in temperature is avoided, it may be isolated from liquid ammonia in high yield as an orange-yellow substance, which is insoluble in benzene but soluble in tetrahydrofuran (yellow color). The compound reduces silver nitrate to silver and is hydrolyzed by wet methanol with the formation of triphenylsilanol, hydrogen, and barium hydroxide. When heated to 160°C, bis(triphenylsilyl)barane decomposes with charring and the formation of polyphenylsilanes.

2.3. Derivatives of Zinc

The reaction of hexaiododisilane with diethylzinc gives hexaethyldisilane and tetraethylsilane simultaneously [700, 702]. The reaction of silane with compounds containing the triethylzincate anion leads to the formation of ethylsilane and diethylsilane [90]. The reaction of diethylzinc with triethylsilane [264] under sufficiently drastic conditions (heating for 8 hr at 160°C) gives a very low yield (4.5%) of compounds containing the Si–Zn–Si group, while the main products of the reaction are metallic zinc, tetraethylsilane, and hexaethyldisilane. The silyl derivative of zinc obtained (the individuality of this compound is doubted by the authors of the paper cited) is very sensitive to light and oxygen, and involatile, and cannot be isolated by distillation or sublimation; it decomposes at 80-90°C with the liberation of zinc. The reaction of triorganosilanes with halomethylzinc halides gives a 55-65% yield of triorganomethylsilanes [1146]:

$$R_3SiH + XCH_2ZnX \xrightarrow{\text{ether}} R_3SiCH_3 + ZnX_2 \qquad (2\text{-}13)$$
$$(R = C_2H_5,\ C_4H_9;\ X = Br,\ I)$$

The mechanism of the process of "methylene insertion", in which triorganosilanes show higher activity than olefins, is not yet clear (see also Eq. 2-55, p. 164).

The hypothesis of Emeleus and his co-workers [677] on the formation of H_3SiZnI in the reaction of iodosilane with metallic zinc was not confirmed by the investigations of Aylett [76].

Wiberg and his co-workers [123] showed that if anhydrous zinc chloride is added gradually in the absence of air and moisture to a solution of triphenylsilylpotassium in liquid ammonia (molar ratio of $ZnCl_2$ to triphenylsilylpotassium = 1 : 2), there is an exothermic reaction and a yellowish precipitate is formed. After removal of the ammonia, the precipitate may be separated by means of dry benzene into KCl and $Zn[Si(C_6H_5)_3]_2$ (more than 85% yield of a powder with the color of elephant hide).

Bis(triphenylsilyl)zincane* is monomeric in benzene solution, forms an ammoniate, reduces silver nitrate to silver, and is hydrolyzed readily by acids and alkalis. Hydolysis by dilute hydrochloric acid forms zinc hydroxide and triphenylsilane. Treatment of bis(triphenylsilyl)zincane with an alcoholic solution of potassium hydroxide results in the liberation of hydrogen (2 moles per mole) and the formation of triphenylsilanol or the product of its anhydrocondensation, i.e., hexaphenyldisiloxane. When heated above 105°C, bis(triphenylsilyl)zincane decomposes with the liberation of metallic zinc. Contrary to expectations, it is not hexaphenyldisilane which is formed together with the metal, but a mixture of tetraphenylsilane and higher polyphenylsilanes. During the "thermolysis," there is apparently disproportionation of the bis(triphenylsilyl)zincane into tetraphenylsilane and the hypothetical "diphenylsilene" $(C_6H_5)_2Si$, which polymerizes with "saturation" of the ends of the chain by triphenylsilyl groups and is converted into polyphenylsilanes:

$$(C_6H_5)_3Si[Si(C_6H_5)_2]_nSi(C_6H_5)_3.$$

2.4. Derivatives of Cadmium

When heated for 5 hr at 110°C, a mixture of triethylsilane and diethylcadmium in a molar ratio of 2 : 1 gives a 27% yield of bis(triethylsilyl)cadmane [261, 441].† Metallic cadium, ethane, and tetraethylsilane are formed together with this. Bis(triethylsilyl)cadmane is a lemon colored liquid, which is readily oxidized and does not distill; when heated at 140°C for 8 hr, it decomposes quantitatively into cadmium and hexaethyldisilane. It reacts exothermally with benzoyl peroxide with the formation of cadmium benzoate and triethylsilyl benzoate. It reacts with bromine in carbon tetrachloride, also with the liberation of heat, to give cadium

* See footnote on p. 138.

† See footnote on p. 138. In the papers cited the names used are based on the name of the element and not its highest hydride.

bromide and triethylbromosilane (scheme 2-14a). The direction of the reaction of bis(triethylsilyl)cadmane with ethyl bromide depends on the temperature (2-14b, 2-14c). The reaction with 1,2-dibromoethane is exothermic and is accompanied by the liberation of cadmium (2-14d):

$$(R_3Si)_2Cd \xrightarrow{+2Br_2\,(CCl_4)} CdBr_2 + 2R_3SiBr \quad (2\text{-}14a)$$

$$(R_3Si)_2Cd \xrightarrow{+2RBr} CdR_2 + 2R_3SiBr \quad (2\text{-}14b)$$

$$(R_3Si)_2Cd \xrightarrow[>100°\,C]{+2RBr} CdBr_2 + 2R_4Si \quad (2\text{-}14c)$$

$$(R_3Si)_2Cd \xrightarrow{+BrCH_2CH_2Br} Cd + C_2H_4 + 2R_3SiBr \quad (2\text{-}14d)$$

$$(R = C_2H_5)$$

2.5. Derivatives of Mercury

The reaction of hydrosilanes, particularly triethylsilane, with mercury salts is in general a reduction reaction [358]. The course of the reduction to monovalent (formally) or metallic mercury is connected with the character of both the salt and the solvent used. It is only possible to talk of the formation of organosilicon derivatives of mercury here from the point of view of the reduction mechanism, which includes complex formation:

$$R_3SiH + HgX_2 \longrightarrow R_3Si\begin{matrix} H\cdots X \\ \\ X-Hg-X \end{matrix}Hg-X \longrightarrow R_3SiX + HX + Hg_2X_2 \quad (2\text{-}15)$$

$$(R = C_2H_5;\ X = Cl)$$

In the case of organomercury halides there is also reduction [359]:

$$R_3SiH + R'HgX \longrightarrow R'H + Hg + R_3SiX \quad (2\text{-}16)$$

(R = ethyl; X = chlorine; R' = phenyl, thienyl, and furyl)

However, during the slow reaction of triethylsilane with diethylmercury (117 hr at 130-140°C), which is also accompanied by the liberation of ethane, compounds with a Si–Hg bond are formed [28, 262, 441]. At first there is apparently monosubstitution:

$$(C_2H_5)_3SiH + (C_2H_5)_2Hg \longrightarrow C_2H_6 + C_2H_5HgSi(C_2H_5)_3 \quad (2\text{-}17)$$

Together with ethyl(triethylsilyl)mercurane (yield ~15%), bis(triethylsilyl)mercurane is obtained but in very low yield (~3%).* This may be due to the reaction of ethyl(triethylsilyl)mercurane with triethylsilane (i.e., the stepwise nature of reaction 2-17) and it also may be explained by "symmetrization." The second variant is preferable since when ethyl(triethylsilyl)mercurane is heated at 170°C for 11 hr, the conversion reaches almost 70%, while a third of the reaction products consist of bis(triethylsilyl)mercurane [263]:

$$2R_3SiHgR \longrightarrow (R_3Si)_2Hg + R_2Hg \qquad (2\text{-}18)$$

The low yields of silyl derivatives of mercury in reaction (2-17) must be explained by secondary decomposition processes, which are also indicated by the quite considerable content of metallic mercury in the reaction products [262].† Decomposition occurs readily under the action of UV light even in evacuated ampoules:

$$2(R_3Si)_n HgR_{2-n} \longrightarrow 2Hg + (2-n)R_2 + n(R_3Si)_2 \qquad (2\text{-}19)$$

Both mono- and bis(triethylsilyl) derivatives of mercury are oxidized readily and in both cases the main products are mercury and hexaethyldisiloxane. However, the oxidation of ethyl(triethylsilyl)mercurane also gives diethylmercury which also indicates the initial symmetrization. The reaction of ethyl(triethylsilyl) mercurane with benzoyl peroxide also has the same character:

$$2R_3SiHgR + (C_6H_5COO)_2 \longrightarrow R_2Hg + Hg + 2R_3SiOCOC_6H_5 \qquad (2\text{-}20)$$

Nonetheless, there is the possibility that ethyl(triethylsilyl)-mercurane undergoes a secondary reaction with triethylsilane since, for example, this compound reacts smoothly [263] with triethylgermane:

$$R_3GeH + RHgR' \xrightarrow{100^\circ C} RH + R_3GeHgR' \qquad (2\text{-}21)$$
$$[R = C_2H_5;\ R' = (C_2H_5)_3Si \text{ or } (C_2H_5)_5Si_2]$$

* The term "silylmercurane" is also more rational for such compounds, but names of the type "bis(triethylsilyl)mercury" are still common.

† At the same time, mercury replaces cadmium in $[(C_2H_5)_3Si]_2Cd$ [441].

In this connection we should point out that pentaethyldisilane reacts with diethylmercury much more vigorously than triethylsilane [263, 441]. The yield of monosubstituted and disubstituted pentaethylbisilyl derivatives formed in accordance with schemes analogous to (2-17) and (2-18) is higher than the yield of triethylsilyl derivatives, while less metallic mercury is obtained; this indicates the high thermal stability of pentaethylbisilyl derivatives of mercury.

Like bis(triethylsilyl)mercurane, a mixed silylgermyl derivative of mercury decomposes in light with the quantitative liberation of metallic mercury [28]. However, symmetrization does not occur in this case and the second product is triethyl(triethylgermyl)silane. The reaction may be regarded as a successful route to the preparation of compounds of this type (see Ch. IV).

The possibility of the formation of compounds of the type H_3SiMI by the reaction of iodosilane with metallic mercury is no higher than the corresponding possibility in reactions with other Group II elements (see Sections 2.1 and 2.3) [106, 677].

No reaction is observed between triethylchlorosilane and furylmercuryl chloride in boiling o-xylene [359]. At the same time, trimethylhalosilanes react with sodium amalgam to give bis(trimethylsilyl)mercurane and not hexamethyldislane [123]:

$$2\,(CH_3)_3SiX + 2Na + Hg \longrightarrow [(CH_3)_3Si]_2Hg + 2NaX \quad (X = Cl, Br) \qquad (2\text{-}22)$$

The beautiful, highly refractive yellow crystals of bis(trimethylsilyl)mercurane sublime in high vacuum at 60°C without decomposition; they are soluble in ether, tetrahydrofuran, benzene, hexane, and carbon disulfide and the molecular weight in ether corresponds to the monomeric form. The IR spectrum contains a strong absorption band at 318 cm^{-1}, which evidently may be ascribed to vibrations of the Si–Hg bond.

Heating bis(trimethylsilyl)mercurane for 24 hr at 100-160°C produces quantitative decomposition in accordance with scheme 2-19.* The decomposition occurs in a few days in light in ether at room temperature. The half-decay period of the compound at 190°C

* Thermal decomposition in the absence of solvents in accordance with scheme (2-19) may be used to analyze compounds of type $(R_3Si)_2Hg$ since both the hexaorganodisilane and mercury are readily determined quantitatively [123].

in cyclohexane is ~ 80 hr [660]. The action of light on the solution accelerates the decomposition, but the photolysis is retarded by a change to aromatic solvents or octene-1. Thermal or photolytic decomposition of bis(trimethylsilyl)mercurane involves the formation of trimethylsilyl radicals. The action of light on solutions of the compound in methyl methacrylate or styrene produces polymerization of the latter [660]. When a solution of bis(trimethylsilyl)-mercurane in toluene is heated for two weeks at 190°C, 90% decomposition occurs, but in addition to metallic mercury and hexamethyldisilane (scheme 2-19), trimethylsilane, benzyltrimethylsilane, and bibenzyl are obtained. Under analogous conditions, benzene and chlorobenzene give phenyltrimethylsilane and trimethylsilane with bis(trimethylsilyl)mercurane, but no biphenyl is formed.

Despite its relative instability, bis(trimethylsilyl)mercurane is more stable than its carbon analog di-tert-butylmercury, which decomposes with the liberation of the metal even at 40°C [123]. Bis(trimethylsilyl)mercurane reacts with zinc at high temperature with the liberation of dimethylmercury. The replacement of mercury by zinc and the formation of bis(trimethylsilyl)zinc do not occur.

Bis(trimethylsilyl)mercurane reacts with hydrochloric acid with the liberation of the metal:

$$(CH_3)_3Si]_2Hg + HCl \longrightarrow (CH_3)_3SiH + Hg + (CH_3)_3SiCl \qquad (2\text{-}23)$$

This reaction may also be used as an analysis method.

If we add together schemes (2-19) and (2-22), as applied to trimethylhalosilanes, the final result will correspond to the synthesis of hexamethyldisilane by a Wurtz reaction, where sodium amalgam acts as the halogen-abstracting agent. On the other hand, the reaction of H_3SiCl with Na/Hg, during which it was not possible to establish the intermediate formation of $Hg(SiH_3)_2$ [1190], leads not to the disilane, but to its disproportionation products SiH_4 and $(SiH_2)_x$ and a small amount of $(SiH)_x$. Therefore Wiberg and his coworkers [123] made an attempt to prepare the partial methylation products, bis(dimethylsilyl)mercurane $Hg[Si(CH_3)_2H]_2$ and bis(methylsilyl)mercurane $Hg[Si(CH_3)H_2]_2$, which may be formed by the reaction of bis(trimethylsilyl)mercurane and disilylmercurane, and to study their decomposition.

Shaking liquid dimethylbromosilane with liquid sodium amalgam for days at room temperature evidently forms bis(dimethylsilyl)mercurane. This compound was not isolated since it is extremely unstable and decomposes even during the preparation process. It was possible to establish with great difficulty that the decomposition proceeds in two directions. The smaller part ($\sim ^1/_5$) of the $Hg[Si(CH_3)_2H]_2$ is converted into tetramethyldisilane in accordance with scheme (2-19); the bulk ($\sim ^4/_5$) loses dimethylsilane with the formation of the hypothetical dimethylsilylenemercury, which decomposes with the formation of metallic mercury and "dimethylsilene", while part is converted into a polymeric form:

$$\begin{array}{l} Hg[Si(CH_3)_2H]_2 \longrightarrow |HgSi(CH_3)_2| + (CH_3)_2SiH_2 \\ \qquad\qquad\qquad\qquad\qquad \downarrow \\ \qquad\qquad\qquad Hg + |Si(CH_3)_2| \longrightarrow [Si(CH_3)_2]_n \end{array} \tag{2-24}$$

The "dimethylsilene" abstracts a $H(CH_3)_2Si$ group from bis(dimethylsilyl)mercurane with the formation of hexamethyltrisilane, octamethyltetrasilane, and decamethylpentasilane. The polymeric silylenemercury, which could also be synthesized from $(CH_3)_2SiBr_2$ by the action of Na/Hg, could be extracted with a solvent from the involatile residue after distillation of the volatile reaction products. Removal of the solvent left a viscous, orange-yellow oil, which liberated mercury under the action of light or on heating above 60°C, and was converted into polydimethylsilylene, a waxy, colorless, involatile substance that was soluble in hexane, benzene, and ether.

In the reaction of sodium amalgam with methylbromosilane, it was also impossible to isolate the expected primary reaction product $Hg[Si(CH_3)H_2]_2$, since the compound is unstable and decomposes immediately. No dimethylsilane was detected in the volatile decomposition products; the decomposition apparently does not proceed in accordance with scheme (2-19) in this case. In accordance with a scheme of type (2-24) one would also expect the formation of hypothetical methylsilylenemercury $HgSi(CH_3)H$ and methylsilane. In actual fact, the latter constituted the bulk of the volatile products, while instead of $HgSi(CH_3)H$, in analogy with scheme (2-24) the polymeric product $[Si(CH_3)H]_n$ was formed rapidly.

The data obtained by Wiberg and his co-workers show that the methylated silyl compounds of mercury $Hg[Si(CH_3)_nH_{3-n}]_2$ (n=1,

2, and 3), which are formed initially from silyl bromides of the type (CH_3) $SiH_{3-n}Br$ and sodium amalgam, may decompose in two ways with the formation of both methylated disilanes $[Si(CH_3)_nH_{3-n}]_2$ and the disproportionation products of the latter $Si(CH_3)_nH_{4-n}$ and $Si(CH_3)_nH_{2-n}$. With an increase in the number of methyl groups the reaction proceeds predominantly by the first mode and with a decrease in the number, by the second mode. Thus, the most methylated compound $Hg[Si(CH_3)_3]_2$ gives exclusively $[(CH_3)_3Si]_2$, while the compound containing the smallest number of the methyl groups $Hg[Si(CH_3)H_2]_2$ gives exclusively a mixture of methylsilane CH_3SiH_3 and methylsilylene CH_3SiH. At the same time, if the intermediate compound is $Hg[Si(CH_3)_2H]_2$, then the decomposition proceeds simultaneously in both ways. Of the mercury derivatives which are the primary reaction products, only the fully methylated compounds $Hg[Si(CH_3)_3]_2$ in the first case and $HgSi(CH_3)_2$ in the second, are quite stable.

Bis(trimethylsilyl)mercurane was used in the synthesis of silyl derivatives of aluminum (see Ch. 3, Section 8).

The reaction of triphenylsilyllithium and mercuric chloride ($HgCl_2$) in equimolar amounts forms [715] triphenylchlorosilane, mercury, mercurous chloride (Hg_2Cl_2), and mainly hexaphenyldisilane (45%). In the reaction with mercuric bromide ($HgBr_2$) in a ratio of 2 : 1 the yield of hexaphenyldisilane is increased to 63%. The reaction of triphenylsilyllithium with Hg_2Cl_2 proceeds analogously. These results were initially explained by an exchange process:

$$\begin{aligned} R_3SiLi + HgX_2 &\longrightarrow R_3SiX + [LiHgX] \\ &\qquad\qquad\qquad \llcorner\!\!\longrightarrow LiX + Hg \\ R_3SiLi + R_3SiX &\longrightarrow R_3SiSiR_3 + LiX \end{aligned} \tag{2-25}$$

However, such an exchange cannot explain the formation of tetraphenylsilane in ~70% yield in the reaction of triphenylsilyllithium with phenylmercuric bromide or with diphenylmercury [715]. Here a different scheme is more probable:

$$\begin{aligned} R_3SiLi + RHgX &\longrightarrow R_3SiHgR + LiX \\ &\qquad\qquad \llcorner\!\!\longrightarrow R_4Si + Hg \\ &(X = Br \text{ or } C_6H_5) \end{aligned} \tag{2-26}$$

This treatment is also suitable for describing the reaction of triphenylsilyllithium with mercury halides:

$$R_3SiLi + HgX_2 \longrightarrow R_3SiHgX + LiX \longrightarrow R_3SiX + Hg \qquad (2\text{-}27)$$

However, in all these schemes the triphenylsilyl compounds of mercury play a purely subsidiary role and the authors of the paper cited [715] did not attempt to isolate and study them. Wiberg and his co-workers investigated the possibility of carrying out reaction (2-13) with $HgCl_2$ for this purpose [123].

The reaction of mercuric chloride with triphenylsilylpotassium in liquid ammonia forms an unstable monosilyl derivative and bis(triphenylsilyl)mercurane, which also decomposes to a considerable extent during the preparation process:

$$\begin{aligned} &HgCl_2 \xrightarrow[-KCl]{+KSiR_3} Hg(SiR_3)Cl \xrightarrow[-KCl]{+KSiR_3} Hg(SiR_3)_2 \\ &Hg(SiR_3)Cl \longrightarrow Hg + R_3SiCl \xrightarrow{(NH_3)} R_3SiNH_2 \qquad (2\text{-}28) \\ &Hg(SiR_3)_2 \longrightarrow Hg + (R_3Si)_2 \qquad (R = C_6H_5) \end{aligned}$$

Thus, the main reaction products are triphenylsilylamine and hexaphenyldisilane. Together with these, a small amount of tetraphenylsilane is obtained, indicating the possibility of the decomposition of bis(triphenylsilyl)mercurane in accordance with the scheme (2-29) (and not 2-28), in analogy with the decomposition of methylated silyl compounds of mercury:

$$Hg(SiR_3)_2 \longrightarrow Hg + R_4Si + [R_2Si] \qquad (2\text{-}29)$$

It was also possible to isolate a very small amount of undecomposed bis(triphenylsilyl)mercurane, which is a yellow substance that is soluble in benzene and sensitive to the action of light and air. Its separation from the triphenylsilylamine, which is obtained simultaneously, is difficult since the mercury derivative has a low decomposition point (55-60°C) and the two compounds are similar in solubility. Like bis(triphenylsilyl)mercurane and in contrast to hexaphenyldisilane, triphenylsilylamine is soluble in benzene so that bis(triphenylsilyl)mercury may be separated by means of benzene from KCl, mercury and hexaphenyldisilane, but not from triphenylsilylamine.

TABLE 19. Organosilyl Derivatives of Mercury

Compound	b.p., °C (mm)	Preparation scheme	Yield, %	Literature
$(C_2H_5)_3SiHgC_2H_5$	—	2-17	15	262
$(C_2H_5)_5Si_2HgC_2H_5$	100–104 (1)*	2-17	20	263
$[(CH_3)_3Si]_2Hg$	—**	2-22	—	123
$[(C_2H_5)_3Si]_2Hg$	98–100 (1)	2-18	34	263
$[(C_6H_5)_3Si]_2Hg$	decomp. 55–60°	2-28	—	123
$[(C_2H_5)_5Si_2]_2Hg$	180–190 (1), decomp.	2-18	8	263
$(C_2H_5)_3SiHgGe(C_2H_5)_3$	130–131 (1.5)	2-21	26	263
$(C_2H_5)_5Si_2HgGe(C_2H_5)_3$	159–163 (1)	2-21	50	263

* $n_D^{20} = 1.5353$.

** m.p., 102-104°C (decomp.).

The physicochemical constants of organosilyl derivatives of mercury are given in Table 19.

3. COMPOUNDS CONTAINING THE $Si-(C)_n-M$ GROUP

3.1. Organomagnesium Compounds Containing Silicon in the Organic Radical and Their Use in Organosilicon Chemistry

Silicon organomagnesium compounds containing the grouping $Si-(C)_n-Mg$ are important intermediates in the synthesis of various carbofunctional organosilicon compounds (alcohols, aldehydes, and acids), polysilyl-substituted hydrocarbons, and many silicon-heteroorganic compounds [78, 89]. The commonest method of preparing silicon organomagnesium compounds is based on the reaction of (haloorgano) silanes with metallic magnesium:

$$\geq Si-\left(\overset{|}{\underset{|}{C}}\right)_n-X + Mg \longrightarrow \geq Si-\left(\overset{|}{\underset{|}{C}}\right)_n-MgX \qquad (2\text{-}30)$$

Both haloalkyl- and haloarylsilanes react in this way. The reaction is usually carried out in ethyl ether or tetrahydrofuran. The latter cannot be used in the case of (haloalkenyl)silanes [367, 374], which contain a halogen atom at a double bond. The use of tetrahydrofuran as the solvent increases the yield of the organomagnesium compound in the case of the reaction of magnesium with (chloroaryl)silanes [808, 1135, 1349]. The rate of formation and the yields of silicon organomagnesium compounds depend on the structure of the organic radicals attached to the silicon atom and also on the nature of the halogen atom and its position relative to the silicon atom.

The presence of electron-acceptor groups attached to the silicon atom lowers the reactivity of chloromethyltriorganosilanes toward magnesium. The reaction rate falls when the methyl groups in chloromethyltrimethylsilane are replaced by phenyl groups [638, 1172] and also with a change from p-bromophenyltriethysilane to p-bromophenyldiphenylsilane and p-bromotriphenylsilane [298]. It is not possible to obtain the corresponding compounds of the type R_3SiCH_2MgCl by the reaction of methyldiethoxy(chloromethyl)silane and dimethylethoxy(chloromethyl)silane with magnesium [704]. In contrast to this, p-bromophenylmethyldiphenoxysilane, p-bromophenylmethyldi-o-cresoxysilane [1036], and p-bromophenoxytrimethylsilane [703] react with magnesium to form normal Grignard reagents. Compounds containing the Si−H bond give organomagnesium derivatives in good yields [808−810, 958, 1062, 1111]. Chloromethylpentamethyldisiloxane reacts readily with magnesium [446, 563, 1035, 1172, 1313, 1543, 1560] and chloromethylheptamethylcyclotetrasiloxane reacts with somewhat more difficulty [307, 1013, 1601, 1615].

Chloroalkyl or bromophenyl derivatives of silanes and siloxanes are usually used to prepare silicon organomagnesium compounds. In some cases it is also convenient to use bromoalkyl derivatives for this purpose [375, 425, 447, 595, 596, 620, 652, 832, 1178, 1179, 1187, 1278, 1562]. As in the organic series, with a change from chloroalkyl- to bromoalkyl- and iodoalkyl- silanes the rate of the reaction with magnesium increases. Thus, while the yield of $(CH_3)_3SiCH_2MgCl$ from the reaction of $(CH_3)_3SiCH_2Cl$ with magnesium for 10 hr at room temperature is 90%, $(CH_3)_3SiCH_2MgI$ is formed by the reaction of $(CH_3)_3SiCH_2I$ with Mg at 35°C in 95% yield after only 20 min [1257]. Replacement of the chlorine atom by iodine in the halomethyl group increases the yield of $(CH_3)_3SiOSi(CH_3)_2CH_2MgX$ from 80 to 93% [1035]. In the reaction of a halomethylheptamethylcyclotetrasiloxane with magnesium, the yields of the organomagnesium compound are 20% in the case of the chloro derivative, 65-67% in the case of the iodo derivative, and 85-90% in the case of the bromo derivative [307]. However, it should be remembered that an increase in the rate of reaction of a haloalkylsilane (siloxane) is accompanied by a simultaneous increase in the rate of side reactions of the Wurtz type [813, 832, 1278].

An attempt at the synthesis of a bis Grignard reagent from dimethylbis(chloromethyl)silane in THF gave exclusively the polymer $[(CH_3)_2SiCH_2CH_2]_n$, where $n \simeq 12$ [813]. In the case of sym-bis-(chloromethyl)tetramethyldisiloxane, the bis-magnesium derivative is formed in ether in 18-21% yield, judging by secondary reactions with CO_2 and chloromethyldimethylchlorosilane, while there is 60% of condensation. In THF or when sym-bis(iodomethyl)tetramethyldisiloxane is used there is solely condensation. In the case of 1,7-bis(chloromethyl)octamethyltetrasiloxane, no signs of condensation are observed in the reaction with magnesium in ether, but no derivatives corresponding to the bis-magnesium derivatives could be isolated either. In the opinion of Greber and Metzinger [813], the Grignard reagent formed "operates on itself" with cleavage of siloxane bonds in the manner of the scheme

$$ClCH_2[Si(CH_3)_2O]_3Si(CH_3)_2CH_2Cl + ClMgCH_2[Si(CH_3)_2O]_3Si(CH_3)_2CH_2Cl$$
$$\longrightarrow ClCH_2Si(CH_3)_2OSi(CH_3)_2CH_2[Si(CH_3)_2O]_3Si(CH_3)_2CH_2Cl +$$
$$+ClMgOSi(CH_3)_2OSi(CH_3)_2CH_2Cl$$
$$\xrightarrow{H_2O} Mg(OH)Cl + HOSi(CH_3)_2OSi(CH_3)_2CH_2Cl \qquad (2\text{-}30a)$$

Silicon organomagnesium derivatives may be obtained from haloalkyl derivatives of silanes containing a halogen atom in the α-position [337, 447, 652, 801, 1003, 1175, 1176, 1179, 1257, 1258], the γ-position [375, 447, 1062, 1178, 1179, 1562], and the δ-position [425, 1278] relative to the silicon atom. β-Haloalkylsilanes also react with magnesium, but the main products from hydrolysis of the compounds formed are the corresponding disiloxanes [425]. It is not possible to obtain C–Mg derivatives by the reaction of trimethyl-β-chloroethoxysilane with magnesium [967]. Instead of it $(CH_3)_3SiOMgCl$ is formed. In contrast to this, 4-chlorobutoxy- and 5-chloropentoxytrialkylsilanes react normally with magnesium [1568]:

$$R_3SiO(CH_2)_nCl + Mg \longrightarrow R_3SiO(CH_2)_nMgCl \qquad (2\text{-}31)$$
$$(n = 4 \text{ and } 5)$$

The second method of synthesizing compounds containing the Si–(C)n–Mg group is based on the replacement of a labile hydrogen at a carbon atom in some organosilanes by a MgX group by the action of a Grignard reagent. This is applicable in the case of cyclopentadienylsilanes [392, 864, 1056] and ethynylsilanes [498, 505, 506]:

$$C_5H_5\text{—}Si\lessdot\ (H) + RMgX \xrightarrow[-RH]{} C_5H_4(MgX)\text{—}Si\lessdot \qquad (2\text{-}32)$$

$$>Si\text{—}C\equiv CH + RMgX \xrightarrow[-RH]{} >Si\text{—}C\equiv C\text{—}MgX \qquad (2\text{-}33)$$

Silicon organomagnesium compounds undergo all the reactions which are characteristic of normal organomagnesium compounds. Their only difference lies in the fact that in many cases in the synthesis of organosilicon compounds containing electronegative groups in the β-position relative to the silicon atom, β-decomposition occurs when they are isolated.

Silicon organomagnesium compounds are oxidized readily by oxygen [637, 671, 1013, 1615]:

$$>Si\text{—}(C)_n\text{—}MgX + \frac{1}{2}O_2 \longrightarrow >Si\text{—}(C)_n\text{—}OMgX \xrightarrow[-Mg(OH)X]{H_2O} >Si\text{—}(C)_n\text{—}OH \qquad (2\text{-}34)$$

Heating p-(trimethylsilyl)phenylmagnesium bromide with sulfur forms p(trimethylsilyl)thiophenol [1051]:

$$(CH_3)_3Si\text{—}C_6H_4\text{—}MgBr + S \xrightarrow[-Mg(OH)Br]{+H_2O} (CH_3)_3Si\text{—}C_6H_4\text{—}SH \qquad (2\text{-}35)$$

The reaction of silicon organomagnesium chlorides or bromides with bromine and iodine is used for the introduction of bromine [504, 598] and iodine [638, 665] into the hydrocarbon part of organosilicon compounds:

$$>Si\text{—}(C)_n\text{—}MgX + I_2 \xrightarrow[-MgXI]{} >Si\text{—}(C)_n\text{—}I \qquad (2\text{-}36)$$

Thus, the reaction of triphenylsilylmethylmagnesium chloride with iodine forms triphenyl(iodomethyl)silane (57% yield). Analogously, p-(trimethylsilyl)phenylmagnesium bromide is converted into trimethyl(p-iodophenyl)silane in 30% yield [665].

Dicyanogen reacts with trimethylsilylmethylmagnesium chloride like halogens:

$$(CH_3)_3SiCH_2MgCl + (CN)_2 \xrightarrow[-Mg(CN)Cl]{} (CH_3)_3SiCH_2CN \qquad (2\text{-}37)$$

Silicon organomagnesium compounds which do not contain reactive groups at the silicon atom are hydrolyzed by the action of water or aqueous solutions of acids [267, 425, 476, 598, 703, 801, 813, 1035, 1111, 1173, 1275, 1349]:

$$\geq Si-(C)_n-MgX \xrightarrow[-Mg(OH)X]{+H_2O} \geq Si-(C)_n-H \qquad (2\text{-}38)$$

In the hydrolysis of a p-(trialkylsiloxy)phenylmagnesium bromide the Si–O bond is also cleaved; as a result of this the reaction products also contain phenol and a hexaalkyldisiloxane in addition to 2 trialkylphenoxysilane [267]. Hydrolysis of the product from the reaction of trimethyl(2-bromopropyl)silane with magnesium forms mainly hexamethyldisiloxane and a very small amount of trimethylpropylsilane [425].

The reaction of silicon organomagnesium compounds with carbon dioxide, together with condensation with malonic ester, is one of the most important methods of synthesizing organosilicon carboxylic acids:

$$\geq Si-(C)_n-MgX + CO_2 \xrightarrow[-MgX_2]{HX} \geq Si-(C)_n-COOH \qquad (2\text{-}39)$$

By this method it has been possible to prepare trimethylsilylacetic acid [567, 1172, 1258] with a maximum yield of 88% [1172], and also triphenylsilylacetic acid (30%) [638], (dimethylphenylsilyl) acetic acid (69%) [1172], pentamethyldisiloxanylacetic acid (85%) [1172], heptamethylcyclotetrasiloxanylacetic acid [307, 1601] (with the reaction carried out under pressure the yield is 65% [307], γ-(trimethylsilyl)butyric acid (74%) [1178, 1562], and δ-(triphenylsilyl)valeric acid (60%) [1278], and also a series of p-(trialkylsilyl)-, p-(dialkylarylsilyl)-, and p-(triarylsilyl)benzoic acids [298, 555, 619, 634, 635, 703, 1028, 1313, 1314, 1560], p-(trialkylsilylmethyl) benzoic acids [479, 567], p-[2-(trialkylsilyl)ethyl]benzoic acids [479], α- and β-(trimethylsilyl)acrylic acids (64%) [374], trimethylsilylpropiolic acid [1630], unstable triethylsilylpropiolic acid [504-

506], and some organosilicon dicarboxylic acids [813, 1313, 1314, 1560]. It was not possible to prepare the corresponding carboxylic acids from p-(trialkylsiloxy)phenylmagnesium bromide [267] or α-(trimethylsilyl)benzylmagnesium bromide [832].

Silicon organomagnesium compounds add at the carbonyl group of aldehydes and ketones:

$$\geq Si-(C)_n-MgX + O=C< \xrightarrow[-Mg(OH)X]{+H_2O} \geq Si-(C)_n-\overset{|}{\underset{|}{C}}-OH \qquad (2\text{-}40)$$

In some cases when n = 1, only the products of β-decomposition are obtained [337, 429], while in reactions with organomagnesium derivatives of trialkylphenoxysilanes, cleavage of the Si–O bond was observed [267].

The products of the reactions of silicon organomagnesium compounds with aldehydes and ketones and their yields are given in Table 20.

In the reaction of silicon organomagnesium compounds with anhydrides [82, 618, 832] and acid chlorides of carboxylic acids [1258], organosilicon ketones are formed:

$$\geq Si-(C)_n-MgX + ClCOR \xrightarrow[-MgXCl]{} \geq Si-(C)_n-CO-R \qquad (2\text{-}41)$$

The reaction of trimethylsilylmethylmagnesium chloride with acetyl chloride does not form trimethylsilylacetone, but only its β-decomposition products [1258]. Organosilicon ketones are also obtained by the reaction of silicon organomagnesium compounds with nitriles [356, 555]:

$$\geq Si-(C)_n-MgX + RCN \xrightarrow[-Mg(OH)X \atop -NH_3]{2H_2O} \geq Si-(C)_n-CO-R \qquad (2\text{-}42)$$

In the reaction of trimethylsilylmethylmagnesium bromide with benzonitrile, only the β-decomposition products were isolated [832]. The reaction of trialkylsilylethynylmagnesium bromides with dimethylformamide leads to the formation of trialkylsilylpropynals [312].

The reaction of trimethylsilylphenylmagnesium bromide with the ethyl esters of formic and orthoformic acids (after hydrolysis of the organosilicon acetal, which is formed in 96% yield) gives

TABLE 20. Products of the Reaction of Silicon Organomagnesium Compounds with Aldehydes and Ketones

Silicon organomagnesium compound	Carbonyl compound	Reaction products	Yield, %	Literature
$(CH_3)_3SiCH_2MgCl$	CH_3CHO	$(CH_3)_3SiCH_2CHOHCH_3$	—	1258
	Cl_3CCHO	$(CH_3)_3SiCH_2CHOHCCl_3$	17	429
	$CH_2{=}CHCHO$	$(CH_3)_3SiCH_2CHOHCH{=}CH_2$	50.5	444, 447
	C_4H_3O—CHO	$(CH_3)_3SiCH_2CHOH$—C_4H_3O	53.3	355
	$CH_3COC_2H_5$	$(CH_3)_3SiCH_2COH(CH_3)C_2H_5$	1.9	429
	$ClCH_2COCH_3$	$(CH_3)_3SiCH_2COH(CH_3)CH_2Cl$	70	446, 448
	$CH_3COC_3H_7$-*n*	$(CH_3)_3SiCH_2(CH_3)C{=}CHC_2H_5$	32	337
	$C_6H_5COCOC_6H_5$	$(CH_3)_3SiCH_2(C_6H_5)COHCOC_6H_5$	69.5	429
$(CH_3)_3SiCH_2MgBr$	CH_3COCH_3	$(CH_3)_3SiCH_2(CH_3)COHCH_3$	52	832
	C_6H_5CHO	$(CH_3)_3SiCH_2CHOHC_6H_5$	14	832
$(CH_3)_2C_2H_5SiCH_2MgCl$	$CH_2{=}CHCHO$	$(CH_3)_2C_2H_5SiCH_2CHOHCH{=}CH_2$	68.7	444, 447
	$CH_3CH{=}CHCHO$	$(CH_3)_2C_2H_5SiCH_2CHOHCH{=}CHCH_3$	54.4	444, 447
	$ClCH_2COCH_3$	$(CH_3)_2C_2H_5SiCH_2(CH_3)COHCH_2Cl$	70	446, 448
$CH_3(C_2H_5)_2SiCH_2MgCl$	$CH_2{=}CHCHO$	$CH_3(C_2H_5)_2SiCH_2CHOHCH{=}CH_2$	51.5	444, 447
	$ClCH_2COCH_3$	$CH_3(C_2H_5)_2SiCH_2(CH_3)COHCH_2Cl$	50	446, 448
$(CH_3)_2(n\text{-}C_4H_9)SiCH_2MgCl$	$CH_2{=}CHCHO$	$(CH_3)_2(n\text{-}C_4H_9)SiCH_2CHOHCH{=}CH_2$	64.7	444, 447
$(C_2H_5)_3SiCH_2MgCl$	$CH_2{=}CHCHO$	$(C_2H_5)_3SiCH_2CHOHCH{=}CH_2$	53.6	444, 447

$(CH_3)_3Si(CH_2)_3MgBr$	$CH_2{=}CHCHO$	$(CH_3)_3Si(CH_2)_3CHOHCH{=}CH_2$	80	447
$(CH_3)_2C_2H_5Si(CH_2)_3MgBr$	$ClCH_2COCH_3$	$(CH_3)_2C_2H_5Si(CH_2)_3COH(CH_3)CH_2Cl$	50	446, 448
$CH_3(C_2H_5)_2Si(CH_2)_3MgBr$	$CH_2{=}CHCHO$ $CH_3CH{=}CHCHO$ $ClCH_2COCH_3$	$CH_3(C_2H_5)_2Si(CH_2)_3CHOHCH{=}CH_2$ $CH_3(C_2H_5)_2Si(CH_2)_3CHOHCH{=}CHCH_3$ $CH_3(C_2H_5)_2Si(CH_2)_3COH(CH_3)CH_2Cl$	64 70 55	447 447 446, 448
$(C_2H_5)_3SiCH_2CH_2CH_2MgCl$	(furan ring)—CHO	$(C_2H_5)_3SiCH_2CH_2CH_2CHOH$—(furan ring)	37.2	357
$(CH_3)_3SiOSi(CH_3)_2CH_2MgCl$	$ClCH_2COCH_3$	$(CH_3)_3SiOSi(CH_3)_2CH_2COH(CH_3)CH_2Cl$	75	446, 448
$[(CH_3)_2SiO]_3(CH_3)SiCH_2MgCl$ (ring closed through O)	CH_2O	$[(CH_3)_2SiO]_3(CH_3)Si{-}CH_2CH_2OH$ (ring closed through O)	13	1013, 1615
$(CH_3)_3SiC{\equiv}CMgBr$	CH_3COCH_3	$(CH_3)_3SiC{\equiv}CCOH(CH_3)_2$	—	1630
$(C_2H_5)_3SiC{\equiv}CMgBr$	$CH_2{=}CHCHO$ (furan ring)—CHO $ClCH_2COCH_3$	$(C_2H_5)_3SiC{\equiv}CCHOHCH{=}CH_2$ $(C_2H_5)_3SiC{\equiv}CCHOH$—(furan ring) $(C_2H_5)_3SiC{\equiv}CCOH(CH_3)CH_2Cl$	60 53 60	505, 506 496 505, 506
$(CH_3)_2Si(C{\equiv}CMgBr)_2$	CH_3COCH_3	$(CH_3)_2Si[C{\equiv}CCOH(CH_3)_2]_2$	—	1630
$(CH_3)_3SiC{\equiv}CCH_2MgCl$	CH_3CHO	$(CH_3)_3SiC{\equiv}CCH_2CHOHCH_3$	16.4	311

TABLE 20. (Cont'd)

Silicon organomagnesium compound	Carbonyl compound	Reaction products	Yield, %	Literature
$(CH_3)_3Si-C_6H_4-MgCl$	CH_3CHO	$(CH_3)_3Si-C_6H_4-CHOHCH_3$	50	1135
$(CH_3)_3Si-C_6H_4-MgBr$	CH_3CHO	$(CH_3)_3Si-C_6H_4-CHOHCH_3$	47.1	78, 118, 434, 890, 1571
$(CH_3)_3Si-C_6H_4-MgBr$	C_6H_5CHO	$(CH_3)_3Si-C_6H_4-CHOHC_6H_5$	20	617
	CH_3COCH_3	$(CH_3)_3Si-C_6H_4-COH(CH_3)_2$	46	434
	CH_3COCH_3	$(CH_3)_3Si-C_6H_4-COH(CH_3)_2$	42	591
	CH_3COCH_3	$(CH_3)_3Si-C_6H_4-COH(CH_3)_2$	49	890
m-$(CH_3)_3Si-C_6H_4-MgBr$	CH_3COCH_3	m-$(CH_3)_3Si-C_6H_4-COH(CH_3)_2$	41	591
$(C_2H_5)_3Si-C_6H_4-MgBr$	CH_3CHO	$(C_2H_5)_3Si-C_6H_4-CHOHCH_3$	57.6	434
	CH_3CHO	$(C_2H_5)_3Si-C_6H_4-CHOHCH_3$	40.6	78, 82, 617, 820
	C_2H_5CHO	$(C_2H_5)_3Si-C_6H_4-CHOHC_2H_5$	37.8	78, 82, 617, 820

	$n\text{-}C_3H_7CHO$	$(C_2H_5)_3Si-C_6H_4-CHOHC_3H_7\text{-}n$	33	78, 82, 617, 820
	$(CH_3)_2CHCHO$	$(C_2H_5)_3Si-C_6H_4-CHOHCH(CH_3)_2$	40,1	78, 82, 617, 820
	$(CH_3)_2CHCHO$	$(C_2H_5)_3Si-C_6H_4-CHOHCH(CH_3)_2$	25,1	82, 617
	$n\text{-}C_6H_{13}CHO$	$(C_2H_5)_3Si-C_6H_4-CHOHC_6H_{13}\text{-}n$	11	82, 617
	$n\text{-}C_7H_{15}CHO$	$(C_2H_5)_3Si-C_6H_4-CHOHC_7H_{15}\text{-}n$	25	82, 617
	C_6H_5CHO	$(C_2H_5)_3Si-C_6H_4-CHOHC_6H_5$	35,4	82, 617
	CH_3COCH_3	$(C_2H_5)_3Si-C_6H_4-COH(CH_3)_2$	34,8	434
$(C_2H_5)_3SiCH_2-C_6H_4-MgBr$	CH_3CHO	$(C_2H_5)_3SiCH_2-C_6H_4-CHOHCH_3$	13	434
$(C_2H_5)_3SiCH_2CH_2-C_6H_4-MgBr$	CH_3CHO	$(C_2H_5)_3SiCH_2CH_2-C_6H_4-CHOHCH_3$	13	434
$(CH_3)_3SiO-C_6H_4-MgBr$	CH_3CHO	$[(CH_3)_3Si]_2O$	—	267
$(C_2H_5)_3SiO-C_6H_4-MgBr$	CH_3CHO	$(C_2H_5)_3SiO-C_6H_5$; $[(C_2H_5)_3Si]_2O$	—	267
$(n\text{-}C_3H_7)_3SiO-C_6H_4-MgBr$	CH_3CHO	$(n\text{-}C_3H_7)_3SiO-C_6H_4-CHOHCH_3$	—	267

trimethylsilylbenzaldehyde [276, 619, 703]:

$$(CH_3)_3Si-C_6H_4-MgBr + HCOOC_2H_5 \xrightarrow{-Mg(OC_2H_5)Br} (CH_3)_3Si-C_6H_4-CHO \quad (2\text{-}43)$$

An analogous reaction with trimethylsilylmethylmagnesium chloride forms bis(trimethylsilylmethyl)carbinol (7.9%) and trimethyl(allyl)silane [429]. The reaction with esters of acetic and butyric acids leads only to the corresponding alkenylsilanes [337]. Ethyl chlorocarbonate reacts smoothly with trimethylsilylmethylmagnesium chloride [801]:

$$(CH_3)_3SiCH_2MgCl + ClCOOC_2H_5 \xrightarrow{-MgCl_2} (CH_3)_3SiCH_2COOC_2H_5 \quad (2\text{-}44)$$

Silicon organomagnesium compounds cleave the three-membered oxide ring in ethylene oxide [425, 502, 597, 617, 1179], propylene oxide [596], and epichlorohydrin [595] and also the four-membered ring in trimethylene oxide [1187]:

$$\equiv Si-(C)_n-MgX + \underset{\diagdown O \diagup}{CH_2CH_2} \xrightarrow[-Mg(OH)X]{H_2O} \equiv Si-(C)_n-CH_2CH_2OH \quad (2\text{-}45)$$

Silicon organomagnesium compounds condense with halohydrocarbons containing a sufficiently labile halogen atom [355, 375, 424, 427, 428, 445, 504, 506, 615, 832, 890, 1278]:

$$\equiv Si-(C)_n-MgX + X-C\equiv \xrightarrow{-MgX_2} \equiv Si-(C)_n-\overset{|}{\underset{|}{C}}- \quad (2\text{-}46)$$

Isomerization occurs in the reaction of a trialkylsilylmethylmagnesium chloride with propargyl bromide [445]:

$$R_3SiCH_2MgCl + BrCH_2C\equiv CH \xrightarrow{-MgClBr} R_3SiCH_2CH=C=CH_2 \quad (2\text{-}47)$$

Chloromethyl methyl ether reacts with trimethylsilylmethylmagnesium chloride in accordance with the scheme

$$(CH_3)_3SiCH_2MgCl + ClCH_2OCH_3 \xrightarrow{-MgCl_2} (CH_3)_3SiCH_2CH_2OCH_3 \quad (2\text{-}48)$$

Reactions of silicon organomagnesium compounds with halosilanes proceed according to the scheme

$$\rangle Si-(C)_n-MgX + X-Si\langle \xrightarrow[-MgX_2]{} \rangle Si-(C)_n-Si\langle \qquad (2\text{-}49)$$

They are used to prepare organosilicon compounds containing two or more silicon atoms, connected to each other through a carbon chain, and also for the synthesis of disilylethylenes and disilylacetylenes. The reaction of silicon organomagnesium compounds with alkoxysilanes proceeds analogously [449, 1154].

In the aromatic series reaction (2-49) is used for the synthesis of disilylbenzenes containing different triorganosilyl groups:

Data on the products from the reaction of silicon organomagnesium compounds with halosilanes are given in Table 21.

Silicon organomagnesium derivatives react with halides of the elements and their organic derivatives according to the general scheme

$$y\rangle Si-(C)_n-MgX + X_kElR_{m-k} \xrightarrow[-yMgX_2]{} \left[\rangle Si-(C)_n-\right]_y ElX_{k-y}R_{m-k} \qquad (2\text{-}50)$$

where m is the valence of the element El, k varies from 1 to m, and y varies from 1 to k.

These reactions are examined in more detail in later chapters in the discussion of methods of synthesizing organosilicon heterocompounds containing the grouping $Si-(C_n)-El$.

3.2. Derivatives of Mercury

By the action of the magnesium derivative of "siliconeopentyl chloride," i.e., trimethylsilylmethylmagnesium chloride, on mercuric chloride, in 1945 Whitmore and Sommer synthesized the first compound containing the grouping $Si-(C)_n-Hg$ [1257]:

$$(CH_3)_3SiCH_2MgCl + HgCl_2 \xrightarrow[ether]{} (CH_3)_3SiCH_2HgCl + MgCl_2 \qquad (2\text{-}51)$$

Mercuric chloride reacts analogously with organomagnesium derivatives of trimethyl-α-chloroethylsilane [1180], phenyldimeth-

TABLE 21. Products of the Reaction of Silicon Organomagnesium Compounds with Halosilanes

Silicon organomagnesium compound	Halosilane	Reaction products	Yield, %	Literature
$(CH_3)_3SiCH_2MgCl$	$SiCl_4$	$[(CH_3)_3SiCH_2]_3SiCl$	—	1311
		$[(CH_3)_3SiCH_2]_4Si$	—	
	$SiCl_4$	$(CH_3)_3SiCH_2SiCl_3$	—	1176
	CH_3SiCl_3	$[(CH_3)_3SiCH_2]_3SiCH_3$	33	1154
	$(CH_3)_2SiCl_2$	$[(CH_3)_3SiCH_2]_2Si(CH_3)_2$	65	1173
	$(CH_3)_3SiCl$	$(CH_3)_3SiCH_2Si(CH_3)_3$	63	1173, 1175
	$(CH_3)_2SiClCH_2Cl$	$(CH_3)_3SiCH_2Si(CH_3)_2CH_2Cl$	32	1173
$(CH_3)_3SiCH_2MgBr$	$HSiCl_3$	$(CH_3)_3SiCH_2SiHCl_2$	64	652
		$[(CH_3)_3SiCH_2]_2SiHCl$	13.3	
		$[(CH_3)_3SiCH_2]_3SiH$	32.7	
		$[(CH_3)_3SiCH_2]_3SiH$	18.6	
$(CH_3)_2SiHCH_2MgCl$	$(CH_3)_3SiCl$	$(CH_3)_2SiHCH_2Si(CH_3)_3$	80	809
	$(CH_3)_3SiCl$	$(CH_3)_2SiHCH_2Si(CH_3)_3$	47	1111
$(CH_3)_2SiHCH_2MgCl$	$(CH_3)_2SiClCH_2Cl$	$(CH_3)_2SiHCH_2Si(CH_3)_2CH_2Cl$	73.1	809
	$(CH_3)_2SiCl_2$	$(CH_3)_2SiHCH_2SiCl(CH_3)_2$	69	
	$[(CH_3)_2SiCl]_2CH_2$	$(CH_3)_2SiHCH_2Si(CH_3)_2CH_2SiCl(CH_3)_2$	52.1	
	$[(CH_3)_2SiClCH_2]_2Si(CH_3)_2$	$(CH_3)_2SiH[CH_2Si(CH_3)_2]_3Cl$	38.3	
$(CH_3)_2C_6H_5SiCH_2MgCl$	$(CH_3)_2SiClCH_2Cl$	$(CH_3)_2C_6H_5SiCH_2Si(CH_3)_2CH_2Cl$	—	705
	$CH_3SiCl_2CH_2Cl$	$[(CH_3)_2C_6H_5SiCH_2]_2Si(CH_3)CH_2Cl$	17	
	$(CH_3)_2SiCl_2$	$[(CH_3)_2C_6H_5SiCH_2]_2Si(CH_3)_2$	16.9	

$(CH_3)_3SiCH{=}CHMgBr$	$(CH_3)_3SiCl$	$(CH_3)_3SiCH{=}CHSi(CH_3)_3$	—	367, 374
$(C_2H_5)_3SiC{\equiv}CMgBr$	$(CH_3)_3SiCl$	$(C_2H_5)_3SiC{\equiv}CSi(CH_3)_3$	—	504
$(n\text{-}C_3H_7)_3SiC{\equiv}CMgBr$	$(CH_3)_3SiCl$	$(n\text{-}C_3H_7)_3SiC{\equiv}CSi(CH_3)_3$	—	504
$CH_3(C_6H_5)SiHC{\equiv}CMgBr$	$(CH_3)_3SiCl$	$CH_3(C_6H_5)SiHC{\equiv}CSi(CH_3)_3$	—	504
$(C_6H_5)_3SiC{\equiv}CMgBr$	$(CH_3)_3SiCl$	$(C_6H_5)_3SiC{\equiv}CSi(CH_3)_3$	—	504
$(CH_3)_2SiHCH_2Si(CH_3)_2CH_2MgCl$	$(CH_3)_3SiCl$	$(CH_3)_2SiHCH_2Si(CH_3)_2CH_2Si(CH_3)_3$	40.8	809
$(CH_3)_2C_6H_5SiCH_2Si(CH_3)_2CH_2MgCl$	$(CH_3)_2SiBrCH_2Si(CH_3)_2CH_2Cl$	$(CH_3)_2C_6H_5Si[CH_2Si(CH_3)_2]_3CH_2Cl$	73	705
$(CH_3)_3SiOSi(CH_3)_2CH_2MgCl$	$(CH_3)_3SiCl$ $(CH_3)_2SiClCH_2Cl$ $(CH_3)_2SiCl_2$	$(CH_3)_3SiOSi(CH_3)_2CH_2Si(CH_3)_3$ $(CH_3)_3SiOSi(CH_3)_2CH_2Si(CH_3)_2CH_2Cl$ $(CH_3)_3SiOSi(CH_3)_2CH_2SiCl(CH_3)_2$ $[(CH_3)_3SiOSi(CH_3)_2CH_2]_2Si(CH_3)_2$	— 73 —	1532 886, 1620 563, 1534, 1543

TABLE 21. (Cont'd)

Silicon organomagnesium compound	Halosilane	Reaction products	Yield, %	Literature
$(CH_3)_2SiH-C_6H_4-MgCl$	$(CH_3)_2SiCl_2$	$(CH_3)_2SiH-C_6H_4-SiCl(CH_3)_2$	65	
	$(CH_3)_2SiCl_2$	$[(CH_3)_2SiH-C_6H_4-]_2Si(CH_3)_2$	40	808
	$(CH_3)_2Si(Cl)-C_6H_4-Si(Cl)(CH_3)_2$	$(CH_3)_2SiH[-C_6H_4-Si(CH_3)_2]_3H$	27	
$(CH_3)_3Si-C_6H_4-MgCl$	$(CH_3)_2SiCl-C_6H_4-Cl$	$(CH_3)_3Si-C_6H_4-Si(CH_3)_2-C_6H_4-Cl$	52	984
$(C_2H_5)_3Si-C_6H_4-MgBr$	$SiCl_4$	$(C_2H_5)_3Si-C_6H_4-SiCl_3$	—	820
$(CH_3)_3SiO-C_6H_4-MgBr$	$(CH_3)_3SiCl$	$(CH_3)_3SiO-C_6H_4-Si(CH_3)_3$	77.4	703, 976
$CH_3(C_6H_5O)_2Si-C_6H_4-MgBr$	$(CH_3)_3SiCl$	$CH_3(C_6H_5O)_2Si-C_6H_4-Si(CH_3)_3$	14	1036
$CH_3(o\text{-}CH_3C_6H_4O)_2Si-C_6H_4-MgBr$	$(CH_3)_3SiCl$	$CH_3(o\text{-}CH_3C_6H_4O)_2Si-C_6H_4-Si(CH_3)_3$	41.5	1036

ylchloromethylsilane [705], pentamethylchloromethyldisilane [913 b], and tert-butyldimethylchloromethylsilane [913 c]. If the reaction is carried out in tetrahydrofuran with the ratio $(CH_3)_3SiCh_2MgCl : HgCl_2 = 2:1$, then bis(trimethylsilylmethyl)mercurane is obtained in almost 50% yield [1147]. The replacement evidently proceeds stepwise. In any case, by organomagnesium synthesis it is possible to obtain from $(CH_3)_3SiCH_2HgCl$ alkyl or aryl derivatives of (trimethylsilylmethyl)mercuranes [1257]. In the case of the germanium analog a compound is obtained in which the mercury is attached to two Group IV elements, namely, silicon and germanium [40, 371].

Mercury derivatives of the type RHgR' have been used as model compounds in the study of the relative electronegativity of different groups in organic compounds.

Thus, it follows from the following reaction that the group $(CH_3)_3GeCH_2$ has a higher electronegativity than the group $(CH_3)_3SiCH_2$:

$$2\,(CH_3)_3SiCH_2HgCH_2Ge(CH_3)_3 \xrightarrow{HCl}$$

$$\longrightarrow \underset{(80\%)}{(CH_3)_4Ge} + \underset{(85\%)}{(CH_3)_3SiCH_2HgCl} + \underset{(20\%)}{(CH_3)_4Si} + \underset{(traces)}{(CH_3)_3GeCH_2HgCl} \qquad (2\text{-}52)$$

On the other hand, the relative electronegativity of radicals is characterized by the cleavage of organo(trimethylsilylmethyl)-mercuranes by means of HCl:

$$R_3SiCH_2HgR' \xrightarrow[C_2H_5OH]{HCl} \begin{cases} R_4Si + R'HgCl & (R'=CH_3,\ C_6H_{13}\text{-n}) \\ R_3SiCH_2HgCl + R'H & (R'=C_6H_5) \end{cases} \qquad (2\text{-}53)$$

In combination with other data, on the basis of this reaction it is possible to arrange radicals in the following series with respect to the electronegativity [1257]:

$$C_6H_5 > (CH_3)_3SiCH_2 > CH_3 > n\text{-}C_6H_{13} > (CH_3)_3CCH_2$$

With regard to its cleaving effect, $HgCl_2$ is analogous to HCl [1257]:

$$R_3SiCH_2HgC_6H_{13} + HgCl_2 \longrightarrow R_3SiCH_2HgCl + C_6H_{13}HgCl \qquad (2\text{-}54)$$

The reaction of bis[(phenyldimethylsilyl)methyl]mercurane with lithium is used to prepare the corresponding lithium derivative, which is not formed by the direct action of lithium on phenyldimethylchloromethylsilane [705]:

$$[C_6H_5(CH_3)_2SiCH_2]_2Hg + 2Li \xrightarrow[N_2]{ether} Hg + 2C_6H_5(CH_3)_2SiCH_2Li \quad (2\text{-}54a)$$

The reaction is 50% complete after boiling for 40 hr and 86% complete after 110 hr.

No organosilicon derivatives of mercury are formed by the reaction of triorganosilanes with mercury compounds of the type $C_6H_5HgCX_2Br$ ($X = H$, Cl or Br). There is no reduction either [1144], but the following reaction occurs:

$$R_3SiH + C_6H_5HgCX_2Br \longrightarrow R_3SiCX_2H + C_6H_5HgBr \quad (2\text{-}55)$$

A detailed description and discussion of such "insertion reactions" is given in [114a].

In the oxymercuration of trimethylvinylsilane, in contrast to its hydroboronation (see Ch. 3, Section 2.1.2), only one isomer is formed [1159] and its reduction with sodium amalgam gives β-(trimethylsilyl)ethyl alcohol.

The properties of compounds containing the grouping Si–C–Hg are given in Table 22.

TABLE 22. Compounds Containing the Grouping Si–C–Hg

Formula	m.p., °C	b.p. °C (mm)	n_D^{20}	d_4^{20}	Yield, %	Literature
$(CH_3)_3SiCH_2HgCl$	74–76	—	—	—	80; 85 *	40, 1257
$(CH_3)_3SiCH(CH_3)HgCl$	97	—	—	—	40 *	1180
$(CH_3)_3CSi(CH_3)_2CH_2HgCl$	82–83	—	—	—	65 *	9136
$(CH_3)_3SiSi(CH_3)_2CH_2HgCl$	69–70	—	—	—	62 *	9138
$(CH_3)_3SiCH_2HgCH_3$	—	—	—	—	— **	1257
$(CH_3)_3SiCH_2HgC_6H_{13}$-n	—	—	—	—	— **	1257
$(CH_3)_3SiCH_2HgC_6H_5$	—	—	—	—	— **	1257
$[(CH_3)_3SiCH_2]_2Hg$	—	49–50 (0,35)	1.4869^{25}	—	49 *	1147
$[C_6H_5(CH_3)_2SiCH_2]_2Hg$	32–34	180–182 (1)	—	—	71 *	705
$(CH_3)_3SiCH_2HgCH_2Ge(CH_3)_3$	—	71 (4)	1.5028	1.6839	58 **	40, 371
	—	95.5 (7)	1.5015	1.6746	81 **	371

* Compound obtained by scheme (2-51).

** Combination of scheme (2-51) with the replacement of a halogen by an organic radical by the Grignard reaction.

4. COMPOUNDS CONTAINING THE Si – N – M GROUP

When sodium bis(trimethylsilyl)amide is treated with $BeCl_2$ [605] or dihalides of elements of the zinc subgroup [604, 606] in ether there is a rapid exothermic reaction:

$$2\,(R_3Si)_2NNa + MX_2 \xrightarrow{\text{ether}} [(R_3Si)_2N]_2M \qquad (2\text{-}56)$$
$$(R = CH_3,\ M = Zn,\ Cd,\ Hg,\ Be;\ X = Cl,\ Br,\ I)$$

The colorless, crystalline, low-melting substances obtained dissolve readily in ether, ligroin, benzene, and carbon tetrachloride. They do not form adducts with pyridine or tetrahydrofuran and are hydrolyzed with the formation of $M(OH)_2$ and hexamethyldisilazan. The interatomic distances r(Zn–N) = 2.06 Å, and r(Cd–N) = r(Hg–N) = 2.23 Å. The frequencies ν_{as} MN_2 lie in the infrared spectra in the region of 400-440 cm^{-1} while for $\nu_s MN_2$ the Raman spectra show strong bands at 385–400 cm^{-1}.

The reaction of hexamethyldisilazan with Grignard reagents [1237] leads after strong cooling of the concentrated ether solutions to crystalline compounds which have the following composition and structure:

$$\{[(CH_3)_3Si]_2N\}_2Mg\langle X_2\rangle Mg[O(C_2H_5)_2]_2 \cdot nO(C_2H_5)_2$$
$$(X = Cl,\ Br,\ I;\ n = 0 \text{ or } 2)$$

These compounds are readily soluble in normal organic solvents and are decomposed rapidly by water. At elevated temperatures, together with elimination of ether there is pyrolysis with the formation of volatile compounds containing the Si–N bond.

Prolonged boiling of $\{[(CH_3)_3Si]_2N\}_2Mg \cdot 2MgI_2 \cdot 4O(C_2H_5)_2$ in benzene converts this compound into the dietherate. Dioxane displaces MgI_2 from it to form bis[di(trimethylsilyl)amino]magnane in the form of a complex with dioxane (1 : 3), which is white, crystalline, strongly hygroscopic, and highly soluble. The reaction of the iodine-containing compound with sodium bis(trimethylsilyl)amide liberates NaI and yields the etherate of bis[di(trimethylsilyl)amino]-magnane, which loses ether during vacuum sublimation to give the magnesium analog of compounds formed by scheme (2-56). The bromine-containing compound does not react with $SiCl_4$ in boiling benzene.

Ammonolysis of bis(triphenylsilyl)barane proceeds in stages [123]:

$$[(C_6H_5)_3Si]_2Ba \xrightarrow[-2H_2]{+4NH_3} 2(C_6H_5)_3SiNH_2 + Ba(NH_2)_2 \xrightarrow[-2NH_3]{} [(C_6H_5)_3SiNH]_2Ba \quad (2\text{-}57)$$

Bis(triphenylsilylamino)barane is a colorless substance, which ignites in air, is hydrolyzed by the action of wet methanol to give barium hydroxide and triphenylsilanol, forms an ammoniate, and does not reduce $AgNO_3$.

When the polymer obtained by the reaction of dimethyldichlorosilane with ethylenediamine $[(CH_3)_2SiNHCH_2CH_2NH]_n$ is heated with anhydrous beryllium chloride in xylene a polymer is formed in which each third nitrogen atom is coordinated to beryllium [963]. It is more resistant to hydrolysis than the starting polymer.

The properties of compounds containing the grouping Si–N–M are given in Table 23.

TABLE 23. Compounds Containing the Grouping Si–N–M

Formula	m.p., °C	b.p., °C (mm)	n_D^{20}	d_4^{20}	Yield, %	Literature
$\{[(CH_3)_3Si]_2N\}_2Be$	–3–1	110 (3)	1.4369	0.8249	68 *	605
$\{[(CH_3)_3Si]_2N\}_2Mg$	—	—	—	—	—	1237
$\{[(CH_3)_3Si]_2N\}_2Mg \cdot (C_2H_5)_2O$	—	subl. 100 (1) decomp.	—	—	—	1237
$\{[(CH_3)_3Si]_2N\}_2Mg \cdot 3OC_4H_8O$	—	—	—	—	—	1237
$\{[(CH_3)_3Si]_2N\}_2Mg \cdot MgCl_2 \cdot 2(C_2H_5)_2O$	106	—	—	—	—	1237
$\{[(CH_3)_3Si]_2N\}_2Mg \cdot MgBr_2 \cdot 2(C_2H_5)_2O$	98	—	—	—	—	1237
$\{[(CH_3)_3Si]_2N\}_2Mg \cdot MgBr_2 \cdot 4(C_2H_5)_2O$	—	—	—	—	—	1237
$\{[(CH_3)_3Si]_2N\}_2Mg \cdot MgI_2 \cdot 4(C_2H_5)_2O$	60	—	—	—	—	1237
$[(C_6H_5)_3SiNH]_2Ba$	—	—	—	—	—	19
$\{[(CH_3)_3Si]_2N\}_2Zn$	12.5	82 (0.5)	1.4506	0.952	83 *	604, 606
$\{[(CH_3)_3Si]_2N\}_2Cd$	8	93 (0.5)	1.4660	1.062	75 *	604, 606
$\{[(CH_3)_3Si]_2N\}_2Hg$	11	78 (0.15)	1.4717	1.288	97 *	604, 606

* Compounds obtained by scheme (2-56).

5. SALTS OF SILICOORGANIC ACIDS

γ-Silicoorganic acids form calcium salts, which, in analogy with the calcium salts of organic acids, form symmetrical silicoorganic ketones in 23-28% yield on dry distillation [129, 338]:

$$(RR'_2SiCH_2CH_2COO)_2Ca \longrightarrow (RR'_2SiCH_2CH_2)_2CO + CaCO_3 \qquad (2\text{-}58)$$
$$(R = CH_3;\ R' = CH_3,\ C_2H_5)$$

Correspondingly, heating a mixture of calcium salts of silicoorganic and isobutyric acids gives a 13-16% yield of unsymmetrical ketones containing a trialkylsilyl group in one of the radicals.

The reactions of α, ω-bis(β-carboxyethyl)-polydimethylsiloxanes with calcium, zinc, and cadmium hydroxides in benzene and its homologs under conditions of azeotropic distillation of the water lead to viscous or viscoelastic hydrolytically stable polymers of the type $\{-MOCOCH_2CH_2[-Si(CH_3)_2O]_m-Si(CH_3)_2CH_2CH_2COO-\}_n$ [289, 290, 1518a] (m=29, 32, 74, 106; yield 63-83%). Their IR spectra contain a band at 1600-1620 cm^{-1}, which belongs to antisymmetrical valence vibrations of M–O in the –COOM–group (see also Ch. 8; scheme 8-54, Table 97).

6. COMPLEXES OF HALOSILANES AND OTHER ORGANOSILICON COMPOUNDS WITH HALIDES OF GROUP II ELEMENTS

Silicon tetrachloride and methylchlorosilanes do not catalyze the polymerization of sylvan (the latter is polymerized exclusively by ionic catalysts). However, the presence of traces of zinc chloride or metallic zinc is sufficient for rapid and quantitative polymerization at a temperature of the order of 50°C [317, 1484].

The authors of the papers cited conclude that complexes of halosilanes with $ZnCl_2$ in a ratio of 1 : 1 participate in the process. It is not possible to remove the silicon from the polymers formed (mol. wt. 8000-10000) by treatment with alcohol. The polymerization of sylvan is also catalyzed by the complex of $CdCl_2$ with phenyltrichlorosilane [472]; its activity is lower than that of complexes of antimony chloride with methylchlorosilanes, but higher than that of the complex of phenyltrichlorosilane with $AlCl_3$ (see Ch. 3, Section 9).

TABLE 24. Complexes of Mercury Halides with Organosilicon Heterocompounds

Compound	m.p., °C	Yield, %	Literature
$[(CH_3)_3SiCH_2]_3P \cdot HgCl_2$	175—177 Decomp.	—	1137
$[(CH_3)_3SiCH_2]_3As \cdot HgCl_2$	176—176.8 Decomp.	—	1137
$p\text{-}[(C_2H_5)_3Si]C_6H_4As(C_6H_5)_2 \cdot HgCl_2$	188	—	821
$[(CH_3)_3SiCH_2SH]_2 \cdot HgCl_2$	142	—	654
$[(C_2H_5)_3SiCH_2SH]_2 \cdot HgCl_2$	105	69	654
$[n\text{-}C_3H_7)_3SiCH_2SH]_2 \cdot HgCl_2$	135	—	654
$[(CH_3)_3SiCH_2P(C_6H_5)_3]Br \cdot HgBr_2$	134—135	—	815, 1139
$p\text{-}[(C_2H_5)_3Si]C_6H_4As(C_6H_5)_2 \cdot HgBr_2$	181	—	821
$\{CH_3[(CH_3)_3SiCH_2]_3As\}I \cdot HgI_2$	134—135	91	1137
$p\text{-}[(C_2H_5)_3Si]C_6H_4As(C_6H_5)_2 \cdot HgI_2$	139.5	—	821
$[(CH_3)_3SiCH_2]_2S \cdot HgI_2$	63	20	654

Stable complexes with mercury halides are formed by triorganosilylmethyl derivatives of phosphorus and arsenic, triorganosilylaryl derivatives of arsines, and also organosilicon mercaptans. The properties of these adducts are given in Table 24.

The adducts of tris(trimethylsilylmethyl)phosphine and tris-(trimethylsilyl)arsine with $HgCl_2$ are soluble in acetone [1137]. Compounds of (triethylsilylphenyl)diphenylarsine with mercury halides are soluble in benzene and have a sharp melting point, which may be used for the identification of silylphenylarsines [821].

Chapter 3

Organosilicon Compounds of Group III Elements

BORON

1. COMPOUNDS CONTAINING THE Si – O – B GROUP

1.1. Preparation Methods

1.1.1. Reactions of Halosilanes with Boric Acid and Its Derivatives

The simplest method of synthesizing tris(trialkylsilyl) borates is based on the reaction of trialkylchlorosilanes with boric acid [251]:

$$3R_3SiCl + B(OH)_3 \longrightarrow (R_3SiO)_3B + HCl \qquad (3\text{-}1)$$

Thus, boiling boric acid with trialkylchlorosilanes for 12-15 hr gives tris(trialkylsilyl) borates in about 25% yield.

The reaction of trialkylchlorosilanes with trialkyl borates proceeds according to the scheme

$$3R_3SiCl + B(OR')_3 \longrightarrow (R_3SiO)_3B + 3R'Cl \qquad (3\text{-}2)$$

The process is catalyzed by ferric chloride, but the yields of tris(trialkylsilyl) borates are still lower than by scheme (3-1) (~10%) [251, 1318, 1563].

The reaction of triethylbromosilane with phenylboric acid at 140°C leads to the formation of very small amounts of bis(triethylsilyl)phenylboronate [929, 930]. At a lower temperature there is only anhydride formation by the phenylboric acid. Cyclic phenylboroxane was also obtained by the reaction of phenylboric acid with diorganodichlorosilanes [65]. It was believed that when a mixture of dimethyldichlorosilane and boric acid (in a molar ratio of 3:2) was heated until the liberation of HCl ceased [251] the composition of the reaction products would correspond to the equation

$$3nR_2SiCl_2 + 2nB(OH)_3 \longrightarrow 6nHCl + (3R_2SiO \cdot B_2O_3)_n \qquad (3\text{-}3)$$

However, the viscous colorless polymer obtained contained 8.9% of boron and 26.0% of silicon, as compared with 7.4 and 28.8%, respectively, for $3(CH_3)_2SiO \cdot B_2O_3$. It rapidly hardened in air, becoming coated with a dense white film, and had the properties of "bouncing putty".*

A more detailed study of the reaction of boric acid with dimethyldichlorosilane in a molar ratio of 2:3 showed [1217] that in the presence of a solvent (tetrahydrofuran) the rate of the reaction increases appreciably, passing through a maximum with an increase in the dilution. Ketones and esters also catalyze the reaction. Under optimal conditions the yield of HCl is 90-95%; no chlorine is detected in the solid polymers obtained, which have viscoelastic properties and molecular weights above 760-940. At the same time, the IR spectra show the presence of terminal hydroxyl groups in the polymers and strong hydrogen bonds between them. As a result of this, despite their low molecular weight, these polymers have the properties of "bouncing putty" which is usually obtained by introducing boric acid derivatives into high-molecular polyorganosiloxanes (see below).

An additional amount of HCl (up to a total yield of 100%) is liberated during the reaction of the solid polymers with excess dimethyldichlorosilane. The liquid borasiloxanes obtained in this way contain terminal chlorine atoms. Their hydrolysis again leads to solid polymers.

The presence of hydroxyl groups in the products from the reaction of dimethyldichlorosilane with boric acid indicates that

* This name is given to peculiar materials, which combine the properties of flow with elastic properties (p. 195).

the equation (3-3) proposed previously describes the reaction only in the first approximation. The following scheme is more accurate:

$$3nR_2SiCl_2 + 2nB(OH)_3 \longrightarrow \left[B_2(OH)_x \cdot (R_2SiO_2)_{\frac{6-x}{2}}\right]_n + (6-x)\,n\,HCl \quad (3\text{-}4)$$

According to this scheme there is the intermediate formation of products of the type $(HO)_2BOSiR_2OB(OH)_2$. On the basis of the fact that in the condensation process there is always more than 2/3 of the theoretically possible amount of HCl liberated, Vale considers that the possibility of the further formation of linear polymers with the fragments $-B(OH)OSiR_2OB(OH)-$ is excluded and the following cyclic structure is more probable:

$$HOB\begin{matrix}\diagup OSiR_2O \diagdown \\ \diagdown OSiR_2O \diagup\end{matrix}BO\left[SiR_2OB\begin{matrix}\diagup OSiR_2O \diagdown \\ \diagdown OSiR_2O \diagup\end{matrix}BO\right]_n SiR_2OB\begin{matrix}\diagup OSiR_2O \diagdown \\ \diagdown OSiR_2O \diagup\end{matrix}BOH \quad (R{=}CH_3) \quad (3\text{-}5)$$

In this connection we should mention a report of the preparation of bicyclic compounds $B[OSi(C_6H_5)_2O]_3B$ by the reaction of diphenyldichlorosilanes with boric acid [523].

In 1948 a patent [1456, 1525] was published which described the preparation of polyalkylborasiloxanes* with the ratio B:Si = 0.16:1.0 by treatment of α,ω-dichloropolydialkylsiloxanes with boric acid or a mixture of it with acetic acid. Andrianov [8] has proposed the stepwise scheme

$$Cl(R_2SiO)_n SiR_2Cl + B(OH)_3 \longrightarrow Cl(R_2SiO)_n SiR_2B(OH)_2 + HCl, \text{ etc.} \quad (3\text{-}6)$$

$$(R = \text{ alkyl ratic } Cl:Si = 0.16\text{—}1.0)$$

In the case of α,ω-dichlorosiloxanes the linear condensation scheme is less debatable than in the case of diorganodihalosilanes. A development of scheme (3-6) is the synthesis of borasiloxane elastomers by "hydrolytic condensation" [269, 270]. Essentially this is a two-stage process, the first stage of which yields α,ω-dichloropolydiorganosiloxanes, which then react with small amounts of boric acid with continuing elimination of HCl.

---

* To denote the replacement of a silicon atom by a boron atom it seems preferable to use the terminology of the "oxa-aza" nomenclature with the use of the prefix <u>bora-</u>. It has been suggested [66] that the prefix <u>boro-</u> should be used for the group > BR' in organosilicon compounds.

There is no definite opinion on the structure of the units of the chain which contain boron and the following different variants are regarded as possible:

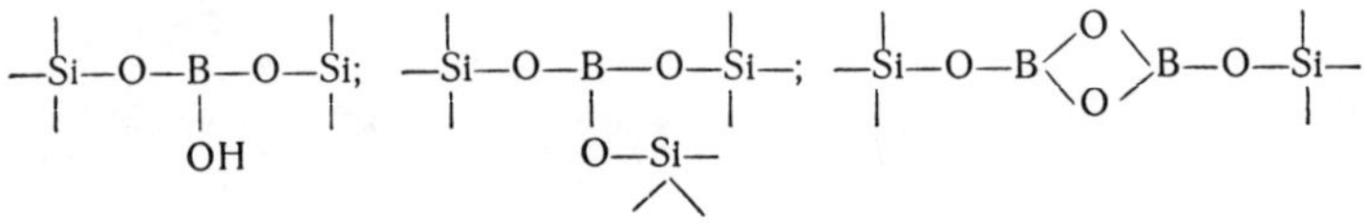

$$\begin{array}{c} | \quad\quad | \\ -Si-O-B-O-Si; \\ | \quad\ | \quad\ | \\ OH \end{array} \quad \begin{array}{c} | \quad\quad\quad | \\ -Si-O-B-O-Si-; \\ | \quad\ | \quad\ | \\ O-Si- \\ \wedge \end{array} \quad -\overset{|}{\underset{|}{Si}}-O-B\langle\!\!\begin{array}{c}O\\ \\O\end{array}\!\!\rangle B-O-\overset{|}{\underset{|}{Si}}- \qquad (3\text{-}7)$$

However, in general, the specific properties of borasiloxane elastomers (see below) are explained by the formation of supermolecular structures as a result of coordination bonds of boron [269].

The use of boric acid and the products from its reaction with halosilanes as catalysts for the arylation of silicochloroform has been reported in patents [1318, 1563]. By studying arylation on the example of the dehydrocondensation of methyldichlorosilane with benzene and its homologs and derivatives, Mikheev, Mal'nova, and others [46, 361, 363, 364, 381-384, 1467, 1474, 1482] established that the best catalyst is the product from the reaction of methyldichlorosilane with boric acid. The latter may be introduced directly into the mixture of starting materials, but from the technological point of view it is preferable to prepare the catalyst by boiling methyldichlorosilane with boric acid (molar ratio 6:1) for 50 hr. The liquid catalyst is soluble both in the mixture of starting materials and in the final arylation products. In the case of benzene, at the optimal concentration of the catalyst (1.35 wt.%) the process is carried out at 240-250°C, while the uncatalyzed reaction proceeds only at 470°C.

The mechanism of dehydrocondensation is treated as a heterolytic (ionic) process:

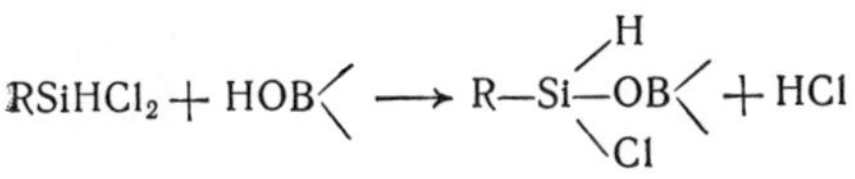

$$RSiHCl_2 + HOB\langle \longrightarrow R-Si\begin{array}{l}\diagup H\\ -OB\langle\\ \diagdown Cl\end{array} + HCl \qquad (3\text{-}7a)$$

The electron-acceptor effect of the boron atom produces a shift in electron density, which weakens the bond of the silicon to hydrogen. Protonization of the hydrogen promotes its coordinations to electrons of the aromatic nucleus:

$$C_6H_5{\to}H{\to}\underset{Cl\ \ CH_3}{Si}{\to}O{\to}B\langle \longrightarrow [C_6H_5(H)H]\ H^+\ \overline{\underset{Cl\ \ CH_3}{Si}}{-}O{-}B\langle \longrightarrow C_6H_5\underset{Cl\ \ CH_3}{Si}{-}O{-}B\langle + H_2 \qquad (3\text{-}7b)$$

As a result there is the possibility of the formation of a carbenium ion – silyl borate anion pair. The recombination of this unstable system leads to an arylsilyl borate. The latter reacts with excess methyldichlorosilane:

$$C_6H_5Si(Cl)(CH_3){-}O{-}B\langle + Cl_2SiHCH_3 \longrightarrow C_6H_5Si(Cl)(CH_3){-}Cl + \rangle B{-}O{-}Si(Cl)(CH_3){-}H \qquad (3\text{-}7c)$$

The use of products from the reaction of boric acid with hydrochlorosilanes as catalysts for arylation of the latter has also been mentioned elsewhere [184, 362, 478].

It has been reported [1260, 1725] that the high-temperature reaction of boric esters with polyfunctional halosilanes ($RSiCl_3$; $SiCl_4$) is accompanied by the formation of highly condensed borasiloxanes. Oligomers with a molecular weight up to 800 may be isolated by distillation. In particular, the preparation of a borasiloxane polymer as a result of boiling phenyltrichlorosilane with butyl borate (3:1) for 20 hr has been described [725].

However, according to the data of Gerrard and his co-workers [623], tributyl borate does not react with $SiCl_4$ at the boiling point, while in the presence of ferric chloride, butyl chloride is liberated and SiO_2 and B_2O_3 are formed.

In a study of the addition of hydrochlorosilanes to butyl esters of alkenylboric acids it was observed [376] that even under mild conditions (chloro-platinic acid as the catalyst) there are side reactions involving the ester groups. In the case of the model system, the butyl ester of propylboric acid – silicochloroform ratio 3:2; boiling for 16 hr), from the reaction products was isolated the butyl ester of propylchloroboric acid $-C_3H_7B(OC_4H_9)Cl$, indicating alkoxyl – halogen exchange. In addition a substance was obtained with 160-166°C (3 mm), $n_D^{20} = 1.4392$, mol. wt. 493, which contained silicon, boron, and chlorine. Its exact composition and structure were not established, but it was evidently formed as a result of condensation processes like reaction (3-1).

1.1.2. Reactions of Alkoxysilanes and Acyloxysilanes with Boric Acid and Its Derivatives

The first report of the preparation of a tris(trialkylsilyl) borate by the reaction of trimethylethoxysilane with boric acid appeared in 1948 [1307, 1528]:

$$3R_3SiOR' + B(OH)_3 \rightleftharpoons (R_3SiO)_3B + 3R'OH \qquad (3\text{-}8)$$

However, as is often the case with patent data, a check on them showed [251] that when trimethylethoxysilane was boiled with boric acid (molar ratio 3:1) for 6-48 hr in the presence of the catalyst given in the patent (p-toluene-sulfonic acid), tris(trimethylsilyl)borate was obtained in a yield of no more than 3-7%.*

The reaction mixture was found to contain ethyl alcohol (b.p. 78°C), hexamethyldisiloxane (b. p. 100°C), and triethyl borate (b.p. 119°C). No silyl borate was formed under analogous conditions in the absence of a catalyst and only the above products were obtained, i. e., the process proceeded mainly in accordance with the scheme

$$\rangle Si{-}OR + HO{-}B\langle \rightleftharpoons \rangle Si{-}OH + RO{-}B\langle \qquad (3\text{-}9)$$

Trimethylsilanol readily undergoes anhydrocondensation to hexamethyldisiloxane. The formation of alcohol is presumably due to hydrolysis by water liberated in the condensation. No substances boiling above 67°C (b. p. of trimethyl borate 67.3°C) were formed during prolonged boiling of boric acid with trimethylmethoxysilane (b. p. 67°C). However, when not less than two methyl groups at the silicon atom in trimethylmethoxysilane were replaced by higher alkyls it was possible to obtain tris(trialkylsilyl) borates in 70-75% yield by simple distillation of trialkylmethoxysilanes with boric acid [251]. Preliminary boiling for 6 hr increased the yield of tris(triethylsilyl)borate to 94%. Products with aryl radi-

* It should be noted that Krieble [1307, 1528] gives a too low a boiling point (90°C) for tris(trimethylsilyl)borate, while tris(tert-butyl)borate has a boiling point of 175°C [1574].

cals at the silicon were obtained in high yield and by this method it was possible to obtain not only derivatives of boric acid such as tris(ethyldiphenyl) borate (50% yield) [413], but also such compounds as the tetrakis(triphenylsilyl) ester of p-phenyleneboric acid [533].

According to the data of Andrianov and his co-workers [216], in the reaction of $CH_3(C_6H_5NHCH_2)Si[OSI(CH_3)_3]OC_2H_5$ with boric acid (3:1) the reaction occurs mainly at the ethoxyl group without touching other potentially reactive groups and tris[phenylaminomethyl(methyl) triethylsiloxysilyl]borate is obtained in 44% yield.

When boric acid is replaced by its anhydride [515], trimethylethoxysilane reacts readily:

$$3R_3SiOR' + B_2O_3 \longrightarrow \underset{(99\%)}{(R'O)_3B} + \underset{(75\%)}{(R_3SiO)_3B} \tag{3-10}$$

The reaction of boric acid with triorganoacetoxysilane proceeds smoothly and requires no catalyst [253]:

$$R_3SiOCOCH_3 + B(OH)_3 \rightleftarrows (R_3SiO)_3B + CH_3COOH \tag{3-11}$$

There are contradictory data on the reaction of boric acid with diorganodialkoxysilanes. In the slow distillation of dimethyldiethoxysilane with boric acid in a molar ratio of 3:1, ethyl ether and triethyl borate distill [251]. The remaining viscous, colorless, readily soluble polymer contains 4.2% boron:

$$6nR_2Si(OR')_2 + 3nB(OH)_3 \longrightarrow [6R_2SiO \cdot B_2O_3]_n + 9nR'OH + nB(OR')_3 \tag{3-12}$$
$$(\text{when } R = CH_3, R' = C_2H_5; \text{when } R = C_2H_5, R' = CH_3)$$

In other papers [64, 324, 471] it is reported that the reaction of dimethy- or diethyldiethoxysilane with boric acid in a ratio of 3:2 is accompanied by the formation of a viscous, sticky polymer with the composition $[3R_2SiO \cdot B_2O_3]_n$ ($n \cong 8$; yield 90-97%) together with the liberation of alcohol and small amounts of triethyl borate. According to the same data, in the case of methylphenyldiethoxysilane, with a ratio of it to H_3BO_3 of 3:1, a polymer is obtained with a com-

* Although this method of writing the compounds does not reflect their structure, it is more convenient than the method adopted in [64, 324]: $[(RR'Si)_3(BO_3)_2]_n$ and $[(CH_3 \cdot C_6H_5SiO_{1.5})_3B]_n$.

position [6RR'SiO · B_2O_3]* (n ≅ 4; yield 95%), but in contrast to scheme 3-12, the reaction products contain diethyl ether in addition to alcohol and triethyl borate. The following structure has been assigned to the products from the reaction of methylphenyldiethoxysilane and methyl(phenylaminomethyl) diethoxysilane with boric acid in a ratio of 3:2 [216, 221]:

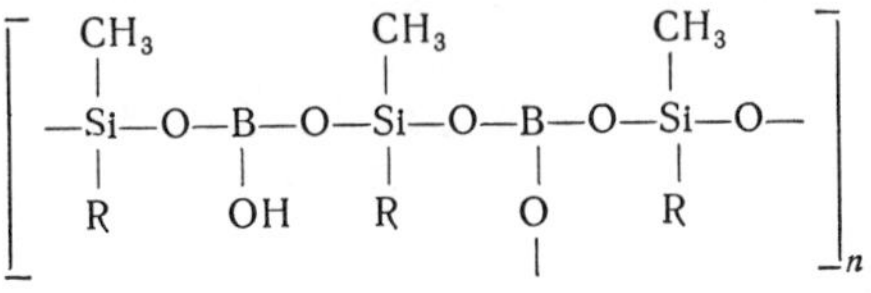

$$\left[-\underset{R}{\overset{CH_3}{Si}}-O-\underset{OH}{B}-O-\underset{R}{\overset{CH_3}{Si}}-O-\underset{\underset{|}{O}}{B}-O-\underset{R}{\overset{CH_3}{Si}}-O- \right]_n \quad (R = C_6H_5 \text{ or } CH_2NHC_6H_5)$$

Alcohol was the only by-product of the reaction. The character of R determined both the yield and the consistency of the polymer obtained (viscous and sticky when R = C_6H_5, yield 87%; solid and transparent when R = $C_6H_5NHCH_2$, yield 71%). The yield was 60-65% with respect to alcohol, but the polymers contained terminal ethoxyl groups. A polymer of analogous structure was obtained by the reaction of boric acid with a mixture of methyl-(phenylaminomethyl) diethoxysilane and dimethyldiethoxysilane in a molar ratio of 2:1:2 [216]. However, in this case the liberation of alcohol was reported, while triethyl borate was not mentioned among the reaction products.

It is reported in a patent [1304] that when dimethyldimethoxysilane is boiled with boric acid, trimethyl borate is liberated and transparent substances are formed, which have the consistency of putty. Together with this it is reported [351] that heating diethyldiethoxysilane with boric acid in ratios of 3:2 - 3:1 at 90-100°C forms triethyl borate, alcohol, cyclic diethylsiloxanes, and a viscous resinous residue, which distills at 250°C (1 mm). The latter constitutes from 25 to 65% of the weight of the high-boiling products, depending on the reagent ratio and the reaction conditions (catalyst and solvent). Boric acid is liberated during the hydrolysis of this residue. The authors consider that the main reaction is as follows:

$$3\,(C_2H_5)_2\,Si\,(OC_2H_5)_2 + B\,(OH)_3 \longrightarrow [(C_2H_5)_2\,SiO]_x + B\,(OC_2H_5)_3 + 3C_2H_5OH \qquad (3\text{-}13)$$

Together with this there is the formation of polymers of the following type, which are quite stable in the reaction medium:

$$\begin{matrix}RO\\RO\end{matrix}\!\!>\!B{-}O\left[\begin{matrix}R\\|\\-Si-O\\|\\R\end{matrix}\right]_x\!\!-B\!<\!\begin{matrix}OR\\OR\end{matrix}\quad\text{and}\quad\left[\begin{matrix} & & R & \\ & & | & \\ -B-O- & & Si-O- & \\ | & & | & \\ OR & & R & \end{matrix}\right]_x$$

While the reaction of boric acid with diorganodialkoxysilanes is interpreted in very different ways, hampering the estimation of the preparative possibilities of the method, an unequivocal interpretation has been put forward for the reaction of boric acid with diorganodiacetoxysilanes [325, 1495]:

$$3nRR'Si(OR'')_2 + 2nB(OH)_3 \longrightarrow [3RR'SiO \cdot B_2O_3]_n + 6nR''OH \qquad (3\text{-}14)$$
$$(R = CH_3;\ R' = CH_2{=}CH;\ R'' = CH_3COO)$$

When boric acid is replaced by phenylboric acid, the reaction proceeds by a stepwise condensation scheme [216, 221]:

$$nRR'Si(OR'')_2 + nC_6H_5B(OH)_2 \longrightarrow R''O\left[\begin{matrix}R & & \\ | & & \\ -Si-O- & B-O- & \\ | & | & \\ R' & C_6H_5 & \end{matrix}\right]_n H + nR''OH \qquad (3\text{-}15)$$
$$(R = CH_3;\ R' = C_6H_5 \text{ or } C_6H_5NHCH_2;\ R'' = C_2H_5)$$

At the same time, the direction of the reaction of phenylboric acid with α,ω-diethoxypolydiorganosiloxanes depends on the character of the substituents at the silicon atom [241]. If in C_2H_5O-$(SiRR'O)_nC_2H_5$, $R = CH_3$, and $R' = C_6H_5$ or $R = R' = C_2H_5$, then, the processes tend toward cyclization with the formation of all the three possible variants $\underline{RR'SiOB(C_6H_5)O}$, $(RR'SiO)_n$, and $(C_6H_5BO)_m$. In the case where $R = R' = CH_3$, the boron appears in a linear structure and the oligomers $C_2H_5OB(C_6H_5)O[Si(CH_3)_2O]_nB(C_6H_5)OC_2H_5$ are formed.

When $R_2Si(OR')_2$ is heated with acids of the type $(HO)_2BR''B$-$(OH)_2$, polymers are formed, whose consistency is determined by the character of the radical at the silicon and the biradical of the acid [533, 1357]. The possibilities for modification of the properties are wide since it is possible to combine in one copolymer units with polymethylene and phenylene groups (Table 25).

TABLE 25. Products of the Reaction of $R_2Si(OR')_2$ (I), with $(HO)_2B(CH_2)_4B(OH_2$ (II), p-$(HO)_2BC_6H_4B(OH)_2$ (III), or a Mixture of (II) and (III) [533, 1357]

R	R′	Molar ratio of I : II : III	Character of copolymer
CH_3	C_2H_5	2:1:0	Very viscous liquid
C_6H_5	CH_3	2:1:0	Elastic, very viscous, solid product; transparent melt at 90°C
CH_3	C_2H_5	2:0:1	Solid polymer, softening point ~40°C
C_6H_5	CH_3	2:0:1	Solid, friable polymer, softening point ~120°C
C_6H_5	CH_3	6:1:2	Elastic solid product, softening point ~100°C
CH_3 and C_6H_5	C_2H_5	6:1:2	Very viscous liquid

The following structures are considered probable for the copolymers:

$$>B{-}R''\!\left\langle\begin{matrix} O{-}SiR_2{-}O \\ O{-}SiR_2{-}O \end{matrix}\right\rangle\!B{-}R''{-}B< \quad \text{or} \quad {-}\underset{R}{\overset{R}{Si}}{-}O{-}\underset{R''}{B}{-}O{-}\underset{R}{\overset{R}{Si}}{-}O{-}\overset{R''}{B}{-}O{-}$$

The reaction of dimethyldiacetoxysilane with triethyl borate [836, 837] and the "metameric" reaction of dimethyldiethoxysilane with triacetoxyboron [149] proceeds with elimination of an ester in accordance with a stepwise condensation scheme. The first stages of these processes may be represented in the following way [5, 8, 149, 722, 836, 837]:

$$R_2Si(OR')_2 + (CH_3COO)_3B \longrightarrow R'OSiR_2OB(OCOCH_3)_2 + CH_3COOR' \qquad (3\text{-}16)$$

$$R_2Si(OCOCH_3)_2 + (R'O)_3B \longrightarrow CH_3COOSiR_2OB(OR')_2 + CH_3COOR' \qquad (3\text{-}17)$$

$$(R{=}CH_3;\ R'{=}C_2H_5)$$

Other diorganodialkoxy- and diorganodiacetoxysilanes, including methyl(phenylaminomethyl)diethoxysilane, react analogously with boric esters [219, 467].

By using trimethyl borate labeled with O^{18} in the reaction with diethyldiacetoxysilane it was possible to establish that the formation of the ester involves rupture of the SiO−C and B−OC bonds [336]. A subsequent study of the model system triacetoxyboron−tetramethoxysilane (labeled with O^{18}) showed that in this case heterofunctional condensation occurs with rupture of the Si−OC and BO−C bonds.

The condensation of heterofunctional intermediates with each other and with excess of the starting monomers leads to the formation of various polyorganoborasiloxanes, whose character is determined by the ratio of the reagents, their nature, and the catalyst (Table 26).

Thus, the reactivity of compounds of the type $RR'Si(OR'')_2$ in condensation with boric esters decreases in the series:

$$R' = CH_3 > C_6H_5 \text{ and } R'' = C_2H_5 > CH_3CO > n\text{-}C_4H_9$$

This is evidently determined by steric factors. In the case of $CH_3(C_6H_5)Si(OR)_2$ the viscosity of the polymer increases more sharply than in the reaction of boric esters with $(CH_3)_2Si(OR)_2$ since in the first case the molecular weight of the polymer is higher at the same degree of condensation (Fig. 1). Andrianov and Volkov [149] assigned a linear structure to the polyorganoborasiloxanes obtained, i. e., they considered that the group I was more probable than II or other units:

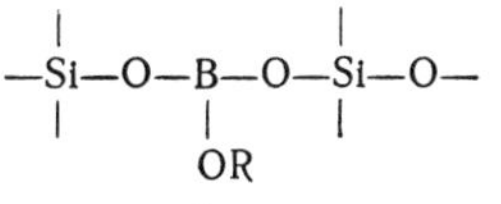

```
  |           |                |           |
—Si—O—B—O—Si—O—          —Si—O—B—O—Si—O—
  |     |     |                |     |     |
        OR                           O
        I                            |
                                   >Si—O—
                                     II
```

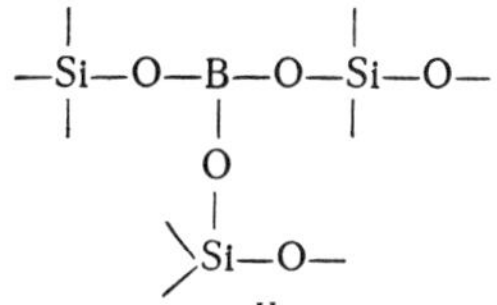

Wick [1267] considers that the reaction of polydiorganosiloxanes containing terminal ethoxyl groups with boroacetic anhy-

TABLE 26. Products of the Reaction of $R_2Si(OR')_2$ with $(R''O)_3B$ [149, 836, 837]

$R_2Si(OR')_2$	$(R''O)_3B$	Molar ratio of re-agents (catalyst)	Degree of reac-tion, % (from ester liberated)	Molecular weight and character of polymer	Viscosity of polymer, centipoise
$(CH_3)_2Si(OC_2H_5)_2$	$(CH_3COO)_3B$	2:1 ($NaOC_2H_5$)	—*	Elastic-plastic mass	—
		3:2	94	Viscous oil	500
		1:1 (HCl)	—	Resinous rigid mass	—
$CH_3(C_6H_5)Si(OC_2H_5)_2$	$(CH_3COO)_3B$	3:2	86	Viscous oil	430
$(CH_3)_2Si(OC_4H_9\text{-}n)_2$	$(CH_3COO)_3B$	3:2	61	Liquid	12
$CH_3C_6H_5Si(OC_4H_9\text{-}n)_2$	$(CH_3COO)_3B$	3:2	56	Highly viscous liquid	650
$(CH_3)_2Si(OCOCH_3)_2$	$(C_2H_5O)_3B$	1:1 ($NaOC_2H_5$)	—	Highly viscous oil	—
		1:3 ($NaOC_2H_5$)	—	Viscous plastic mass	—
$(CH_3)_2Si(OCOCH_3)_2$	$(n\text{-}C_4H_9O)_3B$	3:2	74	Liquid	7
$CH_3(C_6H_5)Si(OCOCH_3)_2$	$(n\text{-}C_4H_9O)_3B$	3:2	76	Highly viscous liquid	700

* 70-80% of the distillable part of the reaction products consisted of ethyl acetate.

dride is one of the ways of preparing borasiloxane elastomers, which give vulcanizates which are self-sealing.

Kreshkov [318-320] reported that the reaction of organotrialkoxysilanes with boric acid at a high temperature forms boric esters and organoalkoxysilanols. However, it was established later [64, 251, 324, 325, 355] that slow heating of organotriethoxysilanes with boric acid in an equimolar ratio forms alcohol and solid, insoluble polyorganoborasiloxanes, which are characterized by the absence of hydroxyl and ethoxyl groups. According to some data [251] they have the composition $(4RSiO_{1.5} \cdot B_2O_3)_n$ (R = CH_3; alcohol yield 97%) while according to other data [64, 324, 325], they have the composition $(RSiO_3B)_n$. Regardless of the character of the radical at the silicon atom (alkyl, aryl, or alkenyl), the polymer yield is 93-97%. However, when R is vinyl, only 88% of the theoretically possible amount of alcohol is obtained. At the same time, the reaction products were found to contain triethyl borate, whose appearance is explained by ethoxylation of boric acid [324].

It was reported [455] that the reactions of mixtures of organotrialkoxysilanes and diorganodialkoxysilanes with boric acid or triacetyl borate (more probably, tetraacetyl diborate) and also the reactions of mixtures of organotriacetoxysilanes and diorganodiacetoxysilanes with boric esters are very similar. In both cases the compounds ROR' (R = alkyl and R' = acetyl) are liberated and polymeric products, which are mainly solids, are formed. When organoacetoxysilanes are used the polymers formed are characterized by quite a high content of acetoxy groups, indicating stepwise polycondensation. These data are similar to the data in patents [1409, 1444, 1616] in which it is reported that the reaction of partly condensed alkylalkoxysilanes with products from the reaction of boric acid with alcohols and phenols (i. e., apparently esters of this acid) form solid thermoplastic resins, which harden in air.

It was believed previously [318-320, 699] that the reaction of boric acid with tetraalkoxysilanes proceeds according to the scheme [3-18a]:

$$3(RO)_4Si + 4(HO)_3B - \begin{cases} \rightarrow 4(RO)_3B + 3SiO_2 + 6H_2O & \text{a)} \\ \rightarrow 12ROH + [3SiO_2 \cdot B_2O_3] & \text{b)} \end{cases} \quad (3\text{-}18)$$

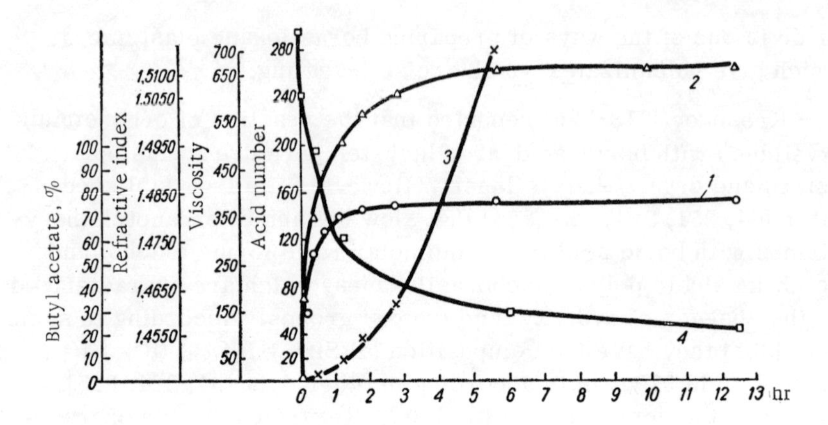

Fig. 1. Reaction of methylphenyldiacetoxysilane with butyl borate [149]: 1) Amount of butyl acetate liberated, %; 2) refractive index; 3) viscosity (centipoise); 4) acid number.

However, it was shown subsequently [251] that the distillation of tetraethoxysilane with boric acid forms predominantly ethanol (90%) and a white solid residue, which contains boron, silicon, and, according to indirect data, ethoxyl groups. The reaction may be represented approximately by the scheme (3-18b).

As regards the reaction of boric acid with polyalkoxysilanes $(RO)_3SiO[(RO)_2SiO]_nSi(OR)_3$, it is believed that they lead to the formation of trialkyl borate and siloxanols

$$(RO_3)_3SiO[(RO)_2SiO]_nSi(OR)_2OH$$

which then undergo anhydrocondensation ("dimerization"). However, since the reactions of tetraalkoxysilanes proceed according to scheme (3-18b), while siloxanediols are quite stable, this interpretation cannot be regarded as well-founded.

Boric anhydride reacts quite readily with tetraalkoxysilanes [251]:

$$3Si(OR)_4 + 2B_2O_3 \longrightarrow 4B(OR)_3 + 3SiO_2 \qquad (3\text{-}19)$$

This reaction is particularly convenient for the synthesis of the lower trialkyl borates ($R = CH_3$ or C_2H_5), which are most difficult to obtain since they form azeotropic mixtures with the corresponding alcohols. It is surprising that three years after the publication of [251], a method of preparing trialkyl borates by scheme (3-19) was patented in the USA [1640].

There is a report [1321] that when a mixture of trimethyl borate and tetramethoxysilane in the stoichiometric ratio is heated to 200-250°C, dimethyl ether is liberated and polyborasiloxanes are formed.*

In the reaction of divinyltetraethoxydisiloxane with boric acid [64, 325] in a ratio of 3:4, the siloxane bonds are not touched; there is the liberation of alcohol and the formation of a white, powdery, insoluble polymer with the composition $\{3(CH_2=CH-SiO)_2O \cdot 2B_2O_3\}_n$.

1.1.3. Reactions of Alkoxysilanes and Acyloxysilanes with Boron Halides

Judging by literature data [650, 718, 816, 949, 1186, 1586, 1708], the reaction of alkoxysilanes with boron halides is well represented by the scheme (3-20)

$$R_nSi(OR')_{4-n} + BX_3 \longrightarrow R_nSiX_{4-n} + B(OR')_3 \qquad (3\text{-}20)$$

(R = hydrogen or alkyl, R′ = alkyl, X = halogen, $n = 0-3$)

In individual cases there is the formation of partial replacement products, i. e., alkoxyhalosilanes or esters of haloboric acids. According to the same information, alkylboron halides react like boron halides with alkoxysilanes.

Nonetheless, Gerrard and his co-workers [698] showed that the direction of the reaction of alkoxysilanes with boron halides depends substantially on the structure of the alkoxyl radical and the character of the other substituents at the silicon atom. While tetrabutoxysilane reacts with boron trichloride by stepwise replacement of butoxyl by chlorine [718], sec-butoxytrichlorosilane and isopropoxytrichlorosilane react with it according to the scheme:

$$Cl_3SiOR + BCl_3 \longrightarrow (Cl_3SiO)_3B + RCl$$

*Under normal conditions no intermolecular reaction is observed in the system tetraethoxysilane–trimethyl borate [273]. Trimethyl borate reacts to a very slight extent with trimethylethoxysilane [249].

Tris(trichlorosilyl)borate is formed in 50% yield. At the same time, trimethyl-sec-butoxysilane is converted into trimethylchlorosilane by the action of boron trichloride. However, it is sufficient for only one electron-acceptor substituent to remain at the silicon atom to produce weakening of the C−OSi bond and to direct the reaction in accordance with the scheme (3-21). Thus, in the case of dimethyl-sec-butoxychlorosilane the corresponding trisilyl borate was obtained.

In the reaction of methoxysilane with boron trifluoride there is the possibility of the formation of an intermediate adduct in a ratio of 1:1 [949, 1186]. At the same time, neither methoxysilane ("methyl silyl ether") nor methyl silyl sulfide react with diborane under the same conditions.

No data have appeared in the literature as yet on the direct reaction of boron halides with acyloxysilanes. However, there is a report [16] of the catalytic effect of boron trifluoride and its etherate on the acylation of thiophene by tetraacetoxysilane. The mechanism of the effect is represented by the scheme

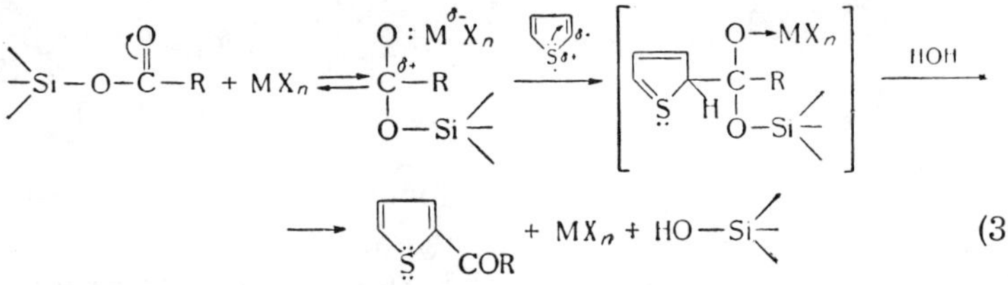

$$\geq Si-O-C(=O)-R + MX_n \rightleftharpoons \overset{\delta-}{O}:MX_n - \overset{\delta+}{C}(R)-O-Si\leq \xrightarrow{C_4H_4S} \left[C_4H_3S(H)-C(R)(O\rightarrow MX_n)-O-Si\leq \right] \xrightarrow{HOH}$$

$$\longrightarrow C_4H_3S-COR + MX_n + HO-Si\leq \qquad (3\text{-}22)$$

1.1.4. Reactions of Silanols with Boric Acid and Its Derivatives

In discussing the different possible routes to the synthesis of borasiloxanes, in 1954 Wiberg and Kruerke [1260] put forward the opinion that the formation of the group $\geq Si-O-B<$ as a result of anhydrocondensation (scheme 3-23)

$$\geq Si-OH + >B-OH \longrightarrow \geq Si-O-B< + H_2O \qquad (3\text{-}23)$$

is improbable as this would require suppression of the tendency to form both siloxanes and boroxanes.

However, this point of view was found to be groundless. A few years later it was shown that the reaction of triorganosilanols with boric acid is a good method of synthesizing tris(triorganosilyl)-borates [412, 413, 515, 1472]:

$$3R_3SiOH + B(OH)_3 \rightleftharpoons (R_3SiO)_3B + 3H_2O \qquad (3\text{-}24)$$

The reaction is carried out by continuous distillation of the water formed with an inert solvent (benzene). As a rule, the yields of tris(trialkylsilyl)borates are 80-90% and only tris(ethyldimethyl-silyl)borate is obtained in a yield of 47%. If we take into account the fact that tris (triphenylsilyl)borate is formed quantitatively [412], it is logical to relate the increase in the yield of silyl borates to the corresponding increase in the stability of the silanols and the suppression of the side reaction of their anhydrocondensation to a hexaorganodisiloxane [51]. In this connection it should be noted that in the reaction of triethylsilanol with boric acid in a molar ratio of 2.5:1 [412], tris(triethylsilyl)borate is obtained in 84% yield, while hexaethyldisiloxane is not formed. With a reagent ratio of 5:1 [515] the yield of the silyl borate is 88%, but hexaethyldisiloxane is obtained in addition to it (60%, calculated on the unreacted triethylsilanol).

Replacement of boric acid by trialkyl borates reduces the yield of tris(triethylsilyl) borate to 38% when R' = C_2H_5 (60% of the unreacted triethylsilanol is converted into hexaethyldisiloxane) [515] and to 24% when R' = n-C_4H_9 [251]. The reaction proceeds according to the scheme:

$$3R_3SiOH + B(OR')_3 \longrightarrow (R_3SiO)_3B + 3R'OH \qquad (3\text{-}25)$$

The process is catalyzed by sodium alcoholates silanolates. At the same time, in reaction of triethylsilanol with tris(trimethyl-silyl) borate, "silanolysis" proceeds to give a 35% yield of tris-(trimethylsilyl)borate [515]:

$$3R'_3SiOH + (R_3SiO)_3B \rightleftharpoons (R'_3SiO_3)_3B + 3R_3SiOH \qquad (3\text{-}26)$$

However, due to the high tendency of trimethylsilanol to undergo anhydrocondensation, hexamethyldisiloxane and water are obtained together with it.

On the example of triethyl borate it was shown that with a decrease in the ratio of triethylsilanol:borate from 3:1 to 1:1 it is not possible to obtain a monosubstituted product, i. e., triethylsilyl diethyl borate, and only the tris derivative is formed [163]. It may be surmised that this is the result of disproportionation of the initially formed partial substitution products during their distillation. With a reagent ratio of 2:1 there is apparently formed a bis(triethylsilyl)alkyl borate, but this was not isolated or characterized. Its formation is quite probable since the addition of a dialkyldiacetoxysilane to the reaction mixture and distillation of ethyl acetate gave compounds of the type $[(C_2H_5)_3SiO]_2BOSiR_2OB \cdot [OSi(C_2H_5)_3]_2$, where R = CH_3 or C_2H_5. The same product is formed when the diethyldiacetoxysilane is replaced by diethylsilanediol.

The reaction of triethylsilanol with boric anhydride is represented by the equation:

$$6R_3SiOH + B_2O_3 \longrightarrow 2\,(R_3SiO)_3B + 3H_2O \tag{3-27}$$

When the reaction is carried out in the presence of anhydrous copper sulfate [509], the yield of tris(triethylsilyl)borate is only 40%. Azeotropic distillation of the water with benzene is much more effective (92%) [51, 163, 412, 413]. Hexaethyldisiloxane was not detected in the reaction products; boric anhydride evidently does not have a condensing effect under the reaction conditions.

The silanolysis of tris(diethylamino) boron is less interesting from the preparative point of view [515]:

$$3R_3SiOH + [(C_2H_5)_2N]_3B \longrightarrow (R_3SiO)_3B + 3\,(C_2H_5)_2NH \tag{3-28}$$

In the case of triethylsilanol, after distillation of the diethylamine, practically pure tris(triethylsilyl) borate (98% yield) remains in the reaction flask. Although the method gives high yields and does not require the use of solvents, it is inconvenient in that it requires the preliminary preparation of tris(diethylamino) boron.

The replacement of boric acid by alkyl(aryl) boric acids does not introduce any complications. Their reaction with triorganosilanols with azeotropic removal of the water readily yields bis(triorganosilyl)alkyl(aryl) boronates [635, 1645]. A change to silanediols yields under the same conditions cyclic compounds of the type

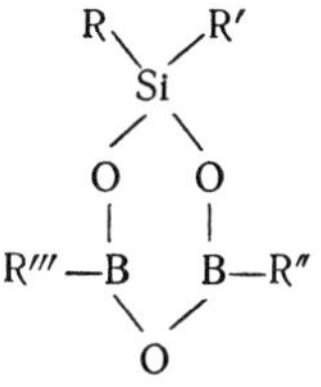

$$\begin{array}{c} R \quad R' \\ \diagdown \diagup \\ Si \\ \diagup \quad \diagdown \\ O \qquad O \\ | \qquad | \\ R'''-B \qquad B-R'' \\ \diagdown \diagup \\ O \end{array}$$

where R, R', R", and R'" are identical or different aryl or cycloalkyl radicals such as phenyls [1642]. At the same time, it is reported that prolonged heating of diphenylsilanediol with phenylboric acid at a temperature above 200°C forms an elastic polymer which is soluble in the usual solvents [1549, 1726].

sym-Tetramethyldisiloxanediol reacts with boric acid at room temperature to form a substance which has the consistency of putty at 50-55°C and is rigid at 25-35°C [1320, 1573].

Rochow [1296, 1402, 1520] considered that during the production of high-molecular polysiloxanes from low-molecular polysiloxanediols, boric acid derivatives act only as water-removing agents, i. e., catalysts of anhydrocondensation.* The following serves as an example: when a mixture of hydrolysis products of dimethyldichlorosilane is heated with ethyl borate (1:1 by volume) at 190°C there is resin formation after only 10 min, while in the absence of the ester, the solidification of the hydrolyzate requires many hours and is accompanied by considerable evolution of voltile products. In the opinion of Andrianov, [4, 5, 8], this example† only indicates the introduction of boron into the siloxane structure:

$$2HOSiR_2[OSiR_2]_n OSiR_2OH + B(OR')_3 \longrightarrow$$
$$\longrightarrow 2R'OH + R'OB\begin{cases} OR_2SiO[R_2SiO]_n R_2SiOH \\ OR_2SiO[R_2SiO]_n R_2SiOH \end{cases} \qquad (3\text{-}29)$$

* Later the use of boron compounds as catalysis for the condensation of siloxanes containing terminal OH or OR groups was the subject of other patents [1382, 1415, 1421, 1439, 1448, 1537, 1665].

† It should be noted that Andrianov mistakenly ascribes this example to Brewer and Haber in J. A. C. S., 70, 3888 (1948), but this paper has no bearing on the given problem.

The validity of this point of view is confirmed both by the course of reactions of trialkylsilanols with alkyl borates and by direct experimental data [167, 171].

When α,ω-polydimethylsiloxanediol (mol. wt. 2500) is heated with tributyl borate (molar ratio 1:1) at 200°C in a stream of CO_2, after only 30 min 43% (of the theoretically possible amount) of the butanol is liberated in accordance with the equation:

$$mHO[Si(CH_3)_2O]_nH + mB(OC_4H_9\text{-n})_3 \longrightarrow$$

$$\longrightarrow HO\left\{[(CH_3)_2SiO]_n\underset{\underset{OC_4H_9\text{-n}}{|}}{B}{-}O{-}\right\}_m C_4H_9\text{-}n + (2m-1)\,n\text{-}C_4H_9OH \qquad (3\text{-}30)$$

Further thermal condensation of the oligomers obtained in the first stage of the process leads apparently to gel formation. However, the high-molecular polymer (mol. wt. 3400) does not have a three-dimensional structure, but is readily soluble in organic solvents, has a glass transition point of −125°C, and has plastic-elastic properties.

The authors of the paper cited explain these properties by the presence of coordination cross links:

$$\begin{array}{c} \diagdown\quad\quad\diagup \\ -Si-\overset{\delta}{O}-Si- \\ \diagup\quad\ \ |\ \ \quad\diagdown \\ \diagdown\quad\quad\downarrow\quad\quad\diagup \\ -Si-O-B-O-Si- \\ \diagup\quad\quad|\quad\quad\diagdown \end{array}$$

According to the same data [171], the condensation according to scheme (3-30) does not occur at room temperature, though tributyl borate forms associates with polydimethylsiloxanediols.

Wick [1267] describes the preparation of boron-containing siloxane elastomers with a molecular weight up to 4000 by condensation of polydimethylsiloxanediols (mol. wt. 5000) with boric acid at 120°C. In his opinion, in processes of this type the boron compounds not only enter the polymer chain, but also act as a catalyst for the condensation; thus, there is no need to use special condensing agents. This hypothesis combines the views of Rochow and Andrianov. However, a recent detailed study of the reaction of α,ω-dihydroxypolydimethylsiloxanes with boric acids over the range of temperatures of 20-130°C [353, 354] shows that it is also accompanied by more extensive processes. The first act is the formation of coordination compounds through the electrophilic boron and the silanol oxygen and these activate polycondensation. In the spectra of the reaction products there is a line at 1590 cm^{-1}, which corresponds to water. Polycondensation does not occur in the presence of compounds with strong electron-donor groups (formamide). Together with condensation, the coordination compounds promote oxidation of methyl groups at silicon atoms and a cross-linked structure develops. In the presence of 4 mol. % H_3BO_3 at 50°C, the absorption of oxygen by the polymer occurs at a rate of 0.2-0.3 ml/hr. The activation character of the oxidation is confirmed by the fact that it does not proceed either in the presence of electron-donor additives or in the absence of silanol groups.

1.1.5. Reactions of Sodium Silanolates with Boron Halides

It is not possible to obtain silyl borates and borasiloxanes by the direct reaction of silanols with boron halides because of the replacement of hydroxyl by halogen which occurs [1262, 1586, 1708]. The use of pyridine makes it possible to obtain only 34% of tris(triethylsilyl)borate [515] though it is well known that trialkyl borates are obtained in good yield from boron trichloride and alcohols in the presence of pyridine. Hexaethyldisiloxane and a complex of pyridine with boron trichloride are formed in considerable amounts during the reaction.

The reaction proceeds readily when the triorganosilanol is replaced by its sodium salt [515, 1102, 1262]:

$$3R_3SiONa + BX_3 \longrightarrow (R_3SiO)_3B + 3NaX \quad (R = CH_3,\ C_2H_5;\ X = Cl,\ Br) \tag{3-31}$$

Alkyl(aryl) boron halides react analogously [515, 1262].

1.1.6. Cleavage of Siloxanes by Boron Halides

The cleavage of alkyl-substituted disiloxanes and polysiloxanes by boron halides in the general form is represented by the scheme [103, 111, 658, 679, 816, 886, 946, 947, 994, 1170, 1263, 1586]:

$$3R_3SiOSiR_3 + 2BX_3 \longrightarrow 6R_3SiX + B_2O_3 \quad (\Gamma = H \text{ or alkyl};\ X = F,\ Cl,\ \text{or Br}) \tag{3-32}$$

In contrast to ethers, whose cleavage by boron halides occurs at 70-80°C, disiloxanes react even at −80°C. At the same time, neither methyl-substituted disiloxanes nor sym-dimethyldisilthiane react with trimethylboron at temperatures from −78 to +170°C [679].

Equation (3-32) characterizes only the final products of the reaction; in actual fact, this reaction has many stages.

In the first stage there is apparently formed on adduct, which is then converted into R_3SiX and an organosiloxyboron dihalide. The stability of the latter, which is determined by the nature of the substitutents at the silicon atom and also at the boron atom*, increases in the following series [415]:

$$1)\ (CH_3)H_2SiOBCl_2 < (CH_3)_2HSiOBCl_2 < (CH_3)_3SiOBCl_2 < (C_2H_5)_3SiOBCl_2$$

$$2)\ (CH_3)_3SiOBF_2 < (CH_3)_3SiOBCl_2 < (CH_3)_3SiOB(CH_3)_2$$

Raising the amount of BX_3 reacting with the disiloxane to the equimolar ratio promotes the formation of triorganosiloxyboron dihalides [658].

In the second stage of the reaction there is decomposition of the R_3SiOBX_2 with the formation of R_3SiX and a boron oxyhalide, which then disproportionates to BX_3 and B_2O_3.†

In the case of excess disiloxane, the second stage of the reaction may be different [103, 1263]:

$$R_3SiOSiR_3 + R_3SiOBX_2 \longrightarrow R_3SiX + (R_3SiO)_2BX \qquad (3\text{-}33)$$

Bis(triethylsiloxy) boron bromide was obtained in this way, for example.

There are reports in the literature [699] that during the cleavage of a siloxane bond by boron trifluoride etherate, in some cases there is simultaneous cleavage of the neighboring Si−C bond, but these are not confirmed by any convincing experimental data.

* $(CH_3)_3SiOBCl_2$ has been synthesized, for example, in 77% yield [679].

† It is interesting to note that hexamethyldisiloxane has found application as an acceptor of BF_3 for removing the excess of it from a reaction mixture in the form of trimethylfluorosilane [1170].

Moreover, it is reported [658] that the reaction of boron trichloride with $(H_3SiSiH_2)_2O$ according to the scheme (3-32) proceeds at $-78°C$ with chlorodisilane formed in 92% yield, while no cleavage of the Si–Si bond is observed.

Scheme (3-32) takes a somewhat different form when BX_3 is replaced by organoboron halides [1263]:

$$(3-n)\,R_3SiOSiR_3 + R'_nBX_{3-n} \longrightarrow (R_3SiO)_{3-n}BR'_n + (3-n)\,R_3SiX \qquad (3\text{-}34)$$

$$(R'{=}CH_3,\ CH_3O;\ X{=}Br;\ n=1 \text{ or } 2)$$

It is interesting to note that sym-dimethyldisilthiane does not react with BX_3 over the range from -132 to $+20°C$ [679], while sym-dimethyldisiloxane gives 91-93% of CH_3SiH_2X at $-78°C$. Heating sym-dimethyldisilthiane with BX_3 at 100°C is accompanied by the liberation of hydrogen and the formation of products, which do not contain boron, but whose charater has not been determined.

We should examine the mechanism of the cleavage of hexaorganodisiloxanes by boron halides. The Si–O bond is known to be considerably more ionic than the C–O bond ($x_0 - x_{Si} = 1.61$, $x_0 - x_C = 1.04$) and consequently it is more sensitive to the action of polar reagents.* Since the siloxane bond is polar (the positive end of the dipole is at the silicon atom, which is less electronegative than the oxygen atom), either end may be the site of primary attack by a reagent. As a result of this, siloxanes readily undergo heterolytic reactions and are cleaved by both nucleophilic and electrophilic reagents. A "four-center" scheme has been proposed for the cleavage of siloxanes by boron halides [679]:

$$\begin{array}{ccc} \diagdown & & \diagup \\ -Si & -O- & Si- \\ \diagup\uparrow & & \downarrow\diagdown \\ Cl & \text{———} & BCl_2 \end{array} \longrightarrow \ \diagdown\!\!\!-SiCl + Cl_2BOSi\!\!\!-\!\!\diagup \qquad (3\text{-}35)$$

However, this scheme is unsuitable for explaining a whole series of facts and, in particular, the identical behavior of siloxanes toward protonic and aprotic acids, the catalytic effect of these acids on the cleavage reaction, and the stability of hexachlorodisiloxane toward electrophilic reagents – BF_3 does not cleave it even at room temperature [816], etc.

* Cleavage reactions of the Si–O–Si group, leading to the formation of organosilicon heterocompounds are also examined in other sections of the book. A general bibliography on the heterolytic cleavage of Si–O bonds is given in [24] and [246].

It is more probable that there is coordination of the siloxane oxygen atom (more accurately, the molecular orbital covering the silicon and oxygen atoms) with the boron halide. The latter is a much more powerful electron acceptor than a silicon atom attached to oxygen and as a result of this it attracts the unshared pair of electrons of oxygen, which are captured by 3d-orbitals of silicon [24, 246]. As a result of this there is weakening (as a result of a decrease in order and an increase in polarity) of the Si−O bond, facilitating its heterolytic breakdown, and also an increase in the electrophilicity of the Si atom.* The latter facilitates breakdown of the complex by a cyclic transfer of electrons.

In the opinion of Voronkov [24], the cleavage of siloxanes by boron halides proceeds as a result of the reaction of an initially formed bimolecular compound, which may be isolated in a number of cases [111, 679, 994, 1263], with a second molecule of BX_3 through an intermediate six-membered active complex (i. e., autocatalytically):

$$\text{cyclic complex } [\geq Si{-}O(\to BX_2{-}X)\cdots Si\leq \cdots X{-}BX_2] \longrightarrow \geq Si{-}O{-}BX_2 + \geq Si{-}X + BX_3 \quad (3\text{-}36)$$

The marked weakening of the electron-donor properties of the siloxane oxygen is one of the indications of an Si−O partial double bond in siloxanes [24, 246, 248]; however, this does not exclude the possibility of the formation of coordination compounds with electrophilic reagents in general since the electron-donor power of the siloxane oxygen is determined by the nature of the substituents at the silicon. Thus, the complex $[(CH_3)_3Si]_2O \cdot BF_3$ is formed at −72°C and is involatile at lower temperatures [679]. The complex $[(CH_3)_3Si]_2O \cdot BBr_3$ is formed at −40°C [1263]. A decrease in the number of methyl groups attached to the silicon atom, which have a +I effect and produce an increase in electron density at the oxygen, reduces the stability of the coordination complexes. In par-

* It was shown recently [248] that organosiloxanes do not show electrophilic or nucleophilic properties toward most reagents. Reaction is observed only with such very powerful electron-donor and electron-acceptor polar reagents as strong protonic and aprotic acids and bases (H_2SO_4, BX_3, AlX_3, KOH, etc.).

ticular, complexes of BF_3 with $(CH_3SiH_2)_2O$ and $[(CH_3)_2SiH]_2O$ could not be isolated and a reaction leading to the formation of the corresponding methylhydrofluorosilane was complete in a few minutes at $-130°C$ [111, 679]. In hexachlorodisiloxane the oxygen atom has hardly any electron-donor properties due to the powerful $-I$ effect of the six chlorine atoms. This explains the inertness of the compound toward BF_3 and other electrophilic agents.

It is quite natural that the cleaving effect of boron halides also extends to linear and cyclic polysiloxanes [946, 947, 1263, 1563]. The reaction of boron halides with cyclodiorganosiloxanes of the type $(R_2SiO)_n$, where $R = CH_3$ or C_2H_5, while $n = 3$ or 4, may be represented, for example, by the following series of conversions [946, 947]:

$$\begin{array}{c} 2(R_2SiO)_3 + 6BX_3 \longrightarrow 2[(R_2SiO)_3 \cdot 3BX_3] \longrightarrow 6R_2SiXOBX_2 \\ \downarrow -4BX_3 \\ B_2O_3 + 3(R_2SiX)_2O \longleftarrow 2(R_2SiXO)_3B \\ \hookrightarrow 3R_2SiX_2 + (R_2SiO)_3 \end{array} \tag{3-37}$$

This gives two final products, namely boric anhydride and a dialkyldihalosilane, and the reaction may be reduced formally to the scheme (3-32).

As in the case of hexaalkyldisiloxanes, the stability of the intermediate compounds formed at the separate stages of the process depends largely on the character of the halogen. The nature of the radicals at the silicon atom and the size of the ring determine mainly the rates of the separate stages and the degree of conversion. When $X = F$, the coordination complexes are quite stable even at $+5°C$, while when $X = Cl$ it was not possible to demonstrate their formation even at low temperatures. On the other hand, dialkylchlorosiloxyboron dichlorides could be isolated by vacuum distillation of the reaction mixture, though there was disproportionation to a large extent with the liberation of BCl_3 and the formation of tris(dialkylchlorosilyl) borates. At the same time, dialkylfluorosiloxyboron difluorides are very short-lived compounds, which readily undergo further conversion to tris(dialkylfluorosilyl)-borates.

The formation of compounds of the type $(R_2SiXO)_3B$ by the decomposition of $R_2SiXOBX_2$ and no others such as $(R_2SiO)_3 \cdot BX_3$, has been established definitely. Tris(dialkylchlorosilyl) borates may be purified by refractionation on a packed column at temperatures of 100-150°C (2-10 mm) with no sign of disproportionation. At the same time, their fluorine analogs disproportionate rapidly according to the scheme (3-37) even at room temperature, to form the corresponding final products in more than 90% yield. The rate of the disproportionation of a sym-dihalotetraalkyldisiloxane when X = F is higher than in the case of analgous chlorine derivatives.

It should be noted that regardless of whether the starting cyclosiloxane is a "trimer" or a "tetramer" the final reaction product is always a "trimer." In the opinion of McCusker and Ostdick [946, 947], this is explained by the fact that disproportionation of a sym-dihalotetraalkyldisiloxane proceeds in stages with the intermediate formation of more complex α,ω-dihalopolydiorganosiloxanes. There then arise chain configurations which favor the formation of six-membered rings (i. e., "trimers").

It is characteristic that tris(diorganochlorosilyl) borates are obtained in good yield only with the exact ratio $(R_2SiO)_n:BCl_3 = n$. Thus, for hexaethylcyclotrisiloxane the yield of tris(diethylchlorosilyl)borate with a molar ratio of reagents of 1:3 is 72%, while with ratios of 1:2 and 1:1 the yields are only 50 and 30%, respectively. The excess siloxane does not undergo any change.. The degree of conversion of cyclosiloxanes is related to the steric accessibility of the siloxane oxygen. It was shown on atomic models that steric hindrance in cyclodimethylsiloxanes is less than in cyclodiethylsiloxanes, while in $(R_2SiO)_3$ it is less than in $(R_2SiO)_4$. The "bulk" of the boron halide also plays a part. Thus, in the case of octaethylcyclotetrasiloxane, in which the oxygen atoms are most sterically hindered, there is only 16% cleavage by BCl_3, while in the case of hexamethyltrisiloxane there is 81% cleavage. At the same time, the "small" molecules of BF_3 quantitatively cleave octaethylcyclotetrasiloxane.

1.1.7. Cleavage of Siloxanes by Boric Acid and Its Anhydride

In the catalytic reaction of hexaalkyldisiloxanes with boric acid the yield of tris(trialkylsilyl) borates does not exceed 10% even though the whole of the boric acid reacts [24, 51]. Due to the capacity of boric acid to cleave the siloxane bond, hexamethyldisiloxane is used as a molecular weight regulator in the production of borasiloxane elastomers by condensation of hydroxyl-containing polydimethylsiloxanes with boric acid [1267]. It is obvious that this compound would be unable to fulfill these functions without participating in the reaction with boric acid.

It has been reported that the reaction of boric acid with hexamethylcyclotrisiloxane [24, 51], particularly in the presence of acid catalysts, proceeds readily with the formation of a polymer with the composition $[9(CH_3)_2SiO \cdot B_2O_3]_n$. Higher polydimethylsiloxanes are also cleaved readily by boric acid.

At temperatures up to 250°C, boric and phenylboric anhydrides hardly react with hexaalkyldisiloxanes [251, 515]. However, at 350°C under pressure it is possible to obtain tris(trimethylsilyl)-borate [251]:

$$3R_3SiOSiR_3 + B_2O_3 \longrightarrow 2(R_3SiO)_3B \quad (3\text{-}38)$$

Attempts to cleave hexachlorodisiloxane with boric anhydride were unsuccessful [251].

1.1.8. "Bouncing Putty"

The possibility of the cleavage of siloxanes by boric acid and boric anhydride with the introduction of boron into the siloxane structure must be borne in mind in examining data on the effect of their addition on the properties of polyorganosiloxanes and, in particular, in the discussion of the problem of so-called "bouncing putty."

The reaction of polyorganosiloxanes with boric anhydride, depending on their viscosity, the amount of anhydride, and the temperature, will give polymers [1399, 1524] of very different consistencies. They all have low freezing points, high boiling points, and good thermal stability. In this way and also by the introduction of boric or pyroboric acids, boric anhydride, or triethyl borate into siloxane elastomers [1366, 1419, 1550, 1561] there was obtained, in particular, a material which combines opposite properties, namely, elasticity like the elasticity of rubber and the fluidity of a highly viscous liquid. Under the slow action of a constant force the material runs, while under the instantaneous action of a force (such as an impact with surface) it shows high elasticity (a rebound resilience of more than 50%). Under a sharp external impact the polymer behaves as a brittle solid. The consistency and the unusual combination of properties of this material is the reason for calling it by the descriptive name "bouncing putty" [109].

Substances with the viscoelastic properties of putty, as has already been mentioned, are formed by the reactions of bifunctional monomeric organosilicon compounds with boric acid [251, 1217, 1304] or its esters [1357] and by other methods [1267]. This confirms the presence in the structure of "bouncing putty" of the units $\rangle$Si—O—B$\langle$. Taking into account the starting materials used for preparing this material, the formation of these units may be the result of the introduction of boron into the siloxane structure both at the point of cleavage of disiloxane bonds and by intermolecular condensation through terminal hydroxyl groups [1366, 1399, 1419, 1524, 1550, 1561]. The first variant is more probable in view of the low tendency of polydimethylsiloxane elastomers to condense through OH groups.

Wright, one of the originators of "bouncing putty" relates the peculiarity of its properties to the two-phase system, which consists in his opinion of 1) a cross-linked polysiloxane network,which is hydrophobic due to the presence of methyl radicals at the silicon atoms, and 2) a viscous hydrophilic product from the reaction of polydimethylsiloxane with boron-containing compounds (i. e., a borasiloxane), which fills the spaces in the network. With the momentary action of a stress the repulsive forces between the hydrophobic and hydrophilic phases are preserved and this is reflected in the high elasticity ("bounce") of the material. During the prolonged action of a stress the internal counter forces are gradually overcome and there occurs deformation and plastic flow of the polymer [1550].

At the present time there is a tendency [1267] to group together as "bouncing putties" boron-containing siloxane polymers, with a molecular weight from several hundreds to ~70,000, containing more than 0.5 mol.% boron, which cannot be vulcanized by the agents usually used for hot or cold vulcanization of polysiloxane rubbers.*

Borasiloxane polymers of low molecular weight show the properties of "bouncing putty" because of interchain hydrogen bonds

* True borasiloxane elastomers will be discussed below.

[1217] and also coordination bonds of siloxane oxygen of one chain with a boron atom of another [1262, 1267].

1.1.9. Other Methods of Preparing Borasiloxanes

A resinous film-forming polyphenylborasiloxane was obtained [1723] by cohydrolysis of dichlorosilane with phenyldifluoroboron with subsequent thermal condensation of the hydrolyzates.

A mixture of dimethyldichlorosilane and the product from the reaction of phenyltrichlorosilane with butyl borate was cohydrolyzed with aqueous methyl ethyl ketone and removal of the solvent by distillation yielded a viscous polymer, which was called a "boron-containing silanol" [1726]. Heating converted it into a resin, which was used in combination with polydimethylsiloxane resin (CH_3:Si = 1.3) to obtain thermally stable coatings.

Careful hydrolysis of tris(dimethylethoxysilyl)borate yielded a polymer with a molecular weight of 660, which had the properties of "bouncing putty" [1267].

Boron hydrides and their complexes with ammonia and amines and also phenylboranes are effective catalysts for the polymerization of cyclodimethylsiloxanes [1323, 1552] and consequently they produce scission of disiloxane bonds. However, diborane does not react with $(H_3Si)_2O$ up to 25°C [1194].

The reaction of boric acid with triorganosilanes [216, 274, 413, 1475] proceeds according to the scheme

$$3R_3SiH + B(OH)_3 \longrightarrow (R_3SiO)_3B + 3H_2 \qquad (3\text{-}39)$$

The reaction is catalyzed by colloidal Group VIII metals, is effected by heating the mixture at 100-130°C, and is readily controlled through the evolution of hydrogen. The yield of tris(triorganosilyl) borates is 90-95% of the theoretical regardless of the structure of the starting trialkylsilane (the reaction rate is related to this factor).*

The reaction of alkyl(aryl) boric acids with triorganosilylamines yields the same products as in the case of the correspond-

* The conditions for preparing tris(triethylsilyl) borate in accordance with scheme (3-39) [412] are evidently not optimal (61% yield).

ing silanes [1635]:

$$2R_3SiNH_2 + R'B(OH)_2 \longrightarrow R'B(OSiR_3)_2 + 2NH_3 \qquad (3\text{-}40)$$
$$(R = n\text{-}C_4H_9,\ C_6H_5;\ R' = n\text{-}C_4H_9,\ C_6H_5,\ C_6H_{11})$$

The reaction between boric acid and a secondary amine, namely, triethyl(phenylamino) silane proceeds analogously [146].

1.2. Physical Properties of Silyl Borates and Borasiloxanes

Table 27 gives the physicochemical constants of organosilicon esters of oxygen-containing acids of boron.

Tris(trialkylsilyl) borates are colorless, oily, high-boiling liquids with a peculiar odor. They distill without decomposition (up to mol. wt. 800) and do not change on heating to 300-350°C [251, 274, 415, 471, 1260]. The intense absorption bands in the IR spectra of silyl borates and borasiloxanes in the region of 1310-1350 cm^{-1} are characteristic of all esters of boric acid and belong to asymmetric valence vibrations of BO_3. In the case of tris[trialkyl(aryl) silyl] borates the frequency ν_{as} BO_3 appears at 1310-1325 cm^{-1} [253] or 1334 ± 5 cm^{-1} [515]. While in the IR spectrum of $[(CH_3)_3SiO]_3B$ the band at 1320 cm^{-1} is intense, the bands at 1308 cm^{-1} for $[(C_2H_5)_3SiO]_3B$ and 1325 cm^{-1} for $[(CH_3)_3SiO]_3B$ in the Raman spectra are weak or inactive [253]. Silyl metaborates are characterized by a maximum at 1380 cm^{-1} [515]. In the spectra of compounds of the type $(RSiO_3B)_3$ there are absorption maxima corresponding to the frequency ν_{as} BO_3 at 1350 cm^{-1} (R = CH_3), 1370 cm^{-1} (R = C_2H_5) [324], and 1400 cm^{-1} R = CH = CH_2 [325]. The symmetrical valence vibrations of BO_3 for tris(triorganosilyl) borates lie in the region of 735-880 cm^{-1}. The spectra of polymers obtained by heterocondensation of triacetoxyboron with diorganodialkoxysilanes [219] are characterized by a band at 1345 cm^{-1}, which corresponds to Si–O–B groups. The IR spectra of borasiloxane rubbers also contain a characteristic absorption band in the region of 7.4-7.5 mμ [1267] or, more accurately, 1340 cm^{-1} [308], which is not present in boric acid, a polydimethylsiloxane rubber, or a mixture of the latter with boric acid. This, in particular, indicates that in the preparation of borasiloxane elastomers by the reaction of polydimethylsiloxanediols with boric acid [1267]

TABLE 27. Organosilicon Esters of Oxygen-Containing Acids of Boron

Formula	b.p., °C (mm)	n_D^{20}	d_4^{20}	Preparation	Yield, %	Literature
$[(CH_3)_3SiO]_3B$	48 (5)	1.3859	0.8299	3-1	26	251
	47 (5)	1.3252[25]	—	3-1	—	1318 1563
	—	1.3848	0.8250	3-1	—	46
	84 (20); 186 (760)	1.3840	—	3-10	75	515
	184.5 (776)	1.3860	0.8225	3-38	20	251
	M.p., 35	—	—	3-31	—	1102
$[(CH_3)_2C_2H_5SiO]_3B$	93—94 (6)	1.4072	0.8598	3-24	47	412, 413
$[CH_3(C_2H_5)_2SiO]_3B$	131—133 (5)	1.4225	0.8751	3-8	73	251
	138—140 (6)	1.4240	0.8773	—	61	412, 413
$[(C_2H_5)_3SiO]_3B$	178—179 (13)	1.4382	—	3-1	24	251
	120—130 (1)	1.4360	—	3-2	10	251
	139—143 (2); 310—312 (780)	1.4379	0.8918	3-8	94	251
	168—170 (5)	1.4378	0.8930	3-11	79	253
	172—175 (6)	1.4378	0.8904	3-24	84	412, 413
	120 (1)	1.4370	—	3-24	89	515
	142—149 (4)	1.4342	—	3-25	23	251
	184 (20)	1.4370	—	3-25	38	515
	152—154 (3)	1.4375[21]	0.8962[18]	3-27	40	509
	165—167 (5)	1.4377	0.8906	3-27	92	412, 413
	183 (17)	1.4370	—	3-31	73	515
	195 (9,5)	1.4380	0.8921	3-39	94	274
	154—155 (4)	1.4373	0.8915	3-39	61	412
	—	1.4378	0.8903	—	52	163
$[CH_3(n\text{-}C_3H_7)_2SiO]_3B$	185 (3)	1.4308	0.8661	3-39	95	274
	157—160 (1)	1.4332	0.8668	3-8	74	251
$[C_2H_5(n\text{-}C_3H_7)_2SiO]_3B$	214 (3)	1.4410	0.8768	3-39	97	274
$[(n\text{-}C_3H_7)_3SiO]_3B$	215—217 (1)	1.4425	0.8662	3-39	91	274
$[CH_3(n\text{-}C_4H_9)_2SiO]_3B$	232 (4)	1.4370	0.8736	3-39	90	274
$[C_2H_5(n\text{-}C_4H_9)_2SiO]_3B$	235 (3)	1.4462	0.8875	3-39	96	274
$[(n\text{-}C_4H_9)_3SiO]_3B$	273—274 (3)	1.4488	0.8753	3-39	89	274
$[CH_3(n\text{-}C_5H_{11})_2SiO]_3B$	273—274 (9)	1.4410	0.8613	3-39	94	274
$[C_2H_5(n\text{-}C_5H_{11})_2SiO]_3B$	297—298 (11)	1.4480	0.8753	3-39	95	274
$[(C_2H_5)_2C_6H_5SiO]_3B$	295—300 (1)	1.5840	—	3-11	56	253
	320—325 (3)	1.5850	1.120	3-24	77	412, 413

TABLE 27. (Cont'd)

Formula	b.p., °C (mm)	n_D^{20}	d_4^{20}	Preparation	Yield, %	Literature
$[CH_3(C_6H_5)_2SiO]_3B$	250—255 (5)	1.5200	1.0125	3-24	85	412, 413
$[C_2H_5(C_6H_5)_2SiO]_3B$	—	—	—	3-8	50	413
	354—358 (8)	1.5976^{50}	1.086	3-39	92	274
$[(C_6H_5)_3SiO]_3B$	M.p. 150°	—	—	3-24	98	412
$[(CH_3)_3SiOSi(CH_3)(C_6H_5NHCH_2)O]_3B$	—	—	—	3-8	44	221
$[Cl(CH_3)_2SiO]_3B$	60 (0.1)	—	—	3-21	—	698
	81.5—82 (2); 98—98.5 (11)	1.4130	1.0966	3-37	81—89	946
$[F(CH_3)_2SiO]_3B$	—	1.3649	1.0797	3-37	—	947
$[Cl(C_2H_5)_2SiO]_3B$	137.5—138 (2)	1.4401	1.0665	3-37	72	946
$[F(C_2H_5)_2SiO]_3B$	—	1.4019	1.0419	3-37	—	947
$(Cl_3SiO)_3B$	98—102 (12); 108—112 (28); 24—28 (35)	—	1.586	3-21	49	698
$[(CH_3)_2C_2H_5OSiO]_3B$	—	—	—	—	—	1318, 1563
$[(C_2H_5)_3SiO]_2BCH_3$	—	—	—	3-31	—	1262
$[(C_2H_5)_3SiO_3SiO]_2BOCH_3$	$p = 2.5$ mm (115°)	—	—	3-34	—	1263
$[(C_2H_5)_3SiO]_2BOC_2H_5$	—	—	—	3-25	—	163
$[(n\text{-}C_4H_9)_3SiO]_2BC_4H_9\text{-}n$	—	—	—	3-40	—	1635
$[(C_6H_5)_3SiO]_2BC_4H_9\text{-}n$	—	—	—	3-24	—	1635
$[C_6H_5(C_6H_{11})_2SiO]_2BC_4H_9\text{-}n$	—	—	—	3-24	—	1635
$[(C_6H_5)_3SiO]_2BC_6H_{11}$	—	—	—	3-40	—	1635
$[(C_2H_5)_3SiO]_2BC_6H_5$	120 (0.2)	1.4730	0.928	3-31	56	515
$[(C_6H_5)_3SiO]_2BC_6H_5$	M.p. 132—133	—	—	3-24; 3-40	71	1635, 1645
$[(m\text{-}ClC_6H_4)_3SiO]_2BC_6H_5$	—	—	—	3-24	65	1635, 1645
$[C_6H_5CH_2(C_6H_5)_2SiO]_2BC_6H_5$	—	—	—	3-24	—	1635
$[(p\text{-}CH_3C_6H_4)_3SiO]_2BC_6H_4Cl\text{-}m$	—	—	—	3-24	70	1635, 1645
$[(C_2H_5)_3SiO]_2BBr$	$p = 0.5$mm(70°)		—	3-33	—	1263
$(C_2H_5)_3SiOB(CH_3)_2$	—	—	—	—	—	1263
$(C_2H_5)_3SiOB(C_6H_5)_2$	135 (0.5)	1.5270	0.974	3-31	97	515
$(CH_3)_3SiOBCl_2$	$p = 6.8$mm(0°)		—	3-32	—	658
	$p = 10.5$mm(0°)		—	3-32	77	679

TABLE 27. (Cont'd)

Formula	b.p., °C (mm)	n_D^{20}	d_4^{20}	Preparation	Yield, %	Literature
$\{[(C_6H_5)_3SiO]_2B\}_2C_6H_4$- p	320 (0,2)	—	—	—	—	533
$\{[(C_2H_5)_3SiO]_2BO\}_2Si(CH_3)_2$	169—175 (8)	1.4320	—	3-25	21	163
$\{[(C_2H_5)_3SiO]_2BO\}_2Si(C_2H_5)_2$	176—179 (82)	1.4390	—	3-16; 3-25	24	163
$[(CH_3)_3SiOBO]_3$	—	1.4101	0.988	3-44	87	515
$[(C_2H_5)_3SiOBO]_3$	—	1.4360	0.955	3-44	84	515
$(C_6H_5)_2SiOB(C_6H_5)OBC_6H_4O$ (cyclic)	M.p. 156—160	—	—	3-24	—	1642

the latter is not a condensing agent or a specific additive, but enters the siloxane chain.

1.3. Chemical Properties of Silyl Borates and Borasiloxanes

Like other borasiloxane compounds, tris(triorganosilyl) borates are hydrolyzed more or less readily in acid, alkaline, and neutral media [251, 274, 515]. According to Wiberg [1260], hydrolysis proceeds with rupture of the B−O bond:

$$\equiv Si-O-B\langle + HOH \longrightarrow \equiv Si-OH + HO-B\langle \qquad (3\text{-}41)$$

The hydrolysis rate is determined by the boron content and the character of the organic radicals at the silicon atoms.

The hydrolysis of polyorganoborasiloxanes is represented by the general scheme

$$x[-SiR_2OB(OR')OSiR_2O-] \xrightarrow[OH^- \text{ or } H^+]{H_2O} \qquad (3\text{-}42)$$

$$\longrightarrow (-SiR_2OSiR_2O-)_x + B(OH_3) + R'OH$$

In principle, the higher their boron content, the lower the resistance of the polymers to hydrolysis.*

A serious drawback which limits the possibility of application of "bouncing putty," is its sensitivity to hydrolysis. On prolonged

* It is reported in [103] that borasiloxanes containing B and Si in a ratio of 1:1 are unsuitable for practical applications because of their hydrolytic instability.

standing in contact with atmospheric moisture, this material loses its elastic properties. Unfilled and unvulcanized borasiloxane elastomers, obtained by polycondensation of dimethylsiloxanediols with boric acid, are as sensitive to hydrolysis as "bouncing putty" [1267]. However, data have been obtained in recent years on the possibility of substantially raising the hydrolytic resistance of borasiloxane polymers by modification of the side and main chains and also the supermolecular structure [216, 219, 269, 413, 467]. Thus, while the introduction of pyridyl groups into the borasiloxane molecule does not improve its hydrolytic stability [522], the latter is increased appreciably when the polymer contains polar phenylaminomethyl radicals at the silicon atoms in an amount which is equal to or greater than the number of boron atoms in the molecule [216, 219, 269, 413, 467]. Polymers obtained by condensation of diorganodialkoxysilanes with an alkylene(arylene) diboric acid have an increased resistance to the action of atmospheric moisture [533]. Good moisture resistance is shown by borasiloxane elastomers obtained by condensation of α,ω-dichloropolydimethylsiloxanes with boric acid [269, 270]. Their properties do not change during storage for a year. This is explained by the ordered structure of the polymers, i. e., the presence of sections with parallel chains, connected by coordination bonds.

When a borasiloxane with mol. wt. 940, which contained 7.9% B and 26.6% Si and was obtained by scheme (3-4) [1217], was treated with high-energy radiation (electrons with an energy of 4 MeV), its melting point was raised from 105 to 145°C with a dose of 10^9 R. The polymer became infusable with higher doses. With an increase in the radiation dose, a polymer with a rebound resilience of 80% gradually became rigid and at doses greather than $5 \cdot 10^8$ R, it became brittle. In comparison with control samples, the irradiated polymer retained its sensitivity to the action of atmospheric moisture up to a dose of $5 \cdot 10^7$ R. Then it gradually showed inertness to water and after a dose of $5 \cdot 10^9$ R, the polymer became insoluble and practically insensitive to the action of water. As a result of heating at 210°C (in a desiccator over P_2O_5), a sample which had received a dose of $5 \cdot 10^8$ R lost about 1% weight in 6 hr, while an unradiated polymer lost 4.5% weight under the same conditions. For a sample which had received a dose of $\sim 5 \cdot 10^9$ R, the loss in weight after 3 hr at 420°C was 4.6%.

Hydrogen halides readily cleave tris(trialkylsilyl) borates and polyorganoborasiloxanes with the formation of boric acid and organohalosilanes [515, 1217].

As has already been pointed out, the reaction between trialkylchlorosilanes and boric esters, which is one of the methods of preparing tris(triorganosilyl) borates, is catalyzed by ferric chloride. The low yields of these compounds are explained by decomposition during distillation. Thus, during the distillation of tris(trialkylsilyl) borates over anhydrous $FeCl_3$ at ~225°C [251] they decompose partly with the formation of a hexaalkyldisiloxane (up to 60% yield):

$$2(R_3SiO)_3B \longrightarrow 3R_3SiOSiR_3 + B_2O_3 \qquad (3\text{-}43)$$

A comparison of schemes (3-43) and (3-38) shows that the reaction between hexaorganodisiloxanes and boric anhydride is an equilibrium reaction and may be directed in one direction or the other, depending on the conditions.

By reaction with boric anhydride, tris(trialkylsilyl) borates are converted into metaborates [515]:

$$(R_3SiO)_3B + B_2O_3 \rightleftarrows (R_3SiOBO)_3 \qquad (3\text{-}44)$$

Triethylsilyl metaborate is trimetric like alkyl metaborates and anhydrides of alkylboric acids [8]. Like them it evidently has a cycloboroxane structure:

```
            OSiR3
              |
              B
            /   \
           O     O
           |     |
   R3SiO—B       B—OSiR3
            \   /
              O
```

In particular, this is confirmed by a characteristic doublet at 720-735 cm^{-1} in the IR spectrum. Silyl metaborates do not undergo changes when heated for 50 hr at 250°C in a sealed tube, but during an attempt at distillation they are reconverted into boric anhydride and silyl orthoborates. The formation of a metaborate

TABLE 28. Phenylation of Methyldichlorosilane [46]

Catalyst*	Optimal reaction temperature, °C	Yield of $CH_3(C_6H_5)SiCl_2$, % on CH_3SiHCl_2 reacted
H_3BO_3 or products of its reaction with methyldichlorosilane	240±5	37–40
BCl_3	245	35
Mixture containing $[(CH_3)_3SiO]_3B$, obtained by reaction of H_3BO_3 with trimethylchlorosilane	260	34
$[(CH_3)_3SiO]_3B$	270	27
$BF_3 \cdot O(C_2H_5)_2$	275	21

* The amount of boric acid added was 0.1% of the weight of the mixture. The other catalysts were present in an amount equivalent to 0.1% of H_3BO_3.

is evidently the reason why the yield of tris(triethylsilyl)borate does not exceed 75% in reaction (3-10) when there is the quantitative formation of triethyl borate [515].

Siloxyboron halides are monomeric and, as has already been pointed out above, their stability is determined by the character of the substitutents at the silicon and boron atoms (p. 190). In particular, H_3SiOBF_2 decomposes with the liberation of SiH_4 and H_3SiF even at room temperature [111].

1.4. Application of Silyl Borates and Borasiloxanes

Tris(trialkylsilyl) borates have found application as plasticizers [1307, 1528], components which give to siloxane rubber mixes improved technological properties and a reduced shrinkage [1369, 1647, 1681], catalysts for the polymerization of cyclodimethylsiloxanes [1307, 1528], and catalysts for the arylation of hydrosilanes [46, 1318, 1563]. The activity of different boron compounds in the arylation of methyldichlorosilane is illustrated by the data in Table 28.

Organoborasiloxane polymers of the "bouncing putty" type*

* There are other curious epithets such as "silly putty" and "crazy clasy" [109].

were long regarded as a peculiar physicochemical curiosity and no practical application was found (Fig. 2). In our day polymers of this type are used relatively widely despite their inadequate hydrolytic stability (see above) and the impossibility of vulcanizing them directly. The elastic properties of polymers of the "bouncing putty" type are improved by the addition of glycerol, zinc hydroxide and oxide or zinc salts of aliphatic acids, but deteriorate when oleic acid is introduced. Polymers filled with titanium dioxide, silica, or lithopone are recommended as adhesives [1366] and high-temperature puttys for vacuum apparatus [1550, 1561]. When used in some applications (for example, as the center of golf balls) these materials have better elastic properties than any known rubber mixture [1561]. The use of "bouncing putty" has been proposed for cleaning movie film, in medicine (as a replacement for paraffin in physiotherapy and as elastic materials for treatment of affected extremities), and in acoustics (as a sound absorbing material) [1550].

A study of the possibility of using boron nitride as a filler for methylsiloxane resins showed that vulcanizates containing about 30% BN stick together during storage [1267, 1294, 1429, 1672].

The hypothesis that this was due to the effect of small amounts of free boric acid present in the boron nitride was one of the reasons for investigating the possibilities of borasiloxanes with a higher molecular weight but a lower boron content than bouncing

Fig. 2. Scientists from General Electric reflect on "Bouncing Putty" (Cartoon from Colliers).

putty. It was established that rubber mixtures based on borasiloxane elastomers begin to be vulcanized by peroxides with a ratio Si:B ≃ 150:1 (optimal mol. wt. 350-500 thousand). The vulcanizates obtained are self-sealing at room temperature. The best combination of properties (vulcanization, tensile strength, adhesion properties, and tendency of the vulcanizates to be self-sealing), is shown by polymers containing one boron atom for every 300-400 silicon atoms in the siloxane chain [132, 1267].

The insensitivity of self-sealing filled peroxide vulcanizates of borasiloxane rubbers to moisture was reported by Wick [1267]. A multilayer composition remains whole after boiling in water for a week. Borasiloxane elastomers were subsequently obtained which retain during storage in contact with atmospheric moisture for a year the capacity to confer on vulzanizates adhesion to metals and other materials, autoadhesion, and good thermal stability [269]. The highest level of thermal stability (390-400°C) is achieved by vulcanization of borasiloxane rubbers with γ-radiation (Co^{60}) [126, 438].* The increased thermal stability is connected with the "blocking effect" of the hetero atoms, which inhibit the development of breakdown processes. On the other hand, to achieve the optimal physicomechanical properties of γ-vulcanizates of borasiloxane elastomers it is necessary to use a higher physical dose of radiation than for a dimethylsiloxane rubber, for example. This is explained by the increase in the polarity of the system when hetero atoms are introduced (and, correspondingly, the capacity for rearrangement) and also the presence of cyclic structures of the following type which are capable of dissipating energy:

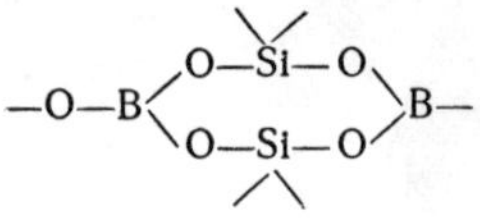

```
           \ /
        O—Si—O
      /        \
—O—B            B—
      \        /
        O—Si—O
           / \
```

Having a high thermal stability and adhesive and autoadhesive properties, borasiloxane elastomers are equivalent in weather resistance, ozone resistance, and dielectric properties to normal siloxane rubbers. Therefore, self-sealing tapes are used primarily in the electrical industry for the insulation of complex

* An even higher level of thermal stability of the vulcanizates is achieved by using heterosiloxane elastomers containing both boron and phosphorus atoms in the chain and phenyl groups at the silicon atoms [438].

circuits, including large size circuits. This gives an electrically insulating, thermally stable protective layer of practically any thickness even at room temperature (moderate heating accelerates the "optimization" of the properties of the coating). There is also the possibility of the use of materials based on borasiloxane rubbers, particularly self-sealing tapes, for the replacement of mica in heat-stable electrical insulation, joints in siloxane rubber cables, the windings of electronic tubes, the production of mandrel-wrapped hose sections, etc. [1267, 1500]. In addition to insulating tapes, it is also possible to make conducting self-sealing tapes, filled with carbon black instead of silicic acid (silica filler) or other mineral fillers. At the present time self-sealing tapes are produced by many firms which make siloxane rubbers. As an example, there is the material of type R27VL of the company "Wacker–Chemie" in West Germany [1267]. A mixture consisting of borasiloxane polymer- 100 parts by weight, synthetic silicic acid of the aerosil type- 40 parts, Hysil-X-303 synthetic silicic acid- 10 parts, iron oxide- 0.5 parts, and 2,4-dichlorobenzoyl peroxide- 1.5 parts is vulcanized for 10 sec at 300°C in the form of a continuously extruded tape. After conversion to a solid sample, the vulcanizate has the following properties:

Density, g/cm^3	1.19
Hardness, Shore	60
Tensile strength, kg/cm^2	70
Relative elongation, %	230
Rebound resilience, %	50
Compression set (22 hr, 175°C), %	12
Volume resistivity, $\Omega \cdot cm$	$1 \cdot 10^{15}$
Electrical strength	30 kV/mm
Tan δ at 10^3 Hz	0.005
Dielectric constant	2.3

To prevent self-sealing during storage, the tapes are wound onto polyethylene or some other material.

According to Wick [1267], the initial "adhesion" of layers of tape occurs as a result of the formation of O–B coordination bonds. There is then partial hydrolysis of the Si–O–B groups by atmospheric moisture with the formation of silanol groups and through them, normal "cold vulcanization," which is catalyzed by components of the rubber mixture (aerosil containing HCl and iron oxide).

However, while these ideas on the mechanism of the process were sound for Wick's polymers, which had a low hydrolytic stability, the concept of cross-linking in thc second stage as a result of promoted oxidation by atmospheric oxygen [353, 354], is more appropriate for hydrolytically stable borasiloxane elastomers produced in the USSR [269].

Various borasiloxane polymers are recommended as hydraulic fluids [1321, 1399, 1534], lubricants, lacquers, antifoaming agents [1321], water-repellant compositions [1456, 1525], electrical insulants, water-proofing agents, sealants [1320, 1399, 1524, 1573], additives to increase the thermal stability of vulcanizates from dimethylsiloxane rubbers [451], and components of heat-resistant adhesives [64, 455]. Thus, adhesive compositions 201 and 202, which are based on borasiloxanes from the condensation of phenylethoxysilanes with boric acid in combination with phenyl-formaldehyde resins, paraformaldehyde, and asbestos, retain quite a high reserve of strength in adhesive joints with steel 30 KhGSA (40-50 kg/m^2) after heating for 2 hr at 400°C, 1 hr at 500°C, and for a minute up to 750-1000°C. At the same time, compositions 201 and 202 have good water resistance [455].

Bis(triorganosilyl) boronates and cyclic silaboroxanes have been proposed for improving the hydrolytic and thermal stability of other organoboron compounds and for modification of polyvinyl chloride, vinylidene chloride, and synthetic resins, particularly for improving their impact strength [1376, 1635, 1642, 1645].

They are also recommended as flamc-proofing coatings, plasticizers, scintillators, active ingredients of bactericides, germicides, fungicides, and pesticides, neutron-absorbing media, and additives for petroleum products [1642, 1645].

The introduction of boric and pyroboric acids, boric esters and anhydride, and also polyborates is valuable for improving the molding and extrusion of resin mixtures and for improving the physicomechanical properties of vulcanizates from siloxane rubbers [1383, 1391, 1591, 1681], hardening, improving, and modifying the properties of organosilicon resins and plastics and combinations of them with organic resins [1297, 1376, 1409, 1439, 1507, 1521, 1616, 1621], stabilization of organosilicon hydraulic fluids [1623] and combinations of organosilicon resins with polymethyl methacrylate [1530], and inhibition of the polymerization of cyclosiloxanes of the type $[(CH_2)_5SiO]_n$ [1538]. Partial condensation of high-molecular polydiorganosiloxanediols and products of the reaction of boric acid with polyhydric alcohols and phenols or the direct introduction of these products into siloxane elastomers with subsequent filling and vulcanization of the mixtures makes it possible to obtain, depending on the recipe, either self-sealing tapes of the type described above [1293, 1371, 1409, 1441, 1616, 1673], or a tape which is not self-sealing, but which adheres well to metallic surfaces [1683]. A result analogous to the latter is obtained by introducing N-triethylborazole into polydimethylsiloxanediols [1382]. Boron hydrides and alkylboranes are recommended as substances for accelerating the vulcanization of rubber mixes based on polydimethylsiloxane rubber and simultaneously improving the thermal stability of the vulcanizates [97, 1325, 1397, 1553] and also as additives which make organosilicon dielectric fluids capable of absorbing oxygen [1348]. It is also reported that a complex of boron trifluoride with piperidine (1:1) produces rapid hardening of methylphenylsiloxane resins [614, 1583].

1.5. Analysis of Silyl Borates and Borasiloxanes

It is generally recommended that boron in tris(triorganosilyl) borates and borasiloxanes is determined by hydrolysis of a sample of the substance with aqueous alcohol [251, 1271] or aqueous dioxane [251, 253] with subsequent titration with 0.1-0.2 N NaOH with phenolphthalein [253, 471, 1217] or Thymol Blue [251] as the indicator in the presence of mannitol. If the substance contains hydrolyzable chlorine or the hydrolysis is carried out with aqueous or alcoholic alkali [324, 326], before the addition of mannitol the solution is neutralized with 0.1 N NaOH or HCl to Methyl Red. The relative error of the determination is 1-3% [326].

Good results are obtained with individual silyl borates and soluble borasiloxanes by nonaqueous potentiometric or visual titration in media with a high dielectric constant [323, 1504a]. For the potentiometric determination of boron (glass indicator electrode and calomel reference electrode) the recommended solvents are methanol, methyl ethyl ketone, nitromethane, and acetonitrile. The nature of the titrant is not of great importance and it is possible to use alkali metal and quaternary ammonium methylates or nitron (diphenylendaminodihydrotriazole). Small amounts of mannitol or glycerol increase the potential jump by a factor of more than two

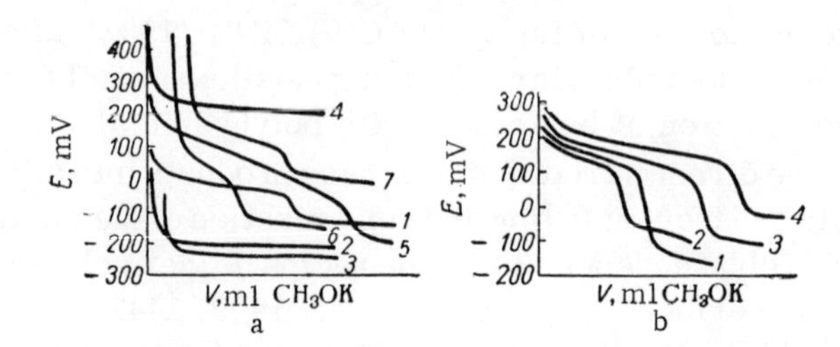

Fig. 3. Curves of the potentiometric titration of $[(CH_3)_3SiO]_3$ B with potassium methylate in various media [164]. a) Without additives: 1) CH_3OH; 2) C_2H_5OH; 3) n-C_3H_7OH; 4) $C_6H_5CH_2OH$; 5) $CH_3COC_2H_5$; 6) CH_3CN; 7) CH_3NO_2; b) in the presence of glycerol: 1) $(CH_3)_2CO$; 2) CH_3OH; 3) CH_3CN; 4) CH_3NO_2.

(Fig. 3).* For visual titration the dyes recommended are 1,4- and 1,5-dihydroxyanthraquinone. The relative error of the determination is ±0.8% for silyl borates and 3 – 5% for borasiloxanes.

The silicon content of silyl borates and borasiloxanes if established from the sum of the boron and silicon oxides formed by mineralization of a sample of the substance with a mixture of oleum and nitric acid (wet combustion) with subsequent calcination of the oxides at 900°C [251].† It is also possible to determine the silicon after decomposition of the substance with alkali in the form of silicomolybdenum blue by photocolorimetry at pH = 4.1 – 4.4 [326].

Emmission spectral analysis with a graphite electrode with a porous base, using an internal standard, is recommended for borasiloxanes containing less than 1% Si [326a]. The reference element is cobalt in the form of $CoCl_2$ and for this purpose the substance analyzed, such as $[3CH_3 \cdot CH_2 = CHSiO \cdot B_2O_3]$, is prepared as a solution in a mixture of o-xylene with 96% ethanol. The spectra are excited in a high-frequency spark (DG-2 generator). The relative error of the method is 0.4%.

Microcombustion in the presence of copper oxide on asbestos is recommended for determination of carbon and hydrogen in borasiloxanes [324].

* The separate determination of silyl borates and chlorosilanes in mixtures of them is possible in the presence of glycerol.

† It has been reported that boron may be determined in the residue from wet combustion by titration with 0.1 N alkali in the presence of glycerol [455].

2. COMPOUNDS CONTAINING THE $S-(C)_n-B$ GROUP

2.1. Preparation Methods

2.1.1. Reaction of Organosilicon Grignard Reagents with Boron Halides

Organosilicon compounds containing carbon bridges between silicon and boron atoms have a shorter history than borasiloxanes. The first communications on the preparation of compounds of this type appeared only in 1958.

By the reaction of organosilicon Grignard reagents with boron trifluoride etherate it was possible to synthesize compounds of the type $[R(CH_3)_2SiCH_2]_3B$, where R = CH_3, C_2H_5, n-C_3H_7, $(CH_3)_3SiCH_2$, and $(CH_3)_3SiO$ [1138, 1613].

When $[(CH_3)_3SiOSi(CH_3)_2CH_2]_3B$ was heated with octamethylcyclotetrasiloxane in the presence of potassium trimethylsilanolate there was polymerization at the siloxane bonds, while the C–B bond was not touched. Polymerization for 16 hr at 170-190°C lead to the formation of $\{(CH_3)_3Si[OSi(CH_3)_2]_nOSi(CH_3)_2CH_2\}_3B$, where n = 16 [1613].

The reaction of the combined Grignard reagent obtained by the action of magnesium on a mixture of $[ClCH_2Si(CH_3)_2]_2O$ and butyl bromide with boron trifluoride etherate or ethyl borate yielded $[(n\text{-}C_4H_9)_2BCH_2Si(CH_3)_2]_2O$ and a polymer, to which was assigned the formula $[n\text{-}C_4H_9BCH_2Si(CH_3)_2OSi(CH_3)_2CH_2BC_4H_9\text{-}n]_x$ without discussion of the structure [1633].

2.1.2. Addition Reactions

The well-known addition of hydrosilanes to unsaturated compounds in the presence of a complex of chloroplatinic acid with isopropyl alcohol ("Speier's catalyst") has also found wide application for the preparation of silylalkylboranes. Thus, in particular, pentamethyldisiloxane and trivinylborane at 100°C yielded $[(CH_3)_3SiOSi(CH_3)_2CH_2CH_2]_3B$ [1613]. The reaction of hydrochlorosilanes with alkenyl boronates [376] may be represented by the general scheme

$$CH_2{=}CH(CH_2)_nB(OR')_2 + RSiHCl_2 \longrightarrow RCl_2Si(CH_2)_{2+n}B(OR')_2 \quad (3\text{-}45)$$

$$(R = CH_3 \text{ or } Cl;\ R' = n\text{-}C_4H_9 \text{ or iso-}C_4H_9;\ n = 0 \text{ or } 1)$$

Trichlorosilane and methyldichlorosilane add to alkenyl boronates under much milder conditions than to olefins and this is explained by the activating effect of the boron on the double bond. However, the addition does not occur in the absence of a catalyst. Only an insignificant amount of the product from the addition of triethylsilane to the butyl ester of allylboric acid is obtained in the presence of chloroplatinic acid, but under more drastic conditions.

The reaction according to scheme (3-45) is complicated by side processes – the products obtained contain more chlorine than the esters which should be formed. It is generally impossible to isolate the butyl ester of (3-trichlorosilylpropyl)boric acid in a pure form. A particularly complex mixture was obtained in an attempt at initiation of addition with γ-radiation [376]. An investigation of the model system trichlorosilane – butyl ester of propylboric acid showed that under the reaction conditions there is the possibility of alkoxyl – halogen exchange and also the formation of high-boiling substances. On these grounds it was suggested that the reaction of trichlorosilane with $CH_2 = CH(CH_2)_nB(OC_4H_9\text{-n})_2$ forms not only esters of the type $Cl_3Si(CH_2)_{2+n}B(OC_4H_9\text{-n})_2$, but also the esters $Cl_3Si(CH_2)_{2+n}B(OC_4H_9\text{-n})Cl$, and possibly the telomerization products $Cl_3Si\{CH_2CH[(CH_2)_nB(OC_4H_9\text{-n})_2]\}_mH$.

The addition of boron derivatives to unsaturated organosilicon compounds has been studied much more. The reaction of olefins with the complex $NaBH_4 - AlCl_3$ in a ratio of 3:1 in 1,2-dimethoxyethane (Brown and Rao's reagent) forms trialkylboranes of the type $(RCH_2CH_2)_3B$ [592, 592a]. At the same time, according to data in [1248], no boronation products are obtained by the reaction of sodium borohydride with vinyltrichlorosilane; reduction occurs with cleavage of the Si–C bond and the liberation of SiH_4.

In studying the possibility of applying the Brown–Rao reaction to alkenylsilanes, Seyferth [1138, 1140, 1161] discovered that with a molar ratio of $(CH_3)_3SiCH = CH_2 : NaBH_4$ equal to 2:1 there is formed a substance with the composition $[(CH_3)_3SiC_2H_4]_3B$ in a yield of 65-75%. Oxidation of this substance with 30% hydrogen peroxide in alcoholic alkali and a study of the nature of the alcohols obtained shows that regardless of the reaction conditions, the addition product has a structure close to $[(CH_3)_3SiCH_2CH_2]_2BCH(CH_3)Si(CH_3)_3$ (ratio of α- and β-trimethylsilylethyl groups ~35:65). An identical product is formed in 85-93% yield by replacing the complex $NaBH_4 - AlCl_3$ by the more accessible reagent $(CH_3)_3N \cdot BH_3$ [1140, 1159]. In addition to its higher efficiency, the latter method is simpler to use.

In the reaction of the Brown–Rao reagent with trimethylvinylsilane (ratio B:Si = 1.6:1), the subsequent hydrolysis of the reaction mixture with dilute hydrochloric acid did not give the expected trimethylsilylethylboric acid. Boiling the reaction products with ethanol gave three products, namely, bis(trimethylsilylethyl)ethylboronate, $\{[(CH_3)_3SiCH_2CH_2]_2B\}_2O$, and also tris(trimethylsilylethyl)borane.

Summation of the oxidation products of all the components of the reaction mixture gives a mean ratio of α- and β-addition of 53:47. As regards individual compounds, the decomposition of tris(trimethylsilylethyl)borane forms α- and β-trimethylsilylethyl alcohols in a ratio of 43:57 in the given case and in a ratio of 70:30 for bis(trimethylsilylethyl)ethylboronate.

The reaction of trimethylvinylsilane with excess of yet another modification of the hydrobororating agent, namely, the product from the reaction of sodium borohydride with boron trifluoride etherate in tetrahydrofuran yields, after treatment of the reaction mixture with methanol, a mixture of isomers of [(trimethylsilyl)ethyl] dimethylborinate, where $\alpha:\beta$ = 59:41 [1161]. On the other hand, oxidation of the products from an analogous reaction in diglyme led to an equimolar ratio of α- and β-isomers mixed with hydroxyalkyl derivatives of silicon [913a]. Thus, depending on the hydroboronation conditions, the ratio of α- and β-structures varies from 2:1 to 1:2. This indicates practically the same probability of the addition of boron to the α- or β-carbon atom of a vinyl group attached to silicon. The presence of trimethylsilyl groups attached to α- and β-carbon atoms in the same compound is evidently determined by steric factors. In this connection, very interesting results were obtained by replacing one of the methyl radicals in $(CH_3)_3SiCH = CH_2$ by the more bulky groups $(CH_3)_3Si$ and $(CH_3)_3C$ [913a]. When the products from the reaction of sodium borohydride with BF_3 etherate were used as the hydroboronating agent in diglyme, oxidation of the reaction mixture in the case of pentamethylbisilylvinylsilane (ratio of compound to $NaBH_4$ 2:1) led to a mixture of organosilicon alcohols with an $\alpha:\beta$ ratio of 25:75. This may be related to both the large steric effect and also the stronger electron-donor character of the group $(CH_3)_3SiSi(CH_3)_2$ in comparison with $(CH_3)_3Si$. However, the ratio of isomeric alcohols obtained from the hydroboronation products of tert-butyldimethylvinylsilane, namely, $\alpha:\beta = 56:44$, was unexpected since the electronic contribution of the group $(CH_3)_3CSi(CH_3)_2$ is practically identical to that of $(CH_3)_3Si$, while the steric effect is substantially greater. This result may obviously be connected with the difficulty of adding a second molecule of an alkenylsilane to a primary adduct which already contains boron. In any case, hydrolysis of the hydroboronation products of tert-butyldimethylvinylsilane with subsequent refluxing of the mixture in ethanol gave two incompletely alkylated boron compounds, namely $(CH_3)_3CSi(CH_3)_2C_2H_4B(OC_2H_5)_2$ and $\{[(CH_3)_3CSi(CH_3)_2C_2H_4]_2B\}_2O$. Their separate oxidation led to a mixture of α- and β-hydroxyethyl derivatives of silicon in a ratio of 89:11 in the first case and only 45:55 in the second. The introduction of a methylene bridge between the silicon atoms and the vinyl group directs the hydroboronation exclusively toward the addition of boron to the terminal C atom. Tris(γ-trimethylsilylpropyl)borane, for example, was obtained in this way [1161].

The processes described above reduce in practice to the reaction of alkenylsilanes with diborane and in the final form they correspond to the scheme

$$6R(CH_3)_2Si(CH_2)_nCH{=}CH_2 + B_2H_6 \longrightarrow 2[R(CH_3)_2Si(CH_2)_{n+2}]_3B \quad (3\text{-}46)$$

$(R = CH_3$ [543, 1161], $(CH_3)_3SiO$ [543]; $(CH_3)_3Si, (CH_3)_3C$[913a]; $n = 0-1)$

The inactivity of $CH_2{=}CHSiCl_3$ in contrast to the activity of $CH_2{=}CHSi(CH_3)_3$ in the reaction with diborane is evidently connected with the opposite I-effects of the substituents at the silicon atom in these compounds which produce strengthening or weakening of the $Si-CH{=}CH_2$ bond. However, when methylvinyldichlorosilane or vinyltrichlorosilane is mixed with pentaborane or decaborane in an inert atmosphere in carbon disulfide in the presence of $AlCl_3$ or HCl addition occurs and the corresponding β- adducts are obtained in up to 75% yield [1697].

When triisobutylboron is heated with alkenylsilanes at 150-200°C transalkylation occurs [377, 378]:

$$3R_3Si(CH_2)_nCH{=}CH_2 + (iso\text{-}C_4H_9)_3B \longrightarrow [R_3Si(CH_2)_{n+2}]_3B + 3\ iso\text{-}C_4H_8 \quad (3\text{-}47)$$

The groups $\geqslant SiCH_2CH_2B\leqslant$ and $\geqslant SiCH_2CH_2CH_2B\leqslant$ being present in the products obtained is assumed only on the basis of the fact that primary trialkylboranes are formed in the Brown–Rao reaction. This hypothesis is not supported by definite proof.

The following chain of conversions may be carried out to prepare tris(methyldiethylsilylethyl)borane in accordance with scheme (3-47) [377]:

$$R_3B \xrightarrow[-RH]{n\text{-}C_4H_9SH} R_2BSC_4H_9\text{-}n \xrightarrow{R'OH} \begin{cases} \xrightarrow{-n\text{-}C_4H_9SH} R_2BOR' \\ \longrightarrow RB\begin{matrix} OR' \\ SC_4H_9\text{-}n \end{matrix} + RH \end{cases} \tag{3-48}$$

$$R_2BSC_4H_9\text{-}n \xrightarrow[-n\text{-}C_4H_9SH]{+H_2O} R_2BOH \qquad RB(OR')(SC_4H_9\text{-}n) \xrightarrow[-n\text{-}C_4H_9SH]{R'OH} RB(OR')_2$$

$$[R = CH_3(C_2H_5)_2SiCH_2CH_2;\ R' = CH_3 \text{ or } n\text{-}C_4H_9]$$

In contrast to dialkylboric acids, bis(methyldiethylsilylethyl)-boric acid (a thick liquid with the odor of camphor) may be vacuum distilled. It was first suggested that the formation of esters of methyldiethylsilylethylboric acid occurs as a result of the action of excess alcohol on the primary reaction products, namely, bis-(methyldiethylsilylethyl) borinates. However, the latter are not converted into methyldiethylsilylethyl boronates by boiling with alcohol. It is therefore believed that in the first stage there are formed complexes of bis(methyldiethylsilylethyl) butylthioborinate with alcohols, which decompose not only with elimination of the mercaptan and the formation of the corresponding ester, but also with cleavage of the B–O bond and the formation of mixed esters of the type RB(OR')(SR"). Under the action of excess alcohol the latter are converted into methyldiethylsilyl boronates.

Methylvinyldiethoxysilane and methylvinyldichlorosilane will also undergo reaction (3-47). Howevever, a more complex mixture of products is formed in this case than in the case of alkenylsilanes which contain no functional groups at the silicon atom.

Tetrabutylmercaptodiborane reacts with methyldiethylvinylsilane at ~50°C in the presence of pyridine with the formation of the dibutyl ester of 2-methylethylsilylethylthioboric acid [379].

It is interesting to note that the reaction of triethylsilane with cyclohexene in the presence of BCl_3 (3:3:1) yields tricyclohexylboron and triethylchlorosilane [1437]. However, this result is understandable in the light of data [503, 1498] on the reduction of boron halides and their amino derivatives by hydrosilanes:

$$\rangle Si{-}H + X{-}B\langle \longrightarrow \rangle Si{-}X + H{-}B\langle \qquad (3\text{-}49)$$

In the second state there is addition of borane to cyclohexene.

2.1.3. Synthesis of Carborane Derivatives Containing Silicon

A series of papers was published in the USA in 1963 describing the results of seven years of investigations by scientists of two chemical corporations of the preparation of a new class of organoboron compound, namely, derivatives of decaborane. Almost simultaneously there began to appear in the USSR analogous work of Zakharkin and his co-workers.*

$$B_{10}H_{14} + RC{\equiv}CR' \xrightarrow[B]{2A} \underset{B_{10}H_{10}}{RC{-}CR'} + 2H_2 \qquad (3\text{-}50)$$

R and R' = H or organic radical; A = CH_3CN or $(C_2H_5)_2S$;

B = C_6H_6, $CH_3C_6H_5$, or $(iso\text{-}C_3H_7)_2O$

The simplest representative of this type of compound is $B_{10}H_{10}$-C_2H_2, that was given the trivial name carborane, which then became the group name for compounds of this type. In the formation of carborane, the decaborane "purse" is given a "zipper" of two acetylene carbon atoms to form a closed structure, which is reminiscent of a bird cage (geometrically it is an icosahedron with a diameter of 8 Å). Hence we have the more complete name for carborane dicarbaclovododecaborane in which the Greek prefix clovo-(cage) reflects the steric structure of the compound.

In principle there is the possibility of the existence of three geometric isomers of the icosahedral carborane system. Deri-

---

*So as not to overload the bibliography, we merely quote the announcement [636] and the review published in June 1965 [59].

†Here we will adhere to the term carboranes which has become more common than the name proposed by Zakharkin and his co-workers, "barenes," that is a combination of the words "boron" and "arenes" and reflects the aromaticity of these compounds.

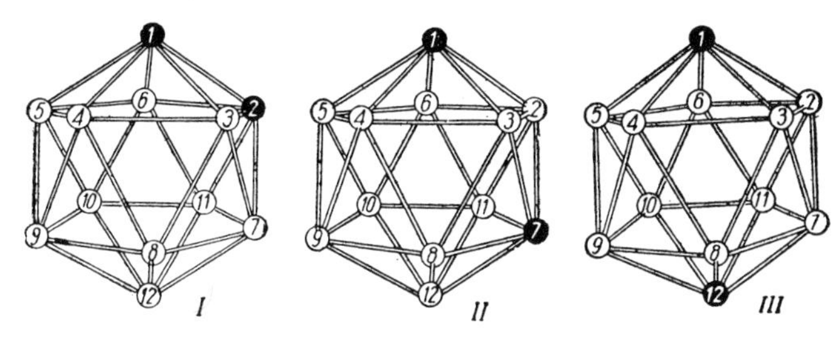

Fig. 4. Isomers of carborane [1301]: I) Ortho-carborane; 1,2-dicarbaclovododecabrorane (12); carborane; II) meta-carborane; 1,7-dicarbaclovododecaborane; neocarborane; III) para-carborane; 1,12-dicarbaclovododecaborane.

vatives of two of them (I and II) are already known (Fig. 4).*

The outstanding properties of carboranes, namely, their high chemical and thermal stability, prompted attempts to incorporate carborane nuclei into polymers. In particular, this stimulated the preparation of derivatives of carboranes containing silicon.

The synthetic possibilities are manifold here. First of all we should point out that silylcarboranes are obtained by a "direct" route, i. e., by the reaction of complexes of decaborane with silyl-

---

*As long as only derivatives of 1,2-dicarbaclovododecaborane were known, only the symbol was used in the chemical literature to denote the carborane (12) nucleus

$-C\underset{B_{10}H_{10}}{\overset{O}{\text{——}}}C-$. With the appearance of derivatives of neocarborane there arose a

need for symbols which reflect the structural differences of the isomers. In an article on the nomenclature of compounds of the type $C_2B_{10}H_{10}R'R''$ [462a] the formulas

$HC\overset{..}{—}CH$ / $B_{10}H_{10}$, (C-H / B_{10} H_{10} / H-C), and (C-H / B_{10}—H_{10} / H-C) were proposed for the ortho-, meta-, and

para- isomers, respectively. The formulas $HC\underset{B_{10}H_{10}}{\overset{O}{\text{——}}}CH$, $HCB_{10}H_{10}CH$, and

(H / C / B_{10} H_{10} / C / H) are given in the review [102]. Unfortunately we were unable to use these

formulas either in the monograph or in its translation.

acetylenes by a scheme of type (3-50) [299, 1128]:

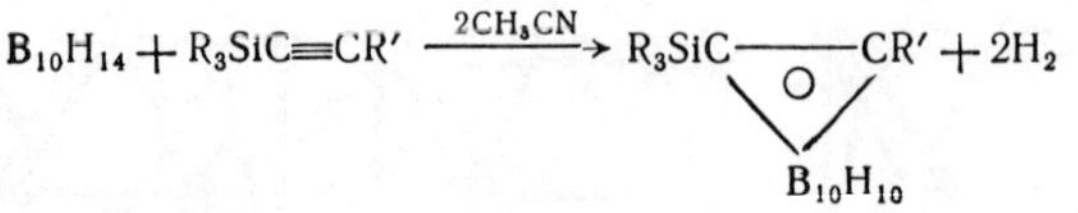

$$B_{10}H_{14} + R_3SiC{\equiv}CR' \xrightarrow{2CH_3CN} R_3SiC\text{—}CR' + 2H_2 \quad (\text{ring: } B_{10}H_{10}) \qquad (3\text{-}51)$$

$(R = CH_3,\ C_6H_5,\ \text{or}\ OC_2H_5;\ R' = H,\ CH_3,\ \text{or}\ C_6H_5)$

However, the yields of compounds are low and this method is only of fundamental and not preparative value. Much better results are obtained by organometallic synthesis. The action of butyllithium on carboranes results in metallation at the carbon and both mono- and dilithium derivatives may be obtained [299, 842, 1001, 1003, 1004, 1693]. Their reactions with chlorosilanes are exothermic and lead to compounds with carboranyl groups at silicon atoms or carboranylene bridges between silicon atoms:

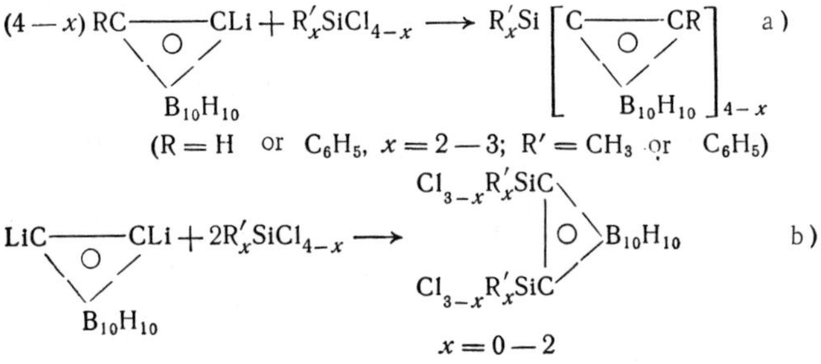

$$(4-x)\,RC\text{—}CLi\ (B_{10}H_{10}) + R'_xSiCl_{4-x} \longrightarrow R'_xSi\left[C\text{—}CR\ (B_{10}H_{10})\right]_{4-x} \qquad a)$$

$(R = H\ \text{or}\ C_6H_5,\ x = 2-3;\ R' = CH_3\ \text{or}\ C_6H_5)$

$$LiC\text{—}CLi\ (B_{10}H_{10}) + 2R'_xSiCl_{4-x} \longrightarrow (Cl_{3-x}R'_xSiC)_2B_{10}H_{10} \qquad b)$$

$x = 0-2$

The synthesis in accordance with scheme (3-52b) is of great interest from the point of view of preparing reactive compounds. However, it is not the presence of potentially active chlorine atoms at the silicon, but the isomerism of the compound which is of decisive importance for reactions involving chlorosilylcarboranes. Thus, the hydrolysis of bis(chlorodimethylsilyl)-ortho-carborane was expected to form a linear carboranylenesiloxane or a macrocyclic system or, in the extreme case, a diol. Contrary to all expectations, exocyclic ring closure occurred quantitatively [636, 1001]:

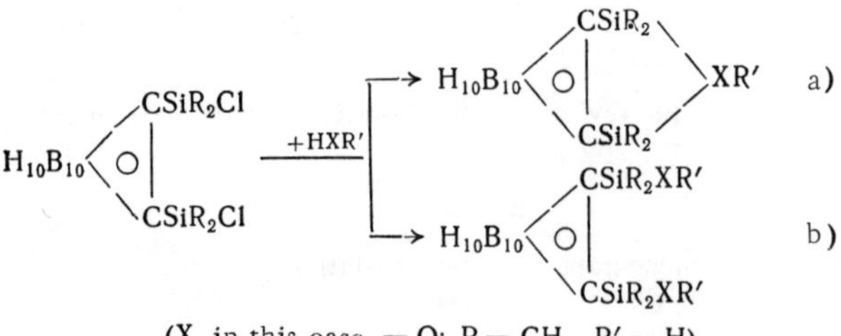

$$H_{10}B_{10}(CSiR_2Cl)_2 \xrightarrow{+HXR'} \begin{cases} H_{10}B_{10}(CSiR_2)_2XR' & a) \\ H_{10}B_{10}(CSiR_2XR')_2 & b) \end{cases} \qquad (3\text{-}53)$$

$(X\ \text{in this case} = O;\ R = CH_3,\ R' = H)$

The same compound is obtained by the reaction of dilithio-ortho-carborane with sym-tetramethyldichlorodisiloxane [1001]. Five-membered exocyclic compounds are also formed by the reaction of bis(chlorodimethylsilyl)-ortho-carborane with ammonia (X = NH; R' = H), methylamine (X = NH; R' = CH_3), aniline (X = NH, R' = C_6H_5) and even hydrazine (X = NH; R' = NH_2) [1001, 1004]. It is characteristic that a six-membered exocycle is not formed in the reaction with hydrazine or when the reaction is carried out with sym-dimethylhydrazine – in this case there is obtained a practically unseparable mixture of acyclic products (if we do not count the carborane nucleus).

At the same time, in the hydrolysis of bis(chlorodimethylsilyl)-meta-carborane* cyclization does not occur at all and a diol is formed quantitatively (scheme 3-53b; X = 0; R' = H) [1002]. A diol is also obtained by hydrolysis of bis(chlorodiphenylsilyl)-meta-carborane, while in the ortho isomer the chlorine at the silicon is hydrolytically stable. Normal substitution products are formed by the action of ammonia or methanol on bis(chlorodiorganosilyl)-meta-carboranes. In the case of bis(dichloromethylsilyl)-meta-carborane the hydrolysis leads to a tetraol [1002]. At the same time, in the reaction of the ortho-isomer with ammonia or methylamine with a reagent ratio of 1:3 there is simultaneously the replacement of one chlorine atom at each silicon atom by an amino(methylamino) group and the formation of an exocycle at the expense of the other two chlorine atoms [1001].

While it is not possible to form a six-membered exocycle by simple methods, this occurs comparatively readily in the reaction of dilithiocarboranes with bis(chlorosilyl) carboranes [1001, 1004]:

$$\mathrm{LiR''_mC\underset{B_{10}H_{10}}{\overset{O}{—}}CLi} + \mathrm{Cl_{x-1}Si(CH_3)_{4-x}R''_nCl} \longrightarrow \mathrm{(CH_3)_{4-x}(Cl_{x-2})Si \langle R''_nC, R''_mC \rangle (O) B_{10}H_{10}} \qquad (3\text{-}54)$$

$$\mathrm{R''} = \mathrm{C\underset{B_{10}H_{10}}{\overset{O}{—}}CSi(CH_3)_{4-x}Cl_{x-2}}$$

a) $m=0,\ n=1,\ x=2-4$

b) $m=1,\ n=0,\ x=2$

* This is obtained in good yield from the corresponding dilithio derivative and dimethyldichlorosilane in hexane [1002]. On the other hand, when the reaction is carried out in ether the ortho-isomer is formed in low yield.

If sym-tetramethyldichlorodisiloxane is used in reaction (3-54b) instead of dimethyldichlorosilane, an eight-membered ring is obtained with two carborane and one siloxane bridge between the silicon atoms [1004].

The customary ideas on the reactivity of chlorine at a silicon atom are upset by experimental data for compounds of the type

```
  X              X
  |              |
—Si—C———C—Si—C———C—
| |   \ O /  |   \ O /  |
| Cl   \/    Cl   \/    |
|    B10H10      B10H10 |
|_______________________|
```

Their geometric form is similar to a dumbbell. In actual fact, in the reaction of the bis-chloro derivative (X = CH_3) with water in a solvent a diol is formed quantitatively [1004]. However, the reaction with ammonia unexpectedly produces cleavage of the ring at the $Si-C_{carb}$ bonds, while the bis-amino derivative is not formed. At the same time, when X = Cl the compound is inert toward water (!) while it is converted quantitatively into the bis-diamino derivative by ammonia. It is not yet clear which structural details are responsible for these anomalies.

When trimethylchlorosilane is replaced by trimethylchloromethylsilane in the reaction of monolithiocarborane, a compound is obtained with a methylene bridge between the silicon and the carborane nucleus [842, 1693]. However, while it is impossible to form an exocyclic compound by the reaction of dilithiocarborane with dimethyldichlorosilane, by using dimethylbis (chloromethyl)-silane it is possible to obtain the following heterocycle in one stage:

```
            CH2—C
           /    | \
(CH3)2 Si<      |O >B10H10
           \    | /
            CH2—C
```

The synthesis of compounds of the type $Si-(C)_n-$carboranyl through organomagnesium derivatives of carborane is complicated by isomerization [842, 1128]. An illustration is provided by scheme

(3-55):

$$\underset{B_{10}H_{10}}{HC\overset{O}{-\!-\!-}CCH_2Br} + Mg + CH_3SiX_3 \longrightarrow \underset{B_{10}H_{10}}{CH_3C\overset{O}{-\!-\!-}CSi(CH_3)X_2} \quad (3\text{-}55)$$

When X = OC_2H_5, instead of the expected (carboranylmethyl)-methyldiethoxysilane, 1-methyl-2-(methyldiethoxysilyl) carborane is formed in 50% yield. However, when X = Cl it is possible to identify (carboranylmethyl)methyldichlorosilane and then its diethoxy derivative.

However, the only result of the experiments is the spectral characteristics since the compounds decomposed during an attempt at distillation. The same occurred during the separation of the products from the reaction of methyltriethoxysilane with the magnesium derivative of 1-methyl-2-bromomethylcarborane, a compound which is incapable of isomerization. Because of these complications,organomagnesium compounds of carboranes cannot yet be used in the synthesis of carborane derivatives containing silicon. Moreover, compounds of the type

$$\underset{B_{10}H_{10}}{RC\overset{O}{-\!-\!-}CSiR'X_2}$$

are not very thermally or hydrolytically stable. In any case, their hydrolysis in an acid or alkaline medium does not lead to cyclosiloxanes with carborane "appendages" at the silicon atom, but to cleavage of the $Si-C_{carb}$ bond [1128]:

$$x\underset{B_{10}H_{10}}{CH_3C\overset{O}{-\!-\!-}CSi(CH_3)(OC_2H_5)_2} \xrightarrow[H^+;\ OH^-]{H_2O} x\underset{B_{10}H_{10}}{CH_3C\overset{O}{-\!-\!-}CH} + (CH_3SiO_{1,5})_x \quad (3\text{-}56)$$

Derivatives of carboranes containing silicon with bridges between the Si and the carborane nucleus, two or more carbon atoms in length, may be obtained by the classical addition of hydrosilanes to unsaturated compounds, in this case, alkenylcarboranes [492,

807, 1128, 1511]:

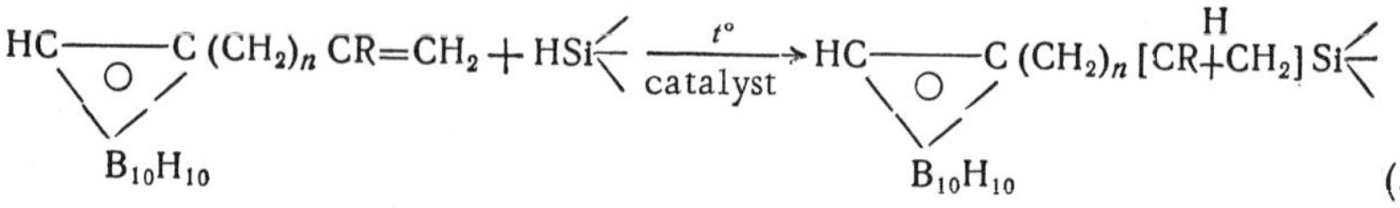

$$\underset{\diagdown O \diagup \atop B_{10}H_{10}}{HC\text{———}C}(CH_2)_n CR{=}CH_2 + HSi\!\lessgtr \xrightarrow[\text{catalyst}]{t^\circ} \underset{\diagdown O \diagup \atop B_{10}H_{10}}{HC\text{———}C}(CH_2)_n[\overset{H}{CR}{-}CH_2]Si\!\lessgtr \qquad (3\text{-}57)$$

Prolonged boiling of the reaction mixture in the presence of platinized charcoal gives products with n = 3 or 4, containing silicon at the terminal atom of the carbon chain. The reaction of vinylcarborane with methyldichlorosilane does not occur under these conditions. At the same time, Shapatin and other investigators [492, 492, 1511] report that the addition of hydrochlorosilanes to alkenylcarboranes, particularly those containing vinyl and isopropenyl groups, occurs in the presence of chloroplatinic acid at ~150°C. Ferric chloride catalyzes the reaction above 200°C, while if the reaction mixture is heated in a steel autoclave at ~250°C, there is no need to specially introduce a catalyst. There are no data on the order of addition which predominates under these conditions (α- or β-) or data on the structure of the adducts obtained.

The use of compounds of the type

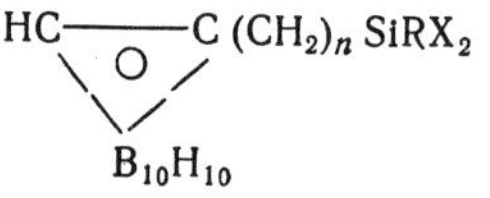

$$\underset{\diagdown O \diagup \atop B_{10}H_{10}}{HC\text{———}C}(CH_2)_n SiRX_2 \quad (X = Cl \text{ or } OC_2H_5;\ n = 3 \text{ or } 4)$$

for preparing siloxane polymers with carboranyl nuclei in the side chain has been reported [636, 801, 814]. In the hydrolysis of these compounds with subsequent condensation, ring formation competes with the formation of a linear chain and the homopolymer has a low molecular weight. A combination of the monomer with dimethyldichlorosilane yields finally a copolymer with mol. wt. ~13,000. Despite the relatively low degree of polymerization (~40) the product is a solid with viscoelastic properties [814].

The large size of the carborane nucleus is a serious hindrance to the polymerization of cyclosiloxanes obtained by hydrolysis of bifunctional (carboranylalkyl) alkylsilanes [636]. The polymerization is facilitated if the ring contains both (carboranylalkyl)-alkylsiloxane and dimethylsiloxane units. Such "mixed" cyclosiloxanes may be obtained both by cohydrolysis of the corresponding monomers and by using in reaction (3-57) cyclosiloxanes containing the unit $-SiH(CH_3)O-$ such as heptamethylcyclotetrasiloxane [641]. Polymerization in the presence of acidic or basic ionic catalysts at 150°C leads to the formation of elastomers with carboranylalkyl radicals at some of the silicon atoms of the siloxane chain [636, 641]. In this connection it should be noted that the action of the alkali on the exocycle

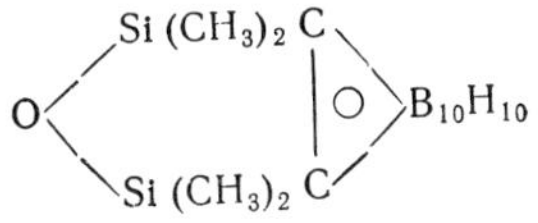

$$O\langle \begin{matrix} Si(CH_3)_2C \\ Si(CH_3)_2C \end{matrix} \bigcirc B_{10}H_{10}$$

also forms a polymer, but this is a polydimethylsiloxane in character since alkaline cleavage is accompanied by the liberation of free carborane [1004]. Sodium carbonate or bicarbonate produces ring opening at the $Si-C_{carb}$ bond with the formation of an oily product*

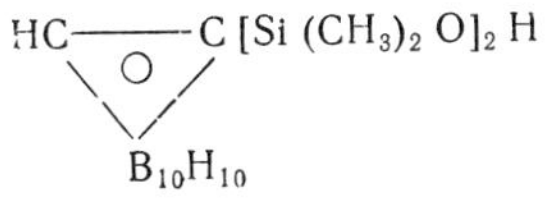

$$HC \underset{B_{10}H_{10}}{\overset{\bigcirc}{—}} C[Si(CH_3)_2O]_2H$$

The explanation of conversions similar in character to the scheme (3-56) evidently lies in the strongly marked electrophilic properties of the carborane nucleus.

Thus, the preparation of linear carboranesiloxane polymers is not an easy task, particularly since when it was finally possible to synthesize neocarborane with silanol groups (see above), it was found to be exceptionally inactive in condensation as a carborane derivative [1002]. On the other hand, it has been reported [641] that polymers of the following type may be prepared by polyaddition:

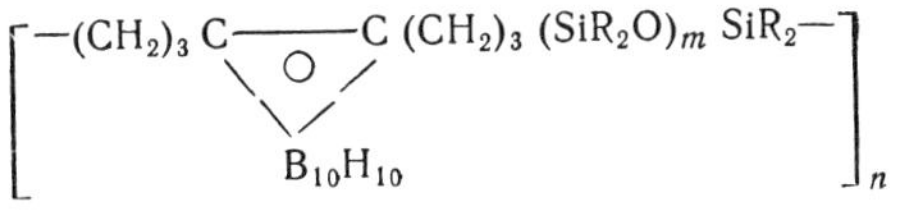

$$\left[-(CH_2)_3C \underset{B_{10}H_{10}}{\overset{\bigcirc}{—}} C(CH_2)_3(SiR_2O)_m SiR_2- \right]_n$$

*(1-Carboranyldimethylsiloxy) dimethylmethoxysilane is formed by the action of absolute methanol [1004].

While carboranylmethylmagnesium bromide is an unreliable reagent, carboranylmethyl alcohol reacts with methylchlorosilanes readily and without any complications even in the absence of HCl acceptors [1128]:

$$(4-x)\,\underset{B_{10}H_{10}}{HC\overset{O}{\text{———}}CCH_2OH} + R_xSiCl_{4-x} \xrightarrow[\text{toluene}]{t} R_xSi\left[OCH_2\underset{B_{10}H_{10}}{C\overset{O}{\text{———}}CH}\right]_{4-x}$$

$$(x=1 \text{ or } 2)$$

An analogous reaction occurs between trimethylchlorosilane and 1,2-bis(hydroxymethyl)carborane in a ratio of 2:1. The hope of forming a linear polymer by the reaction of this glycol with dimethyldichlorosilane was not justified here. Instead of this there was closure of a seven-membered exocycle:

$$(CH_3)_2SiCl_2 + \begin{matrix} HOCH_2C \\ | \\ HOCH_2C \end{matrix}\!\overset{O}{\rangle} B_{10}H_{10} \longrightarrow (CH_3)_2Si\begin{matrix} OCH_2C \\ | \\ OCH_2C \end{matrix}\!\overset{O}{\rangle} B_{10}H_{10}$$

This crystalline compound is the main product of the reaction of the diol with dimethyldiethoxysilane, but at the same time a low-molecular oil (linear polymer ?) is obtained. Under the action of KOH or para-toluenesulfonic acid, the seven-membered exocycle behaves like a five-membered exocycle with a Si–O–Si bridge (see above). No carborane-containing polymer is obtained in this case either and the products are the starting bismethylolcarborane and polydimethylsiloxane [1128].

The thermal stability of silicon derivatives of the diol (decomposition begins below 200°C) is lower than that of derivatives of carboranylmethanol $Si\overset{O}{\diagup\diagdown}Si$. In connection with this and also the failure to prepare linear polymers, a search was undertaken for organosilicon compounds which would be incapable of forming an exocycle, but, on the other hand, would increase the thermal stability of the system. According to preliminary data [807], these conditions are satisfied by bifunctional silicophthalocyanines. The thermal condensation (200-250°C) of silicophthalocyaninediol or silicophthalocyanine diisocyanate with bis-methylolcarborane* yields oligomers containing the structural fragments

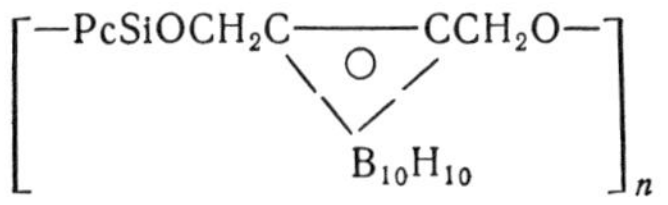

$$\left[-\mathrm{PcSiOCH_2C}\underset{B_{10}H_{10}}{\overset{O}{\diagdown\!\diagup}}\mathrm{CCH_2O}-\right]_n$$

(Pc — phthalocyanine cyclic system)

Sterically this is quite a surprising alternation of phthalocyanine disks perpendicular to the polymer axis and carborane spheres, connected by cylinders of ...$O-CH_2$...CH_2-O.... From copolymers of this type one would expect high thermal stability and other practically valuable properties.

2.2. Physical Properties and Application

The physicochemical constants of compounds containing the group $Si-(C)_n-B$, but which are not carboranes are given in Table 29.

As a rule, tris(triorganosilylalkyl)borane decomposes during liquids which distill in vacuum without decomposition, do not ignite spontaneously in air (slow oxidation without ignition), and are soluble both in organic solvents and polysiloxane fluids. They are recommended as stabilizers of lubricants, hydraulic and dielectric fluids, and additives for absorbing oxygen from liquids [1613, 1633].

Tris(methyldichlorosilylethyl)borane decomposes during distillation. The products from the addition of higher boranes to alkenylchlorosilanes have been proposed [1697] for preparing siloxane polymers with boron-containing "appendages" by hydrolysis of individual compounds such as methyl-β-decaboranylethyldichlorosilane or cohydrolysis of these compounds with other chlorosilanes of different functionality.

The properties of carborane derivatives containing silicon are illustrated by the data in Table 30.

In most cases carboranes containing silicon are white crystalline substances. With a change from the ortho to the meta isomer there is generally a fall in melting point or a change in the stage of aggregation.

* In the case of silicophthalocyanine dichloride the reaction is catalyzed by concentrated sulfuric acid.

TABLE 29. Compounds Containing the Si—(C)$_n$—B Group

Formula	b.p., °C (mm)	n_D^{20}	d_4^{20}	Yield, %	Literature
$[(CH_3)_3SiCH_2]_3B$	55—56 (0.4); 78 (1.6)	—	—	83	1138, 1633
$[C_2H_5(CH_3)_2SiCH_2]_3B$	87—89 (0,3)	—	—	71	1138
$[n\text{-}C_3H_7(CH_3)_2SiCH_2]_3B$	111—113 (0.3)	—	—	80	1138
$[(CH_3)_3SiCH_2Si(CH_3)_2CH_2]_3B$	—	—	—	—	1138
$[(CH_3)_3SiOSi(CH_3)_2CH_2]_3B$	110—113 (22)	—	—	—	1613
$\{(CH_3)_3Si[OSi(CH_3)_2]_{17}CH_2\}_3B$	—	1.4042 (25°)	0.9779 (25°)	—	1613
$(CH_3)_3SiCH(CH_3)B[CH_2CH_2Si(CH_3)_3]_2$	65—72 (0.12)	—	—	75	1140, 1613
$[CH_3(C_2H_5)_2SiCH_2CH_2]_3B$	182—184 (2)	1.4632	0.8462	63	377
$[CH_3(C_2H_5O)_2SiCH_2CH_2]_3B$	115—118 (0.01)	1.4270	0.9409	21	377
$[CH_3\cdot Cl_2SiCH_2CH_2]_3B$	127—130 (0.05) decomp	—	—	25	377
$[(CH_3)_3SiOSi(CH_3)_2CH_2CH_2]_3B$	—	—	—	—	1613
$[(CH_3)_3SiCH_2CH_2CH_2]_3B$	104—106 (0.09)	1.4410	0.8222	62	377
	—	—	—	70	1161
$[Cl_3SiCH_2CH_2CH_2]_3B$	157—159 (0.1)	1.4870	1.3487	69	377
$[(CH_3)_3SiC_2H_4]_2BOC_2H_5$	69—72 (1.2)	—	—	—	1138
$[CH_3(C_2H_5)_2SiCH_2CH_2]_2BOH$	152—160 (0.01)	1.4608	0.8813	94	377
$[CH_3(C_2H_5)_2SiCH_2CH_2]_2BOCH_3$	150—151 (3.5)	1.4558	0.8447 (21°)	48	377
$[CH_3(C_2H_5)_2SiCH_2CH_2]_2BOC_4H_9\text{-}n$	163—166 (3.5)	1.4538	0.8525	—	377
$[CH_3(C_2H_5)_2SiCH_2CH_2]_2BSC_4H_9\text{-}n$	163—166 (0.5)	1.4821	0.8767	83	377
$(CH_3)_3SiC_2H_4B(OCH_3)_2$	65—79 (21)	—	—	41	1161
$(CH_3)_3CSi(CH_3)_2C_2H_2B(OC_2H_5)_2$	120—140 (2.8)		—	—	913c
$(CH_3)_3SiC_2H_4B(NHC_6H_5)_2$	128—130 (0.03)	—	—	—	1161
$(CH_3)_3SiC_2H_4BCl_2$	60—63 (30)	—	—	63	1161
$CH_3(C_2H_5)_2SiCH_2CH_2B(OCH_3)_2$	71—72 (1.5)	1.4335	0.8741 (21°)	30	377
$CH_3(C_2H_5)_2SiCH_2CH_2B(OC_4H_9\text{-}n)_2$	150—154 (5.5)	1.4375	—	—	377
$CH_3(C_2H_5)_2SiCH_2CH_2B(SC_4H_9\text{-}n)_2$	156—158 (1)	1.5737	0.9516	37	379
$CH_3\cdot Cl_2SiCH_2CH_2B(OC_4H_9\text{-}iso)_2$	128—130 (3)	1.4370	0.9915	56	376
$Cl_3SiCH_2CH_2B(OC_4H_9\text{-}iso)_2$	115—118 (3)	1.4385	1.0775	72	376
$CH_3\cdot Cl_2SiCH_2CH_2CH_2B(OC_4H_9\text{-}iso)_2$	77—80 (0.06)	1.4400	0.9884	64	376
$\{[(CH_3)_3SiC_2H_4]_2B\}_2O$	110 (0.12); 120 (0,15)	—	—	—	1138
$\{[(CH_3)_3CSi(CH_3)_2C_2H_4]_2B\}_2O$	165—168 (2.8); M.p. 52—56	—	—	—	913a
$[(n\text{-}C_4H_9)_2BSi(CH_3)_2]_2O$	—	—	—	—	1633
$CH_3(B_{10}H_{13}CH_2CH_2)SiCl_2$	—	—	—	75	1697
$B_{10}H_{13}CH_2CH_2SiCl_3$	—	—	—	—	1697

TABLE 30. Carborane Derivatives Containing Silicon

Note. $\triangledown = -C \overset{\bigcirc}{—} C-$ (with $B_{10}H_{10}$); o - ortho isomer, m - meta isomer

Formula	Isomer	b.p., °C (mm)	m.p., °C	Preparation method (scheme).	Yield, %	Literature
$(CH_3)_3Si\triangledown H$	o	—	80—85	3-51	2,5	1128
		—	94—95	3-52a	—	842, 1693
$(C_6H_5)_3Si\triangledown H$	o	—	165—167	3-52a	—	842, 1693
$CH_3\triangledown Si(CH_3)(OC_2H_5)_2$	o	150—160 (0,05)	—	3-55	54	1128
$CH_3\triangledown Si(OC_2H_5)_3$	o	—	—	3-51	Traces	299
$C_6H_5\triangledown Si(CH_3)_3$	o	—	104—106	3-51	—	299
		—	105—106	3-52a	61	299
$H\triangledown CH_2Si(CH_3)_3$	o	—	42—52	~ 3-52a	—	842, 1693
$H\triangledown C_2H_4Si(CH_3)_2Cl$	o	145—147 (3)	68	3-57	—	492
$H\triangledown C_2H_4Si(CH_3)Cl_2$	o	149—150 (3)	72,5	3-57	—	492
$H\triangledown C_2H_4SiCl_3$	o	138—141 (3)	90	3-57	—	492
$H\triangledown C_2H_3(CH_3)Si(CH_3)_2Cl$	o	147—148 (3)	44	3-57	—	492, 1511
$H\triangledown C_2H_3(CH_3)Si(CH_3)Cl_2$	o	155—156 (3)	52.5	3-57	78	492, 1511
$H\triangledown C_2H_3(CH_3)Si(C_2H_5)Cl_2$	o	169—170 (3)	—	3-57	—	492
$H\triangledown C_2H_3(CH_3)Si(C_6H_5)Cl_2$	o	211 (3)	—	3-57	—	492, 1511
$H\triangledown C_2H_3(CH_3)SiCl_3$	o	151—152 (3)	—	3-57	76	1511
		—	60	3-57	92	492
$CH_3\triangledown CH_2C_2H_4Si(CH_3)Cl_2$	o	180—183 (5)	—	3-57	—	492
$H\triangledown CH_2CH_2CH_2Si(CH_3)(OC_2H_5)_2$	o	134—138 (0,2)	—	3-57	27	1128
$H\triangledown (CH_2)_4Si(CH_3)(OC_2H_5)_2$	o	160—162 (0,1)	—	3-57	53	1128
$H\triangledown CH_2CH_2C_2H_4Si(CH_3)Cl_2$	o	187—188 (4)	—	3-57	—	492

* The tilde denotes that the method of preparing the compound is analogous to this scheme.

† d_4^{20} 1,0930; ‡ d_4^{20} 1,1398.

TABLE 30. (Cont'd)

Formula	Isomer	b.p., °C (mm)	m.p., °C	Preparation method (scheme)	Yield, %	Literature
◁ $(CH_2)_2$ > $Si(CH_3)_2$	o	—	149—150	~3-52	—	842, 1693
◁ $(CH_2O)_2$ > $Si(CH_3)_2$	o	—	98—100	3-59	59	1128
$CH_3OSi(CH_3)_2$▽$Si(CH_3)_2OCH_3$	m	108—110 (0.1)	36—37	3-53 b	—	1002
$HOSi(CH_3)_2$▽$Si(CH_3)_2OH$	m	—	98—99.5	3-53 b	Quant.	1002
$H_2NSi(CH_3)_2$▽$Si(CH_3)_2NH_2$	m	—	41.5—43.5	3-53 b	77	1002
$ClSi(CH_3)_2$▽$Si(CH_3)_2Cl$	o	—	112.5—113.5	3-52 b	88	1001
	m	102—104 (0.1)	—	3-52 b	84	1002
$CH_3OSi(C_6H_5)_2$▽$Si(C_6H_5)_2OCH_3$	o	—	153—155	3-53 b	76	1002
$HOSi(C_6H_5)_2$▽$Si(C_6H_5)_2OH$	m	—	153—155	3-53 b	Quant.	1002
$ClSi(C_6H_5)_2$▽$Si(C_6H_5)_2Cl$	o	—	244—245	3-52 b	23	1001
	m	—	131—133	3-52 b	65	1002
$(HO)_2Si(CH_3)$▽$Si(CH_3)(OH)_2$	o	—	136.5—138.5	~3-53 b	—	1002
$Cl_2Si(CH_3)$▽$Si(CH_3)Cl_2$	o	—	119—120	3-52 b	70	1001
	m	103—105 (0.15)	—	3-52 b	68	1002
Cl_3Si▽$SiCl_3$	o	—	121—122	3-52 b	60	1001
◁ $Si(CH_3)_2$ > O $Si(CH_3)_2$	o	—	—	~3-52 b	75	1001
◁ $Si(CH_3)_2$ > NH $Si(CH_3)_2$			160—161	3-52a	Quant.	1001
	o	—	190—192	~3-53a	Quant.	1001
◁ $Si(NH_2)(CH_3)$ > NH $Si(NH_2)(CH_3)$	o	—	189—191.5	—	Quant.	1001

$Fc[Si(CH_3)_2]_2NCH_3$	o	—	156—158	3-53a	97	1004
$Fc[Si(NHCH_3)(CH_3)]_2NCH_3$	o	—	128—129.5	—	Quant.	1001
$Fc[Si(CH_3)_2]_2NC_6H_5$	o	—	183—185	3-53a	74	1004
$Fc[Si(CH_3)_2]_2NNH_2$	o	—	163—165	3-52a	78	1004
$(CH_3)_2Si(FcH)_2$	o	—	195—196.5	3-52a	—	842, 1693
$(CH_3)_2Si(OCH_2FcH)_2$	o	—	98—100	3-58	65	1128
$-[-(CH_3)_2SiFc-]_2-$ (cyclic)	o	—	309—310	3-54a	23	1001
		—	—	3-54 b	—	1004
$-[-HO(CH_3)SiFc-]_2-$ (cyclic)	o	—	304	—	Quant.	1004
$-[-Cl(CH_3)SiFc-]_2-$ (cyclic)	o	—	281—283	3-54a	51	1004
$-[-(H_2N)_2SiFc-]_2-$ (cyclic)	o	—	347—349	—	Quant.	1004
$-[-Cl_2SiFc-]_2-$ (cyclic)	o	—	271—272	3-54a	52	1004
$-[-(CH_3)_2SiFc-]_2-Si(CH_3)_2O-$ (cyclic)	o	—	278	3-54 b	35	1004
$CH_3Si(OCH_2FcH)_3$	o	—	220	~3-58	—	1128

The simplest silicon derivatives of carborane are recommended as components of solid rocket fuels [1693]. However, most attention has been directed toward introducing carborane nuclei into the structure of organosilicon polymers, particularly elastomers, as units conferring thermal and chemical stability.

Exocycles of the following type are exceptionally stable:

$$X\langle \begin{matrix} Si(CH_3)_2C \\ Si(CH_3)_2C \end{matrix} \bigcirc B_{10}H_{10} \quad \left(X = O,\ NH \text{ or } -C \overset{\bigcirc}{\underset{B_{10}H_{10}}{\text{———}}} C- \right)$$

They do not change when heated to ~490°C [1004]. Slow liberation of methane and hydrogen begins above this temperature, but decomposition occurs in practice above 590°C. In a thermogravimetric investigation of a carboranylalkylsiloxane copolymer with mol. wt. 13,000 in a nitrogen or air atmosphere, data were obtained indicating that the carborane "appendage" was not touched up to 400-450°C. Such polymers have withstood a temperature of 350°C for 100-200 hr in a sealed tube [641, 814].

3. ORGANOSILICON COMPOUNDS CONTAINING BORON AND NITROGEN

3.1. Preparation Methods

3.1.1. Reactions of Silylamines, Aminosilanes, and Azidosilanes with Boron Compounds

The reaction of boron trihalides with tertiary silylamines [84, 512, 600, 666, 1045, 1198] proceeds according to the general scheme

$$RR'_2N + BX_3 \longrightarrow [RR'_2N \cdot BX_3] \longrightarrow RX + R'_2NBX_2 \qquad (3\text{-}60)$$

$(R = SiH_3,\ CH_3SiH_2,\ (CH_3)_3Si;\ R' = \text{alkyl}\ \ SiH_3,\ CH_3SiH_2,\ (CH_3)_3Si;\ X = \text{halogen})$

Regardless of the number and character of the silicon-containing radicals at the nitrogen atom the decomposition of the crystalline 1:1 adduct is accompaned by the liberation of a molecule of a compound with a Si–X bond.

It was originally believed [600] that the reaction of trisilylamine with boron trifluoride at temperatures from −78 to −40°C is reversible and that above −40°C there is decomposition of the adduct with the formation of a complex mixture of products, including SiH_4, SiH_2F_2, etc. However, it was found later [1198] that the reaction of silylamines (including trisilylamine) with boron trifluoride proceeds according to the scheme (3-60), while such compounds as disilylaminoboron difluoride, for example, react with excess boron trifluoride with the formation of a silane and various fluorosilanes and this actually explains the results in [1198].* Moreover, $(CH_3SiH_2)_3N$ does not give an adduct with BF_3 with excess of the latter (1:2) [84].

Since a solid adduct is formed with an equimolecular ratio, it must be assumed that its decomposition involves a second molecule of boron trifluoride. Naturally, when the starting amine contains only one silicon-containing radical the organoboron dihalide obtained does not contain silicon.

The stability of the adducts and silylaminoboron dihalides depends on the number of silyl and methyl groups at the nitrogen (see section 3.2).

Trichlorosilyldimethylamine (dimethylaminotrichlorosilane) also reacts with boron trifluoride in accordance with the scheme (3-60) and is converted into an adduct, which decomposes quantitatively at room temperature with the formation of trichlorofluorosilane [817]. Compounds of the type

$$(CH_3)_m Si [N (CH_3)_2]_n X_{4-(m+n)} \quad (m=1\text{—}2;\ n=1-3;\ X=F \text{ or } Cl)$$

react with BF_3 at −50°C to form adducts, which decompose in the temperature range from −50 to + 90°C with the liberation of $(CH_3)_m SiF_n X_{4-(m+n)}$ [818]. Anomalous behavior is shown only by bis(dimethylamino) dichlorosilane, which does not form such an adduct [817]. The replacement of BX_3 by phenylboron dichloride [512] or dipropyl (butyl) boron chloride [986] does not change

* In the opinion of Nöth [986], the reaction of trimethylsilyldimethylamine with BX_3 is accompanied by the simple replacement of X by $(CH_3)_2N$, while the formation of mono-, bis-, or tris(dimethylamino) derivatives of boron depends on the amount of aminosilane.

the character of the reaction with (dialkylamino)alkylsilanes. The same alkylchlorosilanes and dialkylamino derivatives of boron are obtained ultimately. It is interesting that the reaction also proceeds in this way with silicon-containing heterocycles [514]:

$$C_6H_5BCl_2 + (CH_3)_2Si\begin{matrix}\diagup N(C_2H_5)CH_2 \\ \quad\quad\quad\quad | \\ \diagdown N(C_2H_5)CH_2\end{matrix} \longrightarrow \underset{(83\%)}{(CH_3)_2SiCl_2} + \underset{(65\%)}{C_6H_5B\begin{matrix}\diagup N(C_2H_5)CH_2 \\ \quad\quad\quad\quad | \\ \diagdown N(C_2H_5)CH_2\end{matrix}} \quad (3\text{-}61)$$

At the same time, in the reaction of trisilylamine with dimethylboron bromide [600] it was possible to isolate a viscous liquid with a composition close to $(CH_3)_2BN(SiH_2Br)_2$. The complexity of this reaction is indicated by the fact that the products identified included SiH_4, SiH_3Br, $(CH_3)_3B$, etc.

In comparison with trimethylamine, trisilylamine is a weak electron donor and does not react with diborane [600, 1196]. Disilylmethylamine forms a 1:1 adduct at − 80°C and this both dissociates and decomposes above this temperature while the adduct of diborane with silyldimethylamine decomposes at room temperature with the formation of SiH_4 and dimethylaminoborine [1196]. Trisilylamine undergoes an interesting reaction with bromodiborane, which forms diborane and disilylaminoborine – $(H_3Si)_2NBH_2$ – a volatile substance which exists as a mixture of monomer and dimer (10-15:90-85) [600]. The monomer reacts with diborane at room temperature to form $(H_3Si)_2B_2H_5$ (80% yield over 3 days); this compound, which ignites instantaneously in air, decomposes with the liberation of diborane and the formation of polymeric forms of disilylaminoborine:

$$2(H_3Si)_3N + 2B_2H_5Br \xrightarrow[-2H_3SiBr]{} B_2H_6 + 2(H_3Si)_2NBH_2 \longrightarrow$$
$$\longrightarrow 2(H_3Si)_2NB_2H_5 \xrightarrow[-B_2H_6]{} \text{polymer [Si, N, B]} \quad (3\text{-}62)$$

Methyldisilylamine evidently reacts with bromodiborane analogously to trisilylamine. However, the $(CH_3NSiH_3)BH_2$ formed is a highly active compound, which is capable of further reaction in several directions:

$$\left[\begin{matrix} H_3Si \\ \quad\quad NBH_2 \\ H_3C \end{matrix}\right] \longrightarrow \begin{cases} CH_4 + \text{polymer}[Si, N, B] \\ SiH_4 + (H_3CNBH)_2 \\ \xrightarrow{+B_2H_6} \begin{matrix} H_3Si \\ \quad\quad NB_2H_5 \\ H_3C \end{matrix} \end{cases} \qquad (3\text{-}63)$$

Silyl(methylamino)diborane is more volatile than the disilylamino derivative of diborane, but with trimethylamine it forms an involatile complex, which is stable in vacuum. In this connection, we should point out that despite the increase in the electron donor properties in the series $(SiH_3)_3N-SiH_3N(CH_3)_2$, only silyldimethylamine forms a sufficiently stable complex with trimethylborane* [1197]. Compounds of the series $(CH_3SiH_2)_3N-CH_3SiH_2N(CH_3)_2$ behave analogously. It is possible that steric factors play a part here. Like (methylsilyl)dimethylamine, (trimethylsilyl)dimethylamine gives a solid 1:1 adduct with trimethylborane [666] and also with BH_3 [986]. The latter decomposes at 140°C with the formation of a mixture of products, which includes trimethylsilane, dimethylaminodiborane, the dimer of methylaminoborine, and bis(dimethylamino)borine.

In the reaction of dimethylaminodimethylchlorosilane with lithium borohydride [883, 985, 986] there is reduction, accompanied by the formation of the solid adduct $(CH_3)_2SiHN(CH_3)_2 \cdot BH_3$, which gives dimethylsilane on decomposition.

Neither tris(trimethylsilyl)amine nor diphenyl- bis(dimethylamino)silane undergoes transamination with phenyl-bis(dimethylamino)borane [856].

Hexamethyldisilazan [bis(trimethylsilyl)amine] reacts with boron trihalides analogously to tertiary silylamines [142, 544, 816, 986]:

$$R_3SiNHSiR_3 + BX_3 \longrightarrow R_3SiX + R_3SiNHBX_2 \qquad (3\text{-}64)$$

* The following series of basicities was compiled from the results of a study of the reaction of silyl-containing compounds with various derivatives of boron, aluminum, and gallium [935]:

$$(CH_3)_3N > (CH_3)_2NSiH_3 > CH_3N(SiH_3)_2 > N(SiH_3)_3 < O(SiH_3)_2 < FSiH_3.$$

As in the case of tertiary silylamines, the reaction evidently proceeds through a complex of the disilazan with BX_3. In any case, it was shown that there is the formation of a stable adduct (1:1) of hexamethyldisilazan with boron trifluoride [816, 986]. Under mild conditions (0–20°C) the products of this decomposition correspond to the scheme (3–64). Trimethylfluorosilane and B-trifluoroborazole are obtained at 200°C, but there is quite definitely the intermediate formation of trimethylsilylaminoboron difluoride since substituted borazoles ("cyclotriborazans") are, on the one hand, the final products of the reaction of hexamethyldisilazan with BCl_3 and BBr_3 [986]* and on the other hand, they are obtained by decomposition of individual silylaminoboron dihalides [600, 1198].†

$$3RR'NBX_2 \longrightarrow 3RX + \text{[ring: BX–NR'–BX–NR'–XB–R'N]} \qquad (3\text{–}65)$$

$(R = H_3Si, (CH_3)_3Si;\ R' = H, CH_3;\ X = \text{halogen})$

The most probable mechanism for the cleavage of hexaalkyldisilazans by boron trihalides is one similar to the cleavage of hexaalkyldisiloxanes (scheme 3–6), which includes the intermediate formation of a six-membered cyclic active complex:

$$(R_3Si)_2NH + BX_3 \longrightarrow (R_3Si)_2NH \cdot BX_3 \xrightarrow{+BX_3} \text{[cyclic complex: } R_3Si(SiR_3)NH\text{–}BX_2\text{–}X\text{–}BX_2\text{–}X\text{–}] \longrightarrow$$

$$\longrightarrow R_3SiX + R_3SiNHBX_2 + BX_3 \qquad (3\text{–}66)$$

A similar mechanism is evidently valid in the case of the reaction of tertiary silylamines with BX_3 in accordance with the scheme (3–60). The proposed "four-center" scheme for the reaction [544, 935] is not feasible in the given case.

* In particular, in this way it was possible to obtain B-tribromo(iodo)-borazoles which were previously unknown.

† An exception is disilylaminoboron dichloride, which decomposes slowly at 60–65°C with the formation of a complex mixture of products, which contain hydrogen, silane, and dichlorosilane [600].

In the cleavage of hexamethyldisilazan by organoboron dichlorides, B-organo substituted borazoles with practically any radical at the boron atoms are ultimately formed [986]. When this reaction is carried out with a bis(trimethylsilyl)alkylamine, B-organoborazoles with organic substitution at the nitrogen atoms in addition are obtained [512]:

$$3(R_3Si)_2NR' + 3R''BCl_2 \longrightarrow 6R_3SiCl + (R'NBR'')_3 \qquad (3\text{-}67)$$

$$(R' = H,\ CH_3;\ R'' = CH_3,\ C_2H_5,\ n\text{-}C_3H_7,\ n\text{-}C_4H_9,\ tert\text{-}C_4H_9,\ C_6H_5)$$

In the equimolar reaction of hexamethyldisilazan with dipropyl(butyl)boron chloride, we do not have the conditions for the formation of a borazole and only oily trimethylaminodialkylborines are obtained quantitatively. These compounds are sensitive to air and moisture. Their reaction with R_2BCl (or the direct reaction of hexamethyldisilazan with R_2BCl in a molar ratio of 1:2 gives bis(dialkylboryl)amines [986]:

$$(R_3Si)_2NH + nR'_2BCl \longrightarrow nR_3SiCl + (R'_2B)_nNH(SiR_3)_{2-n} \quad (n = 1 \text{ or } 2) \qquad (3\text{-}68)$$

Using the boron-containing heterocycle $o\text{-}C_6H_4O_2BCl$ in reaction (3-68) or replacement of one organic radical at the boron atom by a dimethylamino group does not change its general character [924]. The only difference is that in the first case it is possible to isolate one product, namely, bis(1,3,2-benzodioxabor-2-yl)amine, while in the reaction of hexamethyldisilazan with phenyl(dimethylamino)boron chloride it is possible to obtain both products from substitution of the R_3Si groups at the silicon atom, depending on the ratio of the reagents.

In the reaction of nonamethyltrisilazan with organoboron dichlorides [512] there is "expulsion" of the silicon from the cyclic structure with the conversion of the silazane into a borazane. This is essentially a development of reaction (3-61):

$$(R_2SiNR')_3 + 3R''BCl_2 \longrightarrow 3R_2SiCl_2 + (R''BNR')_3 \qquad (3\text{-}69)$$

The mechanism of this reaction is difficult to visualize. At the same time, in the reaction of dipropylboron chloride with hexamethylcyclotrisilazan (with a molar ratio of 1:1) the ring is opened with the formation of a linear trisilazan [986]. With a molar ratio of reagents of 3:1 the reaction product is $(n\text{-}C_3H_7)_2BNHSi(CH_3)_2Cl$ while with a larger amount of dipropylboron chloride, dimethyldichlorosilane is formed. This indicates a stepwise development of the reaction:

$$\text{cyclo-}(R_2SiNH)_3 + R_2B\vdots Cl \longrightarrow R_2B(NHSiR_2)_3Cl \xrightarrow[R_2BNHSiR_2Cl]{+R_2BCl} R_2B(NHSiR_2)_2Cl \xrightarrow{+R_2BCl} \text{etc.} \qquad (3\text{-}70)$$

The inner logic of the process ("boron to nitrogen and chlorine to silicon") is preserved to the end since R_2BCl is in excess and there is no need to consider the disproportionation of $R_2BNHSiR_2Cl$.

Heating $[(CH_3)_2SiNH]_n$ (where n = 3 or 4) with tributyl borate at 180-250°C leads to cleavage of the rings with the formation of linear butoxysilazans $n\text{-}C_4H_9O(CH_3)_2Si[NHSi(CH_3)_2]_mOC_4H_9\text{-}n$ (where m = 0−2) [220]. Solid substances of undetermined composition are obtained at the same time and it is not possible to isolate boron-containing products.

Hexamethyldisilazan does not react with trimethylboron [986] or with phenyl-bis(dimethylamino)borane. At the same time, the primary silylaminc $(C_2H_5)_3SiNH_2$ undergoes transamination with dialkylaminoboranes [856]:

$$R_3SiNH_2 + R'_2BNR''_2 \xrightarrow{\text{benzene}} R_3SiNHBR'_2 + R''_2NH \qquad (3\text{-}71)$$

However, scheme (3-71) describes a simple example using a monoamino derivative of boron. In the reaction of R_3SiNH_2 with organo-bis(dialkylamino)boranes and tris(dialkylamino) boranes it is possible to obtain products from both partial and complete replacement of the amino groups at the boron atom, depending on the reagent ratio and the reaction conditions. Therefore, it was unexpected when it was found that heating diphenyldiaminosilane with phenyl-bis(dimethylamino)borane does not form a compound of the borasilazan type, but that the reaction proceeds with the liberation of methylamine and the formation of a substance with m. p. 252-256°C, which is the linear borazan $CH_3NH(C_6H_5BNCH_3)_3H$. It is suggested [856] that in this case there is as usual transamination with subsequent heterocondensation of aminodimethylamino derivatives of boron.

When reaction (3-71) is carried out with a compound with a B−B bond, the latter is not cleaved. In this case transamination leads to the formation of tetrakis(triethylsilylamino)diborane [517].

Compounds containing an amino group in the radical at the silicon atom react with triarylboranes with the liberation of heat and the formation of products whose IR spectra show bands characteristic of $B-H_2N$ complexes [1660].

Triorganosilylazides form 1:1 complexes with boron tribromide and the thermal stability of these increases with an increase in the number of phenyl groups at the silicon atom [119, 1208]. Moisture causes hydrolysis. Triphenylsilazides react with the complex $(C_2H_5)_3OBF_4$ ("triethyloxonium tetrafluoroborate") [1266] with the formation of triphenylfuorosilane, ethyl azide, and boron trifluoride etherate.

It is belived that the reaction proceeds through the intermediate complex (I):

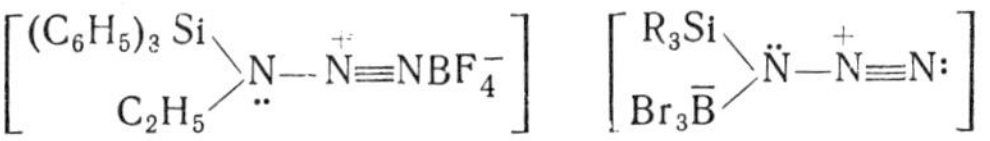

$$\left[\begin{matrix}(C_6H_5)_3Si \\ C_2H_5\end{matrix}\!\!>\ddot{N}-\overset{+}{N}\equiv NBF_4^-\right] \qquad \left[\begin{matrix}R_3Si \\ Br_3\bar{B}\end{matrix}\!\!>\ddot{N}-\overset{+}{N}\equiv N:\right]$$

I II

In this connection it should be noted that complexes of R_3SiN_3 with BBr_3 have been assigned structure (II), which is supported by the presence in the IR spectrum of a strong band at 980 cm^{-1}, which is not present in the spectra of silyl azides themselves, and a shift in the bands corresponding to $Si(N_3)$ by + 60 − 65 cm^{-1} and − 120 -140 cm^{-1} for antisymmetrical and symmetrical valence vibrations, respectively.

3.1.2. Synthesis Using Organometallic Compounds

Triorganosilyllithiums react with bis(dimethylamino)chloroborane to form derivatives in which the silicon is attached to boron by a direct bond. The same result is obtained by the reaction of bis(dimethylamino)chloroborane with a triorganochlorosilane in the presence of metallic potassium [987]:

$$\left.\begin{array}{l} R_3SiLi \xrightarrow{a)} \\ R_3SiCl + 2K \xrightarrow{b)} \end{array}\right| \xrightarrow[-MCl]{+ClB[N(CH_3)_2]_2} R_3SiB[N(CH_3)_2]_2 \quad (3\text{-}72)$$

$$(R = CH_3 \text{ or } C_6H_5;\ M = LiK)$$

Derivatives containing an alkyl at the boron atom may also be obtained by means of triorganosilyllithium compounds. However, it is not possible to synthesize compounds of the type $R_3SiBR'_2$ in this way and this indicates that the S−B bond is only stable in the system Si−B−N. Hydrolysis of the products obtained by scheme (3-72) involves cleavage of both Si−B and B−N bonds. The silicon−boron bond is not touched when the compound is treated with HCl in ether:

$$2HCl + R_3SiB[N(CH_3)_2]_2 \longrightarrow R_3SiB[N(CH_3)_2]Cl + (CH_3)_2NH \cdot HCl \quad (3\text{-}73)$$

Triorganosilylboranes reduce silver salts; when R = CH_3, the compounds inflame spontaneously in air like silylamino derivatives of boron.

In 1961 Wannagat and his co-workers prepared sodium bis(trimethylsilyl)amide, a compound with which it was subsequently possible to synthesize many nitrogen-containing hetero-organic compounds of silicon [122]. In particular, the action of this compound on a boron trihalide etherate gave various organo [bis(silyl)-amino]boron halides [719, 720]:

$$n(R_3Si)_2NM + BX_3 \longrightarrow [(R_3Si)_2N]_nBX_{3-n} + nMX \quad (3\text{-}74)$$

$$(R = CH_3;\ M = Na;\ n = 1 \text{ or } 2 \text{ when } X = Cl,\ 2 \text{ when } X = F)$$

Analogous results are obtained by passing BX_3 through a solution of lithium bis(trimethylsilyl)amide in pentane or hexane [1045].

It is not possible to obtain the completely substituted product (n = 3) by scheme (3-74) and this is evidently explained by steric factors. In any case, bis[bis(trimethylsil)amino]fluoroborane does not react with sodium bis(trimethylsilyl)amide. It is also characteristic that BF_3 does not give a monosubstitution product in reaction (3-74). A monosubstitution product is formed by prolonged heating of bis[bis(trimethylsilyl)amino]fluoroborane with BF_3 at 100°C, when this reaction is irreversible [719, 720]:

$$[(R_3Si)_2N]_2BX + BX_3 \longrightarrow 2(R_3Si)_2NBX_2 \qquad (3\text{-}74a)$$

In contrast to its fluorine analog, bis[bis(trimethylsil)amino]-chloroborane reacts with $NaN(SiR_3)_2$ in boiling xylene [720]. However, instead of the expected tris[bis(trimethylsil)amino]borane, the products obtained are tris(trimethylsilyl)amine and a four-membered substituted cycloborazan, which is a white crystalline substance that sublimes at 110°C (0.1 mm):

$$\begin{array}{l} (R_3Si)_2N-B-N-SiR_3 \\ \qquad\quad\;\; | \quad\; | \\ \quad R_3Si-N-B-N(SiR_3)_2 \end{array} \qquad (R = CH_3)$$

This compound is obtained directly by heating bis[bis(trimethylsilyl)amino]haloboranes [719, 1045]:

$$2[(R_3Si)_2N]_2BX \longrightarrow \left[(R_3Si)_2NB{-}NSiR_3\right]_2 + 2R_3SiX \qquad (3\text{-}75)$$

In this heterocycle the silicon is in the outer sphere, while from the dilithium derivative of octamethyltrisilazan (dimeric in nonpolar solvents) it is possible to obtain a compound with the silicon atom in a four-membered ring [693]:

$$R_2Si\begin{matrix}\diagup N(SiR_3)Li \\ \diagdown N(SiR_3)Li\end{matrix} + \begin{matrix}Cl\diagdown \\ Cl\diagup\end{matrix}BR' \xrightarrow{\text{hexane}} R_2Si\begin{matrix}\diagup N(SiR_3)\diagdown \\ \diagdown N(SiR_3)\diagup\end{matrix}BR' \qquad (3\text{-}76)$$

$$(R = CH_3;\quad R' = C_6H_5)$$

The conversion of lithium reaches 95%, but the compound is isolated in 55% yield.

The action of sodium bis(trimethylsilyl)amide on compounds of the type $C_6H_5BCl(OC_2H_5)$ or $(CH_3O)_2BCl$ [719] results in replacement of only the chlorine atoms and the formation of trimethylsilyl)amino-alkoxyborane.*

* The thermal decomposition of these compounds, like that of bis(trimethylsilyl)-aminodichloroborane, will be discussed below.

The reaction of sodium bis(trimethylsilyl)amide with (dialkylamino)dichloroboranes leads to compounds of the type $R_2NB(Cl)N$-$[Si(CH_3)_3]_2$, which have a higher thermal stability. When diethyl(diisopropyl) derivatives are boiled with excess sodium bis(trimethylsilyl)amide in xylene and also in the case of bis(trimethylsilyl)-aminochloroborane there is not simple substitution, but rearrangement. However, its character is different since a silazan and not a borazan four-membered ring is formed and there is migration of a methyl group from the silicon atom to the boron atom:

$$2R_2'NBCl_2 \xrightarrow[-2NaCl]{+2NaN(SiR_3)_2} 2R_2'NB(Cl)N(SiR_3)_2 \xrightarrow[-2NaCl]{+2NaN(SiR_3)_2} \left(R_3SiN\!-\!SiR_2\right)_2 + R_2'NB(R)N(SiR_3)_2 \qquad (3\text{-}75a)$$

The reaction of dimethyl(dimethylamino)chlorosilane with lithium borohydride proceeds in the following way [893, 985, 986]:

$$R_2Si(NR_2)Cl + LiBH_4 \longrightarrow R_2Si(H)NR_2 \cdot BH_3 + LiCl \xrightarrow{60°} R_2SiH_2 + (R_2NBH_2) \qquad (3\text{-}77)$$

At the same time compounds of the type $(RO)_2SiHCl$ are simply hydrogenated by lithium borohydride to $(RO)_2SiH_2$. At −115°C trimethylamine displaces lithium from the complex $Li[H_3B \cdot SiCl_3]$ (see Section 4) to form the complex $(CH_3)_3N \cdot SiCl_2 \cdot BH_3$, which is converted on prolonged standing into $(CH_3)_3N \cdot SI_2Cl_4 : BH_3$ [95, 985].

3.1.3. Reactions of Cyanosilanes with Boron Compounds

Cyanosilane and trimethylcyanosilane react with boron halides from − 196 to − 78°C according to the following general scheme:

$$R_3SiCN + BX_3 \longrightarrow R_3SiCN \cdot BX_3 \longrightarrow R_3SiX + \frac{1}{n}(BX_2CN)_n \longrightarrow BX_3 + B(C_3N_3) \qquad (3\text{-}78)$$

The vapor pressure of the adduct of trimethylcyanosilane and boron trifluoride at −126°C equals practically zero. On heating to −95°C, the pressure increases to 20 mm, but on cooling to −126°C, it again falls to zero. A rise in temperature to −78°C produces an increase in pressure to 100 mm, but when the adduct is kept at this temperature for 3 hr, the pressure falls to 2 mm. At −32°C there is liberation of trimethylfluorosilane. In the opinion of the authors of [689], this indicates that over the given temperature range there is the following equilibrium:

$$\underset{\text{(solid)}}{(CH_3)_3SiCN \cdot BF_3} \rightleftharpoons \underset{\text{(solid)}}{(CH_3)_3SiCN} + \underset{\text{(gas)}}{BF_3} \qquad (3\text{-}79)$$

At $-78°C$ there is a solid-phase rearrangement with the formation of the complex $(CH_3)_3SiF \cdot BF_2CN$, which facilitates the subsequent decomposition in accordance with the scheme (3-78). The formation of a coordination complex of trimethylcyanosilane with BBr_3 has not been observed experimentally, but the character of their reaction is analogous to the reaction with other boron trihalides. The reaction proceeds quantitatively without any indication of the formation of an intermediate complex [688]:

$$(CH_3)_3SiCN + (n\text{-}C_4H_9)_2BCl \longrightarrow (CH_3)_3SiCl + (n\text{-}C_4H_9)_2BCN \qquad (3\text{-}80)$$

Methylcyanosilane, which is assigned the structure of an isocyanide, reacts with excess boron trifluoride to form a yellow-brown insoluble solid, which decomposes on heating [680]. The reaction products contain 19.5% CN while the starting compound contains 36.6% CN and BF_2CN contains 34.8% CN.

The reaction of silyl cyanides with diborane is very similar to reaction (3-78) [687]:

$$2R_3SiCN + B_2H_6 \xrightarrow{-196°C} 2R_3SiCN \cdot BH_3 \xrightarrow{>25°C} 2R_3SiH + \frac{1}{n}(BH_2CN)_n \qquad (3\text{-}81)$$

$$(R = H \text{ or } CH_3)$$

The solid adducts are stable at room temperature. It is believed that their decomposition proceeds through the formation of an intermediate complex, which promotes the elimination of R_3SiH:

```
R3SiCNBH2
 :    |
 H    H
 |    :
H2BNCSiR3
```

When R = CH_3, hydrolysis of the adduct involves the liberation of hydrogen and the formation of hexamethyldisiloxane, HCN, and H_3BO_3. In the reaction with HCl, at first there is evidently replacement of one of the hydrogen atoms at the boron.* Decomposition then occurs:

$$(CH_3)_3SiCN \cdot BH_3 + HCl \longrightarrow (CH_3)_3SiCN \cdot BH_2Cl + H_2 \qquad (3\text{-}82)$$
$$\longrightarrow (CH_3)_3SiCl + (BH_2CN)$$

Trimethylamine displaces trimethylcyanosilane from the adduct, forming in its turn a complex with BH_3. Decomposing the adduct in a medium of trimethylcyanosilane forms trimethylsilane and a new compound $(CH_3)_3SiCN \cdot BH_2CN$, which is a mobile yellow liquid. This substance may be regarded as the result of replacement of hydrogen at the boron atom, as in the case of the reaction with HCl. However, since the cyanoborine is obtained in a polymeric form (a colorless glassy substance with a very high thermal stability) when the adduct is decomposed in the absence of excess trimethylcyanosilane, it has been suggested [687] that the excess trimethylcyanosilane acts as an acceptor of cyanoborine liberated during the decomposition of the adduct and inhibits its polymerization.

In a study of the reaction of trimethylcyanosilane with trimethylborane, data were obtained [688] indicating addition at the $C \equiv N$ bond and the formation of $(CH_3)_3SiC(CH_3)=NB(CH_3)_2$, which is the only organosilicon compound with a double bond in the bridge connecting the silicon and boron atoms.

3.1.4. Synthesis of Borazole Derivatives Containing Silicon

The term borazole (or the English borazene or borazene) is generally used at the present time. However, it is more correct to name this compound "cyclotriborazan":

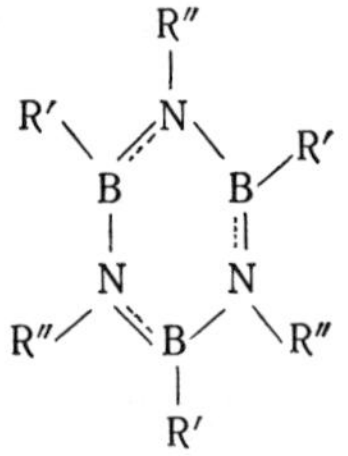

R″
N
R′ B B R′
N N
R″ B R″
R′

* The reaction of $(CH_3)_3N \cdot BH_3$ with HCl forms $(CH_3)_3N \cdot BH_2Cl$.

It has been pointed out above that the decomposition of compounds of the type $RR'NBX_2$ leads to the formation of substituted borazoles. In this case, if R = alkyl and R' = silyl, the reaction according to scheme (3-65) proceeds with elimination of the radical containing silicon in the form R'X, while the organic substituent passes into the borazole structure. The formation of a compound with a borazole ring containing silicon was first observed in the decomposition of $(H_3Si)_2NBF_2$ [1119]. However, the N-trisilyl-B-trifluorocycloborazan obtained is formally an inorganic compound. The methods first used to prepare cycloborazans with or-organosilicon substituents at the boron atoms were organometallic syntheses with organosilicon Grignard reagents [1148, 1151], triphenylsilyllithium [1151], triphenylsilylpotassium [647], and sodium triorganosilanolates [956, 1151].* In the reaction of the organomagnesium reagent obtained from trimethylsilylmethyl chloride with N-trimethyl-B-trichlorocyclotriborazane it is also possible to isolate the partial substitution product and by the reaction of this with a dimagnesium compound it is possible to form a bridge between two silicon-containing borazole rings [1148]:†

$$(RNBCl)_3 \xrightarrow{\text{a) } R'_3SiCH_2MgX;\ \text{b) } R'_3SiM;\ \text{c) } R'_3SiOM} (RNBR'')_3 \qquad R'' = \text{a) } R'_3SiCH_2;\ \text{b) } R'_3Si;\ \text{c) } R'_3SiO \tag{3-83}$$

* When B-trichloroborazole is treated with trimethyl (diethylamino)silane there is exchange with the formation of trimethylchlorosilane and B-tris(diethylamino)-borazole [512].

† One of the experiments on the synthesis of B-tris(trimethylsilylmethyl)-N-trimethyl-borazole, when insufficient trimethylsilylmethylmagnesium chloride was used, unexpectedly gave such a bicyclic compound, but with an oxygen bridge between boron atoms of the two rings [1151].

$$\text{(Cl)}_3\text{B}_3\text{N}_3\text{R}_3 \xrightarrow{2\text{R}_3\text{SiCH}_2\text{MgCl}} (\text{R}_3\text{SiCH}_2)_2\text{ClB}_3\text{N}_3\text{R}_3 \xrightarrow{\text{ClMg}(\text{CH}_2)_4\text{MgCl}} [(\text{R}_3\text{SiCH}_2)_2\text{B}_3\text{N}_3\text{R}_3]\text{CH}_2\text{CH}_2\text{CH}_2\text{CH}_2[\text{B}_3\text{N}_3\text{R}_3(\text{CH}_2\text{SiR}_3)_2] \quad (3\text{-}84)$$

The chlorine remaining in the partial silylation product may also be replaced by an alkyl radical.

When a mixture of trichloroborazole and sym-tetramethyldi-butoxydisiloxane in a molar ratio of 1:8 is heated at 100°C in a nitrogen atmosphere until the liberation of butyl chloride ceases, there is formed a friable, dark brown polymer which melts above 300°C, is resistant to the action of atmospheric moisture, and is insoluble in the usual solvents [315]. The polymer (intrinsic viscosity in cresol 0.16) has a network structure, where the borazole rings are connected by siloxane chains:

$$\left[\begin{array}{c} n\text{-C}_4\text{H}_9(\text{SiR}_2\text{O})_2\text{—B}_3\text{N}_3\text{H}_3\text{—O}(\text{SiR}_2\text{O})_2\text{—C}_4\text{H}_9\text{-}n \\ | \\ \text{O} \\ | \\ (\text{R}_2\text{SiO})_2 \\ | \\ n\text{-C}_4\text{H}_9(\text{SiR}_2\text{O})_2\text{—B}_3\text{N}_3\text{H}_3\text{—O}(\text{SiR}_2\text{O})_2\text{—C}_4\text{H}_9\text{-}n \end{array}\right]_3 \qquad (\text{R} = \text{CH}_3)$$

Seyferth and his co-workers [1156] used the classic addition of hydrosilanes to a double bond to synthesize silicon-containing cyclotriborazanes:

$$(RNBCH{=}CH_2)_3 + 3HSi\lessgtr \longrightarrow \left(RNBCH_2CH_2Si\lessgtr\right)_3 \qquad (3\text{-}85)$$

The addition of alkylhydrochlorosilanes to B-trivinyl-N-triphenylcyclotriborazan does not occur in the presence of radical initiators, but in accordance with Farmer's rule, the reaction proceeds readily when "Speier's catalyst" is used. Compounds with a siloxane chain in the radical at the boron atom (addition of hydrosiloxanes) are also obtained readily by reaction (3-85). Derivatives containing chlorine at the silicon atoms may undergo further conversions (hydrolysis, alkylation, and the reaction with silanolates), considerably extending the range of silicon-containing borazoles and, in particular, offering a route to the preparation of siloxane polymers with cyclic borazan appendages in the side chain.

Hexamethyldisilazan reacts with diborane [986] to form an adduct, which decomposes when heated in a closed system:

$$(R_3Si)_2\,NH + B_2H_6 \longrightarrow (R_3Si)_2\,NH\cdot BH_3 \xrightarrow{120^\circ\,C}$$

$$\longrightarrow {}^1/_3\,(R_3SiNBH)_3 + R_3SiH + H_2 \qquad (3\text{-}86)$$

The bis(trimethylsilyl)aminodichloroborane obtained by reaction (3-74) is also thermally unstable and even in boiling xylene it decomposes with the formation of trimethylchlorosilane and B-trichloro-N-trimethylsilylcyclotriborazan [719, 720]. The fluorine analog of the latter is evidently formed by thermal decomposition of bis(trimethylsilyl)aminodifluoroborane [1045]. At the same time, when bis(trimethylsilyl)aminodimethoxyborane is heated to 200°C, trimethylmethoxysilane is eliminated, but a resinous polymer is formed, which is apparently cross-linked [719]. On the other hand, the thermal decomposition of bis(trimethylsilyl)amino-(phenyl)ethoxyborane proceeds not according to scheme (3-75), but with the formation of a substituted borazole:

$$3\,[(CH_3)_3\,Si]_2\,NB\,(C_6H_5)\,OC_2H_5 \xrightarrow{200^\circ\,C} 3\,(CH_3)_3\,SiOC_2H_5 + [(CH_3)_3\,SiNBC_6H_5]_3 \qquad (3\text{-}85)$$

In principle, cyclization may also be used to prepare a cyclotriborazan with silicon-containing substituents at the boron atoms. Thus, the decomposition of (trimethylsilylethyl) bis (phenylamino)-borane (~330°C) yields B-trimethylsilylethyl-N-phenylcyclotriborazan [814].

Borazoles of the type $(RBNR')_3$, containing Si—B bonds are not sensitive to the action of dry air right up to 160°C, but their hydrolysis occurs very readily and is accompanied by both cleavage of the Si—B bonds and also complete destruction of the ring. In the case of B-tris(triphenylsilyl)-N-trimethylborazole, triphenylsilane, boric acid, and methylamine are formed [1151]. According to data in [647], the bromination of such borazoles in CCl_4 involves elimination of triphenylbromosilane and decomposition of the ring. At the same time, B-tris(triphenylsilyl)-N-trimethylcyclotriborazan decomposes even when dissolved in carbon tetrachloride, but bromination of it forms triphenylbromosilane and a brominated borazole [1151].

B-Tris(trimethylsiloxy)-N-trimethylborazole is hydrolyzed even by atmospheric moisture. Its triphenylsiloxy analog is more stable and is hydrolyzed by aqueous acids and alkalis. Replacement of the methyl radicals at the nitrogen atoms by hydrogen atoms increases the sensitivity of the compound to hydrolysis by moisture. On the other hand, a derivative containing the group N-C_6H_5 is hydrolyzed only by boiling water or alkalis.

Cyclotriborazans containing Si—CH_2—B groups are resistant to oxidation by air and hydrolysis in neutral and weakly alkaline conditions. B-Tris(trimethylsilylmethyl)-N-trimethylborazole is not decomposed by alkaline hydrogen peroxide at room temperature. At the same time, B,B-bis(trimethylsilyl)-B-butyl-N-trimethylborazole is not hydrolytically stable nor is the bicyclic compound with a tetramethylene bridge between the boron atoms (scheme 3-84). This and data presented previously indicate the great importance of steric factors in the hydrolytic stability of the silicon-containing borazoles.

B-Tris(trimethylsilylmethyl)-N-trimethylcyclotriborazan is not changed by heating for 0.5 hr in a sealed tube at 380°C. After heating for 3 hr at 300°C, the melting point is lowered to 54-56°C, but more than 90% of the product with the initial melting point

(59-60°C) can be recovered. At 400°C, these figures are reduced to 43-52°C and > 70%, respectively. After heating at 450°C, the product consists of a suspension of white crystals in a yellowish oil (recovery ~40%) and is completely and irreversibly converted into an oily liquid after heating at 500°C [1148].

B-Tris (triphenylsiloxy) cyclotriborazan is thermally unstable; it gives insoluble products even during recrystallization from dibutyl ether or ligroin and also on sublimation in vacuum [956].

3.2. Physical Properties and Application

The physical properties of various organosilicon compounds containing boron and nitrogen (with the exception of borazole derivatives) are given in Table 31.

For the complexes $R_3N \cdot BF_3$, the replacement of each methyl by a silyl radical raises the heat of formation by 9 kcal. There is a corresponding decrease in the rate of decomposition of the adducts at 25°C [1198]. The complexes of azidosilanes with BX_3 have a considerable stability and the liquid complex of methyldiphenylsilyl azide with BBr_3 does not change when heated for 2 hr at 200°C and distills without decomposition. Under these conditions the complex of trimethylsilyl azide dissociates, but it has a much higher stability than complexes of trimethylsilyl cyanides with BX_3, which decompose below room temperature [1208].

With compounds of the type R_2NBF_2 the vapor pressure rises sharply with a change from methyl to silyl derivatives. Disilylaminodifluoroborane also differs markedly in heat of evaporation and melting point from dimethylaminodifluoroborane [p = 0.2 mm (10°); H_{evap}19.1 kcal/mole, m. p. 162°C; cf. Table 31]. This indicates that disilylaminodifluoroborane is not associated. At the same time, the difference between disilylaminodichloroborane and its dimethyl analog [m. p. 43°C; p = 26 mm (25°)] is appreciably less [666]. Bis(methylsilylamino)difluoroborane forms an adduct (1:1) with trimethylamine and this indicates that boron retains its electron-acceptor properties in this compound [84].

Examples of the practical application of compounds containing Si—N—B or Si—B—N bonds are not yet found in the literature. Products of the reaction of compounds of the type $H_2N(CH_2)_xSi$-$(OC_2H_5)_yR$ (where x = 3 or 4; R = CH_3, OC_2H_5; y = 2 or 3) with

TABLE 31. Organosilicon Compounds Containing Boron and Nitrogen

Formula	m.p., °C	b.p., °C(mm)	Other properties and yields	Literature
	Complexes			
$H_3Si(CH_3)_2N \cdot B(CH_3)_3$	16.5	—	$p = 138$ mm (0°)	1197
$CH_3SiH_2(CH_3)_2N \cdot B(CH_3)_3$	—35	—	$p = 180$ mm (0°)	666
$(CH_3)_3Si(CH_3)_2N \cdot B(CH_3)_3$	—	—	$p = 218$ mm (0°)	666
$H(CH_3)_2Si(CH_3)_2N \cdot BH_3$	75-79, decomp	—	—	986
$(CH_3)_3Si(CH_3)_2N \cdot BH_3$	102—105, decomp.	—	—	986
$H[(CH_3)_3Si]_2N \cdot BH_3$	50-51, decomp	—	—	986
$H_3SiCN \cdot BH_3$	35	—	—	687
$(CH_3)_3SiCN \cdot BH_3$	69—69.3	—	—	687
$(CH_3)_3SiCN \cdot BH_2CN$	—	—	$p = 0.2$ mm (20°)	687
$(CH_3)_3SiCN \cdot BH_2Cl$	—	—	—	687
$(H_3Si)_3N \cdot BF_3$	—	—	H_{form} = 1 kcal/mole	1198
$CH_3(H_3Si)_2N \cdot BF_3$	—	—	H_{form} = 10 kcal/mole	1198
$(CH_3)_3Si(CH_3)_2N \cdot BF_3$	—	—	Involatile at 20°C; evap. at 100°C, decomp.	666
$H_3Si(CH_3)_2N \cdot BF_3$	—	—	H_{form} = 19 kcal/mole	1198
$CH_3SiH_2(CH_3)_2N \cdot BF_3$	—	—	solid	666
$H_3SiCN \cdot BF_3$	—	—	solid	687
$(CH_3)_3SiCN \cdot BF_3$	—	—	solid	687
$H_3SiCN \cdot BCl_3$	—	—	solid	687
$(CH_3)_3SiCN \cdot BCl_3$	—	—	solid	687
$(CH_3)_3SiN_3 \cdot BBr_3$	—	—	solid	1208
$CH_3(C_6H_5)_2SiN_3 \cdot BBr_3$	—	—	Liquid; dist. at 0.5 mm	1208
$(C_6H_5)_3SiN_3 \cdot BBr_3$	—	—	solid	1208

TABLE 31. (Cont'd)

Formula	m.p., °C	b.p., °C (mm)	Other properties and yields	Literature
		Silylaminohaloboranes		
$(H_3Si)_2NBF_2$	< —120	—	p=230 mm (0°); liquid H_{evap} = 7.2 kcal/mole	1198
$H_3Si(CH_3)NBF_2$	—	—	p = 23mm (0°); solid	1198
$(CH_3SiH_2)_2NBF_2$	—	—	Involatile at —96°C; stab. at 20°C; yield 80%	666
$(CH_3)_3SiNHBF_2$	92—104 part	—	—	986
	> 26, decomp.	—	—	544
$[(CH_3)_3Si]_2NBF_2$	—37.6	21 (5.2)	—	1045
$(H_3Si)_2NBCl_2$	< —78	—	$p = 22$ mm (25°)	666
$(CH_3)_3SiNHBCl_2$	> 310, decomp.	—	—	544
	> 300, decomp.	—	—	142
$[(CH_3)_3Si]_2NBCl_2$	5—6	82 (11)	n_D^{20} 1,4554	719, 720
$[(CH_3)_3SiNH]_2BF$	—	136 (11)	—	122
$\{[(CH_3)_3Si]_2N\}_2BF$	18	63 (0.5)	—	1045
	21	76 (0.2); 136 (11)	—	719, 720
$\{[(CH_3)_3Si]_2N\}_2BCl$	36	105 (0.1—2)	—	719, 720
$[(CH_3)_3Si]_2N[(C_2H_5)_2N]BCl$	—	113 (10)	n_D^{20} 1.4551	719
$[(CH_3)_3Si]_2N[(iso\text{-}C_3H_7)_2N]BCl$	—	78 (0.1)	n_D^{20} 1.4592	719
$[(CH_3)_3Si]_2N[(C_6H_{11})_2N]BCl$	77	140 (0.01)	—	719
$[(CH_3)_3Si]_2N[(C_6H_5)_2N]BCl$	101	—	—	719
$(CH_3)_3Si[(CH_3)_2N]BCl$	—	42 (9)	—	987
$(C_6H_5)_3Si[(CH_3)_2N]BCl$	135	150—160 (high vac.)	—	987

TABLE 31. (Cont'd)

Form	m.p., °C	b.p., °C(mm)	Other properties and yields	Literature
	Silylaminoboranes			
$(H_3Si)_2NBH_2$	—68.8—69.4	—	—	600
$H_3Si(CH_3)NB_2H_5$	—39	—	—	600
$(CH_3)_3SiNHB(C_3H_7\text{-n})_2$	—	70—71 (9)	—	986
$(CH_3)_3SiNHB(C_4H_9\text{-n})_2$	—	104—106 (15)	—	986
$(CH_3)_3SiNHBC_6H_5[N(CH_3)_2]$	—	82—83 (3)	Yield 21%	856
$Cl(CH_3)_2SiNHB(C_3H_7\text{-n})_2$	—	77—79 (1)	—	986
$(C_2H_5)_3SiNHB(C_5H_{11}\text{-n})_2$	—	114—117 (3)	Yield 67%	856
$(C_2H_5)_3SiNHBC_6H_5[N(CH_3)_2]$	—	119—122 (3)	Yield 79%	856
$(C_2H_5)_3SiNHB[N(CH_3)_2]_2$	—	83—91 (3)	Yield 84%	856
$Cl[(CH_3)_2SiNH]_3B(C_3H_7]_2$	—	125—130 (1)	—	986
$[(C_2H_5)_3SiNH]_2BC_6H_5$	—	146—154 (3)	Yield 59%	856
$[(C_6H_5)_3SiNH]_2BC_6H_5$	181—4	—	Yield 84%	856
$[(CH_3)_3Si]_2NB(OCH_3)_2$	—	69 (12)	n_D^{20} 1.4220	719
$[(CH_3)_3Si]_2NB(CH_3)N(C_2H_5)_2$	—	118 (10)	—	719
$(CH_3)_3Si]_2NB(CH_3)N(C_3H_7\text{-iso})_2$	—	137 (17)	—	719
$[(CH_3)_3Si]_2NB(C_6H_5)OC_2H_5$	—	82 (0.03)	n_D^{20} 1.4858	719
$(CH_3)_2Si\langle N[Si(CH_3)_3], N[Si(CH_3)_3] \rangle BC_6H_5$ (ring)	20—22	90—91 (1.2)	n_D^{20} 1.4783; Yield 55%	693
$[(CH_3)_3Si]_2NB—NSi(CH_3)_3$ / $(CH_3)_3SiN—BN[Si(CH_3)_3]_2$ (B–N bonded ring)	208	—	Yield 75%	1045
	212	—	subl. 110° (0.01)	719, 720
$[(C_2H_5)_3SiNH]_2BB[NHSi(C_2H_5)_3]_2$	—	—	Yield 20%	517
$[(CH_3)_2N]_2BSi(CH_3)_3$	—	65 (9)	—	987
$n\text{-}C_4H_9[(CH_3)_2N]BSi(C_6H_5)_3$	—	155—165 (high vac.)	—	987
$[(CH_3)_2N]_2BSi(C_6H_5)_3$	50	150—160 (high vac.)	—	987

triarylboranes $(XC_6H_4)_3B$, where X = H, o-Cl, o-CH_3O, are recommended as adhesives for bonding rubber to glass, asbestos, and steel [1660].

Table 32 gives the physicochemical constants of borazole derivatives containing silicon.

Borazoles with triorganosilylalkyl and triorganoalkoxy substituents are readily soluble in ether and toluene and sparingly soluble in hexane (the latter is used for recrystallization).

Data on the solubility of B-triorganosilyl substituted borazoles are contradictory. According to Seyferth [1151], B-tris(triphenylsilyl)-N-trimethylborazole dissolves readily in benzene, chloroform, and methylene chloride, dissolves with difficulty in ether and hexane, and is decomposed in carbon tetrachloride. Cowley and his co-workers [647] first reported that such compounds are soluble only in ether and then stated that they are soluble in ether, benzene, and carbon tetrachloride.

There are as yet no data on the practical use of borazole derivatives containing silicon.

4. OTHER BORON-CONTAINING ORGANOSILICON COMPOUNDS

The reaction of triphenylsilylithium with triphenylboron in tetrahydrofuran (nitrogen atsmosphere) yielded a white crystalline compound, which the authors named lithium triphenylsilyltriphenylborate [1152]:

$$(C_6H_5)_3SiLi + B(C_6H_5)_3 \longrightarrow Li[(C_6H_5)_3SiB(C_6H_5)_3] \qquad (3\text{-}87)$$

This substance was readily hydrolyzed, insoluble in hexane and carbon tetrachloride, but soluble and quite stable in methanol. Replacement of the lithium by other cations such as K, $[(CH_3)_4N]$, and $[(C_6H_5)_3PCH_3]$ led, as in the case of other "tetraarylborane anions," to the formation of insoluble compounds, which did not have a sharp melting point and decomposed over a wide temperature range. If the cation was lithium, decomposition began at about 100°C. The decomposition of a compound containing a quaternary phosphonium cation requires a temperature above 200°C. When the

TABLE 32. Derivatives of Borazoles Containing Silicon

R′	R″	m.p., °C	b.p., °C (mm)	Characteristic frequency in IR spectrum, cm^{-1}	Preparation method (scheme)	Yield, %	Literature
Compounds of the type $(R'BNR'')_3$							
$(C_6H_5)_3Si$	CH_3	248—251	—	1339	3-83 b	—	1151
		Viscous oil	—	1441, 1428, 1375	3-83 b	78 (99)	647
$(C_6H_5)_3Si$	C_6H_5	55—58	—	1495, 1428, 1360	3-83 b	90	647
$(CH_3)_3SiCH_2$	CH_3	64	—	1387	3-83 b	58	1151
		59—60	128—129 (0.45)	—	~3-84 *	5	1148
$C_2H_5(CH_3)_2SiCH_2$	CH_3	—	183—184 (1)	1387	3-83a	60	1151
n-$C_4H_9(CH_3)_2SiCH_2$	CH_3	—	208—217 (0.9)	1387	3-83a	57 (69)	1151
$(CH_3)_3SiOSi(CH_3)_2CH_2$	CH_3	—	175—176 (0.55)	1384	3-83a	50	1151
$(CH_3)_3SiCH_2CH_2$	C_6H_5	157—159	—	1377	3-85 + alkyl	76	1156
		132—133	—	1385	3-65	—	1161
$Cl(CH_3)_2SiCH_2CH_2$	C_6H_5	164—166	—	—	3-85	81	1156
$HO(CH_3)_2SiCH_2CH_2$	C_6H_5	137	—	1377	3-85 + hydrolysis	76	1156
$(CH_3)_3SiOSi(CH_3)_2CH_2CH_2$	C_6H_5	86—87	—	1375	3-85	57	1156
$Cl_2(CH_3)SiCH_2CH_2$	C_6H_5	170—172	—	—	3-85	—	1156
$(HO)_2(CH_3)SiCH_2CH_2$	C_6H_5	—	—	1380	3-85 + hydrolysis	59	1156
$CH_3[(CH_3)_3SiO]_2SiCH_2CH_2$	C_6H_5	—	Oil	—	3-85	—	1156
$(CH_3)_3SiO$	CH_3	22—23	129—131 (0.85)	1407	3-83 c	60	1151
$(C_6H_5)_3SiO$	H	164—7	—	—	3-83 c	Quant.	956
$(C_6H_5)_3SiO$	CH_3	206	—	—	3-83 c	Quant.	956
$(C_6H_5)_3SiO$	C_6H_5	234—235	—	—	3-83 c	Quant.	956
H	$Si(CH_3)_3$	—	104 (0.95)	—	3-86	—	986
F	SiH_3	—	—	—	3-65	—	1198
F	$Si(CH_3)_3$	—	—	—	3-65	—	1045
Cl	$Si(CH_3)_3$	140; subl 95 (0.2)	—	1325	~3-65 *	—	719, 720
C_6H_5	$Si(CH_3)_3$	156	—	—	3-65a	—	719

*See footnote to Table 30.

TABLE 32. (Cont'd)

Formula	m.p., °C	b.p., °C (mm)	Preparation method (scheme)	Yield, %	Literature
Other compounds					
CH_3 | N $(CH_3)_3SiCH_2$—B B—$CH_2Si(CH_3)_3$ | | H_3C—N N—CH_3 B | C_4H_9-n	—	—	3-84 +	—	1148
CH_3 | N $(CH_3)_3SiCH_2$—B B—$CH_2Si(CH_3)_3$ | | H_3C—N N—CH_3 B | Cl	—	105—108 (0.03)	3-84	39; 49	1148
CH_3 CH_3 | | N N R—B B—$(CH_2)_4$—B B—R | | | | H_3C—N N—CH_3 H_3C—N N—CH_3 B B | | R R †	93—94	—	3-84	58	1148
CH_3 CH_3 | | N N R—B B—O—B B—R | | | | H_3C—N N—CH_3 H_3C—N N—CH_3 B B | | R R †	95—96	—	—	—	1151

† R = $(CH_3)_3SiCH_2$.

Li derivative was treated with phenyllithium with the subsequent addition of bromobenzene, tetraphenylsilane was not formed.

The reaction of silane with $KB(C_4H_9\text{-}n)_2$ yielded potassium "dihydrobutylborate" [90]. With bromosilane the products were silane, tributylborane, and an involatile product, which contained a Si–B bond. It was suggested [78] that this was $H_3SiB(C_4H_9\text{-}n)_2$.

Silicon tetrachloride reacts with lithium borohydride in ether mainly with the formation of silane, diborane, and lithium chloride. However, at −115°C the complex $Li_2[(BH_4)_2SiCl_4]$ is also formed. The substance reacts with tetrahydrofuran (THF) to form lithium chloride, dichlorosilane, and the complex $THF \cdot BH_3$.

When heated above −80°C, each mole of adduct liberates one mole of hydrogen and one mole of borine (diborane), and above −60°C one mole of lithium chloride. The new complex $Li[H_3B \cdot SiCl_2 \cdot Cl]$ isolated from the decomposition products as the trietherate, in which the silicon was in the form of the dichloride, lost LiCl and BH_3 in its turn at 20°C and was converted into the complex $Li[H_3B \cdot Si_2Cl_4 \cdot Cl]$, which is soluble in tetrahydrofuran. The reaction of $Li[H_3B \cdot SiCl_3]$ with lithium borohydride at 0°C forms $Li[H_3B \cdot SiH_2Cl]$, which loses LiCl and BH_3 at room temperature to form $H_3B \cdot Si_2H_4$, which is stable in ether (etherate). By means of trimethylamine it is possible to remove from this substance all the BH_3 in its turn at 20°C and was converted into the complex $Li[H_3B \cdot$ silicon tetrachloride behave analogously to silicon tetrachloride in reactions with lithium borohydride [883, 985, 1259].

Fluorosilane and trimethylfluorosilane form 1:1 adducts with BF_3 at −128°C, while trimethylchlorosilane does not form a coordination compound with BCl_3 under the same conditions [689, 1186]. BCl_3 likewise does not react with trimethylchlorosilane or tetramethylsilane at room temperature [649], but in the presence of $\gamma\text{-}Al_2O_3$ or a silicon-copper alloy at room temperature or when the mixture is heated to 350°C there occurs a disproportionation, which has an equilibrium character:

$$(CH_3)_4Si + BCl_3 \rightleftarrows (CH_3)_3SiCl + CH_3BCl_2 \qquad (3\text{-}87a)$$

$$(CH_3)_3SiCl + BCl_3 \rightleftarrows (CH_3)_2SiCl_2 + CH_3BCl_2 \qquad (3\text{-}87b)$$

There are no signs of interaction in the systems $SiCl_4\text{–}BCl_3$ and $SiBr_4\text{–}BBr_3$ [398, 399].

The action of diboron tetrachloride on quartz forms Cl_3SiBCl_2 [943, 944].

ALUMINUM

5. COMPOUNDS CONTAINING THE Si – O – Al GROUP

5.1. Preparation Methods

5.1.1. Hydrolysis of Alkylhalosilanes and Alkylakoxysilanes by Aqueous Solutions of Alkali Metal Aluminates

In the hydrolysis of dimethyldichlorosilane by an aqueous solution of sodium aluminate obtained by dissolving aluminum chloride in aqueous alkali, the aluminum is present in the hydrolyzate exclusively in the alkaline medium [236, 456, 458]. If the amount of alkali is insufficient to neutralize the HCl liberated by hydrolysis, then after the neutral point has been passed the alumasiloxane polymer formed initially is cleaved and the liquid polysiloxanes obtained finally do not contain aluminum. The hydrolysis process forms a three-phase system, namely, a precipitate of NaCl and aqueous alkali and organic layers. The consistency of the organic component depends on the aluminum content of the polymer (as a rule, the Si:Al ratio in the products is somewhat higher than in the starting materials). With a ratio of Si:Al = 10 - 20:1, sticky, white, semisolid products are obtained. Over the range of ratios Si:Al = 30 – 50:1, greasy polymers are formed, which gel in air in the presence of alkali. Despite friability, infusibility, and very little flow, the gelling of these polymers is only relative as they

* There is as yet no single system of nomenclature for organoaluminum compounds. In the latest monograph on this subject [49] the compounds are treated as derivatives of the element (aluminum hydride, trialkylaluminum, etc.). To unify the nomenclature of heteroorganic compounds (see [25]), silicon-containing aluminum compounds are treated here mainly as derivatives of hypothetical AlH_3 and $H_2AlOAlH_2$. We propose to call then alumane (aluane according to [97] and alane according to [114]) and alumoxane, respectively. This agrees well with the use of the "oxa-aza" principle for denoting the replacement of a silicon atom by an aluminum atom in a polymer chain by introducing the prefix aluma-. The suggestion [66] that the group >ALR' should be denoted by the prefix alumo- is not appropriate.

are soluble in benzene. With a Si:Al ratio above 60, oily liquids are obtained which do not gel during prolonged storage.

Alkaline cohydrolysis of aluminum chloride with diphenyldichlorosilane in a ratio of 1:1-4 yields solid, friable, resinous polymers [127, 1370]. Here also the glassiness is not connected with a cross-linked structure. Products based on diphenyldichlorosilane with a molecular weight of about 4000 dissolve in ethanol as well as benzene and carbon tetrachloride. Analogous results are obtained in the case of dibenzyldichlorosilane and mixtures of bifunctional and trifunctional arylchlorosilanes [1370]. Bis(dichlorophenyl)dichlorosilane also gives a solid soluble polymer, but with a higher molecular weight (~6000) and a higher softening point [127].

Fractionation shows the quite considerable polydispersity of these polymers. However, while a polyalumadiphenylsiloxane is homogeneous in composition regardless of the molecular weight, in the case of its chlorinated analog the concentration of the dichlorophenyl component decreases with an increase in the molecular weight.

The alkaline cohydrolysis of a mixture of dimethylchlorosilane, phenyltrichlorosilane, and aluminum chloride in a molar ratio of 8:2:1 in a solvent yields [151] a liquid polymer, which is converted by prolonged heating at 200°C into an infusible, insoluble substance, which has a composition that differs appreciably from the composition of the starting mixture:

$$
-[-OSi(CH_3)_2\,\underset{\substack{| \\ O \\ |}}{OSi}(C_6H_5)-]_4-O-\underset{\substack{| \\ O \\ |}}{Al}-O-
$$

When ethyltrichlorosilane or phenyltrichlorosilane is used, the ratio Si:Al = 1:1 in the gel-like polymers (mol.wt. 1500-3000) formed by alkaline cohydrolysis is reached at pH = 10−12 [130, 177].

An attempt to obtain polyalumaorganosiloxanes in an acid medium by cohydrolysis of methyl(phenyl)trichlorosilane with aluminum sulfate was not successful either [14, p. 456]. Here also the prepration of a heterosiloxane requires an alkaline medium or in any case a neutral medium with an HCl acceptor.

When trimethylmethoxysilane is stirred with a solution of aluminum in 30% NaOH, after 2.5-3 hr there floats on the surface of the reaction mixture white, acicular crystals whose composition corresponds to sodium bis(trimethylsilyl) aluminate [33, 453, 1499]. For the formation of this it is necessary for the Si:Al ratio in the reaction mixture to lie within the range of 1:4 – 1:1. With an increase in the ratio to 2:1 – 3:1 or a rise in the hydrolysis temperature above 50°C, the reaction product is hexamethyldisiloxane. It is believed that the process occurs in two stages:

$$(CH_3)_3SiOSi(CH_3)_3 \xrightarrow[OH^-]{H_2O} (CH_3)_3SiOH \xrightarrow[H_2O,\ OH^-]{NaAlO_2} [(CH_3)_3SiO]_2AlONa \quad (3\text{-}88)$$

Triphenylethoxysilane behaves differently under the same conditions [329, 1505]. The reaction products arc tris(triphenylsiloxy)-silane, a white solid, which floats on the surface of the reaction mixture, and a white powdery precipitate, to which has been assigned the composition:

$$[(C_6H_5)_3SiO]_2AlO-\left[-Al\begin{matrix}\diagup ONa \\ \diagdown O-\end{matrix}\right]_5 Al[OSi(C_6H_5)_3]_2 \cdot 18H_2O$$

A study of the hydrolysis of dimethyldiethoxysilane [474] shows that the reaction may be divided into several stages. When dimethyldiethoxysilane is mixed with an aqueous solution of sodium aluminate, aluminum hydroxide and silanols are formed initially (cf. [329]). There is then homogenization of the system as a result of condensation, which is accompanied by an exothermic effect. The final stage of the process is the separation of the phases, and the formation of a stable system, which consists of solid products (condensation of silanols with aluminum hydroxide) and liquid products, which do not contain silicon or aluminum. An increase in the Si:Al ratio in the reaction mixture from 1:10 to 2:1 sharply reduces the reaction time (from 2 or more hours to a few minutes). Then with a further change in the Si:Al ratio to 10:1 the reaction rate remains constant and high. With high concentrations of aluminate the first separation of the mixture into two phases is not observed and this is evidently due to the stability of the aluminum hydroxide sol and the formation of sodium silanolates with the large amount of alkali in the mixture. Aqueous alkaline solutions of aluminates may be regarded as systems which are intermediate between true solutions and colloids, in which there exists a whole series of equilibria [474, 481]:

$$\underset{\text{true solution}}{NaAlO_2 + 2H_2O} \rightleftarrows 3NaOH + \underset{\text{sol}}{Al(OH)_3} \rightleftarrows 3NaOH + \underset{\text{gel}}{Al(OH)_3} \quad (3\text{-}89)$$

Therefore, dilution of the aluminate solution with water or alkali appreciably affects the hydrolysis of dimethyldiethoxysilane.

In particular, crystalline substances to which are assigned the following structures were isolated from the reaction products:

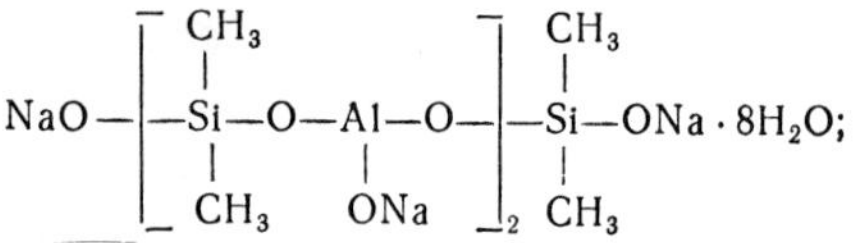

$$NaO-\left[-\underset{CH_3}{\overset{CH_3}{Si}}-O-\underset{ONa}{Al}-O-\right]_2-\underset{CH_3}{\overset{CH_3}{Si}}-ONa \cdot 8H_2O;$$

starting ratio Si:Al = 1

$$Na-\left[-\underset{CH_3}{\overset{CH_3}{Si}}-O-\underset{ONa}{Al}-O-\right]_2-ONa \cdot 3{,}5H_2O$$

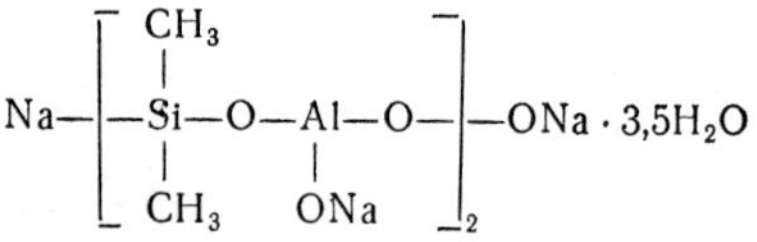

starting ratio Si:Al = 2

With the ratio Al:Si = 1:6, the solid reaction product contained a small number of ethoxyl groups.

The nature of the reaction of methyltriethoxysilane with aqueous solutions of sodium aluminate is very similar to the hydrolysis of dimethyldiethoxysilane [474]. However, the ratio of Si:Al in the hydrolysis system has a different effect on the reaction rate. With a change in the ratio from 1:10 to 1:3 the rate of the process increases, over the range 1:3 – 5:1 it hardly changes, and then it falls. Solid or sticky polymers containing residual ethoxyl groups are obtained depending on the initial Si:Al ratio.

The reaction of solutions of alkali metal aluminates with tetraalkoxysilanes forms alumosilicates and hydroalumosilicates [67, 332, 385]. The reaction rate depends on the size of the hydrocarbon radical in the alkoxyl group and falls with a change from CH_3 to C_4H_9. The hydrolysis of a mixture of tetraethoxysilane and aluminum acetate with aqueous alcohol [32, 321, 322] with a ratio

Si:Al = 3:1 – 6:1 yields polymeric substances containing about 24% of ethoxyl groups, which are not removed by prolonged heating at 120°C. These substances are chemical compounds and not a mixture of products from separate hydrolysis. With a small amount of water in the cohydrolysis system, ethyl acetate is liberated and not acetic acid and ethanol.*

5.1.2. Reactions of Halosilanes, Alkoxysilanes, and Acyloxysilanes with Aluminum Compounds

As early as 1892 Stokes [1191] put forward the following scheme to explain the reaction of alkoxytrichlorosilanes with aluminum chloride:

$$nCl_3SiOR + AlCl_3 \longrightarrow (Cl_3SiO)_n AlCl_{3-n} + nRCl \qquad (n = 1 \text{ or } 2) \qquad (3\text{-}90)$$

The compound obtained in the first stage then reacts with excess alkoxytrichlorosilane with the liberation of aluminum chloride. An alkyl chloride and polydichlorosilane $[SiOCl_2]_n$ are obtained finally.

The reaction of aluminum chloride with the products from the reaction of tetraalkoxysilanes and acetylene ends with the formation of the compound $Si(OAlCl_2)_4$ [508].

To explain the formation of polydiorganosiloxanes by "heterofunctional condensation" of dialkyldialkoxysilanes and dialkyldichlorosilanes in the presence of $AlCl_3$ [823, 825], a mechanism has also been proposed which is analogous to that put forward by Stokes for the case of alkoxytrichlorosilanes.

In discussing the possible routes for the reactions of aluminum halides with tetraalkoxysilanes, Kreshkov and his co-workers [335] postulated the following schemes, which are the most probable and simplest in their opinion:

$$3Si(OR)_4 + 4AlX_3 \rightleftarrows Al_4(SiO_4)_3 + 12RX \qquad (3\text{-}91)$$

* Kreshkov and his co-workers name the products from the cohydrolysis of $(C_2H_5O)_4Si$ with metal acetates organosilicon silicates. From this point of view, most of the substances described above are organoalumosilicates. We are not examining organosilicates especially. The literature data on this subject has been generalized in a review [43] and papers [57, 339].

$$3Si(OR)_4 + 4AlX_3 \rightleftarrows 4(RO)_3Al + 3SiX_4 \qquad (3-92)$$

$$3Si(OR)_4 + 2AlX_3 \longrightarrow 2ROAl\langle O_2 \rangle Si(OR)_2 + (RO)_2SiX_2 + 4RX \qquad (3-93)$$

Unfortunately these schemes were not checked experimentally and remain hypothetical. In any case there are no grounds for regarding them as "classical" as is done by the authors of [335]. In actual fact, according to the data of Stokes, reaction (3-91) (in the case of $AlCl_3$) does not occur in that form since ethers are formed in addition to alkyl chlorides. Likewise, halogen–alkoxyl exchange does not occur in the reaction of tetralkoxysilanes with aluminum halides (scheme 3-92) and the reaction proceeds according to Stokes' scheme with the intermediate formation of compounds of the type $(RO)_3SiOAlX_2$.* Likewise, there has been no experimental confirmation of the "compromise" scheme (3-93).

In the example listed above, the compounds containing the Si–O–Al group are only hypothetical intermediate reaction products. The reality of their existence in the reactions of aluminum halides with alkoxysilanes was first confirmed by the isolation of trimethylsiloxydichloroalumane from the products of the reaction of trimethylethoxysilane with aluminum chloride [116, 1168]:

$$R_3SiOR' + AlX_3 \longrightarrow R_3SiOAlX_2 + R'Cl \quad a) \qquad (3-94)$$
$$R_3SiOAlX_2 \longrightarrow R_3SiX + (AlOX)_n \quad b)$$

The quite stable compound formed in this way undergoes thermal decomposition with the formation of trimethylchlorosilane and polymeric aluminum oxychloride (3-94b).

The mechanism of the cleavage of the alkoxysilanes by aluminum halides is as follows [24]. A dimeric molecule of the halide coordinates with the oxygen atom of the OR group with the subsequent formation of an active cyclic transition complex as a

* The reaction of alkoxysilanes with aluminum halides has been studied in detail by Voronkov and Podchekaeva [54, 245]. According to their data, RO–halogen exchange is observed only when tetraaroxysilanes are treated with aluminum halides. A review [21] is devoted to a detailed discussion of the reaction of AlX_3 with organosilicon compounds, including alkoxysilanes.

result of the donation of electrons of the halogen atom to the most electrophilic center. Depending on the relative polarity of the Si–O and C–O bonds, this center may be the silicon or the carbon atom:

$$\begin{matrix} R & & AlX_2 \\ & O & \curvearrowleft X \\ \equiv Si & & AlX_2 \\ & X & \end{matrix} \longrightarrow ROAlX_2 + AlX_3 + \equiv SiX \qquad (3-95)$$

$$\begin{matrix} \equiv Si & & AlX_2 \\ & O & \curvearrowleft X \\ R & & AlX_2 \\ & X & \end{matrix} \longrightarrow \equiv SiOAlX_2 + AlX_3 + RX \qquad (3-96)$$

The character of the complex and consequently, the composition of its decomposition products are determined by the nature of the radical R and also the other substituent at the silicon. If R = alkyl, then both the Si–O and the C–O bonds may be cleaved. The greater the number of alkoxyl groups at the silicon atom, the stronger is the Si–O bond (as a result of $p_\pi - d_\pi$ bonding) and, consequently, the more readily the C–O bond is cleaved. When $R = C_6H_5$ the interaction of the unshared pairs of electrons of oxygen with the aromatic ring strengthens the C–O bond and cleavage occurs at the siloxane bonds.

It was proposed [318, 328] that the reaction of tetraethoxysilane with aluminum hydroxide proceeds according to the scheme

$$3Si(OC_2H_5)_4 + 4Al(OH)_3 \longrightarrow Al_4(SiO_4)_3 + 12C_2H_5OH \qquad (3-97)$$

Alcohol is actually liberated in the reaction of tetraalkoxysilanes with powdered aluminum hydroxide, but it is not aluminum silicate which is formed, but hydroalumosilicates, close to pyrophyllite ($H_2Al_2Si_4O_{12}$ or $Al_2O_3 \cdot 4SiO_2 \cdot H_2O$). Hydroalumosilicates are also obtained by hydrolysis of the products from the reaction of tetraethoxysilane with aluminum glycerate [480]. The final ratio Si:Al is determined by the pH of the medium.

In the reaction of aluminum alcoholates with organoalkoxysilanes, transesterification occurs [72]:

$$RSi(OR')_3 + (R''O)_3Al \longrightarrow RSi(OR'')_3 + (R'O)_3Al \qquad (3-98)$$

$$(R = C_6H_5;\quad R' = n\text{-}C_4H_9;\quad R'' = C_2H_5 \text{ or iso-}C_3H_7)$$

The reaction of trimethylacetoxysilane with aluminum butylate proceeds entirely according to an analogous scheme [285]:

$$3(CH_3)_3SiOCOCH_3 + (n\text{-}C_4H_9O)_3Al \longrightarrow (CH_3COO)_3Al + 3(CH_3)_3SiOC_4H_9\text{-}n \quad (3\text{-}99)$$

However, using aluminum isopropylate for this reaction completely changes the character of the process. The "acetoxy – alkoxy" exchange is suppressed and heterocondensation occurs instead. Depending on the reagent ratio and the experimental conditions it is possible to obtain the products of partial or complete replacement [75, 314, 955, 973, 996, 1048, 1285, 1678]:

$$nR_3SiOCOCH_3 + (iso\text{-}C_3H_7O)_3Al \longrightarrow (R_3SiO)_nAl(OC_3H_7\text{-}iso)_{3-n} + nCH_3COOC_3H_7\text{-}iso \quad (n = 1-3) \quad (3\text{-}100)$$

When aluminum butylate is heated with diorganodiacetoxysilanes in a molar ratio of 1:1 at 140–160°C, the liberation of butyl acetate is complete in 2–4 hr [167, 285]:

$$RR'Si(OCOCH_3)_2 + (R''O)_3Al \rightarrow CH_3COO[SiRR'OAl(OR'')]_nOR'' + CH_3COOR'' \quad (3\text{-}101)$$

The yield of ester is 63–68% when diethydiacetoxysilane and methylphenyldiacetoxysilane are used, but falls to 33% for dimethyldiacetoxysilane. This is connected with the effect of the radical at the silicon atom on the activity of the acetoxy groups. Together with the formation of butyl acetate there is replacement of acetoxy groups by butoxy groups. In the case of dimethyldiacetoxysilane the yield of dimethyldibutoxysilane reaches 44%. Correspondingly, the acetoxy group content of the alumasiloxane obtained is higher than according to scheme (3-101) in the "pure form." It is suggested [285] that this is connected with the possibility of the formation of different intermediate complexes. [See next page for (3-102)].

The possibility of the formation of a linear polymer containing an alkoxy group at the silicon atom by reaction (3-101) is confirmed indirectly by the fact that only alkoxy groups at aluminum participate in the heterocondensation of dimethyldiacetoxysilane with compounds such as $R_3SiOAl(OR')_2$ and alumasiloxane polymers are obtained with side triorganosiloxy groups with a molecular weight

$$R_2Si(OCOR')_2 + Al(OR'')_3 \rightarrow \text{a) } \xrightarrow{-R'COOR''} \text{cyclic } R_2Si(O)_2[Al(OR'')_2]OC(R')O; \quad \text{b) } R_2Si(OCOR')(OR'') + (R''O)_2AlOCOR' \quad (3\text{-}102)$$

up to 42,000 [404, 1046, 1048, 1132, 1504, 1678]:

$$R_2Si(OCOCH_3)_2 + R'_3SiOAl(OR'')_2 \longrightarrow CH_3COO[SiR_2OAl(OSiR'_3)]_n OR'' \quad (3\text{-}103)$$

The reaction is catalyzed by sodium ethylate. With a molar ratio of 1:2 it is possible to obtain an oligomer [996]:

$$(CH_3)_2Si\{OAl[OSi(CH_3)_3]OCOCH_3\}_2$$

In reaction (3-103) it is possible to replace the diorganodiacetoxysilane by acetic anhydride [1046, 1695]. With an equimolar ratio, triorganosiloxyalumoxanes are obtained:

$$R_3SiOAl(OR')_2 + (CH_3CO)_2O \longrightarrow (R_3SiOAlO)_n + 2CH_3COOR' \quad (3\text{-}104)$$

By varying the ratio of the reagents it is possible to obtain siloxyalumoxane oligomers with terminal alkoxy or acetoxy groups, i. e.:

$$R'O[Al(OSiR_3)O]_nR' \quad \text{or} \quad CH_3COO[Al(OSiR_3)O]_nCOCH_3$$

In the reaction of aluminum diisopropoxycaprylate with dimethyldichlorosilane or sym-dimethyldiphenyldichlorodisiloxane [218], isopropyl chloride is liberated. However, it is not linear polymers which are formed but infusible and insoluble products, evidently due to elimination of capryloxy groups under the action of traces of hydrogen chloride. An attempt to direct the process toward linear heterocondensation by using an internal complex compound, namely, (2-ketopent-3-en-4-oxy) dipropoxyalumane, was unsuccessful likewise. Its reaction with diorganodichlorosilanes is accompanied by "alkoxy–chlorine" exchange and not condensation [1221]. At the same time, the use of diethyldiacetoxysilane in the reaction leads to two products, one insoluble and the other soluble; the later has mol. wt. 1200 and is a linear polymer:

$$R'_2Si(OCOCH_3)_2 + (R''O)_2Al[O{-}C(R){=}CH{-}C(R){=}O\rightarrow] \longrightarrow CH_3COO\left[SiR'_2OAl\left(\overbrace{O{-}C(R){=}CH{-}C(R){=}O}\right)O\right]_n R'' \quad (3\text{-}105)$$

$$(R = CH_3;\ R' = C_2H_5;\ R'' = iso\text{-}C_3H_7)$$

By varying the components it is possible to prepare by this scheme copolymers of different consistencies (resins, waxes, and powders) with molecular weights up to 30,000 – 40,000.

In analogy with $(RO)_4Si$, the reaction of aluminum hydroxide with silicon tetrachloride is accompanied by the formation of products of the hydrosilicate type [328]. The exothermal effect of these processes falls in the series $Cl > CH_3O > C_2H_5O$. On the basis of this data and also ideas on the role of aluminum chloride in heterofunctional condensation reactions it is impossible to exclude the possibility of the formation of alumasiloxanes in the preparation of polyorganosiloxanes by the dry hydrolysis method, i.e., by the reaction of organochlorosilanes with aluminum hydroxide [1300].

5.1.3. Reactions of Silanols with Metallic Aluminum

In Chapter 2 we mentioned the author's certificate of Andrianov on the preparation of various metallasiloxanes by the reaction of silanols with metals [1461]. The following is one example of the use of this method:

$$3HOSiR_2OSiR_2OH + Al \longrightarrow (HOSiR_2OSiR_2O)_3Al + 1.5H_2 \quad (R = C_2H_5) \qquad (3\text{-}106)$$

The degree of conversion under the experimental conditions (1.5 hr at 250°C) according to the hydrogen liberated varies over a range of 44 to 69%. Despite the inadequate reproducibility of the process, this reaction has subsequently been mentioned repeatedly (practically in one context) as a classical example [5, 169, 178, 226, 286].

The reactions of silanols, silanediols, or siloxanediols with metallic aluminum, leading to the formation of compounds containing the Si–O–Al group, may be represented in a general form by the scheme

$$3 \equiv SiOH + Al \longrightarrow Al(OSi\equiv)_3 + 1.5H_2 \qquad (3\text{-}107)$$

In principle, the final result of these reactions is determined by the tendency of the hydroxyl-containing compounds to undergo homocondensation. With an increase in this tendency the possibility of reactions of type (3-107) falls and vice versa. In the case of sym-tetraethyldisiloxanediol, an average of 58% of the silanol groups participate in the reaction (i. e., only 0.38 g-atom of Al participates per mole of diol instead of 0.66 g-atom according to calculation). At the same time, the degree of reaction of triethylsilanol with aluminum is 88%. In the reaction of polyphenylsiloxanes obtained by hydrolysis of trichlorosilane and containing 6-8% OH groups with aluminum, only 5-9% of the total number of these groups participate in the reaction. Data on the reaction of metallic aluminum with diethylsilanediol disagree. Andrianov and his co-workers [169, 226, 286, 287] report that there is mainly condensation with the formation of polydiethylsiloxanes, while only 1-1.5% of the OH groups undergo reaction with the metal in 2 hr at 150°C. According to a patent [1404] in the reaction of diethylsilanediol with activated aluminum* the liberation of hydrogen is complete after 18 hr at room temperature, while the main product is a gray rubbery substance with the composition $(C_2H_5)_2SiO \cdot 0.1\,Al_2O_3$.

* The activators are acids or bases in combination with a solution of $HgCl_2$ [1404]. Tris(triethylsiloxy)alumane was obtained in 88% yield also by using $HgCl_2$ as the catalyst [169, 178, 182, 183, 286].

The reaction of aluminum with triethylsilanol gives a yield of 80-90% of crystalline tris(triethylsiloxy) alumane – $[(C_2H_5)_3SiO]_3$-Al [1404].*

The possibility of reaction (3-107) should be taken into account in explaining the high thermal stability of polysiloxane coatings pigmented with aluminum powder [7, 34, 545, 657, 800, 833, 1183, 1338, 1592, 1597].

Aqueous or aqueous alcohol solutions of so-called "siliconates" of alkali metals $RSi(OH)_2OM$ or $HO[SiR(OH)_x(OM)_{1-x}O]_n$ (x = 0 or 1) are known to be effective water-repellent compositions. However, the high alkalinity of solutions of these compounds (pH not less than 12) interferes with their application in a number of places. According to the data of Mai and his co-workers [45, 345, 347, 1501], when metallic aluminum is dissolved in solutions of "siliconates" hydrogen is liberated and "alumosiliconates" of alkali metals are formed. With a ratio of Si:Al = 3, a snow-white hygroscopic substance is obtained and this corresponds approximately in composition to the compound $[RSi(OH)(OM)O]_3Al$, and the following cyclic structure has been assigned to it:

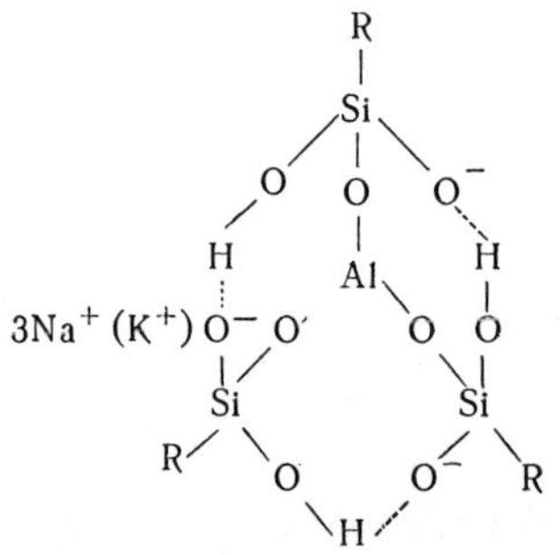

```
                          R
                          |
                          Si
                        / | \
                       O  O  O⁻
                      /   |   :
                     H    Al   H
                     :     \   |
     3Na⁺ (K⁺) O⁻  O       O   O
               |  /         \  |
               Si            Si
              /  \          /  \
             R    O        O⁻   R
                   \      /
                    H ...
```

The optimal regions of stability of solutions of this compound lie at pH values of 3-4 and above 10. By varying the Si:Al ratio in the solutions and by adding acids it is possible to obtain solutions of "alumosiliconates" which are stable over a wide range of pH values and are superior in water-repelling action to compositions based on alkali metal siliconates known previously.

*M. p. 310°C. The lower melting point for this compound (150°C) reported in [183] was incorrect because of the inadequate purity of the product [287]. Andrianov and his co-workers named this compound "nonaethylalumoxytrisiloxane," which gives no idea of its true structure when out of context.

5.1.4. Reactions of Silanols with Aluminum Compounds

When an equimolar mixture of a triorganosilanol and an aluminum alcoholate is heated, alcohol is liberated and vitreous monosubstituted products are obtained in good yield [75, 404, 1285, 1488]:

$$R_3SiOH + Al(OR')_3 \longrightarrow R_3SiOAl(OR')_2 + R'OH \qquad (3\text{-}108)$$
$$(R = C_2H_5,\ C_6H_5;\ R' = n\text{-}C_4H_9,\ iso\text{-}C_3H_7)$$

By increasing the mole fraction of silanol in the mixture it is possible to obtain bis and tris derivatives by this method [50, 404, 407, 651]. There is hardly any homocondensation of silanols, including trimethylsilanol, under the reaction conditions and this may be explained by intermediate complex formation.

The reaction of α,ω-polysiloxanediols with compounds of the type $Al(OR)_3$ may be represented by the general scheme:

$$HO(D)_n H + Al(OR)_2 OR' \longrightarrow [D_n Al(OR')O]_x + 2ROH \qquad (3\text{-}109)$$

$(D = (CH_3)_2SiO$ or $CH_3(C_6H_5)SiO;\ R = n\text{-}C_4H_9,\ iso\text{-}C_3H_7;\ R' = n\text{-}C_4H_9,$

$(C_2H_5)_3SiO,\ C_7H_{15}CO,\ CH_3COCH_2CO,$ or 8-hydroxyquinoline, etc. $n \geqslant 2)$

In the reaction of triethylsiloxydibutoxyalumane with sym-tetramethyldisiloxanediol (1:1), only butoxy groups attached to aluminum react and a waxy polymer is formed [404]:

$$n\text{-}C_4H_9O\left[\begin{array}{l}Al-O-Si(CH_3)_2O-Si(CH_3)_2O\\ |\\ OSi(C_2H_5)_3\end{array}\right]_n H$$

The molecular weight of the polymer is ~200 (n = 6). It is soluble and flows in the cold. Its infrared spectrum shows bands at frequencies characteristic of an eight-membered siloxane ring, but these are not very sharply expressed. These data and also the physical characteristics of the substance indicate that the polymer has a structure which includes intra- or intermolecular coordination bonds:

$$\begin{array}{l}\cdots -Si-O-Al--O-Si-\ddot{O}-\cdots\\ \qquad\qquad\quad \uparrow \qquad\qquad\ \downarrow\\ \cdots -Si-\ddot{O}-Si-O-Al-O-\cdots\end{array}$$

Like sym-tetramethyldisiloxanediol, higher α,ω-polydimethylsiloxanediols will also react with triorganosiloxydialkoxyalumanes [50, 404, 1515]. However, to obtain satisfactory results the triethylsiloxydibutoxyalumane, for example, should be in considerable excess over the stoichiometric amount (a three-fold excess with a diol of mol. wt. 1500-2400 and five-fold with a mol. wt. of 30,000).

Polyalumasiloxanes of the type

$$[Si(CH_3)_2O]_n[AlOSi(C_2H_5)_3O]_m \quad (n = 20 - 400; \quad m = 2 - 5)$$

are rubbery with the rigidity of the polymers and their flow points rising with an increase in the aluminum content. Vulcanizates of a polymer, obtained by using a siloxanediol with mol. wt. ~ 30,000, have physicomechanical properties equivalent to rubbers from a polydimethylsiloxane polymer with a considerably higher molecular weight (500,000); this indicates that the molecular weights of the two elastomers are of the same order. When alumasiloxane polymers are treated with acetylacetone (an agent capable of eliminating aluminum from the siloxane structure) the starting siloxanediols are regenerated.

As the bifunctional aluminum-containing components for condensation with α,ω-polydiorganosiloxanediols it is possible to use so-called "stabilized aluminum alcoholates" [8], i. e., compounds of the chelate type such as:

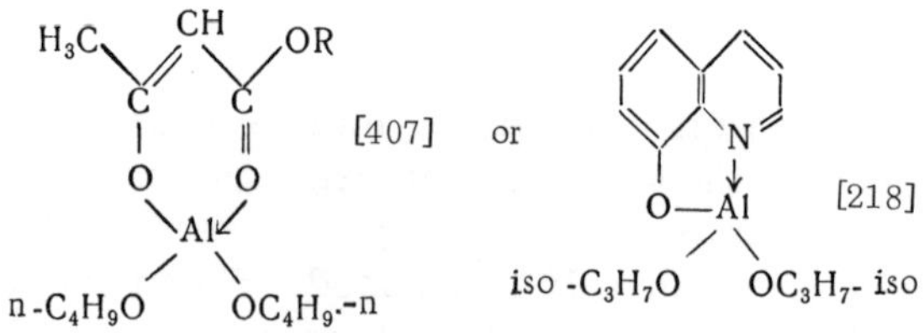

H_3C, CH, OR, C, C, O, O, Al, n-C_4H_9O, OC_4H_9-n [407] or N, O—Al, iso-C_3H_7O, OC_3H_7-iso [218]

In the reaction of aluminum derivatives of acetoacetic esters with polydimethylsiloxanediols the metal content of the rubbery polymer obtained depends only on the molecular weight of the starting diol.

The reaction of the 8-hydroxyquinoline derivative of aluminum with polymethylphenylsiloxanediols containing 2% OH groups for

1.5 hr at 200°C leads to the formation of a viscous, homogeneous, liquid polymer, solutions of which give fast-drying lacquer films. The condensation of diisopropoxycaprooxyalumane with α, ω-polymethylphenylsiloxanediols also forms elastic polymers with Si:Al = 5:1, which have a glass transition point of about −40°C [218]. Although the aluminum atoms in the chain are attached to C_7H_{15}-COO groups, prolonged heating of the polymer at 200°C does not produce cross linking.

The reaction of diethylsilanediol with aluminum isopropoxide yields an involatile product [651].* In the condensation of siloxanediols with aluminum butoxide, as in the reaction of α,ω-polysiloxanediols with boric esters (scheme 3-30) or diacetoxysilanes with aluminum alcholates (scheme 3-101), in the case of an equimolar ratio of the reagents, rubbery linear polymers are formed in the first stage. The aluminum atoms in them, which are incorporated into the siloxane structure, are attached to butoxyl groups [133, 167, 172].

Thus, for example, the condensation of $Al(OC_4H_9\text{-}n)_3$ with a siloxanediol with mol. wt. 2,100 in a ratio of 1:1 yields an alumasiloxane elastomer $\{[(CH_3)_2SiO]_{28}[Al(OC_4H_{9\text{-}n})O]\}_n$ with mol. wt. 30,000. However, because of the residual butoxyl groups the polymer readily cross links in air and gradually loses its solubility. If aluminum butoxide is heated with α,ω-polydimethylsiloxanediol (mol. wt. 5350) in a molar ratio of 1:3, at first there is evidently heterocondensation with the formation of a polymer of the type $Al\{[OSi(CH_3)_2]_{72}OH\}_3$. Subsequent condensation involves the terminal hydroxyl groups and a network structure is formed:

$$
\begin{array}{c}
\cdots -Al-O-D_{144}-\underset{|}{Al}-O-D_{144}-\overset{|}{Al}- \cdots \\
O \\
| \\
D_{144} \\
| \\
\cdots -Al-O-D_{144}-\underset{|}{Al}- \cdots \\
\quad\quad\quad\quad O-
\end{array}
\qquad [D = (CH_3)_2SiO]
$$

* The polycondensation of diphenylsilanediol by means of a mixture of lower alkoxy derivatives of aluminum and magnesium has been patented [1365]. The question of the incorporation of aluminum into the polymer structure was not mentioned in the patents.

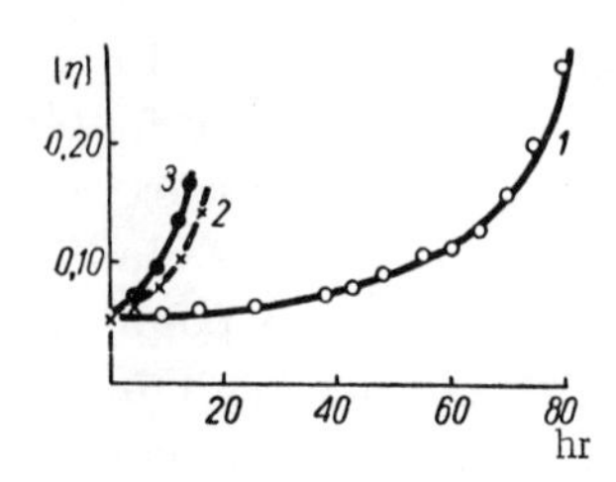

Fig. 5. Reaction of dimethylsiloxandediol (mol. wt. 5350) with aluminum butoxide [172]: 1) Molar ratio 3:1, 200°C; 2) 1:1, 180°C; 3) 1:1, 200°C.

This polymer has elastic properties, is soluble, and has a glass transition point of about −60°C. The retention of the elastic properties is explained [172] by the great length of the flexible chain segments between aluminum atoms. However, during prolonged heating the rigidity of the polymer increases and it loses its solubility (Fig. 5).

The condensation (3-109) may be achieved both by heating and also at room temperature and this accounts for the use of compounds of the type $(RO)_2AlOR'$ for curing liquid and solid organosilicon elastomers under normal conditions ("cold vulcanization") [50, 406, 1486].

According to the data of Andrianov [1461] it is possible to synthesize alumasiloxane polymers by condensation of silanediols with aluminum hydroxide. However, no experimental data on this reaction have appeared in the literature since the publication of the author's certificate (1948). At the same time, the reaction of triphenylsilanol with phthalocyaninoalumanol hydrate with azeotropic distillation of the water [998] yields $PcAlOSiR_3$ (where Pc = a phthalocyanine nucleus and R = C_6H_5). The use of phthalocyaninosilanediol in the process makes it possible to obtain oligomers (Fig. 6) with x = 1 or 2, and according to the data of Joyner and other investigators, in these polymers the flat "wheels" are set on an almost straight axis since the O−Si−O angle in the polymers equals 180° and the Si−O−Si angle is also close to 180°C.

The reaction of triphenylsilanol with trimethylalumane is accompanied by the liberation of methane [1097]:

$$2\,(C_6H_5)_3\,SiOH + 2\,(CH_3)_3\,Al \longrightarrow [(C_6H_5)_3\,SiOAl\,(CH_3)_2]_2 + 2CH_4 \quad (3\text{-}110)$$

The action of triethylalumane on low-molecular siloxanediols with a molar ratio of 1:1 yields oligomers, which are blocked at one end by an alkylalumane group [450]:

$$HO\,[Si\,(CH_3)\,RO]_n\,H + (C_2H_5)_3\,Al \longrightarrow HO\,[Si\,(CH_3)\,RO]_n\,Al\,(C_2H_5)_2 + C_2H_6 \quad (3\text{-}110a)$$

When n = 2.3 and R = CH_3, the yield of oligomers reaches 92-98%; however, it is somewhat lower (83%) when n = 5, R = C_6H_5. With an increase in the length of the chain of the siloxanediol (for example, when n = 37) there is simultaneous condensation with the introduction of aluminum atoms into the siloxane chain and oligomers of the following type are formed:

$$HO\,[Si\,(CH_3)\,RO]_n\,Al\,(C_2H_5)\,[OSi\,(CH_3)\,R]_n\,OAl\,(C_2H_5)_2$$

5.1.5. Reactions of Alkali Metal Silanolates with Aluminum Salts

Aluminum halides react with silanols analogously to boron halides. It is not possible to synthesize compounds with Si−O−M groups in this way. The analogy extends further and aluminum halides readily undergo the following reaction [8, 174, 722, 1075, 1083, 1102]:

$$3R_3SiOM + AlCl_3 \longrightarrow (R_3SiO)_3\,Al + 3MCl \quad (3\text{-}111)$$

In contrast to its boron analog, tris(trimethylxiloxy)alumane is a dimer with a planar four-membered coordination alumoxane

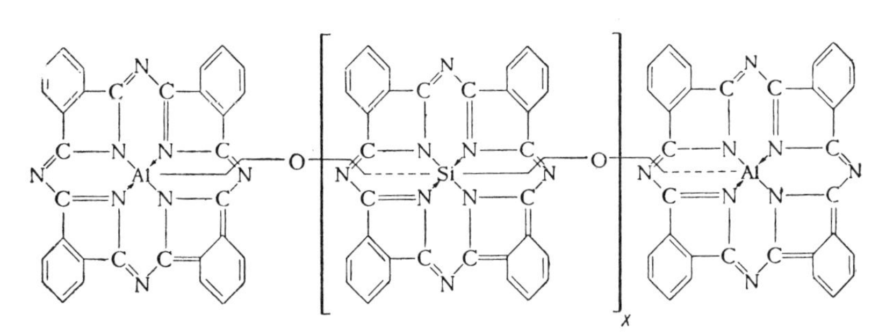

Fig. 6. Structure of phthalocyaninoalumasiloxanes [999].

ring [1075, 1102]: *

$$\begin{array}{c} \quad\quad\quad SiR_3 \\ R_3SiO\quad O\quad OSiR_3 \\ \quad Al\quad\quad Al \\ R_3SiO\quad O\quad OSiR_3 \\ R_3Si \end{array}$$

The elucidation of this structure was facilitated by the NMR spectra, in which two signals with an area ratio of 2:1 correspond to the $(CH_3)_3SiO$ groups. This indicates two types of group, namely, one in which the oxygen does not form part of a "bridge" and another in which it is a member of a ring and coordinated to an aluminum atom. At the same time, the NMR spectra of tris(trimethylsiloxy)borane show only one signal corresponding to the group $(CH_3)_3SiO$. The dimeric form of compounds of the type $R_3SiOAlX_2$ is specific to alumasiloxanes. Determination of the molecular weight of trimethyl(phenyl)siloxydibutoxyalumanes (see above) cryoscopically also showed that they are dimeric in solutions [50]. Trialkylsiloxydihaloalumanes are also dimers. It is not possible to obtain them by scheme (3-111),† but they may be used as intermediates in the preparation of tris(triorganosiloxy) alumanes [1075, 1099]:

$$[R_3SiOAlX_2]_2 + 4R_3SiOM \longrightarrow [(R_3SiO)_3Al]_2 + 4MX \qquad (3\text{-}112)$$
$$(M = Li,\ Na,\ \text{and}\ X = Cl)$$

By replacing aluminum chloride (which is known to be a dimer itself) by alkylaluminum halides it is possible to obtain mixed siloxyalkylalumanes [182] (cf. scheme 3-110):

$$R_3SiOM + XAlR_2 \longrightarrow R_3SiOAlR_2 + MX \qquad (3\text{-}113)$$

This equation has a formal character since both the starting R_2AlX and the compound obtained with a Si−O−Al bond are dimeric.

* Data on the melting point of tris(trimethylsiloxy) alumane disagree: 99°C [174] and 238°C [1075]. In [1102] the melting point is mistakenly reported as 208°C, which refers to tris(trimethylsiloxy) gallane; m. p. 101°C is reported in [973]. Since the melting point of tris(triethylsiloxy) alumane is above 300°C [1404], the most probable value for the m. p. is 238°C. The deviations may be connected with the compound not being dimeric under certain conditions.

† The preparation, structure, and properties of triorganosiloxydihaloalumanes will be examined in detail below (p. 27).

The reaction of potassium diphenylsilanediolate with aluminum chloride in a molar ratio of 3:2 yields an insoluble powdery polymer [995]. However, the reaction of the Na analog with (2-ketopent-2-en-4-oxy)dichloroalumane in benzene [1221] leads to the formation of a resinous soluble product with mol. wt. ~600, i. e.,

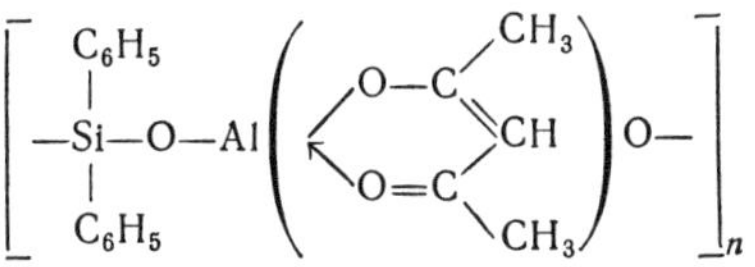

$$\left[-\underset{C_6H_5}{\overset{C_6H_5}{\underset{|}{\overset{|}{Si}}}}-O-Al\left(\begin{matrix} O-C(CH_3) \\ \quad \quad \; CH \\ O=C(CH_3) \end{matrix} \right)O- \right]_n \qquad (n \sim 2)$$

This is evidently a cyclic compound.

The reaction of aluminum halides with sodium siloxanediolates $NaO[RR'SIO]_nNa \cdot xH_2O$ (where $R = CH_2$; $R' = CH_3$ or C_6H_5; $n = 2-4$) [166, 199, 226] leads to the formation of polydimethylalumasiloxanes, whose properties are determined primarily by their aluminum content. When the reaction is carried out in a solvent the polymer solutions formed are stable, but after removal of the solvent the products with a ratio of Si:Al = 0.8−1.3 are converted into an insoluble state in which they do not flow or soften over the temperature range of 20-650°C, indicating cross linking. A polymer with a ratio Si:Al = 3.2 dissolves partially in polar solvents, but is insoluble in nonpolar solvents. Products with a ratio Si:Al $\geqslant$ 3.8 do not lose their solubility, even after heating for 6 hr at 150°C.

The reaction of aluminum salts with monosodium salts of silanetriols is of great interest for practical purposes. As has already been reported, dissolving polysiloxanes of the type $(RSiO_{1.5})_n$ in alkalis leads, in particular, to salt-like compounds of the general form $RSi(OH)_2ONa \cdot xH_2O$, where R = alkyl or aryl. Their reaction with solutions of aluminum salts [$AlCl_3$, $Al_2(SO_4)_3$, $KAl(SO_4)_2$] gives the same result. In the first stage aluminum silanolates of the type $[RSi(OH)_2O]_3Al$ are formed [8, 12, 14, 166, 175]. They then undergo anhydrocondensation or cocondensation with organosilanetriols, formed by partial hydrolysis of the monosodium salt of the triol by water present in the system. A rise in temperature improves the polymer yield. In this case the heating time is of no importance. The most favorable medium is a mixture of toluene and butanol (1:1) and a concentration of the starting polyphenylsiloxane of 15%. In this way it has been possible to obtain polyalumaethylsiloxanes with a Si:Al ratio from 4 to 56 and polyalumaphenylsiloxanes with a Si:Al ratio from 1 to 70 with a molecular weight from 2000 to 40,000. It was observed that the molecular weight of the polymers increases with an increase in their aluminum content. It is higher for ethyl substituted polymers than for polyalumaphenylsiloxane and this may be explained by steric factors. In principle the synthesis of polyalumaorganosiloxanes may also be carried out without the intermediate solution of the sodium salts of organosilanetriols.

5.1.6. Cleavage of Siloxanes by Halides and Other Compounds of Aluminum

The cleavage of hexaalkyldisiloxanes by aluminum halides, which was discovered in 1952 [250, 1462], may be represented in the general form by the following scheme:

$$R_3SiOSiR_3 \xrightarrow{AlX_3} R_3SiOAlX_2 + R_3SiX \quad a) \qquad (3\text{-}114)$$

$$\xrightarrow{AlX_3} R_3SiX + \frac{1}{n}[AlOX]_n \quad b)$$

$$(R = \text{alkyl}; \quad X = Cl, \ Br, \ I)$$

The corresponding trialkylhalosilane distills smoothly from a mixture of hexaalkyldisiloxane and the aluminum halide (yield 76-87%). It is advantageous to replace aluminum iodide by a mixture of aluminum and iodine and this increases the yield of trialkyliodosilanes from 76-80% to 88-93% [256]. The reaction of hexaorganodisiloxanes with anhydrous aluminum halides is a simple and convenient method of synthesizing trialkylhalosilanes.

The intermediate formation of a triorganosiloxydihaloalumane (scheme 3-114a) was first only a working hypothesis [250]. However, it was soon established [51, 408, 414, 415, 1468] that by distilling about 80% of the trialkylhalosilane formed or stopping the reaction with the temperature of the mixture 20-30°C above the boiling point of the starting hexaalkyldisiloxane, compounds of the type $R_3SiOAlX_2$ could be obtained in 60-85% yield.

It was shown subsequently that the reaction proceeds under even milder conditions, namely, when AlX_3 is dissolved in a hexaalkyldisiloxane at 40°C [646] and when a reaction mixture in hexane is heated [1093, 1098]. In this case it is possible to achieve an almost quantitative yield (96%) of the triorganosiloxydihaloalumane.

Originally it was possible to stop the reaction at stage (3-114a) only in the case of aluminum chloride and bromide [51, 646]. Then iodine derivatives were isolated [1093] (reaction in the dark in hexane at 0°C). However, triorganosiloxydifluoroalumanes have not been obtained up to now.

When sym-tetramethyldiethyldisiloxane was distilled over aluminum chloride, in addition to the expected ethyldimethylchlorosilane, trimethylchlorosilane and methyldiethylchlorosilane were detected among the reaction products [408]. This must be ascribed to the secondary process of disproportionation of ethyldimethylchlorosilane under the action of aluminum chloride [21].

In the reaction of aluminum halides with disiloxane – $H_3SiOSiH_3$ – the reaction proceeds mainly according to scheme (3-114a) [902], but the reaction products contain H_2SiX_2 and SiH_4 together with H_3SiX (72-91% yield). Their formation cannot be explained by disproportionation since H_3SiCl, for example, does not undergo changes under the reaction conditions under the action of $AlCl_3$. It may be surmised therefore that in this case scheme (3-114) is modified:

$$H_3SiOSiH_3 \xrightarrow{AlX_3} H_3SiOAlX_2 + H_3SiX \longrightarrow SiH_4 + H_2SiX_2 + \frac{1}{n}[AlOX]_n \qquad (3\text{-}115)$$

The reaction of disiloxane with dimethylaluminum bromide forms $H_3SiOAl(CH_3)_2$.

The mechanism of the cleavage of siloxanes by aluminum halides may be treated on the same basis as in the case of boron halides or ether electrophilic reagents [24, 246]. In this case the dimeric state of aluminum halides favors the explanation from the point of view of the possibility of the intermediate formation of active cyclic 6-membered transition complexes. Coordination compounds are evidently formed initially and these are quite stable adducts, which may be identified, in particular, in the reaction of disiloxane with aluminum halides at −78°C [902]. They undergo subsequent cyclization and decomposition:

$$\text{[cyclic complex: } {>}Si{-}O({\to}AlX_2){-}Si{<} \text{ bridged by } X \text{ to } AlX_2\text{]} \longrightarrow {>}Si{-}O{-}AlX_2 + AlX_3 + {>}Si{-}X$$

In the light of data presented in the examination of the reaction of disiloxanes with boron halides it is not suprising that hexachlorodisiloxane does not change during prolonged boiling with an equimolar amount of aluminum chloride [51].

Interesting results are obtained by studying the cleavage by aluminum halides of Si−O−M bonds in heterosiloxanes of Group IV elements [1092, 1093, 1105]. When trimethylsiloxytrimethylgermane is treated with aluminum chloride at 0°C, cleavage occurs at the Ge−O bond [1105]:

$$R_3SiOMR_3 + AlX_3 \longrightarrow R_3MX + R_3SiOAlX_2 \quad (3\text{-}116)$$

Since the polarity of Sn−O and Pb−O bonds is even higher, it was expected that the cleavage by aluminum chloride of trimethylsiloxytrimethylstannane and trimethylsiloxytrimethylplumbane would correspond even more closely to scheme (3-116). However, in actual fact [1092, 1093], the reaction is more complex. The first stage is cleavage according to scheme (3-116), but the high reactivity of the Sn−O and Pb−O bonds stimulates the secondary process:

$$R_3SiOMR_3 + R_3SiOAlX_2 \longrightarrow R_3MX + (R_3SiO)_2\,AlX \quad (3\text{-}117)$$

Unstable bis(trimethylsiloxy)chloroalumane decomposes with the formation of $(AlOCl)_n$ and hexamethyldisiloxane, which reacts in accordance with scheme (3-114a). As a result, trimethylchlorosilane appears among the products of a reaction which is carried out under conditions most favorable to scheme (3-116).*

Data on the action of AlX_3 on cyclosiloxanes are contradictory. According to the data of Hyde [1317, 1396, 1574, 1656] the reaction of $AlCl_3$ with $(R_2SiO)_n$ ($R = CH_3$, C_2H_5 or C_6H_5; $n = 3-4$, in a ratio of 1:1 or 4:3 (heating at 110-120°C for 12-48 hr) forms crystalline cyclodialumatetrasiloxanes:

$$\begin{array}{ccc} ClAl-O-SiR_2 & & ClAl-O-AlCl \\ | \qquad\quad | & \text{or} & | \qquad\quad | \\ O \qquad\quad O & & O \qquad\quad O \\ | \qquad\quad | & & | \qquad\quad | \\ R_2Si-O-AlCl & & R_2Si-O-SiR_2 \end{array}$$

* Strong cooling of the reaction mixture when M = Sn [1093] and a solvent and strong cooling when M = Pb [1092, 1093]. Without a solvent the reaction of trimethylsiloxytrimethylplumbane with $AlCl_3$ proceeds explosively and a lead mirror is formed [1092].

Linear oligomers $Cl(R_2SiO)_nSiR_2Cl$ (n = 1 – 4) are obtained at the same time.

Andrianov and his co-workers [288] state, however, that under analogous conditions (heating at 120°C for 10 hr) the reaction of aluminum chloride with $(R_2SiO)_n$ ($R = CH_3$ or C_2H_5) in a molar ratio of 4:3 (n = 4) or 3:2 (n = 3, regardless of the nature of the radical at silicon or the number of units in the ring, led to the crystalline compounds, which distill in vacuum, are sensitive to the action of moisture, and have the composition $R_8O_6Cl_5Si_4Al_3$.* Treatment of them with acetylacetone extracts only a third of the aluminum which is present according analysis. This led to the conclusion that the compounds have a complex structure in which two molecules of 1-halo-3,3,5,5-tetraalkyl-1-alumatrisiloxane are connected to a AlX_3 molecule:

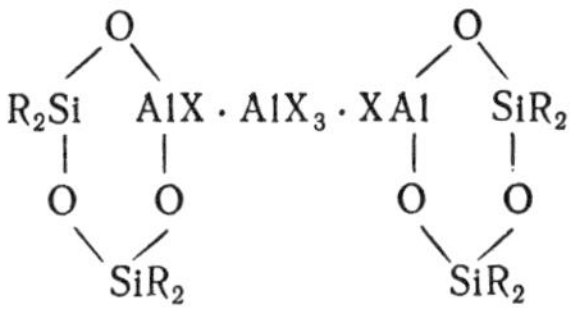

$$\underbrace{R_2Si\overset{O}{\frown}AlX}_{\text{ring: }R_2Si-O-AlX,\ R_2Si-O-SiR_2-O-AlX} \cdot AlX_3 \cdot \underbrace{XAl\overset{O}{\frown}SiR_2}_{\text{ring: }XAl-O-SiR_2,\ XAl-O-SiR_2-O-SiR_2}$$

In addition to these compounds, α,ω-dihalopolydialkylsiloxanes $X[R_2SiO]_m SiR_2X$ are formed and in the case when X = Cl and n = 4, dichlorides are obtained with m = 1 – 3, while in the case where n = 3, m = 0 or 1, i. e., dialkyldichlorosilanes are also isolated. The following scheme has been proposed:

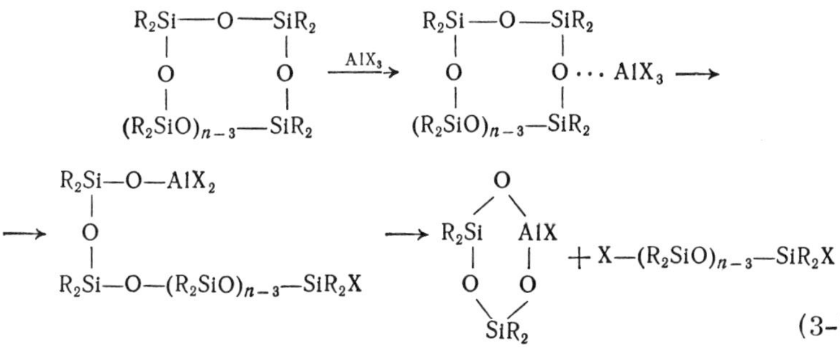

$$\begin{array}{c} R_2Si-O-SiR_2 \\ | \qquad\quad | \\ O \qquad\quad O \\ | \qquad\quad | \\ (R_2SiO)_{n-3}-SiR_2 \end{array} \xrightarrow{AlX_3} \begin{array}{c} R_2Si-O-SiR_2 \\ | \qquad\quad | \\ O \qquad\quad O\cdots AlX_3 \\ | \qquad\quad | \\ (R_2SiO)_{n-3}-SiR_2 \end{array} \longrightarrow$$

$$\longrightarrow \begin{array}{c} R_2Si-O-AlX_2 \\ | \\ O \\ | \\ R_2Si-O-(R_2SiO)_{n-3}-SiR_2X \end{array} \longrightarrow \begin{array}{c} O \\ R_2Si \quad AlX \\ | \qquad | \\ O \qquad O \\ SiR_2 \end{array} + X-(R_2SiO)_{n-3}-SiR_2X \qquad (3\text{-}118)$$

Alumacyclosilanes then form complexes with AlX_3, while sym-dihalotetralkyldisiloxanes, which are obtained when n = 4, undergo

*Analogous results are obtained by cleavage with $AlBr_3$ and AlI_3 [189, 685, 1087].

rearrangements under the action of aluminum halides and these lead to linear α,ω-dihalopolydiorganosiloxanes.

On this basis we can also explain the formation of α,ω-dichlorosiloxanes in the cleavage of cyclosiloxanes by organochlorosilanes under the action of metal halides, particularly aluminum chloride [212].

Scheme (3-118) satisfactorily explains the formation of six-membered alumatrisiloxane rings both from trisiloxane and from tetrasiloxane and also the origin of dialkyldichlorosilanes, which are obtained in the reaction of aluminum chloride with $(R_2SiO)_3$. However, the views of Andrianov and his co-workers both on the character of the process and on the structure of the final compounds have encountered objections, especially in recent Italian investigations [613, 645, 682, 683, 685]:

1) Scheme (3-118) gives no explanation why α,ω-dichlorosiloxanes are formed in the latter case in addition to R_2SiCl_2 and why the proposed rearrangement of sym-dichlorotetraalkyldisiloxanes does not reduce to simple disproportionation under the action of $AlCl_3$ with the formation of not only higher, but also lower members of the homologous series, i. e., R_2SiCl_2;

2) The compound $(CH_3)_8O_6Cl_5Si_4Al_3$ may be sublimed repeatedly in vacuum without change and this is impossible for a complex containing $AlCl_3$;

3) This compound is also obtained by the reaction of octamethylcyclotetrasiloxane with $(CH_3)_3SiOAlCl_2$ (dimer), i. e., in a process in which $AlCl_3$ does not participate;

4) The formation of aluminum acetylacetonate is not a direct demonstration of the presence of molecularly bound $AlCl_3$; the same result may be obtained with the grouping $>SiOAlCl_2$.

Moreover, the Italian chemists showed that the nature of the products from the reaction of $[(CH_3)_2SiO]_4$ and AlX_3 depends substantially on their molar ratio and the temperature of the process. Thus, with a ratio $R_2SiO:AlCl_3 = 1:1$ there are formed dimethyldichlorosilane, dimeric chlorodimethylsiloxydichloroalumane, and a glassy soluble polymer with a molecular weight of about 1000, corresponding to $[(CH_3)_2SiOAl(Cl)O]_n$.

This product and also its bromine-containing analog (see below) are unchanged by prolonged heating at 250-300°C and this disproves the hypothesis that they contain AlX_3 in any coordinated form. At the same time, part of the aluminum is extracted in the form of the tris-acetylacetonate when the polymers are treated with sodium acetylacetonate (see below).

With excess octamethylcyclotetrasiloxane (a molar ratio of 3:1) the results of the interaction coincide with the data in [288]. The so-called "universal compound" $(CH_3)_8O_6Cl_5Si_4Al_3$ (I) and α,ω-dichloropolydimethylsiloxanes are formed. Analogously, the reaction of $[(CH_3)_2SiO]_4$ and $AlBr_3$ in a molar ratio of 4:3 yields (28 hr, 150°C), α,ω-dibromopolydimethylsiloxanes $Br(R_2SiO)_nSiR_2Br$ (n = 1 – 3) and $CH_3)_8O_6Br_5Si_4Al_3$ (Ia).

The "universal compound" is also obtained by the reaction of $[(CH_3)_2SiO]_n$ (n = 3 or 4) with $Cl(CH_3)_2SiOAlCl_2$ [645] and with $(CH_3)_3SiOAlCl_2$ [613, 682, 683]. However, here the character of the products depends on the reagent ratio and the reaction temperature. The following schemes may be put forward (where R = CH_3):

$$[R_2SiO]_4 + 3R_3SiOAlCl_2 \longrightarrow C_8H_{24}O_6Cl_5Si_4Al_3 + R_3SiCl + (R_3Si)_2O \qquad (3\text{-}119a)$$

$$n\,[R_2SiO]_4 + 4nR_3SiOAlCl_2 \longrightarrow 4nR_3SiCl + [R_2SiOAl(Cl)O]_{4n} \qquad (3\text{-}119b)$$

These equations are not realized exactly as written, but only illustrate the direction which corresponds most closely to the composition of the products. When the mixture is heated at 160-170°C for 25 hr the interaction corresponds to the greatest degree to scheme (3-119a),* but a polymer of the type $[R_2SiOAl(Cl)O]_n$ is also formed. With molar ratios of $[(CH_3)_2SiO]_4 : (CH_3)_3SiOAlCl_2$ = 1:2, 1:3, and 1:4 and a temperature of 230-250°C, the results of the process may correspond to the scheme (3-119b), though there is also a tendency for enrichment of the polymer in aluminum at the expense of the silicon content [683].

With a ratio of 1:4 the composition of the polymer is close to the formula $[(CH_3)_6Al_4Cl_4O_7Si_3]_n$ with n equal to ~10 (mol. wt. 5240). With any ratio of the starting materials the polymers obtained are solid glassy amorphous substances, which do not melt below 300°C, are readily soluble in cold benzene, and are hydrolyzed readily by atmospheric moisture.

* The same as when R = C_2H_5 [682].

With a molar ratio $[(CH_3)_2SiO]_4:(CH_3)_3SiOAlBr_2 = 1:4$, heating the mixture for 25 hr at 160–170°C leads to the formation of trimethylbromosilane, compound Ia, and a solid glassy polymer with a molecular weight of 2540, which is close in composition to $[(CH_3)_2SiOAl(Br)O]_n$ [683]. Under more drastic conditions (10 hr at 170–190°C and 10 hr at 210–230°C) a polymer is formed which contains Al, Br, and Si in a ratio of 1.7:2:1 with mol. wt. 2810 [685].

X-ray and chemical investigations of the compounds $(CH_3)_8$-$O_6X_5Si_4Al_3$ (X = Cl or Br) indicate, in the opinion of the Italian chemists [564a, 613, 685], a structure with a central pentacoordinate aluminum atom with two four-membered alumoxane rings, analogous to those found in dimeric triorganosiloxydihaloalumanes:

```
                 O
               /   \
          R2Si       SiR2
            |    X    |
 X          O    |    O          X
   \      /   \  |  /   \      /
     Al ←       Al ←       → Al
   /      \   ↗     \   ↗      \
 X          O         O          X
            |         |
          R2Si       SiR2
               \   /
                 O
```

In the processes described above, compounds of type I evidently play the part of a transition complex and this is confirmed, on the one hand (when X = Cl), by the conversion into the polymer $[(CH_3)_2SiOAl(Cl)O]_n$ with the liberation of dimethyldichlorosilane at 225°C and on the other hand, by the formation of chlorodimethylsiloxydichloroalumane by the action of aluminum chloride [645].

An examination of the reactions of cyclodialkylsiloxanes with aluminum halides leads to the conclusion that there is mainly limited polymerization (telomerization) with the participation of AlX_3 with subsequent intramolecular formation of a ring, intermolecular rearrangements, disproportionation, etc. The polymerization of cyclosiloxanes under the action of AlX_3 has been described in the literature [1538, 1564]. On the other hand, in the reactions of α,ω-dichloropolydimethylsiloxanes with aluminum chloride it was possible to demonstrate the formation of $Cl(CH_3)_2$-$SiOAlCl_2$, a compound which, as it shown by the data above, plays an appreciable part in the cleavage of cyclodiorganosiloxanes by aluminum chloride:

$$(ClR_2Si)_2O \xrightarrow{AlCl_3} R_2SiCl_2 + ClR_2SiOAlCl_2 \xrightarrow{170^\circ C} R_2SiCl_2 + AlCl_3 + [R_2SiOAl(Cl)O]_n \quad (3\text{-}120)$$

In presenting data on the reaction of cyclodimethylsiloxanes with aluminum chloride we should point out that their slow distillation over $AlCl_3$ with a molar ratio $[(CH_3)_2SiO]_n : AlCl_3 = 4-6$ leads to the formation of dimethyldichlorosilane in 60% yield [3]. The patents of Jack are interesting in this respect. One of them [1638] recommends the use of waste dimethylsiloxane rubber and mixtures of it by heating them with 5% $AlCl_3$ at 300°C. This yields up to 80% of a mixture of cyclodimethylsiloxanes. In another patent by this author [1361, 1434] it is reported, on the contrary, that heating jelly-like hydrogen-containing siloxanes with aluminum chloride leads to the formation of methylhydrochlorosilanes.

Prolonged contact of hexaphenylcyclotrisiloxane with excess aluminum chloride in chloroform yielded [690] a distillate which fumed in air and was hydrolyzed with the formation of silicic acid, and a sticky polymer, containing phenyl groups at silicon atoms. All these data again indicate the complex character of the reaction of aluminum halides with siloxanes.

It was expected that the reaction of hexaalkyldisiloxanes with triethylaluminum would proceed by a scheme of type (3-114a). A trialkylsiloxydiethylalumane is actually formed in this reaction. However, ethylene and a trialkylsilane are obtained instead of a trialkylethylsilane [855, 1414]:

$$R_3SiOSiR_3 + Al(C_2H_5)_3 \longrightarrow R_3SiOAl(C_2H_5)_2 + CH_2{=}CH_2 + R_3SiH \quad (3\text{-}121)$$

When $R = CH_3$, bis(trimethylsiloxy)ethylalumane is formed together with trimethylsiloxydiethylalumane. Trimethylalumane does not undergo reaction (3-121) [855], but the reaction of hexamethyldisiloxane with $(iso\text{-}C_4H_9)_2AlH$ [1417] is described by the equation:

$$R_3SiOSiR_3 + R'_2AlH \longrightarrow R_3SiOAlR'_2 + R_3SiH$$

Therefore, the hypothesis was put forward that under the conditions studied there is first decomposition of the triethylaluminum, while the more active diethylalumane formed cleaves the quite stable molecule of the hexaalkyldisiloxane. Alkylhaloalumanes react with siloxane analogously [855, 1355, 1360, 1655, 1735]:

$$R_3SiOSiR_3 + R_3Al_2Cl_3 \longrightarrow R_4Si + R_3SiOAlCl_2 \qquad (3\text{-}123)$$

$$R_3SiOSiR_3 + R'_2AlCl \longrightarrow R_3SiOAlR'Cl + R_3SiH + R'_{-H} \qquad (3\text{-}124)$$

$$(R = CH_3;\ R' = C_2H_5;\ R'_{-H} = CH_2 = CH_2)$$

Depending on the molar ratio of the reagents, in the reaction of trialkylalumanes with octamethylcyclotetrasiloxane there is either simple opening of the ring or more extensive breakdown [855, 1006]:

$$nR'_3Al + (R_2SiO)_4 \longrightarrow nR'_2[-OSiR_2-]_{4/n}R' \qquad (3\text{-}125)$$

$$(R = CH_3;\ R' = CH_3,\ C_2H_5)$$

The oligomers obtained are readily hydrolyzed with the formation of ethane, aluminum hydroxide, and an alkylsil(ox)anol, which condenses to dimers.

Heating triethylaluminum with high-molecular polysiloxanes of the series $[CH_3SiO_{1.5}]_n$ and $[(CH_3)_2SiO]_n$ is also accompanied by the formation of products of the alumasiloxane type [855, 1408, 1420]. Their hydrolysis leads in the first case to methylethylpolysiloxanes and in the second, to dimethylpolysiloxanes of lower molecular weight than the starting polymers.

In the light of the data presented, it is not surprising that it has been reported [704] that the action of lithium aluminum hydride on polydialkyl(hydro) siloxanes forms silane or a dialkylsilane and solid alumasiloxanes of the type $>AlOSiX_3$ (where X = H, R, OR). By further reaction with lithium aluminum hydride, the latter are partly cleaved with the liberation of SiH_4 or a trialkylsilane.

5.2. Physical Properties

Table 33 gives the physicochemical constants of individual compounds containing the Si–O–Al group.

Tris(triorganosiloxy) alumanes are colorless crystalline substances with anomalously high melting points, which are apparently due to their dimeric state. Tris(trimethylsiloxy)alumane may be purified by sublimation and dissolves readily in aprotic solvents in amounts considerably greater than 30 wt.% [1075].

Triorganosiloxydichloroalumanes and their bromine analogs are colorless crystalline substances, which are stable in light and dry air; they dissolve readily in organic solvents and distill in vacuum without decomposition, but are very sensitive to moisture. The report of Simmler [1168] that trimethylsiloxydichloroalumane cannot be distilled in vacuum was not confirmed subsequently.

The melting point reported by Cowley and his co-workers [646] for trimethylsiloxydibromoalumane (100°C with decomposition) probably refers to an inadequately pure substance, which evidently contains some $AlBr_3$ that catalyzed decomposition [51]. According to a more reliable report, this compound melts at 110-115°C, while decomposition begins at a much higher temperature.* In general, the thermal stability of $R_3SiOAlX_2$ falls with an increase in the atomic weight of the halogen (X). Triorganosiloxydiiodoalumanes are least stable and these decompose on heating to 90-100°C and also in the light. When stored in ampoules in the light, triorganosiloxydibromoalumanes acquire a brown color, evidently due to decomposition with the liberation of bromine, while preparations stored in the dark for a year showed no changes.

Siloxydimethylalumane is stable at −78°C, but it decomposes slowly at room temperature (rapidly above 100°C) with the formation of silane and a viscous liquid [902]. At the same time, trimethylsiloxydimethylalumane boils at 200°C practically without decomposition [1108].

Simmler and Wiberg [1168] first observed the dimeric structure of compounds of the type $R_3SiOAlX_2$. The presence of an inorganic alumoxane ring in the structure of these compounds may now be regarded as firmly established. Table 34 summarizes data obtained by investigating the proton magnetic resonance spectra of trialkylsiloxyalumanes.

Without giving a detailed interpretation of NMR spectral data, we should note that the negative value of the chemical shift of sil-

*According to data in [1093], decomposition occurs at 150°C, but Orlov reports [51] that at this temperature the compound does not undergo obvious changes in 13 hr.

TABLE 33. Compounds Containing the Si−O−Al Group (for simplicity the dimeric state of the compound is ignored here)

Formula	m.p., °C	b.p., °C (mm)	Preparation method (scheme)	Yield, %	Literature
$[(CH_3)_3SiO]_3Al$	98—100	—	3-111	—	174, 179
	101	—	3-100	—	973
	—	Subl. 125 (0.25)	3-100	—	955
	238	Subl. 155 (1)	3-111	81	13, 955
	—	—	3-112	88	955
$[(C_2H_5)_3SiO]_3Al$	159	—	3-107	—	174, 182, 183
	310	—	3-107	—	1404
	327	—	3-107	34	179
$[(C_6H_5)_3SiO]_3Al$	—	—	3-88	35	329
	485	—	3-111	—	722
$[(CH_3)_3SiO]_2AlC_2H_5$	—	136—142 (18)	3-121	—	1414
$[(CH_3)_3SiO]_2AlOC_3H_7$-iso)	—	98—105 (0.4)	3-100	—	955
	82	140 (3)	3-100	—	973
$[(CH_3)_3SiO_2]AlOC_4H_9$-n	—	Subl. 145 (0.1)	3-100	—	955
$[(C_2H_5)_3SiO]_2AlOC_4H_9$-n	—	—	3-108	—	404
$H_3SiOAl(CH_3)_2$	41—42	$p = 3$ mm (0°); 109—112 extrapol. decomp. 105°	3-114a	—	902
$(CH_3)_3SiOAlH_2$	80—82	Subl. 40 (1)	—	—	113, 114
$(CH_3)_3SiOAl(CH_3)_2$	45.5	81.5 (10); Subl. 35 (1) *	3-113	84	1082
	—	200 (760)	3-129	68	1082, 1099, 1108
$(CH_3)_3SiOAl(C_2H_5)_2$	—	135—140 (18)	3-121	—	855
	105—113	Subl. 115 (1)	3-113	73	1082
$(CH_3)_3SiOAl(C_4H_9\text{-iso})_2$	—	165—170 (1,5)	3-122	—	855
$(C_2H_5)_3SiOAl(CH_3)_2$	14—16	110—114 (1)	—	—	113, 114
$(C_2H_5)_3SiOAl(C_2H_5)_2$	—	193—197 (16)	3-121	—	855

$(CH_3)_3SiOAl(OC_3H_7\text{-iso})_2$	—	95—97 (0.3)	3-100	—	955, 1285
	123	155 (2)	3-100	76	973, 996
$CF_3CH_2CH_2(CH_3)_2SiOAl(OC_3H_7\text{-iso})_2$	—	145—148 (8) **	3-100	—	314
$CH_3(CF_2CH_2CH_2)_2SiOAl(OC_3H_7\text{-iso})_2$	—	149—150 (6) 3*	3-100	—	314
$(CF_3CH_2CH_2)_3SiOAl(OC_3H_7\text{-iso})_2$	—	174.5 (2) 4*	3-100	—	314
m-$CF_3C_6H_4(CH_3)_2SiOAl(OC_3H_7\text{-iso})_2$	—	137—143 (2) 5*	3-100	—	314
$CH_3(m^2\text{-}CF_3C_6H_4)_2SiOAl(OC_3H_7\text{-iso})_2$	—	128.5 (0.5) 6*	3-100	—	314
$(C_6H_5)_3SiOAl(OC_3H_7)\text{-iso})_2$	(glassy)	—	3-100	—	1285
$(C_2H_5)_3SiOAl(OC_4H_9\text{-n})_2$	—	> 250 (1)	3-108	95	404
$(C_2H_5)_3SiOAl(C_2H_5)Cl$	—	90—101 (1,5)	3-122	—	855
$(CH_3)_3SiOAlHCl$	20,5	63—65 (1)	3-130	48 (67)	1098, 1099
$(CH_3)_3SiOAlCl_2$	88	102 (4)	3-114a	—	408
	87—88	Subl. 145—150	3-114a	—	646
	82—88	226—236	3-123	—	1655, 1735
	88—89	102 (4); decomp. 195	3-114a	72	1093
	—	100 (4)	3-114a	—	682, 684
$Cl(CH_3)_2SiOAlCl_2$	—	—	Ring opening	—	613
$C_2H_5(CH_3)_2SiOAlCl_2$	46.5	125 (4)	3-114a	—	408
$CH_3(C_2H_5)_2SiOAlCl_2$	33	143 (4)	3-114a	—	408
$(C_2H_5)_3SiOAlCl_2$	43	173 (4)	3-114a	—	408
	41—42	140-142(2); decomp. 200	3-114a	96	1093
$(CH_3)_3SiOAlHBr$	7—8	54—57 (1)	3-131	—	113, 114
$(CH_3)_3SiOAlBr_2$	111.5—113	127 (4)	3-114a	—	408
	100 decomp.	—	3-114a	—	646
	113.5—114.5	127—128, 5(4,5); decomp. 150	3-114a	91	1093
	—	110 (1)	3-114a	—	683, 684
$C_2H_5(CH_3)_2SiOAlBr_2$	66	150 (2,5)	3-114a	—	408
$CH_3(C_2H_5)_2SiOAlBr_2$	55	173 (3,5)	3-114a	—	408
$(C_2H_5)_3SiOAlBr_2$	73	181 (3)	3-114a	—	408
	76—78	172-175(2); decomp. 145	3-114a	90	1093

TABLE 33. (Cont'd)

Formula	m.p., °C	b.p., CC (mm)	Preparation method (scheme)	Yield, %	Literature
$(CH_3)_3SiOAlHI$	decomp. 0—2	69—75 (1) decomp.	3-131	—	113, 114
$(CH_3)_3SiOAlI_2$	decomp. 90	—	3-114a	82	1093
$(C_2H_5)_3SiOAlI_2$	decomp. 100	—	3-114a	73	1093
$[(CH_3)_2SiO_2AlCl]_2$	—	165—167 (4)	—	—	1317, 1396, 1574
$[(C_2H_5)_2SiO_2AlCl]_2$	—	184—191 (1)	—	—	1317, 1396, 1574
$(CH_3)_8O_6Cl_5Si_4Al_3$	153	180—182 (5)	—	—	288
	147—160	Subl. 160 (2)	—	—	682
$(CH_3)_8O_6Br_5Si_4Al_3$	154—155 (sealed tube)	Subl. 150—5 (0.16)	3-119a	—	685
$(CH_3)_8O_6(Br, I)_5Si_4Al_3$	159—160 decomp.	—	—	—	1087
$(C_2H_5)_8O_6Cl_5Si_4Al_3$	112	225—233 (10)	—	—	288
$Na\{Al(CH_3)_2[OSi(CH_3)_3]_2\}$	215	—	3-126	—	1096
$K\{Al(CH_3)_2[OSi(CH_3)_3]_2\}$	125	—	3-126	—	1096
$Li\{Al[OSi(CH_3)_3]_4\}$	—	—	3-126	98	1071, 1077, 1083
$Na\{Al[OSi(CH_3)_3]_4\}$	—	—	3-126	93	1071, 1077, 1083
$K\{Al[OSi(CH_3)_3]_4\}$	—	—	3-126	97	1071, 1077, 1083
$(CH_3)_3NH\{Al[OSi(CH_3)_3]_4\}$	112	—	3-127	—	1083
$(CH_3)_4N\{Al[OSi(CH_3)_3]_4\}$	201	—	3-127	—	1083
$(CH_3)_4Sb\{Al[OSi(CH_3)_3]_4\}$	180.5	—[7*]	3-128	—	1080, 1082

* d_4^{20} 0.98; ** n_D^{20} 1.4126; d_4^{20} 1.0513; 3* n_D^{20} 1.4099; d_4^{20} 1.0714; 4* n_D^{20} 1.4100; d_4^{20} 1.0862; 5* n_D^{20} 1.4392; d_4^{20} 1.5018; 6* n_D^{20} 1.4628; d_4^{20} 1.1874;
7* d_4^{25} 1.170.

oxyalumanes is explained by the weakening of the d_π-p_π interaction in the system Si–O–M as a result of oxygen–aluminum coordination bonds, which is promoted by the presence of halogen atoms at the aluminum.

The results of x-ray investigation of the structure of some dimeric siloxyalumanes are given in Table 35.

The solid products obtained by hydrolysis of organoethoxysilanes with aqueous alkaline solutions of sodium aluminate [474] are amorphous in most cases and have refractive indices from 1.460 – 1.474 (methyltriethoxysilane) to 1.501 – 1.519 (dimethyldiethoxysilane). In polarized light the crystalline products obtained from dimethyldiethoxysilane appear as needles and small prisms with well-expressed cleavage, direct extinction, and no pleochroism. The refractive index is ~1.52. These products are insoluble in organic solvents, but dissolve in cold concentrated alkali and hot water and melt in their own water of crystallization at 70°C.

Thermal analysis of the product from the reaction of methyltriethoxysilane with a solution of aluminate (Al:Si = 1:1; aluminate: water = 4:1) showed that up to 430-450°C there is the loss of water and alcohol, then up to 800°C the organic component burns up, in the range of 860°-870°C there is an endothermic sintering process, and finally, at 940°C there is an exothermic effect, which corresponds to the formation of the mineral carnegieite [474, 475].

Data on the thermal stability of tris(triphenylsiloxy)alumane are somewhat contradictory. On the one hand, m. p. ~ 485°C is reported, but it is also reported that even at 300°C decomposition begins with rupture of Si–C bonds (mass spectrometry) [722]; according to other information, on a thermogram of the substance, exothermic effects corresponding to the three phenyl groups at Si are observed at 500, 600, and 700°C [329].

Sodium bis(trimethylsilyl)aluminate dissolves in oxygen-containing solvents, (ether, alcohol, acetone, etc.), but is insoluble in benzene, carbon tetrachloride, carbon disulfide, acetonitrile, and other solvents. It forms acicular, infusible, birefringent crystals with a refractive index of ~1.555. An exothermic effect corresponding to burn-up of the organic part is observed only at 700°C [330].

Polyalumaorganosiloxanes obtained from aluminum silanolates of the type $[RSi(OH)_2O]_3Al$, are colorless, friable glassy polymers,

TABLE 34. Characteristics of Proton Magnetic Resonance Spectra of Trialkysiloxyalumanes (60 MHz; c = 2 wt.% in CCl_4; internal standard—tetramethylsilane; all the values are given in Hz; the accuracy of the determination of δ and $J(H^1-C^{13})$ is ± 1 Hz; $J(H^1-Si^{29})$ ± 0.05 Hz)

Compound	$\delta_{R(Si)}$	$\delta_{R(Al)}$	$J(H^1-C^{13})$	$J(H^1-Si^{29})$	Literature
$(CH_3)_3SiCH_3$	0	—	118—118.5	6.78	1078, 1081, 1093
$(CH_3)_3SiOSi(CH_3)_3$	—3.0	—	—	—	1075
	—3.5	—	118	6.86	1078, 1081
$\{[(CH_3)_3SiO]_3Al\}_2$	—3.5;	—	117.3;		
	—18.5 *		119 *	—	1075
	—4.0;	—	116;	6.65;	1081†
	—19 *	—	116 *	7.0 *	
$\{[(CH_3)_3SiO]_4Al\}Li$	—7.5	—	—	—	1103
$[(CH_3)_3SiOAl(CH_3)_2]_2$	—	—	118.5	7.05	1078
	—12	+49.2	—	—	1108
	—12	—	119	7.08	1093
$[((CH_3)_3SiOAlClH]_2$	—19.5	—221	—	—	1099
$[(CH_3)_3SiOAlCl_2]_2$	—25.5	—	119.2	7.08	1099
	—27.6	—	—	—	1108
$[(CH_3)_3SiOAlBr_2]_2$	—31.5	—	120	7.10	1093
$[(CH_3)_3SiOAlI_2]_2$	—36.8	—	120.3	7.10	1093
Compound	δ_{CH_3}	δ_{CH_2}	Δ	$J\frac{(H^1-Si^{29})}{\Delta}$	**Literature**
$(C_2H_5)_3SiC_2H_5$	—56	—30	—26	0.308	1093
$(C_2H_5)_3SiOSi(C_2H_5)_3$	—56	—31	—25	0.320	1093
$[(C_2H_5)_3SiOAlCl_2]_2$	—64.3	—55	—9.3	0.860	1093
$[(C_2H_5)_3SiOAlBr_2]_2$	—65	—58.3	—6.7	1.19	1093

*The second figure refers to CH_3 groups in "bridge" $(CH_3)_3SiO$ groups.
†Data given in [1102] are incorrect and actually refer to tris(trimethylsiloxy)gallane (private communication).

[14, 166, 175, 176] which do not melt, flow, or sinter on heating to 500°C. At the same time, these polymers will flow even at 150°C when plasticizers such as transformer oil are introduced [215]. The infusibility of alumasiloxane polymers is explained by the high glass transition points which are connected with the high rigidity of the polymer skeleton due to dense branching and the open molecular packing. It should be noted that in the differential thermal analysis of polymethylalumasiloxane, no endothermic decomposition effect is observed at 550-600°C, while the effect is observed in the range of 550-630°C in the case of polyethylalumasiloxane [198]. A characteristic of polymers of this type is the combination of infusibility at high temperatures with solubility in most nonpolar and weakly polar solvents with the exception of petroleum ether [8, 14, 137, 176]. Films formed after evaporation of the solvent readily redissolve. The solubility is completely retained even after heating of the polymers at 150°C for 10 hr. The degree to which the solubility is retained after heating at higher temperatures depends in general both on the character of the polymer and on the nature of the solvent. Thus, after heating at 200°C for 2 hr, polyphenylalumasiloxane is completely soluble in acetone, 90% soluble in ethanol, 47% in chlorobenzene, and only 11% in a nonpolar solvent, toluene. At the same time, the solubility of polyethylalumasiloxane in acetone after similar thermal treatment is only 4%. Thermomechanical curves give a picture which is typical of cross-linked polymers.

The peculiarities of the properties of polyorganoalumasiloxanes obtained from $[RSi(OH)_2O]_3Al$ was first explained by the specific cyclolinear structure of their molecules. However, a study of the behavior of polyalumaphenylsiloxanes in solution [61, 464] shows that the introduction of aluminum atoms into the siloxane structure conveys on the molecules a structure, which differs fundametally from the structure of both linear dimethylsiloxanes and cyclolinear phenylsiloxane polymers. In an investigation of 2 polymer with $T_g - 120°C$ and $M_w/M_n = 1.82$, data were obtained which indicated a cyclic network structure with the main unit:

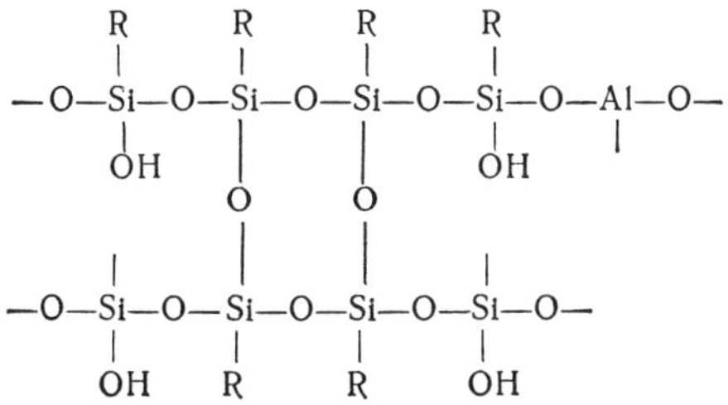

```
      R      R      R      R
      |      |      |      |
 —O—Si—O—Si—O—Si—O—Si—O—Al—O—
      |      |      |      |      |
      OH     |      |      OH
             O      O
             |      |
      |      |      |      |
 —O—Si—O—Si—O—Si—O—Si—O—
      |      |      |      |
      OH     R      R      OH
```

The density of the network is inversely proportional to the number of free OH groups in the elementary unit. The viscosity of the solutions decreases with a decrease in this number, but the solubility is preserved only if there is ordered ring closure. The weak dependence between the intrinsic viscosity and the molecular weight indicates that the form of the particles is close to

TABLE 35. Data from X-ray Investigation of Some Siloxyalumanes

Crystal parameter	Compound		
	$[(CH_3)_3SiOAl(CH_3)_2]$ [1255]	$[(CH_3)_3SiOAlBr_2]_2$ [565]	$(CH_3)_3O_6Br_5Si_4Al_3$ [564a]
β	104°49′	131°7′	93°52′± 10′
a (Å)	7.00_8	10.206±0.01	11.22± 0.03
b (Å)	13.22_0	9.681±0.005	13.48± 0.01
c (Å)	11.02_2	13.718±0.01	18.76± 0.03
U (Å)^3	987	—	2832
Z	2	4	4
D_x	0.984	—	
D_m	0.98	—	1.91 (D_c)
Space group	$P2_1/n$	$P2_1/C$	$P2_1/n$

spherical like Einstein's rigid spheres. The value of the coefficient a in the Mark–Houwink equation varies from 0.17 to 0.488 (at 20°C, 0.345 for benzene and 0.17 for chlorobenzene), depending on the nature of the solvent and the temperature.

The problem of the spectral characteristics of the Al–O bond in the Si–O–Al group is not at all clear. In the infrared spectra of products from the reaction of alkoxysilanes with solutions of aluminates, the Al–O bonds have been assigned absorption bands with a frequency of 970 [330] and 982 cm^{-1} [327], which are characteristic of Si–O–Al groups in albite and anorthoclase. Andrianov and his co-workers [154] have reported the results of studying the infrared spectra of compounds containing the Si–O–Al group and these are given in Table 36.

In analyzing these data the authors assigned the vibration frequencies of the Al–O bonds in Si–O–Al groups for alumasiloxane polymers to the range of 1050-1060 cm^{-1}. At the same time, for monomeric tris(triorganosiloxy) alumanes the band at 1055-1065 cm^{-1} is assigned to vibration of the Si–O bond. The intensity of the band at 1060-1100 cm^{-1} falls with a decrease in the aluminum content of the polymer, though its position hardly changes. In this connection we should note (as was also found previously [199]) that an increase in the aluminum content of polydimethylalumasiloxanes from Si:Al = 23.3 to Si:Al = 0.8 produces a shift in the characteristic band of these polymers in the region of 980-1100 cm^{-1} from 1070 to 1100 cm^{-1}.

The data presented in Table 36 showed that the spectra of polyalumaphenylsiloxane and polyphenylsiloxane are practically identical. Both compounds have a band at 1060 cm^{-1}, though one of these polymers contains no aluminum. The authors assign a band in the region of 945-950 cm^{-1} to an OH group. However, in hydrosilicates a band at 3333 cm^{-1} is characteristics of this group [327]. Moreover, it is not clear how tris(triethylsiloxy)alumane can show the frequency of an OH group if an individual compound with m. p. 327°C was actually investigated in [154]. On the basis of a study of the infrared spectra of polymers obtained according to scheme (3-103), Nudel'man and his co-authors [404] assigned to Si—O—Al bonds a broad band at 1027-1090 cm^{-1}, dividing it into two parts. The part at 1050-1090 cm^{-1} belongs to Si—O valence vibrations, while the part at 1027-1040 cm^{-1} belongs "to a considerable extent" to the Al—O bond. There are other examples of the assignment of infrared bands in the region of 980-1050 cm^{-1} [955] and 1050-1075 cm^{-1} [1075, 1093] to the system Si—O—Al; however, in some cases a band at 700-800 cm^{-1} is also assigned to vibrations of the Si—O bond in this system [995, 1093].

TABLE 36. Infrared Spectra of Compounds Containing the Si—O—Al Group

Compound	Si : Al ratio	Frequency, cm^{-1}	
		Si—O—Si Si—O—Al	OH group
Tris(trimethylsiloxy)alumane	3	1064 982	945
Tris(triethylsiloxy)alumane	3	1055 1002	951
Polyalumamethylsiloxane	4	1118 1050 1036	946
Polymethylsiloxane	—	1122 1036	—
Polyalumaethylsiloxane	5	1109 1060 1036	945
Polyethylsiloxane	—	1120 1036	950 946
Polyalumaphenylsiloxane	4	1099 1057	—
Polyphenylsiloxane	—	1104 1059	950

On the whole the impression is created that vibrations of Al–O bonds in alumasiloxanes appear in the same region as the frequency ν_{as} of Si–O bonds: they depend somewhat on the structure of the polymers, the nature of the intermolecular interaction, and the relative number of Si–O–Al bonds in the polymer. It should be noted that the region from 3 to 15 μ has been investigated mainly; in the longer wave region, bands in the range 540-640 cm^{-1} are assigned to Al–O bonds in alumasiloxanes [443, 535]. In particular, tris(triethylsiloxy)alumane shows absorption bands at 549, 585, and 638 cm^{-1} in this region [535].

5.3. Chemical Properties

As has already been pointed out, tris(trimethylsiloxy)alumane is a dimer with two sp^2-hybridized bridging oxygen atoms. The dimerization decreases the fraction of π-bonding in the bonds of the oxygen with silicon and aluminum and gives the dimers a coordinationally saturated character. This led to the hypothesis that addition reactions involving tris(trimethylsiloxy)alumane are possible. In actual fact, the following exothermic reaction proceeds almost quantitatively at room temperature in carbon tetrachloride [1017, 1077, 1083, 1103]:

$$(R_3SiO)_3Al + MOSiR_3 \longrightarrow M[Al(OSiR_3)_4] \quad (R = CH_3) \tag{3-126}$$

When M = Na or K, the compounds obtained are colorless, crystalline, dimeric substances, which are practically insoluble, do not sublime without decomposition, and decompose without melting when heated above 250°C with the evolution of a gas. The Li derivative differs from them in properties. Thus, it is soluble in nonpolar solvents and sublimes in vacuum (practically without decomposition) at 185-210°C; according to a cryoscopic determination of the molecular weight in benzene, it is a trimer and according to an ebullioscopic determination, a dimer. In the infrared spectra ν_{as} Si–O–Al lies at 1077 for the Li derivative, 985 for the Na derivative, and 971 cm^{-1} for the K derivative. Thus, there is a shift toward lower frequencies with an increase in the radius of the alkali metal. This indicates coordination bonding of M to the oxygen atom of the system Si–O–Al, which disrupts the symmetry of the anion and weakens the bridging bonds in the dimer.

Trimethylsiloxydimethylalumane is also capable of reactions analogous to scheme (3-126) [1096]. The reaction of this compound with alkali metal trimethylsilanolates yields the compounds $M[AlR_2(OSiR_3)]_2$ and it is interesting that when M = Na this compound is a monomer (in benzene) and when M = K, a dimer.

The possibilities of synthesizing compounds with a tetrakis-(trimethylsilyl)aluminate anion are not limited to reactions (3-126) [1072, 1080, 1083]:

$$(R_3SiO)_3Al \rightleftarrows \{H[Al(OSiR_3)_4]\} \xrightarrow{NR_3'} NR_3'H[Al(OSiR_3)_4] \longrightarrow R_4'N[Al(OSiR_3)_4] \quad (3\text{-}127)$$

$$(R = CH_3;\ R' = H,\ CH_3,\ C_2H_5)$$

$$[(R_3SiO)_3Al]_2 + R_4SbOSiR_3 \longrightarrow 2R_4Sb[Al(OSiR_3)_4] \quad (3\text{-}128)$$

These compounds will be examined in more detail in the section on organosilicon derivatives of antimony. The dimeric state of tris(trimethylsiloxy)alumane is emphasized particularly here since the antimony derivative obtained is monomeric.

When dimers of trimethylsiloxydichloro(bromo)alumane are treated with sodium acetylacetonate, the whole of the halogen is bound in the form of NaX. At the same time, 54-55% of the aluminum present in the starting compound is converted into the acetylacetonate [613, 684]. When the reaction is carried out in absolute ethanol, trimethylethoxysilane and hexamethyldisiloxane are also formed, while the remainder of the aluminum is bound in the form of an amorphous solid coordination polymer with a mean molecular weight of 570 (X = Br) − 900 (X = Cl), in which there are three chelating groups to every four aluminum atoms. It has been suggested that the first stage of the process is substitution, in which the structure of the dimer is preserved. The compound formed, which contains pentacoordinate aluminum, is unstable and decomposes with subsequent disproportionation of the new product:

$$[R_3SiOAlX_2]_2 + 4NaOC(CH_3){=}CHCOCH_3 \xrightarrow[-4NaX]{} \left[R_3SiOAl\left(\text{acac}\right)_2\right]_2 \longrightarrow$$

$$\longrightarrow 2R_3SiOAl\left(\text{acac}\right)_2 \longrightarrow Al\left(\text{acac}\right)_3 + (R_3SiO)_2Al\,\text{acac} \quad (3\text{-}129)$$

$$\left[\text{acac} = \begin{matrix} O{-}C(CH_3) \\ CH \\ O{=}C(CH_3) \end{matrix}\right]$$

There is subsequently alcoholysis of the bis(trimethylsiloxy)-(2-ketopent-3-en-4-oxy)alumane and water appears in the system as a result of condensation of the trimethylsilanol formed. Successive processes of solvolysis and condensation lead finally to an alumoxane polymer, whose composition should correspond to formula (I), but in actual fact, because of side reactions it corresponds more closely to formula (II) (4-5% excess of aluminum acetylacetonate):

$$\left[\mathrm{Al}\left(\text{acac}\right)\mathrm{O}\right]_n \qquad \left\{-\underset{\substack{|\\ O_{0.5}}}{\mathrm{Al}}-\mathrm{O}-\left[\mathrm{Al}\left(\text{acac}\right)\mathrm{O}-\right]_3\right\}_n$$

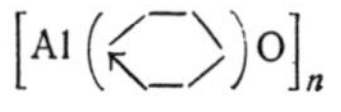

I 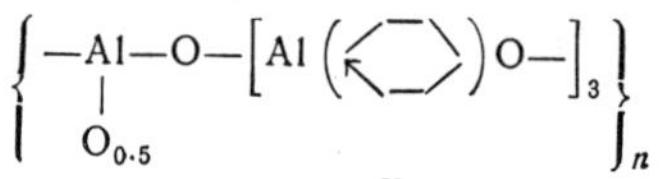 II

When $(CH_3)_8O_6X_5Si_4Al_3$ (where X = Cl, Br) is treated with sodium acetylacetonate in ethanol [613, 685] 41-44% of the total amount of aluminum in the "universal compound" is bound in the form of the acetylacetonate. Together with the trisacetylacetonate, a mixture of liquid and solid polymeric products with mol. wt. 500–2000 is formed. The amount of NaX obtained corresponds to the amount of halogen atoms present in the compound. It is therefore suggested that as in the case of $R_3SiOAlX_2$, the first stage of the process is a replacement reaction. The unstable five-membered ring rearranges into the linear compound

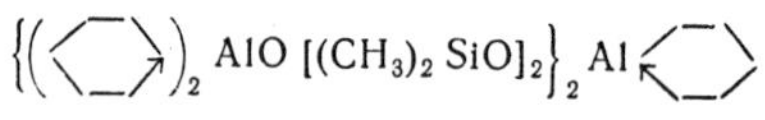

$$\left\{\left(\text{acac}\right)_2 \mathrm{AlO}\left[(CH_3)_2SiO\right]_2\right\}_2 \mathrm{Al}\left(\text{acac}\right)$$

whose chain grows as a result of condensation with the elimination of aluminum acetylacetonate. The amount of the latter is somewhat greater than the calculated amount for the binding of one aluminum atom in three and this must also be due to side processes.

A study of the reactions of sodium acetylacetonate with polymers of the type $[R_2SiOAl(X)O]_n$ leads to the conclusion [613, 683] that this formula reflects only the composition and not the structure of the substance. When X = Cl there is mainly replacement of the halogen atom and the alumasiloxanes formed (mol. wt. 1100-1600), which contain chelate groups at the aluminum atoms, are much more hydrolytically stable than the starting polymers. However, at the same time 17-18% of the aluminum is extracted from the polymer in the form of the tris-acetylacetonate. Comparison of these data with results obtained by studying the action of sodium acetylacetonate on trimethylsiloxydichloroalumane and $(CH_3)_8O_6Cl_5Si_4Al_3$ compelled the authors to draw the conclusion that one of the structural groups $\geq SiOAlX_2$, present in the polymers is eliminated in the form of aluminum acetylacetonate. They therefore assigned to the polymers structure (III) in which one out of every three aluminum atoms is attached to two chlorine atoms, the second is attached to one chlorine atom while the third is attached to only oxygen atoms:

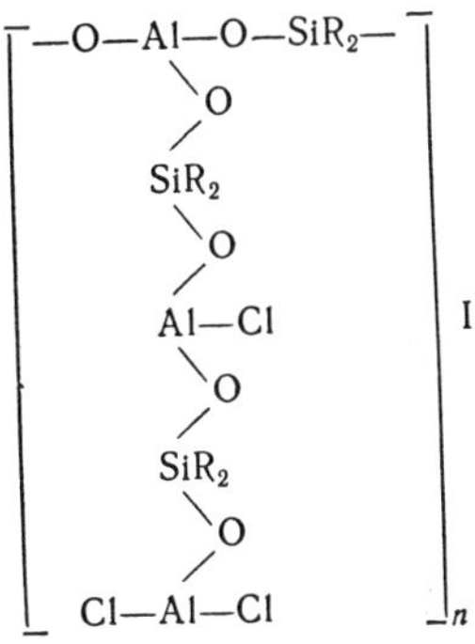

```
 ┌ —O—Al—O—SiR2— ┐
 │      \        │
 │       O       │
 │      /        │
 │    SiR2       │
 │      \        │
 │       O       │
 │      /        │
 │    Al—Cl      │   III
 │      \        │
 │       O       │
 │      /        │
 │    SiR2       │
 │      \        │
 │       O       │
 │      /        │
 └  Cl—Al—Cl     ┘n
```

Through the $\geq Si—O—AlCl_2$ groups there is the possibility of interchain coordination with the formation of four-membered rings. It is proposed that Al–O–Al bonds are present in addition to Si–O–Al bonds in a polymer with a ratio Al : Si = 4:3.

In the case of X = Br, treatment of $(R_2SiO_2AlBr)_n$ with sodium acetylacetonate leads to a polymer with a mean molecular weight of 470. Up to 35% of the aluminum is bound in the form of the trisacetylacetonate and this makes it probable that there is a larger number of $\geq Si—O—AlX_2$ groups in the polymer in comparison with the chlorine-containing analog [683].

It has been reported [1653] that crystalline cyclic alumasiloxanes of the type $[(CH_3)_2SiOAl(Cl)O]_2$ react with the solid disodium salt of sym-tetramethyldisiloxanediol in a benzene-ether mixture with the formation of a straw yellow, mobile liquid, whose structure was not investigated.

Compounds of the type $R_3SiOAlR'_2$ may be synthesized by alkylation of trialkylsiloxydihaloalumanes [1099] as well as by scheme (3-113):

$$R_3SiOAlX_2 + 2CH_3Li \longrightarrow R_3SiOAl(CH_3)_2 + 2LiX \qquad (3\text{-}130)$$

The cyclic structure is not broken down by the action of methyllithium, but the reactivity of $R_3SiOAlX_2$ is not high in methylation.

Trimethylsiloxydimethylalumane is inert toward trimethylamine [1108].

It was hoped that by treating $R_3SiOAlX_2$, with lithium aluminum hydride it would be possible to obtain the simplest trimethylsiloxy derivative of aluminum [113, 114, 1096, 1098, 1099]. In actual fact, it was found that the reduction proceeds in stages. Compounds of the type $R_3SiOAl(X)H$ are formed initially and, for example, when X = Cl, a twofold excess of lithium aluminum hydride only increases the yield of trimethylsiloxychloroalumane. By reducing individual triorganosiloxyhaloalumanes it is possible to obtain the dihydro drivative:

$$R_3SiOAlX_2 \xrightarrow{LiAlH_4} R_3SiOAl(X)H \xrightarrow{LiAlH_4} R_3SiOAlH_2 \qquad (3\text{-}131)$$

Compounds with two hydrogen atoms at the aluminum may be distilled. Their decomposition occurs at quite a high temperature with the liberation of a trialkylsilane and the formation of polymeric $(AlXO)_n$ (where X = H or halogen). However, trimethylsiloxyalumane has a tendency to polymerize and does not exist for long in a dimeric form [113, 114].

$R_3SiOAl(X)H$ will undergo reaction (3-130). Like other compounds of this series, trimethylsiloxychloroalumane is dimeric, while the like substituents at the aluminum atoms (H and Cl) are in the trans-position relative to each other, lying outside the plane of the alumoxane ring [1099]. The colorless crystals are readily soluble in solvents without active protons. It is paradoxical that the compound is hydrolyzed slowly in excess water, while it inflames almost explosively if a small amount of water is added to a considerable amount of trimethylsiloxychloroalumane. Rapid alcoholysis accompanied by the liberation of hydrogen occurs with methanol. In this connection it should be remembered that when bis(trimethylsiloxy)isopropoxyalumane is treated with butanol there is only normal transalkoxylation, while the trimethylsiloxy groups are not touched [955]. Trialkylsiloxydihaloalumanes react vigorously with water with the formation of aluminum hydroxide and a hexaalkyldisiloxane [408]. The compounds $M[Al(OSiR_3)_4]$ are also hydrolytically unstable. At the same time, sodium bis-(trimethylsilyl)aluminate is hydrolyzed by water completely only after boiling for 2-3 days, but is cleaved by acids [330]. Tris-(trialkylsiloxy) alumanes undergo rapid acid hydrolysis [151, 1075]; tris(triphenylsiloxy)alumane is more stable hydrolytically [329].

The mechanism of the acid hydrolysis of the group Si−O−Al may be represented in the following way [178]:

$$\equiv Si-O-Al\lt + H^+ \longrightarrow \equiv Si-\overset{+}{\underset{H}{O}}-Al\lt \xrightarrow{H_2O} \equiv Si-\overset{+}{\underset{H}{O}}-\overset{\uparrow OH_2}{Al}\lt \longrightarrow$$

$$\longrightarrow \equiv SiOH + \gt AlOH + H^+ \text{ etc.} \qquad (3\text{-}132)$$

An oxonium compound is formed initially and this is then converted into a coordination hydrate complex, which decomposes with cleavage of the Si−O−Al bond. However, it is more probable that the hydrolysis proceeds through the intermediate formation of a cyclic active complex:

$$[\text{cyclic complex of } \equiv Si-O-Al\lt \text{ with } H-O \text{ and } H_2O] \longrightarrow \equiv SiOH + \gt AlOH + H_3O^+ \qquad (3\text{-}133)$$

It might have seemed that because of the presence of semi-organic groups at the aluminum atom, which reduce the nucleophilicity of the oxygen in Si−O−Al, alumasiloxanes should show a lower resistance to acid hydrolysis than kaolin. However, in the acid hydrolysis of tris(triethylsiloxy)alumane the triethylsilanol formed initially is then converted into a simple compound, namely, hexaethyldisiloxane. At the same time, the hydrolysis of high-molecular polyorganoalumasiloxanes may form complex branched products through the hydroxyl-containing "fragments" obtained initially; this complicates the hydrolysis and finally is responsible for the somewhat higher hydrolytic stability of alumasiloxane copolymers as compared with alumosilicates.

For soluble polymers obtained from $[RSi(OH)_2O]_3Al$ it was established that polymers with $R = C_6H_5$ show the highest resistance to acid hydrolysis [140]. Polymers with R = C_2H_5 are less resistant, while polymers with R = CH_3 show the lowest hydrolysis resistance.

The Si–O–Al group in phthalocyaninotriphenylsiloxyalumane is not touched when the substance is boiled for 2 hr in concentrated NH_4OH or 12 N H_2SO_4 and is cleaved only by concentrated sulfuric acid [998]. This indicates the screening effect of substituents at the silicon atoms on the hydrolytic resistance of alumasiloxane bonds. While partial hydrolysis of phenoxydiisopropoxyalumane still leads to an insoluble and infusible product, in the case of triphenylsiloxydiisopropoxyalumane a toluene-soluble, low-molecular polymer is obtained [1285].

When moist air is blown through tris(triethylsiloxy)alumane at 170°C, this crystalline product is gradually converted into a polymer, whose viscosity increases with time [5, 135, 164, 169, 174, 179, 182, 284, 286, 291, 1464]*. The volatile products contain hexaethyldisiloxane, while the soluble, glassy polymer with a molecular weight of 4,100 contains silicon and oxygen in amounts intermediate between the values calculated for polymers of the type:

$$(C_2H_5)_3SiO[(C_2H_5)_3SiOAlO]_nSi(C_2H_5)_3 \text{ and } [(C_2H_5)_3SiOAlO]_n$$

These data indicate stepwise hydrolytic condensation:

$$\begin{aligned}
&Al(OSiR_3)_3 + H_2O \longrightarrow (R_3SiO)_2AlOH + R_3SiOH \longrightarrow \\
&\longrightarrow (R_3SiO)_2AlOAl(OSiR_3)_2 + (R_3Si)_2O + H_2O \longrightarrow \\
&\longrightarrow \ldots -\underset{\displaystyle OSiR_3}{\underset{|}{Al}}-O-\underset{\displaystyle OSiR_3}{\underset{|}{Al}}-O-\ldots + R_3SiOH \quad \text{etc.}
\end{aligned} \qquad (3\text{–}134)$$

Initially there is elimination of a triethylsiloxy group with the formation of two hydroxyl-containing compounds, which are capable of condensing with dimerization of the molecule. However, while the condensation of triethylsilanol leads to hydrolytically resistant hexaethyldisiloxane, the product of anhydrocondensation of bis-(triethylsiloxy)alumanol is capable of further elimination of triethylsiloxy groups, etc. With small amounts of water it is in a cycle and in principle, 1.5 mole of H_2O per mole of tris(triethylsiloxy)alumane is sufficient for complete conversion of the latter

* The compound is not changed by prolonged heating at 220°C without air blown through.

TABLE 37. Hydrolysis of Tris(triethylsiloxy) alumane

Molar ratio H_2O: $[(C_2H_5)_3SiO]_3Al$	Gel time of solution, min
2.32	12960
2.80	320
3.27	100
3.74	6

into aluminum hydroxide and hexaethyldisiloxane. However, due to the increasing viscosity of the products of hydrolytic condensation the process proceeds slowly. An increase in the amount of water introduced into the reaction accelerates the hydrolysis and promotes the development of processes forming branched and cross-linked polymers. Instead of linear, soluble polytriethylsiloxyalumoxanes (a new class of semiorganic high-molecular compounds), hydrolysis of dilute solutions yields gels (Table 37).

An attempt at the polymerization of oily alumaxilosanes with sulfuric acid was unsuccessful [456]. The process was accompanied by cleavage of Si—O—Al groups with elimination of the aluminum as the sulfate and the formation of normal polydimethylsiloxane elastomers. In the presence of 0.1-0.2 wt. % KOH, polymerization of alumasiloxane oils with mol. wt. ~1000 and an aluminu content of 0.3-0.6% occurs at an adequate rate and ~100°C. One of the polymers obtained in this way contained 0.3% Al and had a mol. wt. of ~1700, but was superficially equivalent to high-molecular rubbery polysiloxanes and had T_g - 121°C. However, when a filler was introduced on a mill, the alumasiloxane "elastomer" underwent irreversible decomposition.

In the catalytic polymerization of alumasiloxanes of the type $(RSiO_2)_3Al$ (where R = C_2H_5 or C_6H_5) or $[CH_3\,C_6H_5)SiO]_3Al$, they gradually lose their solubility and are converted into a gel state. This can be followed visually as insoluble parts of the gel precipitate from solution [160]. In polymerization with 1% NaOH at 120°C, the polymerization rate falls in the series:

$$[CH_3\,(C_6H_5)\,SiO]_3\,Al > (C_6H_5SiO_2)_3\,Al > (C_2H_5SiO_2)_3\,Al$$

In the presence of ethylsulfuric acid the polymerization occurs more rapidly than with alkali, but the authors did not report whether or not there is elimination of aluminum.

The solubility of the starting aluminum polyorganosilanolates of Andrianov and his co-workers [11, 134, 160, 173, and etc.] is explained by their cyclolinear structure. In the catalytic polymerization, planar rings connected through oxygen are opened with the formation of cross-linked, insoluble molecules. It has already been pointed out above that a study of the behavior of such polymers in solutions made it possible to devise the concept of a cyclic lattice structure. Since the polymers do not correspond completely in composition to the formula $(RSiO_2)_3Al$, but also contain hydroxyl groups, the thermocatalytic process may also include opening of cyclic sections of the structure and condensation with a fall in the number of hydroxyls and the formation of a denser disordered lattice. Moreover, differences are observed in the chemical composition of the starting and final alumasiloxanes. This indicates that the process cannot be reduced exclusively to catalytic rearrangement and there are more complex conversions.

In a series of literature communications it is reported that alumasiloxanes of the type $[-(RSiO_{1.5})_nAlO-]_x$ (R = C_2H_5 or C_6H_5; n = 3 – 5) [52, 165, 266, 416, 419, 420] and $\{[CH_3(C_6H_5)SiO]_3Al\}_n$ [266] and also the alumasiloxanes $Al\{[OSi(CH_3)_2]_{21}OH\}_3$ [242] may themselves act as catalysts for the polymerization of relatively low-molecular polymethyl(dimethyl)phenylsiloxanes [52, 165, 266, 416, 419], their graft copolymerization with epoxide resins [420], and also alkaline polymerization of bicyclic siloxanes [242]. Thus, for example, the time for converting the cohydrolysis products of methyltrichloro(triacetoxy)silanes and phenyltrichloro(triacetoxy)-silanes into an infusible state at 150°C is reduced markedly even when 0.5% of polyethylalumasiloxane with n = 5 is introduced into the mixture [165]. An investigation of the effect of tertiary amines on this process showed that the conversion of the polymer into an infusible state is connected with the formation of coordination bonds between siloxane oxygen of the organosilicon component and aluminum of the polyethylalumasiloxane. Since the starting organosiloxanes and alumasiloxanes contain up to 10% OH groups, as a rule, it is more correct here to talk of polycondensation and hardening processes rather than polymerization.

In this connection we should mention a whole series of references which report the use of aluminum halides [1604, 1618], mixtures of them with iron oxide [1301, 1302, 1526, 1533], saturated [63, 1309, 1702], and unsaturated [1574] aluminum alcoholates, aluminum salts of aliphatic [21, 1406, 1576] and naphthenic [1513] acids, compounds of the aluminum acetylacetonate type [21, 1332, 1439], triethylalumane, and triethylsiloxydiethylalumane [855, 1345] as catalysts for the low-temperature hardening of liquid and resinous polysiloxanes. On the other hand, there are reports of the use of alumosilicates (the acid clay "kill") in the catalytic rearrangement of ethylsiloxane liquid [41]. The possibility of the formation of compounds of the alumasiloxane type cannot be excluded in these reactions.

Some data on the thermal stability of siloxyalumanes and alumasiloxanes have already been given in Section 5.2. The pyrolysis of pure trimethylsiloxydichloroalumane in the absence of aluminum chloride corresponds only partly to the equation (3-114b). Together with trimethylchlorosilane (which is obtained in only 30% of theoretical yield under these conditions), gaseous products are formed, while the solid residue contains not only aluminum oxychloride, but also polymeric substances containing silicon [51].

Crystalline trimethylsiloxydiisopropoxyalumane decomposes at 260°C with the formation of propylene, isopropyl alcohol, trimethylisopropoxysilane, and insoluble products [75, 1286]. The decomposition of tris(trimethylsiloxy)alumane occurs at 260-280°C. Its products are hexamethyldisiloxane and a polymer with a molecular weight of 1200, which is insoluble in benzene [945, 1286].

A study of the thermooxidative breakdown of polyorganosiloxane resins modified with polyalumaorganosiloxanes showed [52, 137, 165, 225, 266, 416-418] that the rate of weight loss of polydimethylphenylsiloxanes containing 0.05-0.5% Al at 400°C increases with an increase in the aluminum content of the modified polymer; however, both the rate and absolute value of the loss in weight were substantially less than for polymers which contained no aluminum. It is not surprising that at 500°C the weight losses were generally higher, but the situation remained the same with respect to the rate, while the nature of the volatile decomposition products was different. Oxidation of radicals, i. e., cross linking, is more characteristic of alumasiloxane polymers, while the formation of cyclosiloxanes is suppressed. At the same time, when purely organosilicon polymers are subjected to a high temperature the losses in weight are connected with their breakdown and their elimination of silicon in the form of cyclosiloxanes.

A study of the thermooxidative breakdown of combinations of polyphenylalumasiloxanes with polymethyl(phenyl)siloxanes confirmed that the absolute weight loss on heating is independent of the aluminum content – this is reflected only in the rate of thermo-oxidative processes. As regards the relative stability of radicals, in the presence of alumasiloxanes the rate of elimination of phenyl radicals is higher, increasing with an increase in the aluminum content of the mixture. The same occurs in the case of modification of polyvinylsiloxanes, though the rate of elimination of vinyl groups is lower than that of phenyl groups. These phenomena may be connected with the fact that in this case there is elimination of radicals during heat aging and not their gradual breakdown, while with alumasiloxane polymers the tendency for cross linking is more marked. At the same time, the thermo-oxidative stability of methyl- and ethylsiloxane polymers increases when alumasiloxanes are introduced into them and this effect is greater for the former than for the latter, as was to be expected, As an illustration, the whole of the organic part was burnt away when a polymethylsiloxane was heated at 400°C for 24 hr, while a polymer containing 0.5% aluminum lost 6% of the radicals in 5 days. An increase in the mean functionality of the copolymers increases the thermo-oxidative stability and this is evidently connected with the favorable effect of the more developed network structure. In a nitrogen atmosphere the breakdown – crosslinking processes proceed at a much lower rate and to a much lesser extent than in air.

5.4. Application

Polyalumaorganosiloxanes are used primarily for preparing modified siloxane resins and plastics based on them with an increased thermal stability. The plastic compositions are effective at 400-420°C [1, 137, 1491]. Electrical insulating compounds containing binders based on alumasiloxane resins have a high thermal stability, arc resistance, and stable dielectric properties under conditions of high humidity and high temperatures [266, 1479, 1494]. After heat aging for 15 days at 400°C or 48 hr at 500°C, flexible glass-micanite of grade GS40(25) K-60, which is produced in the USSR and based on polydimethylphenylsiloxanes modified with polyalumamethylphenylsiloxanes, has breakdown voltages at 20°C and the aging temperature of 13-18 and 10-12 kV/mm, respectively, and a bulk resistivity of 10^{11} and 10^{10} ohm·cm [52]. Testing glass micanite tape based on the same material as a continuous insulation on steel rods showed that after aging at 500°C for 15 days, the insulation remained whole and mechanically strong, while the breakdown voltage had not fallen. These materials are recommended as the main insulation of low-voltage electrical machines with an operating temperature of 300°C and for electrical equipment with a limited life at a working temperature up to 500°C. The modification of a fiberglass laminate based on phenol-formaldehyde resin with a polyalumaphenylsiloxane (0.5-1%) increases the water resistance of the material, its hardness, and strength and electrical insulating properties, while the Martens yield temperature is raised from 215°C to more than 250°C [14, 222, 1413].

The addition of alumasiloxanes resins improves the thermal stability, elasticity, and bonding power of electrical insulating varnishes [1487].

Polyalumaethylsiloxanes and polyalumaphenylsiloxanes, plastics, compounds, varnishes and materials based on them of various grades are produced industrially in the USSR [7, 14, 52, 58]. They include the varnish K-16 (a solution of polyalumaethylsiloxane resin in toluene and butanol), which is used as a hardener for organosilicon resins in the preparation of fiber-glass laminates STK-71 and other electrical insulating materials, the varnish K-37 (a solution of polyalumamethylphenylsiloxane in toluene), which is introduced into other varnishes, K-38 (a solution of polyalumanaphthenatophenylsiloxane resin in toluene) – a hardener for varnish K-71 and a binder, K-39 (a solution of polyalumaphenylsiloxane resin), and others.

Aluminum-pigmented anticorrosion siloxane coatings protect ordinary steel at temperatures up to 540°C and stainless steel up to 760-870°C (briefly at 900°C). Under these conditions the organic part of the polymer is burned up, but the remaining alumasilicate skeleton continues to protect the surface from corrosion [545, 657, 734, 800, 833, 1183, 1338, 1532, 1597].

When alumasiloxanes obtained by condensation of siloxanediols with aluminum alcoholates (scheme 3-109) are introduced into dimethylsiloxane rubber, they have a plasticizing effect and raise the thermal stability of the vulcanizates at 300°C (particularly alumaphenylsiloxanes) to a degree which is proportional to the aluminum content of the mixture [451]. It was found that the distribution of the additive through the elastomer is of great importance. It is therefore quite understandable that alumasiloxane elastomers obtained by heterofunctional condensation have an even better set of properties [132, 269, 438]; γ-vulcanizates based on them have higher thermal stability and better adhesive properties than vulcanizates of vinylsiloxane rubbers,* but they are inferior to rubbers from other heterosiloxane elastomers, particularly, boron and phosphorus-containing elastomers.

On the other hand, polymers obtained by heterocondensation of dimethyldiacetoxysilane with triethylsiloxydibutoxyalumane are [1503, 1504] effective vulcanizing agents of rubber mixtures from carboxyl-containing rubbers. The use of alumasiloxanes eliminates scorching of the mixtures at the same time.

Polymers obtained by the hydrolysis of organoalkoxysilanes by solutions of aluminates are recommended as binders [1470] and electrical insulating resins [1370]. In the opinion of Rast and Takimoto [1678, 1695], triorganosiloxyalumoxanes (scheme 3-104) may have wide application. Depending on the molecular weight and the nature of the terminal groups, they may be used both as thermostable and hydrolytically resistant fluids (hydraulics, heat transfer agents, and lubricants), adhesives, thermoplastic and thermosetting plastics (cast, extrusion, and laminating materials), and modifying additives for organosilicon, alkyd, phenol-aldehyde, and epoxide resins.

Alumasiloxanes with phthalocyanine groups have a high thermal stability, subliming at temperatures above 500°C [999]; this indicates the possibility of using them as heat transfer agents.

We have already mentioned above the strong water repellant effect of solutions of alkali metal "alumosiliconates." They are particularly effective in waterproofing constructional materials with a neutral reaction, namely, plaster, gypsum concrete, and gypsum pozzuolanic cement parts and limestones and also constructional and artistic ceramics [45, 346, 347]. An experiment has been carried out on the waterproofing of the facade of an administrative building (in Moldavia) with an area of 13,000 m^2 with a dilute solution of alumomethylsiliconate (0.7% on silicon, 0.4–0.5 liter/m^2), which is marketed in the form of a 6% solution as grade AMSR. Sodium alumomethylsiliconate is also an effective dressing for

* The optical vulcanization dose is also higher (14.5 MR) [438].

glass fiber and a water repellant for cloth, particularly canvas [45] used for making protective clothing.

According to data in [1653], products from the reaction of cyclochloralumadimethylsiloxanes with sodium sym-tetramethyldisiloxanediolate are good water-repellant preparations and emulsifiers for siloxane oil-water systems.

Triorganosiloxyalkylalumanes are recommended as additives for fuels, reducing and alkylating agents, complex formers, and catalysts for the polymerization of unsaturated compounds [815, 1414, 1417].

There are reports [1387, 1714] that the addition of alkylalkoxysilanes increases the activity of organometallic complexes which are catalysts for the polymerization and copolymerization of olefins, based on ethylaluminum chlorides and titanium trichloride or vanadium halides (VCl_4, $VOCl_3$). In the latter case the most effective systems are those in which the atomic ratios Al:V:Si are in the range of 2–4:1:0.25–0.3. On the basis of data presented in Section 5.1.2. and the conditions for preparing these catalysts, it may be surmised that these systems include compounds of the alumasiloxane type though naturally there is the possibility of the formation of organosilicon derivatives of titanium vanadium (see Chapters 4 and 5).

5.5. Analysis

The aluminum in organoalumasiloxanes may be determined by various methods. A method similar to that used for rapid analysis of silicates consists of wet combustion of the substance with a mixture of oleum and nitric acid with subsequent separation of the silicic acid.

The aluminum in the filtrate is determined by precipitation with ammonia in the presence of Methyl Red and then the precipitate of aluminum hydroxide is calcined to Al_2O_3 at $\sim$1200°C [127, 408].

It has also been recommended that the substance be decomposed with concentrated sulfuric acid (in the presence of ammonium sulfate according to data in [1477]) with the aluminum in the filtrate determined by precipitation with tartaric acid [1477] or by titration with Complexon III [465]. In the analysis of compounds of the type $M[Al(OSiR_3)_4]$ ($R = CH_3$; M – alkali metal) their hydrolytic stability is used [1077]. A sample of the substance is dissolved in excess 0.1 N HCl, the solution is evaporated (the trimethylsilanol is removed) and the content of the alkali metal determined by back titration with 0.1 N NaOH and the aluminum determined gravimetrically or complexometrically.

6. COMPOUNDS CONTAINING THE $Si-(C)_n-Al$ GROUP

The following addition occurs in the reaction of dialkylalumanes with trialkylalkenylsilanes under mild conditions (4–10 hr at 75-90°C without a catalyst) [300, 671]:

$$R_3Si(CH_2)_n{-}CH{=}CH_2 + R'_2AlH \longrightarrow R_3Si(CH_2)_{n+2}AlR'_2 \quad (3\text{-}135)$$

($R = CH_3$ or C_2H_5; $R' = n\text{-}C_4H_9$ or iso-C_4H_9; $n = 0$ or 1)

It is not possible to isolate the primary reaction products. Disproportionation occurs on heating, even in high vacuum (10^{-5} mm):

$$R_3Si(CH_2)_{n+2}AlR'_2 \longrightarrow {}^2/_3R'_3Al + {}^1/_3[R_3Si(CH_2)_{n+2}]_3Al \quad (3\text{-}136)$$

In the case of trimethyl(allyl)silane the aluminum adds to the terminal carbon atom and tris(3-trimethylsilylpropyl) alumane is obtained. Its formation was demonstrated by the isolation of 3-trimethylsilylpropanol from the products of oxidative hydrolysis. Heating this compound under a pressure of ethylene leads to lengthening of the carbon chain between the silicon and the aluminum (in particular, the product of oxidative hydrolysis of the substance obtained is 5-trimethylsilylpentanol) [300]. Determination of the order of addition of $R^1{}_2AlH$ to trialkylvinylsilanes is complicated by side reactions during oxidation. When $R = CH_3$ and R' = iso-C_4H_9, it is not possible to isolate an alcohol at all. In the case where $R = C_2H_5$ it was established that in analogy with hydroboronation, "hydroalumination" of vinylsilanes leads to a mixture of α- and β-adducts in a ratio of 70:30. However, the tendency for the formation of triethylsilylethanols appears to a slight extent,* while hydrolytic oxidation of the reaction mixture yields mainly triethylsilanol (hexaethyldisiloxane).

*Normal and not oxidative hydrolysis leads to the formation of tetraethylsilane [671].

Triphenylvinylsilane reacts with triisobutylaluminum with the liberation of isobutylene and the formation of tris(triphenylsilylethyl)alumane [858], whose structure was not established. Heating it at 200°C with excess triphenylvinylsilane is accompanied by dimerization of the latter:

$$3R_3SiCH{=}CH_2 + Al(C_4H_9\text{-}iso)_3 \longrightarrow (R_3SiC_2H_4)_3Al + 3(CH_3)_2C{=}CH_2$$

$$(R_3SiC_2H_4)_3Al + R_3SiCH{=}CH_2 \longrightarrow (R_3SiC_2H_4)_2AlCH\langle{}^{(CH_2)_3SiR_3}_{SiR_3} \xrightarrow{R_3SiCH=CH_2}$$

$$\longrightarrow (R_3SiC_2H_4)_3Al + R_3SiCH{=}CHCH_2CH_2SiR_3 \quad (R = C_6H_5) \qquad (3\text{-}137)$$

The structure of the dimer was demonstrated by its "hydroalumination" with subsequent alcoholysis which yielded, 1,4-bis-(triphenylsilyl)butane.

A peculiar route was selected by Goubeau and Mayer [806] for the preparation of dimethyl-bis(hydroxymethyl)silane:

$$R_2Si(CH_2OCOCH_3)_2 \xrightarrow{LiAlH_4} [R_2Si(CH_2O)_2Al(OC_2H_5)_2]Li \xrightarrow[H_2O]{HCl}$$

$$\longrightarrow R_2Si(CH_2OH)_2 + C_2H_5OH + LiCl + AlCl_3 \cdot 6H_2O \qquad (3\text{-}138)$$

No individual organosilicon derivatives of aluminum containing the grouping $Si{-}(C)_n{-}Al$ have been isolated up to now.

7. COMPOUNDS CONTAINING THE Si – N – Al GROUP

Distillation of a mixture of hexamethyldisilazan and aluminum chloride or bromide gives about 75% yield of trimethylhalosilanes [143]. Under milder conditions there is the formation of trimethylsilylaminodihaloalumanes [544, 1101]:

$$R_3SiNHSiR_3 + AlX_3 \longrightarrow R_3SiNHAlX_2 + R_3SiX \qquad (3\text{-}139)$$

$$R_3SiNHAlX_2 \longrightarrow R_3SiX + 1/n\,[AlNHX]_n$$

These are crystalline compounds, which are readily soluble in dichloroethane and acetonitrile, moderately soluble in nitrobenzene, and insoluble in petroleum ether. The dichloro derivative is insoluble in benzene, fumes in air, and is readily hydrolyzed by water. The bromine analog dissolves slightly in benzene, but reacts explosively with water. Like trialkylsiloxydihaloalumanes, compounds of the type $R_3SiNHAlX_2$ are dimeric and have a structure which includes a four-membered coodination alumazan ring [1101]:

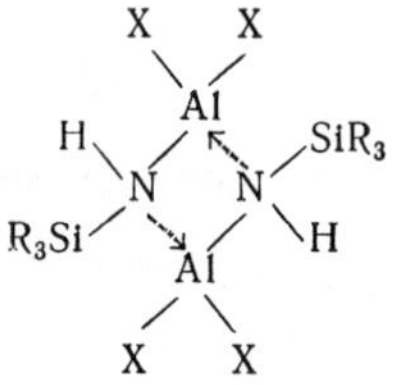

```
          X   X
           \ /
            Al
     H\    / ↖   /SiR3
       N        N
R3Si/    ↘   /   \H
            Al
           / \
          X   X
```

The structure is not broken down in the reaction with methyllithium in ether and this gives dimeric crystalline trimethylsilylaminodimethylalumane.*

According to the data of Andrianov and his co-workers [185, 227], after hexamethyldisilazan has been heated with aluminum chloride (0.3 wt.%) for 5 hr at 140°C the reaction mixture contained 8% of octamethylcyclotetrasilazan. At 160°C the content of it is increased to 14% and the reaction product also contains 17% of liquid oligomers. Increasing the amount of aluminum chloride to 3.8% and the temperature to 240°C increases the content of octamethylcyclotetrasilazan to 20%, while the yield of polymer is 60%. On the other hand, heating octamethyltetrasiloxane for 5 hr with 0.5% $AlCl_3$ led to the formation of a trimer (28% in the mixture). The authors did not find any compounds containing silicon and aluminum. At the same time, when aluminum chloride was added to a benzene solution of hexamethylcyclotrisilazan (molar ratio 1:1) [1225], the solution acquired a black color after two days. A precipitate formed and the following structure was assigned to it:

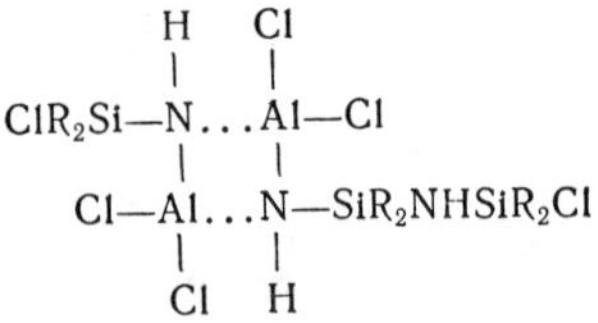

```
          H   Cl
          |   |
ClR2Si—N...Al—Cl
          |   |
     Cl—Al...N—SiR2NHSiR2Cl
          |   |
          Cl  H
```

*Parameters of the crystal according to X-ray data [1255]: β 104°23', a 6.75_9 Å; b $13,18_1$ Å; c $11,22_5$, Å; Z 2; U 968,7 Å, D_X 0,997; D_m 0,998; space group $P\,2^1/n$ (cf. Table 35).

The product was readily hydrolyzed. It dissolved partly in ether. Distillation of the ether solution yielded a colorless solid, which corresponded in analysis to the etherate of chlorodimethylsilylaminodichloroalumane. The ether-insoluble part after evaporation of the solvent was a black product, which was soluble in ether and corresponded closely in composition to $(CH_3)_2Si(NHAlCl_2)_2$.

With trimethylalumane, trisilylamine forms a 1:1 adduct at −46°C, which decomposes at 0°C with the formation of SiH_4 [1196]. Dimethyl(silyl)amine and methyl(disilyl)amine react analogously with trimethylalumane with the former giving the most stable adduct. However, both of them dissociate reversibly.

In the reaction of hexaalkyldisilazans and triorganosilylamines with R_3Al [294] there is the liberation of alkanes and the formation of compounds with Si−N−Al bonds:

$$\gtrdot Si-\overset{|}{N}-H + R_3Al \longrightarrow \gtrdot Si-\overset{|}{N}-AlR_2 + RH$$

$$\gtrdot Si-\overset{|}{N}-AlR_2 + H-\overset{|}{N}-Si\lessdot \longrightarrow \left(\gtrdot Si-\overset{|}{N}\right)_2 AlR + RH \qquad (3\text{-}140)$$

Replacements of the first alkyl in R_3Al by the group $(R'_3Si)_2N$ proceeds with the liberation of heat. The reaction of hexamethyldisilazan with triethylaluminum to introduce a second bis(trimethylsilyl) amino group requires heating, while a third bis(triethylsilyl) amino group cannot be introduced even with prolonged heating of the mixture at 150°C. Since the reaction with $(iso\text{-}C_4H_9)_3Al$ proceeds much more slowly and in lower yields than in the case of triethylalumane, this must be due to steric factors. The compounds obtained according to the scheme (3-140) (Table 38) are high-boiling liquids, which are soluble in hydrocarbon liquids; they are readily oxidized and hydrolyzed in air. In combination with $TiCl_4$, such compounds are active catalysts for the polymerization of ethylene.

The reaction of hexamethylcyclotrisilazan with triethylalumane begins at room temperature with the formation of a 1:1 adduct [295, 1502]. Subsequent conversions are shown by the scheme:

$$R_2Si\begin{matrix}\diagup NH-SiR_2\diagdown \\ \diagdown NH-SiR_2\diagup\end{matrix}NH\cdot AlR'_3 \xrightarrow{-R'H} R_2Si\begin{matrix}\diagup NH-SiR_2\diagdown \\ \diagdown NH-SiR_2\diagup\end{matrix}NAlR'_2 \xrightarrow[\sim 250^\circ C]{+(R_2SiNH)_3}$$

$$\longrightarrow \left[R_2Si\begin{matrix}\diagup NH-SiR_2\diagdown \\ \diagdown \underset{|}{N}-SiR_2\diagup\end{matrix}N\underset{\underset{R'}{|}}{Al}- \right]_n \qquad (3\text{-}140a)$$

Depending on the ratio of the reagents and the reaction conditions, viscous liquids or solid polymers are obtained with molecular weights from 420 to 1840; they are converted into powders by atmospheric moisture and oxygen. With a ratio $[(CH_3)_2SiNH]_3$: $Al(C_2H_5)_3$ = 1:3, heating the mixture to 260°C leads to the cyclic "alumylsilazan" $[(CH_3)_2SiNAl(C_2H_5)_2]_3$. Thus, under these conditions the reaction proceeds according to scheme (3-140) with complete replacement of the silazan hydrogen atoms in $(R_2SiNH)_3$ by R'_2Al groups.

In the reaction of hexamethyldisilazan with lithium aluminum hydride there is double metallation [1015]:

$$4\,(R_3Si)_2\,NH + LiAlH_4 \longrightarrow LiN\,(SiR_3)_2 + Al\,[N\,(SiR_3)_2]_3 + 4H_2 \qquad (3\text{-}141)$$

Lithium bis(trimethylsilyl)amide is readily separated by recrystallization. The colorless crystalline tris[bis(trimethylsilyl)-amino]alumane sublimes in vacuum and is soluble in nonpolar solvents. The compound has a highly symmetrical structure and with 54 hydrogen atoms, in the NMR spectrum there is only one sharp signal. The spherical form of the molecule with screening of the central aluminum atom makes dimerization impossible. However, water cleaves the compound with the liberation of heat and hydrolysis occurs at the Al–N bonds.

According to data in [1097] triphenylalumane adds to (triphenylphosphinimino)trimethylsilane with the formation of a stable compound, which loses benzene above the melting point with the formation of a heterocyclic compound, whose structure is provisionally represented by the formula

$$(CH_3)_3\,SiN\begin{matrix} \diagup Al\,(C_6H_5)_2 \diagdown \\ \diagdown P\,(C_6H_5)_2 \diagup \end{matrix} C_6H_4$$

The properties of nitrogen-containing organosilicon derivatives of aluminum are given in Table 38.

8. COMPOUNDS CONTAINING THE Si – Al BOND

In the reaction of the complex $NaHAl(OC_2H_5)_3$ [1124] with triphenylchlorosilane, sodium chloride precipitates and the carbon content of the reaction mixture is 58-69% [68.6% C is the calculated balue for $(C_6H_5)_3SiHAl(OC_2H_5)_3$. However, the authors were unable to isolate any individual compound or to determine whether products with a direct Si−Al bond were formed in the reaction.

The only paper in which silylalumanes have been described as yet is the review [123], which generalizes the investigations of Wiberg and his students on the chemistry of metal silyls. The reaction of bis(trimethylsilyl)mercurane (see Ch. 2) with excess activated aluminum begins at 60°C and proceeds at an adequate rate at 80°C. Metallic mercury is liberated quantitatively in 1-3 days to form a complex mixture of volatile liquid and involatile solid products. Bis(trimethylsilyl)mercurane evidently "silylates" aluminum, but the unstable tris(trimethylsilyl)alumane undergoes decomposition:

$$Hg(SiR_3)_2 + Al \longrightarrow Hg + Al(SiR_3)_3 \longrightarrow AlR_3 + [R_2Si] \quad (R = CH_3) \qquad (3-142)$$

Under the reaction conditions the bis(trimethylsilyl)mercurane partly decomposes and partly silylates the "dimethylsilene," leading to the formation of compounds of the series $R_3Si[R_2Si]_nR$ (n = 1 − 3). To eliminate side processes connected with the decomposition of bis(trimethylsilyl)mercurane at 80°C, dimethylalumane was used in the reaction. About a day was required for complete conversion of bis(trimethylsilyl)mercurane at room temperature. Its decomposition was eliminated, but the reaction was by no means confined to the simple formation of trimethylsilyldimethylalumane. This compound is also unstable:

$$Hg(SiR_3)_2 + R_2AlH \longrightarrow Hg + R_2AlSiR_3 + R_3SiH \qquad (3-143)$$
$$R_2AlSiR_3 \longrightarrow R_3Al + [R_2Si]$$

As side reactions there is silylation of $[R_2Si]$ and also the formation of the hypothetical tetramethylalumane. The latter disproportionates to trimethylalumane and aluminum, as a result of which there is the possibility of a new chain of conversions in accordance with scheme (3-142).

The instability of alkylsilyl derivatives of aluminum stimulated attempts at the synthesis of aryl derivatives. However, if $AlCl_3 \cdot NH_3$ is added to a solution of triphenylsilylpotassium in liquid ammonia in a molar ratio of 1:3, the solid product formed is not tris-(triphenylsilyl) alumane, but a mixture of HCl, triphenylsilane,

TABLE 38. Organosilicon Compounds Containing the Si–N–Al Group

Formula	m.p., °C	b.p., °C	n_D^{20}	d_4^{20}	δ, Hz*	Yield, %	Preparation method (scheme)	Literature
$(CH_3)_3SiNHAl(CH_3)_2$	78—79 **	—	—	—	—12 (CH_3—Si); +52.2(CH_3—Al)	—	Methylation	1101
	38.5	85 (7)	—	—	—	—	—	1255
$(C_2H_5)_3SiNHAl(C_2H_5)_2$	—	77—79 (12)	1.4800	0.9110	—	—	3-140	294
$[(CH_3)_3Si]_2NAl(C_2H_5)_2$	—	78—79 (2)	—	—	—	—	3-140	294
$[C_2H_5)_3Si]_2NAl(C_2H_5)_2$	—	164—165 (8)	1.4927	0.8889	—	—	3-140	294
$[(CH_3)_3Si)_2NAl(C_4H_9\text{-iso})_2$	—	107—109 (5)	1.4690	0.8549	—	25	3-140	294
$(CH_3)_3SiNHAlCl_2$	272 **	—	—	—	—	52	3-139	858
	270 **	—	—	—	—	83	3-139	143
	163 decomp	—	—	—	—25.2 (CH_3—Si)	—	—	1101
$Cl(CH_3)_2SiNHAlCl_2 \cdot (C_2H_5)_2O$	—	94—95 (1)	—	—	—	—	—	1225
$(CH_3)_3SiNHAlBr_2$	152 **	—	—	—	—	21	3-139	858
	151 **	--	—	—	—	81	3-139	143
$\{[(CH_3)_3Si]_2N\}_2AlC_2H_5$	—	125—126 (3)	1.4725	0.9063	—	Quantitative	3-140	294
$\{[(CH_3)_3Si]_2N\}_2AlC_4H_9\text{-iso}$	—	151—152 (5)	1.4730	0.8992	—	50	3-140	294
$\{[(CH_3)_3Si]_2N\}_3Al$	230 decomp***	—	—	—	14.6 (CH_3—Si)	—	3-141	1015
$(CH_3)_2Si(NHAlCl_2)_2$	—	—	—	—	—	—	—	1225
$[(CH_3)_2SiNAl(C_2H_5)_2]_3$	—	—	—	—	—	—	~3-140	295

* In CCl_4, 60 MHz, internal standard Si $(CH_3)_4$.

** In vacuum.

*** The melting point given in [1015] (> 500°C) is incorrect (private communication).

hexaphenyldisilazan, and compounds with an Al–N bond:

$$3R_3SiK + AlCl_3 \longrightarrow 3KCl + (R_3Si)_3Al \xrightarrow{NH_3}$$
$$\longrightarrow Al(NH_2)_3 + R_3SiH + R_3SiNHSiR_3 + H_2 \quad (R = C_6H_5) \tag{3-144}$$

A study of the reaction products showed that the group $(C_6H_5)_3$-Si migrates without changing. Since ammonia was found to be an unsuitable medium, the reaction was repeated at −20°C with a non-solvating solvent, namely, tetrahydrofuran. The $AlCl_3$ then reacted quantitatively, but ~80% of free KCl was formed. After removal of this and distillation of the tetrahydrofuran, there remained a yellowish residue, which was partly soluble in benzene. The insoluble part contained all the "missing" KCland corresponded in analysis to the compound $KCl \cdot Al[Si(C_6H_5)_3]_3$ (> 50% yield). The product dissolved in tetrahydrofuran, had reducing properties , and on treatment with alcoholic alkali, unexpectedly liberated two and not three moles of hydrogen per mole of complex. The authors of [123] explained this result by isomerization of tris(triphenylsilyl)alumane (I):

$$3(C_6H_5)_3SiK + AlCl_3 \xrightarrow{-KCl} \underset{\text{I}}{[(C_6H_5)_3Si]_2Al\text{—}Si(C_6H_5)_3} \longrightarrow$$

$$\longrightarrow \underset{\text{II}}{(C_6H_5)_3Si\text{—}Al(C_6H_5)\text{—}Si(C_6H_5)_2Si(C_6H_5)_3} \tag{3-145}$$

The isomer (II) forms a complex with KCl and on hydrolysis it yields two moles of hydrogen in accordance with the two Al–Si bonds in it since the Si–Si bond is not cleaved hydrolytically under the conditions adopted. A study of the cleavage of $KCl \cdot Al[Si(C_6H_5)_3]_3$ by gaseous hydrogen chloride led to new unexpected results. There was formed 1 mole of $AlCl_3$ per mole of complex, but the reaction consumed 6 moles of HCl and not 3 moles (calculated for the cleavage of one Al–Cl bond and two Al–Si bonds) to yield 4 moles and not 1 mole of benzene. Under the catalytic effect of $AlCl_3$ the HCl evidently cleaved phenyl groups from silicon. The over-all process is represented by equation (3-147) and not the proposed scheme (3-146):

$$KCl \cdot AlR(SiR_3)(SiR_2SiR_3) + 3HCl \not\longrightarrow KCl + AlCl_3 + RH + R_3SiH + HSi_2R_5$$

$$KCl \cdot AlR(SiR_3)(SiR_2SiR_3) + (3+n+m)\,HCl \longrightarrow$$

$$\longrightarrow KCl + AlCl_3 + (1+n+m)\,RH + HSiCl_nR_{3-n} + HSi_2Cl_mR_{5-m} \quad (3\text{-}147)$$

The benzene-soluble part of the products from the reaction of aluminum chloride with triphenylsilylpotassium in tetrahydrofuran contained tris(triphenylsilyl)alumane, free from KCl according to the material balance. Whether or not this product corresponded to formula I or isomer II was not investigated. The first hypothesis is more probable. If we take as a basis the structure $Al[Si(C_6H_6)_3]_3$, the aluminum is so screened by the triphenylsilyl group that the compound must be free from KCl. The isomeric form (II) favors the formation of a complex with KCl since the aluminum in it is more accessible.

9. COMPLEXES OF HALOSILANES WITH ALUMINUM HALIDES

It is generally accepted nowadays that the rearrangement of organohalosilanes which occurs in the presence of AlX_3 involves the intermediate formation of complexes [21]. It is pertinent to point out the difference in opinions on the nature of these complexes.

In an examination of the reactions of catalytic substitutive chlorination of phenylchlorosilanes it was first suggested [510] that the main role here is played by ionized complexes of the type

$$C_6H_5SiCl_3 + AlCl_3 \longrightarrow C_6H_5SiCl_3 \cdot AlCl_3 \longrightarrow C_6H_5\overset{Cl\ \ Cl}{\underset{Cl}{Si}}\ldots Cl^- \ldots \overset{+}{A}lCl_2 \rightleftarrows$$

$$\rightleftarrows [C_6H_5SiCl_4]^- [AlCl_2]^+ \quad (3\text{-}148a)$$

However, a different treatment was proposed later [47]:

$$Cl_2 + AlCl_3 \rightleftarrows \overset{+}{Cl}\ldots Cl \ldots \overset{-}{Al}Cl_3 \rightleftarrows \overset{+}{Cl} + [AlCl_4]^- \xrightarrow{C_6H_5SiCl_3}$$

$$\rightarrow [C_6H_5(Cl)SiCl_3]^+ [AlCl_4]^- \quad (3\text{-}148b)$$

At the same time, Alpatova and Kessler [2] disputed the possibility of the complexes having an ionic structure as they considered that it was improbable that the silicon compound in them would act as an acceptor.* In their opinion the interaction is limited to strong polarization of the Si–X bonds, which is responsible for the increased reactivity of the complexes.

As a counterbalance to this we should mention work [472] on the polymerization of sylvan (2-methylfuran) in the presence of catalysts based on organochlorosilanes and metal chlorides. Sylvan is known to be polymerized only by ionic compounds and is inert toward radical initiators and catalysts of the Ziegler type. At the same time, a benzine-insoluble complex with the composition $C_6H_5SiCl_3 \cdot AlCl_3$, which was isolated by the authors, is one of the initiators (though not the most active) of the polymerization of sylvan. With a concentration of it of 6 mol.% the yield of the polymer after 100 min at 18°C (sylvan:benzene = 30:70) was 18% and increased to 25% with an increase in the complex concentration to 8 mol. %.

The complex of aluminum chloride with trimethylchlorosilane is most stable. In the opinion of Lengyel and his co-workers [925], the interaction of these reagents forms through a common chlorine atom a three-dimensional structure which consists of a trigonal bipyramid of $(CH_3)_3SiCl_2$, connected through the Cl–Cl edge to a $AlCl_4$ tetrahedron. Believing the following structure to be possible

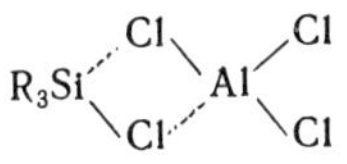

$$R_3Si\begin{matrix} \cdot\cdot Cl \diagdown \\ \diagdown Cl \cdot\cdot \end{matrix}Al\begin{matrix} \diagup Cl \\ \diagdown Cl \end{matrix}$$

the authors extend these concepts to the mechanism of the methylation of chlorosilanes with methylaluminum sesquichlorides:

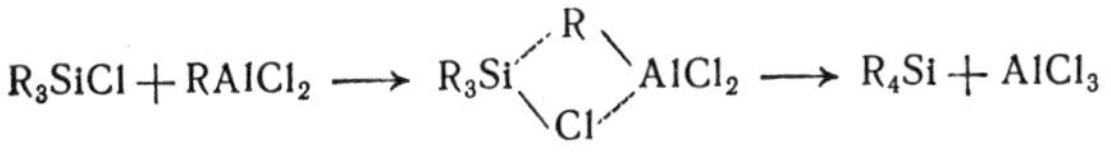

$$R_3SiCl + RAlCl_2 \longrightarrow R_3Si\begin{matrix} \cdot\cdot R \diagdown \\ \diagdown Cl \cdot\cdot \end{matrix}AlCl_2 \longrightarrow R_4Si + AlCl_3 \qquad (3\text{-}149)$$

*Typographical errors in the text of [2] distorted the essence of the viewpoint of the authors.

10. ORGANOSILICON COMPOUNDS OF GALLIUM, INDIUM, AND THALLIUM

Until 1962 the literature on organosilicon compounds containing gallium, indium, and thallium was limited to reports that, for example, trisilylamine and trifluorosilane do not form adducts with trimethylgallane at −78°C [1195, 1196]. Methyldisilylamine gives a 1:1 adduct at 0°C, which decomposes slowly in the solid state and at a high rate on melting (12°C). The adduct of trimethylgallane and dimethyl(silyl) amine melts at a higher temperature and decomposes at a lower rate.

Disiloxane forms an unstable 1:1 adduct with trimethylgallane at −80°C [1194, 1196].

The first true organosilicon derivative of gallium was obtained by a scheme analogous to (3-111) [1075, 1102]:

$$3\,(CH_3)_3\,SiONa + GaCl_3 \longrightarrow [(CH_3)_3\,SiO]_3\,Ga \qquad (3\text{-}150)$$

This colorless crystalline substance has much in common with its Al analog – from the dimeric structure to the capacity to form salt-like isomorphous compounds with trimethylsilanolates of alkali metals [1077, 1103] and trimethylsiloxytetramethylstilbane [1072, 1080]; the rates of their reactions are even higher than in the case of tris(trimethylsiloxy)alumane.

However, the analogy ends at this point. The action of gallium trichloride on hexamethyldisiloxane does not yield trimethylsiloxydichlorogallane, but unexpectedly, methylation occurs [87, 1086]:

$$n\,R_3SiOSiR_3 + n\,GaCl_3 \longrightarrow n\,R_3SiCl + \underset{85\%}{n\,RGaCl_2} + [R_2SiO]_n \qquad (3\text{-}151)$$

In its turn, the reaction of gallium trichloride with dimethylsiloxanes also forms methyldichlorogallane and $(CH_3SiClO)_n$. Indium trichloride does not react with siloxanes even at temperatures above 200°C. Di- and trimethyl derivatives of gallium are not formed by reaction (3-151).

Cleavage of methyl radicals from silicon is not a specific reaction of gallium trichloride with methylsiloxanes. Methyldichlorogallane is also formed by the reaction with tetramethylsilane [87,

1073]:

$$R_4Si + GaCl_3 \longrightarrow R_3Si\begin{matrix} R \\ \quad \\ Cl \end{matrix}GaCl_2 \longrightarrow R_3SiCl + \underset{>90\%}{RGaCl_2} \qquad (3\text{-}152)$$

The reaction with tetraethylsilane proceeds analogously. It is interesting that in the case of $(CH_3)_3SiCH_2Cl$ a methyl radical is also eliminated while the chloromethyl radical is not touched. At the same time, in the reaction with trimethylsilane at −20°C there is quantitative reduction of $GaCl_3$ to $HGaCl_2$ [87, 1088].

While the reaction with tetramethylsilane proceeds according to scheme (3-152) at room temperature, tetramethylsilane does not react with indium trichloride even at 220°C.

It was possible to obtain organosilicon derivatives of gallium and indium of the type R_3SiOMX_2 by a completely different method [87, 1074]:

$$(CH_3)_3SiOH + (CH_3)_3M \cdot O(C_2H_5)_2 \xrightarrow{<20^\circ C} CH_4 + (CH_3)_3SiOM(CH_3)_2 \qquad (3\text{-}153)$$

The compounds are dimeric (like the Al analog) and are distinguished by high hydrolytic and oxidative resistance.

The properties of organosilicon derivatives of gallium and indium are given in Table 39.

The reaction of triethylsilane with triethylthallium (3:1) in a sealed evacuated ampoule for 4-5 hr at 100°C forms metallic thallium (84%), ethane (53%), and a dark cherry red liquid ($d_4^{20} \sim 1.19$), which was found to be the first organosilicon derivative of thallium $[(C_2H_5)_3Si]_3Tl$ (yield 8%) [260, 335a].

11. ORGANOSILICON DERIVATIVES OF ELEMENTS OF THE SCANDIUM SUBGROUP *

According to Andrianov [1461], it is possible to introduce cerium into a siloxane structure by a high-temperature reaction of hy-

* To be systematic we will also examine here derivatives of uranium, as a member of the actinide group.

TABLE 39. Organosilicon Derivatives

Formula	m.p., °C	δ, Hz
$(CH_3)_3SiOGa(CH_3)_2$	16,5; Subl. 34—35 (3 mm)	—5 (CH_3—Si) +17 (CH_3—Ga)
$[(CH_3)_3SiO]_3Ga$	208; Subl. 135 (1 mm)	—4.2; —17.0 *
$Li\{Ga[OSi(CH_3)_3]_4\}$	—	—9.5; —9.9
$Na\{Ga[OSi(CH_3)_3]_4\}$	—	—7.0
$K\{Ga[OSi(CH_3)_3]_4\}$	—	—
$(CH_3)_4Sb\{Ga[OSi(CH_3)_3]_4\}$	190,5; Subl. 190 (1 mm)	—5.0
$H_3Si(CH_3)_2N \cdot Ga(CH_3)_3$	50	—
$CH_3(H_3Si)_2N \cdot Ga(CH_3)_3$	12	—
$(CH_3)_3SiOIn(CH_3)_2$	16; Subl. 57—58 (3 mm)	—0.8 (CH_3—Si) +7.5 (CH_3—In)

* This belongs to the $(CH_3)_3Si$ group attached to the "bridging" oxygen.

† The symbol ~ indicates that the method of preparing the compound is analogous to the given scheme.

‡ As was reported in [1102], the constants of this compound refer to the aluminum analog and vice versa.

§ d_4^{25} 1.260.

droxyl-containing organosilicon compounds with finely dispersed metal or cerium hydroxide. However, no communication describing these reactions in detail or the properties of cerasiloxanes has appeared up to now. At the same time, soluble cerium compounds have been used as catalysts for the hardening of siloxane resins [1342, 1612, 1623] and as stabilizers of siloxane lubricants and fluids [529-532, 970, 1447]. It has also been reported that the addition of 0.02-0.03 wt.% of cerium derivatives to dimethylsiloxane or methylphenylsiloxane fluids [970] retards both destructive processes involving the formation of rings and also thermooxidative processes at 400°C. In the case of siloxane lubricants based on methyl-, methylphenyl-, and methylchlorophenylsiloxanes the most effective compound is cerium toluate or combinations of it with o-phenylenediamine disalicylate (i. e., a chelate compound of cerium). The addition of yttrium and lanthanum compounds also increases the efficiency of o-phenylenediamine disalicylate. Derivatives of praseodymium and europium are less effective, but also increase the thermal stability of siloxane lubricants [532]. Somewhat larger of Gallium and Indium

$J(H^1-C^{13})$, Hz	$J(H^1-Si^{29})$, Hz	Preparation method (scheme)†	Yield, %	Literature
—	—	3-153	Quant.	1074
117.5; 120.2 *	—	3-150	77	1075 ‡
116.3	6.40	~ 3-126	98	1077, 1103
115.8	6.40	~ 3-126	97	1077
—	—	~ 3-126	93	1077
117	6.65	~ 3-128	92	1072, 1080 §
—	—	—	—	1195
—	—	—	—	1195
—	—	3-153	Quant.	1074

amounts of cerium compounds, for example, 1% of its naphthenate, are required to stabilize liquids with silanol groups [1447]. The addition of oxides, hydroxides, and carboxylates of yttrium and rare earth elements [1356, 1374, 1433, 1652] increases the thermal stability of dimethylsiloxane elastomers. For example, the addition of 0.08 wt.% of a mixture of octanoates (caprylates) of rare earth elements in the proportions in which these elements are present in monazite sand are equivalent in action to 5% of red iron oxide, which is well-known for its thermostabilizing effect and is used widely in recipes for rubber mixtures from siloxane elastomers.

The first organosilicon derivative of uranium was prepared in 1961 [570]. An attempt to prepare a tetrakis(triorganosiloxy)uranium by the action of sodium methyldiethylsilanolate on uranium tetrachloride was unsuccessful, but this yielded a bis-siloxy derivative:

$$nR_3SiONa + UCl_4 \longrightarrow (R_3SiO)_2UCl_2 \qquad (3\text{-}154)$$

It was a solid green substance, which was oxidized in air.

Subsequently it was possible to prepare completely substituted siloxy derivatives of penta- and hexavalent uranium by the addition of a triorganosilanol to the corresponding ethoxy derivative of ur-

TABLE 40. Organosilicon Derivatives of Uranium

Formula	Color	Subl. p., °C (mm)	Degree of association †	Preparation method (scheme)	Yield, %
$[CH_3(C_2H_5)_2SiO]_2UCl_2$ *	Green	B.p. 260 (0.01)	—	3-154	—
$[(CH_3)_3SiO]_5U$	Yellow-brown	140—50 (0.1)	1.77	3-155	Quant.
$[C_2H_5(CH_3)_2SiO]_5U$	Yellow-brown	155 (0.1)	1.31	3-155	90
$[CH_3(C_2H_5)_2SiO]_5U$	Orange	160 (0.1)	1.13	3-155	93
$[(C_2H_5)_3SiO]_5U$	Green-brown	170—80 (0.1)	1.0	3-155	98
	The same	170—80 (0.1)	1.0	3-156	95
$[(CH_3)_3SiO]_6U$	Orange	145—50 (0.1)	1.05	3-155	85
$[C_2H_5(CH_3)_2SiO]_6U$	Orange	160—5 (0.1)	0.98	3-155	98
$[CH_3(C_2H_5)_2SiO]_6U$	Red	175 (0.1)	1.0	3-155	95
$[(C_2H_5)_3SiO]_6U$	Red	195 (0.05)	0.97	3-155	92

* Data from [570]. For the other compounds the data are from [569].
† Ebullioscopic determination of the molecular weight in benzene.

anium in boiling benzene [569]:

$$nR_3SiOH + U(OC_2H_5)_n \longrightarrow (R_3SiO)_n U + nC_2H_5OH \quad (n = 5 \text{ or } 6) \qquad (3\text{-}155)$$

The ethanol formed was removed as the azeotrope with benzene to give a yield of the organosilicon compounds of uranium close to quantitative. By using pentakis(trimethylxiloxy) uranium it is possible to prepare higher derivatives by the "transsiloxylation" method:

$$[(CH_3)_3SiO]_5U + 5(C_2H_5)_3SiOH \longrightarrow [(C_2H_5)_3SiO]_5U + 5(CH_3)_3SiOH \qquad (3\text{-}156)$$

In this case the trimethylsilanol is also removed by distillation with a solvent. Siloxy derivatives of penta- and hexavalent uranium are solid colored substances which are often associated. Their physicochemical constants are given in Table 40.

It is recommended [569] that uranium is determined analytically by dissolving a sample of the substance in ethanol in a platinum crucible. Ammonia is added to the solution, which is evaporated under an infrared lamp to remove volatile silanols. The residue is ignited to U_3O_8. The valence of the uranium in the compound is determined by double oxidation with a solution of cerium sulfate. For determination of the silicon a sample of the substance is dissolved in concentrated sulfuric acid and treated with a mixture of ammonium nitrate and sulfate. The solution is evaporated to dryness, water added, and the solution filtered through a fine filter paper, which is then burned in a platinum crucible and the silicon determined as SiO_2.

Chapter 4

Organosilicon Compounds of Group IV Elements

GERMANIUM*

1. COMPOUNDS CONTAINING THE Si – O – Ge, Si – S – Ge, AND Si – Se – Ge GROUPS

1.1. Preparation Methods

1.1.1. Cohydrolysis of Halosilanes and Halogermanes

The cohydrolysis of a mixture of dimethyldichlorosilane and dimethyldichloro(dibromo)germane in an acid medium yields low-molecular oily polydimethylgermasiloxanes, mainly of cyclic structure, containing 2-7.5% germanium [457, 1478]. The inorganic chain contains only about 40% of the amount of germanium introduced into the reaction. This is due to the higher stability (lower tendency for condensation) of dimethylgermanediol in an acid medium in comparison with dimethylsilanediol.

1.1.2. Reactions of Alkoxysilanes and Silanols with Germanium Compounds

Condensation of methyldipropylmethoxysilane with germanium tetrachloride in the presence of traces of α-picoline was the

---

* See also Sections 2.5. (Ch. 2) and 1.1.6. (Ch. 3). The terminology used in this chapter is based on the principles given in [25]. In [66] it is suggested that the prefix germa- is used for the group $>GeR_2$. From the principles of "oxa-aza" nomenclature it is more correct to use the prefix germa- to denote any replacement of a Si atom by a Ge atom.

first example of the synthesis of tetrakis(triorganosiloxy)germane [68]:

$$4RR'_2SiOR + GeCl_4 \longrightarrow (RR'_2SiO)_4Ge + 4RCl \qquad (4\text{-}1)$$
$$(R = CH_3;\ R' = n\text{-}C_3H_7)$$

The reaction of chloromethyldimethylsilanol with dimethyldichlorogermane in the presence of an acceptor of HCl (triethylamine) also leads to a triorganosiloxy derivative of germanium [1268]:

$$2RR'_2SiOH + R'_2GeCl_2 \xrightarrow{R''_3N} (RR'_2SiO)_2GeR'_2 + 2R''_3N \cdot HCl \qquad (4\text{-}2)$$
$$(R = ClCH_2;\ R' = CH_3;\ R'' = C_2H_5)$$

Heating triphenylsilanol with phthalocyaninogermanediol with azeotropic distillation of the water [860] gives a quantitative yield of a compound in which the triphenylsiloxy groups are separated by a planar phthalocyanine ring from the central Ge atom, which is bound by valences and coordinated. When triphenylsilanol is replaced by diphenylsilanediol [861, 862] the reaction under the same conditions gives fine platelets of crystalline $PcGe[OSi(C_6H_5)_2OH]_2$ (Pc represents a phthalocyanine group; see Fig. 6 on p. 271), which are greenish blue in transmitted light and reddish in reflected light.

1.1.3. Synthesis Using Compounds of the Alkali Metal Silanolate Type

A convenient and flexible method of preparing germansiloxanes is the reaction [98, 487, 488, 826, 1040, 1042, 1090, 1103, 1104]

$$R_nGeX_{4-n} + (4-n)\,R'_3SiOM \longrightarrow R_nGe(OSiR'_3)_{4-n} + (4-n)\,MX \qquad (4\text{-}3)$$

(R = alkyl; R' = alkyl, aryl, or aralkyl; n = 0–3; X = halogen, mainly chlorine; M = alkali metal)

The reaction of trimethylsilanolates of alkali metals with dimethyldichlorogermane in a molar ratio of 2 : 1 gives a ~ 70% yield of bis(trimethylsiloxy)dimethylgermane. If the same reagents are used in a molar ratio of 1:1, it is impossible to isolate trimethylsiloxydimethylchlorogermane from the reaction products in a pure state. A difficultly separable mixture of dimethyldichlorogermane and products from replacement of chlorine in it by one or two trimethylsiloxy groups is formed [1104]. At the same time, the reaction of sodium triphenylsilanolate with germanium tetrachloride gives tris(triphenylsiloxy)chlorogermane [488].

Scheme (4-3) is well supplemented by the following reaction [1040, 1041, 1143]:

$$R_3GeOLi + R'_3SiCl \longrightarrow R_3GeOSiR'_3 + LiCl \quad (4\text{-}4)$$

A further development of such synthesis methods is the use in reactions (4-3) and (4-4) of lithium sil(germ)anethio(seleno)lates [1040, 1041, 1043, 1044]:

$$R_3SiSLi + R_3GeCl \longrightarrow R_3SiSGeR_3 + LiCl \quad (4\text{-}5)$$

$$R_3GeElLi + R_3SiCl \longrightarrow R_3SiElGeR_3 + LiCl \quad (4\text{-}6)$$

$$(El = S \text{ or } Se)$$

Trimethylsiloxytrimethylgermane may also be obtained by means of the original reaction of "germyl displacement" [1106]:

$$[(CH_3)_3GeO]_2SO_2 + 2LiOSi(CH_3)_3 \longrightarrow Li_2SO_4 + 2(CH_3)_3SiOGe(CH_3)_3 \quad (4\text{-}7)$$

1.2. Physical Properties

The physicochemical constants of compounds in which silicon and germanium are connected through a Group VI element are given in Table 41.

In the change from the system $(CH_3)_3Si-O-Si(CH_3)_3$ to $(CH_3)_3Ge-O-Ge(CH_3)_3$ the replacement of a silicon atom by germanium is accompanied by a rise in the boiling point of ~18°C, the refractive index by ~0.27, and the density by 0.23. It is characteristic that the melting point of trimethylsiloxytrimethylgermane is lower than that of hexamethyldisiloxane or hexamethyldigermoxane. The considerable discrepancy in the data on the melting point of tetrakis(triphenylsiloxy)germane is noteworthy. Comparison of data on the melting points of triorganosiloxy derivatives of aluminum and titanium indicate that the higher value is more probable.

TABLE 41. Compounds Containing the Group Si–El–Ge (El = O, S, Se, SO_2, CrO_2)

Formula	m.p., °C	b.p., °C (mm)	Preparation method (scheme)	Yield, %	Literature
$(CH_3)_3SiOGe(CH_3)_3$	—68	117 (725) *	4-3; 4-4	70	1040, 1104
	—	118—122	4-4	67	1143
	—	116—117.5	4-7	65	1106
$(CH_3)_3SiOGe(CH_3)_2Cl$	—	—	4-3	—	1104
$(C_2H_5)_3SiOGe(CH_3)_3$	—	34 (1)	4-3	78	1090
$(CH_3)_3SiOGe(C_2H_5)_3$	—	33 (1)	4-3	77	1090
$[(CH_3)_3SiO]_2Ge(CH_3)_2$	—61	166 (725); ** 54.5 (10.5)	4-3	70	1104
$[Cl(CH_3)_2SiO]_2Ge(CH_3)_2$	—	95—98 (2)3*	4-2	—	1268
$[(CH_3)_3SiO]_3GeCH_3$	—	77 (10)	4-3	69	1090
$[(C_6H_5)_3SiO]_3GeCl$	204—205	—	4-3	—	487, 488
$[(CH_3)_3SiO]_4Ge$	—59	198 (725): 59 (1)	4-3	81	1072, 1090
$[CH_3(C_3H_7)_2SiO]_4Ge$	—	—	4-1	—	68
$[(C_6H_5)_3SiO]_4Ge$	236—237	—	4-3	—	487, 488
	472 decomp	—	4-3	—	826
$[(C_6H_5CH_2)_3SiO]_4Ge$	203—204	—	4-3	—	487, 488
$(CH_3)_3SiSGe(CH_3)_3$	—27	63 (12)	4-6	—	1040
	—27	63 (10)	4-6	57	1043
$(CH_3)_3SiOSO_2OGe(CH_3)_3$	92—98	—	4-12	99	1106
$(CH_3)_3SiSeGe(CH_3)_3$	—	74—76 (12)	4-6	—	1040
	—19 to —17	79 (12)	4-6	63	1044
$(CH_3)_3SiOCrO_2Ge(CH_3)_3$	—3 to +2	—	4-12	92	1106

* n_D^{18} 1.4038; d_4^{20} 0,99.

** n_D^{20} 1.4064.

3* n_D^{20} 1.4495.

All trialkylsiloxygermanes, which are colorless mobile liquids with an unpleasant musty odor, except in the case of tetrakis derivatives, crystallize on strong cooling and dissolve readily in the usual organic solvents.

In the infrared spectra of compounds with one triorganosiloxy group attached to the germanium atom there is a characteristic line ν_{as} of Si–O–Ge at 995 cm^{-1}.* Thus, replacement of a silicon atom in the system Si–O–Si by germanium produces a strong shift in the line ν_{as} toward the longwave region of the spectrum. With a

*The vibration spectra of organic compounds containing elements of the carbon subgroup have been examined in detail in the review [70].

TABLE 42. Value of Chemical Shift in NMR Spectra of Trimethylsiloxy Derivatives of Germanes 60 MHz; CCl_4; internal standard $(CH_3)_4Si$

Compound	δ_{CH_3} (Si), Hz	δ_{CH_3} (Ge), Hz
$(CH_3)_3Si—O—Si(CH_3)_3$	—3.5	—
$(CH_3)_3Si—O—Ge(CH_3)_3$	0.0	—21.0
$[(CH_3)_3SiO]_2Ge(CH_3)_2$	—3.5	—29.0
$[(CH_3)_3SiO]_3GeCH_3$	—5.7	—36.2
$[(CH_3)_3SiO]_4Ge$	—8.3	—

change to the system (R_3SiO) Ge, where $n \geq 2$, the line at 995 cm^{-1} is split into two at 1020 cm^{-1} and $\sim$950 cm^{-1} with the number of triorganosiloxy groups having hardly any effect on the position of the lines – it is only important that there should be no less than two. Trimethylsiloxytrimethylgermane has a characteristic broad band at 813-833 cm^{-1}, which is absent from the spectrum of both siloxanes and germoxanes and is also split with a change from the system Si–O–Ge to Si–O–Ge–O–Si. It is connected with the appearance of new symmetry of the molecules when silicon is replaced by germanium [98, 1090, 1104].

The results of a study of the NMR spectra of trimethylsiloxy derivatives of methylgermanes [1081, 1084, 1090] are given in Table 42.

Overall, IR and NMR data indicate that the replacement of silicon by germanium in a disiloxane or, what is equivalent, replacement of germanium by silicon in a digermoxane produces a shift in the electrons of oxygens toward the silicon with an increase in the fraction of π-bonding in Si–O and an increase in the polarization of the Ge–O bond.

1.3. Chemical Properties

Compounds of the type $(CH_3)_3SiOGe(CH_3)_2R$ distill in vacuum without decomposition. However, when heated at normal pressure these substances, which potentially are capable of the elimination of dimethylgermoxane units, disproportionate slowly [1104]:

$$n\,(CH_3)_3\,SiOGe\,(CH_3)_2\,R \longrightarrow [(CH_3)_2\,GeO]_n + n\,(CH_3)_3\,SiR \qquad (4\text{-}8)$$
$$[R = (CH_3)_3\,SiO \text{ or } Cl]$$

Trimethylsilthio(seleno)trimethylgermanes (yellowish liquids with a disgusting odor) boil in vacuum without decomposition, but disproportionate readily when distilled under normal pressure [1044]:

$$2R_3SiElGeR_3 \longrightarrow (R_3Si)_2\,El + (R_3Ge)_2\,El \qquad (4\text{-}9)$$
$$(El = S \text{ or } Se)$$

Trimethylsiloxyorganogermanes of the type $[(CH_3)_3SiO]_nGeR_{4-n}$ (n = 1 or 2) are hydrolyzed even by warm water (more rapidly in the presence of basic agents) [1104]:

$$(R_3SiO)_n\,GeR_{4-n} + n\,H_2O \xrightarrow{OH^-} nR_3SiOH + R_{4-n}Ge\,(OH)_n \qquad (4\text{-}10)$$

Phthalocyaninogermasiloxanes show the highest thermal and hydrolytic stability. Heating $PcGe[OSi(C_6H_5)_2OH]_2$ at 385°C in vacuum leads to dehydration with the formation of a polymer of the type $[Si(C_6H_5)_2OGePcOSi(C_6H_5)_2O]_n$ [861]. Boiling in benzyl alcohol results in "blocking" of the terminal silanol groups to give

$$PcGe[OSi\,(C_6H_5)_2OCH_2C_6H_5]_2$$

Cleavage of trimethylsiloxytrimethylgermane by phenyllithium occurs at the Ge–O bond [1143]:

$$(CH_3)_3\,SiOGe\,(CH_3)_3 + C_6H_5Li \longrightarrow (CH_3)_3\,SiOLi + (CH_3)_3GeC_6H_5 \qquad (4\text{-}11)$$

As has already been pointed out, cleavage at the Ge–O bond also occurs with aluminum chloride (scheme 3-116) [1105, 1112]. This course of reaction with Lewis acids and bases is due to the lower electronegativity of germanium than silicon.

The polymerization of low-molecular cohydrolysis products of dimethyldichlorosilane and dimethyldihalogermanes by means of sulfuric acids proceeds more slowly than in the case of cyclodimethylsiloxanes [457]. Soon after the introduction of the sulfuric acid the system forms two phases and a white flocculent precipitate forms. It is then homogenized again and is gradually converted into a clear rubbery polymer containing germanium. This may be explained by cleavage of Ge–O bonds in the system Si–O–Ge with the formation of cyclodimethylgermoxanes (the more polar Ge–O bonds are weaker than the Si–O bonds), which are insoluble in siloxanes, but are then copolymerized with cyclodimethylsiloxanes. Germasiloxane elastomers containing up to 7.5% Ge do not differ in properties from normal organosilicon rubber.

In principle, the action of sulfur trioxide on trimethylsiloxytrimethylgermane may form both silyl germyl sulfate and a mixture of disilyl and digermyl esters of sulfuric acid. At -35°C a mixed ester is obtained quantitatively [1106]:

$$R_3SiOGeR_3 + SO_3 \longrightarrow R_3SiOSO_2OGeR_3 \qquad (4\text{-}12)$$

Its sublimation at 100-120°C is accompanied by disproportionation:

$$2R_3SiOSO_2OGeR_3 \longrightarrow (R_3SiO)_2SO_2 + (R_3GeO)_2SO_2 \qquad (4\text{-}13)$$

Chromium trioxide reacts analogously with trimethylsiloxytrimethylgermane (see scheme 6-23).

Mixed silyl germyl esters are readily hydrolyzed:

$$R_3SiOElO_2OGeR_3 \xrightarrow{H_2O} R_3SiOH + R_3GeOH + H_2ElO_4 \qquad (4\text{-}14)$$
$$(El = S,\ Cr)$$

The reaction of trimethylsiloxytrimethylgermane with phosphorus oxychloride [1121] results in 80% cleavage of the Ge–O bond and 20% cleavage of the Si–O bond:

$$R_3SiOGeR_3 + POCl_3 \longrightarrow \begin{cases} R_3GeCl + R_3SiOPOCl_2 & (4\text{-}14a) \\ R_3SiCl + R_3GeOPOCl_2 & (4\text{-}14b) \end{cases}$$

In the case of the acid chloride of pyrophosphoric acid ($P_2O_3Cl_4$), trimethylsilyl and trimethylgermyl phosphochloridates are formed in equimolar amounts.

2. COMPOUNDS CONTAINING THE Si – $(C)_n$ – Ge GROUP

2.1. Preparation Methods

2.1.1. Organometallic Synthesis

Although the reaction of trimethylsilylmethylmagnesium chloride with trimethylhalogermanes in ether proceeds slowly and is complicated by side reactions [40, 1076, 1111], it makes it possible to obtain compounds in which silicon and germanium atoms are connected by a methylene bridge:

$$RR'_2SiCH_2MgCl + R'_3GeX \longrightarrow RR'_2SiCH_2GeR'_3 + MgClX \quad (4-15)$$
$$(R = H, CH_3;\ R' = CH_3)$$

By using trimethylsilylphenylmagnesium bromide it is possible to introduce a p-phenylene bridge into the structure [40].

In the "reverse" reaction, i.e., the action of trimethylgermylmethylmagnesium chloride on methyltrimethoxysilane (1 : 1), refluxing for 15 hr gives a ~75% yield of methyl(trimethylgermylmethyl)dimethoxysilane; this compound, which is bifunctional through the methoxy groups, is capable of further conversions [1154, 1155].

The use of organic derivatives of lithium in the synthesis of organosilicogermanium compounds gives good results [661, 953]:

$$4\,(CH_3)_3\,SiC_6H_4Li + GeX_4 \longrightarrow [(CH_3)_3\,SiC_6H_4]_4\,Ge + 4LiX \quad (4-16)$$

When triethylperfluorovinylsilane is treated with triphenylgermyllithium, only one fluorine atom in the β-position is replaced [1157]:

$$(C_2H_5)_3\,SiCF{=}CF_2 + (C_6H_5)_3\,GeLi \longrightarrow (CH_3)_3\,SiCF{=}CFGe\,(C_6H_5)_3 + LiF \quad (4-16a)$$

Compounds with unsaturated or aliphatic-aromatic bridges between the silicon and germanium atoms may also be obtained by the Wurtz reaction [24, 29–31]:

$$(CH_3)_3\,Ge\,|\!\!-\!\!-\!\!|\,Cl + (CH_3)_3\,SiCl \xrightarrow{Na} (CH_3)_3\,Si\,|\!\!-\!\!-\!\!|\,Ge\,(CH_3)_3 \quad (4-17a)$$

$$(CH_3)_3\,Si\,|\!\!-\!\!|\,X + R_3GeX \xrightarrow{Na} (CH_3)_3\,Si\,|\!\!-\!\!|\,GeR_3 \qquad (4\text{-}17b)$$

($|\!\!-\!\!|$ denotes the biradical $-CH{=}CH-$, $-\underset{\underset{CH_2}{\|}}{C}-$, $-CH_2C_6H_4-$, or $-OC_6H_4-$, while $R = CH_3$, C_2H_5)

2.1.2. Addition Reactions

Germanochloroform adds to vinyltrichlorosilane in the absence of any catalysts [426]:

$$Cl_3SiCH{=}CH_2 + HGeCl_3 \longrightarrow Cl_3SiCH_2CH_2GeCl_3 \qquad (4\text{-}18)$$

When methylvinyldichlorosilane is used in the reaction the yield of the adduct is somewhat lower (~68%) [370]. The introduction of a chlorine atom into the vinyl group hampers the addition, but does not change its character [426]. Ethynyltrimethylsilane adds only one equivalent of trichlorogermane* [40]:

$$(CH_3)_3\,SiC{\equiv}CH + HGeCl_3 \longrightarrow (CH_3)_3\,SiCH{=}CHGeCl_3 \qquad (4\text{-}19)$$

With a change to allylsilanes [370, 426, 954], acryloxysilanes [29], or triorganogermanes [29, 954], benzoylperoxide or chloroplantinic acid is recommended as a catalyst. In this case the addition is also contrary to Markovnikov's rule:

$$R_3Si\,(CH_2)_x\,CH{=}CH_2 + HGeR'_3 \xrightarrow{cat} R_3Si\,(CH_2)_x\,CH_2CH_2GeR'_3 \qquad (4\text{-}20)$$

$$(R = Cl,\ CH_3,\ C_6H_5;\ R' = Cl,\ C_6H_5;\ x = 0 \text{ or } 1)$$

In the case of α, β-dichlorovinyltrichlorosilane there is only 20% addition of trichlorogermane, even in the presence of chloroplatinic acid, due to steric hindrance [369].

The addition of dialkyldialkenylgermanes to tetraalkyldisiloxanes leads to polymers with a peculiar structure of the main chain [316]:

$$R_2Ge\,[(CH_2)_xCH{=}CH_2]_2 + HSiR'R''OSiR'R''H \longrightarrow$$

$$\longrightarrow \left[-\underset{R''}{\overset{R'}{Si}}O\underset{R''}{\overset{R'}{Si}}(CH_2)_{x+2}-\underset{R}{\overset{R}{Ge}}(CH_2)_{x+2}- \right]_n \qquad (4\text{-}21)$$

Viscous oily polymers are obtained in 70-80% yield by heating the reaction mixture in the presence of chloroplatinic acid. In the case where $R = C_2H_5$, $R' = CH_3$, $R'' = C_2H_5$ and $x = 0$ after 6 hr at 120°C a polymer is formed with mol. wt. of 6200 (n = 15). When $R = CH_3$ and $x = 1$, the degree of polymerization reaches 13 (mol. wt. 4450). It should be noted that in the series of compounds $(C_2H_5)_2M(CH=CH_2)_2$ the highest molecular weight of the polymer for the conditions chosen (6200) is observed with M = Ge. No polymer at all is formed with the organotin compound, while the reaction with the lead derivative is accompanied by decomposition with the liberation of the free metal. Diethyldivinylsilane gives a polymer with mol. wt. 2500.

In analogy with reactions according to scheme (4-21), polyaddition of diorganodihydrometallanes to diethynyl derivatives of Group IV elements in the presence of peroxide or platinum catalysts gives polymers with unsaturated groups in the chain [360]:

$$R_2MH_2 + HC{\equiv}CR'C{\equiv}CH \longrightarrow [R_2MCH{=}CHR'CH{=}CH]_n \qquad (4\text{-}22)$$

When $R_2M = CH_3(C_6H_5)Si$ or $(C_6H_5)_2Ge$ and R' is $C_2H_5(C_6H_5)Ge$ or $CH_2Si(CH_3)_2CH_2$, respectively, insoluble brown polymers are obtained. The product from the addition of diphenylgermane to bis-(ethynyldimethylsilyl)benzene is yellow and soluble in benzene and toluene; its molecular weight is 2700. A series of data indicate that the introduction of silicon and germanium atoms into a polymer chain containing conjugated double bonds does not disrupt the conjugation. This is connected with the possibility of $d_\pi - p_\pi$ interaction.

The interaction of germanochloroform with a series of organic halides leads to compounds of the type $RGeCl_3$ [368]. However, β-chloroethyltrichlorosilane does not react and it is not possible to obtain compounds with an alkylene bridge between Si and Ge atoms in this way.

2.2. Physical Properties

Table 43 gives the properties of compounds containing carbon bridges between germanium and silicon atoms.

In the infrared spectrum of trimethylsilyltrimethylgermylmethane, a very strong band at 1050 cm^{-1} is assigned to ν_{as} of Si–C–Ge [1076]. Data on the NMR spectra of some silicon-containing derivatives of germanium are given in Table 44.

TABLE 43. Compounds Containing the $Si-(C)_n-Ge$ Group

Compound	m.p., °C	b.p., °C (mm)	n_D^{20}	d_4^{20}	Preparation method (scheme)	Yield, %	Literature
$H(CH_3)_2SiCH_2Ge(CH_3)_3$	—	132—134	—	—	4-15	51	1111
$(CH_3)_3SiCH_2Ge(CH_3)_3$	—74	38—42 (20); 139 (740)	—	—	4-15	42	1076
$CH_3(CH_3O)_2SiCH_2Ge(CH_3)_3$	—	—	—	—	~4-15	75	1154, 1155
$(CH_3)_3SiC(=CH_2)Ge(CH_3)_3$	—	—	—	—	4-17 b	—	40
$(CH_3)_3SiCH_2CH_2Ge(CH_3)_3$	—	49.5 (13)	1.4363	0.9433	Methylation	80	370
$(C_2H_5)_3SiCH_2CH_2Ge(CH_3)_3$	—	—	—	—	Methylation	—	40
$(C_2H_5)_3SiCH_2CH_2GeCl_3$	—	—	—	—	~4-18	—	40
$(CH_3)Cl_2SiCH_2CH_2GeCl_3$	—	121.5 (19.5)	1.5025	1.5883	~4-18	68	370
$Cl_3SiCH_2CH_2GeCl_3$	35	120 (15)	—	—	4-18	83	426
$Cl_3SiCH_2CHClGeCl_3$	—	129 (13)	1.5268	1.8350	4-18	52	426
$Cl_3SiCHClCH_2GeCl_3$	—	—	—	—	4-18	—	33
$Cl_3SiCHClCHClGeCl_3$	—	137 (13)	1.5345	1.8609	~4-18	20	33, 369
$(CH_3)_3SiCH=CHGe(CH_3)_3$	—	158—160 (748)	1.4460	0.9558	4-17a	—	33, 369
$(CH_3)_3SiCH=CHGeCl_3$	—	—	—	—	4-19	—	40
$(C_2H_5)_3SiCF=CFGe(C_6H_5)_3$	64—66	—	—	—	4-16a	55	1157
$(C_6H_5)_3SiCH_2CH_2CH_2Ge(C_6H_5)_3$	134—135	—	—	—	4-20	76 (93)	954
$Cl_3SiCH_2CH_2CH_2GeCl_3$	—	107 (4)	1.5043	1.6538	4-20	—	370
$(CH_3)_3SiOOCCH_2CH_2Ge(C_2H_5)_3$	—	—	—	—	~4-20	—	29
$(CH_3)_3SiOC_6H_4Ge(C_2H_5)_3$	—	126—127 (2)	1.5050	—	4-17 b	42	661
$(CH_3)_3SiC_6H_4Ge(CH_3)_3$	—	—	—	—	4-15	—	40
$(CH_3)_3SiC_6H_4CH_2Ge(CH_3)_3$	—	108 (5)	1.5067	—	4-17a	80	567
$(CH_3)_3SiCH_2C_6H_4Ge(C_2H_5)_3$	—	135—136 (3—4)	1.5088	—	~4-16	85	661
$[(CH_3)_3SiC_6H_4]_4Ge$	351—354	—	—	—	4-16	49	953

*See the second footnote to Table 39.

TABLE 44. Characteristics of NMR Spectra of Compounds Containing the Group Si—(C)—Ge [283] 40 MHz; 1 : 1 with cyclohexane (τ = 8.56)

Compound	$\tau_{CH_3(Si)}$	$\tau_{CH_3(Ge)}$	τ_{CH_2}
$(CH_3)_3SiCH_2Ge(CH_3)_3$	10.02	9.86	10.18
$(CH_3)_3SiCH_2CH_2Ge(CH_3)_3$	10.04	9.93	9.45
$(CH_3)_3SiC(=CH_2)Ge(CH_3)_3$	9.78	9.65	4.30

2.3. Chemical Properties

A study of the cleavage of n-$(C_2H_5)_3GeC_6H_4CH_2Si(CH_3)_3$ by hydrogen chloride at 50°C in aqueous dioxane showed [566] that $K_{H_2O} \cdot 10^3$ for this compound equals 29.6 min^{-1}, in comparison with 140 for 4-trimethylgermylanisole and 38.5 for 4-trimethylsilylanisole. The hydrolysis of $(CH_3)_3GeCH_2Si(CH_3)(OCH_3)_2$ leads to cyclosiloxanes containing trimethylgermyl groups in the side chain (b.p. 90–95°C at 0.2–0.45 mm). Alkaline copolymerization of the latter with cyclodimethylsiloxanes and hexamethyldisiloxane makes it possible to obtain products of linear structure with trimethylgermyl "appendages," which have higher activation energies for viscous flow than siloxane polymers of similar structure and the same degree of polymerization.

In the reaction of aluminum bromide with a compound of the type $R_3SiR'GeR_3$ (where R = CH_3; R' = CH_2, CH_2CH_2, CH = CH, C = CH_2, C_6H_4), the cleavage corresponds to the Vdovin reaction [40, 372] (Table 45):

$$R_3MR'M'R_3 \xrightarrow{AlBr_3} R_3M'R + [-R_2MR'-]_n \qquad (4\text{-}23)$$

TABLE 45. Cleavage of $(CH_3)SiR'Ge(CH_3)_3$ by Aluminum Bromide [372]

R'	$(CH_3)_4Si$, % in mixture	$(CH_3)_4Ge$, % in mixture
CH_2	9	85
CH_2CH_2	21	75
CH=CH	13	80
$C=CH_2$	20	75
C_6H_4	6	90
C≡C	Resinification	

In addition to this we should point out that the NMR spectrum of (trimethylsilyl)trimethylgermylmethane (60 MHz) shows three sharp signals with an intensity ratio of 9 : 9 : 2. Relative to $(CH_3)_4Si$ the chemical shift according to data in [1067] is -0.65 Hz for CH_3(Si), -9.4 for CH_3(Ge), and +11.2 Hz for CH_2.

For the isomeric 1-trimethylsilyl-1(2)-trimethylgermylethylenes the difference between the total and the electronic polarization is close to the values of P_{at} of the symmetrical trans-Si analog, while the dipole moment of both compounds equals zero [306].

These data indicate the high tendency of Ge–R' bonds (in comparison with Si–R') to undergo heterolytic decomposition. The mechanism of methyl migration from the second heteroatom which accompanies the cleavage has not been elucidated as yet.

Compounds containing bridges of the type $-CH_2CHCl-$, between Si and Ge atoms are dehydrochlorinated more readily than their purely organosilicon analogs. However, β-decomposition predominates over this reaction. Depending on the position of the halogen atom in the hydrocarbon bridge either the silicon atom or the germanium atom participates in the β-decomposition. The decomposition of compounds containing the bridge $-CHClCHCl-$ indicates the high tendency of the C–Ge bond to undergo cleavage [33, 369]:

$$Cl_3GeCHClCH_2SiCl_3 \longrightarrow Cl_3GeCH{=}CH_2 + SiCl_4 \quad (4\text{-}24a)$$
$$Cl_3GeCH_2CHClSiCl_3 \longrightarrow GeCl_4 + CH_2{=}CHSiCl_3 \quad (4\text{-}24b)$$
$$Cl_3GeCHClCHClSiCl_3 \longrightarrow GeCl_4 + ClCH{=}CHSiCl_3 \quad (4\text{-}24c)$$

3. COMPOUNDS CONTAINING THE Si – N – Ge GROUP

Reports of the preparation of two compounds of this class were published in 1963-64 [1060, 1064]:

$$(CH_3)_n GeCl_{4-n} + (4-n)[(CH_3)_3 Si]_2 NNa \longrightarrow (CH_3)_n Ge \{N [Si (CH_3)_3]_2\}_{4-n} \quad (4\text{-}25)$$

A new route was reported recently [1112]:

$$(CH_3)_3 GeN (CH_3) Li + (CH_3)_3 SiCl \longrightarrow (CH_3)_3 GeN (CH_3) Si (CH_3)_3 \quad (4\text{-}25a)$$

TABLE 46. Compounds Containing the Si–N–Ge Group

Formula	m.p., °C	b.p., °C (mm)	Yield, % (scheme)	ν_{as}^{SiNGe}, cm⁻¹	Literature
$(CH_3)_3GeN(CH_3)Si(CH_3)_3$	—	159 (735)	— (4-25a)	—	1112
$(CH_3)_3GeN[Si(CH_3)_3]_2$	29—32	54—56 (1)	76 (4-25)	890	1060, 1064
$(CH_3)_2Ge\{N[Si(CH_3)_3]_2\}_2$	—	~135 (1)	30—40 (4-25)	—	1060

The properties of compounds containing the Si–N–Ge group are given in Table 46.

Bis[(trimethylsilyl)amino]trimethylgermane is a crystalline waxy substance, which unexpectedly is exceptionally resistant to hydrolysis.

4. COMPOUNDS CONTAINING THE Si – Ge BOND

4.1. Preparation Methods

The simplest compounds containing the Si–Ge bond, namely, silylgermane, was prepared in 1962 by MacDiarmid and his coworkers [948, 1181] from an equimolar mixture of silane and germane in a silent electrical discharge. It is characteristic that neither disilane nor digermane was formed. Subsequently different mixed hydrides of silicon and germanium, containing the atoms of the elements in ratios right up to Si_5Ge or $SiGe_3$ were synthesized by pyrolysis of mixtures of simple and higher hydrides of silicon and germanium [1210], decomposition by hydrochloric acid of calcium [1038] or magnesium "silicide-germanides" [1210], by the action of hydrofluoric acid on a mixture of silicon and germanium monoxides [1038], and by the reaction of silylpotassium with chlorogermane (synthesis of silylgermane) [1218]. The latter reaction is a particular case of organometallic synthesis:

$$R_3SiM + XGeR'_3 \longrightarrow R_3SiGeR'_3 + MX \quad (4\text{–}26)$$

$$R_3SiX + M'GeR'_3 \longrightarrow R_3SiGeR'_3 + M'X \quad (4\text{–}27)$$

Thus, for example, the first compound containing the Si–Ge bond, namely, triethylsilyltriphenylgermane was obtained from triphenylgermylsodium and triethylbromosilane in accordance with scheme (4–27) in 1933 [898]. It was shown later that triethylsilyltriphenylgermane may be obtained in 54-64% yield from triethylchlorosilane by replacing the sodium derivative by the potassium derivative [746]. However, when triphenylchlorosilane is used a practically inseparable mixture of reaction products due to secondary processes of exchange and condensation with the participation of diphenylchlorosilane [747]. Therefore it is only possible to obtain triphenylsilyltriphenylgermane in accordance with scheme (4-26) by the reaction of triphenylsilylpotassium with triphenylchlorogermane or triphenylbromogermane (yields of 43 and 63%, respectively). A higher yield of triphenylsilyltriphenylgermane (84-87%) is achieved by the reaction [745, 747]

$$(C_6H_5)_3SiLi + (C_6H_5)_3GeCOOCH_3 \longrightarrow (C_6H_5)_3SiGe(C_6H_5)_3 \qquad (4\text{-}28)$$

In the reaction of triphenylgermylsodium with trichlorosilane [959] all three chlorine atoms are replaced by germyl groups (in the case of chloroform only two chlorine atoms are replaced, while the third is reduced). The reaction of triphenylgermylsodium with silicon tetrachloride proceeds unusually and forms hexaphenyldigermane and a powdery substance, which becomes viscous on heating [959]. It is suggested that this compound has the structure

$$[(C_6H_5)_3Ge]_2Si{=}Si[Ge(C_6H_5)_3]_2$$

However, this is not supported by convincing experimental data.

The only completely alkylated compound containing the Si–Ge bond was obtained recently [1162] by the scheme

$$R_3SiX + XGeR_3 + 2Na \longrightarrow R_3SiGeR_3 + 2NaX \qquad (4\text{-}29)$$
$$(R = C_2H_5;\ X = Br)$$

4.2. Physical and Chemical Properties

The physicochemical constants of compounds containing Si–Ge bonds are given in Table 47.

Mixed hydrides of silicon and germanium have a persistent unpleasant odor and are unstable in air. They react with alkali with the liberation of hydrogen. Thermal decomposition occurs at 290-350°C, when hydrides of simpler composition are formed. Silicon and germanium trichlorides and also HCl are obtained with gold chloride (AuCl) [1210].

TABLE 47. Compounds Containing Si–Ge Bonds

Formula	m.p., °C	Other properties	Literature
H_3SiGeH_3	—119.7 ± 0,2	b. p. 7° (760 mm)	948
	~—115	Yield ~20%	1218
Si_2GeH_8	—113.4	$p = 39{,}6$ mm (0°)	1210
Si_3GeH_{10}	—87.4	$p = 4{,}7$ mm (0°)	1210
Si_4GeH_{12}	—71.5	—	1210
$SiGe_2H_8$	—108.5	$p = 19{,}3$ mm (0°)	1210
$(C_2H_5)_3SiGe(C_2H_5)_3$	—	b. p. 254.5° (760 mm) n_4^{20} 1,4860: d_4^{20} 0,9791	1162
$(C_2H_5)_3SiGe(C_6H_5)_3$	93.5	—	898
	95—96.5	Yield 64%	746
$(C_6H_5)_3SiGe(C_6H_5)_3$	357—359	Subl. 500° decomp. Yield 87%	747
$HSi[Ge(C_6H_5)_3]_3$	188, α-form 170,5, β-form	—	959
$C_2H_5Si[Ge(C_6H_5)_3]_3$	283.5	Yield 69%	959
$HOSi[Ge(C_6H_5)_3]_3$	197	—	959
$H_2NSi[Ge(C_6H_5)_3]_3$	206	—	959
$ClSi[Ge(C_6H_5)_3]_3$	230	—	959
$BrSi[Ge(C_6H_5)_3]_3$	242	—	959
$(C_6H_5)_3SnSi[Ge(C_3H_5)_3]_3$	—	—	804

Triphenylsilyltriphenylgermane is inert toward oxygen in boiling xylene and toward iodine in boiling chloroform; 30% decomposition is produced by iodine in xylene [747]. The compound is not cleaved by the action of Na|K alloy (1 : 5) in ether, benzene, xylene, etc., but when tetrahydrofuran, bromobenzene, or tetraphenylsilane is added to the ether, cleavage occurs readily with the formation of the triphenylsilyl or triphenylgermyl derivative of potassium [746, 747].

The Ge–Si bond in tris(triphenylgermyl)silane is not cleaved by alkali metals or bromine and this makes possible a whole series of conversions [$R = (C_6H_5)_3Ge$] [959]:

$$R_3SiH \xrightarrow{Li} R_3SiLi \xrightarrow{C_2H_5Br} R_3SiC_2H_5 \qquad (4\text{-}30)$$

$$R_3SiH \xrightarrow{Br_2} R_3SiBr \begin{cases} \xrightarrow{NH_3\ liq.} R_3SiNH_2 \xrightarrow{HCl\ gas} R_3SiCl \\ \xrightarrow{NH_4OH} R_3SiOH \xleftarrow{NH_4OH} R_3SiCl \end{cases} \qquad (4\text{-}31)$$

Tris(triphenylgermyl)halosilanes are not hydrolyzed by water, but form the corresponding silanol when treated with alkalis in the cold. When the compound is boiled with alkalis the Ge–Si are cleaved. An interesting characteristic of tris(triphenylgermyl)-bromosilane is its capacity to form stable adducts with sym-dichloroethane (2 : 1; dissociation energy 11 kcal/mole) and with benzene (1 : 3). Tris(triphenylgermyl)aminosilane also forms a 1 : 1 adduct with benzene.

The action of tris(triphenylgermyl)silyllithium or triphenylchlorostannane gave a unique compound, containing three elements of Group IV: $[(C_6H_5)_3Ge]_3SiSn(C_6H_5)_3$ [804].

TIN

5. COMPOUNDS CONTAINING THE Si – O – Sn AND Si – S – Sn GROUPS*

5.1. Preparation Methods

5.1.1. Cohydrolysis of Halosilanes and Halostannanes

Cohydrolysis of a mixture of trimethylchlorosilane with dialkyldichlorostannanes by aqueous ammonia in a solvent leads to the formation of crystalline compounds of the type $(CH_3)_3SiO(R_2SnO)_2Si(CH_3)$ (where R = CH_3, C_2H_5, n-C_3H_7, n-C_4H_9) [991, 993]. The yield of these depends on the ratio $(CH_3)_3SiCl : R_2SnCl_2$. Thus, when R = CH_3 and Si : Sn = 1 : 1 the yield equals only 30%, but it increases to 80% with a ratio of 2 : 1 and even up to 95% when Si : Sn = 5 : 1.

The structure of the stannasiloxanes obtained was demonstrated in confirmatory synthesis by cohydrolysis of trimethylchlorosilane with sym-dichlorotetraalkyldistannoxanes. These results refer to hydrolysis by aqueous ammonia with twice the amount of the latter required for neutralization of the HCl liberated. With an equivalent amount, the nature of the cohydrolysis products changes and the compounds $(CH_3)_3SiOSnR_2OSi(CH_3)_3$ are formed. These compounds are not sufficiently stable (though they were identified) and slowly disproportionate on standing:

$$2R_3SiOSnR'_2OSiR_3 \longrightarrow R_3SiOSiR_3 + R_3SiO(R'_2SnO)_2SiR_3 \qquad (4\text{-}32)$$

*The basic nomenclature of organic tin compounds is given in [25]. The prefix stanna- conveniently denotes replacement of a silicon atom by a tin atom in organosilicon heteropolymers. It has been suggested that the prefix stano- should be used for the group $> SnR_2$ [66].

In their turn sym-bis(trimethylsiloxy)tetraalkyldistannoxanes are also capable of disproportionating slowly with the formation of polydialkylstannoxanes with a higher molecular weight, containing terminal trimethylsiloxy groups.

The cohydrolysis of trimethylchlorosilane with stannous chloride ($SnCl_2$) leads to $(CH_3)_3SiO(SnO)_2Si(CH_3)_3$ (34% yield) which shows an even higher tendency to disproportionate than analogous derivatives of tetravalent tin. Changing from trimethylchlorosilane to monofunctional oligomers of the type $(CH_3)_3Si[OSi(CH_3)_2]_nCl$ (where n=30, 50, 150) makes it possible to obtain linear block polydimethylstannasiloxanes $\{(CH_3)_3Si[OSi(CH_3)_2]_nO\}_2Sn(CH_3)_2$ [224]. The use of stannic chloride under analogous conditions yields "cross-shaped" polymers of the type $\{(CH_3)_3Si[OSi(CH_3)_2]_nO\}_4Sn$.

Cohydrolysis of polyfunctional halosilanes and halostannanes leads to polymeric stannasiloxanes, whose character depends on the functionality of each of the components, the ratio Si : Sn, and the nature of the radicals at the silicon and tin atoms [14, 153, 178, 239, 458, 648, 992] (see Section 5.2). As a rule the hydrolysis is carried out in an alkaline medium, mainly with ammonia, in the presence of an inert solvent [178] or without it [239, 458]. On the example of the cohydrolysis of dimethyldichlorosilane with dimethyldibromostannane it was shown [239] that with insufficient ammonia for neutralization of the hydrogen halides liberated the tin was not incorporated into the siloxane structure. In cohydrolysis in an alkaline medium with a high initial ratio of Si : Sn, the composition of the final polymer depends to a considerable extent on the hydrolysis rate. However, with an increase in the mole fraction of the organotin compound in the hydrolysis mixture this factor ceases to be determining. The yield of dry neutral cohydrolyzates is 55-70% with a starting ratio Si : Sn = 9 − 19 and increases to 85-90% when Si : Sn = 49 − 99. At the same time, it has been reported [993] that a 5% yield of stannasiloxanes was obtained in the cohydrolysis of dimethyldichlorosilane with dimethyldichlorostannane. The latter dissolves with difficulty in dimethyldichlorosilane and is hydrolyzed with more difficulty than dimethyldibromostannane [239] and this evidently explains these results.

The cohydrolysis of a ternary mixture of diethyldichlorosilane–phenyltrichlorosilane–diethyldichlorostannane in a molar ratio of 6 : 3 : 1 leads to a polymer with the ratio Si : Sn = 20.3. Successive extraction of it with toluene, acetone, and alcohol made it possible to isolate fractions with Si : Sn = 11.5; 9.3 and 4.1, respectively. When heated to 105-150°C the cohydrolyzates changed into an infusible, insoluble state [178]. In a study of the cohydrolysis of phenyltrichlorosilane and stannic chloride in a molar ratio of 1 : 1 it was found [14] that polystannaphenylsiloxane is not obtained in either an acid or alkaline medium, while in a neutral medium (alkali as an HCl acceptor) a polymer is formed which contains silicon and tin in a ratio of 128 : 1 (23% yield).

It is also possible to introduce tin into a siloxane structure by cohydrolysis of tetraethoxysilane by weakly alkaline solutions of potassium stannate [67]. Depending on the starting ratio of Si : Sn, this process yields insoluble sticky or crystalline substances, to which has been assigned the structure

$$\cdots-\left[\begin{array}{c} R \\ | \\ -Si-O- \\ | \\ OK \end{array}\right]_m-\left[\begin{array}{c} OK \\ | \\ -Si-O- \\ | \\ OK \end{array}\right]_n-\begin{array}{c} OK \\ | \\ Sn-O-\cdots \\ | \\ OK \end{array}$$

$(m = 1-5;\ n = 1-30;\ R = OC_2H_5$ or $CH_3)$

The content of organic radicals in such polymers is low and actually they are silicates.

5.1.2. Heterofunctional Condensation

An attempt to prepare trimethylsiloxytrimethylstannane or bis(trimethylsiloxy)dimethylstannane by heterocondensation of trimethylethoxysilane with trimethylformoxy- or dimethyldiformoxystannane respectively, was unsuccessful [993]. However, it was shown subsequently that this failure to obtain stannasiloxanes by heterocondensation of the "alkoxy-acyloxy" type is a particular case and is evidently due to the reaction conditions and characteristics of the compounds used. In any case, the following reactions occur at 160-170°C though they are quite slow and complicated by secondary processes [1209]:

$$(4-n)\,R_3SnOCOCH_3 + R_nSi\,(OC_2H_5)_{4-n} \longrightarrow R_nSi\,(OSnR_3)_{4-n} + (4-n)\,CH_3COOC_2H_5$$

$$(R = C_6H_5;\ n = 2 \text{ or } 3) \qquad (4\text{-}33)$$

$$2R_3SiOC_2H_5 + R'_2Sn\,(OCOCH_3)_2 \longrightarrow R'_2Sn\,(OSiR_3)_2 + 2CH_3COOC_2H_5$$

$$(R = C_6H_5;\ R' = n\text{-}C_4H_9) \qquad (4\text{-}34)$$

When a mixture of dimethyldiethoxysilane is heated with $Sn(OCOCH_3)_2$ at 120°C for 104 hr only ⅔ of the reaction occurs (calculated from the ethyl acetate liberated), while the product obtained contains Sn and Si in a ratio of 2 : 1 compared with the expected 1 : 1 despite the excess of dimethyldiethoxysilane [839]. The solid yellow polymer has terminal acetoxy groups (making it possible to determine its molecular weight as ~7000) and has a considerable hydrolytic stability. The interaction of dimethyldiethoxysilane with diisobutyldiacetoxystannane is complex. At 140°C the reaction is complete in two days, but the yield of ethyl acetate is only 47%. In addition to crystalline $C_2H_5O[(CH_3)_2SiO(iso\text{-}C_4H_9)_2SnO]_2COCH_3$, the reaction products contain sym-diacetoxytetraisobutyldistannoxane and an insoluble waxy polymer, which softens at 210°C, has a ratio Sn : Si = 4.7, and is hydrolyzed by 50% sulfuric acid only after heating for 1 hr at 80°C.*

Oily stannasiloxanes are formed by condensation of dibutyldiacetoxystannane with a mixture of sym-tetraethoxydiphenyldisiloxane and $C_2H_5O[(CH_3)_2SiO]_{10}C_2H_5$ in the presence of acetyl chloride [1430]. The polymer obtained contains residual ethoxy groups so that it will undergo further condensation with polyalkylene glycols and their monoethers. A viscous syrupy condensation product containing 33.6% Sn was obtained by heating dibutyldiacetoxysilane with tetraethoxysilane in a ratio of approximately 5 : 4 [1731].

Data on the possibility of preparing stannasiloxanes by condensation of the "alkoxy–halogen" type are contradictory. On one hand, it is reported that friable, glassy polymers are formed by the condensation of diphenyldimethoxystannane with diphenyldichlorosilane [1650],† and also tin-containing products are formed by the reaction of dibutyldichlorostannane with polymethyl silicate [1686]. At the same time, according to other data [993] dimethyldichlorostannane does not react with trimethylethoxysilane, while the reaction of trimethylchlorosilane with dimethyldiformoxystannane is an exchange and leads to the formation of dimethyldichlorosilane. The opinion [335] that the reaction of a tetraalkoxysilane with stannic chloride should lead to complete exchange of alkoxyl and halogen was not confirmed. It was shown, for example, that tetraethoxysi-

* The possibility of preparing stannasiloxanes by heterocondensation of the "alkoxy–acetoxy" type has also been reported in [990, 1132].

† The pair is also called diphenyldimethoxysilane–dibutyldimethoxystannane (?).

lane forms complexes with stannous chloride [54, 116, 243, 245, 309b] which decompose, depending on the conditions, with the evolution of ethyl chloride [116, 243] or ethylene [54, 245] and the formation of stannasiloxane polymers [116.*

Triethyl(methylthio)stannane does not react with diphenyldifluorosilane [521].

The formation of stannasiloxanes was observed in the reaction of low-molecular polydimethylsiloxanediols with dibutyldiacetoxystannane [1732]. This may be the basis of the use of bifunctional organotin compounds, particularly diacyloxystannanes, as "cold vulcanization" agents for organosilicon rubbers and liquid siloxane polymers ("elastogens") [18, 20, 50, 231-233, 400, 401, 403, 405, 406, 459, 461, 462, 972, 980, 1029, 1368, 1372, 1373, 1385, 1389, 1421, 1428, 1436, 1442, 1446, 1450, 1453, 1469, 1481, 1486, 1666, 1682, 1689, 1690] and also as hardeners for siloxane resins and varnishes [292, 1303, 1310, 1324, 1328, 1330, 1334, 1347, 1403, 1416, 1439, 1457, 1544, 1593, 1607, 1611, 1617, 1669, 1676, 1696, 1705, 1729].

The possibility that the hardening of organosilicon resins by organotin compounds is a condensation process with the incorporation of tin into the siloxane structure has not been disproved by anyone but also no one has demonstrated it. There is no doubt as to the condensation nature of "cold vulcanization" but the treatment of the mechanism by different investigators differs somewhat. There is a common opinion that in this case the organotin compounds act only as catalysts for a condensation, which involves silanol ends of the organosilicon polymer and alkoxysilanes that are unused . units of the "vulcanizing group."

After studying the kinetics of "cold vulcanization" of polydimethylsiloxanes in bulk by solutions of $R_2Sn(OCOR')_2$ in tetraethoxysilane by measuring the sheer strength and elastic recovery on a torsion plastimeter, Baranovskaya and her co-authors [231-233] proposed a scheme which includes the formation of an active complex. According to their data, the cold vulcanization process is characterized by an inductive effect, which is connected with the formation of the active complex, with a subsequent increase in the reaction rate due to the growth of molecular chains and their

*More details will be given on complexes containing silicon and tin in Section 9.

branching. The rate of the process is also connected with the nature of the alkyl and aryl radicals of the organotin compound:

$$\sim SiR_2{-}OH + R_2Sn(OAc)_2 + Si(OR')_4 \longrightarrow$$

$$\longrightarrow \sim SiR_2{-}\overset{\overset{H}{|}}{O} \dashrightarrow \overset{R\diagdown\ \diagup R}{\underset{AcO\diagup\ \diagdown OAc}{Sn}} \dashleftarrow \overset{\overset{R'}{|}}{O}{-}Si(OR')_3 \xrightarrow[-R_2Sn(OAc)_2]{-R'OH} \sim SiR_2OSi(OR')_3$$

$$\sim SiR_2OSi(OR')_3 + R_2Sn(OAc)_2 + HOSiR_2 \sim \longrightarrow$$

$$\longrightarrow \sim SiR_2OSi(OR')_2 \dashrightarrow \overset{R\diagdown\ \diagup R}{\underset{AcO\diagup\ \diagdown OAc}{Sn}} \dashleftarrow \overset{\overset{H}{|}}{O}{-}SiR_2 \sim \text{ etc.} \qquad (4\text{-}35)$$

On the basis of a study of the kinetics of cold vulcanization by measuring the viscosity of a solution of polydimethylsiloxane in benzene in the presence of an analogous catalytic system, Nudel'man and other investigators [50, 400, 401, 403, 405, 406] came to the conclusion that there is no induction period (Fig. 7).* Structure formation begins at the moment when the components are mixed and proceeds at a rate which increases with time. The reaction is first-order with respect to the catalyst. In the opinion of the authors, a transition donor-acceptor complex is formed first:

$$\begin{matrix} \geqslant SiOH \\ (C_2H_5O)_4Si \end{matrix} + \begin{matrix} OCOR \\ | \\ SnR'_2 \\ | \\ OCOR \end{matrix} \longrightarrow \begin{matrix} \geqslant Si{-}\overset{H}{O} \quad OCOR \\ \quad \downarrow \qquad | \\ (C_2H_5O)_3Si \quad SnR'_2 \\ \diagdown O \nearrow \ \diagdown OCOR \\ C_2H_5 \diagup \end{matrix} \longrightarrow$$

$$\longrightarrow \begin{matrix} \geqslant SiO \\ | \\ Si(OC_2H_5)_3 \end{matrix} + \begin{matrix} R'_2Sn(OCOR)_2 \\ C_2H_5OH \end{matrix} \qquad (4\text{-}36)$$

* The authors consider that viscosimetric determinations are more legitimate since the appearance and hence the measurement of viscoelastic properties is only possible when a cross-linked structure is formed. As a result of this, processes which belong to the earlier phase of the reaction are not observed in an investigation with a plastimeter.

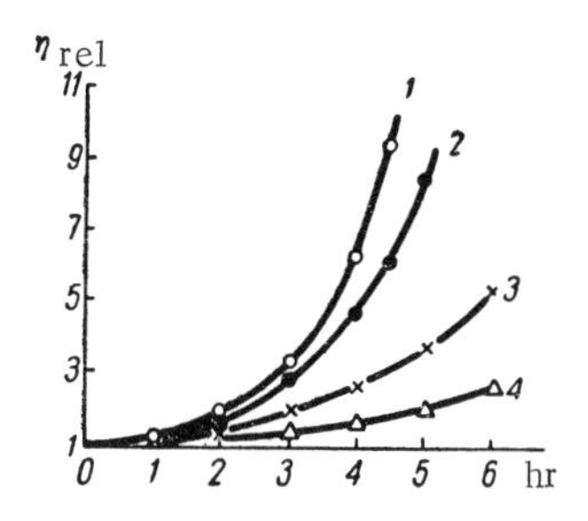

Fig. 7. Relation of relative viscosity to time [403]. System: Polydimethylsiloxane (500,000, 15% solution, 50 ml) + tetraethoxysilane or phenyltriethoxysilane (0.5 g) + catalyst (0.5 g). Catalysts: 1) dibutyldicapryloxystannane + tetraethoxysilane; 2) dibutyldicaproxystannane + tetraethoxysilane; 3) dibutyldistearoxystannane + tetraethoxysilane; 4) dibutyldicapryloxystannane + phenyltriethoxysilane.

The rearrangement by the cyclic transfer schemes leads to the liberation of alcohol in the form of a complex with a dialkyldiacyloxystannane and the formation of a polymer with terminal triethoxysiloxy groups, which then undergoes further condensation reactions analogously.* Infrared spectroscopic data show the disappearance of the absorption band of silanol groups (~3300 cm^{-1}) after the addition of the organotin compound and the cross linking agent to the siloxane polymer.

Nagy and his co-workers [972] regard the "cold vulcanization" of α, ω-polydimethylsiloxanes as a second-order reaction. The reaction rate (mole/min) is represented by the equation

$$V = -\frac{dOH}{dt} = k_0 c^a (n_{OH} - x)(n_{OR} - x)$$

where c is the molar concentration of the initiator (organotin compound), n_{OH} and n_{OR} are the numbers of moles of hydroxyl and ethoxyl groups, respectively, and x is the number of OH and OR groups which react in time t.

After solving the differential equation and rearranging we obtain the equation log M = Kt + A, which gives the relation between the mean molecular weight of the siloxanediol and the vulcanization time, where

$$K = \tan\alpha = \frac{1}{2303} k_0 c^a (n_{OR} - n_{OH})$$
$$(a = 0.5)$$

The values of k_0 characterize the efficiency of different initiators, i.e., diacyloxystannanes. The authors consider that the vulcanization is a condensation between the siloxanediol and alkoxysilane with the liberation of alcohol. They believe that this is con-

*Scheme (4-36) [50] is clearly inaccurate since it distorts the essence of the views of the authors.

firmed by the fact that the reaction of triethylsilanol with tetraethoxysilane in a molar ratio of 4 : 1 yields tetrakis(triethylsiloxy) silane. However, the hardening of siloxanediols by tetraethoxysilane does not occur in the absence of an initiator. This led to the conclusion that the organotin compound forms an intermediate product with tetraethoxysilane, which then reacts with α, ω-siloxanediols. In actual fact, dibutyldiacetoxystannane reacts with tetraethoxysilane (2 : 1) with the liberation of ethyl acetate and the formation of a crystalline compound with an eight-membered ring $[-(C_2H_5O)_2 \cdot SiOSn(C_4H_9)_2O-]_2$.

In the reaction of the oligomers $(CH_3)_3Si[OSi(CH_3)_2]_nOH$ with tetraethoxystannane [224] there is not condensation with the formation of a stannasiloxane, but an exchange reaction with the formation of $(CH_3)_3Si[OSi(CH_3)_2]_nOC_2H_5$. In the reaction of such siloxanols with dimethyldichlorosilane in dry benzene in the presence of an HCl acceptor there occurs the condensation

$$2R_3Si\,(OSiR_2)_n\,OH + R_2SnCl_2 \xrightarrow[3-5^\circ]{(C_6H_5)_3N} R_2Sn\,[O\,(SiR_2O)_n\,SiR_3]_2 \quad (R = CH_3) \qquad (4\text{-}37)$$

The following complex was isolated from the products of the reaction of triphenylsilanol with $SnCl_4$ in benzene in the presence of pyridine [628]:

$$\{[(C_6H_5)_3\,SiO]\,Sn\,(C_5H_5N)_4\,Cl\}\,SnCl_6$$

At the same time, chloromethyldimethylsilanol does not react with $(CH_3)_2SnCl_2$ but only undergoes homocondensation [1268], though the corresponding bis(chloromethyldimethylsiloxy) derivatives are formed smoothly in reactions with dimethyldichlorosilane and dimethyldichlorogermane.

Triethylsilane reacts with triethylbutoxysilane in the presence of iron pentacarbonyl with the formation of hexaethyldisiloxane [490]. It was therefore suggested that compounds containing the Si–O–Sn group could be obtained analogously. However, the reaction of triethylsilane with triethylbutoxystannane forms triethylbutoxysilane in 86% yield.

It is reported that it is possible to prepare stannasiloxane polymers by the catalytic reaction of diphenylsilanediol with diphenylstannane [1650].

5.1.3. Reaction of Silan(stann)olates of Alkali Metals with Halostann(sil)anes

US patent 2998407 [1650] covers the reaction of diorganodichlorostannanes with silanediols and silanedithiols in the presence of alkali metals or their hydroxides amongst various methods of preparing stannasiloxanes and stannasilthianes. Essentially we are concerned here with reactions of compounds of the type $\gt SiOM$ and $\gt SiSM$ (M is an alkali metal) with halostannanes in a masked form. In its pure form the reaction of alkali metal silanolates with halostannanes is a convenient method of introducing trialkyl(aryl)-siloxy groups into tin compounds [98, 168, 627, 628, 826, 1003, 1091, 1107, 1143, 1203, 1209]:

$$nR_3SiOM + R'_{4-n}SnX_n \longrightarrow (R_3SiO)_n SnR'_{4-n} + nMX \qquad (4\text{-}38)$$

$$(R \text{ and } R' = CH_3,\ C_2H_5,\ C_6H_5;\ M = Li,\ Na,\ K;\ X = Cl,\ Br;\ n = 1-4)$$

The solvents recommended for the reaction are ethyl ether, tetrahydrofuran, dioxane, and benzene. By scheme (4-38) it is also possible to prepare triorganosiloxy derivatives of divalent tin such as bis(trimethylsiloxy)stannine $(CH_3)_3SiOSnOSi(CH_3)_3$ [1206].*

As a rule the yield of stannasiloxanes is no less than 60-80%. However, a detailed investigation of the synthesis of triphenylsiloxy derivatives showed [1209] that the conversion on NaX, calculated on the compound reacting, was at a level of 96-98%, while the yield of useful products was 91-99%. The reaction is exothermic and proceeds at a high rate even at room temperature—the conversion on NaX reaches 93-96% after 1 min and then increases by only 1% after 1 hr.

An interesting variant on scheme (4-38) is the use of pseudo-silanolates [1091]:

$$R_3SiOSbR_4 + ClSnR_3 \longrightarrow R_3SiOSnR_3 + R_4SbCl \quad (R = CH_3) \qquad (4\text{-}39)$$

The preparation of tetrakis-siloxy derivatives of tin by the use of sodium monosiloxanolates was described recently [224]:

$$4R_3Si(OSiR_2)_n ONa + SnCl_4 \longrightarrow [R_3Si(OSiR_2)_n O]_4 Sn + 4NaCl \qquad (4\text{-}40)$$

$$(R = CH_3;\ n = 10,\ 34,\ 80)$$

* Sodium triphenylsilanolate does not react with $SnCl_2$ [826].

Scheme (4-38) may also be extended to the synthesis of stannasilthia(selena)nes. In particular $(CH_3)_3SiSSn(CH_3)_3$ was obtained in this way [1095].

The possibilities of synthesizing stannasiloxanes by condensations of the "alkali metal–halogen" type are not limited to the system silanolate–halostannane. The reverse system may also be used effectively [1041, 1089, 1091]:

$$(4-n)\,R_3SnOLi + R_nSiX_{4-n} \longrightarrow (R_3SnO)_{4-n}\,SiR_n + (4-n)\,LiX \quad (4\text{-}41)$$

The reaction of diorganodichlorostannanes with sodium diorganosil(ox)anediolates yields linear soluble polydiorganostannasiloxanes [148, 648] with mol. wt. 1000-5000. In particular, the polymers $-(CH_3)_2SnO-[-(CH_3)_2SiO-]_n-$ were prepared with mol. wt. 2300-2700 and with values of n lying in the range from 2 to 20. If diethyldichlorostannane is used, the polymers are capable of forming films from solutions. Changing the ratio of Si : Sn from 10 : 1 to 50 : 1 does not affect the glass transition point of the polymers, which remains at the level of -100°C [148], but as the silicon content increases, the elasticity of these polymers increases and the same is true when methyl radicals at the silicon are replaced by ethyl radicals. The stability of phenylstannasiloxanes is higher than that of methylstannasiloxanes [648]. A powdery polymer with the composition $[(C_6H_5)_2SiO]_5[SnO]_4$ was obtained by the reaction of sodium diphenylsilanediolate with stannous chloride in acetone [847].

By the reaction of stannic chloride with monosodium salts of alkyl(aryl)silanetriols [14, 228, 1489] it is possible to obtain polymers of the type $[(RSiO_{1.5})_4Sn]_n$, whose consistency, solubility, and thermomechanical properties depend on the ratio of Si : Sn. Thus, with $R = C_2H_5$ and Si : Sn = 1.25–4, glassy, colorless, clear, brittle substances are obtained, while with Si : Sn = 7.75–19.5, light yellow viscous resins are obtained. Ethylstannasiloxanes and phenylstannasiloxanes with Si : Sn $\simeq$ 1.2 are soluble only in acetone, but with a higher silicon content they dissolve in practically all organic solvents (with the exception of petroleum ether when $R = C_6H_5$). These polymers retain their solubility after isolation from solutions, while methylstannasiloxanes are stable only in solutions and lose their solubility and fusibility after removal of the solvent, even at room temperature. Ethyl- and phenylstannasiloxanes lose their solubility when heated above 200°C. The fall in solubility progresses with a change from $R = C_6H_5$ to $R = C_2H_5$ and a decrease in the ratio Si : Sn. The effect of these factors on the thermomechanical properties may be illustrated by the fact that when $R = C_6H_5$ and Si : Sn = 4, the polymer does not flow at 400°C, while with a ratio of 14 : 1, the polymer flows at 130-140°C. At the same time, the product with $R = C_2H_5$ and Si : Sn = 4 flows at 80°C, but does not flow at 400°C when Si : Sn = 1.25. The reaction of monosodium salts of silanetriols with stannic chloride gives a high yield (~ 90%) of polystannaorganosiloxanes and preserves the given ratio of Si : Sn when the process is carried out in an alkaline medium (ending in a neutral medium) or in a mixture of toluene with an alcohol (ethanol or butanol; 15% solution of the reagent).

5.1.4. Reactions of Siloxanes with Tin Compounds and Reaction of Stannoxanes with Organosilicon Compounds

Heating a mixture of octamethylcyclotetrasiloxane and stannic chloride in a molar ratio of 1 : 0.058 at 152°C for 30 hr gave a 86% yield of a polymer with a molecular weight of 350,000 which did not contain tin [223]. With a decrease in the molar ratio to 1 : 1 a high polymer (330,000) was also formed under similar conditions.* It was not possible to isolate compounds of the type $Cl(R_2SiO)\,SnCl_3$, though the organosilicon polymer was isolated in only 53% yield. It is believed that ring cleavage occurred through the formation of the coordination compound:

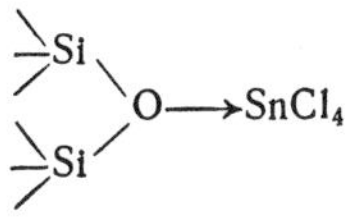

$$\begin{matrix} >Si \diagdown \\ \quad O \longrightarrow SnCl_4 \\ >Si \diagup \end{matrix}$$

Simmler [116] associates the cleavage of hexamethyldisiloxane by stannic chloride both with the formation of an adduct and with the formation of $>Si{-}O{-}SnCl_3$:

* When a mixture of octamethylcyclotetrasiloxane and $SnCl_4$ is treated with γ-radiation at room temperature there is simultaneously polymerization of the "tetramer" and chlorination of it [439, 1492]. At doses of ~ 30 MR, 3 mol. % of chlorine enters the polymer as a result of the conversion of $SnCl_4$ into $SnCl_2$.

$$\begin{array}{c} >Si\quad Si< \\ \bar{O} \\ Cl\quad Cl \\ Sn \\ Cl\quad Cl \\ \bar{O} \\ >Si\quad Si< \end{array} \longrightarrow \begin{array}{c} >Si\quad Si< \\ O \\ \downarrow \\ Cl\quad Cl \\ Sn \\ Cl\quad Cl \\ \uparrow \\ O \\ >Si\quad Si< \end{array} \longrightarrow \begin{array}{c} >Si\quad Si< \\ O \\ \downarrow \\ Cl\quad Cl \\ Sn \\ Cl\quad Cl \\ \uparrow \\ O \\ >Si\quad Si< \end{array} \longrightarrow \begin{array}{c} >Si-\bar{O}-SnCl_3 \\ + \\ Cl-Si< \\ + \\ >Si-\bar{O}-Si< \end{array} \qquad (4\text{-}42)$$

The catalytic effect of stannic chloride in the breakdown of cyclosiloxanes by chlorosilanes [212] is also explained by the formation of compounds of the type $Si-O-MCl_{n-1}$, i.e.,:

$$>Si-O-Si< \xrightarrow{SnCl_4} >Si-O-SnCl_3 \xrightarrow{>SiCl} >Si-O-Si< + SnCl_4 \text{ etc.} \qquad (4\text{-}43)$$

However, the authors presented no proof of the existence of such compounds (as opposed to derivatives of other elements). At the same time, it is stated [440] on the basis of infrared spectroscopic data that neither cyclic nor linear, nor α,ω-(dichloro)methylsiloxanes nor α,ω-(dichloro)ethylsiloxanes show even traces of interaction with stannic chloride.

Heating a suspension of a polydiorganostannoxane in cyclodimethylsiloxanes for several hours at temperatures from 80°C [1733] to 230°C [1431] forms stannasiloxanes:

$$[(CH_3)_2SiO]_4 + (R_2SnO)_n \longrightarrow \{[(CH_3)_2SiO]_4(R_2SnO)_n\}_x \qquad (4\text{-}44)$$
$$(R = n\text{-}C_4H_9;\ R = n\text{-}C_8H_{17})$$

The process is accelerated in the presence of alkalis (neutralization of CO_2 after the end of the polymerization). By adding hexamethyldisiloxane it is possible to vary the consistency of the polymer from viscous liquids to resinous or rubbery products.

At the same time, treatment of solutions of low-molecular polydimethylstannasiloxanes containing 9-19% Sn in octamethylcyclotetrasiloxane [239, 458] with concentrated sulfuric acid leads to the formation of sulfates, which occupy an intermediate position between R_2SnSO_4 and $HOSnR_2OSO_3H$ with respect to composition. Elimination of tin does not occur in the presence of KOH, but rubbery polymers cannot be obtained either under conditions where cyclodimethylsiloxanes are converted into elastomers or under more drastic conditions.

Polydimethylphenylsiloxane oligomers are polymerized on heating in the presence of 0.5-2% of polystannaphenylsiloxanes, obtained from monosodium salts of silanetriols and stannic chloride, with the formation of high-molecular products [52, 416, 419].

The copolymerization of polydiorganostannoxanes with diorganosilanediols may be regarded as a modification of scheme (4-44) [648, 1650, 1651]. Thus, in particular, the insoluble "oxide" gradually dissolves when an equimolar mixture of diphenylsilanediol and polydiphenylstannoxane is refluxed for 6 hours in dioxane. Removal of the solvent in vacuum leaves a clear, brittle glassy, soluble polymer, with mol. mt. 1000-5000 and m.p. 50-70°C.

The interaction of polydimethylsiloxanediols or siloxane rubbers with hexabutyldistannoxane yields oily stannasiloxanes, which contain R_3SiO terminal groups [975, 1734]. On the other hand, bis-(triphenylsiloxy)dimethylstannane was isolated from the reaction products of triphenylsilanol and cyclic polydimethylstannoxane ("dimethyltin oxide"), indicating complete cleavage of the ring [888]. In the case of cyclodiphenylstannoxane, high-molecular stannasiloxanes with terminal R_3SiO groups were obtained [648, 887]. In the simpler case of the reaction of hexaethyldistannoxane with triethylsilanol, transesterification actually occurs [302, 499]:

$$R_3SnOSnR_3 + R'_3SiOH \longrightarrow R'_3SiOSnR_3 + R_3SnOH \qquad (4-45)$$

At the same time, reactions of hexaethyldistannoxane with compounds containing functional groups at the silicon atom of the type R_nSiX_{4-x} (where R is organic; n=1-3; X=F, Cl, I, $OCOCH_3$, $OCOCF_3$, NCO, NCS, CN) and $SiBr_4$ lead in all cases only to the formation of $(C_2H_5)_3SnX$ and disiloxanes (SiO_2 in the case of stannic bromide) [521].* The reaction of trimethylchlorosilane and polydimethylstannoxane likewise did not lead to the formation of stannasiloxanes [993].

* Hexaethyldistanthiane undergoes analogous reactions only with $SiBr_4$ and dodecyltriiodosilane [521]. It is pertinent to mention here that heating an equimolecular mixture of $(C_2H_5)_4Sn$ and $C_6H_5SiCl_3$ (24 hr at ~190°C) forms triethylchlorostannane and diethyldichlorostannane and also resinous substances [501].

TABLE 48. Compounds Containing the Si–O–Sn Group

Compound	m.p., °C	b.p., °C (mm)	n_D^{20}	d_4^{20}	Preparation method	Yield, %	Literature
$(CH_3)_3SiOSn(CH_3)_3$	—59	141 (720); 38 (11)	—	—	4-38	78	1091, 1107
	—	144	—	—	4-38	61	1143
	—	—	—	—	4-39	85	1091
	—	—	—	—	4-41	88	1091
$(C_2H_5)_3SiOSn(CH_3)_3$	—	49 (1)	—	—	4-38	80	1091
	—	48—49 (1,5)	—	—	4-41	78	1089, 1091
$(CH_3)_3SiOSn(C_2H_5)_3$	—	99 (20)	—	—	Cohydrolysis	—	992
	—	52 (1)	—	—	4-38	81	1091
$(C_2H_5)_3SiOSn(C_2H_5)_3$	—	—	—	—	4-45	—	499
$(CH_3)_3SiOSn(C_3H_7\text{-n})_3$	—	120 (20)	1.4575	1.093	Cohydrolysis	—	992
$(CH_3)_3SiOSn(C_4H_9\text{-n})_3$	—	142 (2)	1.4582	1.059	Cohydrolysis	—	992
$(C_6H_5)_3SiOSn(C_6H_5)_3$	138—139	—	—	—	4-38	67	1003
	—	—	—	—	4-38	60	1209
	139—140	—	—	—	4-33	24	1209
$[(CH_3)_3SiO]_2Sn$	—	—	—	—	4-38	—	1206
$[(CH_3)_3SiO]_2Sn(CH_3)_2$	48	160 (720) decomp 75 (11)	—	—	4-38	81	1091, 1107
	—	—	—	1.56^{25}	4-38	—	1206
$[(C_6H_5)_3SiO]_2Sn(CH_3)_2$	155—156	—	—	—	4-38	65	1209
	151—153	—	—	—	4-38	—	628
$\{(CH_3)_3SiO[Si(CH_3)_2O]_{30}\}_2Sn(CH_3)_2$	Tg – 120	—	1.4080	0.9910	Cohydrolysis	80	224
$\{(CH_3)_3SiO[Si(CH_3)_2O]_{34}\}_2Sn(CH_3)_2$	Tg – 115	—	1.4069	0.9900	4-37	—	224

$\{(CH_3)_3SiO[Si(CH_3)_2O]_{50}\}_2Sn(CH_3)_2$	Tg—120	—	1.4061	0.9834	Cohydrolysis	—	224
$\{(CH_3)_3SiO[Si(CH_3)_2O]_{150}\}_2Sn(CH_3)_2$	Tg—100	—	1.4057	0.9681	Cohydrolysis	—	224
$[(C_6H_5)_3SiO]_2Sn(C_4H_9\text{-}n)_2$	69—70	—	—	—	4-34	63	1209
	—	—	—	—	4-38	80	1209
$[(C_6H_5)_3SiO]_2Sn(C_6H_5)_2$	148.5—149.5	—	—	—	4-38	40	1209
$[(C_6H_5)_3SiO]_2Sn(CH_2C_6H_5)_2$	122—123	—	—	—	4-38	78	1209
$[(CH_3)_3SiO]_2Sn[OCOC(CH_3)=CH_2]_2$	67—69	—	—	—	4-49	80	162
$[(CH_3)_3SiO]_3SnCH_3$	34	155 (720) decomp 49 (11)	—	—	4-38	69	1091, 1107
$[(CH_3)_3SiO]_3SnOCOC(CH_3)=CH_2$	—	—	—	—	4-48	—	162
$[(CH_3)_3SiO]_4Sn$	—	—	—	$1{,}41^{25}$	4-38	—	1206
	64	Su61 60 (1)	—	—	4-38	—	1091
	47—49	83—85 (1)	—	—	4-38	41	162
$[(C_2H_5)_3SiO]_4Sn$	—	200—202 (4)	—	—	4-38	54	168
	85—87	—	—	—	4-38	87	162
$[(C_6H_5)_3SiO]_4Sn$	322 разл.	—	—	—	4-38	—	826
$\{(CH_3)_3SiO[Si(CH_3)_2O]_{10}\}_4Sn$	Tg—110	—	1.4057	0.9714	4-40	70	224
$\{(CH_3)_3SiO[Si(CH_3)_2O]_{34}\}_4Sn$	Tg—100	—	1.4039	0.9711	4-40	—	224
$\{(CH_3)_3SiO[Si(CH_3)_2O]_{80}\}_4Sn$	Tg—110	—	1.4090	—	4-40	—	224
$(CH_3)_2Si[OSn(CH_3)_3]_2$	—	77 (1)	—	—	4-41	86	1089, 1091
$(C_6H_5)_2Si[OSn(C_6H_5)_3]_2$	94,5—96,5	—	—	—	4-33	35	1209
$[(CH_3)_3SiOSn]_2O$	—	—	—	—	Cohydrolysis	34	993
$[(CH_3)_3SiOSn(CH_3)_2]_2O$	167—168	—	—	—	Cohydrolysis	95	993
$[(CH_3)_3SiOSn(C_2H_5)_2]_2O$	126—130	—	—	—	Cohydrolysis	90	993
$[(CH_3)_3SiOSn(C_3H_7\text{-}n)_2]_2O$	107—108	—	—	—	Cohydrolysis	85	993
$[(CH_3)_3SiOSn(C_4H_9]_2O$	108—109	—	—	—	Cohydrolysis	95	993
$(CH_3)_3SiSSn(CH_3)_3$	—	—	—	—	~4-38	—	1095

5.2. Physical Properties

The physicochemical constants of compounds containing the Si–O–Sn group are given in Table 48

Low-molecular alkylstannasiloxanes are mainly colorless, readily soluble, volatile liquids with a sharp unpleasant irritating odor and high toxicity, especially the methyl derivatives [992, 1091, 1107]; they are all monomeric. At the same time, the introduction of Sn–O–Sn groups into the structure makes the compounds (siloxystannoxanes) capable of dimerization with the formation of four-membered coordination rings [991, 993]:

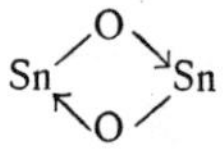

Sn(–O→)Sn (four-membered ring: Sn–O→Sn, Sn←O–Sn)

Such compounds are solids and their melting points are higher, the shorter the carbon chain of the radical at the tin atom. The dimeric state of the compounds is preserved in an inert solvent (benzene, cyclohexane, and carbon tetrachloride). The strength of the Sn→O bond in the dimer of bis(trimethylsiloxy)tetramethyldistannoxane has been estimated at 4.5 kcal/mole [644].

Polydimethylstannasiloxanes obtained by cohydrolysis of dimethyldichlorosilane with dimethyldibromostannane and containing up to 3% tin are colorless oils with $n_D^{20} \sim 1.405$, $d_4^{20} \sim 0.975$, and mol. wt. 800-1000 [229]. With a tin content of 8-10% they are already like honey and with 18-19% tin they have a resinous consistency. The density of these substances is greater than 1; they are soluble in organic solvents and cyclodimethylsiloxanes. This confirms their stannasiloxane structure since the compounds $(R_2SnO)_n$ are insoluble in $(R_2SiO)_n$. In the case of the cohydrolysis of dimethyldichlorosilane and diethyldichlorostannane, the polymer, which contains 11% tin, is resinous, but flows less than polydimethylstannasiloxanes with a higher tin content. Liquid oligomers of the type $\{(CH_3)_3SiO[(CH_3)_2SiO]_n\}_4Sn$ have a lower viscosity and density than the oligomers $\{(CH_3)_3SiO[(CH_3)_2SiO]_n\}_2Sn(CH_3)_2$ similar to them in molecular weight [224].

Triphenylsiloxy derivatives of tin are white crystalline substances with a sharp melting point, which dissolve in hydrocarbons on heating (bis(triphenylsiloxy)diphenylstannane dissolves only in hot xylene) [1209].

In the infrared absorption spectra of monomeric compounds containing the Si–O–Sn group in which the silicon and tin atoms bear alkyl radicals, bands with frequencies of 980 ± 5 cm^{-1} (10.2 ± 0.05 mμ) and 910 ± 5 cm^{-1} belong to the vibrations of the Sn–O bond in Si–O–Sn [443, 992, 993, 1091]. The later band does not appear in the spectra of compounds of the type $R_3SnOSiR_3$, but is characteristic of only substances containing the group

$$\geq Si-O-\overset{|}{\underset{|}{Sn}}-O-Si\leq$$

If the tin in the Si–O–Sn group is in the lower valence state (2), the infrared spectrum contains a sharp band at 870 cm^{-1} [993]. Characteristic bands in the region of 10.2-10.6 mμ are also observed in the IR spectra of triphenylsiloxy derivatives of tin [1209]. In the IR spectra of polymeric organostannasiloxanes the tin introduces bands with frequencies of 950-970 and 530-560 cm^{-1} [443], whose intensity increases with an increase in the content of the heteroatom in the molecule. The bands at ~ 550 cm^{-1} are well resolved, while bands in the region of ~ 960 cm^{-1} merge with bands characteristic of the valence vibrations of the Si–O bond.

A study of the NMR spectra of stannasiloxanes and stannasilthianes was carried out mainly by Schmidbaur and his co-workers [1081, 1084, 1091, 1095] (Table 49).

The data presented in Table 49 show that the screening of the protons in the trimethylsilyl groups decreases with an increase in the number of them in the molecule, i.e., with a change from $(CH_3)_3SiOSn(CH_3)_3$ to $[(CH_3)_3SiO]_4Sn$. In the system Si–O–Sn the Sn–O bond has a higher polarity $\left(\overset{\oplus}{Sn}\rightarrow\overset{\ominus}{O}\right)$, than in stannoxanes, while the fraction of π-bonding for Si–O ($d_\pi - p_\pi$) is reduced in comparison with siloxanes. There is an increase in the s-character of the Sn–C bonds ($sp^3 \rightarrow sp^2$). It may be stated that stannasiloxanes are stannyl silanolates and not silyl stannolates as might have been expected from literature data on the electronegativities of the elements.

The pattern of the shifts in the proton signals when one of the atoms in the systems Si–O–Si and Sn–O–Sn are replaced by a tin or a silicon atom, respectively, are preserved with a change from "oxanes" to "thianes."

TABLE 49. NMR Spectra of Compounds Containing Si–O–Sn and Si–S–Sn Groups*

Compound	Si			Sn				Literature
	δ_{CH_3}	J		δ_{CH_3}	J			
		H–C[13]	H–Si[29]		H–C[13]	H–Sn[117]	H–Sn[119]	
$(CH_3)_3SiOSi(CH_3)_3$	–3.5	118	6.86	—	—	—	—	1081
$(CH_3)_3SnOSn(CH_3)_3$	—	—	—	–14.2	128.8	53.6	56.0	1091, 1095
$(CH_3)_3SiOSn(CH_3)_3$	+2.5	116.8	6.72	–20.5	128.5	54.9	57.4	1081, 1084, 1091
$[(CH_3)_3SiO]_2Sn(CH_3)_2$	–1.2	117.0	—	–33.5	—	68.0	71.4	1091
$[(CH_3)_3SiO]_3SnCH_3$	–3.6	—	—	–39.0	—	—	—	1091
$[(CH_3)_3SiO]_4Sn$	–8.5	118.2	—	—	—	—	—	1091
$(C_2H_5)_3SiOSi(C_2H_5)_3$	–57	—	—	—	—	—	—	1091
$(C_2H_5)_3SiOSn(C_2H_5)_3$	+2.0	—	—	–74.0	—	HCSn	54.0	1091
$(C_2H_5)_3SiOSn(CH_3)_3$	–57.2	—	—	–24.0	—	—	—	1091
$(CH_3)_2Si[OSn(CH_3)_3]_2$	+7.0	—	—	–22.7	—	57.6	60.0	1091
$(CH_3)_3SiSSi(CH_3)_3$	–19.5	119.5	6.9	—	—	—	—	1095
$(CH_3)_3SnSSn(CH_3)_3$	—	—	—	–23.8	130.8	53.9	56.3	1095
$(CH_3)_3SiSSn(CH_3)_3$	–17.4	120.0	6.8	–26.6	131.0	57.0†	53.8 †	1095

* All the measurements were made at room temperature at 60 MHz; internal standard $(CH_3)_4Si$; c = 3 ÷ 5 ± 1%; all the values are in hertz.

† These figures are confused in [1095] (communication from the authors themselves).

5.3. Chemical Properties

Methylstannasiloxanes of the series $(R_3SiO)_nSnR_{4-n}$ (where $R = CH_3$; n = 1–4) are less thermally stable as a whole than polysiloxanes or polystannoxanes,* but differ appreciably between themselves [1091, 1107]. The extreme members of the series, which distill without decomposition, are most stable, and symmetrical tetrakis(trimethylsiloxy)stannane has an even higher thermal stability than trimethylsiloxytrimethylstannane. The bis derivative begins to decompose at 150°C like its germane analog:

$$n\,(R_3SiO)_2\,SnR_2 \longrightarrow nR_3SiOSiR_3 + [R_2SnO]_n \quad (R = CH_3) \qquad (4\text{–}46)$$

Scheme (4-32) ($R' = CH_3$) may be regarded as an intermediate stage. Tris(trimethylsiloxy)methylstannane disproportionates un-

* Correspondingly, polyphenylstannasiloxanes do not reach the level of thermal stability of $[(C_6H_5)_2SiO]_n$, but stand prolonged heating better than linear methyl- or vinylsiloxanes [648]. The thermal stability increases with a decrease in the tin content.

der the same conditions, giving hexamethyldisiloxane and a polymethylstannoxane, which contains a certain number of trimethylsiloxy groups attached to tin atoms.

As regards compounds of the type $(CH_3)_3SiOSnR_3$, trimethylsiloxytripropyl(butyl)stannane does not change when stored in air for more than 50 days. However, the compound with $R = C_2H_5$ forms a solid substance on standing or during distillation, evidently due to the absorption of CO_2 [992]. In actual fact, when CO_2 is passed through a solution of trimethylsiloxytriethylstannane in acetone, bis-(triethylstannyl)carbonate is formed. The tendency of substances of the type $(R_3SiOSnR'_2)_2O$ to disproportionate has already been discussed.

The lability of these compounds characterizes their behavior toward oxygen-containing organic compounds. When they are dissolved in ethanol, dioxane, and acetone there is slow decomposition with the separation of insoluble polydialkylstannoxanes ("oxides of dialkyltin"). Moisture promotes the decomposition, though the starting compounds are insoluble in water [992].

Some properties of polyorganostannasiloxanes of the type $[(RSiO_{1.5})_4Sn]_n$ (where $R = CH_3$, C_2H_5, C_6H_5) have already been described above. There are data [198] indicating that the thermal decomposition of such stannasiloxanes differs substantially in character from that of other metallasiloxanes.* Their decomposition is accompanied by an exothermic effect, unlike the endothermic effect which is observed in the case of alumasiloxanes, for example.

The decomposition point (°C) for different forms of R is 480 for CH_3, 462 for C_2H_5, and 460 for C_6H_5. A study was also made of the thermooxidative breakdown at 400°C of products obtained by polymerization of siloxane oligomers under the action of polyorganostannasiloxanes of the above type with a ratio of Sn : Si = 1–1.5 : 100 [52, 416, 418]. It was found that the introduction of tin into the structure increases the thermal stability of methyl- and ethyl-sub-

*The data in [198] do not refer directly to polymers of this type since in the article itself both the reference to the preparation method and the following formula were given incorrectly:

```
┌ —Si— O—Si—O— ┐
   |      |
   O      O
   |      |
└ —SiR—O—SnR—O— ┘x
```

TABLE 50. Hydrolysis of Stannasiloxane Polymers $[(RSiO_{1.5})_xM]_n$*

M	R		
	CH_3	C_2H_5	C_6H_5
	Metal passing into silicon		
Al	100	87 (0)	69
Sn	100	93 (10)	95
Ti	—	1.5—2 (0)	1.5—2

*Boiling for 10 hr with 10% HCl; in brackets – boiling with water.

stituted polymers (this is more marked for CH_3 than for C_2H_5) and this has a greater effect than the introduction of aluminum or titanium. On the other hand, the thermooxidative breakdown of phenyl- and vinylsiloxane polymers proceeds more rapidly when there are heteroatoms present in the chain (this is greater for C_6H_5 than for $CH=CH_2$). Tin promotes the development of this process to a lesser extent than aluminum, but to a greater extent than titanium.

Stannasiloxanes are more reactive than siloxanes toward electrophilic and nucleophilic agents. According to Schmidbaur [1091] this is due both to the greater polarity of the Sn–O bond and the lower fraction of π-bonding of the Si–O bond and to the greater capacity of tin atoms for coordination and the formation of transition compounds. In particular, trimethylsiloxytrimethylstannane is readily cleaved by water, alcohols, amines, mercaptans, and Lewis acids [1091]. The reaction with phenyllithium proceeds in the following way [1143]:

$$(CH_3)_3SiOSn(CH_3)_3 + C_6H_5Li \longrightarrow (CH_3)_3SiOLi + (CH_3)_3SnC_6H_5 \quad (4\text{-}47)$$
$$\xrightarrow{H_2O} (CH_3)_3SiOSi(CH_3)_3$$

No trimethylphenylsilane was detected among the reaction products.

Tetrakis(triethylsiloxy)stannane is hydrolyzed in an acid medium ~82 times faster than tris(triethylsiloxy)alumane [135, 164, 225, 291]. When a benzene solution of tetrakis(triphenylsiloxy)-stannane is treated with dilute hydrochloric acid there is 72% cleavage with the formation of a mixture of triphenylsilanol and hexaphenyldisiloxane [826]. Polymers of the type $[(RSiO_{1.5})_4Sn]_n$ are inferior to both titana- and alumasiloxanes with respect to hydrolytic resistance to water and hydrochloric acid [14, 140] (Table 50).

When tetrakis(trimethylsiloxy)stannane is treated with methacrylic acid for 1 hr at 40°C, the following reaction occurs quantitatively [162]:

$$(R_3SiO)_4Sn + CH_2{=}CRCOOH \longrightarrow (R_3SiO)_3SnOCOC(R){=}CH_2 + R_3SiOH \quad (R = CH_3) \qquad (4\text{-}48)$$

The reaction with tetramethacryloxystannane at 60°C gives an 80% yield of the symmetrization product:

$$(R_3SiO)_4Sn + (CH_2{=}CRCOO)_4Sn \longrightarrow 2\,(R_3SiO)_2Sn(OCOCR{=}CH_2)_2 \qquad (4\text{-}49)$$

When bis(trimethylsiloxy)dimethacryloxystannane* (30% solution in petroleum ether) is heated at 65°C in the presence of benzoyl peroxide (0.1%) for 2 hours no polymerization occurs; then, in a few minutes, a solid insoluble polymer is formed, which does not flow in the range of 20–500°C. For tris(trimethylsiloxy)methacryloxystannane (80°C, 50% solution in benzene) the induction period is 12 hr and then an insoluble product is obtained equally rapidly. Copolymerization with methyl methacrylate (2 : 1, 65°C) is complete in 25 min. The clear polymer, which is insoluble in organic solvents, has T_g 120°C and T_{flow} ~300°C. Copolymerization with tyrene is complete in 50 min and the polymer has T_g 80°C and T_{flow} 280°C.

5.4. Application

Descriptions have been given of the use of various polyorganostannasiloxanes as lubricating materials [975, 1430, 1650, 1651, 1733, 1734], hydraulic and damping fluids [1650, 1651, 1733], heat transfer agents [1650, 1651], binders for electrical insulation, cementing, and coloring compositions [1494, 1733], heat-resistant coatings [1650, 1651, 1686], oxidation inhibitors for siloxane fluids [1430], biocides [1430], and glues and adhesives for glass, metals, asbestos, cloth, and abrasives [481].

Mention has already been made of the use of a wide range of organotin compounds as hardeners for siloxane resins, films, and varnishes [p. 341]. On the other hand, there is a report of their use as stabilizers for polysiloxane fluids (liquid dielectrics, hy-

*The name "tin bis(trimethylsiloxymethacrylate)" given by the authors of [162] is incorrect.

draulic, polishing compositions, and lubricants) [1322, 1336, 1405, 1596, 1607], additives for siloxane rubber mixtures to improve the residual deformation and relative elongation of vulcanizates [1339], and to give them self-sealing properties after vulcanization [1713] and also additives to improve the water-proofing of fibrous materials by hydrosiloxanes [1352].

Polysiloxane compositions which are vulcanized at room temperature by systems which include organotin compounds are used as elastic, heat-resistant, protective sealing, damping, and electrically insulating materials in aviation, electronics, ship building, electrical engineering, and other branches of industry and also for producing precision-cast parts from plastics, and exact copies in stomatology, criminology, etc. [20]. It is also reported that diorganodiacyloxystannanes have been used as catalyst-hardeners in the reactions of organosilicon alcohols and silanediols with isocyanates [1364, 1715].

6. COMPOUNDS CONTAINING THE $Si-(C)_n-Sn$ AND $Si-(C)_n-O-Sn$ GROUPS

6.1. Preparation Methods

6.1.1. Organometallic Synthesis

Organosilicon derivatives of tin containing the group $Si(CH_2)_nSn$ or SiC_6H_4Sn may be obtained readily by the scheme [499, 821, 958, 1003, 1111, 1136, 1367, 1636, 1641, 1671]:

$$(4-n)\,RR'_2SiAMgX + R''_nSnX'_{4-n} \longrightarrow R''_nSn\left(ASiR'_2R\right)_{4-n} \qquad (4\text{-}50)$$

[R = H or an organic radical; A = a biradical (alkylene or arylene)]

For the preparation of compounds of the type $R_3SiCH_2SnR_3$, a change from a Grignard reagent to a Normant reagent, i.e., when the reaction is carried out in tetrahydrofuran, increases the yield of the required product from 50-60% to 85-95%.*

Reacting the dimagnesium derivative obtained from sym-di-(chloromethyl)tetramethyldisloxane with dialkyldichlorostannanes

* When a compound of the type $HSiR_2(CH_2)_nCl$ is used the yield of the tin derivatives reaches 67-99% in an ether medium also (see below).

with subsequent heating of the products with alkali forms six-membered cyclic compounds [958]:

$$O\begin{matrix} \diagup SiR_2CH_2MgX \\ \diagdown SiR_2CH_2MgX \end{matrix} + R'_2SnX_2 \longrightarrow O\begin{matrix} \diagup SiR_2CH_2 \diagdown \\ \diagdown SiR_2CH_2 \diagup \end{matrix}SnR'_2 \qquad (4\text{-}51)$$

However, this method is of no preparative interest because of the abundance of by-products. A more effective method of preparing such rings is based on the hydrolysis of appropriate silicon organotin compounds containing Si–H or Si–OC_2H_5 groups (yield 87-92%; see 4-62).

Stannasilcarbanes are also obtained by the following route [1154]:

$$(CH_3)_3\,SnCH_2MgCl + CH_3SiX_3 \longrightarrow (CH_3)_3\,SnCH_2Si\,(CH_3)\,X_2 \qquad (4\text{-}52)$$
$$(X = Cl \text{ or } OCH_3)$$

The hydrolysis of such compounds leads to the formation of siloxanes, containing trialkylstannyl groups in the side chain [1155] (see below). Products with terminal R_3SnCH_2 groups are formed by the reaction [1542]

$$2R_3SnCH_2SiR'_2ONa + Cl\left(SiR''_2O\right)_n SiR''_2Cl \longrightarrow \left[R_3SnCH_2SiR'_2\left(OSiR''_2\right)_{\frac{n+1}{2}}\right]_2 O \qquad (4\text{-}53)$$

By using the disilanolates $(NaOSiR'_2CH_2)_2SnR_2$ it is possible to obtain linear polymers in which siloxane sections of the chains alternate with the groups $-CH_2SnR_2CH_2-$.

To prepare compounds containing the groups $Si(CH_2)_nSn$ or SiC_6H_4Sn it is also possible to use organolithium synthesis [599, 953, 1076]:

$$(4-n)\,R_3SiALi + R'_nSnX_{4-n} \longrightarrow R'_nSn\,(ASiR_3)_{4-n} \qquad (4\text{-}54)$$

p-Trimethylsilylbenzyltrimethylstannane was synthesized by the Wurtz reaction [567]:

$$R_3SnCH_2C_6H_4X + 2Na + XSiR_3 \longrightarrow R_3SnCH_2C_6H_4SiR_3 \qquad (4\text{-}55)$$
$$(R = CH_3)$$

6.1.2. Addition Reactions

Triphenylstannane adds to triphenylvinylsilane on heating in vacuum with the formation of a symmetrical disubstituted ethane [840]:

$$R_3SnH + CH_2{=}CHSiR_3 \xrightarrow{t^\circ} R_3SnCH_2CH_2SiR_3 \quad (R = C_6H_5) \tag{4-56}$$

The analogous hydrostannylation of triphenylallylsilane does not occur though triphenylgermane is known to form an addition product quantitatively [954]. The same is observed in the reaction of triphenylstannane with dialkenylsilanes (2:1). It does not react with diphenyldiallylsilane, but adds well to diphenyldivinylsilane, forming bis (β-triphenylstannylethyl) diphenylsilane. It has been suggested that a compound with a central tin atom is obtained in the reaction of diphenylstannane with triphenylvinylsilane in a ratio of 1:2. However, due either to disproportionation of diphenylsilane under the experimental conditions or to exchange reactions, a symmetrical disubstituted ethane was obtained unexpectedly in this case. Metallic tin was also detected among the reaction products.

The addition of diphenylstannane to diphenyldivinylsilane led to a new surprise—the reaction product was not a polymer, but a crystalline derivative of 1-stanna-4-silacyclohexane [841]:

$$R_2Si(CH{=}CH_2)_2 + R_2SnH_2 \longrightarrow R_2Si\begin{matrix} \diagup CH_2CH_2 \diagdown \\ \diagdown CH_2CH_2 \diagup \end{matrix}SnR_2 \tag{4-57}$$

In this connection it should be pointed out that a photochemical reaction of diethylstannane with sym-divinyltetramethyldisiloxane leads to the formation of a clear viscous oil [1427]; however, it was not possible to add sym-dimethyldiethyldisiloxane to diethyldivinylstannane in the presence of chloroplatinic acid [316].

Triethylvinylstannane is much less active in the reaction with silicochloroform than triethylvinylsilane. With benzoyl peroxide as the initiator, the yield of the addition product was only 31% after the reaction mixture had been heated at 95°C for 17 days in comparison with 71% after 6 days in the case of triethylvinylsilane [1141].

6.1.3. Other Reactions

The reaction of hexaalkyldistannoxanes with organosilicon alcohols is essentially analogous to scheme (4-45). The product from the reaction of hexaethyldistannoxane with γ-(dimethylethylsilyl)propanol is γ-(dimethylethylsilyl)propoxytriethylstannane [499]:

$$R_3SnOSnR_3 + R'_3Si(CH_2)_nOH \longrightarrow R'_3Si(CH_2)_nOSnR_3 + R_3SnOH \qquad (4\text{-}58)$$
$$(R = C_2H_5;\ R' = CH_3;\ n = 3)$$

With hexaethyldistannoxane, acetylene alcohols of the type $R_2Si[C{\equiv}CC(CH_3)_2OH]_2$ yield the extremely peculiar compounds $[(C_2H_5)_3SnOC(CH_3)_2C{\equiv}C]_2SiR_2$.

The reaction of trialkylhalosilanes with methyl(trialkylstannyl)acetate does not lead to the formation of a substance containing silicon and tin [234]. Instead of this there is cleavage of the $Sn-CH_2$ bond and a rearrangement of the carbonyl part of the compound with the formation of silyl vinyl ethers. The reaction of R_3SiX with stannyl alkyl ketones proceeds analogously so that we may speak of the general reaction

$$R_3SnCH_2COR' + R''_3SiX \longrightarrow R_3SnX + R''_3SiOC(R'){=}CH_2 \qquad (4\text{-}59)$$
$$(R' = CH_3,\ OCH_3\ \text{etc.})$$

6.2. Physical Properties

The physicochemical constants of compounds containing the groups $Si-(C)_n-Sn$ and $Si-(C)_n-O-Sn$ are given in Table 51.

Mixtures of compounds of the type $[(CH_3)_3SiC_6H_4]_4M$, where M = S, Ge, or Sn, do not give depression of the melting point (~345°C), evidently because of their isomorphism [953].

In the infrared spectrum of trimethylsilylmethyltrimethylstannane a very strong band with a frequency of 1006 cm^{-1} is assigned to ν_{as} Si – C – Sn [1076].

Data on the proton NMR spectra of compounds of the type $R_3SiCH_2MR_3$, where M is a Group IV element will be given in this chapter in the section on lead.

TABLE 51. Compounds Containing the Groups Si – (C)$_n$ – Sn and Si – (C)$_n$ – O – Sn

Formula	m.p. °C	b.p., °C (mm)	n_D^{25}	d_4^{25}	Preparation method (scheme)	Yield, %	Literature
$H(CH_3)_2SiCH_2Sn(CH_3)_3$	—	153—154	—	—	4-60	77	1111
$(CH_3)_3SiCH_2Sn(CH_3)_3$	—	165—166 (760)	1.4569	1.1244	4-50	58	1003
	—	64—65 (24)	1.4594	1.136	4-50	86	1136
	—78	161 (740); 59 (10)	—	—	4-54	—	1076
$(CH_3O)_2(CH_3)SiCH_2Sn(CH_3)_3$	—	77.5—81 (18)	1.4523	1.248	4-52	74	1154
$Cl_2(CH_3)SiCH_2Sn(CH_3)_3$	—	58—59 (3.8—4)	1.4824	1.415	4-52	50	1154
$(CH_3)_2(C_2H_5)SiCH_2Sn(C_2H_5)_3$	—	99—100 (3.5)	1.4800[20]	1.1135[20]	4-50	58	499
$H(CH_3)_2SiCH_2Sn(C_4H_9\text{-}n)_3$	—	133 (5)	1.4764	1.047	4-60	99	958, 1367, 1636, 1671
$C_2H_5O(CH_3)_2SiCH_2Sn(C_4H_9\text{-}n)_3$	—	147 (5)	1.4682	1.053	—	97	958, 1367, 1636
$Cl(CH_3)_2SiCH_2CH_2Si(CH_3)_2CH_2Sn(C_4H_9\text{-}n)_3$	—	180 (30)	1.4836	1.066	4-63	32	958, 1636
$(CH_3)_3SiCH_2Sn(C_6H_5)_3$	—	162—163 (2)	1.5943	1.2369	4-50	62	1003
$(CH_3)_3SiCH_2Sn(CH_3)_2Br$	—	50 (0.35)	1.5101	—	Table 53	31—52	1136
$(CH_3)_3SiCH_2Sn(C_4H_9\text{-}n)_2Br$	—	91 (0.2)	1.5037	—	Table 53	97	1136
$(CH_3)_3SiCH_2Sn(C_4H_9\text{-}n)_2OCOCH_2F$	62—63.4	—	—	—	Table 53	55	1136
$[H(CH_3)_2SiCH_2]_2Sn(CH_3)_2$	—	101 (20)	1.4743	1.108	4-60	82	958, 1367, 1636, 1671
$[H(CH_3)(C_6H_5)SiCH_2]_2Sn(CH_3)_2$	—	—	1.4810	1.043	4-60	—	1671
$[(CH_3)_3SiCH_2]_2Sn(CH_3)_2$	—	146.5—147.5 (65)	1.4644	1.0559	4-50	51	1003
	—	49—52 (0.35); 55 (0.7); 62 (1.3)	1.4702	1.073	4-50	94	1003
$[(CH_3)_3SiCH_2]_2Sn(CH=CH_2)_2$	—	93—94 (3.4)	1.4826	1.078	4-50	80	1136
$[H(CH_3)_2SiCH_2]_2Sn(C_4H_9\text{-}n)_2$	—	130 (5)	1.4810	1.043	4-60	67	958, 1367, 1636, 1671
$[(CH_3)_3SiCH_2]_2Sn(C_4H_9\text{-}n)_2$	—	98 (0.45)	1.4777	1.027	4-50	91	1136

$[C_2H_5O(CH_3)_2SiCH_2]_2Sn(C_4H_9\text{-}n)_2$	—	186 (15)	1.4655	1.056	—	83	958, 1367, 1636, 1671
$[(CH_3)_3SiCH_2]_2Sn(CH_3)Br$	—	71—73 (0.18—0.20)	1.5018	—	Table 53	50—56	1136
$[(CH_3)_3SiCH_2]_2Sn(CH_3)I$	—	85—86 (0.35)	1.5268	—	Table 53	81	1136
$[(CH_3)_3SiCH_2]_2Sn(C_4H_9\text{-}n)I$	—	105 (0.4)	1.5257	—	Table 53	94	1136
$[(CH_3)_3SiCH_2]_2Sn(C_6H_5)_2$	—	138—140 (1.5)	1.5425	1.1404	4-50	57	1003
	—	137 (0.35): 130—132 (0.2)	1.5499	1.149	4-60	88	1136
$[(CH_3)_3SiCH_2]_2SnBr_2$	38.6—39.8	—	—	—	Table 53	98	1136
$[(CH_3)_3SiCH_2]_2SnI_2$	34.6—35.4	—	—	—	Table 53	83	1136
$[(CH_3)_3SiCH_2]_3SnI$	—	105 (0.3)	1.5237	—	Table 53	77	1136
$[(CH_3)_3SiCH_2]_4Sn$	—	94 (0.2)	1.4839	1.018	4-50	84	1136
$(CH_3O)_3SiCH_2CH_2Sn(C_2H_5)_3$	—	78 (0.4)	1.4638	1.209	—	66	1141
$Cl_3SiCH_2CH_2Sn(C_2H_5)_3$	—	85 (0.6)	—	—	—	31	1141
$(C_6H_5)_3SiCH_2CH_2Sn(C_6H_5)_3$	207—208	—	—	—	4-56	54	840
$[H(CH_3)_2SiCH_2CH_2CH_2]_2Sn(CH_3)_2$	—	146 (15)	1.4730	1.052	4-60	76	958, 1367, 1636, 1671
$H(CH_3)_2SiCH_2CH(CH_3)CH_2Sn(C_4H_9\text{-}n)_3$	—	—	—	—	4-60	—	1671
$[H(CH_3)_2SiCH_2CH(CH_3)CH_2]_2Sn(CH_3)_2$	—	—	—	—	4-60	—	1671
$(CH_3)_3Si-C_6H_4-CH_2Sn(CH_3)_3$	—	133—135 (7—8)	1.5251^{20}	—	4-55	32	567
$(CH_3)_3Si-C_6H_4-Sn(CH_3)_3$	103—104	—	—	—	4-54	—	599
$(C_2H_5)_3Si-C_6H_4-Sn(C_2H_5)_3$	—	214 (18)	$1.5276^{21.2}$	$1.1216^{21.2}$	4-50	—	821
$(CH_3)_3Si-C_6H_4-Sn(C_6H_5)_3$	132.5—133.5	—	—	—	4-50	72	1003
$[(CH_3)_3Si-C_6H_4-]_2Sn(C_6H_5)_2$	95—96	—	—	—	4-50	24	1003

TABLE 51. (Cont'd)

Formula	m.p., °C	b.p., °C (mm)	n_D^{25}	d_4^{25}	Preparation method (scheme)	Yield, %	Literature
$[(CH_3)_3Si-C_6H_4]_4Sn$	343—345	—	—	—	4-54	58	953
$(C_6H_5)_2Si[CH_2CH_2Sn(C_6H_5)_3]_2$	143—144	—	—	—	~4-56	86	840
$O[(CH_3)_2SiCH_2Sn(C_4H_9\text{-}n)_3]_2$	—	240 (< 0,1)	1.4852	1.102	—	70	958, 1367, 1636, 1641
$O[(CH_3)_2SiCH_2CH_2Si(CH_3)_2CH_2Sn(C_4H_9\text{-}n)_3]_2$	—	—	1.4850	1.045	4-63	—	958, 1641
$O\langle Si(CH_3)_2-CH_2 \rangle_2 Sn(CH_3)_2$ (cyclic)	—	95 (21)	1.4743	1.203	4-62	87	958, 1641
	—	—	—	—	4-51	37	958
$O\langle Si(CH_3)_2-CH_2 \rangle_2 Sn(C_4H_9\text{-}n)_2$ (cyclic)	—	170 (25)	1.4805	1.116	4-62	92	958, 1641
$(C_6H_5)_2Si\langle CH_2CH_2 \rangle_2 Sn(C_6H_5)_2$ (cyclic)	134—135	—	—	—	4-57	23	841
$O\langle Si(CH_3)_2-CH_2CH_2CH_2 \rangle_2 Sn(CH_3)_2$ (cyclic)	—	148 (16)	1.4840	1.157	—	20	958, 1641
$C_2H_5(CH_3)_2SiCH_2CH_2CH_2OSn(C_2H_5)_3$	—	131—133 (5)	1.4769[20]	1.1208[20]	4-58	73	499

6.3. Chemical Properties

Alcoholic and aqueous alkali do not cleave the group Si – C – Sn. As a result of this the hydrolysis of compounds of the type $R_3SnCH_2SiRX_2$ (where $R = CH_3$; $X = Cl$ or OCH_3) yields siloxanes with stannylmethyl "appendages" [1155]. Copolymerization of these compounds with hexamethyldisiloxane and octamethylcyclotetrasiloxane leads to linear block polymers, which distill in vacuum up to mol. wt. 600 (when Si : Sn = 2 – 3 : 1) (Table 52).

It is possible to carry out reaction (4-50) with compounds containing an active hydrogen at the silicon atom and this is apparently untouched during the Grignard synthesis (the yield of the required compounds varies from 70-80% to quantitative) [958, 1111, 1671]:

$$nH(CH_3)_2Si(CH_2)_mMgCl + R_{4-n}SnCl_n \xrightarrow[\text{ether}]{} [H(CH_3)_2Si(CH_2)_m]_nSnR_{4-n} \quad (4\text{-}60)$$

The presence of Si – H bonds in compounds of this type makes it possible to prepare a series of their derivatives. The action of absolute ethanol in the presence of sodium ethylate forms the corresponding ethoxyl derivatives [958, 1367, 1671]. Aqueous ethanol in an alcoholic medium leads to the formation of siloxanes. Hydrolysis of compounds with n = m = 1, containing one Si – H bond, gives tin-containing disiloxanes:

$$R_3SnCH_2SiR'_2H \xrightarrow[OH^-]{C_2H_5OH} (R_3SnCH_2SiR'_2)_2O \quad (4\text{-}61)$$

When two such bonds are present, six-membered rings are formed:

$$R_2Sn(CH_2SiR'_2H)_2 \longrightarrow R_2Si\begin{matrix} \diagup CH_2\text{—}SiR'_2 \diagdown \\ \diagdown CH_2\text{—}SiR'_2 \diagup \end{matrix}O \quad (4\text{-}62)$$

The same compounds are obtained by hydrolysis of ethoxyl derivatives which have first been synthesized from hydrosilanes [958, 1641]. A bifunctional compound with trimethylene bridges between Si and Sn atoms (n = 2; m = 3) is hydrolyzed with the formation of linear polymers, while a ten-membered ring is obtained only by alkaline cracking (yield 20%).

TABLE 52. Hydrolysis of Monomers of the Type $R_3MCH_2Si(CH_3)$-

M	Hydrolysis products (1 :1 mixture of H_2O + ether in the presence of conc. HCl)				
	Composition	b.p., °C (mm)	n_D^{25}	d_4^{25}	OH groups, %
C	$(CH_3)_3CCH_2Si(CH_3)(OH)_2$	m.p. 98—99	—	—	—
Si	$[(CH_3)_3SiCH_2Si(CH_3)(OH)]_2O$	127—132 (0.1)	Crystallizes		
	$[(CH_3)_3SiCH_2Si(CH_3)O]_n$	140—170 (0.1)	1.4432	0.929	—
	$[(CH_3)_3SiCH_2Si(CH_3)O]_4$	170 (0.1)	1.4422^{28}	—	—
Ge	$[(CH_3)_3GeCH_2Si(CH_3)O]_n$ + oil	90 (0.45)— 95 (0.2)	1.4040	1.194	5.78
Sn	$[(CH_3)_3SnCH_2Si(CH_3)O]_n$ + oil	Not distilled	1.4993	1.448	3.57 *

* 11.9% Si; 50.0% Sn.

** Obtained by heating a mixture of hydrolysis products with $(CH_3)_3SiOSi(CH_3)_3$ in iso-C_3H_7OH in the presence of KOH; we do not give the properties of products of higher molecular weight here.

*** Obtained by heating an equimolecular mixture of $(CH_3)(CH_3)_3MCH_2Si[OSi(CH_3)_3]_2$ and $[(CH_3)_2SiO]_4$ in iso-C_3H_7OH in the presence of KOH.

Chloroplatinic acid promotes the addition of a silicon organotin compound with an active hydrogen atom at the silicon to vinyldimethylchlorosilane [958, 1636]. Since the addition product contains a chlorine atom at the silicon, it is readily "dimerized" by hydrolysis:

$$R_3SnCH_2SiR'_2H + CH_2{=}CHSiR'_2Cl \xrightarrow{H_2PtCl_6} R_3SnCH_2SiR'_2CH_2CH_2SiR'_2Cl \xrightarrow{\text{hydrolysis}}$$
$$\longrightarrow {}^1/_2\,(R_3SnCH_2SiR'_2CH_2CH_2SiR'_2)_2O \qquad (4\text{-}63)$$

Seyferth [1136] studied the cleavage of bis(trimethylsilylmethyl)dialkylstannanes by various aggressive agents. The nature of their action is described in Table 53.

The data presented in this table indicate that the grouping Si−C−Sn in the compounds studied has a high resistance to clea-

$(OCH_3)_2$ and Polymerization of the Hydrolysis Products [1155]

$(CH_3)_3MCH_2Si(CH_3)[OSi(CH_3)_3]_2$ **					$(CH_3)_3SiOSi(CH_3)[CH_2M(CH_3)_3]O[Si(CH_3)_2O]_mSi(CH_3)_3$ ***					
b.p., °C (mm)	n_D^{25}	d_4^{25}	η, centistokes 0/100°C	E, kcal	m	n_D^{25}	d_4^{25}	mol. wt.	η, centistokes 0/100°C	E, kcal
60–64 (0.7–9)	1.4040	0.835	3.33/0.83	2.83	8.2	1.4101	0.930	900	33.9/40	4.30
77–78 (1.5–6)	1.4068	0.838	3.19/0.83	2.74	4.9	1.4163	0.945	671	62.5/2.2	6.27
56–58 (0.3)	1.4200	0.972	3.69/0.79	3.05	5.6	1.4268	1.051	766	93.7/2.5	7.19
60 (0.1)	1.4350	1.078	3.72/0.83	3.06	7.8	1.4322	1.110	974	44.4/3.1	5.42

vage, particularly by iodine and bromine. Thus, when $R = CH_3$, n-C_4H_9 and $(CH_3)_3SiCH_2$, boiling the substance for many hours with an equimolar amount of iodine in ether has no result and the same is true when the ether is replaced by benzene or xylene.* Only after removal of the xylene by distillation and heating the residue to ~175°C is there elimination of one of the radicals and in the case where $R = CH_3$ or n-C_4H_9 it is the alkyl which is eliminated and not the trimethylsilylmethyl group.

Alkyl radicals are usually eliminated from organotin compounds by bromine at temperatures from -30 to -40°C. At the same time, compounds of the type $[(CH_3)_3SiCH_2]_2SnR_2$ ($R = CH_3$, n-C_4H_9) are not cleaved by bromine at 0°C and only react slowly with it at room temperature. The nature of the other substituents at the tin atom has a substantial effect in the cleavage of the Si–CH_2–Sn group at the Sn–C bond. When R = n–C_4H_9, elimination of the trimethylsilylmethyl group predominates. When $R = CH_3$, it is

* In the reaction of iodine with R_3SnR' (where $R = CH_3$, C_2H_5, $R' =$ n-C_5H_{11}, n-$C_{12}H_{25}$) boiling in ether is sufficient to obtain R_3SnI in good yield.

TABLE 53. Cleavage of $[(CH_3)_3SiCH_2]_2SnR_2$ By Various Agents [1136]

R	Cleaving agent, temperature (°C), solvent	Reaction products isolated and their yield, %
CH_3	HBr; —78; $HCCl_3$	$(CH_3)_3SiCH_2SnR_2Br$ (I), 52; $[(CH_3)_3SiCH_2]_2SnRBr$ (II), 43; $(CH_3)_4Si$ (III), 54
	Br_2; 20; CCl_4	I, 31; II, 56
	$HgBr_2$; 80; C_2H_5OH	CH_3HgBr, 94
	I_2; 175	$[(CH_3)_3SiCH_2]_2SnRI$ (IV), 81
C_4H_9	HBr; —78; $HCCl_3$	I, 97; III, 98
	Br_2; 20; CCl_4	Inseparable equimolar mixture of I and II
	I_2; 145; $C_6H_4(CH_3)_2$	IV, 94
	FCH_2COOH; 20—80	$(CH_3)_3SiCH_2SnR_2OOCCH_2F$, 55; III, 26
	HBr; —78; CH_2Cl_2	$[(CH_3)_3SiCH_2]_2SnBr_2$, 98
	I_2; 80; C_6H_6	$[(CH_3)_3SiCH_2]_2SnI_2$, 83; RI 96
$(CH_3)_3SiCH_2$	I_2; 175	R_3SnI, 77; RI, 79

accompanied by elimination of the methyl, while with a change to a compound with $R = C_6H_5$ there is solely cleavage of the $Sn-C_{ar}$ bond while the "siliconeopentyl" group is not touched.* It has been suggested that the cleavage of these silicon-organotin compounds by iodine is a free radical process, while the reactions of HBr and $HgBr_2$ are electrophilic substitutions.

The bis(trimethylsilylmethyl)dihalostannanes obtained in the cleavage reactions described above are crystalline, low-viscosity substances, which are readily soluble in organic solvents, but insoluble in water. In contrast to dialkyldibromostannanes, which instantaneously give a complex with pyridine, bis(trimethylsilylmethyl)dibromostannane does not give an adduct with pyridine because of the steric effect of the "siliconeopentyl" groups.

* When $(CH_3)_3SnCH_2C_6H_4Si(CH_3)_3$ is treated with $HClO_4$ in methanol there is more cleavage of the $Sn-CH_2$ bond than of the $Si-C_{ar}$ bond [567]. In the case of $(CH_3)_3\cdot SnC_6H_4CH_2M(CH_3)_3$, cleavage of the $Sn-C_{ar}$ bond under analogous conditions occurs to a lesser extent when M = Si [relative rate of the reaction, which is pseudo-first order (1.0)] than when M = Ge (1.36) and particularly when M = Sn (3.21) [567a].

Treatment of the corresponding diiodide with hydrogen sulfide in alcohol with subsequent treatment with $Na_2S \cdot 9H_2O$ and again with H_2S leads to crystalline $\{[(CH_3)_3SiCH_2]_2SnS\}_n$ with n=1, 3, and 4.

It is not possible to obtain the corresponding stannanediol by hydrolysis of bis(trimethylsilylmethyl)dibromostannane in an alkaline medium, though di-tert-butyl- and di-tert-amylstannanediols are comparatively stable. In this case the steric hindrance is evidently inadequate to prevent condensation of the bis(trimethylsilylmethyl)stannanediol to a polystannoxane $\{(CH_3)_3SiCH_2]_2SnO\}_x$, which is an amorphous white substance that softens and melts over the range 145-160°C.

6.4. Application

Polymeric products containing the group $Si-(C)_n-Sn$ have been proposed as lubricants, hydraulic fluids, plasticizers, flameproofing agents, coatings, stabilizing additives, fibers, and elastomers [1427, 1636, 1641]. However, the last two recommended applications are from patents and are not actually in use at present. Monomeric (silylmethyl)stannanes with butyl radicals at the Sn atoms are recommended as bactericides and fungicides [1367, 1671].

7. COMPOUNDS CONTAINING THE Si – N – Sn GROUP

7.1. Preparation Methods

A patent was published in 1961 [1650] in which, in particular, the preparation of linear stannasilazans by the following route was described:

$$nR_2Sn(OR')_2 + nR_2Si(NH_2)_2 \longrightarrow [-R_2SiNHSnR_2NH-]_n + 2nR'OH \quad (4\text{-}64)$$
$$(R = C_6H_5;\ R' = CH_3)$$

In the same patent there is also a report of the preparation of linear stannasilazans by the reaction of dimethyldiaminostannane with octamethylcyclotetrasilazan, which evidently has a breakdown (total or partial) condensation mechanism of the type $NR_2-H(NR)$ (here R=H).

This patent publication is unclear on many points and in practice, the chemistry of organosilicon derivatives of tin containing a bridging nitrogen really began with the appearance in 1963-1965 of a series of papers on this subject [931, 1060, 1064, 1065, 1069, 1223, 1238].

The action of Wannagat's reagent on halostannanes forms bis-(trimethylsilyl)amino derivatives of tin:

$$(4-n)\,(R_3Si)_2NNa + R'_nSnCl_{4-n} \longrightarrow R'_nSn[N(SiR_3)_2]_{4-n} \quad (R = CH_3;\ R' = CH_3 \text{ or } n\text{-}C_4H_9) \tag{4-65}$$

In the case where n = 3, the corresponding monosubstituted products are obtained smoothly [1060, 1064]. However, with $SnCl_4$ (R' = Cl) the reaction proceeds according to scheme (4-65) only qualitatively and not quantitatively [1223, 1238]. The reaction yields $(R_3Si)_2NSnCl_3$, $[(R_3Si)_2N]_2SnCl_2$, and also the polymer $[R_3SiNSnCl_2]_n$, which is evidently formed as a result of the condensation of the first of these compounds. The absence of products of a higher degree of substitution is evidently connected with steric hindrance to the introduction of the bulky groups $[(CH_3)_3Si]_2N$. This is confirmed by the fact that it was possible to synthesize trimethylstannylbis-(trimethylsilyl)amine without the use of Wannagat's reagent [931]:

$$R_3SnNR'_2 + HN(SiR_3)_2 \longrightarrow R_3SnN(SiR_3)_2 + HNR'_2 \quad (R = CH_3;\ R' = C_2H_5) \tag{4-66}$$

However, when a bis- or tris-aminostannane was used, even with a large excess of hexamethyldisilazan and prolonged heating (60 hr) the reaction was limited to partial substitution:

$$R_nSn(NR_2)_{4-n} + HN(SiR_3)_2 \longrightarrow R_nSn[N(SiR_3)_2](NR_2)_{3-n} \quad (R = CH_3;\ n = 1 \text{ or } 2) \tag{4-67}$$

It is characteristic that the reaction of $(CH_3)_3SiNHNa$ (suspension in benzene) with trimethylchlorostannane yields trimethyl-silyl-bis(trimethylstannyl)amine i.e., the direction of (4-65) is combined with a reaction of type (4-66) [1067].

By using organolithium synthesis it was possible to obtain silicon-organotin compounds containing several silylamino groups, but less bulky groups than $[(CH_3)_3Si]_2N$ [1065, 1067]:

$$R_nSnCl_{4-n} + (4-n)\,R_3SiN(R)Li \longrightarrow R_nSn[N(R)SiR_3]_{4-n} \quad (n = 1-4) \tag{4-68}$$

Therefore the preparation of $\{[(CH_3)_3Si]_2N\}_2SnCl_2$ by Wannagat should be regarded as an exception to the general rule of steric prohibition. This is evidently connected with the fact that the other substitutents at the tin atom in this substance are not organic groups.

At the same time, there is a reaction [1069] which is a peculiar development of scheme (4-67):

$$(CH_3)_2Sn[N(C_2H_5)_2]_2 + \begin{matrix} HNSi(CH_3)_3 \\ | \\ [C]_n \\ | \\ HNSi(CH_3)_3 \end{matrix} \xrightarrow{-2(C_2H_5)_2NH} (CH_3)_2Sn\begin{matrix} N(Si(CH_3)_3) \\ \quad [C]_n \\ N(Si(CH_3)_3) \end{matrix} \qquad (4\text{-}69)$$

where $[C]_n$ is a biradical: $[-CH_2-]_m$ $(m = 2-4)$ or $o\text{-}C_6H_4\langle$

Triorganosilyl azides form solid, thermally stable 1 : 1 complexes with $SnCl_4$ [119, 1208].

Heating hexamethylcyclotrisilazan in the presence of ~1% of $SnCl_4$ for 8 hr at 195°C leads to ~30% conversion of the "trimer" into a "tetramer" [185, 227]. There is no doubt that this is connected with the intermediate formation of silicon, tin, and nitrogen-containing compounds, but the mechanism of this interesting conversion has not been studied as yet.

7.2. Physical and Chemical Properties

The physicochemical constants of compounds containing the Si–N–Sn group and some of their spectral characteristics are given in Table 54.

In the IR apectra of bis(trimethylsilyl)amino substituted stannanes, the wave number of the antisymmetrical valence vibration of the Si–N–Si bond is shifted by +25 to 30 cm^{-1} in comparison with ν_{as}SiNSi of tris(trimethylsilyl)amine (916 cm^{-1}) [1060, 1064]. Complexes of silyl azides with $SnCl_4$ are characterized by shifts in the frequencies ν_{as} and ν_s SiN_3 in comparison with the individual silyl azides of -20 to 30 and +15 to 20 cm^{-1}, respectively [1208].

The replacement of an organic by a silyl radical at a nitrogen atom attached to a tin atom weakens the Sn–N bond. This is reflected in the hydrolytic sensitivity of all compounds containing the

TABLE 54. Compounds

Formula	m.p., °C	b.p., °C (mm)	Preparation method (scheme)	Yield, %
$(CH_3)_3SiN(CH_3)Sn(CH_3)_3$	—	79—81 (30)	4-68	68
$(CH_3)_3SiN[Sn(CH_3)_3]_2$	—	—	—	—
$[(CH_3)_3Si]_2NSn(CH_3)_3$	20—22	58—59 (1)	4-65	80
	—	55 (1)	4-66	—
$[(CH_3)_3Si]_2NSn(C_4H_9\text{-}n)_3$	—	140—145 (1)	4-65	—
$[(CH_3)_3Si]_2NSnCl_3$	—	—	~4-65	—
$[(CH_3)_3SiN(CH_3)]_2Sn(CH_3)_2$	—	61—63 (0.5)	4-68	60
$[(CH_3)_3Si]_2N[(CH_3)_2N]Sn(CH_3)_2$	—	58 (0.1)	4-67	—
$\{[(CH_3)_3Si]_2N\}_2SnCl_2$	—	—	~4-65	—
$[(CH_3)_3Si]_2N[(CH_3)_2N]_2SnCH_3$	—	78 (0,1)	4-67	80
$CH_2N[Si(CH_3)_3]$ | $CH_2N[Si(CH_3)_3]$ > $Sn(CH_3)_2$ (ring)	14—16	121—123 (18)	4-69	77
$(CH_2)_3$ < $N[Si(CH_3)_3]$, $N[Si(CH_3)_3]$ > $Sn(CH_3)_2$ (ring)	—2	136—138 (10)	4-69	65
$(CH_2)_4$ < $N[Si(CH_3)_3]$, $N[Si(CH_3)_3]$ > $Sn(CH_3)_2$ (ring)	—	70—71 (0.1)	4-69	63
C_6H_4 < $N[Si(CH_3)_3]$, $N[Si(CH_3)_3]$ > $Sn(CH_3)_2$ (ring)	~75	114—117 (0.1)	4-69	70
$(CH_3)_3SiN_3 \cdot SnCl_4$	—	—	—	38
$(C_6H_5)_3SiN_3 \cdot SnCl_4$	—	—	—	—

group Si–N–Sn.* The fact that hydrolysis involves cleavage of the Sn–N bond is well illustrated by the example [1064]:

$$(R_3Si)_2NSnR_3 \xrightarrow{HOH} (R_3Si)_2NH + HOSnR_3 \qquad (R = CH_3) \qquad (4\text{-}70)$$

The same substance is readily hydrogenated at room temperature by lithium aluminum hydride, with which trimethyl(diethylamino)stannane does not react [931].

* Complexes of silyl azides with $SnCl_4$ are also hydrolytically unstable [1208].

Containing the Si–N–Sn Group

NMR spectra [60 MHz, $(CH_3)_4Si$, Hz]				Characteristic frequencies of IR spectra, cm^{-1}	Literature
δCH_3 (Si)	δCH_3 (Sn)	J (HC–Sn^{117})	J (HC–Sn^{119})		
±0	—18.0	52.5	55.0	—	1065
—	—	—	—	—	1067
—5.0	—18.8	53.4	55.8	879 (ν_{as}SiNSn); 950 (ν_{as}SiNSi)	1060, 1064, 1081
—	—	—	—	—	931
—	—	—	—	941 (ν_{as}SiNSi)	1060
—	—	—	—	—	1223, 1238
—1.5	—20.0	57.0	59.5	—	1067
—	—	—	—	—	931
—	—	—	—	—	1223, 1238
—	—	—	—	—	931
—	—23.0	58.0	60.5	—	1069
±0	—21.5	58.0	61.0	—	1069
±0	—21.0	58.0	60.5	—	1069
—11.5	—39.5	59.5	62.0	—	1069
—	—	—	—	2110 ($\nu_{as}SiN_3$); 1347 (ν_sSiN_3)	1208
—	—	—	—	2116 ($\nu_{as}SiN_3$); 1326 (ν_sSiN_3)	119, 1208

8. COMPOUNDS CONTAINING THE Si–Sn BOND

In 1933 a study of the reaction of triphenylsilyllithium with trimethylchlorostannane in liquid ammonia yielded [897] a heavy oily liquid, which did not crystallize at liquid ammonia temperatures, but corresponded in analysis to triphenylsilyltrimethylstannane. However, determination of the molecular weight gave values (599 and 616) which were one and a half times as great as the calculated value (423). No satisfactory explanation of this was found.

Trimethylsilyltributylstannane and trimethylsilyltriphenylstannane are formed in 60-70% yield by the reaction of trimethylchlorosilane with R_3SnLi in ether at -20°C [123]:

$$R_3SnM + ClSiR_3 \longrightarrow R_3SnSiR_3 + MCl \quad (4\text{-}71)$$

If the reaction is carried out at +24°C instead of -20°C, the yield of silylstannanes is substantially reduced because of the formation of the distannanes R_3SnSnR_3. Thus, the reaction of $(C_6H_5)_3 \cdot SnLi$ with $ClSi(CH_3)_3$ in ether at 24°C gives only 23% of trimethylsilyltriphenylstannane, but 54% of hexaphenyldistannane (calculated on the triphenylstannyllithium used). The yield of hexaphenyldistannane in tetrahydrofuran reaches 69%, while it is not possible to isolate a compound containing the Si–Sn bond at all [1205]. Tributylstannyllithium reacts with $(CH_3)_3SiCl$ under normal conditions as if it were a mixture of $(C_4H_9)_2Sn$ and butyllithium. Trimethylbutylsilane is formed in 54% yield and also red-orange dibutyltin, which is then converted into $(C_4H_9)_4Sn$ [562a].

With a rise in the reaction temperature the condensation (4-71) is evidently displaced by the competing metal–halogen exchange, which promotes the formation of R_3SnSnR_3:

$$R_3SnLi + ClSi(CH_3)_3 \rightleftarrows R_3SnCl + LiSi(CH_3)_3 \quad (4\text{-}72)$$
$$R_3SnLi + ClSnR_3 \longrightarrow R_3SnSnR_3 + LiCl$$

Triphenylsilyltriphenylstannane could be obtained both by the reaction of triphenylstannyllithium with triphenylchlorosilane [562a, 775], i.e., by scheme (4-71) and by the reaction of triphenylsilylpotassium with triphenylchlorostannane [796, 1212]:

$$R_3SiM + ClSnR_3 \longrightarrow R_3SiSnR_3 + MCl \quad (4\text{-}73)$$

However, here the reaction is complicated by side reactions which are close to scheme (4-72) or are more complex. As an example, we should point out that hydrolysis of the products from the reaction of triphenylstannyllithium with triphenylchlorosilane in THF yielded triphenylsilyltriphenylstannane (~ 30%), hexaphenyldisilane, hexaphenyldistannane, triphenylsilanol, tetraphenylstannane, and triphenylsiloxytriphenylstannane [1205].

Compounds of the type $R_3SnSiR'_2SnR_3$ may be synthesized by the reaction [123]:

$$2R_3SnLi + R'_2SnCl_2 \xrightarrow[-20^\circ C]{ether} R_3SnSiR'_2SnR_3 + 2LiCl \qquad (4\text{-}74)$$
$$(R = C_2H_5,\ n\text{-}C_4H_9,\ C_6H_5;\ R' = CH_3,\ C_6H_5)$$

The yields of definite compounds are low (15% for N-alkyl compounds and 30-35% for Sn-aryl compounds). This is due to metal–halogen exchange reactions analogous to (4-72), which lead to the formation of distannanes (25-30% in the preparation of alkyl derivatives and 15-20% for aryl derivatives) and higher, chain compounds, containing Sn–Si and Si–Si bonds, whose stability falls with an increase in the chain length. In the reactions of tributylstannyllithium with dimethyldichlorosilane or triphenylstannyllithium with diphenyldichlorosilane, as in the preparation of the compounds R_3SnSiR_3, with a rise in temperature the competing formation of $(R_3Sn)_2$ predominates; the compounds $R_3SnSiR'_2SnR_3$ are obtained in an insignificant amount under these conditions. Thus, in the reaction of tributylstannyllithium with dimethyldichlorosilane at room temperature, 48% of hexabutyldisilane is formed and the second reaction product is tetrabutylstannane (21%), which was not detected when the process was carried out at -20°C. The reaction of triphenylstannyllithium with diphenyldichlorosilane under analogous conditions leads to the formation of hexaphenyldistannane together with tetraphenylstannane.

The stannylsilanes $R_3SnSiR'_3$ and $(R_3Sn)_2SiR'_2$ are readily soluble in nonpolar solvents such as benzene and are monomeric in them. They dissolve less readily in ether and are insoluble in alcohol. These compounds are relatively resistant to the action of atmospheric oxygen, particularly the sterically shielded phenyl derivatives. The liquid substances have the unpleasant odor which is characteristic of many tin compounds. A noteworthy property of stannylsilanes is their reducing power, which, in particular, may be used to prepare $TiCl_3$ that is active in polymerization. If a toluene solution of $TiCl_4$ is added gradually to a solution of bis(triethylstannyl)dimethylsilane in toluene with stirring (molar ratio of silane : $TiCl_4$ = 1 : 2.5), during the exothermic reaction there precipitates a brown material, which becomes brown-violet after heating for 2 hr in an autoclave at 160°C (82% yield on $TiCl_3$):

$$[(C_2H_5)_3Sn]_2Si(CH_3)_2 + 4TiCl_4 \longrightarrow 2(C_2H_5)_3SnCl + (CH_3)_2SiCl_2 + 4TiCl_3 \qquad (4\text{-}75)$$

TABLE 55. Compounds Containing the Si–Sn Bond

Formula	Consistency	m.p., °C	b.p., °C (mm)	Preparation method	Yield, %	Literature
$(C_6H_5)_3SiSn(CH_3)_3$	Heavy oily liquid	—	—	4-73	—	897
$(CH_3)_3SiSn(C_4H_9)_3$	Orange oily liquid	—	94 (1)	4-71	60—70	123
$(CH_3)_3SiSn(C_6H_5)_3$	Colorless crystals	119	—	4-71	60—70	123
$(C_6H_5)_3SiSn(C_6H_5)_3$	The same	289—291	—	4-71	71	775
		296—298	—	4-73	76 (86)	796, 1212
		299—303	—	4-71	30 (19)	1205
$(CH_3)_2Si[Sn(C_2H_5)_3]_2$	Colorless oily liquid	—	135 (1)	4-74	15	123
$(CH_3)_2Si[Sn(C_4H_9\text{-}n)_3]_2$	The same	—	154 (high vacuum)	4-74	15	123
$(CH_3)_2Si[Sn(C_6H_5)_3]_2$	Colorless oily liquid	172	—	4-74	30—35	123
$(C_6H_5)_2Si[Sn(C_6H_5)_3]_2$	The same	199	—	4-74	30—35	123

There are reports that $(C_6H_5)_2SnNa_2$ reacts with dimethyldichlorosilane in the dimethyl ether of glycol [686], but there are no data on the nature of the reaction products.

The reaction of triphenylsilyllithium with tin tetrachloride does not form compounds with the Si–Sn bond, but triphenylchlorosilane (25%) [715]. Since one of the products from the reaction of triphenylsilyllithium with stannous chloride is hexaphenyldisilane, it may be surmised that the reaction of the silicolithium compound with tin chloride proceeds preferentially by the "reverse" scheme (4-72) (first stage) with the subsequent formation of the phenylated disilane:

$$\begin{aligned} R_3SiLi + Sn^{n}Cl_n &\longrightarrow R_3SiCl \\ R_3SiLi + R_3SiCl &\longrightarrow R_3SiSiR_3 \end{aligned} \tag{4-76}$$

The constants of compounds containing the Si–Sn bond are given in Table 55.

9. COMPLEXES OF TIN COMPOUNDS WITH ORGANOSILICON COMPOUNDS

Some data have already been given on this type of complex. In addition, we should also point out that in the systems $SiX_4 - SnX_4$ (X is halogen) hardly any interaction is observed [2].

In a study of the donor-acceptor interaction of chloroethoxysilanes $SiCl_x(OC_2H_5)_{4-x}$ ($x = 0-3$) with $SnCl_4$, measurement of the dielectric constant of isomolar solutions of ethoxychlorosilane–stannic chloride pairs in benzene revealed the formation of the complexes $(RO)_4Si \cdot SnCl_4$ and $2(RO)_3SiCl \cdot SnCl_4$ [309b]. Additional cryoscopic investigations showed:

a) the existence of a 1 : 1 complex for the second system in the region of excess $SnCl_4$;

b) the existence of a weak complex $2(RO)_2SiCl_2 \cdot SnCl_4$;

c) the absence of the interaction in the system $ROSiCl_3 \cdot SnCl_4$. On the example of complexes of tetraethoxysilane and triethoxychlorosilane it was established that they are partly dissociated and that the degree of dissociation increases with dilution, reaching 60-80%. Their IR spectra show a shift in the absorption bands of Si–O–C by 40 to 60 cm^{-1}.

The table below gives the true dipole moments of the complexes $nSiCl_x(OR)_{4-x} \cdot SnCl_4$ and their heats of formation according to data in [309b].

Complex		M	ΔM	Q, kcal/mole	q, (kcal per M ← O bond)
n	x				
1	0	7.30	5.57	9.54	4.77
1	1	3.62	1.84	1.96	1.96
2	1	6.13	3.61	3.11	1.55
2	2	–	–	0.32	0.16
1	3	–	–	0.0	0.0

On the basis of these and spectral data the authors write with certainty of the formation of O → Sn bonds in the interaction of $SnCl_4$ with tetraethoxy- and triethoxychlorosilane. In the case of $(RO)_4Si \cdot SnCl_4$ an octahedral structure is proposed in which the tetraethoxysilane behaves as a bidentate ligand, participating in complex formation with two ethoxyl groups (the coordination number of tin is 6 — sp^3d^2 hydridization). In the series $(RO)_{4-x}SiCl_x$ the donor activity of the compounds falls with an increase in x. Triethoxysilane is a monodentate ligand toward $SnCl_4$ and then the tendency toward complex formation falls still more sharply. Tin tetrabromide and tetrachloride do not form stable adducts with tetraethoxysilane [54, 243, 245].

2-Hydroxypyridine-N-oxide forms a white, crystalline, cationic silicic acid complex, which is characterized in particular in the form of the hexachlorostannate [1249]. The complex is soluble in water, but is not hydrolytically stable and decomposes in the range of 204-206°C.

Chlorostannic acid is a catalyst for the telomerization of olefins with hydrochlorosilanes [69]. There is the possibility that this is connected with the intermediate formation of silicon and tin-containing complexes.

10. ANALYSIS OF ORGANOSILICON DERIVATIVES OF TIN

In the analysis of substances containing the group Si–M–Sn (M = O, C, etc.), regardless of the class to which the compound belongs, it is converted into a mixture of oxides by "wet combustion" of a sample with a mixture of concentrated sulfuric and nitric acids [253, 1003] or H_2SO_4 with KIO_3 [53]. The following methods are recommended for the subsequent determination of tin.

1. The mixture of oxides is boiled with dilute hydrochloric acid and the precipitated SiO_2 separated and then calcined for determination of Si. The filtrate is neutralized with ammonia and the solution boiled in the presence of ammonium nitrate. The filter with the SnO_2 precipitate is ignited in an oxidizing flame [153].

2. The mixture of oxides is treated with a mixture of sodium hydroxide and sulfur, leading to the formation of soluble sodium thiostannate. The solution is treated with acetic acid and then again with NaOH and sulfur to precipitate the tin as the sulfide. Then the thiostannate is boiled with hydrogen peroxide to convert it to sodium stannate, the solution neutralized to Methyl Orange with nitric acid, and the tin precipitated with ammonium nitrate and then determined as the dioxide [1003].

3. The oxides are dissolved in sodium hydroxide, the solution treated with excess 6N HCl with conversion of the sodium stannate into stannic chloride. Then by means of nickel wire in a CO_2 atmosphere the Sn^{4+} is converted into Sn^{2+} and titrated with standard iodine solution [1003].

4. The mixture of oxides is heated with an 8-10 fold amount of ammonium iodide; the tin is removed as the iodide and determined indirectly by difference [616].

A hydrolytic analysis method is used for complexes of silyl azides with tin tetrachloride [1208]:

$$2R_3SiN_3 \cdot SnCl_4 + 5H_2O \longrightarrow (R_3Si)_2O + 2HN_3 + 2SnO_2 + 8HCl \qquad (4\text{-}77)$$

LEAD

11. COMPOUNDS CONTAINING THE Si – O – Pb GROUP

11.1. Preparation Methods

11.1.1. Cohydrolysis and Heterofunctional Condensation Reactions

Khananashvili [67] investigated the reaction of alkoxysilanes of the type $(CH_3)_nSi(OC_2H_5)_{4-n}$ (n = 0 – 2) with aqueous alkali solutions of sodium plumbite with Pb : Si ratios from 3 : 1 to 1 : 30. When n = 0 and Pb : Si = 3 : 1 – 1 : 16, the system forms two phases. When Pb : Si = 1 : 6, 1 : 10, and 1 : 16, during prolonged standing the lower phase deposits crystalline products which are insoluble in organic solvents. They have a structure analogous to the products from the reaction of alkoxysilanes with alkali metal stannates, i.e.,:

$$\left[\cdots-\underset{ONa}{\overset{R}{Si}}-O-\right]_m\left[-\underset{ONa}{\overset{ONa}{Si}}-O-\right]_n-\underset{ONa}{\overset{ONa}{Pb}}-O-\cdots \quad (R = CH_3,\ OC_2H_5)$$

With a Pb : Si ratio from 1 : 16 to 1 : 20 sticky substances are obtained, which liberate a liquid phase on standing. They are insoluble in organic solvents, but dissolve in hot water and may be precipitated from solution by methanol. When Pb : Si = 1 : 20, gels are formed which undergo syneresis. In the hydrolysis of dimethyldiethoxysilane or methyltriethoxysilane by sodium plumbite solutions with a ratio Pb : Si = 1 : 1 – 1 : 25, mainly liquid products are formed together with some solid precipitates.

The reaction of aqueous alkali solutions of sodium plumbite and plumbate with trimethylmethoxysilane [453] forms solid crystalline substances, which are soluble in some organic solvents and whose composition was described by the authors by the following formulas:

$$[(CH_3)_3SiO]_2Pb \cdot 2(CH_3)_3SiONa \cdot 7NaOH \cdot 15H_2O;$$

$$[(CH_3)_3SiO]_2Pb \cdot (CH_3)_3SiONa \cdot C_2H_5OC_2H_5 \cdot 10NaOH \cdot 15H_2O;$$

$$[(CH_3)_3SiO]_2Pb(ONa)_2 \cdot 4NaOH \cdot 10H_2O$$

When mixtures of tetraethoxysilane are heated with an alcohol solution of diacetoxyplumbane $(CH_3COO)_2Pb$ for 1 hr on a water bath, white precipitates are formed, corresponding to the empirical formula $xPb \cdot ySiO_2 \cdot z(C_2H_5)_2O \cdot pH_2O$ [32, 272, 322, 335, 475]. The molar ratio PbO : SiO_2, in the reaction products is close to the value set (1:1.7; 1:3.8; 1:7; 1 : 12 with initial values of 1 : 1; 1 : 3; 1 : 6; 1 : 12, respectively). With careful heating to 500°C, these substances darken due to carbonization of the organic part and then they are converted into fluffy white powders, which sinter at 700°C. Since heating $(CH_3COO)_2Pb$ yields a yellow powder of PbO, which is then converted into dark brown PbO_2, the behavior on heating of the polymers, whose chemical individuality was completely established, indicates that their structure contains the groups Si–O–Pb–O–Si. The thermograms of the substances obtained confirm that they are chemical compounds and not mechanical mixtures.

Under analogous conditions the reaction of lead diacetate with methyl(phenyl)triethoxysilane (with a ratio Si : Pb = 5 – 9) leads to the formation of products of a noncrystalline nature, which are partly soluble in acetone. The following structures are considered possible for them:

```
  [         R     R          ]          [  R     O—Pb—O     R  ]
  [         |     |          ]          [   \Si/        \Si/   ]
  [ ...—Pb—O—Si—O—Si—...     ]          [   /  \        /  \   ]
  [         |     |          ]          [ .   O   O   .     .  ]
  [   R     O     O          ]    or    [  .  |   |  .         ]
  [   |     |     |          ]          [   \Si   Si/          ]
  [ ...—Si—O—Si—O—Si—...     ]          [  R/  \O/  \R         ]
  [   |     |     |          ]          [                      ]n
  [   OR    R     R          ]n
```

At the same time, dialkyldiethoxy- and trialkyl(aryl)alkoxysilanes do not react with diacetoxyplumbine under the conditions described.

11.1.2. Reaction of Silanols, Alkali Metal Silanolates, and Siloxanes with Lead Compounds

In 1945 there appeared a report [1005] of the synthesis of bis(trimethylsiloxy)plumbine:*

$$2\,(CH_3)_3SiOH + PbO \longrightarrow [(CH_3)_3SiO]_2Pb + H_2O \tag{4-78}$$

According to the data of the authors, this compound is a white crystalline substance, which dissolves in the normal solvents and is readily hydrolyzed by dilute H_2SO_4. However, analysis data on the lead content of the product (59.1% instead of the calculated value of 53.7%) raise doubts on the individuality of the compound obtained.

At the same time, compounds of the type $(R_3SiO)_2Pb$ have been patented [1536] as heat and light stabilizers of plastics, but the synthesis route was not reported in the patents.

An attempt to prepare bis(triethylsiloxy)plumbine by the reaction of triethylsilanol with $Pb(OH)_2$ in the presence of a dehydrating agent (freshly calcined sodium sulfate) [180, 181] led to the formation of the benzene-soluble adduct $2[(C_2H_5)_3SiO]_2Pb \cdot Pb(OH)_2$.

In the literature, particularly in patents, one often finds reports of the use of various lead derivatives as additives, which markedly accelerate the hardening of polysiloxane resins, the drying of varnish films, the gelatinization and "cold vulcanization" of

* The principles of the nomenclature of organic compounds of lead, including compounds containing silicon, were presented in [25].

liquid siloxane polymers, and the waterproofing of materials with polyalkylhydrosiloxanes. There have been reports of the use for these purposes of lead salts of various carboxylic acids (naphthenates, resinates, salts of saturated and unsaturated $C_6 - C_{18}$ acids, etc.) [63, 110, 614, 980, 981, 988, 1303, 1328, 1330, 1334, 1368, 1398, 1401, 1403, 1439, 1459, 1460, 1558, 1567, 1569, 1570, 1582, 1646, 1674, 1689, 1696, 1700, 1701, 1705, 1722], tetraphenylplumbane [1319, 1559], various oxides and hydroxides of lead [63, 614, 834, 980, 981, 988, 1522, 1622], its complex compounds [252, 1518], and also finely dispersed metallic lead [1727].* It is also reported that lead salts of carboxylic acids are the most effective of the organic derivatives of metals in these processes [110].

Many organosilicon polymers, toward which lead compounds show a catalytic action, contain OH groups at silicon atoms. Taking account of this and also scheme (4-78) and the possibility of the reaction of dispersed metals with hydroxyl-containing compounds of silicon [1461], it may be surmised that the hardening–gel formation processes develop largely as a result of the condensing action of lead compounds with the formation of plumbasiloxanes as intermediate products. However, Fromberg [63], who made the most complete investigation of the effect of lead derivatives on the properties of polysiloxanes, established that, for example, a 60% toluene solution of polymethylphenylsiloxane, obtained by cohydrolysis of di- and trifunctional chlorosilanes, is converted into a gel in the presence of 0.3% lead naphthenate in 4 hr at 20°C, but the hydroxyl group content of the gelled polymer remains the same (1.3%). This led to the conclusion that the hardening is connected with the rearrangement of the polymer chains, apparently as a result of the closure of internal rings. The question of the incorporation of lead into the siloxane structure in this case remains open. There is the opinion [525, 971] that at an elevated temperature lead oxide reacts with linear block polymethylsiloxanes:

$$R_3SiO(R_2SiO)_n SiR_3 + PbO \longrightarrow R_3SiO(R_2SiO)_m PbO(R_2SiO)_p SiR_3 \quad (p = n - m) \qquad (4\text{-}79)$$

* There has also been a description of the use of lead salts of carboxylic and thiocarboxylic acids as stabilizers of siloxane liquids [1340] and as additives which improve the properties of vulcanizates of siloxane rubbers [1587, 1707].

When a solution of trialkylbromoplumbane in ether is treated with sodium trialkylsilanolate, a precipitate of NaBr forms and trialkylsiloxytrialkylplumbanes are obtained in good yield [1092, 1107]:

$$R_3SiONa + BrPbR'_3 \longrightarrow R_3SiOPbR'_3 \quad (4\text{-}80)$$

A lithium trialkylsilanolate does not react in this way under analogous conditions, evidently due to the solubility of LiBr in ether. In any case, the reaction with trialkylchloroplumbanes proceeds normally:

$$R_3SiOLi + ClPbR'_3 \longrightarrow R_3SiOPbR'_3 \quad (4\text{-}81)$$

Sodium triphenylsilanolate does not react with lead dichloride [826]. Sodium diphenylsilanediolate reacts. With an equimolar ratio of the reagents a mixture of plumbasiloxanes $[-Si(C_6H_5)_2O]_x$ $[-Pb-O-]_y$, diphenylcyclosiloxanes, and lead oxide was obtained in 8 hr at 80°C [847]. The formation of the last two components is evidently due to partial decomposition of polymers formed by condensation of the diolate with the dichloride.

11.2. Physical and Chemical Properties

The physicochemical constants of compounds containing the Si–O–Pb group are given in Table 56.

According to the data of Schmidbaur [1092, 1107], trialkylsiloxytrialkylplumbanes are colorless, mobile, unassociated liquids, which distill in vacuum without decomposition and have compara-

TABLE 56. Compounds Containing the Si–O–Pb Group

Formula	m.p., °C	b.p., °C (mm)	Preparation method	Yield, %	Literature
$(CH_3)_3SiOPb(CH_3)_3$	–1	172 (720); 73 (10)	4-80	76	1092, 1107
$(C_2H_5)_3SiOPb(CH_3)_3$	—	65 (1)	4-81	79	1092
$(CH_3)_3SiOPb(C_2H_5)_3$	—	111.5 (10)	4-80	84	1092
$(C_2H_5)_3SiOPb(C_2H_5)_3$	—	100 (1)	4-80	85	1092
$[(CH_3)_3SiO]_2Pb$	Decomp., 180	—	4-78	—	1005
$2[(C_2H_5)_3SiO]_2Pb \cdot Pb(OH)_2$	Decomp., before melting	—	—	73	180, 181

tively low boiling points relative to the molecular weight. The substances have an extremely unpleasant odor, are very volatile, are readily soluble in organic compounds, including, for example, lipoids, and are highly toxic. Trimethylsiloxytrimethylplumbane does not decompose when distilled under normal pressure (an atmosphere of an inert gas) and under normal conditions it is resistant to the action of dry air and light for a long time. However, at 140°C contact with oxygen products an explosion in which elementary lead is formed. In contrast to their high thermal stability, hexaalkylplumbasiloxanes are highly reactive in heterolytic processes. The compounds are readily solvolyzed by water, alcohols, and other reagents with active protons (trimethylsiloxytrimethylplumbane is decomposed by the action of atmospheric moisture) with the formation of R_3PbOR' ($R = H$, alkyl, etc.), i.e., with cleavage of the Pb–O bond. Acid anhydrides and chlorides react vigorously with hexaalkylplumbasiloxanes to give the corresponding lead salts and acyloxy- or halosilanes. Anhydrous aluminum chloride reacts with trimethylsiloxytrimethylplumbane explosively with the liberation of a lead mirror. At very low temperatures it is possible to limit the reaction almost completely to the stage:

$$2R_3SiOPbR_3 + 2AlCl_3 \longrightarrow 2R_3PbCl + [R_3SiOAlCl_2]_2 \qquad (4\text{-}82)$$

In this connection we should point out that the adduct of bis-(triethylsiloxy)plumbane with $Pb(OH)_2$ acts as a carrier of a trial-

TABLE 57. NMR Spectra of Compounds Containing the Si–O–Pb Group

Compound	Si			Pb			δ_{CH_2}	Literature
	δ_{CH_3}	J		δ_{CH_3}	J			
		H–C[13]	H–Si[29]		H–C[13]	H–Pb[207]		
$(CH_3)_3SiOSi(CH_3)_3$	−3.5	118.0	6.86	—	—	—	—	1081
$(CH_3)_3SiOPb(CH_3)_3$	+3.5	115.5	6.69	−73.0	136.0	69.5	—	1081 1092
$(C_2H_5)_3SiOPb(CH_3)_3$	−54.5	—	—	−71.0	—	69.0	−22.0 (Si)	1092
$(CH_3)_3SiOPb(C_2H_5)_3$	+5.0	—	—	−97.0	—	41.9	−110.5 (Pb)	1092
$(C_2H_5)_3SiOPb(C_2H_5)_3$	−53.8	—	—	−95.7	—	42.8	−21.7 (Si); −109.7 (Pb)	1092
$(C_2H_5)_3SiOSi(C_2H_5)_3$	—	—	—	—	—	—	−32.8 (Si)	1092

TABLE 58. Properties of the Compounds $(CH_3)_3Si-O-M(CH_3)_3$ [1081, 1084, 1107]

M	Freezing p., °C	b.p., °C (mm)	ν_{as} SiOM, cm^{-1}	Si: δ_{CH_3}	Si: $J(H-C^{13})$	Si: $J(HC-Si^{29})$	M: δ_{CH_3}	M: $J(H-C^{13})$	M: $J(HC-M)$
C	—91	103.5 (760)	1052	—	—	—	—	—	—
Si	—57	100.5 (760)	1055	—3.5	118.0	6.86	—3.5	118.0	—
Ge	—68	117 (725)	990	0.0	117.8	6.80	—21.0	126.0	—
Sn	—59	141 (720)	980	+2.5	116.8	6.72	—20.5	128.5	54.0 (Sn^{117}) 57.4 (Sn^{119})
Pb	—1	172 (720)	959	+3.5	115.5	6.69	—73.0	136.0	69.5 (Pb^{207})

kylsiloxy group in the reaction with titanium tetrachloride [180, 181]:

$$\{2Pb[OSi(C_2H_5)_3]_2 \cdot Pb(OH)_2\} + 1.5TiCl_4 \longrightarrow [(C_2H_5)_3SiO]_4Ti + 3PbCl_2 + 0.5Ti(OH)_4 \quad (4\text{-}83)$$

The adduct behaves analogously in the reaction with vanadium oxychloride (see scheme 5-64 on p. 4).

In the IR spectra of $R_3SiOPbR_3$ there is a band of ν_{as} SiOPb at 957-961 cm^{-1} [1092]. Data on the NMR spectra of compounds of this type are given in Table 57 (cf. Table 49, p. 354).

All the characteristics of the bonds in stannasiloxanes (the increase in the s-character of the M–C bond, the polarity of the M–O bond, etc.) are expressed to an even greater extent with plumbasiloxanes. Plumbasiloxanes are clearly plumbyl silanolates in their character.

To conclude this chapter it is interesting to compare the properties of compounds of the type R_3SiOMR_3, where M is an atom of a Group IVa element (C, Si, Ge, Sn, Pb) (Table 58).

The data in Table 58 show that with the replacement of a silicon atom by atoms of other Group IV elements with an increase in the atomic radius there is a regular increase in the chemical screening and a decrease in the spin–spin interaction constants for the $(CH_3)_3Si$ groups and, consequently, an increase in the double-bond character of Si–O and the p-character of the valences of the Si atom. The silanolate character of the compounds also increases more markedly. It is characteristic that no alternation in the properties of the compounds is observed in the series Si–Ge–Sn–Pb,

TABLE 59. Compounds Containing the Si – $(C)_n$ – Pb Group

Formula	m.p., °C	b.p., °C (mm)	Preparation method (scheme)	Yield, %	Literature
$H(CH_3)_2SiCH_2Pb(CH_3)_3$	—	26—27 (1)	4-86	86	1111
$(CH_3)_3SiCH_2Pb(CH_3)_3$	—68	179 (740); 36 (1.5)	4-86	80	1076
$[(CH_3)_3SiCH_2]_3PbCl$	214—216	—	4-85	55	1147
$[(CH_3)_3SiCH_2]_4Pb$	—	104—106 (0.01)	4-84	37	1147
$(C_2H_5)_3Si—C_6H_4—Pb(CH_3)_3$	—	191 (17) *	4-86	—	821

* $n_D^{23,8} = 1.5494$; $d_4^{20} = 1{,}4032$.

though the alternation of the electronegativities of the atoms in this series is well known.

12. OTHER ORGANOSILICON DERIVATIVES OF LEAD

12.1. Compounds Containing the Si – $(C)_n$ – Pb and Si – $(C)_n$ – O – Pb Groups

Compounds of this type are few as yet. When trimethylsilylmethylmagnesium chloride is treated with lead dichloride in tetrahydrofuran the reaction mixture acquires a red-brown color, which changes to grey-green after five hours [1147]. After hydrolysis, metallic lead and tetrakis(trimethylsilylmethyl)plumbane were isolated from the reaction products. The bis(trimethylsilylmethyl)-plumbine formed initially is evidently unstable and disproportionates:

$$4R_3SiCH_2MgX + 2PbX_2 \longrightarrow 2\,(R_3SiCH_2)_2Pb \longrightarrow (R_3SiCH_2)_4Pb + Pb \qquad (4\text{-}84)$$

When the tetrakis derivative is treated with PCl_3, one of the "siliconeopentyl" groups is eliminated:

$$(R_3SiCH_2)_4Pb + PCl_3 \longrightarrow R_3SiCH_2PCl_2 + (R_3SiCH_2)_3PbCl \qquad (4\text{-}85)$$

The reaction of trimethylsilylmethylmagnesium chloride with trimethylchloroplumbane gives a good yield of trimethylsilylmethyltrimethylplumbane [1076]. This colorless liquid with a "sweet" odor is extremely toxic. It has a stability which is unusual for organic compounds of lead. It is not hydrolyzed and not oxidized by air below 100°C. $H(CH_3)_2SiCH_2Pb(CH_3)_3$ was obtained analogously [1111].

As early as 1917 the only compound known then with an arylene bridge between silicon and lead atoms was synthesized by the action of the Grignard reagent from p-triethylsilylbromobenzene on trimethylbromoplumbane [821].

These syntheses may be represented by the general scheme

$$R_3SiAMgX + XPbR'_3 \longrightarrow R_3SiAPbR'_3 \quad (4\text{-}86)$$

$R = CH_3$ or C_2H_5; $R' = CH_3$; $X = Cl, Br$; $A = CH_2$ or C_6H_4)

An attempt to add sym-dimethyldiethyldisiloxane to diethyldivinylplumbane in the presence of chloroplatinic acid was unsuccessful and decomposition with the liberation of lead occurred instead of the formation of a polymer [316].

The properties of compounds containing the $Si-(C)_n-Pb$ group are given in Table 59

It seemed interesting to compare the properties of compounds of the type $(CH_3)_3SiCH_2M(CH_3)_3$ and $H(CH_3)_2SiCH_2M(CH_3)_3$ (where M = Si, Ge, Sn, Pb) (Tables 60 and 61).

Without giving a detailed interpretation of these data we should point out that the trimethylsilyl group has a constant inductive effect in the derivatives of Ge, Sn, and Pb $\left(\Delta\frac{\delta_{CH_3\,(M)}}{\delta_{CH_2}} = 20.3 \pm 0.3\right)$.

In analogy with other carboxylate organosilicon heteropolymers (Ch. 2, Section 5) a polymer with the structure $\{-PbOCOH_2CH_2 \cdot [-Si(CH_3)_2O]_mSi(CH_3)_2CH_2CH_2COO-\}_n$ with n = 29 and $\eta_{toluene} = 0.068$ was prepared recently [289, 290, 1518a].

12.2. Compounds Containing the Si – N – Pb Group

Only two compounds of this type are known as yet and these were synthesized by schemes analogous to (4-65) and (4-68) using trimethylchloroplumbane [1064, 1065, 1067]. Their constants are given in Table 62.

TABLE 60. Properties of the Compounds $(CH_3)_3SiCH_2M(CH_3)_3$ [1076]

M	m.p., °C	b.p., °C (mm)	ν_{as}SiCM, cm^{-1}	Si						M	
				δ_{CH_3}	$J(H_3^1-C^{13})$	$J(H_3^1-Si^{29})$	$J(H_2^1C-Si^{29})$	δ_{CH_2}	$J(H_2^1-C^{13})$	δ_{CH_3}	$J(H_3^1-C^{13})$ *
Si	—	125	1054	—1.8	118.2	6.7	8.8	+15.6	108.4	—1.8	118.2
Ge	—74	139 (740)	1050	—0.65	118.0	6.80	7.40	+11.2		—9.4	124.4
Sn	—78	161 (740)	1006	0.00	118.2	6.79	7.02	+15.5	117.2	—5.1	127.4
Pb	—68	179 (740)	991	—0.3	119.0	6.73	6.00	—25.0		—45.0	136.0

* $J(H_3^1-C-Sn^{117,119})$ 51,5; 53,8; $J(H_3^1-C-Pb^{207})$ 60,5; $J(H_2^1-C-Sn^{117,119})$ 69,4; 72,2; $J(H_2^1-C-Pb^{207})$ 89,0.

TABLE 61. Properties of the Compounds $H(CH_3)_2SiCH_2M(CH_3)_3$ [1111]

M	b.p., °C (mm)	Yield, %	$\delta_{(CH_3)_2Si}$	δ_{SiH}	$J(H_3^1-C-Si-H^1)$	$J(H_2^1-C-Si-H^1)$	δ_{CH_2}	δ_{CH_3M}	$J(H_3^1-O-M)$	$J(H_2^1-C-X)$	$J(H^1-Si-C-M)$
Si	119 (740)	47	—5.5	—242	3.8	4.05	+15.7	—1.6	—	—	—
Ge	133 (740)	51	—5.2	—243	3.8	4.00	+10.1	—10.2	—	—	—
Sn	153.5 (740)	77	—4.8	—247	3.75	4.15	+14.8	—6.2	54.8/52.3 *	71.5/68.0 *	(31.0) **
Pb	26.5 (1)	86	—4.5	—252	3.65	4.05	—24.5	—47.0	61.2 ***	84.5 ***	64.0 ***

* Sn^{119}/Sn^{117}.
** Mean for Sn^{119} and Sn^{117}.
*** Pb^{207}.

TABLE 62. Compounds Containing the Si–N–Pb Group

Formula	b.p., °C (mm)	Yield, %	ν_{as}SiNPb, cm^{-1}	$\delta_{CH_3(Si)}$	$\delta_{CH_3(Pb)}$	Literature
$(CH_3)_3SiN(CH_3)Pb(CH_3)_3$	42—43 (0,5)	52	—	+2.5	—56.0	1065
$[(CH_3)_3Si]_2NPb(CH_3)_3$	85—87 (3)	61	875	±0.0	—	1064

[Bis(trimethylsilyl)amino]trimethylplumbane (a colorless liquid) is hydrolyzed readily with cleavage of the Pb–N bond.

The properties of compounds of the type $[(R_3Si)_2N]_2MR_3$, where $R = CH_3$; M = Si, Ge, Sn, Pb, are compared in Table 63.

The data presented in Table 63 indicate strengthening of Si–N bonds on introduction into the structure of the tertiary silylamine the groups $-MR_3$, whose +I-effect falls in the series Pb > Sn > Ge > Si [1064].

12.3. Compounds Containing the Si–Pb Bond

These compounds are unknown at the present time. Unsuccessful attempts have been made to prepare compounds with a direct bond between silicon and lead both by the reaction of triphenylsilylpotassium with triphenylchloroplumbane [796, 1212] and by the action of triphenylsilyllithium on $PbCl_2$ [715].

TABLE 63. Properties of the Compounds $\{[(CH_3)_3Si]_2N\}_2M(CH_3)_3$ [1064, 1081]

M	m.p., °C	b. p., °C (mm)	ν_{as}, cm^{-1}		Si			M	
			SiNSi	SiNM	δ_{CH_3}	J (H–C[13])	J (HC–Si[29])	δ_{CH_3}	J (H–C[13])
Si	67—69	78—80 (13)	916	916	—10.7	118.0	6.76	—10.7	118.0
Ge	29—32	54—56 (1)	931	890	—7.2	118.0	6.75	—23.5	125.4
Sn	20—22	58—59 (1)	950	879	—5.0	118.0	6.60	—18.8	130.0
Pb	—	85—87 (3)	970	875	±0.0	—	—	—	—

TITANIUM

13. COMPOUNDS CONTAINING THE Si – O – Ti GROUP

13.1. Preparation Methods

13.1.1. Cohydrolysis of Silicon and Titanium Compounds

Alkaline cohydrolysis of organochlorosilanes with titanium tetrachloride [12] or esters of orthotitanic acid such as tetrabutoxytitane* in toluene leads to liquid polymers, which are soluble in the usual organic solvents [153, 178]. Heating at 200°C converts them into a glassy state and the solid polymers are infusible, but soluble. Fractionation shows that they have considerable polydispersity with respect to the titanium content. The solubility of the polymers in a given solvent depends on their titanium content and also the nature of the starting monomers, evidently in connection with differences in the structure of the polymerization products. As a rule, the titanium content of a copolymer is considerably less than the titanium content of the starting mixture and this indicates competing processes of alkaline hydrolysis (or hydrolytic condensation) of tetrabutoxytitane. The structure of the cohydrolysis products has not been established definitely. It may be surmised that there is the initial formation of linear or lightly branched titanasiloxanes, containing unreactive butoxyl groups at titanium atoms and hydroxyl groups at silicon. There is then thermal condensation through these groups to form polymers of cyclolinear character and this explains their solubility.†

According to patent data [1292, 1329, 1589], viscous liquid titanasiloxanes may also be obtained by cohydrolysis of alkylalkoxysilanes with alkyl titanates or amino derivatives in an organic solvent. To obtain impregnating coatings based on titanasiloxanes it is recommended that the cohydrolysis is carried out in situ by impregnating, for example, glass cloth with a mixture of phenyltri-

* This form of nomenclature has been adopted in accordance with the authors' system. See the section on ziconium and hafnium—Translator.

† We will return to the problem of the structure of products from the combination of polyfunctional compounds of silicon and titanium somewhat later.

TABLE 64. Products of Cohydrolysis of $RR'SiCl_2$ with an Equimolar Amount of Bis(2-ketopent-3-en-4-oxy)dichlorotitane [204]

R	R′	Hydrolysis in the absence of an HCl acceptor		Hydrolysis in the presence of pyridine		
		Polymer yield, %	Si : Ti	Polymer yield, %	Si : Ti	Tg, °C
CH_3	CH_3	30	9 : 1	57,6	1 : 1	—50
CH_3	$CH_2=CH$	—	—	62	2 : 1	—20
CH_3	C_6H_5	38	8 : 1	63,8	2 : 1	+45
C_2H_5	C_2H_5	40	5 : 1	70,5	1 : 1	—25

methoxysilane and tetraisopropoxytitane and then treating the cloth with water and drying [1381].

The cohydrolysis of diorganodichlorosilanes with bis(cyclopentadienyl)dichlorotitane in the presence of an HCl acceptor gives a 75-80% yield [52, 201] of dark red viscous polymers, which are soluble in toluene and xylene and in which the molar ratio of Si : Ti practically coincides with the initial ratio (10 : 1). Spectral analysis established that their structure includes Si–O–Ti groups. At the same time, chemical analysis showed that the titanium atoms in the polymer chain are "bare" and instead of the expected structure $[-(R_2SiO)_nTi(C_5H_5)_2O-]_x$, they have a structure which should be represented schematically as $[-(R_2SiO)_nTi(O_{0.5})_2O-]_x$. Thus, the titanocene group is broken down during hydrolysis, but this does not lead to the formation of cross-linked polymers. In an acid medium the cohydrolysis reaction does not occur at all. In the case of dimethyldichlorosilane the products are octamethylcyclotetrasiloxane and titanium dioxide.

The hydrolytic instability of tetraalkoxytitanes and titanocene derivatives led to attempts to use for the construction of linear polytitanasiloxanes "stabilized" alkoxides, i.e., internal complexes of titanium. In the cohydrolysis of diorganodichlorosilanes with bis(2-ketopent-3-en-4-oxy)dichlorotitane in the absence of an HCl acceptor there is a considerable amount of "individual" hydrolysis and 60-70% of the titanium compound undergoes complete hydrolytic cleavage [53, 204]. Correspondingly, the yield of polymers containing the Si–O–Ti group does not exceed 30-40%. Moreover, with an equimolar ratio of the starting monomers the value of Ti : Si in the polymer is considerably less than the initial value. The use of pyridine appreciably improves the degree of participation of the bis(2-ketopent-3-en-4-oxy)dichlorotitane in the cohydrolysis reaction without elimination of the chelate groups and the final ratio Ti : Si in the copolymer aproaches the initial value. Table 64 gives some idea of the effect of the substitutents at the silicon on the yield of the polymer and the Ti : Si ratio in it.

With a change from bifunctional organochlorosilanes to trifunctional compounds, Andrianov and Pichkhadze [53, 205] observed that with an equimolar ratio of $RSiCl_3 : (C_5H_7O_2)_2TiCl_2$ the yield of the copolymer (hydrolysis in toluene in the presence of pyridine) with $R = CH_3$, C_2H_5, and C_6H_5 was 40, 48, 62%, while the ratio Si : Ti = 5 : 1, 4 : 1 and 3 : 1, respectively. There is no doubt that acetylacetonate groups are present at the titanium atoms in the titanosiloxanes obtained in this way. Taking into account the high solubility of these polymers and the bifunctionality of the titanium derivative in the cohydrolysis process, the authors came to the conclusion that the third function at the silicon atom does not result in the formation of a cross-linked structure, but cyclolinear chains of the type:

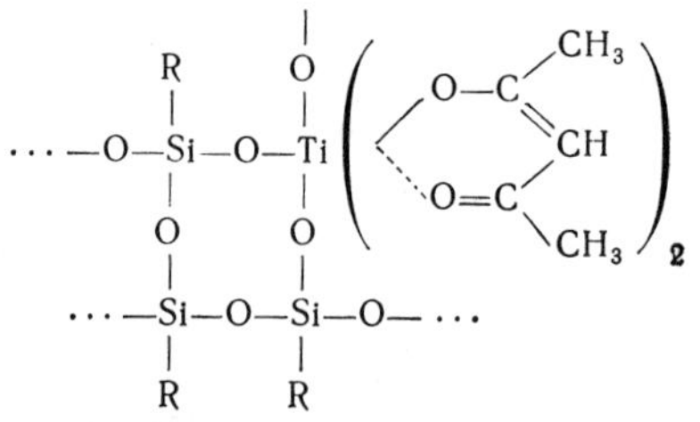

```
                     |          /CH3
         R           O    / O—C      \
         |           |   / /   \\     \
... —O—Si—O—Ti   (  <      CH      )
         |           |   \ \   /      /
         O           O    \ O=C       /
         |           |          \CH3 /2
   ... —Si—O—Si—O— ...
         |       |
         R       R
```

These polytitanasiloxanes lose their solubility on heating and show the properties of cross-linked polymers.

Cohydrolysis of methylphenyldiethoxysilane with bis(2-ketopent-3-en-4-oxy)dibutoxytitane in toluene [202] yields two polymers, namely, one which is soluble in water with a titanium content of more than 37% [for $(C_5H_7O_2)TiO$ the titanium content is 18%] and toluene-soluble, dark red poly [bis(2-ketopent-3-en-4-oxy)titana]-methylphenylsiloxane with a ratio of Si:Ti = 2:1.

In the cohydrolysis of another internal complex compound, namely, bis(8-quinolyloxy)dibutoxytitane with dimethyldiacetoxysilane, the higher hydrolysis rate of the latter promotes its individual hydrolysis with the formation of octamethylcyclotetrasiloxane [53, 203]. As a result, the yield of the titanasiloxane copolymer is 75-80%, while the Si : Ti ratio in it is less than the initial value (3 : 8 and 9 : 1 with initial ratios of 5 : 10 and 12 : 1, respectively). At the same time, the hydroxyquinoline groups are not touched in the hydrolysis processes. This is also confirmed by the formation of bis(8-quinolyloxy)titanediol in the hydrolysis of the dibutoxy derivative [53]. With a change to methylphenyldiacetoxysilane the rates are equalled out and the Si : Ti ratio in the cohydrolysis products, which are obtained in quantitative yield, equals the initial value.

In 1958 Andrianov put forward the opinion [8, 136] that the cohydrolysis of monomers of the type R_2SiCl_2 or $(R_3SiO)_2SiCl_2$ with $(R_3SiO)_2TiCl_2$ could be used to prepare titanasiloxane copolymers with a triorganosiloxy framework about the inorganic chain, which would have a high thermal stability. Selecting the conditions to preserve the trimethylsiloxy groups during hydrolysis is quite complex. Thus, for example, the action of water on bis(trimethylsiloxy)dichlorotitane even with a molar ratio of 1 : 100 liberates 66% of the theoretically possible trimethylchlorosilane and forms an insoluble powdery polymer containing 15.7% Cl and 55% of the mineral part [191]. At the same time, stirring a mixture of bis(trimethylsiloxy)diisopropoxytitane with an equimolar amount of water in isopropanol for 3 hr [1661] and removal of the alcohol by distillation yields a viscous polymer, which hardens in air and loses its solubility only after prolonged heating.

13.1.2. Reactions of Alkoxysilanes and Acyloxysilanes with Esters of Orthotitanic Acid

As yet there is only one paper [281] in which there is a mention (without discussion of the quantitative side of the problem) of heterocondensation of alkoxysilanes with acyloxytitanes. Bis(cyclopentadienyl)bis(trifluoroacetoxy)titane reacts with dialkyldiethoxysilanes with the quantitative liberation of ethyl trifluoroacetate and the formation of polymers of the titanasiloxane type.

A paper [188] which describes the reaction of triethylbutoxysilane with titanium derivatives of organophosphinic acids is also unique [188]. When an α-styryl radical is present at the phosphorus atom, butanol is liberated with the formation of a viscous oil:

$$Ti\{OP(O)C(C_6H_5){=}CH_2[OSi(C_2H_5)_3]\}_4$$

An analogous result is obtained by carrying out the reaction with triethylchlorosilane. This substance reacts with tetrabutoxytitane with replacement of the $(C_2H_5)_3Si$ group by $(C_4H_9O)_3Ti$ and "regeneration" of triethylbutoxysilane. In the reaction of the latter with a derivative of methylphosphinic acid the condensation silylation is accompanied by partial rupture of the Ti–O–P bonds; bis(triethylsilyl)methylphosphinate and a product with the composition $C_2H_6O_6P_2Ti$ are formed (see also scheme 4-102).

An overwhelming majority of communications in this field describe the reaction of acyloxysilanes with alkoxytitanes by the general scheme

$$\geq SiOCOR + R'O{-}Ti\leq \longrightarrow \geq SiOTi\leq + RCOOR' \qquad (4\text{-}87)$$

Thus, for example, the condensation of trimethylacetoxysilane with tetraethoxytitane leads to the formation of tetrakis(trimethylsiloxy)titane [1335, 1411]. The latter is formed in almost quantitative yield by adding a solution of trimethylacetoxysilane in cyclohexane to a boiling solution of tetraisopropoxytitane (molar ratio of 4 : 1) in the same solvent with subsequent slow distillation of the isopropyl acetate–cyclohexane azeotrope [574, 575]. The analogous reaction with triethylacetoxysilane is only complete in 5.5 days (instead of 6 hr in the case of trimethylacetoxysilane), but the yield of tetrakis(triethylsiloxy)titane reaches 93%. In an attempt to accelerate still further the formation of tetrakis(trimethylsiloxy)-titane by carrying out the condensation of trimethylacetoxysilane with tetraisopropoxytitane without a solvent, but with an excess of the first component (~6.5 : 1), it was not possible to obtain tetrakis-(trimethylsiloxy)titane at all. The solid reaction product formed contained 36.3% CH_3COO, 24.0% Ti and 3.7% Si, but did not contain isopropoxyl groups.

The communication of Bradley and Thomas cited above [574] was published in "Chemistry and Industry" on September 20, 1958 only two months after receipt by the editor, while the work of Andrianov and Ganina [150], was submitted to the editor of the "Zhurnal Obshchei Khimii" on December 26, 1957, but was not published until the February, 1959 issue. Intending to prepare tetrakis(trimethylsiloxy)titane, these authors added tetrabutoxytitane to hot excess trimethylacetoxysilane (7.5-8 mole per mole of tetrabutoxytitane). As a result they obtained not tetrakis(trimethylsiloxy)titane, but the solid condensation products:

$$\begin{array}{ccc} (CH_3COO)_2\,Ti{-}O{-}Ti\,(OCOCH_3)_2 & & (CH_3COO)_2\,Ti{-}O{-}Ti\,(OCOCH_3)_2 \\ \quad\;\; | \qquad\;\; | & \text{and} & \quad\;\; | \qquad\;\; | \\ \quad\;\; O \qquad\; O & & \quad\;\; O \qquad\; O \quad\; OCOCH_3 \\ \quad\;\; | \qquad\;\; | & & \quad\;\; | \qquad\;\; | \;\diagup \\ (CH_3COO)_2\,Ti{-}O{-}Ti\,(OCOCH_3)_2 & & (CH_3COO)_2\,Ti{-}O{-}Ti \\ & & \qquad\qquad\qquad\quad \diagdown OSi\,(CH_3)_3 \end{array}$$

In silicon (4.0%) and titanium (25.0%) contents the second substance was quite close to that obtained by the British chemists. The compounds were soluble only in hot alcohols and did not melt up to 340°C, but decomposed at this temperature.

On the basis of these data it was concluded [150] that under certain conditions reaction (4-88), which is a particular case of scheme (4-87), is of secondary importance:

$$4\,(CH_3)_3\,SiOCOCH_3 + Ti\,(OC_4H_9\text{-}n)_4 \longrightarrow [(CH_3)_3\,SiO]_4\,Ti + 4CH_3COOC_4H_9\text{-}n \qquad (4\text{-}88)$$

As was shown subsequently [973, 1047, 1048], by varying the molar ratios of trimethylacetoxysilane and tetraisopropoxytitane and strictly observing a set of conditions in the reaction procedure (the total absence of moisture, the addition of the acetoxysilane to the alkyl titanate and not the reverse, the purity of the starting materials, cooling the reaction mixture, and the minimal period of heating after mixing of the components) it was possible to synthesize all the compounds of the type $[(CH_3)_3SiO]_nTi(OC_3H_{7\text{-}iso})_{4-n}$.

To determine whether the transfer of the acetoxy groups to the titanium in the reaction of esters of orthotitanic acid with excess trimethylacetoxysilane occurs directly according to the scheme

$$\geq TiOR + R'COOSi\leq \longrightarrow \geq TiOCOR' + ROSi\leq \qquad (4\text{-}89)$$

or as a result of the secondary process

$$\geq TiOSiR_3 + R'COOSi\leq \longrightarrow \geq TiOCOR' + R_3SiOSi\leq \qquad (4\text{-}90)$$

let us examine the following facts:

1) In the reaction of excess trimethylacetoxysilane with $(iso\text{-}C_3H_7O)_4Ti$, isopropyl acetate was detected among the volatile

products [574]; in the case of $(n\text{-}C_4H_9O)_4Ti$, butyl acetate was obtained together with the solid condensation products [150];

2) boiling tetrakis(trimethylsiloxy)titane with butyl acetate for six hours did not give even traces of trimethylacetoxysilane [574, 575];

3) the action of acetic anhydride on tetrakis(trimethylsiloxy)-titane yields trimethylacetoxysilane and a solid product which contains acetoxy groups [1047];

4) boiling tetrakis(trimethylsiloxy)titane with trimethylacetoxysilane gives a solid condensation product, which is similar to that obtained by the reaction of $(iso\text{-}C_3H_7O)_4Ti$ with excess trimethylacetoxysilane [574, 575].

On correlating these data it is possible to draw only one conclusion, namely, that there is specific replacement of trimethylsiloxy groups at a titanium atom by acetoxy groups, which proceeds according to scheme (4-90) or (4-91):

$$\gt TiOSiR_3 + (CH_3CO)_2O \longrightarrow \gt TiOCOCH_3 + CH_3COOSiR_3 \qquad (4\text{-}91)$$

The reaction of tetrakis(trimethylsiloxy)titane and trimethylsiloxyisopropoxytitane with "chelating" compounds (β-diketones or 8-hydroxyquinoline) is described, for example, by the following scheme [577, 1202, 1203]:

$$[(CH_3)_3SiO]_n Ti(OC_3H_7\text{-iso})_{4-n} + 2(RCO)_2CH_2 \longrightarrow [(CH_3)_3SiO]_2Ti\left[\begin{matrix} O-C(R) \\ \quad\;\; \backslash\!\backslash CH \\ O=C(R) \end{matrix}\right]_2$$

$$(n = 2-4;\ R = CH_3,\ C_6H_5) \qquad (4\text{-}92)$$

This indicates that in mixed trimethylsilyl alkyl esters of orthotitanic acid the isopropoxy groups are more reactive toward the "chelating" agent than the trimethylsiloxy groups. The replacement of only two of the four trimethylsiloxy groups in tetrakis(trimethylsiloxy)titane by the chelating group is explained by the coordination number 6 of titanium.

The reactivity of the alkoxyl groups in internal complex compounds of titanium is not sufficiently clear. According to the data of Takimoto and Rust [1203], bis(2-ketopent-3-en-4-oxy)diisopropoxytitane does not react with excess trimethylacetoxysilane. The interpretation of this is that the mechanism of the condensation according to scheme (4-89) includes the formation of a coordination complex, which cannot exist for a dichelate derivative, in which the titanium atom is coordinationally saturated. Steric hindrance is put forward as another possible reason.

The work of Breed and Haggerty [577, 578] indicates the second version. In actual fact, refluxing a mixture of bis(8-quinolyloxy)diisopropoxytitane with dimethylphenylacetoxysilane in toluene for 3 hr and subsequently removing the solvent by heating for 6 hr at 160-186°C and keeping the mixture in vacuum it is possible to obtain a quantitative yield of bis(8-quinolyloxy)bis(dimethylphenylsiloxy)titane. In the case of the bis-acetylacetone derivative it was possible to observe the liberation of isopropyl acetate, contrary to the data in [1047]. However, it was not formed as a result of condensation, but as a result of the thermal decomposition of the starting titanium compound.

The reactions of bis(triorganosiloxy)dialkoxytitanes with diorganodiacetoxysilanes proceed according to the heterofunctional condensation scheme [1048, 1132, 1504, 1658]:

$$n\,(R_3SiO)_2\,Ti\,(OR')_2 + nR_2Si\,(OCOCH_3)_2 \longrightarrow$$
$$\longrightarrow R'O\,[Ti\,(OSiR_3)_2\,OSiR_2O]_n\,COCH_3 + (2n-1)\,CH_3COOR' \qquad (4\text{-}93)$$
$$(R = CH_3,\ C_6H_5;\ R' = \text{iso-}C_3H_7,\ \text{n-}C_4H_9)$$

The condensation is catalyzed by sodium ethylate [1048]. Depending on the nature of the substituents at the silicon atoms and the degree of condensation, the process yields viscous liquid [1658] or waxy [1504] soluble polymers, which lose their solubility and fusibility on heating, evidently due to the development of condensation involving the R_3SiO groups.

If one of the compounds participating in the reaction (4-93) is replaced by a trifunctional reagent, i.e., a triorganosiloxytrialkoxytitane or organotriacetoxysilane, respectively, then the condensation is accompanied by crosslinking and brittle polymers are formed with a high softening point and a high thermal stability [1658]. The reaction of bis(trimethylsiloxy)diisopropoxytitane with

TABLE 65. Character of Polymers Formed by Condensation of $Ti(OR)_4$ with $CH_3(R')Si(OCOCH_3)_2$ [145]

R	R′	Polymer			Yield of CH_3COOR, %
		Si, %	Ti, %	mol. wt. (cryoscopically in C_6H_6)	
C_2H_5	CH_3	5.42	29.87	980	90.9
C_2H_5	C_6H_6	6.88	25.80	1278	93.8
C_4H_9-*n*	C_6H_5	7.08	24.50	3318	91.0

acetic anhydride yields polybis(trimethylsiloxy)titoxane* which is a solid polymer melting above 300°C without decomposition and which has fiber- and film-forming properties [1048, 1132, 1460]:

$$n\,(R_3SiO)_2\,Ti\,(OR')_2 + n\,(RCO)_2\,O \longrightarrow R'O\,[Ti\,(OSiR_3)_2\,O]_n\,COCH_3 + (2n-1)\,CH_3COOR' \quad (4\text{-}94)$$

Adding to the reaction mixture tris(trimethylsiloxy)isopropoxytitane (molar ratio 9 : 10 : 1) yields a viscous polymer "blocked" with terminal trimethylsiloxy groups [1661].

Segal and his co-authors [1132] also reported the possibility of preparing polymers (with consistencies of viscous liquids to rigid solids) with the structure

$$\text{iso-}C_3H_7O\left[-\overset{\cap}{\underset{\cup}{Ti}}-O-(-SiR_2O-)_m-\right]_n COCH_3$$

[R = CH_3 or n-C_4H_9; m = 1 or 0; ∩ is a chelate group (a derivative of 8-hydroxyquinoline, acetylacetone, or benzoylacetophenone)]

Heterofunctional polycondensation with the formation of compounds of this type was described in more detail in the work of Andrianov and Pichkhadze [53, 202]. Heating bis(2-ketopent-3-en-4-oxy)dibutoxytitane with methylphenyldiacetoxysilane (equimolar ratio) at 165°C for 2 hr liberates 95% of the theoretically possible amount of butyl acetate and gives a 95% yield of a solid brown polymer which is readily soluble. The condensation of bis(8-quinolyloxy)dibutoxytitane with $CH_3(R)Si(OCOCH_3)_2$ proceeds analogously.

* The basic principles of a rational nomenclature for organic compounds of titanium were presented in [25].

The yield of butyl acetate exceeds 90% in all cases, but when R = C_6H_5 the degree of conversion and the relative viscosity of the copolymer are somewhat lower than when R = CH_3 [203].

Distillation of the products from the reaction of an equimolar mixture of $Ti(OR)_4$ (R = CH = CH_2, n-C_4H_9) with $CH_3(R')Si(OCOCH_3)_2$ (R', = Ch_3, C_6H_5) yields 90-94% of the theoretically possible amount of CH_3COOR. At the same time, $CH_3(R')Si(OR)_2$ and a viscous soluble polymer are formed (Table 65).

An analogous reaction of compounds with R = n- C_4H_9 and R' = CH_3 (molar ratio 1 : 1) was reported in 1961 [463]. However, the mixture thickened during the removal of the volatile condensation products by distillation and the final polymer was solid (its composition was not described). With an increase in the molar ratio of $Ti(OC_4H_9\text{-n})$: $(CH_3)_2Si(OCOCH_3)_2$ to 2 : 1 – 4 : 1, in addition to butyl acetate and dimethyldibutoxysilane, viscous orange liquids were formed, which consisted of polytitoxanes of the type $(n\text{-}C_4H_9O)_{2n+2}Ti_nO_{n-1}$.

According to patent data [1720], the reaction of a mixture of dibutoxydiphenoxytitane and dimethyldiethoxysilane with glycol (molar ratio 7 : 3 : 4) in xylene and subsequent removal of the solvent and alcohols in vacuum yields a light yellow, readily soluble resin. It has a structure which includes the fragments $TiOCH_2CH_2OSiO$.

A patent [1423] describes the preparation of varnishes and air-drying paints based on combinations of organoalkoxysilanes $R_nSi(OR')_{4-n}$(R = CH_3, C_2H_5, C_6H_5; n = 1 – 3) with esters of orthotitanic acid, including "stabilized" esters. On the other hand, such mixtures (essentially, condensation titanasiloxanes) are recommended as catalysts for the cold hardening of hydroxyl-containing siloxane oligomers [1384].

13.1.3. Reaction of Chloro Derivatives of Titanium with Alkoxysilanes and Acyloxysilanes

According to the data of Andrianov and his co-workers [144], the exothermic reaction of trialkylalkoxysilanes with titanium tetrachloride may be described by the equation

$$n\,(CH_3)_3\,SiOR + TiCl_4 \longrightarrow n\,(CH_3)_3\,SiCl + (RO)_n\,TiCl_{4-n} \qquad (4\text{-}95)$$

The nature of the radical R is not of decisive importance here. However, n never exceeds 2, i.e., it is not possible to replace more than two chlorine atoms at titanium by alkoxyl groups, even with prolonged boiling of a mixture of reagents in a molar ratio of 3 : 1. The yield of mono- and dihalo esters of orthotitanic acid is 80-100%. The formation of titanasiloxanes by the reaction of trialkylalkoxysilanes with titanium tetrachloride has not been observed.

Bradley and Hill [568] report that in the reaction of ethoxychlorosilanes $Cl_xSi(OC_2H_5)_{4-x}$ (x = 0 – 3) with titanium tetrachloride at 0°C, ethoxyl – chlorine exchange proceeds to a lesser extent, the higher the value of x. There is predominantly the formation of ethoxytrichlorotitane. It is surmised that the process is connected with the formation of the complex

$$\overset{\delta+}{Ti}\cdots\overset{\delta-}{Cl}\cdots Si,\quad Ti\cdots \underset{\underset{R}{|}}{O}\cdots Si$$

with subsequent exchange due to the high affinity of the titanium atom for nucleophilic atoms and, correspondingly, the higher electrophilicity of the silicon atom. At the same time, the reaction of titanium tetrachloride with tetraethoxysilane (molar ratio 1 : 1) is also accompanied by the formation of polymeric products. The following explanation is given for this:

$$\gt\overset{\delta+}{Ti}-\overset{\delta-}{Cl} \longrightarrow \underset{CH_2}{\overset{CH_3}{|}}-\overset{\delta+}{O}\cdots\overset{\delta-}{Si}\lt \longrightarrow \gt\overset{+}{Ti}\cdots Cl-\underset{CH_2}{\overset{CH_3}{|}}\cdots O=\overset{-}{Si}\lt \qquad (4\text{-}96)$$

Thus, the attack of the chlorine atom is directed at the α-carbon atom and not the silicon atom. In the opinion of the authors, there is then liberation of ethyl chloride, while the positive charge of the titanium is "neutralized" as a result of the capture of a chlorine atom or ethoxyl group from the negatively charged silicon part of the complex, which rearranges in its turn into a siloxane polymer. However, the formation of titanasiloxane polymers may equally probably be the result of the decomposition of the intermediate complex (4-96). In this connection we should mention patent data on the formation of titanasiloxanes in the reactions of polyfunctional alkylalkoxysilanes with $TiCl_4$ [74, 1333, 1631]:

$$4\,(RO)_3\,SiR' + TiCl_4 \longrightarrow [R'\,(RO)_2\,SiO]_4\,Ti + 4RCl \quad (4\text{-}97)$$

$$(R = C_2H_5;\ R' = C_2H_5\ [1333],\ C_6H_{13},\ C_{10}H_{21},\ C_2H_5O\ [74])$$

The condensation of bis(trimethylsiloxy)dichlorotitane with $RR'Si(OR'')_2$ (where $R = CH_3$; $R' = C_6H_5$; $R'' = C_2H_5$, n-C_4H_9) at 150°C is accompanied by the liberation of trimethylchlorosilane and a trimethylalkoxysilane and the formation of a viscous polymer [191]. In analyzing the process on the example of methylphenyldiethoxysilane, the authors divided it into a series of reactions:

1) The action of atmospheric moisture on bis(trimethylsiloxy)-dichlorotitane forms HCl and a partial hydrolysis product:

$$(R_3SiO)_2\,TiCl_2 \xrightarrow[-HCl]{+H_2O} (R_3SiO)_2\,Ti\,(OH)\,Cl \quad (4\text{-}98a)$$

2) This product condenses with methylphenyldiethoxysilane:

$$\geq TiOH + C_2H_5OSi\leq \longrightarrow\ \geq TiOSi\leq + C_2H_5OH \quad (4\text{-}98b)$$

3) The hydrogen chloride splits off a $(CH_3)_3SiO$ group from the initial condensation product and this gives a OH group at the titanium atom:

$$(R_3SiO)_2\,Ti\,(Cl){-}O{-}SiRR'\,(OC_2H_5) + HCl \longrightarrow R_3SiO\overset{Cl}{\underset{OH}{Ti}}{-}O{-}\overset{R}{\underset{R'}{Si}}OC_2H_5 + R_3SiCl \quad (4\text{-}98c)$$

4) The alcohol reacts with trimethylchlorosilane:

$$R_3SiCl + C_2H_5OH \longrightarrow R_3Si{-}O{-}C_2H_5 + HCl \quad (4\text{-}98d)$$

5) The alcohol also reacts with the primary condensation product:

$$(R_3SiO)_2\,Ti\,(Cl){-}O{-}SiRR'\,(OC_2H_5) + C_2H_5OH \longrightarrow$$
$$\longrightarrow C_2H_5O\,(R_3SiO)_2\,Ti{-}O{-}SiRR'\,(OC_2H_5) + HCl \quad (4\text{-}98e)$$

6) The primary condensation products, modified by replacement of trimethylsiloxy groups by OH or chlorine by C_2H_5O, then condense by a scheme of type (4-98b) with growth of the titanasiloxane chain.

In addition, we have already pointed out above (p. 395) that the reaction of bis(trimethylsiloxy)dichlorotitane with water is accompanied by the liberation of 66% of the theoretical amount of trimethylchlorosilane and the formation of an insoluble powdery substance. When dry HCl is passed through bis(trimethylsiloxy)dichlorotitane for four hours, 45% of trimethylchlorosilane is liberated. These facts still do not make the proposed condensation scheme sufficiently convincing. Likewise, it is confirmed by the structure of the polymer formed by condensation of bis(trimethylsiloxy)dichlorotitane with methyl(phenyl)diethoxysilane, which was put forward by the authors [191] only on the basis of partial analysis data:

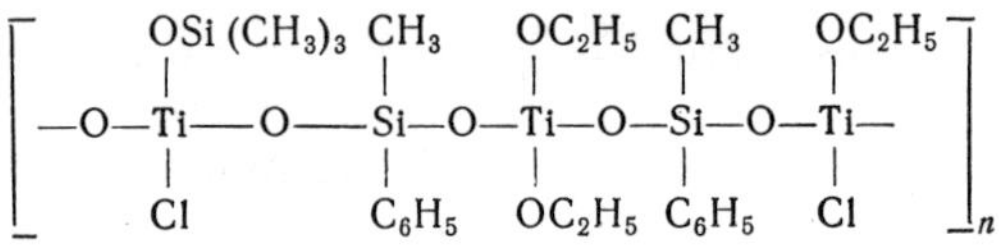

$$\left[-O-\underset{Cl}{\overset{OSi(CH_3)_3}{Ti}}-O-\underset{C_6H_5}{\overset{CH_3}{Si}}-O-\underset{OC_2H_5}{\overset{OC_2H_5}{Ti}}-O-\underset{C_6H_5}{\overset{CH_3}{Si}}-O-\underset{Cl}{\overset{OC_2H_5}{Ti}}- \right]_n$$

Such a structure must be confirmed not only by determination of the ethoxyl groups and data on the separate determination of titanium and silicon (which were not given in the paper), but also by demonstration of the nature of the substitutents at the silicon and titanium atoms and the order of alternation of the chains. The data in [191] only indicate the probability of the titanosiloxane character of the copolymers formed and the high reactivity of the trimethylsiloxy groups in bis(trimethylsiloxy)dichlorotitane under the reaction conditions.

When titanium tetrachloride is added to excess trimethylacetoxysilane [150], a white precipitate is formed, which is soluble only in hot alcohol and contains titanium, silicon, hydrolyzable chlorine, and acetoxy groups. On the basis of analytical data, the two following structures are apparently equally probable:

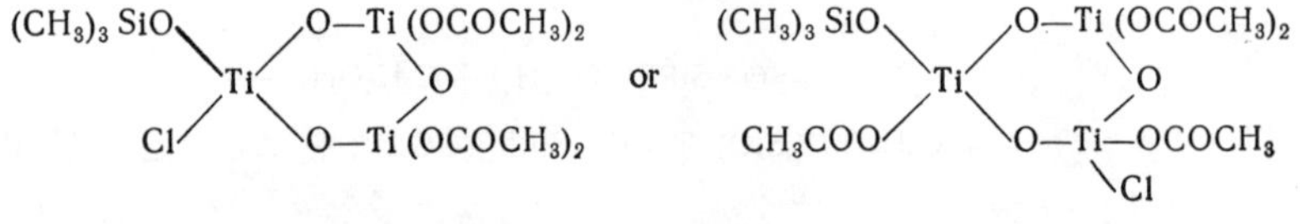

$$\begin{matrix} (CH_3)_3SiO & & O-Ti(OCOCH_3)_2 \\ & Ti & \quad O \\ Cl & & O-Ti(OCOCH_3)_2 \end{matrix} \quad \text{or} \quad \begin{matrix} (CH_3)_3SiO & & O-Ti(OCOCH_3)_2 \\ & Ti & \quad O \\ CH_3COO & & O-Ti-OCOCH_3 \\ & & \quad Cl \end{matrix}$$

Other products of the reaction are acetyl chloride and trimethylchlorosilane. This clearly indicates the presence of competing exchange and condensation reactions with the formation of Ti–O–Si groups and possibly Si–O–Si.

13.1.4. Reactions of Silanols with Esters of Orthotitanic Acid

The reaction of triorganosilanols with tetraalkoxytitanes may be represented by the general scheme

$$nR_3SiOH + Ti(OR')_4 \longrightarrow (R_3SiO)_n Ti(OR')_{4-n} + nR'OH$$
$$(R = \text{alkyl or aryl}; R' = CH_3, C_2H_5, \text{iso-}C_3H_7, \text{n-}C_4H_9; n = 1-4) \quad (4\text{-}99)$$

Tetrakis(triorganosiloxy)titanes (n = 4) were apparently prepared for the first time in this way by Yakovlev [509], according to whose data the yield of $[(C_2H_5)_3SiO]_4Ti$ in the reaction of excess triethylsilanol with $Ti(OC_2H_5)_4$ reaches 60%. Although the author carried out his investigations at the beginning of the 50s, the work was published only in February 1959. At the same time there appeared in the press not only the results of several analogous searches [141, 277, 413, 414, 573, 575, 651, 1289], but also an author's certificate [1466]. The number of publications on this subject has subsequently increased still further [50, 51, 415].

The reactions (4-99) may be accompanied by side processes of homocondensation of the trialkylsilanol and its reaction with the alcohol liberated. To suppress these processes and raise the yields of tetrakis(trialkylsiloxy)titane it is recommended, for example, that a solution of the silanol in benzene be introduced slowly into a benzene solution of the alkyl titanate under conditions such that there is continuous removal of the alcohol from the reaction zone by azeotropic distillation.* The reaction may be carried out in the absence of a solvent, but in this case the process proceeds more slowly and does not go to completion.

In the case of triphenylsilanol it is not essential to remove the alcohol from the reaction zone: a quantitative yield of tetrakis-(triphenylsiloxy)titane [1289] is aided by its low solubility in organic solvents and the low capacity of triphenylsilanol for homocondensation.

The mechanism of reaction (4-99), which is catalyzed by the corresponding sodium alcoholate (silanolate), is treated as a bimolecular nucleophilic substitution (S_N2) [51, 413–415]:

* It should be noted that the reaction of tetrakis(trimethylsiloxy)titane with butanol leads to the formation of $(n\text{-}C_4H_9O)_4Ti$ [575].

$$R'O-\overset{OR'}{\underset{OR'}{Ti}}-OR' \;\; (R_3SiO^- \rightarrow) \rightleftharpoons R_3SiOTi(OR')_3 + R'O^- \text{ etc.} \qquad (4\text{-}100)$$

The steric factor is apparently of great importance in this reaction. This is indicated by the fact that tris(trimethylsiloxy)-silanol does not react with $(C_4H_9O)_4Ti$ on prolonged heating [159].

Scheme (4-99) is supplemented by the reaction [575]:

$$4R_3SiOH + [(CH_3)_3SiO]_4Ti \xrightarrow{(CH_3)_3SiOSi(CH_3)_3} (R_3SiO)_4Ti + 4(CH_3)_3SiOH \qquad (4\text{-}99a)$$

The use of hexamethyldisiloxane as the solvent suppresses the condensation of the trimethylsilanol formed.

By carrying out the reaction of triphenylsilanol with tetrabutoxytitane in a molar ratio of 4 : 1, Zeitler and Brown [1289] observed the formation of triphenylbutoxysilane. With a ratio lower than 4 : 1 no triphenylsiloxyalkoxytitanes were detected in the reaction products [51, 415, 1289]. This is apparently explained again by the insolubility of tetrakis(triphenylsiloxy)titane. By varying the ratio of the components and the character of the alkyl radicals in them, by means of reaction (4-99) it was possible to obtain a whole series of mixed silyl alkyl esters of orthotitanic acid [50, 141, 404, 651, 1488].

The mechanism of the formation of bis(triethylsiloxy)dibutoxytitane from triethylsilanol and tetrabutoxytitane in a molar ratio of 2 : 1 (the reaction was carried out without a catalyst) has been explained [404] by the intermediate formation of a donor-acceptor complex, containing a hexavalent titanium atom:

$$\begin{array}{ccccc} H & R'O & & OR' & H \\ | & \diagdown & & \diagup & | \\ R_3SiO & \longrightarrow & Ti & \longleftarrow & OSiR_3 \\ & \diagup & & \diagdown & \\ & R'O & & OR' & \end{array}$$

Since triethylsilanol is part of the complex, its capacity for homocondensation appears only if the ratio $(C_2H_5)_3SiOH : Ti(OC_4H_9)_4$ is > 2. In the opinion of the authors, hexaethyldisiloxane may form an unstable complex with bis(triethylsiloxy)dibutoxytitane, which disproportionates to triethylbutoxysilane and tetrakis(triethylsiloxy)titane. It is regarded as confirmation of this that tetrakis(triethylsiloxy)titane cannot be obtained by the reaction of triethylsilanol with butyl titanate in a ratio of 4 : 1. This point of view is unwieldy and unjustified, particularly as Zeitler and Brown [1289] obtained under analogous conditions a 56% yield of tetrakis(trimethylsiloxy)titanium using trimethylsilanol, which is less stable than triethylsilanol and shows a higher tendency for complex formation.

In the reaction of triphenylsilanol with the product from hydrolytic condensation of butyl titanate $[TiO_{1.33}(OC_4H_9\text{-}n)_{1.33}]_{13.9}$ (refluxing in toluene), the replacement of the butoxyl groups is accompanied by cleavage of the Ti−O−Ti groups with the result that tetrakis(triphenylsiloxy)titane is obtained [1290]. It is also formed by the reaction of chlorobutoxytitane $(n\text{-}C_4H_9O)_{1.55}TiCl_{2.45}$ with triphenylsilanol.

In this case there is not only replacement of butoxyl groups and chlorine atoms at titanium atoms, but also more complex processes since, for example, the tetrakis(triphenylsiloxy)titane obtained contains some titanium dioxides.

Triorganosilanols are similar to triorganoacetoxysilanes in their action on "stabilized" esters of orthotitanic acid [209, 577, 578]:

$$(\subset)_2 Ti(OR)_2 + 2R'_3SiOH \xrightarrow{50-150^\circ} (\subset)_2 Ti(OSiR'_3)_2 + 2ROH \qquad (4\text{-}101)$$

where R = iso-C_3H_7, n-C_4H_9; R' = CH_3, C_2H_5, C_6H_5; the sign $\subset$ corresponds to

O, N; or HC, C=O, C—O, C_6H_5, C_6H_5

At the same time in the reaction of triphenylsilanol with bis-(2-ketopent-3-en-4-oxy)diisopropoxytitane there is replacement not only of isopropoxyl radicals, but also the acetylacetonate groups and the formation of tetrakis(triphenylsiloxy)titane [577, 578]. On the other hand, the reaction of triorganosilanols with a mixed ester of orthotitanic and dimethylphosphinic acids proceeds peculiarly [186]:

$$[(CH_3)_2P(O)O]_2Ti(OC_4H_9\text{-n})_2 + 2R_3SiOH \longrightarrow$$

$$\longrightarrow \{[(CH_3)_2P(O)O]_2TiO\}_n + R_3SiOC_4H_9\text{-n} + \text{n-}C_4H_9OH + R_3SiOSiR_3 \quad (4\text{-}102)$$

The polymer obtained is close in character to the product of the hydrolysis of bis(dimethylphosphonyl)dibutoxytitane. The fact that it is the dimethylphosphinic acid group which is mainly responsible for this course of the reaction is confirmed by the following example:

$$[(C_2H_5)_3SiO]_2Ti(OC_4H_9\text{-n})_2 + 2(CH_3)_2P(O)OH \longrightarrow \{[(CH_3)_2P(O)O]_2TiO\}_n +$$

$$+ (C_2H_5)_3SiOC_4H_9\text{-n} + \text{n-}C_4H_9OH + (C_2H_5)_3SiOSi(C_2H_5)_3 \quad (4\text{-}102a)$$

The selectivity of triorganosilanols toward alkoxyl groups at a titanium atom is illustrated by a reaction described in a patent [1668]:

$$[-(RO)Ti(OOCR')O-]_n + nR''_3SiOH \longrightarrow [-(R''_3SiO)Ti(OOCR')-]_n + nROH$$

(R = for example, iso-C_3H_7; R' is a long-chain aliphatic radical such as stearyl)

When an equimolar mixture of poly(isopropoxy) (stearoxy)titoxane and diphenylsilanediol is heated in cyclohexane at 69°C, isopropanol–cyclohexane azeotrope distills and a viscous oil, which thickens on standing, is formed.

The reaction of diphenylsilanediol with tetrabutoxytitane (2 : 1) in ether finally leads to the formation of a crystalline substance with m.p. 314°C and a molecular weight of 1320-1510, which contains 3.2% Ti and 13.7% Si. On the basis of these and spectroscopic data the authors [1290] favored a spirocyclic structure:

```
R2Si—O—SiR2—O\        /O—SiR2—O—SiR2
 |             \      /              |
 O              Ti                   O     (R = C6H5)
 |             /      \              |
R2Si—O—SiR2—O/        \O—SiR2—O—SiR2
```

Calculated values for this structure: 2.9% Ti, 13.5% Si, mol. wt. 1666.

From the work of Andrianov and his co-workers [11, 133, 167, 213] it follows that in the condensation of butyl titanate with α, ω-polyorganosiloxanediols the number of butoxyl groups participating in the reaction is regulated by the reagent ratio. In this case it is possible to obtain both polymers with reactive groups of type (I) [167] and also tetrakis derivatives (II) [133, 193, 213]:

$$HO\{[Si(CH_3)_2O]_n Ti(OC_4H_9\text{-n})_2O\}_m C_4H_9\text{-n} \quad [HO(RR'SiO)_x]_4 Ti$$

I II

In the first case, starting from a siloxanediol with mol. wt. 2600 it was possible to obtain an elastic polymer with mol. wt ~2700, which was soluble in aromatic hydrocarbons. The synthesis of such a polymer requires relatively drastic conditions (heating for 3 hr at 150°C in a nitrogen atmosphere and then for 30 min at 200°C and 15 mm). A substance with x=2, $R=CH_3$, $R'=C_6H_5$, and mol. wt. 1256 (calculated 1202) is formed as a result of stirring a mixture of tetrabutoxytitane with $HO(CH_3)(C_6H_5)SiOSi(C_6H_5)(CH_3)OH$ in a ratio of 1 : 4 for 18 hr at 40°C (1 mm). When x=3 the discrepancy between the molecular weight calculated for the condensation product (1888) and that found (1751) is quite high. When $R=R'=CH_3$, oligomers are obtained with values of x of 6, 7, 8, 10, and 13. It is necessary to remove the alcohol from the reaction medium as otherwise there is more extensive conversion, including the liberation of TiO_2.

The condensation of siloxanediols with esters of orthotitanic acid of the type $R_2Ti(OC_4H_9)_2$, where R is a triorganosiloxy or a chelate group has been described [50, 53, 206, 404]. It was found that in the condensation of α, ω-polydimethylsiloxanediols with bis(triethylsiloxy)dibutoxytitane there must be five times the calculated amount of the latter for complete binding of all the OH groups [404]. This is explained by the fact that water which is inevitably present in the system produces hydrolysis of the mixed ester and the siloxanediol condenses not with this ester, but with $R'O[(R_3SiO)_2TiO]_nR'$ (n=2−3). Therefore, the condensation product has a structure, which consists of alternating siloxane and titoxane units with connecting Ti−O−Si groups. A copolymer based on a polydimethylsiloxanediol with mol. wt. 30,000 has a ratio of Si : Ti = 400 : 3. In the case of a diol with mol. wt. 2400, this ratio was 32 : 2.

Bis-chelate derivatives react with $HO[(CH_3)_2SiO]_nH$ or with $(CH_3)[(CH_3)_2SiO]_mH$ (100−200°C in vacuum) as bifunctional compounds with the formation of reactive or blocked linear polymers [44, 53, 206]. In contrast to the reactions with triphenylsilanol, the chelate group is not eliminated in this case either with the 8-hydroxyquinoline derivative or with the acetylacetone stabilized ester. The reaction of 8-quinolyloxytributoxytitane with polyorganosilan(di)ols is analogous, but the process occurs under much milder conditions (70°C, benzene medium) [44, 196, 197]:

$$\left(C_9H_6NO\right)_x Ti(OR)_{4-x} + (4-x)\,HO[(CH_3)_2SiO]_pR' \longrightarrow$$

$$\longrightarrow (4-x)\,ROH + \left(C_9H_6NO\right)_x Ti\{O[(CH_3)_2SiO]_pR'\}_{4-x} \quad (4\text{-}101a)$$

There are many reports, largely in patents [1165, 1166, 1167, 1327, 1378, 1410, 1413, 1439, 1584, 1590, 1608, 1646, 1675, 1696, 1702, 1706, 1710, 1712] that the addition of various esters of orthotitanic acid (including mixed esters) to polysiloxane resins and varnishes markedly accelerates their hardening and drying, making it possible to lower the "baking" temperature from the normal range of 200-250°C to 50-140°C. At the same time there is an increase in the mechanical strength of the coatings. As a rule, the amount of the titanium-containing compound is quite high (10-50 wt.%) therefore, even in the first communications on this subject [1706, 1710], the improved properties of the combined products are explained by the incorporation of titanium into the siloxane structure through the condensation reaction:

$$\equiv Si{-}O\,\boxed{H + RO}{-}Ti\equiv \longrightarrow \equiv Si{-}O{-}Ti\equiv + ROH$$

A direct confirmation of this point of view is provided by the formation of titanasiloxane polymers by heating the products from the hydrolysis of alkyl(aryl)chlorosilanes with tetrabutyl titanate [153, 178]. In the initial stage, soluble liquid polymers are obtained in toluene at 80°C in 4-5 hr. Heating at 200°C promotes the further condensation of these polymers, but the glassy polymers obtained retain their solubility. An analogous reaction for $(C_2H_5O)_4Ti$ is described in the patent [1378].

Esters of orthotitanic acid, particularly tetraethoxytitane are excellent catalysts in the waterproofing of cloth, fibers, paper, leather and other materials of organic origin with nonaqueous solutions of many organosilicon water-proofing agents [22, 23, 247,

252, 303, 304, 365, 1331, 1333, 1341, 1463, 1578, 1603, 1626, 1637, 1703, 1704]. In this case the consumption of catalyst is usually low (down to 0.01%). Apparently it not only accelerates the coupling of silanol groups, but also their bonding to hydroxyl groups of the material treated.

The process of "cold vulcanization," i.e., the conversion of linear liquid and elastomeric polysiloxanes into rubbery materials under the action of $(RO)_4Ti$ [231–233, 400, 403, 405, 406, 1350, 1486, 1689, 1711, 1716], proceeds in room temperature at a considerable rate and does not require the presence of alkoxy derivatives of silicon, as in the case of organotin hardeners. In the case of polydiorganosiloxanes containing $OSiR_3$ terminal groups, as might have been expected, this reaction does not proceed in the cold, though compounds containing R_3SiO groups at titanium atoms participate in it [1716].

From the point of view of Baranovskaya and other investigators [231, 233], the process consists of condensation between terminal $\gt Si{-}OH$ groups of siloxane polymers and $\gt Ti{-}OH$ groups formed as a result of hydrolysis of esters of orthotitanic acid by moisture present in the system or in the atmosphere. The titanium derivative participates in the reaction as a polyfunctional compound with the result that cross-linking processes (more accurately, interchain interaction) develop rapidly.

In the opinion of Nudel'man [405] this scheme is inaccurate since soluble polymers are formed by the condensation of α,ω-polydiorganosiloxanediols with tetrabutyl titanate. Thus, $(RO)_4Ti$ reacts as a bifunctional compound in the first stages of the process, increasing the weight of the linear polymers. It is surmised that the condensation proceeds with the intermediate formation of complexes of two molecules of siloxanediols with $(RO)_4Ti$ with the titanium showing the coordination number 6.

13.1.5. Reactions of Silanols with Titanium Tetrachloride

The synthesis of tetrakis(trimethylsiloxy)titane from trimethylsilanol and titanium tetrachloride in the presence of ammonia, which was achieved in 1954 [681], was the first example of the preparation of a monomeric compound containing the Ti–O–Si group:

$$4R_3SiOH + TiCl_4 \xrightarrow[-4HCl]{\text{acceptor}} (R_3SiO)_4Ti \quad (4\text{-}103)$$

However, the yield of tetrakis(trimethylsiloxy)titane was only 18% and therefore no further attempts were made to prepare it by the method of English and Sommer. Tetrakis(methyldiphenylsiloxy)-titane is obtained in 36% yield under the same conditions [157] and tetrakis(triphenylsiloxy)titane is obtained almost quantitatively [1289]. This indicates a direct connection between the yield of the tetrakis derivative and the capacity of the triorganosilanol for homocondensation. The nature of the HCl acceptor is also important. Thus, for example, when pyridine is used the yield of tetrakis(methyldiethylsiloxy)titane is only 39%. The addition to the reaction of more stable triethylsilanol raised the degree of conversion to 45%. At the same time, in the presence of dimethylaniline the yield of tetrakis(triethylsiloxy)titane increases to 89% [51, 278, 413].*

Up to now it has not been possible to obtain products of partial replacement of chlorine at the titanium atom by the reaction of triorganosilanols with titanium tetrachloride, evidently due to the fact that in $R_3SiOTiCl_3$ the Si–O–Ti group is quite vulnerable [158]. Thus, for example, the action of butyl alcohol and NH_3 on trimethylsiloxytrichlorotitane even in the cold yields tetrabutoxytitane (!). When the alcohol is replaced by a silanol the same picture is observed and in the case of methyldiphenylsilanol, tetrakis(methyldiphenylsiloxy)titane was isolated from the reaction products.

Reaction (4-103) may be extended to siloxanols of the type $(CH_3)_3Si[OSi(CH_3)_2]_nOH$ [193]. Tetrakis derivatives have been prepared with values of n of 23, 27, 47, 66, and 72. If the titanium tetrachloride in this reaction is replaced by bis(trimethylsiloxy)-dichlorotitane, then linear blocked oligomers are obtained with the central grouping $[(CH_3)_3SiO]_2Ti\langle$.

According to the data of Andrianov and his co-workers [131, 133, 155, 192] under mild conditions and in the presence of an HCl acceptor polydimethylsiloxanediols may participate in the reaction (4-103) as monofunctional compounds with the formation of viscous liquid oligomers $\{HO[(CH_3)_2SiO]_n\}_4Ti$, where n = 5 – 104. At the same

* It is reported in a patent [1724] that oily polymeric products are obtained by the reaction of $TiCl_4$ with triethylsilanol without an acceptor.

time, the reaction of titanium tetrachloride with diphenylsilanediol yields solid resinous products [1714].

The reaction of dimethylphenylsilanol with bis(8-quinolyloxy)-dichlorotitane proceeds in the presence of pyridine without touching the chelate groups [577, 578]:

$$\left(C_9H_6NO\right)_2 TiCl_2 + 2RR'_2SiOH \xrightarrow{C_5H_5N}$$

$$\longrightarrow \left(C_9H_6NO\right)_2 Ti\left(OSiR'_2R\right)_2 + 2C_5H_5N \cdot HCl \qquad (4\text{-}104)$$

13.1.6. Reactions of Halogen Derivatives of Titanium with Sodium Silanolates

The low yield of tetrakis(trimethylsiloxy)titane in its preparation by scheme (4-103) provided the first impetus for a search for more productive synthesis methods. It was found that when the sodium salt, i.e., sodium trimethylsilanolate, is used instead of the silanol the yield of tetrakis(trimethylsiloxy)titane is 60-70% [161, 179, 182]. At the same time [161] it was established that the reaction has a general character:

$$nR_3SiONa + TiCl_4 \longrightarrow (R_3SiO)_n TiCl_{4-n} + nNaCl \qquad (4\text{-}105)$$

With the ratios $(CH_3)_3SiONa : TiCl_4 = 3:1$ and $1:1$, reaction (4-105) gives partly substituted titanium chlorides – $[(CH_3)_3SiO]_n \cdot TiCl_{4-n}$ with n = 3 or 1. With a decrease in n the yield of useful compounds falls from 67% when n = 4 to 53% when n = 3 and 27% when n = 1. It is strange that it is not possible to obtain bis(trimethylsiloxy)dichlorotitane by the same method; this is ascribed [161] to its disproportionation under the action of impurities since this compound has been obtained by other methods (see below).

When n = 4 it is possible to carry out reaction (4-105) with other triorganosilanolates such as $(C_2H_5)_3SiONa$ [168, 179], $(CH_3)_2C_6H_5SiONa$ [157] and even $[(CH_3)_3SiO]_3SiONa$ [159]. The last example is most interesting since the reactivity of the OH group in tris(trimethylsiloxy)silanol is extremely low; for example, it does not react with tetrabutoxytitane, does not condense to the corresponding disiloxanes when heated in the presence of HCl, and does not react with titanium tetrachloride. Its reaction with sodium begins only above 150°C and the silanolate is formed in 18% yield, while the main product of the reaction is tetrakis(trimethylsiloxy)silane. However, the reaction of this silanolate with titanium tetrachloride gives a very bulky molecule, namely, tetrakis[tris(trimethylsiloxy)-siloxy]titane comparatively readily (57% yield).

It has already been pointed out above that the cohydrolysis of diorganodichlorosilanes with bis(cyclopentadienyl)dichlorotitane is accompanied by elimination of cyclopentadienyl groups [201]. To introduce bis(cyclopentadienyl)titoxane units into a siloxane structure the conditions were changed and the reaction was carried out between bis(cyclopentadienyl)dichlorotitane and sodium salts of siloxanediols, namely, $NaO[(CH_3)_2SiO]_nNa$, (n=2 or 4) [53, 201].* The yellow polymers obtained dissolved readily in toluene and xylene. However, it was found that as in the case of cohydrolysis, the condensation was accompanied by partial elimination of cyclopentadienyl groups from titanium atoms. The solubility of these polymers is explained only by the specific characteristics of their structure:

$$R_2Si\begin{matrix} O-[SiR_2-O]_{n-1} \\ O \end{matrix}Ti(-O-)-O-(SiR_2O)_{n-1}SiR_2O-Ti(C_5H_5)_2-$$

Gutmann and Meller [826] obtained tetrakis(triphenylsiloxy)-titane by the reaction of bis(cyclopentadienyl)dichlorotitane with sodium triphenylsilanolate in a ratio of 1 : 2 in toluene. However, Gidins [1664] found that elimination of cyclopentadienyl groups from a titanium atom is not the only result of the reaction: after separation of the difficultly soluble tetrakis(triphenylsiloxy)titane, he isolated from the filtrate the product of "normal" siloxylation:

$$(C_5H_5)_2TiCl_2 + R_3SiONa \rightarrow \begin{cases} (C_5H_5)_2Ti(OSiR_3)_2 \\ (R_3SiO)_4Ti \end{cases} \qquad (4\text{-}106)$$

* The compound $NaO[(CH_3)_2SiO]_4Na$ was named incorrectly in [201] as the disodium salt of 1,5- instead of 1,7-dihydroxyoctamethyltetrasiloxane.

Later Noltes and van der Kerk [983] selected conditions under which the yield of bis(triphenylsiloxy)bis(cyclopentadienyl)titane exceeded 60% (heating the mixture in toluene for 30 min at 110°C) while only 8% of tetrakis(triphenylsiloxy)titane was formed. Moreover; brief heating of the mixture at 70°C made it possible to obtain the product from the replacement of only one chlorine atom out of the two, i.e., bis(cyclopentadienyl)(triphenylsiloxy)chlorotitane.

The reaction of bis(8-quinolyloxy)dichlorotitane with sodium dimethylphenylsilanolate proceeds without complications and the yield of the mixed esters is 84% [577, 578].

The reaction of sodium salts of organosilanetriols (isolated beforehand or without special isolation) regardless of the reaction medium leads to solid glassy titanasiloxanes [14, 139, 1480] of the type

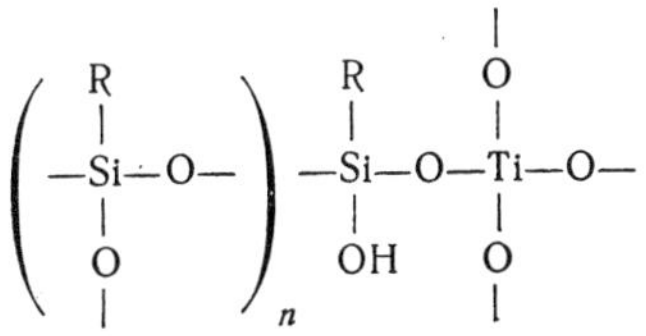

$$\left(\begin{array}{c} R \\ | \\ -Si-O- \\ | \\ O \\ | \end{array}\right)_n \begin{array}{c} R \\ | \\ -Si-O- \\ | \\ OH \end{array}\begin{array}{c} | \\ O \\ | \\ Ti-O- \\ | \\ O \\ | \end{array} \quad (R = CH_3,\ C_2H_5,\ C_6H_5;\ n = 0-19)$$

The best yield (80-90%) and a given Si : Ti ratio are attained when at the end of the synthesis the pH is 7 when the reaction is carried out in toluene or a mixture of it with butanol. The polymers obtained have a high solubility in various organic solvents, but they lose their solubility on heating and do not melt up to 500°C. In accordance with the properties the authors assign to these polymers the cyclolinear structure

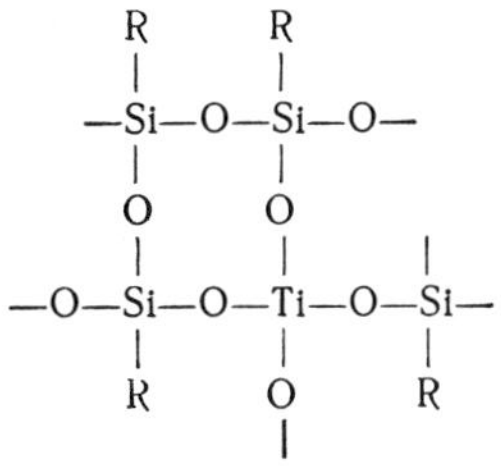

$$\begin{array}{ccccccc} & & R & & R & & \\ & & | & & | & & \\ & & -Si & -O- & Si & -O- & \\ & & | & & | & & \\ & & O & & O & & \\ & & | & & | & & | \\ -O- & & Si & -O- & Ti & -O- & Si- \\ & & | & & | & & | \\ & & R & & O & & R \\ & & & & | & & \end{array}$$

A description is given in a patent [1724] of a combined method of synthesizing solid resinous titanosiloxanes by the action of sodium triethylsilanolate and diphenylsilanediol on ethoxytrichlorotitane.

As has already been pointed out, triorganosiloxy derivatives of lead may "supply" R_3SiO groups to other elements. By scheme (4-107) [180, 181], which may be extended to reactions of silanolates with sufficient justification, tetrakis(triethylsiloxy)titane is obtained in 50% yield:

$$2\{2[(C_2H_5)_3SiO]_2Pb \cdot Pb(OH)_2\} + 3TiCl_4 \longrightarrow 2[(C_2H_5)_3SiO]_4Ti + 6PbCl_2 + Ti(OH)_4 \quad (4\text{-}107)$$

13.1.7. Cleavage of Siloxanes by Titanium Halides

When a mixture of hexamethyldisiloxane and titanium tetrachloride (molar ratio 2 : 1) is heated in the presence of anhydrous ferric chloride, cleavage of both Si–O bonds occurs [24, 51, 279, 414]:

$$2R_3SiOSiR_3 + TiCl_4 \xrightarrow{FeCl_3} 4R_3SiCl + TiO_2 \quad (4\text{-}108)$$

The yield of the trimethylchlorosilane formed exceeds 85%. However, in the absence of a catalyst prolonged heating of a mixture of $R_3SiOSiR_3$ with titanium tetrahalides makes it possible to replace in the latter only some of the halogen atoms by trialkysiloxy groups [24, 51, 190, 279, 414, 1465, 1485]:*

$$nR_3SiOSiR_3 + TiX_4 \longrightarrow (R_3SiO)_nTiX_{4-n} + R_3SiX \quad (n = 1-2) \quad (4\text{-}109)$$

It is preferable to carry out this reaction in the presence of AlX_3. In this case the yield of trimethylsiloxyhalotitanes is quantitative. With a molar ratio $R_3SiOSiR_3 : TiCl_4 = 1 : 1$ the main products are trialkylsiloxytrichlorotitane (n=1). With a ratio of 2 : 1, bis(trialkylsiloxy)dichlorotitanes (n=2) are obtained. The replacement of the second halogen atom by a trimethylsiloxy group is much more difficult than that of the first. Thus, when a mixture of $(CH_3)_3SiOSi(CH_3)_3$ and $TiCl_4$ (1 : 1) is heated for 3 hr, 94% of the theoretically possible amount of trimethylchlorosilane distills. The temperature of the reaction mixture thereupon rises to 155°C and vacuum distillation yields 73% of trimethylsiloxytrichlorotitane.

* Work published much earlier [279, 414] was not cited in [190] though the authors undoubtedly knew of this work since by a happy chance the list of literature cited in [190] included an incorrect reference to the authors certificate [1465].

At the same time, bis(trimethylsiloxy)dichlorotitane was obtained in 76% yield only by heating a mixture of the components in a ratio of 2 : 1 for 40 hr. The temperature of the reaction mixture rises to 190°C during the distillation of the trimethylchlorosilane.

When a mixture of vapor of hexamethyldisiloxane and titanium tetrachloride was passed through a quartz tube at 350°C the yield of bis(trimethylsiloxy)dichlorotitane was ~35%. Heating an equimolar mixture of the reagents for 1 hr at 120°C in the absence of any catalysts made it possible to obtain ~ 70% of trimethylsiloxytrichlorotitane [190].

It must be assumed that bis(trimethylsiloxy)dichlorotitane is obtained as a result of the reaction:

$$R_3SiOTiCl_3 + R_3SiOSiR_3 \longrightarrow (R_3SiO)_2TiCl_2 + R_3SiCl \qquad (4\text{–}110)$$

This point of view is well confirmed by the 43% yield of $[(CH_3)_3SiO]_2TiCl_2$ which is obtained by passing an equimolar mixture of $(CH_3)_3SiOTiCl_3$ and $(CH_3)_3SiOSi(CH_3)_3$ through a quartz tube at 320°C [190].

In the opinion of Andrianov and Kurasheva [190], the mechanism of the cleavage of hexamethyldisiloxane by titanium tetrachloride is connected with the formation of a transition coordination complex:

$$(\rangle Si)_2O + TiCl_4 \longrightarrow \left[(\rangle Si)_2O \cdots TiCl_4 \right] \longrightarrow (CH_3)_3SiCl + (CH_3)_3SiOTiCl_3 \qquad (4\text{–}110a)$$

According to other data [24], taking into account the catalytic effect of Lewis acids, the reaction proceeds through the intermediate formation of a cyclic active complex:

$$\rangle Si\text{–}O(\text{–}Si)\cdots TiX_3 \cdots X \cdots MX_2 \cdots X \longrightarrow \rangle SiX + \rangle SiOTiX_3 + MX_3 \qquad (4\text{–}110b)$$

$(M = Al, Fe)$

Instead of an FeX_3 or AlX_3 molecule, the complex may contain a second molecule of TiX_4 so that there are no grounds for assuming that the noncatalytic decomposition proceeds in a different way.

In reaction (4-109) when n=1 and X=Cl the yield of trialkylsiloxytrichlorotitanes falls substantially with a change from methyl to ethyl radicals. On the other hand, the yield of the bromo derivative with n=1 appreciably exceeds the yield of the chlorine-containing analog (96 and 72%, respectively) [51, 414].

Trimethylsiloxytrichlorotitane decomposes with prolonged heating [24, 51, 279, 414]:

$$2R_3SiOTiCl_3 \longrightarrow 2TiOCl_2 + 2R_3SiCl; \quad 2TiOCl_2 \longrightarrow TiCl_4 + TiO_2 \qquad (4\text{-}111)$$

Under milder conditions (repeated vacuum distillation) the pyrolysis has a different character and is evidently a disproportionation reaction:

$$2R_3SiOTiCl_3 \longrightarrow (R_3SiO)_2TiCl_2 + TiCl_4 \qquad (4\text{-}111a)$$

In this connection we should point out the formation of bis(trimethylsiloxy)dichlorotitane by the symmetrization reaction [161]:

$$(R_3SiO)_4Ti + TiCl_4 \longrightarrow 2(R_3SiO)_2TiCl_2 \qquad (4\text{-}112)$$

Polydimethylsiloxanes are known to be depolymerized by titanium tetrachloride [1579]. Thus, the reaction of TiX_4 with hexaalkyldisiloxanes or with linear siloxane polymers leads to breakdown of the molecule due to rupture of Si—O bonds. In the case of cyclic siloxanes, ring opening occurs:

$$\left[\begin{matrix} \text{Si}< \\ | \\ \text{O} \\ | \\ \text{Si}< \end{matrix}\right. + \begin{matrix} \text{X} \\ | \\ \text{TiX}_3 \end{matrix} \longrightarrow \left[\begin{matrix} \text{Si}\lessgtr\text{X} \\ \\ \\ \\ \text{Si}\lessgtr\text{O}-\text{TiX}_3 \end{matrix}\right. \qquad (4\text{-}113)$$

The reaction proceeds at a temperature of the order of 170°C in the absence of catalysts [147, 200]. With a molar ratio $(R_2SiO)_n$: $TiCl_4$ = 1 : 1 (when $R = CH_3$, n=4 and when $R = C_2H_5$, n=3) the yield of the cleavage products $Cl(R_2SiO)_nTiCl_3$ is 20-25%. When this ratio

TABLE 66. Reaction of 1,3,5-Trimethyl-1,3,5-triphenylcyclotrisiloxane with $TiCl_4$ (molar ratio 1 : 1) [147]

Temperature, °C	Reaction time, hr	$TiCl_4$ conversion, % of initial	Yield of reaction products, %	
			$[CH_3(C_6H_5)SiO]_4$	Polymer
150	3.5	4	44	38.8
170	2.5	26	23.4	59.4
170	4.0	30	6.1	71.4

is changed to 2 : 1, in the case of octamethylcyclotetrasiloxane it is possible to isolate from the reaction mixture $Cl[(CH_3)_2SiO]_4TiCl_3$ (42%) and a substance which has the composition $C_{16}H_{48}O_8Si_8TiCl_4$ (8%). Two structures may be assigned to the latter:

$$Cl[(CH_3)_2SiO]_8TiCl_3 \text{ and } Cl[(CH_3)_2SiO]_4TiCl_2[OSi(CH_3)_2]_4Cl$$

The hydrolysis of this compound, which leads to the formation of a cyclic tetramer, indicates that it is bis(ω-chlorooctamethyltetrasiloxanoxy)dichlorotitane. Thus, the process proceeds in stages:

$$[(CH_3)_2SiO]_4 + TiCl_4 \longrightarrow Cl[(CH_3)_2SiO]_4TiCl_3 \xrightarrow{[(CH_3)_2SiO]_4} \{Cl[(CH_3)_2SiO]_4\}_2TiCl_2 \quad (4\text{-}114)$$

The character of the reaction of titanium tetrachloride with sym-trimethyltriphenylcyclotrisiloxane (1 : 1) is illustrated by the data in Table 66. The polymer obtained at 150°C corresponds in composition to $[CH_3(C_6H_5)SiO]_n$. At 170°C, regardless of the reaction time, a brown elastic polymer $C_{63}H_{72}O_{10.5}Si_9TiCl_4$ is formed.

In the reaction of methylhydrosiloxanes with titanium tetrachloride [979] a brownish red precipitate forms, indicating the conversion of the titanium into a lower valence state.* When olefins are introduced into this system they are polymerized. The nature of the products from the reaction of $TiCl_4$ with $(CH_3SiHO)_n$ has not been determined exactly, but it has been established that they contain trivalent titanium, chlorine, siloxane units, and Si–H bonds. There is either cleavage of siloxane bonds with the inser-

* Trialkylsilanes readily reduce $TiCl_4$ to $TiCl_3$ and $TiCl_2$ [254]. At the same time, methyldichlorosilane does not reduce titanium tetrachloride [979].

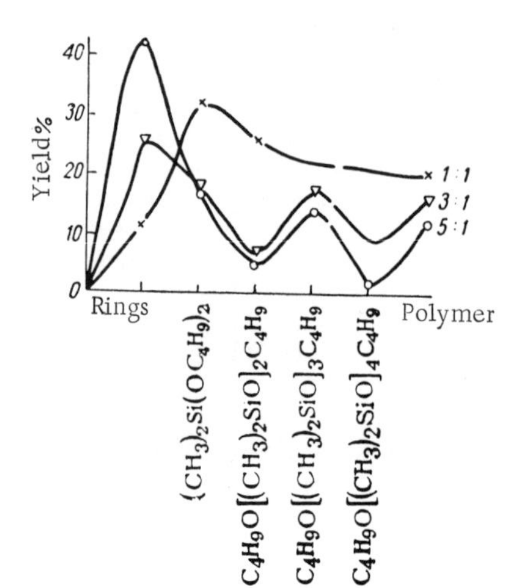

Fig. 8. Effect of the molar ratio of octamethylcyclotetrasiloxane and tetrabutoxytitane on the yield of reaction products [208].

tion of titanium into the polymer structure or there is complex formation and it is difficult to say which at present.

13.1.8. Reactions of Siloxanes with Esters of Orthotitanic Acid

The reaction of methylhydropolysiloxanes with tetrabutoxytitane is complex, like the reaction with titanium tetrachloride [979]. The reaction yields a dark blue solution, which also polymerizes olefins.

The reaction of cyclodialkylsiloxanes with tetrabutoxytitane is represented by the scheme [207, 208]:

$$(RR'SiO)_n + Ti(OR'')_4 \xrightarrow{180-200^\circ C} R''O(-RR'SiO-)_{n-3}R'' + R''O(-RR''SiO-)_{n-2}R'' + \text{polymer} \quad (4\text{-}115)$$

($n = 3$ or 4; $R = R' = CH_3$, C_2H_5 or $R = CH_3$ and $R' = CH{=}CH_2$; $R'' = C_4H_9\text{-n}$)

In the case of hexamethylcyclotrisiloxane the chemical composition of the polymer formed corresponds to the ratio Ti : Si = 1.78. The reaction with octamethylcyclotrisiloxane yields a polymer with Ti : Si = 1. This value is also maintained with an increase in the starting ratio of $[(CH_3)_2SiO]_4 : Ti)OC_4H_9\text{-n})_4$ to 3–5 : 1. However, in this case two molecules of octamethylcyclotetrasiloxane participate in the reaction with tetrabutoxytitane and a range of oligomers

$n-C_4H_9O[(CH_3)_2SiO]_mC_4H_9-n$ is formed with values of m = 1−4 (Fig. 8). The polymeric products are soluble in benzene and have $T_n \sim 100°C$.

In the opinion of Andrianov and his co-workers [208], the first stage of ring opening by tetrabutoxytitane is analogous to the action of titanium tetrachloride. The cleavage products of the type $n-C_4H_9O(RR'SiO)_nTi(OC_4H_9-n)_3$ obtained initially disproportionate with the formation of $RR'Si(OC_4H_9-n)_2$, α,ω-dibutoxypolysiloxanes, and titanasiloxane polymers of the composition $[-RR'SiOTi(O_{0.5})_2O-]_x$.

Cyclodialkylsiloxanes may be arranged in the following series with respect to their ease of cleavage by tetrabutoxytitane:

$$[(CH_3)_2SiO]_3 \gg [(CH_3)_2SiO]_4 > [CH_3(CH_2{=}CH)SiO]_4 > [(C_2H_5)_2SiO]_4$$

13.1.9. Reaction of Alkoxysilanes and Chlorosilanes with Dialkoxytitanones

In investigating compounds of the type $(RO)_2TiO$, Nesmeyanov and Nogina [394, 402] established that at least in dilute solutions they are monomeric, i.e., they have a double bond between titanium and oxygen atoms and are dialkoxytitanones. Addition occurs when these compounds are heated briefly with tetraalkoxysilanes:

$$m(RO)_2TiO + (RO)_4Si \longrightarrow [(RO)_3TiO]_mSi(OR)_{4-m} \qquad (4\text{-}116)$$
$$(R = n\text{-}C_3H_7,\ iso\text{-}C_4H_9;\ m = 1 \text{ or } 2)$$

The (triisobutoxysiloxy)triisobutoxytitane obtained by scheme (4-116) decomposed during an attempt to distill it at 1mm with the formation of tetraisobutoxytitane. The character of the second decomposition product, which had m.p. 194°C, was not established.

Compounds of the type $[(RO)_3TiO]_mSi(OR)_{4-m}$ are evidently strongly associated since determination of their molecular weights even by the isopiestic method, which makes it possible to work with very dilute solutions, gives values of m 1.5 times the calculated values.

It was found subsequently [395] that dialkyldialkoxysilanes may also be used in reaction (4-116). Thus, boiling an equimolar mixture of dipropoxytitanone with dimethyldipropoxysilane for 4 hr and distillation of the reaction products at 10^{-5} mm yielded

$$(n\text{-}C_3H_7O)_3Ti{-}O{-}Si(CH_3)_2OC_3H_7\text{-}n \quad \text{and} \quad [(n\text{-}C_3H_7O)_3TiO]_2Si(CH_3)_2$$

When an attempt was made to distill the first substance at 1 mm it decomposed with the formation of tetrapropoxytitane, apparently according to the scheme

$$n\,(RO)_3\,TiOSiR'_2OR \longrightarrow n\,(RO)_4\,Ti + (R'_2SiO)_n \qquad (4\text{-}117)$$

Treatment of dipropoxy(isobutoxy)titanone with chlorine formed titanium oxychloride, which could be isolated in the form of a complex with the composition $Cl_2TiO \cdot 2ROH$ [395, 402]. It was first suggested [402] that the nature of the addition of tetraalkoxysilanes to Cl_2TiO is the same as in the case of $(RO)_2Ti{=}O$.

However, it was found [395] that the reaction has a different character:

$$Cl_2TiO \cdot 2ROH + Si\,(OR)_4 \longrightarrow Cl_2Ti\,(OR)_2 \cdot ROH + ROH + 1/n\,[OSi\,(OR)_2]_n \qquad (4\text{-}118)$$
$$(R = n\text{-}C_3H_7)$$

According to the data of Nogina, Nesmeyanov, and Kudryavstev [402], silicon tetrachloride will also undergo a reaction by scheme (4-116)*

$$(n\text{-}C_3H_7O)_2\,TiO + SiCl_4 \longrightarrow Cl\,(n\text{-}C_3H_7O)_2\,TiOSiCl_3 \qquad (4\text{-}119)$$

The reaction of $(RO)_2TiO$ with diorganodichlorosilanes is similar to reactions of R_2SiCl_2 with ketones and oxides of divalent metals [1199, 1201]:

$$nR_mMO + nR'_2SiCl_2 \longrightarrow (R'_2SiO)_n + nR_mMCl_2 \qquad (4\text{-}120)$$
(when M = C, m = 2 and R = CH_3; when M = Zn or Mg, m = 0; when M = Ti, m = 2 and R = alkoxyl)

However, while low-molecular cyclosiloxanes (mainly with n=3 or 4) are formed by the reaction of diorganodichlorosilanes with ketones and metal oxides, the reaction of dipropoxytitanone with dimethyldichlorosilane yielded polydimethylsiloxanes with a mean molecular weight of 1800 ($n \simeq 24$), while in the case of $CH_3 \cdot (C_2H_5)SiCl_2$ the degree of polymerization was ~9.5.

* The reaction of dialkoxytitanones with $TiCl_4$ [395] yields polyhalogen-substituted titanones $Cl(RO)Ti{=}O$ and alkoxytrichlorotitanes.

An attempt to carry out reaction (4-116) with "stabilized" alkoxytitanones, i.e., with $(\subset)_2TiO$ † was unsuccessful. Breed and Haggerty [577, 578] are inclined to associate this with the different structures of dialkoxy and dichelate derivatives, believing, in particular, that chelates have the structure of dimers $[(\subset)_2TiO]_2$. Nonetheless, triphenylsilanol reacts with bis(8-quinolyloxy)titanone with the quantitative formation of

$$(C_9H_7NO)_2Ti[OSi(C_6H_5)_3]_2$$

Moreover, by prolonged boiling of trimethylethoxysilane with bis-(2-ketopent-3-en-4-oxy)titanone it was possible to synthesize $(C_5H_7O)_2Ti[OSi(CH_3)_3]_2$, though in low yield (10-20%).

13.2. Physical Properties

The physicochemical constants of individual compounds containing the Si–O–Ti group are given in Table 67.

An overwhelming majority of tetrakis(triorganosiloxy)titanes are colorless liquids with a weak camphor-like odor, which dissolve readily in organic solvents. but are insoluble in water [161, 278, 681]. The only crystalline compounds are tetrakis(triethylsiloxy)titane and its completely phenylated analog, which has an unusually high melting point because it is not monomeric [826, 1289]. Moreover, tetrakis(triphenylsiloxy)titane, in contrast to trialkylsilyl derivatives, is insoluble in alcohols, acetone, carbon disulfide, and carbon tetrachloride, while it reacts with camphor and dimethylformamide. In benzene and its derivatives (toluene and nitrobenzene) it dissolves sparingly and only at the boiling point. Therefore, it was only possible to determine the molecular weight of $[(C_6H_5)_3SiO]_4Ti$ ebullioscopically. The value obtained for the molecular weight (~1200) is 150 units higher than the value calculated for the monomer. According to the data of Bradley and his co-workers [535, 575] tetrakis(trimethylsiloxy)titane has a degree of assocation of 1.2, while all the other alkylsilyl esters of orthotitanic acid are monomeric. The dielectric constant of $[(CH_3)_3SiO]_4Ti$ at 20°C and 10^3 Hz equals 2.18 [509].

Extrapolation of literature data on the boiling points of tetrakis(trialkylsiloxy)titanes to normal pressure leads to the value

† See p. 400.

TABLE 67. Compounds Containing the Si–O–Ti Group

Formula	m.p., °C	b.p., °C (mm)	n_D^{20}	d_4^{20}	Preparation method (scheme)	Yield, %	Literature
$(CH_3)_3SiOTiCl_3$	Cryst.	67 (9)	—	—	4-105	27	161
	32—33	82—88 (17)	—	—	4-109	72	279, 414
	—	62—63 (8)	—	—	4-109	70	190
$(CH_3)_2C_2H_5SiOTiCl_3$	—	94—98 (17)	—	1.305	4-109	46	279, 414
$CH_3(C_2H_5)_2SiOTiCl_3$	—	122—126 (9)	—	1.217	4-109	35	279, 414
$Cl[(CH_3)_2SiO]_4TiCl_3$	—	97—98 (2)	—	—	4-114	20	200
	—	—	—	—	4-114	42	147
$Cl[(C_2H_5)_2SiO]_4TiCl_3$	—	141—143 (2)	—	—	4-113	24	200
$(CH_3)_3SiOTiBr_3$	32—34	102—104 (10)	—	1.807^{40}	4-109	96	279, 414
$(CH_3)_3SiOTi(OC_3H_7\text{-iso})_3$	—	91 (5)	1.4509^{25}	—	4-87	93	1047
	—	114 (13)	1.4490^{22}	—	4-99	50—70	651
$n\text{-}C_3H_7(CH_3)_2SiOTi(OC_3H_7\text{-n})_3$	—	—	1.5135	—	4-116	—	392
$(n\text{-}C_3H_7O)_3SiOTi(OC_3H_7\text{-n})_3$	—	66—69 (10^{-5}); 125—126 (1)	1.4647	—	4-116	26	395
$(\text{iso-}C_4H_9O)_3SiOTi(OC_4H_9\text{-iso})_3$	—	75—78 (10^{-5})	1.4610	—	4-116	—	394
$(C_6H_5)_3SiOTi(C_5H_5)_2Cl$	210—212	—	—	—	~ 4-106a	76	983
$[(C_6H_5)_3SiO]_2Ti(C_5H_5)_2$	202—203	—	—	—	4-106a	—	1664
	208—210	—	—	—	4-106a	63	983
$[(CH_3)_3SiO]_2TiCl_2$	Cryst.	72—73 (3)	—	—	4-112	81	161
	—	106—110 (18)	—	1.155	4-109	76	279, 414
	—	87—88 (8)	—	—	4-109	35; 43	190
$[(CH_3)_2C_2H_5SiO]_2TiCl_2$	—	114—118 (9)	—	1.111	4-109	—	279, 414
$[CH_3(C_2H_5)_2SiO]_2TiCl_2$	—	127—135 (2)	—	1.068	4-109	90	279, 414
$\{Cl[(CH_3)_2SiO]_4\}_2TiCl_2$	—	133—135 (0.014)	—	—	4-114	8	147

$[(CH_3)_3SiO]_2Ti(OC_3H_7\text{-iso})_2$	—	103 (9)	1.4378^{25}	—	4-87	76	1047
	—	120 (14)	1.4408^{22}	—	4-99	50—70	651
	—	92—95 (6)	—	—	4-87	—	973
$[(C_2H_5)_3SiO]_2Ti(OC_3H_7\text{-n})_2$	—	144—146 (1,5)	1.4800	0.9680	4-99	—	141
$[(C_2H_5)_3SiO]_2Ti(OC_4H_9\text{-n})_2$	—	182 (4)	1.4758	0.9517	4-99	34	141
	—	186—188 (6); 145—150 (1)	1.4679	—	4-99	70	404, 1480
	—	165 (1)	1.4740	0.9524	4-99	—	186
$[(CH_3)_3SiO]_2Ti[O{-}C(CH_3){=}CH{-}C(CH_3){=}O]_2$ 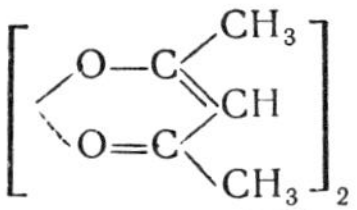	56—57; 54—55	—	—	—	4-92	84; 88	1203
	49—50	110 (0,06); 113—116 (0,1)	—	—	4-116	9; 19	578
$[(CH_3)_3SiO]_2Ti[O{-}C(C_6H_5){=}CH{-}C(C_6H_5){=}O]_2$ 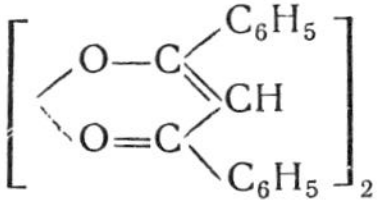	159—165; 161—170; 160—167	—	—	—	4-92	51; 83; 93	1203
$[(C_6H_5)_3SiO]_2Ti[O{-}C(C_6H_5){=}CH{-}C(C_6H_5){=}O]_2$ 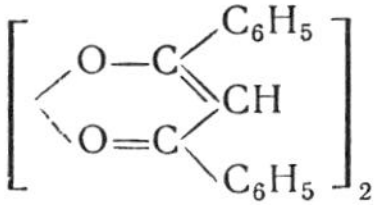	233—238	—	—	—	4-101	100	578
$[(CH_3)_3SiO]_2Ti[OC_9H_6N]_2$ 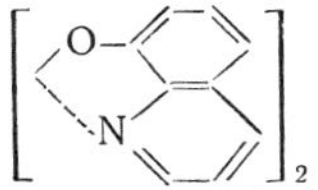	144—148; 145—148	—	—	—	4-92	85; 91	1203
	143—144	—	—	—	4-101	78	209
$[(CH_3)_2C_6H_5SiO]_2Ti[OC_9H_6N]_2$ 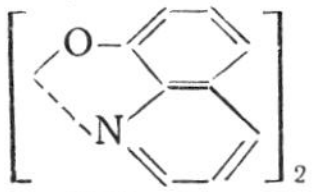	137—139	—	—	—	4-87	100	578
	139—140	—	—	—	4-101	100	578
	138—140	—	—	—	4-104	99	578
$[(C_2H_5)_3SiO]_2Ti[OC_9H_6N]_2$ 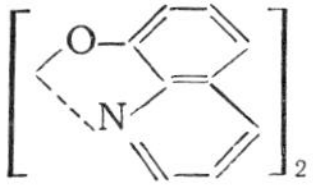	162—164	—	—	—	4-101	83	209

TABLE 67. (Cont'd)

Formula	m.p., °C	b.p., °C (mm)	n_D^{20}	d_4^{20}	Preparation method (scheme)	Yield, %	Literature
$[(C_6H_5)_3SiO]_2Ti[OC_9H_6N]_2$	189—191	—	—	—	4-101	99	578
	190—191	—	—	—	4-116	97	578
	188	—	—	—	4-101	68	209
$[(CH_3)_3SiO]_3TiCl$	Cryst.	103—105 (10)	—	—	4-105	53	161
$[(CH_3)_3SiO]_3TiOC_3H_7$-iso	—	107 (8)	1.4321	—	4-87	65	1047
	—	91—93 (8)	—	—	4-87	—	973
$[(CH_3)_3SiO]_3TiOC_4H_9$-n	—	76—77 (2)	1.4317	0.9070	—	46	161
$[(C_2H_5)_3SiO]_3TiOC_4H_9$-n	—	174—177 (3)	1.4687	0.9378	4-99	46	141
$[(CH_3)_3SiO]_4Ti$	—	69—74 (1—2)	—	—	4-87	—	1335
	—	62 (0.1)	—	—	4-87	95	574, 575
	—	125 (8)	1.4283[25]	—	4-87	82	1047
	—	83 (3)	—	—	4-87	—	973
	—	60 (0.1)	—	—	4-99	100	573, 575
	—	69 (0.5)	—	—	4-99	47	575
	—	114 (11)	1.4300	0.9038	4-99	50	277, 413
	—	102—106 (5—7)	—	—	4-99	56	1289
	—	100 (2)	1.4300	—	4-99	50—70	651
	—	134 (21)	—	—	4-99	—	901
	—	106 (7)	1.4278	0.9078	4-103	18	681
	—	110 (10)	—	—	4-105	—	182
	—	110 (10)	1.4292	—	4-105	58	179
	—	112 (11)	1.4277	0.9051	4-105	66	161, 174
$[C_2H_5(CH_3)_2SiO]_4Ti$	—	86 (0.1)	—	—	4-87	97	575
	—	86 (0.1)	—	—	4-99	94	575
	—	86 (0.1)	—	—	4-99a	100	575
	—	140—142 (5)	1.4461	0.9237	4-99	87	413
$[CH_3(C_2H_5)_2SiO]_4Ti$	—	120 (0.1)	—	—	4-87	96	575
	—	154 (3)	1.4545	0.9244	4-99	90	277, 413
	—	120 (0.1)	—	—	4-99	98	575
	—	120 (0.1)	—	—	4-99a	90	575
	—	186 (6)	1.4565	0.9248	4-103	39	278, 414

$[(C_2H_5)_3SiO]_4Ti$	110	150 (0,1)	—	—	4-87	93	514, 575
	—	199—202 (3.5)	—	$0.917^{20.5}$	4-99	60	509
	99—101	195 (3.5)	—	—	4-99	98	277
	110	150 (0.1)	—	—	4-99; 4-99a	94	575
	93—95; 95—97	227—230 (7); 195—198 (4)	—	—	4-103	45 (89)	278; 1466
	—	204 (4)	1.4689	0.9408	4-105	49	179
	96	221—222 (9)	—	—	4-105	56	1664
	96	176—178 (2,5)	—	—	4-107	50	181
$[(CH_3)_2n\text{-}C_3H_7SiO]_4Ti$	—	112 (0.2)	—	—	4-99	99	575
$[(CH_3)_2iso\text{-}C_3H_7SiO]_4Ti$	—	115 (0.2)	—	—	4-99	98	575
$[CH_3(n\text{-}C_3H_7)_2SiO]_4Ti$	—	190 (3.5)	1.4582	0.9056	4-99	83	277
$[(CH_3)_2C_6H_5SiO]_4Ti$	—	274—275 (7)	1.5392	1.0533	4-105	52	157
$[CH_3(C_6H_5)_2SiO]_4Ti$	—	370—374 (6)	1.5960	1.1248	4-103	36	157
	—	346—348 (3)	1.5988^{21}	1.1394	—	25	158
$[(C_2H_5)_2C_6H_5SiO]_4Ti$	—	300—305 (5)	1.5455	1.0459	4-99	37	413
$[(C_6H_5)_3SiO]_4Ti$	~480	—	—	—	4-99	85	413
	370 (1) Subl.	—	$1.6488^{114.5}$	1.215^{29}	4-99	100	1289
	460—470 decomp.						
	501—505 sealed tube decomp.						
	505, sealed tube decomp.	—	—	—	4-101	—	578
	480	—	—	—	4-103	96	278, 413
	502	Subl. 360 (10^{-4})	—	—	4-106	76	826
$\{[(CH_3)_3SiO]_3SiO\}_4Ti$	—	223—225 (1.5)	1.4201	0.9623	4-105	57	159
$(CH_3)_2Si[OTi(OC_3H_7\text{-}n)_3]_2$	—	—	1.5310	—	4-116	—	395
$(n\text{-}C_3H_7O)_2Si[OTi(C_3H_7\text{-}n)_3]_2$	—	78—81 (10^{-5})	1.4910	—	4-116	36	394
$(CH_3)_3SiO$, CH_3COO >Ti—O[Ti(OCOCH$_3$)$_3$O]— (cyclic)	340 decomp.	—	—	—	—	—	150

TABLE 68. Properties of Titanasiloxane Oligomers and Polymers

Formula	Value of n	T_g, °C	n_D^{20}	d_4^{20}	mol. wt., thous.	Preparation method (scheme)	Literature
$\{HO[(CH_3)_2SiO]_n\}_4Ti$	5	—60	—	—	1.65	~4-99	133, 192
	9	—110	1.4123	0.9870	2.8: 3,0		133, 195
	13	—120	—	—	3.54		133, 195
	18	—100	1.4110	0.9949	5.0		155
	25	—100	1.4082	0.9900	7.68		155
	34	—105	1.4070	0.9808	—		155
	42	—105	1.4065	—	—		155
	52	—100	1.4060	0.9801	14.2		155
	80	—100	1.4055	0.9783	22.47		155
	104	—105	1.4049	0.9773	31.6		155
$\{HO[(CH_3)(C_6H_5)SiO]_n\}_4Ti$	2	—	1.5465	1.1370	—	~4-99	214
	3	—	1.5440	1.1260	—		
	4	—	1.5374	1.112	—		
$\{CH_3[(CH_3)_2SiO]_{n+1}\}_4Ti$	23	—120	1.4031	0.9691	—	~4-99	155
	66	—120	1.4042	0.9747	—		
	72	—120	1.4044	0.9763	—		
$\{(CH_3)_3SiO[(CH_3)(C_6H_5)SiO]_n\}_4Ti$	2	—60	—	—	—	4-121	214
	3	—44	1.5186	1.087	—		
$\{(C_6H_5)_3SiO[(CH_3)(C_6H_6)SiO]_n\}_4Ti$	2	m.p. 268—270	—	—	—	4-121	214
	3	m.p. 46—47	—	—	—		
$\{HO[(CH_3)_2SiO]_n\}_3Ti$ (O—, N— quinoline ring structure)	7	—53	1.4438	1.012	1.65	4-101a	44, 196
	15	—	1.4210	—	3.31; 3.84		44
	22	—110	1.4181	0.994	6.3		44, 196
	66	—115	1.4118	0.980	10.97		44, 196
	100	—105	1.4089	0.979	—		44, 196

$\{HO[(CH_3)(C_6H_5)SiO]_n\}_3 Ti$ (O, N)	12 21	—36 —25	1.5550 1.5553	— —	5,1 8,8	4-101a	44
$\{CH_3[(CH_3)_2SiO]_{n+1}\}_3 Ti$ (O, N)	10 15 30 98 136 170	—102 —115 —105 —110 —118 —	1.4287 1.4207 1.4111 1.4084 1.4070 1.4052	1.0150 0.9832 — 0.9801 0.9793 0.9790	2.17 4.8 12.2 — 24.22 41.6	4-101a	44, 197 44, 197 44, 197 44, 197 44, 197 44
$\{(CH_3)_3SiO[(CH_3)(C_6H_5)SiO]_n\}_3 Ti$ (O, N)	12 21	—36 —23	1.5393 1.5439	1.1121 1.1232	— —	4-101a	44
$\{HO[(CH_3)_2SiO]_n\}_2 Ti [\;]_2$ (O, N)	7 20 74	—76 —76 Cryst. p. —47	1.4863 1.4371 1,4142	— — —	1.35 2.41: 3.1 13.62	4-101a	44
$\{CH_3[(CH_3)_2SiO]_{n+1}\}_2 Ti [\;]_2$ (O, N)	15 60 98 170 350	—85 — —103 —110 —33 *	1.4408 1.4091 1.4089 1.4077 1.4050	1.018 0.984 0.983 0.980 0.979	2.07 7.5 10.72 23.17 54.2	4-101a	44
$\{[(CH_3)_2SiO]_n Ti [\;]_2 O\}_m$ (O, N)	6 17 60 100 200	+6 —10 —110 —120 —60 *	— — — — —	— — — — —	18.0 34.0 98.0 128.0 140.0	—	53 53 53, 201 53 53

TABLE 68. (Cont'd)

Formula	Value of n	T_g, °C	n_D^{20}	d_4^{20}	mol. wt., thous.	Preparation method (scheme)	Literature
$\{[(CH_3)_2SiO]_n Ti[O-C(CH_3)=CH-C(CH_3)=O]_2 O\}_m$	60	—55 (—75)	—	—	58.0	—	53 (201)
	200	—50 *	—	—	138.0		53
$[HO[(CH_3)_2SiO]_n]_3 Ti[OSi(CH_3)_2]_{2n} OTi[[OSi(CH_3)_2]_n OH]_3$	9	—	—	—	—	4-122	155
	18	—	—	—	69.0		
	25	m.p. —42 – 46; Cryst. p. —75 – 87	—	—	37.6		
	34	m.p. —42 – 52; Cryst. p. —72 – 85	1,4095	—	185.0		
	42	m.p. —47 – 57; Cryst. p. —83 – 99	1,4040	—	58.3		
	52	m.p. —42 – 47	—	—	62,4		
	80	m.p. —37 – 47	—	—	82,0		
	104	m.p. —44 – 58; Cryst. p. —72 – 85	—	—	200.0		

* Data on T_g of these copolymers are evidently incorrect since dimethylsiloxane rubber (m = 5-7 thousand) has T_g of the order −130°C. The data given are close to the crystallization points of $[(CH_3)_2SiO]_n$ (from − 35 to − 55°C).

242°C for $[(CH_3)_3SiO]_4Ti$ and 393°C for $[(C_2H_5)_3SiO]_4Ti$. Comparison of the values found and calculated for the molecular refraction of $(R_3SiO)_4Ti$ on the basis of literature data for d_4^{20} and n_D^{20} and also the refined values of the bond refractions and group refractions for organosilicon compounds [373] shows that the refraction for the Ti–O bond of 4.02, derived by English and Sommer [681] is low and should equal 4.32.

Trimethylsiloxychlorotitanes crystallized on standing [161] and change color during storage [279, 414]. The high sensitivity of the compounds to hydrolysis hampers their isolation and investigation. It must be assumed that like $ROAlCl_2$ they are dimeric.

Tris(trimethylsiloxy)butoxytitane is a colorless liquid with an ether-like odor [161]. Mixed trimethylsiloxyisopropoxy derivatives of titanium are also colorless, but they acquire a bright blue color after storage for several months in a sealed ampoule [651]. Organosilicon derivatives of titanium containing chelate groups are crystalline substances and have a yellow color regardless of the nature of the chelate substituent [578, 1203]. There are two possible diastereoisomeric forms for derivatives of β-diketones [1203]:

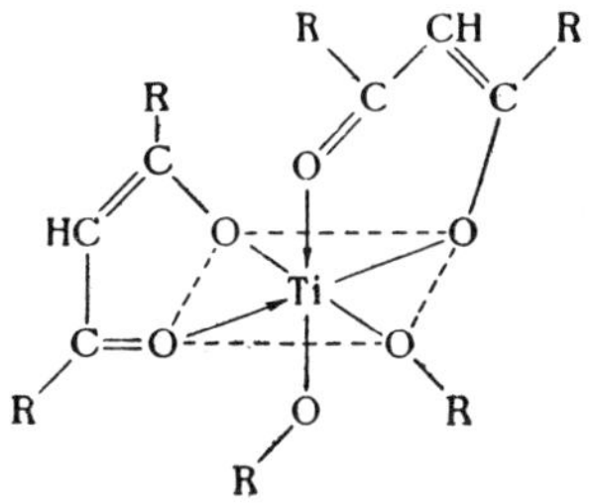

cis

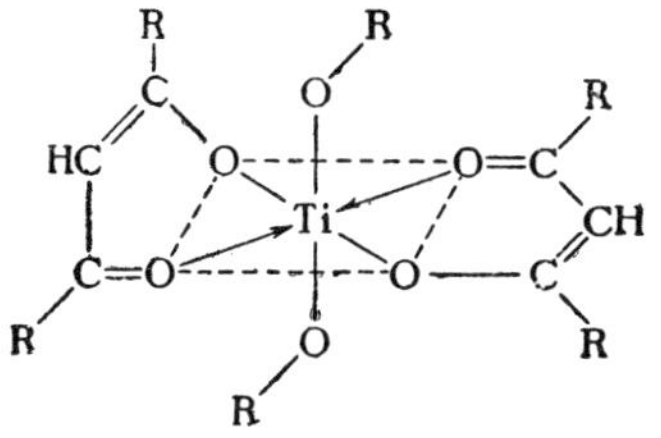

trans

However, the steric configuration of bis(triorganosiloxy) derivatives isolated up to now has not been established.

Triphenylsiloxy derivatives obtained from bis(cyclopentadienyl)dichlorotitane [983] (orange-yellow plates) sublime at 230°C (0.05 mm) without decomposition, are stable in air, but are cleaved readily in an acid or alkali medium, and dissolve readily in hot organic solvents (chloroform, benzene, acetone, and ligroin) apart from methanol.

The constants of a series of titanasiloxane oligomers and polymers are given in Table 68.

An x-ray investigation of "cross-shaped" oligomers containing terminal $HOSiR_2$–groups, including oligomers with polydimethylsiloxane bridges between titanium atoms, showed [155, 156] that they are all amorphous at room temperature. With cooling to -120°C a crystalline phase appears at values of $n > 35$-40. In all cases the structure of the crystalline phase is the same and its heat of fusion is 5.5. cal/g in comparison with 5.9 cal/g for polydimethylsiloxanes. It is believed that the oligomers have "bundle" packing of the molecules. Oligomers of the type $(C_9H_6NO)Ti\{[OSi \cdot (CH_3)_2]_{n+1}CH_3\}_3$ and $(C_9H_6NO)_2Ti\{[OSi(CH_3)_2]_{n+1}CH_3\}_2$ are transparent, viscous, red liquids, which have a linear relation between $\log\eta$ and $1/T$, indicating that they are normal liquids [44, 197]. The activation energy of viscous flow falls from ~4.6 kcal/mole for $n=15$ to 3.7-4.0 kcal/mole for $n=170$. Thus, the magnitude of the intermolecular interaction is determined by the hydroxyquinoline groups (when $n=10$ the oligomer does not flow). The viscosity of linear oligomers is higher than that of branched oligomers.

Comparison of the properties of titanasiloxane copolymers obtained by cohydrolysis of bis(8-quinolyloxy)dibutoxytitane and $RR'SiCl_2$ [53, 203] shows that with the same Si : Ti ratio the polymers containing methylphenylsiloxane units have higher values of T_g, but surpass the dimethylsiloxane analogs in thermal stability (retention of solubility on heating). On the other hand, an increase in the content of the titanium component in the polymer raises the glass transition point and reduces the thermal stability somewhat.

The solid products from the cohydrolysis of bis(2-ketopent-3-en-4-oxy)-dichlorotitane with $RR'SiCl_2$ are readily soluble in alcohol, benzene, acetone, carbon tetrachloride, and other organic solvents [204].

Heating for ½ hr at 200°C sharply changes the thermomechanical properties: a polymer with $R = R' = CH_3$ begins to flow at ~100°C, while the polymer with $R = CH_3$ and $R' = C_2H_5$ begins to flow at 400°C. Polymers with $R = CH_3$ and $R' = C_6H_5$ and also with $R = R' = C_2H_5$ do not flow at 500°C.

A comparison of the properties of poly[bis(2-ketopent-3-en-4-oxy)titana]methylphenylsiloxanes with different Si:Ti ratios, obtained by cohydrolysis [204] and heterofunctional condensation [53, 202], showed that an increase in their titanium content leads to a rise in the glass transition point and acceleration of the loss of the power of the polymer to dissolve after heating.

Poly[bis(8-quinolyloxy)titana]methylphenylsiloxanes are appreciably superior to acetylacetone derivatives in their power to retain their solubility after heating (see below). From the data presented in Table 68 it follows that the introduction of titanium atoms bearing 8-hydroxyquinoline and acetylacetonate groups into a polydimethylsiloxane structure appreciably reduces the flexibility of the chain. The introduction of acetylacetonate groupings gives a higher rigidity.

It should also be pointed out that an investigation of the solution behavior of a polymer obtained by cohydrolysis of methylphenyldichlorosilane and bis[2-ketopent-3-en-4-oxy]dichlorotitane (molar ratio 3 : 1) in the presence of pyridine [422] confirmed that it was a copolymer and was chemically homogeneous. The molecular weight of the first fraction (11,200), determined by light scattering, corresponds to a degree of polymerization of 17. Other fractions have a lower molecular weight. A study of the behavior of solution of the copolymer in various solvents confirmed that its molecule may be regarded as linear. The glass transition point of a titanasiloxane polymer is higher and its intrinsic viscosity lower than those of a polymethylphenylsiloxane with the same molecular weight. In contrast to polydimethylsiloxanes, the viscosity of solutions of a titanasiloxane with chelate groups falls with a rise in temperature, while the intrinsic viscosity depends on the time of flow of the solution through the viscometer capillary.

An increase in the titanium content of copolymers containing cyclopentadienyl groups at the Ti atoms also leads to a rise in the glass transition point [53, 201].

13.3. Thermal Stability

We have already given some data above on the behavior of polymers of the type $[-OSiRR'OTi(\supset)_2-]_n$ on heating. When $R = CH_3$, $R' = C_6H_5$, and the chelate group $\supset$ is 8-hydroxyquinoline, their solubility is retained completely after heating at 200°C for more than 20 hr regardless of the Si : Ti ratio in the polymer and the nature of the solvent [203]. However, even with replacement of

$R' = C_6H_5$ by CH_3 the capacity of the polymer for thermal cross linking begins to depend on these factors and the heating time. Thus, for example, poly[bis(8-quinolyloxy)titana]methylphenylsiloxane with Si : Ti = 1 is 90% soluble in benzene and acetone after heating at 200°C for 20 hr, 92-95% soluble after 8 hr, and completely soluble after heating for 4 hr. At the same time, poly[bis(8-quinolyloxy)titana]dimethylsiloxane with the same Si : Ti ratio is 78% soluble in benzene and 70% soluble in acetone after heating at 200°C for 4 hr. After 8 hr the solubility falls to 70 and 55% respectively, and after 20 hr, to 40 and 15%. With increase in the Si : Ti ratio to 9 the solubility in acetone and benzene after the polymer has been heated at 200°C for 20 hr reaches 69 and 60%, respectively.

Investigation of the thermooxidative breakdown of linear and branched oligomers, obtained by condensation of (8-quinolyloxy)-butoxytitanes by scheme (4-102), by thermogravimetric analysis showed [44] that their thermal stability depends on the concentration of the 8-quinolyloxy groups and not on the structure. The minimum losses in weight on heating were observed for a branched polymer with n = 98. The thermooxidative stability of oligomers at 300°C in air with a chelate group content of 0.25% on titanium was 20 times higher than that of PMS-400 fluid. 8-Quinolyloxy groups inhibit the oxidation of organic groups at silicon (i.e., they inhibit radical processes), but they do not prevent depolymerization processes (i.e., ionic reactions). The activation energy of thermal decomposition of 8-quinolyloxytitanasiloxane oligomers is 22-26 kcal/mole.

Thermal cross linking of polymers containing acetylacetone groups proceeds at a higher rate than for their 8-hydroxyquinoline analogs. Of the cyclolinear polymers obtained by cohydrolysis of bis[2-ketopent-3-en-4-oxy]-dichlorotitane with $RSiCl_3$ [205], only the polymer $R = C_6H_5$ and Si : Ti = 3 is 79% soluble in toluene and 33% soluble in acetone after heating at 200°C for 4 hr. Its ethyl (Si : Ti = 4) and methyl (Si : Ti = 5) analogs completely lose their solubility after heating at 160°C for 2 hr. Heating sharply changes the thermomechanical properties of such polymers.

Interesting results were obtained by studying the properties of polymers obtained by the reaction of monosodium salts of organosilanetriols with titanium tetrachloride [12, 14, 138, 139]. Polymers prepared by combining trifunctional compounds of silicon with tetrafunctional compounds of titanium have a high solubility; on these grounds they may be assigned a cyclolinear structure:

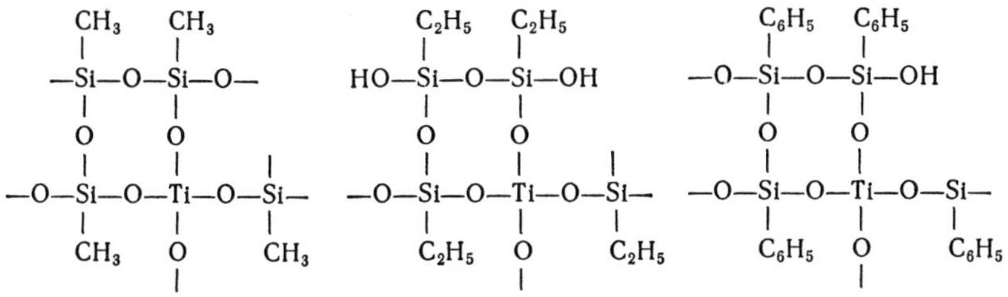

```
    CH3   CH3                C2H5  C2H5                 C6H5  C6H5
    |     |                  |     |                    |     |
  —Si—O—Si—O—           HO—Si—O—Si—OH             —O—Si—O—Si—OH
    |     |                  |     |                    |     |
    O     O                  O     O                    O     O
    |     |     |            |     |     |              |     |     |
—O—Si—O—Ti—O—Si—      —O—Si—O—Ti—O—Si—        —O—Si—O—Ti—O—Si—
    |     |     |            |     |     |              |     |     |
    CH3   O     CH3          C2H5  O     C2H5           C6H5  O     C6H5
          |                        |                          |
          I                        II                         III
```

Data on the solubility of these polymers after heating are given in Table 69.

TABLE 69. Solubility of Polytitanaorganosiloxanes of Types I, II, and III [12, 138, 139]

Polymer	Heating		Solubility, %	
	Temperature, °C	Duration, hr	In benzene (III) or in toluene (I, II)	In carbon tetrachloride
I	150	10	100	—
	200	2	90	—
		4	24	—
	300	0.25	Insoluble	
II	150	10	100	100
	200	2	95	99
		4	14	25
	300	0.25	Insoluble	
III	200	15	100	100
	300	0.25	80	47
	400	0.25	Insoluble	

Polymers I and II have no highly elastic or viscoelastic regions on the thermomechanical curve right up to 500°C, regardless of whether they are in the soluble or insoluble form. The soluble form of polymer III and heated polymers, regardless of the heating temperature, have a region of high elasticity, but do not show flow up to 700°C.

There are data which indicate that the breakdown of the main chains of polymers of types I-III with an inadequate supply of air occurs at ~610°C with the absorption of heat [12, 198]. The combination of these data with the fact that under analogous conditions the elimination of methyl groups from methylphenylsiloxane polymers occurs at ~350°C and elimination of phenyl radicals at ~500°C [198] and also data on the thermomechanical properties of titanasiloxanes indicate the possibility of making coatings from titanaorganosiloxane polymers with a high thermal stability and good service properties, as is confirmed in practice. Thus, for example, "baking" partly hydrolyzed tetrakis(trimethylsiloxy)titane on a steel surface forms a varnish film, which does not crack after the plate has been heated many times to red heat [681].

Replacing a small number of the silicon atoms in the main chains of siloxane elastomers by titanium atoms raises the thermal stability of the vulcanizates based on them without affecting the hydrolytic stability of the polymers [269]. This is explained by the formation of an ordered structure through the coordination bonds formed by the heteroatoms.

13.4. Vibration Spectra

An intense band in the region of 915-925 cm^{-1} is most characteristic of the infrared and Raman spectra of compounds containing titanium atoms in a siloxane structure [12, 134, 139, 154, 158, 192, 196, 197, 203, 204, 205, 443, 535, 571, 901, 1291]. There is no doubt about the assignment of this band to the Si–O–Ti group. However, Andrianov and his co-workers assign it to the Ti–O bond in the Ti–O–Si group, while other investigators assign this band to antisymmetrical valence vibrations of the Si–O bond in the Si–O–Ti group, believing that the Ti–O bond corresponds primarily to the band at 510-525 cm^{-1}. This question is not yet completely clear. The assignment of the bands in the spectrum of tetrakis-(trimethylsiloxy)titane according to the data of Kriegsmann and Licht [901] is given below (Table 70).

According to the data of Rodionov and his co-workers [443], in the IR spectra of titanasiloxane polymers the band of symmetrical valence vibrations of the Ti–O bond in Si–O–Ti lies in the region of 530-580 cm^{-1} and its intensity depends on the Si–Ti ratio. In the IR spectra of the hydrolysis products of tetrakis(trimethylsiloxy)titane (see below) the intensity of the band at 918 cm^{-1} falls

TABLE 70. Assignment of Frequencies in IR and Raman Spectra of $[(CH_3)_3SiO]_4Ti$ [901]

Band, cm^{-1} (intensity)		Type of vibration	Comments
IR	Raman		
—	415 (6)	ν_s Ti—O	485 (M) in IR according to [443]
521 (M)	520 (1)	ν_s Ti—O	423 (S) in IR according to [443]
646 (W)	631 (8)	ν_s SiC_3	—
687 (W)	687 (4)	ν_{as} SiC_3	—
773 (W)	776 (0)	ν_{as} Ti—O	—
918 (VS)	917 (1)	ν_{as} Si—O—Ti	919 (VS) in IR according to [443]
950 (M)	958 (1)	ν_{as} Ti—O—Si	948 (W) in IR according to [443]
1060 (W)	1068 (6)	ν_s Si—O (Ti)	—

as the R_3SiO : Ti ratio decreases and simultaneously there appears a new band at 975 cm^{-1} [571].

In the spectrum of $(C_5H_5)_2Ti[OSi(C_6H_5)_3]Cl$ a band at 959 cm^{-1} belongs to vibrations of Si–O–Ti [983]. In the spectrum of $(C_5H_5)_2Ti[OSi(C_6H_5)_3]_2$ it is split into two bands at 984 and 917 cm^{-1} and this is connected with the symmetrical and unsymmetrical vibrations of the $Ti(OSi)_2$ system.

13.5. Chemical Properties

The chemical properties of compounds containing the Si–O–Ti group have been described already to a considerable extent in previous sections of this chapter. Therefore we will examine here in somewhat more detail only some conversions of oligomers and problems of hydrolytic cleavage of Ti–O–Si groups.

As has already been pointed out, oligomers of the type $Ti[(OSiR_2)_nR]_4$ are obtained by heterocondensation of monosiloxanols $R(R_2SiO)_nH$ with tetrafunctional titanium compounds. However, there is the possibility of "blocking" of prepared oligomers [214]:

$$Ti\{[OSi(CH_3)(C_6H_5)]_n OH\}_4 + 4R_3SiCl \xrightarrow{C_5H_5N} Ti\{[OSi(CH_3)(C_6H_5)]_n OSiR_3\}_4$$
$$(R = CH_3 \text{ or } C_6H_5;\ n = 2 \text{ or } 3) \qquad (4\text{-}121)$$

When heated to 150-200°C, "cross-shaped" oligomers $Ti\{[OSi(CH_3)_2]_nOH\}_4$ undergo intermolecular condensation [131, 133, 155, 192–194]. The initial increase in molecular weight does not lead to loss of solubility by the polymer, i.e., linear polymers are

TABLE 71. Relation of the Rate of Cross Linking of Tetrakis(ω-hydroxypolydimethylsiloxanyl)titanes to the Length of the Chain of the Starting Oligomer

n	2	3	23	34	128
η	0.25	0.36	0.75	0.96	1.07
Cross linking time at 200°C, hr	14	24	105	221	215
Insoluble part of polymer, %	23.8	22.5	21.3	23.0	7.7
$K \cdot 10^{-2}$, 1/wt. %, hr.	3.0	2.83	2.06	0.77	0.71

formed. Thus, for example, when an oligomer with n = 52 (mol. wt. ~15,500) is heated at 200°C for 25 hr in an inert atmosphere the molecular weight is increased to 25,000 and after 250 hr it reaches 62,400. However, the tetrafunctionality of the oligomers gradually has an effect and cross linking occurs with the formation of insoluble polymers:

$$Ti[(OSiR_2)_nOH]_4 \xrightarrow[-H_2O]{t^\circ} [HO(SiR_2O)_n]_3Ti(OSiR_2)_nO(SiR_2O)_nTi[(OSiR_2)_nOH]_3 \longrightarrow$$

$$\longrightarrow \begin{array}{c} | \\ -Ti-(OSiR_2)_{2n}-O-Ti- \ldots \text{etc.} \\ | \qquad\qquad\qquad\qquad | \\ O \qquad\qquad\qquad\qquad O \\ | \qquad\qquad\qquad\qquad | \\ \left(\begin{array}{c} SiR_2 \\ | \\ O \end{array}\right)_{2n} \qquad \left(\begin{array}{c} SiR_2 \\ | \\ O \end{array}\right)_{2n} \\ | \qquad\qquad\qquad\qquad | \\ -Ti-(OSiR_2)_{2n}-O-Ti- \\ | \qquad\qquad\qquad\qquad | \end{array} \quad (R = CH_3) \qquad (4\text{-}122)$$

Data on the relation of the rate of this process to the length of the chain of the starting oligomers are presented in Table 71 [194].

A characteristic of the final polymers is the order of the steric structure – all the titanium atoms in the molecule of the substance are separated by dimethylsiloxane bridges of the same length (2n). When n = 9 – 13 these titanasiloxanes have elastic properties which are paradoxically combined with their insolubility. With a lower value of n the polymers are brittle and resinous. The length of the bridge has a definite effect on their thermal stability. When n = 10 – 25 the temperature at which decomposition begins in an inert atmosphere is 390°C; when n = 52 – 80 it equals 410-415°C, while when n = 104 it is close to the temperature at which the decomposition of pure polydimethylsiloxane is observed (~330°C).

Heating oligomers of the type $C_9H_6NOTi\{O[Si(CH_3)_2O]_nH\}_3$ to 200°C initially yields soluble polymers, whose rate of formation is inversely proportional to the concentration of functional groups (i.e., the value n) in the starting substance [195]. Subsequently cross-linked polymers with an ordered structure are formed. The heterocondensation of these oligomers with (8-quinolyloxy)tributoxytitane in equimolar amounts proceeds similarly.

Trimethylsiloxytrichlorotitane, like other trialkylsiloxyhalotitanes, fumes in air and reacts vigorously with water with the formation of hexamethyl disiloxane [279].

In the reaction of tris(trimethylsiloxy)chlorotitane with butanol in the presence of pyridine, the trimethylsiloxy groups are not eliminated and tris(trimethylsiloxy)butoxytitane is obtained [161]. Historically this compound was the first mixed silyl alkyl ester of orthotitanic acid.

In synthesizing tetrakis(trimethylsiloxy)titane, English and Sommer [681] observed that it showed no signs of change for a long time at the surface of water, 0.1 N NaOH solutions, or dilute HCl. However, hydrolysis occurred when the mixture was stirred and in the case of water and alkali, a curdy white precipitate was formed, while with acid, an oily layer was formed. Later [161] it was reported that treatment of tetrakis(trimethylsiloxy)titane with dilute hydrochloric or acetic acid finally yields solid polymers.

Tetrakis(triorganosiloxy)titanes are much more resistant to hydrolysis than tetraalkoxytitanes. The rate of hydrolytic cleavage of compounds of the type $(R_3SiO)_4Ti$ is connected to a considerable extent with the nature of the alkyl radical R and decreases with an increase in the length of the chain or if R is an aryl group [277, 287]. Thus, for example, tetrakis(triphenylsiloxy)titane does not undergo changes when heated with 4% NaOH, or dilute hydrochloric and sulfuric acids, and also does not react with hot concentrated HCl [287, 826, 1289]. The Ti–O bonds screened by triphenylsilyl groups are broken only by concentrated nitric and sulfuric acids. As another example of the effect of the nature of the substitutents we can point out that diisopropoxy-bis(8-quinolyloxy)titane is readily hydrolyzed at room temperature to give a high-melting insoluble substance $[(\supset)_2TiO]_n$. At the same time, bis(trimethylsiloxy)bis(8-quinolyloxy)titane does not undergo any changes under these conditions [1203].

The hydrolytic cleavage of tetrakis(trialkylsiloxy)titanes was studied in more detail by Andrianov and his co-workers [5, 8, 135, 164, 174, 179, 182, 225, 291] and Bradley [571]. However, their data on the conditions and the character of the products of hydrolysis differ appreciably. Thus, Andrianov and his co-workers report that heating tetrakis(trimethylsiloxy)titane with water in a ratio of 2 : 1 in acetone at 50°C for 3 hr does not change the compound. With a molar ratio $(R_3SiO)_4Ti : H_2O > 1$ and in the presence of acid catalysts there is elimination of trimethylsiloxy groups and polymeric products are formed. The first stage of the process yields linear polymers, which are soluble in toluene, ethyl acetate, cello-solve, etc. Subsequently, branched and cross-linked insoluble solid condensation products are formed [5, 8, 174, 179, 182]:

$$2(R_3SiO)_4Ti + 2H_2O \longrightarrow 2(R_3SiO)_3TiOH + 2R_3SiOH;$$

$$2(R_3SiO)_3TiOH \longrightarrow (R_3SiO)_3TiOTi(OSiR_3)_3 + H_2O;$$

$$(R_3SiO)_3TiOTi(OSiR_3)_3 + H_2O \longrightarrow (R_3SiO)_3TiOTi(OSiR_3)_2OH;\ \text{etc.} \quad (4\text{-}123)$$

At the same time, according to the data of Bradley [571], when tetrakis(trimethylsiloxy)titane is stirred vigorously with water (2 : 1) in dioxane at room temperature for 1 hr there is 74% of hydrolysis corresponding to the scheme

$$2(R_3SiO)_4Ti + H_2O \longrightarrow [(R_3SiO)_3Ti]_2O + 2R_3SiOH \quad (4\text{-}124)$$

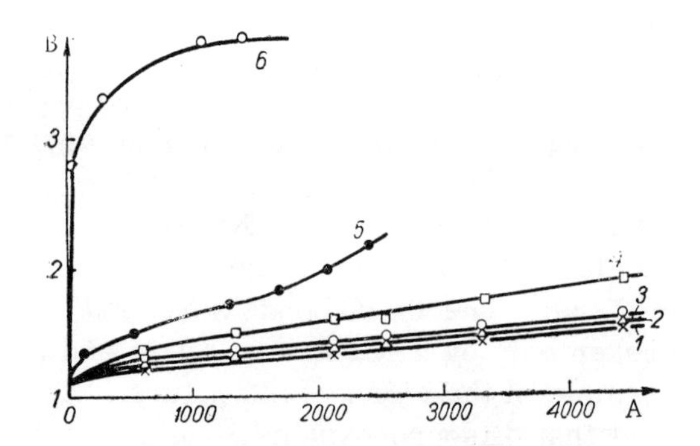

Fig. 9. Rate of hydrolysis of tetrakis(triethylsiloxy) titane [291]. A) Hydrolysis time, hr; B) relative viscosity of 10% solution. Molar ratio H_2O: $Ti[OSi(CH_3)_3]_4$: 1–2.33; 2–2.80; 3–3.27; 4–3.74; 5–7.50; 6–15.

TABLE 72. Hydrolysis of Tetrakis (trimethylsiloxy)-titane (A) [571]*

H_2O: A ratio (h), mole	Degree of hydrolysis, %	$\{TiO_x[OSi(CH_3)_3]_{4-2x}\}_n$		
		x	n	
			Found	Calculated
0.1	—	1.505	8.1	8.26
0.3	—	1.52	9.2	9.09
0.5	74.2	1.56	12.0	11.63
0.75	—	1.59	14.8	14.71
1.0	53.3	1.60	18.4	16.13
1.5	48.6	1.645	26.5	28.57
2.0	47.3	1.68	Difficultly soluble	
2.5	44.2	1.68	Difficultly soluble	

* Mixing with aqueous dioxane at 1 hr at 23°C, evaporation of the solution in vacuum, distillation of the unreacted $[(CH_3)_3SiO]_4Ti$ at 160°C in vacuum.

The unstable dititoxane obtained disproportionates (slowly at room temperature and rapidly at 120°C and above):

$$3\,[(R_3SiO)_3\,Ti]_2\,O \longrightarrow 4\,(R_3SiO)_4\,Ti + 1/n\,[Ti_2O_3\,(OSiR_3)_2]_n \qquad (4\text{-}125)$$

According to a cryoscopic determination of the molecular weight in cyclohexane, in the polymer obtained n = 12. Data on the effect of the amount of water on the character of the hydrolysis products are presented in Table 72.

By introduction of trimethylsilanol into the system it was shown that the hydrolysis process is reversible. By calculation $1/n = 1.028 - 0.604x$ (accuracy ±6%). When h = 0.1 the final product corresponds to the formula $Ti_8O_{12}[OSi(CH_3)_3]_8$. It is believed that the compound may have the cubic structure

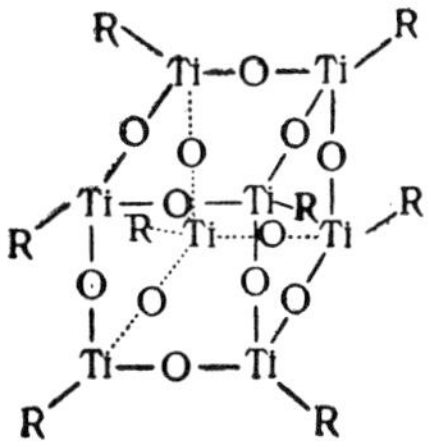

$R = (CH_3)_3SiO$

Polymers corresponding to values of $x > 1.5$ are evidently "cubes" connected by oxygen bridges.

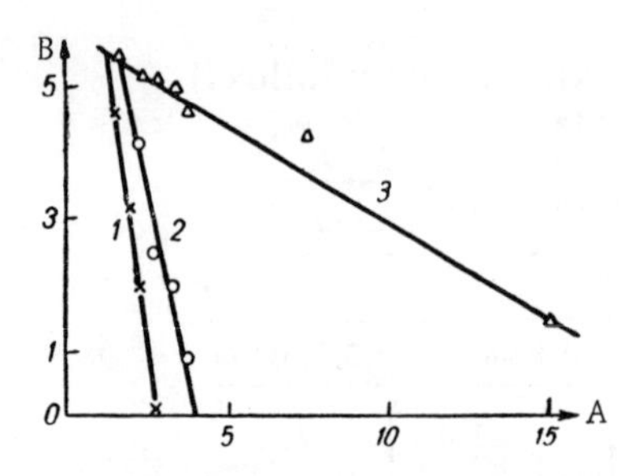

Fig. 10. Relation of the logarithm of gelation time (B) to the number of moles of water per mole of substance hydrolyzed (A) [291]. 1 – $Sn[OSi(C_2H_5)_3]_4$; 2 – $Al[OSi\cdot(C_2H_5)_3]_3$; 3 – $Ti[OSi(C_2H_5)_3]_4$.

In a study of the kinetics of hydrolysis of compounds of the type $[(C_2H_5)_3SiO]_mM$ (where M = Al, Sn, Ti; m is the highest valence of M), it was found [139, 164, 291] that the hydrolysis rates lie on the scale 2220 (Sn) : 27.2 (Al) : 1 (Ti). These values were calculated from the measurement of the viscosity of 10% solutions of these substances in a benzene – alcohol mixture at 20°C in the presence of various amount of water (Fig. 9). The lower rate of hydrolysis of tetrakis(triethylsiloxy)titane in comparison with triethylsiloxy derivatives of other metals is characterized by the lowest value of the proportionality coefficient, which reflects the change in time of gelation (or time to reach a definite viscosity) of the solution in relation to the amount of water used for hydrolysis [164, 291]. If the gelation time is denoted by T and the water concentration by [M], then

$$-\frac{dT}{d[M]} = kT \text{ and } \ln T = -k[M] + C \quad \text{or} \quad \log T = A - \frac{k}{2.3}[M] = A - B[M]$$

For compounds of the type $[(C_2H_5)_3SiO]_nM$ the following values were obtained for k = 2.3 B (Fig. 10): 8.51 (Sn); 5.52 (Al); 1.33 (Ti).

Because of the exceptionally slow hydrolysis of tetrakis(triethylsiloxy)titane the gelation time could not be established and calculations were carried out using the time to reach a definite viscosity (0.4).

The reaction of tetrakis(triethylsiloxy)titane with dimethylphosphinic acid proceeds in the following way [187]:

$$(R_3SiO)_4Ti + 2R'_2P(O)OH \longrightarrow R_3SiO\{[R'_2P(O)O]_2TiO\}_nSiR_3 + 2R_3SiOH(R_3SiOSiR_3)$$

$$(R = C_2H_5;\ R' = CH_3) \qquad (4\text{-}126)$$

Depending on the conditions, oligomers are obtained with mol. wt. 800-1770 (n = 3 – 6). It is believed that the reaction (4-126) proceeds in stages in analogy with scheme (4-123) with gradual replacement of triethylsiloxy groups by radicals from dimethylphosphinic acid.

A study of hydrolytic cleavage in hydrochloric acid solutions of a polytitanaphenylsiloxane with a mean molecular weight of 1500 and containing 16.8% Si and 7.21% of Ti (Si : Ti = 4) showed that heating for 10 hr at 95°C in 10% hydrochloric acid (100 : 1) decomposed only 1.5% of the Ti–O bonds. In a similar polyalumaphenylsiloxane, 87% of the Si–O–Al groups were destroyed under the same conditions [138, 140]. In the same period 20% and 30% solutions of hydrochloric acid cleaved 15% and 45% of the Ti–O bonds in the first polymer, respectively, and 99.8 and 100% of the Al–O bonds in the second.

13.6. Analysis

Wet combustion methods are recommended for the simultaneous determination of silicon and titanium in compounds containing the Si–O–Ti group. A sample of the substance for analysis is treated with a mixture of concentrated sulfuric acid or oleum with concentrated nitric acid [277],* potassium iodate [153], or ammonium nitrate or sulfate [575, 1289]. The mixture is evaporated, diluted with water, and filtered. The precipitate of silicic acid is calcined at ~900°C and weighed and the titanium isolated from the filtrate by precipitation as $Ti(OH)_4$ by treatment with ammonia and then calcined and weighed as TiO_2 [153, 277, 1289].

A wider range of methods is used for the determination of titanium alone. For example, it is recommended that $Ti(OH)_4$ is precipitated with ammonia directly from a solution of the compound for analysis in alcohol or dioxane. In this case, the silicon is removed in the form of diorganosilanols, which remain in the filtrate [279] or are volatilized by evaporation of the solution in a platinum crucible under an infrared lamp [575].

For colorimetric determination of titanium [138, 575, 901, 1289] it is most effective to burn a sample of the substance with a mixture of sodium peroxide and finally powdered carbon (12 : 1) in a nickel crucible [1289]. The ash is dissolved in water and acidi-

* When there are phenyl radicals attached to the silicon atom it is recommended that nitric acid not be used so as to avoid tar formation [1289].

fied with sulfuric acid; the titanium content is determined by means of a calibration curve plotted at 420 mμ, using 30% H_2O_2 and standard solutions.

A polarographic method is also recommended [466]. The substance is decomposed with hydrofluoric acid and subsequently concentrated with excess concentrated sulfuric acid, which provides a strongly acid medium and guarantees the stability of the titanium solution. The polarographic determination of Ti^{4+} is carried out with a mercury dropping electrode in a nitrogen atomsphere with 0.01% gelatin added to the solution. The reduction wave of $Ti^{4+} \rightarrow Ti^{3+} \varepsilon_{1/2} = -0.37$ V over the range of concentrations of $1 \cdot 10^{-3} - 1 \cdot 10^{-4}$ mole/liter with 70% H_2SO_4 as the base electrolyte has a well expressed form.

It has also been suggested that organosilicotitanium compounds are broken down with hydrofluoric acid and the solution treated with benzidine or other amines which form complex salts with H_2TiF_6 [42]. By measuring the electrical conductivity of the solutions and the ratio of concentrations of oxidized and reduced forms of the amine conductometrically or potentiometrically it is possible to determine titanium rapidly and quantitively.

13.7. Application

Protective coatings based on titanasiloxane resins have high thermal stability and mechanical strength, good adhesion to metals, and electrical and moisture insulating properties. Varnish films are smooth, nontacky, and are characterized by an exceptional luster and impermeability to ultraviolet rays. Coatings withstand operating temperatures from -60 to +550°C and their pigmentation with aluminum powder raises the thermal stability to temperatures above 600°C [138, 1165–1167, 1661, 1696].

Titanasiloxane copolymers may be used successfully as enamels for refrigerators [1548], backings ("primers") and cements for attaching polysiloxane resins [1594, 1598] and Teflon [1595] to plastic and metal surfaces (aircraft fuselages, car bodies, and membranes of pressure transducers). It has also been suggested that they are used as impregnating (waterproofing) agents and coatings for cotton cloth, fibers, paper, cellulose, photographic films, leather, and asbestos [1292, 1329, 1335, 1354, 1411, 1589,

1603, 1631, 1668, 1728], as effective catalysts for the condensation of urea-, phenol-, and melamine-formaldehyde, alkyd [1292, 1329, 1589, 1664], and also butadiene [1628] resins, agents for "cold vulcanization" of siloxane elastomers [1504, 1716], and additives to improve the thermal stability of siloxane rubbers [132, 541], catalysts for the polymerization of alkenylsiloxanes [1433] and siloxane oligomers [419], stabilizers for polyvinyl resins, additives which improve the quality of asphalt [1333], antistick coatings [852, 1341], and particularly coatings for tooth enamel to prevent caries [852].

The addition of 8-quinolyloxytitanadimethylsiloxane oligomers to PMS fluids sharply raises their thermooxidative stability at 250-300°C even when the titanium content of the mixture is 0.2-0.3% while the lubricating, viscosity, and low-temperature characteristics of the PMS are retained [44, 1509]. The addition of 5% of poly-[bis(8-quinolyloxy)titana]dimethylsiloxanes with Si : Ti = 160 [53] or siloxane oligomers containing 8-quinolyloxytitanium groups [44] to dimethylsiloxane rubbers improves the physicomechanical properties of the vulcanizates and their thermal stability at 300°C.

14. COMPOUNDS CONTAINING THE Si – N – Ti GROUP

In 1961 Andrianov and his co-workers report [143] that simple distillation of an equimolar mixture of hexamethyldisilazan with titanium tetrachloride splits off both trimethylsilyl groups:

$$R_3SiNHSiR_3 + TiCl_4 \longrightarrow 2R_3SiCl + [NHTiCl_2] \qquad (4\text{-}127)$$

The yield of trimethylchlorosilane reaches 88%. However, Burger and Wannagat established later [601, 608] that if hexamethyldisilazan is added in a ratio $< 0.6:1$ to a solution of titanium tetrachloride in toluene or carbon tetrachloride cooled to 0–10°C, then the colorless solution deposits yellow crystals, whose composition corresponds to the adduct $[(CH_3)_3Si]_2NH \cdot 2TiCl_4$. The low solubility and relative stability of the substance (crystals melt exactly at 83°C with an orange coloration) make the following salt-like structures probable:

$$\{[(CH_3)_3Si]_2NHTiCl_3\}^+[TiCl_5]^- \quad \text{or} \quad \{[(CH_3)_3Si]_2NH \cdot TiCl_5\}^-[TiCl_3]^+$$

TABLE 73. Compounds Containing the Si–N–Ti Group

Formula	m.p., °C	b.p., °C (mm)	Consistency, color	Preparation method (scheme)	Yield, %	Literature
$[(CH_3)_3Si]_2NTiCl_3$	75—77	104—105 (1)	Waxy, dark yellow	4-131	62	122, 604, 608
$[(CH_3)_3Si]_2NTi(OC_3H_7\text{-iso})_3$	—78—80	53 (1)	Colorless liquid*	4-129	57	602, 604, 608
$[(\text{iso-}C_3H_7O)_3Si]_2NTi(OC_3H_7\text{-iso})_3$	180 decomp.	160 (1) subl.	Colorless liquid	4-129	82	608
$\{[(CH_3)_3Si]_2N\}_2Ti(OC_3H_7\text{-iso})_2$	189 decomp.	140 (1) subl.	Colorless liquid	4-129	71	602, 608
$[(CH_3)_3Si]_2NH \cdot TiCl_4$	83	—	Yellow crystals	—	—	601
$(CH_3)_3SiNTiCl_2 \cdot 2C_5H_5N$	—	—	Orange crystals	4-132	—	601, 604, 608

* $n_D^{20} = 1.4708$; $d_4^{20} = 0.937$.

Incidently, we cannot exclude a structure with bridges of chlorine atoms:

$$\begin{array}{c} R_3Si \quad Cl \qquad Cl \qquad Cl \\ \quad \diagdown^{\oplus} \quad \diagdown^{\ominus} \; \diagup \quad \diagdown \quad \diagup \\ H{-}N{-}{-}Ti \qquad\quad Ti{-}Cl \\ \quad \diagup \quad \diagup \,|\, \diagdown \quad \diagup \quad \diagdown \\ R_3Si \quad Cl \;\; Cl \;\; Cl \qquad Cl \end{array}$$

Infrared spectroscopic data indicate that the hexamethyldisilazan is bound through the nitrogen to the titanium atom as a ligand.

At temperature above 30°C and with a molar ratio $(R_3Si)_2NH : TiCl_4 = 0.6-2$, in all cases the solutions deposit orange-red crystals, but their composition depends markedly on the reaction conditions. With a ratio of 1 : 1 over the temperature range of 20-50°C the composition of the products corresponds closely to the reaction

$$R_3SiNHSiR_3 + TiCl_4 \longrightarrow R_3SiNHTiCl_3 + R_3SiCl \qquad (4\text{-}128)$$
$$(R = CH_3)$$

The results of analysis of the proposed (trimethylsilylamino)-trichlorotitane are difficult to reproduce: Burger and Wannagat [608] explained this by the possible presence of the adduct or product of disubstitution. With an increase in the reaction time the composition of the substance containing the Si–N–Ti bond gradually changes from $C_3H_{10}Cl_3NSiTi$ to $C_5H_{13}Cl_2NSiTi$.

Hexamethyldisilazan forms no adduct with triethoxychlorotitane [601, 602]. At the same time, when alkoxychlorotitanes are treated with compounds such as Wannagat's reagent, replacement occurs smoothly [601, 604, 608]:

$$(RO)_n TiCl_{4-n} + (4-n)\, NaN\left(SiR'_3\right)_2 \longrightarrow (RO)_n Ti\left[N\left(SiR'_3\right)_2\right]_{4-n} \qquad (4\text{-}129)$$
$$(R = \text{iso-}C_3H_7;\ R' = CH_3 \text{ or iso-}C_3H_7O;\ n = 2-3)$$

An attempt to replace all four chlorine atoms in $TiCl_4$ by the group $N[Si(CH_3)_3]_2$ was unsuccessful and tris-substituted halogen-containing polymeric substances whose structure was not established were obtained. A monosubstitution product was obtained quite readily, but the reaction could not be described by the simple equation

$$TiCl_4 + NaN[Si(CH_3)_3]_2 \longrightarrow Cl_3TiN[Si(CH_3)_3]_2 + NaCl \qquad (4\text{-}130)$$

The reaction proceeds by a more complex route:

$$3TiCl_4 + 2NaN[Si(CH_3)_3]_2 \longrightarrow Na_2TiCl_6 + 2Cl_3TiN[Si(CH_3)_3]_2 \qquad (4\text{-}131)$$

The physicochemical constants of compounds containing the Si–N–Ti group are given in Table 73.

When stored in ampoules under nitrogen, compounds of the type $(RO)_3TiN(SiR'_3)_2$ and $(RO)_2Ti[N(SiR'_3)_2]_2$ do not change for an indefinite time. In the case where R = iso-C_3H_7 and R' = CH_3 the substances decompose at 130 and 165°C (1 mm), respectively, with the formation of tetraisopropoxytitane, hexamethyldisilazan, and a dark brown residue. Both derivatives of orthotitanic acid and [bis(trimethylsilyl)amino]trichlorotitane are hydrolyzed readily and react with alcohols, ammonia, orthotitanates, and orthosilicates.

Bis(trimethylsilyl)aminotrichlorotitane is the least stable of the nitrogen-containing organosilicon compounds of titanium and it decomposes during storage and reacts with pyridine, with which the derivatives $(RO)_4Ti$ do not react [604, 608]:

$$(R_3Si)_2NTiCl_3 + 2C_5H_5N \longrightarrow R_3SiCl + R_3SiN{=}TiCl_2 \cdot 2C_5H_5N \qquad (4\text{-}132)$$

Thus, a component of the complex formed is trimethylsilylaminodichlorotitane.

All the compounds known up to now containing the Si–N–Ti group are monomeric in solution. In the IR spectra a band at 785-810 cm^{-1} is assigned to ν_{as} SiNTi [608].

15. OTHER ORGANOSILICON DERIVATIVES OF TITANIUM

The action of the Na-derivative of a substituted cyclopentadiene on a carbonyl compound of titanium with the subsequent addition of R_2SiCl_2 to the reaction mixture yielded bis[(didodecylcyclopentadienyl)tetracarbonyltitana]dibenzylsilane [1680]. In this compound the titanium and silicon atoms may be connected directly.

No interaction is observed in the systems $SiX_4 - TiX_4$ [2]. At the same time, titanium tetrachloride forms stable 1 : 1 complexes with methylchlorosilanes and phenyltrichlorosilanes, which are soluble in toluene and active as catalysts for the polymerization of silvan [473]. Mixing $TiCl_4$ with tetraethylsilane leads to a vigorous reaction, as a result of which the titanium is reduced to the trivalent state and the ethylene liberated is polymerized by the $TiCl_3$ [35]. However, in benzene the complexes $(C_2H_5)_4Si \cdot 2TiX_4$ are formed and these are stable for 1-3 hr; their heats of formation and dipole moments are 3.16 and 2.11 kcal/mole and 2.38 and 2.98 D when X = Cl and Br, respectively [35, 494].

The concept of complex formation has already been introduced above to explain the nature of the interaction of chloro derivatives of titanium with alkoxysilanes (see Section 13.1.3). A study of the system $(C_2H_5O)_4Si-TiCl_4$ showed the presence of quite a stable 1 : 1 complex (dipole moment 6.30 D and degree of dissociation in dilute solutions 0.75) and an unstable 2 : 3 complex [309a].

In the system $(C_4H_9O)_4Ti \cdot SiCl_4$ it was found that 1 : 1, 2 : 1, and 4 : 1 complexes are formed. The existence of a 1 : 1 complex with such a weak acceptor as $SiCl_4$ led the authors [309] to suggest that tetrabutoxytitane acts as a bidentate ligand in this extremely unstable compound. The existence of the latter is evidently connected with decomposition of the 2 : 1 complex in which only one butoxyl group out of four participates in complex formation. The 2 : 1 and 4 : 1 complexes are current conducting, but the electrical conductivity decreases with time due to complex stepwise dissociation and electrolytic decomposition processes.

Compounds of the type $R_3SiOTi(OC_2H_4)_3N$ (where R = CH_3, C_2H_5, $C_6H_5CH_2$, C_6H_5) form complexes with iso-$C_3H_7OTi(OC_2H_4)_3N$ and $[N(C_2H_4O)_3Ti]_2O$ in which, in the opinion of the author of [639], the titanium is in a pentacoordinate state.

ZIRCONIUM AND HAFNIUM

16. ORGANOSILICON DERIVATIVES OF ZIRCONIUM AND HAFNIUM

In the author's certificate of Andrianov [1461] it was reported that zirconasiloxanes may be obtained by the reactions of organosilicon compounds containing silanol groups with metallic zirconium or its inorganic derivatives. However, no single concrete example of the preparation of such compounds was given. The first organosilicon derivative of zirconium, namely, tetrakis(trimethylsiloxy)zircane,* was described 10 years ago [574, 575]. It was obtained in 56% yield from trimethylacetoxysilane and tetra-

*The principles of the nomenclature of organic derivatives of elements of the titanium subgroup were described in [25].

isopropoxyzircane under conditions which had proved successful in the synthesis of analogous titanium derivatives:

$$(RO)_4Zr + 4R'_3SiOCOCH_3 \longrightarrow (R'_3SiO)_4Zr + 4ROCOCH_3 \quad (4\text{-}133)$$

The replacement of methyl by ethyl radicals raises the yield. Thus, $[C_2H_5(CH_3)_2SiO]_4Zr$ is obtained in 80% yield and $[CH_3(C_2H_5)_2 \cdot SiO]_4Zr$ in 92% yield. At the same time, in the preparation of tetrakis(triorganosiloxy)titanes by an analgous scheme the yield of these compounds is independent of the nature of the alkyl radical in $R_3SiOCOCH_3$ ($R = CH_3$ or C_2H_5) and is 93-97% in all cases [575].

Tetrakis(triorganosiloxy)zircanes may also be synthesized by means of triorganosilanols [575]:

$$4R_3SiOH + (R'O)_4Zr \longrightarrow (R_3SiO)_4Zr + 4R'OH \quad (4\text{-}133a)$$

In contrast to titanium derivatives, the dependence of the yield of $(R_3SiO)_4M$ on the character of the radical attached to the silicon atom is also preserved in this case.

Triphenylsiloxy derivatives of zirconium and hafnium have been synthesized [826] by a scheme which is a general method of preparing compounds of the type $[(C_6H_5)_3SiO]_4M$, where M is a Group IV element:

$$4R_3SiONa + MCl_4 \longrightarrow (R_3SiO)_4M + 4NaCl \quad (4\text{-}134)$$

In this case $M = Zr$ or Hf.

In addition to these methods of synthesizing organosilicon derivatives there is also the silanolysis reaction [575]:

$$4R_3SiOH + [(CH_3)_3SiO]_4Zr \longrightarrow (R_3SiO)_4Zr + 4(CH_3)_3SiOH \quad (4\text{-}135)$$

A mention of tetrakis(dimethylethylsiloxy)hafnane [535] was not supported by any information on its properties or the preparation method.

The constants of organosilicon compounds containing Si–O–Zr and Si–O–Hf groups are given in Table 74.

In a comparison of the properties of tetrakis(triorganosiloxy)-derivatives of titanium and zirconium the tendency of the latter for association is particularly noteworthy. Tetrakis(trimethylsiloxy)zircane is evidently a dimer, though no band corresponding to a Si–O–Zr coordination bond has been detected in its spectrum [535]. As a rule the degree of polymerization decreases when methyl radicals are replaced by ethyl radicals, evidently due to steric effects. Trimethylsilyl and dimethylethylsilyl derivatives of

TABLE 74. Organosilicon Derivatives of Zirconium and Hafnium

Formula	m.p., °C	b.p., °C (mm)	Preparation method (scheme)	Yield, %	Degree of association†	Absorption band in IR spectrum, cm^{-1}	
						for Si–O	for Mn–O
$[(CH_3)_3SiO]_4Zr$	152	135° (0.1) subl.	4-132 4-133	56 73	2.05	916	521
$[C_2H_5(CH_3)_2SiO]_4Zr$	105	105 (0.1)	4-132 4-133 4-135	80 79 * 97	1.20	913	521
$[n\text{-}C_3H_7(CH_3)_2SiO]_4Zr$	60	103 (0.05)	4-133	81 *	1.11	—	—
$[iso\text{-}C_3H_7(CH_3)_2SiO]_4Zr$	—	110 (0.1)	4-133	91	1.01	916	515
$[CH_3(C_2H_5)_2SiO]_4Zr$	30	120 (0.1)	4-132 4-133	92 95	1.05	912	521
$[(C_2H_5)_3SiO]_4Zr$	—	147 (0.1)	4-132 4-133 4-135	90 99 91	0.98	914	—
$[(C_6H_5)_3SiO]_4Zr$	410, decomp.	360 (10^{-4}) subl.	4-134	—	—	—	—
$[C_2H_5(CH_3)_2SiO]_4Hf$	—	—	?	?	1.0	932	—
$[(C_6H_5)_3SiO]_4Hf$	383, decomp.	360 (10^{-4}) subl. decomp.	4-134	—	—	—	—

* Obtained from $(tert\text{-}C_4H_9O)_4Zr$; in all other cases $(iso\text{-}C_3H_7O)_4Zr$ was used.
† On the basis of an ebullioscopic determination of the molecular weight.

zirconium boil at a higher temperature than their titanium analogs. It is interesting that compounds of the type $(R_3SiO)_4M$ (R = alkyl) with M = Ti and Zr are "antipodes" with respect to physical state. Compounds of zirconium analogous in structure and composition to liquid compounds of titanium are solid and vice versa. The IR spectra of triorganosiloxy derivatives of titanium and zirconium are very similar, primarily in the position of the bands corresponding to vibrations of the Si–O and M–O bonds in the Si–O–M group.

In all cases the yield of $(R_3SiO)_4Zr$ is lower than for the titanium analogs. This is evidently connected with the great strength of tetraalkoxyzircanes as Lewis acids and, simultaneously, their high sensitivity to hydrolysis [575].

According to the data of Bradley [572], the hydrolysis of tetrakis(trimethylsiloxy)zircane under the same conditions as its titanium analog [571] proceeds similarly. Polymeric products of the series $\{ZrO_x[OSi(CH_3)_3]_{4-2x}\}_n$ are formed initially. When h [the molar ratio $H_2O:(R_3SiO)_4Zr$] is 0.1–0.5, the value $x > h$; this is explained by partial condensation of the trimethylsilanol formed by hydrolysis:

$$2Zr(OSiR_3)_4 + H_2O \longrightarrow 2R_3SiOH + [(R_3SiO)_3Zr]_2O \quad (4\text{-}136)$$
$$\quad\quad\quad\quad \hookrightarrow (R_3Si)_2O + H_2O$$

When $h = 0.75-3.0$ the value $h > x$ and this is connected with the increasing resistance of the compounds to hydrolysis with an increase in x or the reversibility of the process (possibly with both factors).

When the hydrolysis product is heated to 150°C (0.1 mm) disproportionation occurs with the liberation of $(R_3SiO)_4Zr$ and the formation of "secondary" polymers $\{ZrO_y[OSi(CH_3)_3]_{4-2y}\}_m$, which lose their solubility when $y \cong 1.6$ (Table 75).

It is believed that polymers of the first type consist of repeating octahedra, connected by chains of four Zr–O–Zr groups. The structure of polymers of the second type is based on a cubic octamer.

The zirconium content of $(R_3SiO)_4Zr$ is determined by precipitation with ammonia from an alcohol solution as in the case of analogous titanium derivatives [575].

TABLE 75. Hydrolysis Products of $[(CH_3)_3SiO]_4Zr$ [572]

Hydrolysis time, hr	$\{ZrO_x[OSi(CH_3)_3]_{4-2x}\}_n$			$\{ZrO_y[OSi(CH_3)_3]_{4-2y}\}_m$		
	x	molecular weight*	n	y	molecular weight*	m
0	0	916	2.05	—	—	—
0.5	0.32	1150	2.85	1.505	3460	17.34
1.0	0.675	1850	5.46	1.560	4130	26.97
2.0	1.107	2730	10.16	1.611	9211	51.16
3.0	1.377	2790	12.36	1.632	Insoluble	

* Cryoscopic determination in cyclohexane.

The preparation of bis[(methylethylcyclopentadienyl)tetracarbonylzircana]dibutylsilane has been described [1680].

Organic salts of zirconium (naphthenate, stearate, etc.), tetraalkoxyzircanes, complexes of caprolactam with $ZrOCl_2$,* and products of the reaction of $ZrCl_4$ with glycols are used as catalysts for the hardening of siloxane resins [1445, 1582, 1607] and water proofing compositions based on organosilicon compounds [252, 1354, 1627], agents for "cold vulcanization" of siloxane elastomers [1416], and catalysts for the copolymerization of siloxane resins with alkyd [1659].

Compositions based on methylphenylsiloxane resins, tetraethoxysilane, and tetrabutoxyzircane are recommended as separating compositions [1440].

Zirconates and fluorozirconates of Group II metals, zirconium silicates [1407, 1577], and $Zr(OH)_4$ [1433] are used as additives to improve the thermal stability of siloxane rubbers.

* And also with $TiCl_3$ [1445].

Chapter 5

Organosilicon Compounds of Group V Elements

ARSENIC

1. COMPOUNDS CONTAINING THE Si – O – As GROUP

1.1. Preparation Methods

1.1.1. Cohydrolysis Reactions

The reaction of arsenic trichloride with triphenylchlorosilane in benzene in the presence of the calculated amount of 30% ammonium hydroxide forms crystalline tris(triphenylsilyl) arsenite [631]:

$$3\,(C_6H_5)_3\,SiCl + AsCl_3 + 3H_2O \xrightarrow[-NH_4Cl]{+NH_3} [(C_6H_5)_3\,SiO]_3\,As \qquad (5\text{-}1)$$

The yield of this compound is substantially affected by even slight variations in the rate of addition of $AsCl_3$ and the stirring rate. The maximum yield (50%) was obtained by adding $AsCl_3$ (in the form of a 0.2 M solution in benzene) to a mixture of $(C_6H_5)_3SiCl$, benzene, and NH_4OH at a rate of 0.02 mole/hr. The best reproducibility is obtained by adding ammonium hydroxide to a mixture of arsenic trichloride and triphenylchlorosilane in benzene. However, in this case the yields are lower. Applying an analogous procedure to an equimolar mixture of $AsCl_3$ and diphenyldichlorosilane led to the separation of a crystalline product that corresponded according to analysis data and determination of the molecular weight to the compound $C_{36}H_{30}As_2Si_3O_6$ which was assigned a

bicyclic structure

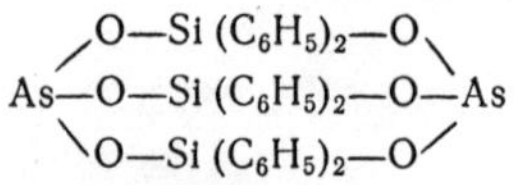

$$\mathrm{As}\begin{matrix}\diagup \mathrm{O{-}Si(C_6H_5)_2{-}O} \diagdown \\ \mathrm{{-}O{-}Si(C_6H_5)_2{-}O{-}} \\ \diagdown \mathrm{O{-}Si(C_6H_5)_2{-}O} \diagup\end{matrix}\mathrm{As}$$

In addition to this substance, a very viscous liquid, which contained silicon and arsenic in a ratio of 10 : 1, was obtained as a residue.

Cohydrolysis of arsenic trichloride with phenyltrichlorosilane in the presence of ammonia forms a polymer which is close in composition to $[-(C_6H_5SiO_{1.5})_8AsO_{1.5}H-]_n$ [630, 632]. It is soluble in many organic solvents, but does not soften or melt on heating. On this basis it was assigned a cyclolinear structure, analogous to the structure of alumasiloxanes obtained from trifunctional organosilicon compounds [11, 173]:

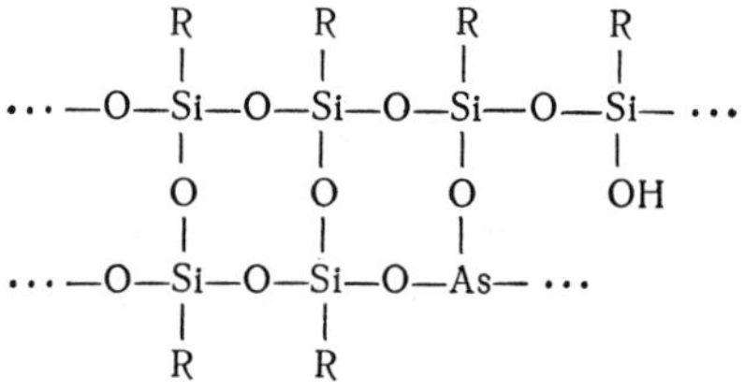

$$\begin{array}{ccccccccc}
 & R & & R & & R & & R & \\
 & | & & | & & | & & | & \\
\cdots\text{—O—} & \text{Si} & \text{—O—} & \text{Si} & \text{—O—} & \text{Si} & \text{—O—} & \text{Si} & \text{—}\cdots \\
 & | & & | & & | & & | & \\
 & O & & O & & O & & OH & \\
 & | & & | & & | & & & \\
\cdots\text{—O—} & \text{Si} & \text{—O—} & \text{Si} & \text{—O—} & \text{As} & \text{—}\cdots & & \\
 & | & & | & & & & & \\
 & R & & R & & & & &
\end{array}$$

1.1.2. Reaction of Arsenic Halides with Silanols

The reaction of triphenylsilanol with $AsCl_3$ in the presence of ammonia as a hydrogen halide acceptor yields tris(triphenylsilyl)-arsenite in ~60% yield in benzene or ~40% yield in ether [631]:

$$3\,(C_6H_5)_3\,SiOH + AsCl_3 \xrightarrow[-NH_4Cl]{+NH_3} [(C_6H_5)_3\,SiO]_3\,As \qquad (5\text{-}2)$$

The product from the reaction of equimolar amounts of diphenylsilanediol and phenyldichloroarsine [630, 632] has an eight-membered cyclic structure:

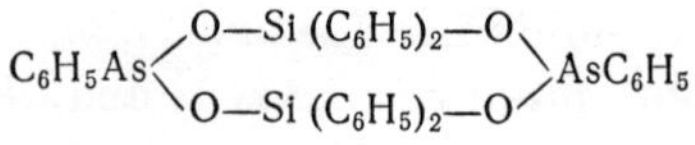

$$C_6H_5As\begin{matrix}\diagup O{-}Si\,(C_6H_5)_2{-}O \diagdown \\ \diagdown O{-}Si\,(C_6H_5)_2{-}O \diagup\end{matrix}AsC_6H_5$$

It may be regarded as a dimer of the series $[-OSi(C_6H_5)_2 \cdot OAsC_6H_5-]_n$, particularly as there is the simultaneous formation of oligomers with $n > 2$.

Blowing anhydrous ammonia through a solution of a mixture of $AsCl_3$ and diphenylsilanediol (2 : 3) in ether led to the elimination of 82% of the HCl as NH_4Cl, but the yield of $As[OSi(C_6H_5)_2O]_3As$ was only 30% [631]. When triethylamine was used as the HCl acceptor, 95% of the hydrogen chloride was bound, but the yield of the bicyclic compound was only 42%. We will examine the possible reasons for this somewhat later.

1.1.3. Reactions of Halosilanes with Arsenic Acids and Their Salts

Another variant of the formation of arsenasiloxane groups by condensation with intermolecular elimination of hydrogen halide is the reaction of organoarsonic acids with alkylchlorosilanes [867, 1625]. With a molar ratio of 1 : 2 the reaction proceeds mainly according to the scheme

$$2R_nSiX_{4-n} + R'As(O)(OH)_2 \xrightarrow[-HCl]{} R'As[OSi(R_n)X_{3-n}]_2 \qquad (5\text{-}3)$$

$$(n = 1\text{—}3;\ R \text{ and } R' = CH_3 \text{ or } C_6H_5;\ X = \text{preferably Cl})$$

In a benzene medium the reaction proceeds vigorously in the cold, but for the maximum elimination of HCl it is recommended that the mixture be boiled under reflux.* For isolation of the primary products in a sufficiently pure form the reaction must be carried out in the absence of moisture. Hydrolytic conversion of oligomers to polymeric arsenasiloxanes may be observed subsequently, but this need not occur. However, it should be noted that the hydrolysis is accompanied by partial cleavage of Si–O–As groups and the composition of the final copolymers is represented by the formula $\{-(R_2SiO)_{An}-[R'As(O)O]_n-\}_x$ when n = 2 and by the formula $\{-(RSiO_{1.5})_{An}-[R'As(O)O]_n-\}_x$ when n = 1.

Thus, in particular, when reaction (5-3) is carried out with dimethyldichlorosilane, hydrolysis of the bis(chlorodimethylsilyl)-methylarsenate formed in CCl_4 with water with subsequent removal of the HCl and distillation of the solvent leads to the formation of a semitransparent rubbery material with a composition close to $C_{25}H_{75}O_{14}Si_{12}As$. In the case of the pair methylarsonic acid–methyltrichlorosilane, hydrolysis of the primary reaction product with

* When n = 3 it is recommended that the reaction be carried out in the presence of pyridine.

boiling water finally gives a white powdery polymer, which is close in composition to $C_{13}H_{39}O_{20}Si_{12}As$. Thus, for $R = R' = CH_3$ and n = 1 or 2 the value of A is practically the same and equal to 12.

The reaction of triphenylchlorosilane with KH_2AsO_4 (3 : 1) in ether proceeds actively even at room temperature despite the insolubility of KH_2AsO_4 in ether [631]. The main product of the reaction is tris(triphenylsilyl)arsenate. Together with this, KCl and free arsenic acid are formed, but not HCl as might have been expected. Therefore, Chamberland and MacDiarmid believe that the process consists of two stages:

$$R_3SiCl + KOAs(O)(OH)_2 \longrightarrow R_3SiOAs(O)(OH)_2$$
$$3R_3SiOAs(O)(OH)_2 \longrightarrow (R_3SiO)_3AsO + 2(HO)_3AsO \quad (5\text{-}4)$$
$$(R = C_6H_5)$$

However, the possibility of the second stage (disproportionation) is doubtful even to the authors themselves, who believe that it is probable that there is stepwise condensation of the primary products with intermediate active complexes involving the unshared electron pair of the oxygen in As = O and the 3d orbital of silicon:

$$R_3Si{-}O{-}As(OH)(OH){=}O \dashrightarrow SiR_3{-}O{-}As(=O)(OH)_2$$

It was not possible to isolate intermediate products of reaction (5-4) even with a twofold excess of KH_2AsO_4 – in this case there was only a fourfold reduction in the yield of tris(triphenylsilyl)arsente (from 87 to 20%).

The action of excess KH_2AsO_4 on diphenyldichlorosilane (stirring in ether for 100 hr at room temperature) leads to the formation of a polymer [631]:

$$4xKH_2AsO_4 + 2x(C_6H_5)_2SiCl_2 \xrightarrow{\text{ether}}$$
$$\longrightarrow HO{-}\underset{OH}{\overset{O}{\overset{\|}{As}}}\left[{-}O{-}\underset{C_6H_5}{\overset{C_6H_5}{Si}}{-}O{-}\underset{OH}{\overset{O}{\overset{\|}{As}}}{-}\right]_{2x}OH + (2x - 1)H_3AsO_4 + 4xKCl \quad (5\text{-}5)$$

The degree of conversion in reaction (5-5) is 99% according to the KCl and 96% according to the arsenic acid. With a slow distillation of the ether from the filtrate there first separates a small amount of hexaphenylcyclotrisiloxane (!). With further evaporation of the ether the reaction product assumes the consistency of a clear viscous sticky liquid, which forms an elastic resinous film on glass. Further vacuum treatment yields a solid spongy polymer, in which x=1 according to analysis. However, it was not possible to determine its molecular weight due to the fact that careful drying was accompanied by partial decomposition and the loss of solubility in benzene. On the other hand, if it were left as a viscous liquid, it continued to deposit H_3AsO_4 gradually. In this case the value of x rises and the final polymer approaches the composition $[-OSi(C_6H_5)_2OAs(O)OH-]_n$. It melts in the range of 150-180°C. The authors are inclined to ascribe the formation of hexaphenylcyclotrisiloxane to the elimination of diphenylsiloxano fragments from the arsenasiloxane polymer, which is accompanied by the formation of polyarsenates.

The reaction of triorganochlorosilanes with silver arsenate is essentially a development of scheme (5-4) [630, 632, 1116, 1119]:

$$3R_3SiCl + Ag_3AsO_4 \longrightarrow (R_3SiO)_3AsO + 3AgCl \qquad (5-6)$$

1.1.4. Other Methods

The reaction of silanolates of alkali metals with halides or organohalides of trivalent arsenic proceeds in the following way [630-632, 1085, 1100]:

$$(3-n)R_3SiOM + R_nAsX_{3-n} \longrightarrow (R_3SiO)_{3-n}AsR_n + (3-n)MX \qquad (5-7)$$

$$(R = CH_3,\ C_6H_5;\ M = Li,\ Na;\ X = Cl,\ I;\ n = 0-2)$$

In this way it is possible to obtain derivatives containing from one to three triorganosiloxy groups at the As^{III} atom.

It is not possible to obtain an arsenasiloxane by the action of sodium trimethylsilanolate on R_4AsCl ($R=CH_3$ or C_6H_5) [1085]. In particular, this is connected with the insolubility of R_4AsCl in organic solvents. On the other hand, the formation of the compounds $R_3SiOAsR'_4$ from the fragments R'_4As and $OSiR_3$ is not favored energetically. Compounds of the type R'_3AsCl_2 are soluble, but their reaction with sodium trimethylsilanolate proceeds in the following way:

$$R'_3AsCl_2 + 2NaOSiR_3 \longrightarrow R_3SiOSiR_3 + R'_3AsO + 2NaCl \quad (5\text{-}8)$$

Since the reaction proceeds in this direction even under mild conditions it is believed that the arsenasiloxane formed initially decomposes readily:

$$R_3Si \vdots \overset{\frown}{O} - \underset{R' \quad R'}{\overset{R'}{As}} \vdots \overset{\frown}{O} - SiR_3$$

In the case where $R = R' = CH_3$, with an equimolar ratio trimethylchlorosilane and $(CH_3)_3AsO$ are obtained.

The reaction of arylarsonic acids with triorganosilanols with azeotropic distillation of the water with an inert solvent [24, 51, 1516] and with triorganomethoxysilane or triorganoacetoxysilanes [24] leads to the formation of bis(triorganosilyl)arylarsonates:

$$2R_3SiOR' + ArAs(O)(OH)_2 \longrightarrow ArAs(O)(OSiR_3)_2 + 2R'OH \quad (5\text{-}9)$$
$$(R = C_2H_5,\ C_6H_5;\ R' = H,\ CH_3,\ COCH_3;\ Ar = C_6H_5,\ o\text{-}NH_2C_6H_4)$$

One might have expected the formation of tris(triorganosilyl) arsenates by the action of arsenic pentoxide on hexaorganodisiloxanes. However, in an investigation of the properties of tris(trimethylsilyl)arsenate it was found that in contrast to its stable phosphorus analog, the arsenic derivative decomposes even at room temperature (and also during distillation in vacuum) with the formation of hexamethyldisiloxane and polymeric trimethylsilyl metaarsenate. It is precisely for this reason that the main product of the reaction of arsenic pentoxide with hexamethyldisiloxane is trimethylsilyl metaarsenate, while the yield of tris(trimethylsilyl)arsenate is less than 5% [1116, 1119]:

$$3xR_3SiOSiR_3 + As_2O_5 \longrightarrow 2[R_3SiOAsO_2]_x + 2xR_3SiOSiR_3 \rightleftarrows 2x(R_3SiO)_3AsO$$
$$(R = CH_3) \quad (5\text{-}10)$$

1.2. Physical Properties

The physicochemical constants of organosilicon esters of oxygen-containing acids of arsenic are given in Table 76.

TABLE 76. Organosilicon Esters of Oxygen-Containing Acids of Arsenic

Formula	m.p., °C	b.p., °C (mm)	n_D^{20}	Preparation method	Yield, %	Literature
$[(CH_3)_3SiO]_3AsO$	−1.5	76—77 (1.2)	1.4249	5-6	29	1116, 1119
$[(C_6H_5)_3SiO]_3AsO$	236—240; 240—242*	—	—	5-4	87	631
$[Cl(CH_3)_2SiO]_2As(O)CH_3$	—	—	—	5-3	68	867
$[Cl_2(CH_3)SiO]_2As(O)CH_3$	—	—	—	5-3	84	867
$[(C_2H_5)_3SiO]_2As(O)C_6H_5$	—	195	1.4969†	5-9	71	1516
$[Cl(C_6H_5)_2SiO]_2As(O)C_6H_5$	—	—	—	5-3	80	867
$[Cl_2(C_6H_5)SiO]_2As(O)C_6H_5$	—	—	—	5-3	78	867
$[(C_6H_5)_3SiO]_2As(O)C_6H_5$	—	—	—	5-3	91	867
$[(C_6H_5)_3SiO]_2As(O)C_6H_4NH_2$-*o*	193—195 (from hexane); 195–196,5 (from benzene)	—	—	5-9	66	1516
$[(C_6H_5)_3SiO]_3As$	190	—	—	5-1	50	631
	188—191; 189.5	—	—	5-2	62; 39	631
	189—190	—	—	5-7	49	631
$[(C_6H_5)_3SiO]_2AsC_6H_5$	119—120	—	—	5-7	—	632
$(CH_3)_3SiOAs(CH_3)_2$	< −70	56.5 (90); 116 (720—725)	—	5-7	84	1085, 1100
$As[OSi(C_6H_5)_2O]_3As$	194—195	—	—	~5-1	40	631
	194	—	—	~5-2	42; 30	631
$C_6H_5As—O—Si(C_6H_5)_2$ / O, O / $(C_6H_5)_2Si—O—AsC_6H_5$ (cyclic)	145	—	—	—	—	630, 632

* After additional purification, m.p. 238.5–239.8°C.
† d_4^{20} 1.1117.

Tris(triphenylsilyl)arsenate is moderately soluble in benzene, carbon tetrachloride, and acetone, but is insoluble in methanol and boiling ether [631]. The bis(triphenylsilyl) ester of o-aminophenylarsonic acid is readily soluble in benzene and xylene, but difficulty soluble in ether and hexane [1516]. Trimethylsiloxydimethylarsine is very soluble in organic solvents and solidifies at a temperature below -70°C (colorless crystals) [1085, 1100].

In the IR spectra of compounds containing the Si–O–As group, a line with a wave number of 912 cm^{-1} is assigned to vibrations of the Si–O–As^V group [631, 632]. As regards the Si–O–As^{III} group, both a band at 882-887 cm^{-1} [631, 632] and one at 934 cm^{-1} [1085] are assigned to this group. The NMR spectrum of tris(trimethylsilyl)arsenate is characterized by a sharp signal at 18.5 Hz (CCl_4, 60 MHz, standard tetramethylsilane) [1084].

1.3. Chemical Properties

We have already mentioned the instability of tris(trimethylsilyl)arsenate. According to the data of Schmidt and his co-workers [1116] scheme (5-10) does not fully cover its thermal conversions. The first stage is a condensation:

$$2\,(R_3SiO)_3\,AsO \longrightarrow (R_3SiO)_2As(=O){-}O{-}As(=O)(OSiR_3)_2 + R_3SiOSiR_3 \quad (R = CH_3) \tag{5-11}$$

Tetrakis(trimethylsilyl)diarsenate then decomposes with the liberation of hexamethyldisiloxane and the formation of polymeric trimethylsilyl metaarsenate.

The thermal instability of tris(trimethylsilyl)arsenate is combined with sensitivity to solvolysis:

$$(R_3SiO)_3\,AsO + HOH \longrightarrow H_3AsO_4 + R_3SiOSiR_3 \tag{5-12a}$$

$$(R_3SiO)_3\,AsO + R'OH \longrightarrow H_3AsO_4 + R_3SiOR' \tag{5-12b}$$

Tris(triphenylsilyl)arsenate cannot be distilled in vacuum. It decomposes at an appreciable rate at 290°C. After the material has been heated at 300-360°C for several hours, solid products are collected in a trap and these melt or soften at ~142°C, but do not contain arsenic. The residue in the flask consists of a sticky viscous substance [631]. It is believed that the decomposition initially proceeds according to scheme (5-11), i.e., with the liberation of hexaphenyldisiloxane and the formation of a compound containing the As–O–As bond.

The action of aqueous acetone on tris(trimethylsilyl)arsenite for 2 hr produces hardly any changes. In general the hydrolytic resistance to pure water at room temperature rises in the following series [632]:

$$(HO)_2\,As(O)[OSiR_2OAs(O)(OH)]_2OH < (R_3SiO)_3\,AsO <$$

$$< (R_3SiO)_3\,As < RAs\,[OSiR_2O]_2\,AsR < As\,[OSiR_2O]_3\,As \quad (R = C_6H_5)$$

Thus, the bicyclic compound shows the least tendency for hydrolysis. However, after treatment with 5% aqueous acetone at room temperature for 1.5 hr with subsequent removal of the solvent in vacuum in the cold, this compound, though it does not lose weight, melts at a lower temperature (at 176-180°C instead of 194°C) and does not crystallize on cooling, but changes into a glassy state. If the solvent is removed by normal distillation, the melting point is lowered to 140-143°C. In the opinion of Chamberland and MacDiarmid [632], this may be connected with partial conversion of the cyclic into a linear structure. We should remember that in the reaction of arsenic trichloride with diphenylsilanediol the yield of bicyclic $As[OSi(C_6H_5)_2O]_3As$ is low. However, it is possible to isolate from the reaction mixture another compound, to which has been assigned the structure $As[OSi(C_6H_5)_2OSi(C_6H_5)_2O]_3As$ in analogy, though this substance does not have sharp physicochemical constants, but softens and melts over quite a wide range (35-50°C). It is difficult to say whether this compound, which may equally be cyclic or cyclolinear, is the product of a conversion of the primary bicyclic compound (disproportionation with the separation of As_2O_3) or whether it is formed directly in the condensation process. Another cyclic compound $[C_6H_5AsOSi(C_6H_5)_2O]_2$ is characterized by the fact that at 350°C in vacuum it is converted into a substance of the same composition, though with a higher molecular weight (this is believed to be the trimer).

Tris(triphenylsilyl)arsenite has a much higher thermal stability than tris(triphenylsilyl)arsenate. It distills in vacuum at 351°C and changes substantially only when heated to 420°C for several hours.

Trimethylsiloxydimethylarsine [1085, 1100] is a highly toxic colorless liquid with an unpleasant odor; it is stable below 150°C and distills without decomposition in pure dry nitrogen. The compound is highly sensitive to oxidation and hydrolysis. In contact with atmospheric oxygen at 60°C the oxidation is explosive. The hydrolysis is accompanied by the formation of trimethylsilanol (hexamethyldisiloxane) and $(CH_3)_2AsOH$ (cacodyl oxide).

The reaction with thionyl chloride, which is accompanied by the elimination of siloxane oxygen, may be used for analytical purposes:

$$(CH_3)_3SiOAs(CH_3)_2 + SOCl_2 \longrightarrow (CH_3)_3SiCl + (CH_3)_2AsCl + SO_2 \quad (5\text{-}13)$$

1.4. Application

Up to the present time the only possible practical use described [867] and patented [1625] is for polyorganoarsenasiloxanes obtained by hydrolysis of the products from the reaction of organoarsonic acids with polyfunctional organochlorosilanes (see scheme 5-3). According to the data of Kary and Frisch these compounds combine the usual properties of polysiloxanes with the capacity to resist the action of various fungi and microorganisms. This is the reason why they are recommended for use as water-repellent impregnating agents for cloths and papers and also as pesticides and fungicides.

2. COMPOUNDS CONTAINING THE Si – (C)$_n$ – As GROUP

2.1. Preparation Methods

The first organosilicon compound of arsenic, namely, [p-(triethylsilyl)phenyl] diphenylarsine, was synthesized in 1917 [821]:

$$R_3SiC_6H_4Cl + 2Na + ClAsR'_2 \longrightarrow R_3SiC_6H_4AsR'_2 + 2NaCl \qquad (5\text{-}14)$$

Later Seyferth used organomagnesium derivatives of organic compounds of silicon [1137, 1644] and arsenic [1154] to prepare substances containing the Si – (C)$_n$ – As group:

$$(C_6H_5)_2As{-}C_6H_4{-}MgBr + (CH_3O)_3SiCH_3 \xrightarrow{\text{ether}}$$

$$\longrightarrow (C_6H_5)_2As{-}C_6H_4{-}SiCH_3(OCH_3)_2 \quad (60\%) \qquad (5\text{-}15)$$

$$3\,(CH_3)_3SiCH_2MgCl + AsCl_3 \xrightarrow{\text{THF}} [(CH_3)_3SiCH_2]_3As \quad (80\%) \qquad (5\text{-}16)$$

These reactions show no unusual features and evidently they may be used equally well for the synthesis of organosilicon compounds of arsenic with various organic bridges between Si and As atoms and different hydrocarbon substituents or functional groups at these atoms.*

* There is a reference in [89] to the synthesis of Grutner's compound by the organomagnesium method from $(C_2H_5)_3SiC_6H_4Br$. However, the corresponding description is absent from the abstract of the patent [1519].

A third of the methods known at the present time for preparing organosilicon derivatives of arsenic of this type involve the reactions of triphenylarsinemethylene [815, 1145]:

$$(C_6H_5)_3\,AsCH_2 + (CH_3)_3\,SiBr \xrightarrow[\text{nitrogen}]{\text{ether}} [(C_6H_5)_3\,AsCH_2Si\,(CH_3)_3]\,Br \qquad (5\text{-}17)$$

The formation of a compound with a (trimethylsilylmethyl)-triphenylarsonium cation is demonstrated by its conversion into an analytically pure picrate, which has a sharp melting point.

2.2. Properties

Like its phosphorus analog, crystalline tris(trimethylsilylmethyl)arsine has a sharp unpleasant odor. However, it is much more resistant to the action of atmospheric oxygen and may be stored for several months in a dark bottle with a hermetical seal. The substance readily adds bromine and iodine to form derivatives of pentavalent arsenic of the type $(R_3SiCH_2)_3AsX_2$, which melt with decomposition at ~120°C. The reaction with organic iodides leads to quaternary arsonium derivatives:

$$[(CH_3)_3\,SiCH_2]_3\,As + RI \xrightarrow{N_2} \{R\,[(CH_3)_3\,SiCH_2]_3\,As\}\,I \qquad (5\text{-}18)$$

$$(R = CH_3,\ C_2H_5,\ (CH_3)_3\,SiCH_2)$$

Both tris(trimethylsilylmethyl)arsine and [p-(triethylsilyl)-phenyl]diphenylarsine form crystalline 1 : 1 adducts with mercury halides and these are soluble in organic solvents. When methyl-tris(trimethylsilylmethyl)arsonium iodide is boiled in alcohol with mercurous iodide, the latter gradually dissolves. When the solution is cooled there is an almost quantitative precipitate of a new "onium" compound $\{CH_3[(CH_3)_3SiCH_2]_3As\}HgI_3$.

The physicochemical constants of arsenic derivatives containing silicon atoms in an organic radical are given in Table 77.

3. COMPOUNDS CONTAINING THE Si – N – As GROUP

3.1. Preparation Methods

Until recently the only compound of this class known was trimethylsilylaminodichloroarsine [544]:

$$(CH_3)_3\,SiNHSi\,(CH_3)_3 + AsCl_3 \longrightarrow (CH_3)_3\,SiNHAsCl_2 + (CH_3)_3\,SiCl \qquad (5\text{-}19)$$

TABLE 77. Organosilicon Derivatives of Arsenic Containing the Si–(C)–As Group

Formula	m.p., °C	Preparation method (scheme)	Yield, %	Literature
$[(CH_3)_3SiCH_2]_3As$	67–68,5	5-15	80	1137, 1644
$[n\text{-}C_4H_9(CH_3)_2SiCH_2]_3As$	—	5-15	—	1644
$[C_6H_{11}(CH_3)_2SiCH_2]_3As$	—	5-15	—	1644
$[(CH_3)_3SiCH_2As(C_6H_5)_2]Br$	177–180 decomp. 160–165	5-17	—	1145
$[(CH_3)_3SiCH_2As(C_6H_5)_2][OC_6H_2(NO_2)_3\text{-}2,4,6]$	113–115	—	—	1145
$\{CH_3[(CH_3)_3SiCH_2]_3As\}I$	193–195 decomp.	5-18	—	1137
$\{C_2H_5[(CH_3)_3SiCH_2]_3As\}I$	112–115	5-18	—	1137
$[(CH_3)_3SiCH_2]_3AsBr_2$	118–120 decomp.	—	—	1137
$[(CH_3)_3SiCH_2]_3AsI_2$	118–120 decomp.	—	—	1137
$\{(CH_3)_3SiCH_2]_3As \cdot HgCl_2$	176–176.8 decomp	—	—	1137
$\{CH_3[(CH_3)_3SiCH_2]_3As\}HgI_3$	134–135	—	91	1137
$\{[(CH_3)_3SiCH_2]_4As\}I$	143.5–145	5-18	—	1137
$CH_3(CH_3O)_2Si-C_6H_4-As(C_6H_5)_2$	—*	5-14	60	1154
$(C_2H_5)_3Si-C_6H_4-As(C_6H_5)_2$	—†	5-14	Quant.	821
$(C_2H_5)_3Si-C_6H_4-As(C_6H_5)_2 \cdot HgCl_2$	188	—	—	821
$(C_2H_5)_3Si-C_6H_4-As(C_6H_5)_2 \cdot HgBr_2$	181	—	—	821
$(C_2H_5)_3Si-C_6H_4-As(C_6H_5)_2 \cdot HgI_2$	139.5	—	—	821

* B. p. 180-200°C(0.45 mm); n_D^{25} = 1.6111; d_4^{25} = 1.243.
† B. p. 279-281°C(17mm); $n_D^{21.3}$ = 1.6146; d_4^{20} = 1.1673.

It was shown recently [1058] that when arsenic trichloride is replaced by (dimethylamino)dichloroarsine the reaction proceeds differently:

$$(CH_3)_3SiNHSi(CH_3)_3 + (CH_3)_2NAsCl_2 \longrightarrow (CH_3)_3SiCl + (CH_3)_2NH + \\ + 1/n\left[(CH_3)_3Si\overset{|}{N} - \underset{|}{As}Cl\right]_n \qquad (5\text{-}20)$$

The first product is evidently $(CH_3)_3SiNHAs[N(CH_3)_2]Cl$, which then decomposes with the liberation of dimethylamine and a silicon-substituted arsenazan. At the same time, heating hexamethyldisilazan with excess dimethylchloroarsine for 6 hr at 100-120°C without a solvent [1062] leads to the formation of tris(dimethylarsino)amine.

The reaction of phenyldichloroarsine with the heterocycle (I) proceeds analogously to scheme (3-61) to form dimethyldichlorosilane in 75% yield and compound (II) in 42% yield [514]:

$$\begin{array}{ccc} H_2C & — & CH_2 \\ | & & | \\ H_5C_2—N & & N—C_2H_5 \\ & \diagdown\ \diagup & \\ & Si & \\ H_3C\diagup & & \diagdown CH_3 \\ & \text{I} & \end{array} \qquad \begin{array}{ccc} H_2C & — & CH_2 \\ | & & | \\ H_5C_2—N & & N—C_2H_5 \\ & \diagdown\ \diagup & \\ & As & \\ & | & \\ & C_6H_5 & \\ & \text{II} & \end{array}$$

The action of arsenic trichloride on trimethyl(ethylamino)silane [513] yields $(CH_3)_3SiCl$, and polymeric $[C_2H_5N-AsCl]_n$ and liberates HCl and this may be explained by the decomposition of the intermediate $C_2H_5NHAsCl_2$.

The reaction of dimethylchloroarsine with Wannagat's reagent [1062] in ether proceeds without any complications:

$$(R_3Si)_2NNa + ClAsR_2 \longrightarrow (R_3Si)_2NAsR_2 + NaCl \qquad (5\text{-}21)$$

Contrary to expectations, excess $(CH_3)_2AsCl$ does not direct the process toward the formation of $[(CH_3)_2As]_3N$. The reaction of sodium bis(trimethylsilyl)amide with cacodyl disulfide leads to similar results [1058]:

$$(R_3Si)_2NNa + R_2AsSAs(S)R_2 \longrightarrow (R_3Si)_2NAsR_2 + R_2As(S)SNa \qquad (5\text{-}22)$$

It is also possible to use organolithium compounds of arsenic for synthesizing arsenic derivatives containing silicon and nitrogen [1058, 1063]:

$$R_2AsN(R)Li + ClSiR_3 \longrightarrow R_2AsN(R)SiR_3 + LiCl \quad (R = CH_3) \qquad (5\text{-}23)$$

The reaction of an analogous derivative of silicon $[R_3SiN(R)Li]$ with arsenic trichloride gives a complex mixture of products, which contain the Si–N–As group. At the same time, with triphenoxyarsine the tris derivatives are formed smoothly [1067]:

$$3(CH_3)_3SiN(CH_3)Li + (C_6H_5O)_3As \longrightarrow [(CH_3)_3SiN(CH_3)]_3As + 3LiOC_6H_5 \qquad (5\text{-}24)$$

TABLE 78. Compounds Containing the Si–N–As Group

Formula	m.p., °C	b.p., °C (mm)	Preparation method (scheme)	Yield, %	Literature
$(CH_3)_3SiNHAsCl_2$	> 310	—	5-9	75	544
$(CH_3)_3SiN(CH_3)As(CH_3)_2$	—50	66—68 (14)	5-23	—	1058, 1063
$[(CH_3)_3Si]_2NAs(CH_3)_2$	—44—46 freezing	45—46 (1)	5-21	75	1062
	—	—	5-22	—	1058
$[(CH_3)_3SiN(CH_3)]_3As$	—	—	5-23	—	1067
$\left[(CH_3)_3SiNAsCl\right]_n$	~190	—	5-20	5	1058
$[p\text{-}CH_3OC_6H_4)_2Si[N{=}As(C_4H_9\text{-}n)_3]_2$	—	—	5	—	1691

A method of preparing silicon- and nitrogen-containing derivatives of pentavalent arsenic has been patented [1691]:

$$R_2Si(N_3)_2 + 2R'_3As \xrightarrow{\text{pyridine}} 2N_2 + R_2Si(N{=}AsR'_3)_2 \qquad (5\text{-}25)$$
$$(R = p\text{-}CH_3OC_6H_4;\ R' = n\text{-}C_4H_9)$$

3.2. Properties

The physicochemical constants of compounds containing the Si–N–As group are given in Table 78.

Trimethylsilylaminodichloroarsine is a colorless crystalline substance, which does not melt or decompose up to 310°C, is difficultly soluble in most organic solvents (like its phosphorus analog), and is very sensitive to moisture [544]:

$$(CH_3)_3SiNHAsCl_2 \xrightarrow{H_2O} H_3AsO_3 + [(CH_3)_3Si]_2O + NH_4Cl \qquad (5\text{-}26)$$

Bis(trimethylsilyl)aminodimethylarsine is a colorless liquid, which is sensitive to moisture and oxygen. When this compound is treated with methyllithium (65°C, nitrogen) there is cleavage of the As–N bond and the formation of trimethylarsine and a lithium Wannagat reagent [1062]. In the NMR spectrum there are two signals with an area ratio of 3 : 1 with chemical shifts of -10 (Si–CH_3) and -77.8 (As–CH_3) Hz, respectively. Cleavage of (trimethylsilyl)-(methyl)aminodimethylarsine by methyllithium proceeds analogously with the formation of $(CH_3)_3SiN(CH_3)Li$ [1065, 1067]. The NMR spectrum of $(CH_3)_3Si(CH_3)NAs(CH_3)_2$ is characterized by three signals at -4 (Si–CH_3), -59 (As–CH_3), and -150.5 (N–CH_3) Hz with an area ratio of 3 : 2 : 1 [1058].

4. COMPOUNDS CONTAINING THE Si – As BOND

4.1. Preparation Methods

The action of an electric discharge (of the type used in ozonizers) on a mixture of SiH_4 and AsH_3 at -78°C and a pressure of 0.25-0.5 atm. yields mixed silylarsines [655]. Mass spectrometric measurements indicated the formation of H_3SiAsH_2, Si_2AsH_7, and evidently $SiAs_2H_6$, though the structure of the last two products has not yet been established.

Iodosilane does not react with arsine even at 100°C [76]. With powdered arsenic at room temperature there is a slow reaction with the formation of a mixture of products, which includes compounds containing the Si – As bond such as: H_3SiAsH_2, H_3SiAsI_2 and $As(SiH_3)_3$. Of these it was possible to isolate only silyldiiodoarsine H_3SiAsI_2 in a pure form [526].

The reaction of iodosilane with potassium and sodium arsenides evidently forms trisilylarsine [76, 107, 526]. In the case of mercury arsenide a single substance was obtained and this was nongaseous and close in composition to $[(H_3Si)_4As]I$. Since H_3SiI forms an unstable 1 : 1 adduct with trimethylarsine at certain temperatures it may be surmised that the primary product of the reaction of iodosilane with mercury arsenide is also trisilylarsine, which then forms an arsonium compound with iodosilane.

Trisilylarsine could be isolated from the products of the reaction of bromosilane with potassium arsenide at a temperature below -100°C in methyl ether [518]:

$$3H_3SiBr + KAsH_2 \xrightarrow[-3KBr]{} (H_3Si)_3As + 2AsH_3 \quad (5\text{-}27)$$

Although the reaction evidently proceeds in stages, mono- and disilyl derivatives of arsine are not formed under these conditions.

The final result of the reaction of trimethylchlorosilane with potassium arsenide is represented by the following scheme according to the authors of [240]:

$$5(CH_3)_3SiF + 5KAsH_2 \longrightarrow [(CH_3)_3Si]_2AsH + [(CH_3)_3Si]_3As + 5KF + 3AsH_3 \quad (5\text{-}28)$$

However, although the coefficients in this equation are arranged with a schoolboy's care they do not reveal the inner meaning of the reaction. At the same time, it may be surmised that the $(CH_3)_3SiAsH_2$ formed initially then disproportionates successively and is finally converted into the given products.

Gilman and his co-workers observed [713] that the reaction of triphenylsilyllithium with arsenic trichloride yields hexaphenyldisilane (44%) and metallic arsenic. To explain this it is necessary to assume that there is a transition compound which contains an Si–As bond (we will discuss the interpretation of this process again in examining organosilicon derivatives of bismuth). At the same time, triphenylsilyl derivatives of arsenic are formed readily by the action of triphenylsilyllithium on organobromoarsines [933]:*

$$nR_3SiLi + Br_nAsR'_{3-n} \xrightarrow{THF} (R_3Si)_n AsR'_{3-n} + nLiBr \qquad (5\text{-}29)$$
$$(R = C_6H_5;\ R' = CH_3 \text{ or } C_2H_5;\ n = 1 \text{ or } 2)$$

MacDiarmid and his co-workers [932] used dimethylarsyllithium to prepare compounds containing the Si–As bond:

$$R_2AsLi + ClSiR_3 \longrightarrow R_2AsSiR_3 + LiCl \quad (R = CH_3) \qquad (5\text{-}30)$$

This exothermic reaction begins at room temperature.

According to the data of Wittenberg and Gilman [1280], lithium cleaves triphenylarsine in tetrahydrofuran with the formation of diphenylarsyllithium and phenyllithium. The reaction of the cleavage product with trimethylchlorosilane with subsequent hydrolysis of the mixture (contact with air) gives trimethylphenylsilane, tetraphenyldiarsoxane, and diphenylarsinic acid. In this case there is no need to talk of intermediate formation of a compound with a Si–As bond. Here it should be noted that in the reaction of the cleavage products of triphenylphosphine by lithium with triphenylchlorosilane under the same conditions, together with $(C_6H_5)_4Si$ and $(C_6H_5)_2P(O)OH$, $(C_6H_5)_2PH$ and $(C_6H_5)_3SiOH$ are formed and these are reasonably regarded as the cleavage products of an intermediate product containing an Si–P bond.

* In [932] it was reported incorrectly that Mayer had prepared triphenylsilyldiphenylarsine. In actual fact, the compound in question was triethylsilyldiethylarsine.

TABLE 79. Properties of Compounds Containing the Si–As Bond

Formula	m.p., °C	b.p., °C (mm)	Other properties	Preparation method (scheme)	Yield, %	Literature
H_3SiAsH_2	—	—	ν_{as} SiAs 890 cm^{-1}	—	—	655, 859
H_3SiAsI_2	-4.0 ± 0.5	210 ± 5 (extrap.)	H_{vapor} 9.2 kcal/mole, Trouton's constant 19.3 19.3	—	—	526
$H_3SiAs(CH_3)_3I$	8.1—9.6	—	$p = 665$ mm (21 °C)	—	—	526
$(H_3Si)_3As$	—	120 (extrap.)	d_4^{20} 1.201; $p = 17$ mm (0° C)	5-27	49	518, 526
$(H_3Si)_4AsI$	—	—	$p = 16$ mm (21° C)	—	—	526
$(CH_3)_3SiAs(CH_3)_2$	—89	136 (extrap.)	—	5-30	76	932
$[(CH_3)_3Si]_2AsH$	—	55 (15)	—	5-28	7	240
$[(CH_3)_3Si]_3As$	—	82—84 (4)	d_4^{20} 0,9939	5-28	~18	240
$(C_6H_5)_3SiAs(CH_3)_2$	53	—	—	5-29	—	933
$(C_6H_5)_3SiAs(C_2H_5)_2$	127	—	—	5-29	—	933
$[(C_6H_5)_3Si]_2AsCH_3$	166	—	—	5-29	—	933
$SiCl_4 \cdot AsCl_3$	—29.5	—	—	—	—	1169

Investigation of the freezing curve of the binary system silicon tetrachloride–arsenic trichloride [1169] indicates the formation of a 1 : 1 adduct, which does not have a very high stability. It has been suggested that 1 : 2–1 : 3 adducts may be formed at a lower temperature though the evidence in favor of their existence is not sufficiently convincing. At -69.5°C a eutectic is observed with a $SiCl_4$ content of 92–93 mol.%.

To conclude this section we should mention a patent which refers to compounds of the type $R_3MMn(CO)_4XR_3'$, where M = Si, Ge, Sn, Pb; X = P, As, Sb [1667]. However, no actual example of the preparation of a substance with M = Si and X = As is given in this patent.

4.2. Properties

The physicochemical constants of compounds containing the Si–As bond are given in Table 79.

Trisilylarsine decomposes at room temperature with the liberation of hydrogen [526]. The vapor pressure of the products from the reaction of iodosilane with trimethylarsine (at low temperatures the product consists of a solid, salt-like, compound which conducts electricity in acetonitrile) is 100 mm at 0°C. Above 5°C the vapor pressure corresponds to values calculated for an equimolar mixture of iodosilane and trimethylarsine. However, with a fall in temperature the vapor pressure found departs increasingly from that calculated for the mixture. The heat of dissociation of $[H_3SiAs(CH_3)_3]I$ is 23.4 kcal/mole at 0°C. Silyldiiodoarsine decomposes slowly at room temperature (rapidly at 80°C) with the formation of a complex mixture of products. Solid, volatile $[(H_3Si)_4As]I$ decomposes at room temperature with the liberation of iodosilane and hydrogen.

Until recently it was only known that compounds containing the group $(CH_3)_3Si$ at an arsenic atom are less stable than their phosphorus analogs (though they do not ignite spontaneously in air) while the Si–As bond in them is hydrolytically unstable [240]. Therefore, the investigation of the properties of trimethylsilyldimethylarsine, which was carried out recently by MacDiarmid and his co-workers [932] is of great value.

Water produces instantaneous quantitative decomposition:

$$R_2AsSiR_3 + H_2O \longrightarrow 2R_2AsH + R_3SiOSiR_3 \qquad (5\text{-}31)$$

Rapid cleavage with a high yield of dimethylarsine is produced by lower aliphatic alcohols. An adduct is formed with HBr at -96°C and this decomposes quantitatively at -78°C:

$$R_2AsSiR_3 + HBr \longrightarrow [R_2As(H)SiR_3]Br \longrightarrow R_2AsH + R_3SiBr \qquad (5\text{-}32)$$

Methyliodide reacts quantitatively with trimethylsilyldimethylarsine at room temperature according to the scheme

$$R_2AsSiR_3 + RI \longrightarrow [R_3AsSiR_3]I \longrightarrow R_3As + R_3SiI \qquad (5\text{-}33)$$
$$\xrightarrow{\;RI\;} [R_4As]I$$

The following reaction proceeds quantitatively with BF_3 over the temperature range of -50 to +20°C:

$$2R_2AsSiR_3 + BF_3 \longrightarrow (R_2As)_2BF + 2R_3SiF \qquad (5\text{-}34)$$

At the same time, the products from the reaction of trimethylsilyldimethylarsine with phosphorus pentafluoride include the adduct of the latter with tetramethyldiarsine. The following scheme has been proposed:

$$R_2AsSiR_3 \xrightarrow[-R_3SiF]{+PF_5} R_2AsPF_4 \xrightarrow[-PF_3]{} R_2AsF \xrightarrow[-R_3SiF]{+R_2AsSiR_3}$$
$$\longrightarrow R_2AsAsR_2 \xrightarrow{+PF_5} R_2AsAsR_2 \cdot PF_5 \qquad (5\text{-}35)$$

The reaction of trimethylsilyldimethylarsine with carbon disulfide forms a liquid which is involatile even at 90°C. However, prolonged heating of it at this temperature produces decomposition with complete recovery of the starting substances. The reversible nature of the reaction made it impossible to isolate the product formed. This was apparently $(CH_3)_2AsC(S)SSi(CH_3)_3$.

Trimethylsilyldimethylarsine is resistant toward ammonia and dimethylamine.

ANTIMONY

5. COMPOUNDS CONTAINING THE Si – O – Sb GROUP

5.1. Preparation Methods

In starting the study of the reaction of dimethyldiacetoxysilane with triethyl antimonite in 1955, Henglein and his co-workers [838] considered that in the case of trialkyl borates (see Ch. 3), the process would proceed according to the equation:

$$3n\,(CH_3)_2Si(OCOCH_3)_2 + 2n\,Sb(OC_2H_5)_3 \rightleftarrows$$

$$\rightleftarrows \left[\begin{array}{c} \qquad\qquad\qquad\qquad\qquad | \\ CH_3 \qquad\qquad\quad CH_3 \quad O \\ | \qquad\qquad\qquad\quad | \qquad | \\ -Si-O-Sb-O-Si-O-Sb-O- \\ | \qquad\quad | \qquad\qquad | \\ CH_3 \quad O \qquad\quad CH_3 \\ | \\ CH_3-Si-CH_3 \\ | \end{array} \right]_n + 6nCH_3COOC_2H_5 \qquad (5\text{-}36)$$

The authors considered that the formation of a theoretically possible compound of the type $Sb[OSi(CH_3)_2O]_3Sb$ was sterically unreal. In this connection it should be noted that there is in [5] the statement that Henglein and other investigators assigned this structure to one of the products obtained by this reaction– mistakenly. In the same place and also in [8] Henglein and his co-workers were quite groundlessly stated to have achieved the following reaction:

$$2n\,(CH_3)_2\,Si\,(OCOCH_3)_2 + n Sb\,(OC_2H_5)_3 \longrightarrow$$
$$\longrightarrow -Si\,(CH_3)_2-O-Sb\,(OC_2H_5)-O-Si\,(CH_3)_2\,O- + CH_3COOC_2H_5 \quad (5\text{-}37)$$

To create favorable conditions for reaction (5-36) the reagents were used in a molar ratio of 2 : 1 and not in a stoichiometric ratio. The reaction begins with simple heating of the mixture and below 100°C it proceeds steadily with the slow liberation of ethyl acetate. However, above 100°C the liberation of the latter proceeds vigorously and a solid insoluble polymer is formed, which contains antimony in a ratio of 3 : 2 and also ~15% of acetoxy groups. The yield of ethyl acetate is about 90% of theoretical. Purification of the polymer by heating in vacuum yielded a small amount of an oily liquid with the ratio Si : Sb = 80.

The preparation of tris(triorganosilyl) antimonites from trialkylalkoxysilanes and antimony trichloride requires quite drastic conditions [24, 56]:

$$3R_3SiOR' + SbX_3 \longrightarrow (R_3SiO)_3\,Sb + 3R'X \quad (5\text{-}38)$$

When $R = CH_3$, the formation of tris(trimethylsilyl)antimonite was observed when a mixture of trimethylmethoxysilane and antimony trichloride was heated for 12 hr in an autoclave at 240°C in the presence of $ZnCl_2$.

The first tris(triorganosilyl) antimonites were described in the literature in 1959. They were synthesized simultaneously and independently in Leningrad [51, 410, 411, 1483] and Moscow [180] and immediately there were put forward three methods, which had been used previously for preparing organosilicon esters of other inorganic acids:

$$3R_3SiOH + Sb\,(OR')_3 \rightleftharpoons (R_3SiO)_3\,Sb + 3R'OH \quad (5\text{-}39)$$

$$3R_3SiONa + SbCl_3 \longrightarrow (R_3SiO)_3\,Sb + 3NaCl \quad (5\text{-}40)$$

$$6R_3SiOH + Sb_2O_3 \rightleftharpoons 2\,(R_3SiO)_3\,Sb + 3H_2O \quad (5\text{-}41)$$

Reaction (5-39) is essentially silanolysis of a triorgano antimonite. It proceeds with simple heating of a mixture of the components (in which a catalytic amount of metallic sodium is dissolved) with gradual introduction of the silanol and distillation of the alcohol formed. The side reaction of hydroxyl–alkoxyl exchange occurs only with a large excess of the alkoxyl derivative, while condensation of the silanols occurs to an appreciable extent under the experimental conditions only in the case of trimethylsilanol and $R(CH_3)_2SiOH$. A development of this method is the reaction of a triorganosilanol with triacetoxyantimony in xylene, with which the acetic acid is distilled in the form of an azeotrope [1517]:

$$3R_3SiOH + Sb(OCOCH_3)_3 \longrightarrow (R_3SiO)_3Sb + 3CH_3COOH \quad (R = C_2H_5 \text{ or } C_6H_5) \qquad (5\text{-}42)$$

The synthesis of triorganosilyl antimonites by scheme (5-41) is carried out with azeotropic distillation of the water formed with benzene or another solvent. The reaction product is separated from unreacted antimony trioxide by filtration and isolated by vacuum distillation. It is recommended that reaction (5-40) is carried out by adding a solution of antimony trichloride in dry benzene to a benzene solution of the silanolate. In 1961 this method was extended to the preparation of trimethylsiloxydimethylstibine (which was called "pentamethylstibiosiloxane" by Schmidbaur) [1085, 1100] and other trimethylsiloxy derivatives of tri- and pentavalent antimony, containing organic substituents at the Sb^{III} and Sb^{V} atoms:

$$(n-m)R_3SiONa + R'_mSbX_{n-m} \longrightarrow R'_mSb(OSiR_3)_{n-m} \qquad (5\text{-}43)$$

$$(n = 3 \text{ or } 5;\ m = 2-4;\ R = CH_3;\ R' = CH_3,\ C_6H_5;\ X = Cl \text{ or } Br)$$

The reaction of sodium trimethylsilanolate with an organohalostibine (stibane) is carried out in a stream of nitrogen in a dry solvent, free from dissolved oxygen.

5.2. Properties

The physicochemical constants of compounds containing the Si–O–Sb group are given in Table 80.

Trimethylsiloxydimethylstibine (a colorless evil-smelling liquid) is analogous in many ways to its arsenic analog, decomposes when distilled at normal pressure in a nitrogen atmosphere, and is oxidized explosively by atmospheric oxygen at 40°C. At the same time, the completely methylated derivative of pentavalent antimony [1085, 1100] $(CH_3)_3SiOSb(CH_3)_4$ (trimethylsiloxytetramethylstibane*) obtained by scheme (5-43) is much more resistant to heating and

*Unsystematically named "heptamethylstibiosiloxane."

TABLE 80. Compounds Containing the Si–O–Sb Group

Formula	m.p., °C	b.p., °C (mm)	n_D^{20}	d_4^{20}	Preparation method (scheme)	Yield, %	Literature
$(CH_3)_3SiOSb(CH_3)_2$	—	48 (24); 144 (720)	—	—	5-43	79	1085, 1100
$(CH_3)_3SiOSb(C_6H_5)_2$	—	70 (1)	—	—	5-43	49	1085
$[(C_2H_5)_3SiO]_2SbSi(C_2H_5)_3$		134–139(1)	1.4655	1.093		58	264a; 210
							215(762)
$[(CH_3)_3SiO]_3Sb$	—	80 (3)	1.4374	1.1448	5-39	67	410
$[C_2H_5(CH_3)_2SiO]_3Sb$	—	132 (6)	1.4508	1.1318	5-39	90	410
$[CH_3(C_2H_5)_2SiO]_3Sb$	—	150 (3.5); 160 (5)	1.4588	1.1132	5-41	62	410, 411 1483
$[(C_2H_5)_3SiO]_3Sb$	—	160—162 (1.5)	1.4681	1.1037	5-40	56	180
	—	203 (30)	1.4684	1.1038	5-39	91	410
	—	170 (3)	1.4675	1.1041	5-41	63	410
	—	151 (2)	1.4682	1.1023	5-42	94	1517
$[(C_6H_5)_3SiO]_3Sb$	214	—	—	—	5-39	—	410
	112	—	—	—	5-42	81	1517
$(CH_3)_3SiOSb(CH_3)_4$	—32	59,5—61 (9); 164—165 (725)	—	1.238	5-43	88	1085, 1100
$(CH_3)_3SiOSb(CH_3)_3Cl$	59	—	—	—	5-43	91	1085
$(CH_3)_3SiOSb(C_6H_5)_4$	76—78	—	—	—	5-43	81	1085
$(CH_3)_3SiOSb(C_6H_5)_3Cl$	—	180 (1) (decomp.)	—	—	5-43	27	1085
$[(CH_3)_3SiO]_2Sb(CH_3)_3$	21	46 (1); 89 (10,5) 112 (30); 210 (725)	—	—	5-43	70	1085
$[(CH_3)_3SiO]_2Sb(C_6H_5)_3$	89	161 (1)	—	—	5-43	56	1085
$\{[(CH_3)_3SiO]_4Al\}^-[Sb(CH_3)_4]^+$	180,5	180 (1) (subl.)	—	1.170^{25}	5-45	91	1072, 1080
$\{[(CH_3)_3SiO]_4Ga\}^-[Sb(CH_3)_4]^+$	190,5	190 (1) (subl.)	—	1.260^{25}	5-45	92	1072, 1080
$\{[(CH_3)_3SiO]_4Fe\}^-[Sb(CH_3)_4]^+$	193	190 (1) (subl.)	—	1.246^{25}	5-45	68	1072, 1080

oxidation (the reaction with oxygen begins considerably higher than 100°C). This compound is very hygroscopic and is hydrolyzed rapidly to give a strongly alkaline reaction due to $[(CH_3)_4Sb]OH$, namely, tetramethylstibonium hydroxide, which may be titrated acidimetrically.

In chemical behavior trimethylsiloxytetramethylstibane is not like heterosiloxanes, but resembles alkali metal silanolates. This makes it possible to use it, for example, for the synthesis of other heterosiloxanes:

$$(CH_3)_3SiOSb(CH_3)_4 + (CH_3)_3SnCl \longrightarrow (CH_3)_3SiOSn(CH_3)_3 + [(CH_3)_4Sb]^+Cl^- \quad (5\text{-}44)$$

Another interesting property of trimethylsiloxytetramethylstibane is its capacity to form salt-like compounds with tris(trimethylsiloxy) derivatives of aluminum, gallium, and iron [1072,

1080, 1083]:

$$2\,(CH_3)_4\,SbOSi\,(CH_3)_3 + \{[(CH_3)_3\,SiO]_3\,M\}_2 \longrightarrow 2[(CH_3)_4\,Sb]^+\,\{M\,[OSi\,(CH_3)_3]_4\}^- \quad (M = Al,\ Ga,\ Fe) \tag{5-45}$$

The exothermic reaction begins even at room temperature and a solution of an equimolar mixture of the reagents in CCl_4 or cyclohexane deposits acicular crystals of the tetramethylstibonium salts, which are stable in air and are readily soluble in many organic solvents.

According to cryoscopic data, these compounds are monomeric in benzene. The NMR spectra indicate a purely ionic structure for these peculiar substances and the equivalence of all four methyl groups at the antimony atom and likewise all four trimethylsiloxy groups at the metal atom. All three complexes are isomorphous. Crystallographic analysis of the aluminium derivative showed that this compound is a combination of an almost ideally tetrahedral anion $\{[(CH_3)_3SiO]_4Al\}^-$ and cation $[(CH_3)_4Sb]^+$. All the angles in the molecule are tetrahedral apart from the Al–O–Si angle, which equals 147°. The bond lengths calculated from the lattice parameters were as follows (in Å): Si–C 1.88; Si–O 1.56; Al–O 1.79; Sb–C 2.20. The closest approach of the ions, namely 3.30 Å, corresponds to the distance between a CH_3 group of the cation and an oxygen atom of the anion. Spectral and x-ray structural data are given in Table 81 [1072, 1080, 1256].

Compounds of this type are highly thermally stable, but are very reactive chemically. The following hydrolysis proceeds readily [1080]:

$$[R_4Sb]\,[M\,(OSiR_3)_4] \xrightarrow{+4H_2O} R_4SbOH + M\,(OH)_3 + 4R_3SiOH \quad (R = CH_3) \tag{5-46}$$

The highest thermal and chemical stability in the series of trimethylsiloxy derivatives of pentavalent antimony are shown by compounds of the type $R_3Sb[OSi(CH_3)_3]_2$ [1085, 1100]. When $R = CH_3$ the substance distills in dry nitrogen at 210°C without decomposition. When $R = C_6H_5$ slow decomposition with the evolution of hexamethyldisiloxane begins only at 280°C. The relative stability of compounds of this type is associated with the symmetry of their structure.

TABLE 81. Some Characteristics of Compounds of the Type $[(CH_3)_4Sb]^+\{M[OSi(CH_3)_3]_4\}^-$

Properties	Metal (M)		
	Al	Ga	Fe
Crystal class	Orthorhombic		
Space group	P_{mmn} (Z=2)		
a, *b*, *c* (Å) *U* (Å³) *d*, calculated/found	13.41; 11.88; 9.90 1578 1.191/1.17	13.45; 11.94; 9.87 1584 1.275/1.26	13.51; 11.95; 9.82 1585 1.246/1.25
Characteristic band in IR spectra, cm^{-1}	1087, 1020, 990	1069, 990, 948	1107, 954, 927
Magnetic properties of compounds	Diamagnetic	Diamagnetic	Paramagnetic
Characteristic signals in NMR spectra, $\delta(Sb-CH_3)/\delta(Si-CH_3)$ Hz	—104/—1.0	—109/—5.0	—

In the $R_3Sb[OSi(CH_3)_3]_2$ molecule the $(CH_3)_3SiO$ groups occupy the apices of a trigonal bipyramid, one opposite the other, while the organic radicals R lie at the corners of the base.

Hydrolytic cleavage of triorganobis(trimethylsiloxy)stibanes is produced by boiling water, acids, and alkalis, and is accompanied by the formation of silanols and organic compounds of antimony. The action of HCl on bis(trimethylsiloxy)trimethylstibane yields hexamethyldisiloxane and $(CH_3)_3SbCl_2$, indicating cleavage at the Sb–O bond. The reaction with thionyl chloride is analogous to scheme (5-13):

$$[(CH_3)_3SiO]_2Sb(CH_3)_3 + 2SOCl_2 \longrightarrow 2(CH_3)_3SiCl + (CH_3)_3SbCl_2 + 2SO_2 \quad (5\text{-}47)$$

In the IR spectrum of trimethylsiloxydimethylstibine, a line with a frequency of 941 cm^{-1} belongs to the antisymmetric valence vibration [1085]. The replacement of methyl groups at the silicon atom in this compound by phenyl groups produces a shift in the line toward the long wave region (ν_{as} 926 cm^{-1}). At the same time,

$(CH_3)_3SiOSb(CH_3)_4$ is characterized by a line at 978 cm^{-1}, whose position changes little with the introduction of a second $(CH_3)_3SiO$ group instead of the CH_3 group (969 cm^{-1}). However, a change to compounds containing the C_6H_5(Sb) group leads to a shift in the ν_{as} SiOSb lines by -35-40 cm^{-1}.

In the NMR spectra of trimethylsiloxytetramethylstibane and bis(trimethylsiloxy)trimethylstibane [1084, 1085], the chemical shifts for CH_3(Si) and CH_3(Sb) are +10; +7.2, and -61; -89 Hz, respectively.

The molecular refraction increment for the Sb–O bond in tris(triorganosilyl) antimonites equals 5.25 ± 0.05 [410].

The polymeric stibasiloxane obtained by Henglein and his co-workers is stable to approximately 170°C and is very sensitive to hydrolysis, decomposing in water with the liberation of SbOOH as powder. After a few days in hydrochloric acid it forms an elastic silicopolymer, which essentially contains no antimony.

5.3. Application

Tris(triorganosilyl) antimonites are recommended in the literature as catalysts for the polymerization of organosiloxanes and also as additives to make the polymers flame resistant [1483, 1517].

Cohydrolyzates of methylchlorosilanes with $CH_3 : Si = 1.6$, which have undergone partial condensation, are hardened at room temperature in the presence of 1-5 wt.% $SbCl_5$ [441, 519, 1402].

6. COMPOUNDS CONTAINING THE $Si-(C)_n-Sb$ GROUP

Tris(trimethylsilylmethyl)stibine is obtained in about 75% yield by the Grignard reaction [1137, 1644]:

$$3\,(CH_3)_3SiCH_2MgCl + SbCl_3 \longrightarrow [(CH_3)_3SiCH_2]_3Sb + 3MgCl_2 \qquad (5\text{-}48)$$

The analogy with the similar compound of arsenic extends still further since this compound readily adds methyl iodide and bromine. However, it is more sensitive to oxidation. Other similar derivatives of antimony may be obtained by using reaction (5-48) [1644].

TABLE 82. Compounds Containing the Si – $(C)_n$ – Sb Group

Formula	m.p., °C	Literature
$[(CH_3)_3SiCH_2]_3Sb$	64—65 *	1137, 1644
$[p\text{-}CH_3C_6H_4(CH_3)_2SiCH_2]_3Sb$	—	1644
$[(CH_3)_3SiCH_2]_3SbBr_2$	158—160	1137
$\{CH_3[(CH_3)_3SiCH_2]_3Sb\}I$	147—148	1137

*Yield 74%.

All the compounds described containing the grouping Si – $(C)_n$ – Sb in the molecules are solids (Table 82).

7. COMPOUNDS CONTAINING THE Si – N – Sb GROUP

At present only one compound of this class is known and this is trimethylsilylaminodichlorostibine, $(CH_3)_3SiNHSbCl_2$, which is obtained by cleavage of hexamethyldisilazan with antimony trichloride [544]. This is a colorless crystalline substance with m.p. 192-194°C (in vacuum), which is insoluble in the usual organic solvents, very hygroscopic, and readily hydrolyzed.

8. COMPOUNDS CONTAINING THE Si – Sb BOND

In contrast to potassium arsenide, $KSbH_2$ (potassium stibide) reacts with bromosilane not according to scheme (5-27), but with the formation of silane and other products [518]:

$$2H_3SiBr + 2KSbH_2 \xrightarrow[-2KBr]{} 2SiH_4 + 0{,}5H_2 + (Sb_2H)\ (?) \tag{5-49}$$

At the same time, trisilylstibine is obtained by the scheme:

$$3H_3SiBr + Li_3Sb \xrightarrow[-3LiBr]{} (H_3Si)_3\,Sb \tag{5-50}$$

The properties of this compound as compared with the properties of its phosphorus and arsenic analogs are characterized by the data in Table 83.

TABLE 83. Properties of Compounds of the Type $(H_3Si)_3$ El [518]

Parameters		Elements (El)		
		Sb	As	P
$\log p = -A/T + B$	A	1670.3	2142.6	1901.8
	B	6.04	8.39	7.79
	T, °C	–6 +15	–16 +15	–30 +11
Q, cal/mole		7638	9798	8697
b.p., °C (extrap.)		255	120	114

Like its As and P analogs, trisilylstibine is a colorless liquid, which ignites spontaneously in air. However, its decomposition with 33% NaOH does not liberate SbH_3, but forms free antimony, while more hydrogen is liberated than in the case of trisilylphosphine or trisilylarsine. Decomposition with the liberation of antimony and hydrogen also occurs during storage of trisilylstibine.

The method of preparing trilithioantimony is important for the synthesis of compounds containing the Si–Sb bond by a scheme of the type (5-50) [519]. If the reagent is prepared by fusion of lithium with antimony with subsequent reduction of the alloy to a fine powder, then the reaction with compounds of the type $(CH_3)_3MCl$ (M = Si, Ge, Sn) does not occur. An active reagent is obtained by addition of powdered antimony to a solution of lithium in liquid ammonia. After removal of the ammonia (the system is heated to 80°C in high vacuum), the reagent reacts with trimethylchlorosilane in ether over the temperature range of -30 to +40°C. With a reaction time of 30-40 hr the yield of tris(trimethylsilyl)stibine is ~80%. This compound (m.p. from -1 to +1°C) may be stored without change for several weeks at room temperature in vacuum and in the dark. It bursts into flame instantly in air.* Measurement of the dipole moment in benzene (1.45-1.50 D) led the authors of [519] to conclude that the system Si_3Sb has a pyramidal structure with class C_{3v} symmetry.

*An attempt to plot the Raman spectrum was unsuccessful due to rapid decomposition of the substance with the liberation of metallic antimony during irradiation with a mercury lamp at + 16°C.

The action of trimethylchlorosilane on the product from cleavage of triphenylstibine with lithium and subsequent hydrolysis yielded trimethylphenylsilane and diphenylstibinic acid [1280]. The difference between this reaction and the similar reaction involving triphenylarsine lies in the predominance of the acid and not the dimetalloxane. However, in this case we cannot postulate the possibility of the intermediate formation of compounds containing the Si – Sb bond. An attempt to prepare such a compound by the reaction of triphenylsilyllithium with antimony trichloride was also unsuccessful [713]. Nonetheless, the nature of a series of products of this reaction isolated after hydrolysis of the reaction mixture, namely, hexaphenyldisilane (50%) and metallic antimony (90%), led the authors to the conclusion that in this case, as in the reaction of $(C_6H_5)_3SiLi$ with $AsCl_3$, there must be the intermediate formation of such a compound.

Dibutylbromostibine reacts with magnesium in THF to form $[(C_4H_9)_2Sb]_2Mg$, which, in its turn, reacts with trimethylchlorosilane to give a 22% yield of trimethylsilyldibutylstibine. The latter is a colorless liquid, which ignites spontaneously in air [841a].

In 1965 Razuvaev and his co-workers discovered the general reaction [259c, 264a, 441]:

$$3R_3MH + R_3M' \rightarrow 3RH + (R_3M)_3M' \qquad (5\text{-}51)$$

$$(M = Si, Ge, Sn;\ M' = Sb, Bi;\ R = C_2H_5)$$

Tris(triethylsilyl)stibine is obtained by this scheme in 37% yield by heating a stoichiometric mixture of the reagents in a sealed evacuated ampoule for 20 hr at 230°C. Ethane is formed in 95.3% of the theoretical amount. Tripropylsilane reacts with $[(C_2H_5)_3Si]_3Sb$ with displacement of the triethylsilyl groups, but this reaction is reversible. In the case of triphenylsilane the reaction is irreversible. Analogously the reaction of tris(triethylsilyl)stibine with triethylgermane leads to replacement of silyl groups at the antimony atom by germyl groups so that the following reaction is evidently general:

$$(R_3Si)_3Sb + 3R'_3MH \longrightarrow 3R_3SiH + (R'_3M)_3Sb \qquad (5\text{-}52)$$

$$(R = C_2H_5;\ \text{when } M = Si,\ R' = n\text{-}C_3H_7,\ C_6H_5;\ \text{when } M = Ge,\ R' = C_2H_5)$$

A study of the behavior of compounds of the type R_3MH in reactions (5-51) and (5-52) enabled the authors of [264a] to construct the following series of activities:

$$(C_2H_5)SnH > (C_2H_5)GeH > (C_6H_5)_3SiH > (C_3H_7)_3SiH \simeq (C_2H_5)_3SiH$$

Tris(trialkylsilyl)stibines are pale yellow viscous liquids, which will distill under nitrogen and are quite thermally stable. In the absence of oxygen, thermal decomposition, which involves homolytic cleavage of the Si–Sb–Si bonds, is complete in 22.5 hr at 300°C in the case of the ethyl derivative. The decomposition may be represented by the scheme:

$$2(R_3Si)_3M \rightarrow 2M + 3R_6Si_2 \qquad (5\text{-}53a)$$

However, the reaction is practically quantitative only with respect to antimony (this may be used for analytical purposes). The other products are triethylsilane and a complex mixture of high-boiling organosilicon compounds:

$$(R_3Si)_3M \rightarrow M + 3R_3Si. \qquad (5\text{-}53b)$$

$$R_3SiH, \text{ etc. (n-donor).}$$

Tris(triethylsilyl)stilbine is oxidized by molecular oxygen at room temperature to form bis(triethylsiloxy)triethylsilylstibine [364a]. The following exothermic reaction with benzoyl peroxide proceeds vigorously [259c, 264a, 441]:

$$2(R_3Si)_3M + 3(C_6H_5COO)_2 \rightarrow 6R_3SiOCOC_6H_5 + 2M \qquad (5\text{-}54)$$

In benzene at 5–7°C the yield of triethylbenzoxysilane reaches 87% in 5–7 min with a quantitative yield of antimony. The authors believe that the reaction proceeds through a cyclic intermediate complex, which decomposes homolytically. This also applies to the reaction of tris(triethylsilyl)stibine with sym-dibromoethane:

$$2(R_3Si)_3M + 3BrCH_2CH_2Br \xrightarrow{\sim 100°C} 2M + 3C_2H_4 + 6R_3SiBr \qquad (5\text{-}55a)$$

The composition of the actual products agrees with the scheme. With an equimolar ratio, apart from ethylene and triethylbromosilane, there is formed a viscous liquid, which is readily oxidized with the appearance of a cherry red color, and which decomposes at 180°C with the liberation of antimony and the starting stibine:

$$n(R_3Si)_3Sb + nBrCH_2CH_2Br \xrightarrow{\sim 90°C} nC_2H_4 + 2n\,R_3SiBr +$$

$$+ (R_3SiSb)n \xrightarrow{180°C} (n/3)\,(R_3Si)_3Sb + (2n/3)Sb \qquad (5\text{-}55b)$$

TABLE 84. Properties of Compounds Containing the R_3Si-Sb Group

Formula	b.p., °C (mm)	n_D^{20}	d_4^{20}	Preparation method (scheme)	Yield, %	Literature
$(CH_3)_3SiSb(C_4H_9\text{-}n)_2$	48-50 (0.025)	–	–	–	22	841a
$[(C_2H_5)_3Si]_3Sb$	148-153 (I)	–	1.099	5-51	37	259a
$[(n\text{-}C_3H_7)_3Si]_3Sb$	161-165 (I)	–	1.008	5-52	36	264a
$(C_2H_5)_3SiSb[OSi(C_2H_5)_3]_2$	134-139 (I)	1.4655	1.093	–	58	264a

The physicochemical constants of organosilyl derivatives of antimony are given in Table 84.

9. COMPLEX COMPOUNDS CONTAINING Si AND Sb ATOMS

Yakubovich and Motsarev [510] established that phenylchlorosilanes form adducts with antimony halides, which are used as catalysts for halogenation. This is observed through the change in color when SbX_3 and SbX_5 are dissolved in phenylchlorosilane. In the latter these complexes are stable for many days and may be isolated. The green crystalline complex $C_6H_5SiCl_3 \cdot SbCl_5$ and the orange liquid $C_6H_5SiCl_3 \cdot SbCl_3$ (m.p. -30°C) were obtained in this way, for example.

The most stable adduct $C_6H_5SiCl_3 \cdot SbCl_3$ has a low electrical conductivity in solutions of chlorosilanes, but during prolonged electrolysis, metallic antimony is liberated at the cathode. A solution of $(ClC_6H_4)_2SiCl_2 \cdot SbCl_5$ with a concentration of 0.25 M has $\varkappa = 2.4 \cdot 10^{-2}$, $ohm^{-1} \cdot cm^{-1}$, indicating a considerable degree of dissociation of this complex. It cannot be isolated in an individual form as there is decomposition with rupture of the Si–C bond. These conversions are treated in the following way by the authors:

$$RSiCl_3 + SbCl_3 \rightleftarrows R\text{—}\overset{\delta-}{Si}Cl_3 \cdots\cdots Cl\text{—}\overset{\delta+}{Sb}Cl_2 \rightleftarrows [Cl_4SiR]^- \, [SbCl_2]^+ \longrightarrow$$

$$\longrightarrow SiCl_4 + RSbCl_2$$

Alpatova and Kessler [2] disputed the ionic nature of complexes of this type, believing that the interaction is limited to strong polarization of the Si–Cl bond, which is responsible for the high reactivity of the complexes.

Adducts with $SbCl_3$ are also formed by dichlorophenyltrichlorosilane and $SiCl_4$ [510]. When a solution of $SbCl_3$ in $SiCl_4$ is cooled, the colorless crystalline adduct separates at -65°C, though its melting point is considerably higher (-25°C). According to the data of Yakubovich and Motsarev, the introduction of organic solvents produces decomposition of the complexes to the starting components. However, Khaidarov and other investigators [472] reported that complexes of $SbCl_3$ with methylchlorosilanes, namely, $2CH_3SiCl_3 \cdot SbCl_3$, $2(CH_3)_2SiCl_2 \cdot SbCl_3$, $2(CH_3)_3SiCl \cdot SbCl_3$ and phenyltrichlorosilane, $2C_6H_5SiCl_3 \cdot SbCl_3$, do not change in ligroin. The adducts produce the polymerization of sylvan (in bulk or ligroin solution): this confirms the ionic nature of the complexes as sylvan is polymerized only in the presence of ionic catalysts. The complexes are formed immediately on mixing of the components. The adducts of antimony trichloride with methyltrichlorosilane is most active. The yield of polymer depends on the complex concentration, increasing, for example, from 35% at a concentration of $2CH_3SiCl_3 \cdot SbCl_3$ of 0.5% to 82% when its concentration is 3% (18°C, 100 min, sylvan : benzene = 30 : 70). The organosilicon part of the complex remains in the polymer when the polysylvan is reprecipitated.

Triorganosilyl azides also form complexes with $SbCl_5$. The solid, pale yellow adduct with trimethylsilyl azide (1 : 1) is unstable, but it may be isolated from CCl_4 solution [1208]. Triphenylsilyl azide interacts with $SbCl_5$ to form a solid 1 : 1 complex, which is quite stable thermally, but is decomposed by moisture [119]. In the opinion of Wiberg and Schmid [1265], the formation of the complex and its decomposition proceed in the following way:

$$R_3SiN_3 + SbCl_5 \longrightarrow \left[\begin{matrix} R_3Si \\ \overset{\ominus}{Cl_5Sb} \end{matrix} \!\!>\! \bar{N}-\overset{\oplus}{N}\equiv N| \right] \longrightarrow R_3SiCl + [Cl_4Sb-\overset{\ominus}{\underline{\bar{N}}}-\overset{\oplus}{N}\equiv N|] \quad (5\text{-}56)$$

This scheme is obviously incomplete since it is reported in the literature that methyldiphenylsilyl azide reacts with antimony pentachloride explosively with the liberation of gaseous products and a solid black residue [1208].

BISMUTH

10. ORGANOSILICON DERIVATIVES OF BISMUTH

Compounds containing the Si–O–Bi group have remained unknown up to now. A substance in which bismuth is attached to silicon through nitrogen, namely, (trimethylsilylamino)dichlorobismuthine, $(CH_3)_3SiNHBiCl_2$, is obtained by a reaction which is general for elements of the phosphorus subgroup, namely, cleavage of hexamethyldisilazan by the trichloride of the element (a scheme of type 5-19). This compound (colorless crystals with m.p. 224-226°C) dissolves with difficulty only in benzene. On hydrolysis it forms basic bismuth chloride [544].

Reaction (5-16) is also general. The reaction of $BiCl_3$ with $(CH_3)_3SiCH_2MgCl$ yields tris(trimethylsilylmethyl)bismuthine $[(CH_3)_3SiCH_2]_3Bi$ [1137, 1644]. In the series of analogous derivatives of As, Sb, and Bi, this compound is the least stable and, in particular, this is reflected in its yield (35-62%). It has not been possible to isolate this substance in an analytically pure form. Even when stored in a nitrogen atmosphere the white crystalline tris(trimethylsilylmethyl)bismuthine darkens slowly, leaving a mirror of metallic bismuth on the glass. It fumes strongly in air and then ignites; when heated it melts at 107-109°C, while it decomposes even at 140°C. In contrast to its analogs, it does not form quaternary compounds with methyl iodide or iodomethyltrimethylsilane and likewise, it does not add $HgCl_2$. In this connection it is pertinent to mention that trimethylbismuthine lacks donor properties.

In discussing the nature of the products from the reaction of triphenylsilyllithium with bismuth trichloride, which include hexaphenyldisilane (79%) and metallic bismuth (81%), Gilman and his co-workers [88, 713] put forward two possible routes for their formation:

$$\begin{aligned} 1.\quad & R_3SiLi + BiCl_3 \longrightarrow [R_3SiBiCl_2] \longrightarrow R_3SiCl + [BiCl] \\ & R_3SiCl + LiSiR_3 \longrightarrow R_3SiSiR_3 \\ & 3\,[BiCl] \longrightarrow BiCl_3 + 2Bi \end{aligned} \qquad (5\text{-}57)$$

$$2.\quad 3R_3SiLi + BiCl_3 \longrightarrow \underset{(R = C_6H_5)}{[(R_3Si)_3\,Bi]} \longrightarrow Bi + 1.5R_3SiSiR_3 \qquad (5\text{-}58)$$

It is difficult as yet to show preference for either of these schemes. However, the need to postulate the intermediate formation of a compound containing the Si–Bi bond to explain the results obtained is obvious.

While Gilman and his co-workers merely put forward hypotheses on the existence of the compounds with Si–Bi bonds, the investigators of G. A. Razuvaev's school demonstrated that they actually exist by synthesizing tris(triethylsilyl)bismuthine [259c, 264a, 441]. In reaction (5-51) triethylbismuthine is superior to the Sb analog in activity and the reaction is complete in 15 hr at ~170°C. On the other hand, the decomposition of tris(triethylsilyl)-bismuthine (a liquid with b.p. 145-146°C at 1 mm and d_4^{20} 1.273) is complete in 8 hr at 290°C (provisionally according to scheme 5-53a). Likewise, the "displacement" reaction with triethylgermane (a scheme of type 5-52), the reaction with benzoyl peroxide (scheme 5-54), and reaction (5-55a) proceed more readily than with tris-(triethylsilyl)stibine.

Bismuth dialkyldithiocarbamates have been described as components of siloxane rubber mixtures, which improve the compression set and thermal stability of the vulcanizates [1587, 1707].

VANADIUM

11. COMPOUNDS CONTAINING THE Si – O – V GROUP

11.1. Preparation Methods

11.1.1. Reaction of Silanols and Alkali Metal Silanolates with Vandium Compounds

It is not possible to prepare triorganosilyl phosphates by the reaction of silanols with phosphorus oxychloride even in the presence of HCl acceptors. In contrast to this, the reaction of silanols with vanadium oxychloride is a convenient method of preparing tris-(triorganosilyl) vanadates [51, 409, 413, 628, 629, 1473]:

$$3R_3SiOH + VOCl_3 \xrightarrow[3-8^\circ C]{\text{acceptor}} (R_3SiO)_3VO + \text{acceptor} \cdot HCl \qquad (5\text{-}59)$$
$$(R = C_2H_5 \text{ or } C_6H_5)$$

A possible hydrogen chloride acceptor is dry ammonia, passed through the reaction mixture simultaneously with the introduction of vanadium oxychloride solution to a benzene solution of the silanol. To simplify the procedure, it is possible to use pyridine and this also increases the yield of the reaction products.

The reaction of a sodium triorganosilanolate with $VOCl_3$ [51, 180, 409, 413, 1473, 1629] proceeds according to the scheme:

$$3R_3SiOM + VOCl_3 \longrightarrow 3MCl + (R_3SiO)_3VO \qquad (5\text{-}60)$$

The extension of this method to vanadium tetrachloride made it possible to obtain compounds of the type $(R_3SiO)_4V$ [628, 629, 640, 1685]. Tetrakis(triphenylsiloxy)vanadine is obtained from $(C_6H_5)_3SiOK$ and VCl_4 in dry benzene in the cold in about 60% yield in a nitrogen atmosphere and ~90% yield in a natural gas atmosphere. With a change to sodium triphenylsilanolate the absence of an inert gas gives an unexpected result, namely, on contact of the reaction mixture with air tris(triphenylsilyl) vanadate is formed in almost quantitative yield [629]. As a result of this observation the reaction was carried out in an inert atmosphere, but in the presence of sulfur. The attempt was successful and the reaction product was the expected $[(C_6H_5)_3SiO]_3VS$.

On the other hand, the reaction of sodium triphenylsilanolate with vanadium oxychloride in the presence of metallic sodium gives $[(C_6H_5)_3SiO]_4V$ [628]. The reaction of VCl_4 with diphenylsilanediol forms sticky products, which are evidently polymeric.

It is not possible to use a reaction of type (5-60) for the synthesis of vanadium derivatives in which this element has a valence lower than 4.

The transesterification of boric and orthotitanic esters with triorganosilanols is one of the convenient methods of synthesizing triorganosiloxy derivatives of these elements (see Ch. 3 and 4). In the case of phosphoric esters serious difficulties are encountered with this method. Nonetheless, transesterification (silanolysis) is carried out readily with trialkyl vanadates [24, 51, 409, 413, 1476]:

$$3R'_3SiOH + (RO)_3VO \xrightarrow{Na} (R'_3SiO)_3VO + 3ROH \qquad (5\text{-}61)$$

$$(R=C_2H_5;\ n\text{-}C_4H_9;\ R'=CH_3,\ C_2H_5,\ C_6H_5)$$

Distillation of the alcohol from the mixture of reagents, in which a catalytic amount of sodium is dissolved, leads to the formation of tris(triorganosilyl) vanadates in up to 90% yield.

The reaction of silanols with V_2O_5 proceeds as in the case of P_2O_5 and Sb_2O_3 [24, 51, 409, 413, 1471, 1624]:

$$6R_3SiOH + V_2O_5 \longrightarrow 2\,(R_3SiO)_3VO + 3H_2O \qquad (5\text{-}62)$$

The only condition necessary to obtain a good yield of tris-(triorganosilyl) vanadates by this method is azeotropic distillation of the water liberated with a suitable solvent.

11.1.2. Cleavage of Siloxanes by Vanadium Compounds

Like polyvalent acid anhydrides of other Group V elements, vanadium pentoxide cleaves hexamethyldisiloxane with the formation of tris(trimethylsilyl)vanadate [1118]. However, this route to the preparation of the compound is ineffective as refluxing the reaction mixture (100°C) for 60 hr gave a yield of $[(CH_3)_3SiO]_3VO$ of less than 10%. At the same time, hexamethyldisiloxane reacts more readily and more extensively with $VOCl_3$ than with phosphorus oxychloride:

$$(CH_3)_3\,SiOSi\,(CH_3)_3 + VOCl_3 \longrightarrow (CH_3)_3\,SiCl + (CH_3)_3\,SiOV\,(O)\,Cl_2 \qquad (5\text{-}63)$$
$$(CH_3)_3\,SiOV\,(O)\,Cl_2 + (CH_3)_3\,SiOSi\,(CH_3)_3 \rightarrow (CH_3)_3\,SiCl + [(CH_3)_3\,SiO]_2\,V\,(O)\,Cl$$

The first stage is complete even with boiling of a stoichiometric mixture of the reagents in the absence of moisture. With excess hexamethyldisiloxane there is ~50% of the second reaction after heating for a day. It is not possible to prepare tris(trimethylsilyl)vanadate by scheme (5-63). However, it is formed by the reaction [180]

$$3\,\{[(C_2H_5)_3\,SiO]_2\,Pb \cdot Pb\,(OH)_2\} + 2VOCl_3 \longrightarrow$$
$$\longrightarrow 2\,[(C_2H_5)_3\,SiO]_3\,VO + 3PbCl_2 + 3Pb\,(OH)_2 \qquad (5\text{-}64)$$

11.1.3. Other Reactions

An attempt to prepare tris(trimethylsilyl)vanadate by the reaction of trimethylethoxysilane with $VOCl_3$ was as unsuccessful as when $POCl_3$ was used for the synthesis of the corresponding phosphorus analog. Trialkylbromosilanes react with trialkyl vanadates but it is impossible to isolate tris(trialkylsilyl) vanadates in a pure form in this case. At the same time, tris(trimethylsilyl)vanadate is formed by slowly adding trimethylchlorosilane to a suspension of silver orthovanadate in methylene chloride or benzene at 8-10°C [1118]. The amount of AgCl which separates is 90% of theoretical.

11.2. Physical and Chemical Properties

Table 85 gives the physicochemical constants of triorganosiloxy derivatives of vanadium.

Tris(trialkylsilyl) vanadates are colorless mobile liquids, which are readily soluble in organic solvents. They do not dissolve in water, but are hydrolyzed with the formation of a trialkylsilanol and also vanadic acid [409]. The hydrolysis becomes more difficult as the radicals at the silicon atom becomes more complex. Tris(trimethylsilyl)vanadate turns red in contact with moist air, indicating its hydrolysis with the formation of polyvanadic acids. Crystalline triphenylsiloxy derivatives of vanadium are hydrolytically stable. Compounds of the type $[(CH_3)_3SiO]_n VOCl_{3-n}$ acquire an intense red color even in the presence of traces of moisture. In addition to polyvanadic acid, HCl and hexamethyldisiloxane are formed during their decomposition [1118]. At the same time, $[(C_6H_5)_3SiO]_3VS$ is not decomposed by cold water, but is converted by repeated recrystallization from wet acetone into $[(C_6H_5)_3SiO]_3VO$ [629]. The reaction of tris(trimethylsilyl)vanadate with alcolhols gives trimethylalkoxysilanes and polyvanadic acids [1118].

According to the data of Orlov [415], among the triorganosilyl esters of B, Al, Ti, P, and V acids, the lowest thermal stability is shown by tris(triorganosilyl) vanadates. Tris(phenyldiethylsilyl)-vanadate decomposes appreciably when distilled in vacuum (2 mm) with the formation of lower oxides of vanadium. Bis(trimethylsilyl)-chlorovanadate is generally difficult to obtain in a pure form since even above 60°C its decomposition occurs with the formation of trimethylchlorosilane, hexamethyldisiloxane, $VOCl_3$, and a red-brown polymer [1118].

Andrianov and his co-workers [180] considered that in the IR spectrum of tris(triethylsilyl)vanadate the bands at 909 and 876 cm^{-1} may be assigned to vibrations of the V–O–Si system.

TABLE 85. Properties of Triorganosiloxy Derivatives of Vanadium

Formula	b.p., °C	m.p., °C (mm)	n_D^{20}	d_4^{20}	Preparation method (scheme)	Yield, %	Literature
$(CH_3)_3SiOVOCl_2$	—39	35—36 (1)	1.5186	—	5-63	—	1118
$[(CH_3)_3SiO]_2VOCl$	—	53 (1.1)	—	—	5-63	—	1118
$[(CH_3)_3SiO]_3VO$	—	65—66 (0,35)	$1.4515^{25.9}$	—	5-60	—	1649
	—	118—120 (18)	1.4542	0.9812	5-62	43	409, 413
	—18	100 (9.5)	1.4578^{18}	—	—	—	1118
$[C_2H_5(CH_3)_2SiO]_3VO$	—	124—126 (7)	1.4670	0.9802	5-61	71	409, 413, 1476
$[CH_3(C_2H_5)_2SiO]_3VO$	—	169—170 (5)	1.4730	0.9816	5-62	39	409, 413
	—	169—170 (5)	1.4730	0.9813	5-62	40	1471
$[(C_2H_5)_3SiO]_3VO$	—	192—194 (7)	1.4812	0.9816	5-59 (ammonia)	54	409, 413, 1473
	—	198—201 (13)	1.4820	0.9825	5-59 (pyridine)	76	409, 413, 1473
	—	170—173 (4)	1.4818	0.9837	5-60	85	409, 413
	—	170—173 (4); 188—194 (7)	1.4818	0.9837	5-60	87	1473
	—	186.5 (3.5)	1.4820	0.9816	5-60	60	180 *
	—	182—185 (5)	1.4828	0.9835	5-61	94	409, 413, 1476
	—	189—192 (6)	1.4808	0,9830	5-62	81	409, 413, 1471
	—	169—171 (1.5)	1.4812	0.9809	5-64	—	180 *
$[C_6H_5(C_2H_5)_2SiO]_3VO$	—	263—268 (1,5)	1.5485	1.0767	5-62	46	409, 413
$[(C_6H_5)_3SiO]_3VO$	225—226	—	—	—	5-59	95	629
	228	—	1.689N_g † 1.664N_p	—	5-61	82—90	409, 413, 1476
	228—229	—	—	—	5-62	30—40	1624
$[(C_6H_5)_3SiO]_3VS$	200—202	—	—	—	—	—	629
$[(CH_3)_3SiO]_4V$	—	—	—	—	~5-60	—	1685
$[(C_6H_5)_3SiO]_4V$	260	—	—	—	~5-60	—	640, 1685
	264—268	—	—	—	~5-60	63—86	629

*The compound was incorrectly named tris(triethylsiloxy) vanadate.

†Immersion method.

11.3. Analysis

For the determination of vanadium in lower tris(trialkylsilyl) vanadates it is recommended that the compound is hydrolyzed with acid aqueous alcohols with subsequent evaporation of the solution and calcination of the V_2O_5 precipitate at 900°C. When wet combustion of the substance with a mixture of oleum and nitric acid is used, the precipitate of silicic acid is removed by filtration and calcined at 900°C, while the filtrate is evaporated and the V_2O_5 then calcined. A more meticulous method has been proposed [629] for triphenylsiloxy derivatives of vanadium. A sample of the substance (0.3-0.4 g) is mixed with 50 ml of 6N NaOH and kept at 90°C for 3 hr. The cold solution is acidified to pH 9 with sulfuric acid and again heated at 90°C for 10-15 min. The hot solution is filtered on a Gooch filter, which is then dried to constant weight at 110°C. The silicon is thus determined as hexaphenyldisiloxane. The best results are obtained with a sample containing 0.03-0.06 g of silicon. The hydrolysis conditions must be observed strictly since more prolonged heating leads to partial decomposition of the $(C_6H_5)_3SiOSi(C_6H_5)_3$ and hence to a low silicon content. An inadequate hydrolysis time leads to incomplete cleavage of the group Si–O–V.

The vanadium in the acidified filtrate is determined. To ensure that it is in the pentavalent state, a small amount of hydrogen peroxide is added to the solution, which is boiled. The solution is buffered to pH 4.5 (20 ml of 20% ammonium acetate + dilute ammonia or acetic acid). Then 15 ml of a 2% solution of 8-hydroxyquinoline in 5% acetic acid is added rapidly with stirring. A dark green precipitate forms rapidly and this turns black when the suspension is heated to boiling. The solution is filtered and the precipitate of the hydroxyquinoline complex washed twice with 10 ml of 5% ammonium acetate and dried at 120°C.

11.4. Application

The only recommendation as yet on the practical application of organosilicon derivatives of vanadium are given in patents of Cohen [649, 1685]. It is suggested that trimethylsilyl vanadates are used as catalysts in petroleum cracking and the polymerization of olefins, corrosion inhibitors, siccatives, and fuel additives.

12. COMPOUNDS CONTAINING THE Si – N – V GROUP

Nitrogen-containing organosilicon derivatives of vanadium were described only in the papers of Wannagat and his co-workers, which appeared in 1964 [122, 604, 607]. The reaction of a sodium bis(triorganosilyl)amide with isopropyl chlorovanadates proceeds in petroleum ether even at room temperature:

$$(RO)_n V(O)Cl_{3-n} + (3-n) NaN(SiR'_3)_2 \longrightarrow (3-n) NaCl + (RO)_n V(O)[N(SiR'_3)_2]_{3-n}$$
$$(n = 1 \text{ or } 2;\ R = \text{iso-}C_3H_7;\ R' = CH_3,\ \text{iso-}C_3H_7O) \quad (5\text{-}65)$$

Attempts to prepare monosubstitution products by the action of Wannagat's reagent on $VOCl_3$ were unsuccessful. This is connected with the fact that in addition to the possibility of secondary reactions involving vanadium oxychloride in nonpolar solvents, complications arise in connection with the formation of the complex salt $NaCl \cdot xVOCl_3$, and in the case of polar solvents (tetrahydrofuran and ether), the reaction products form adducts with them. Since unstable $[(CH_3)_3Si]_2NTiCl_3$ could be stabilized by conversion into the complex with pyridine (see Ch. 4), this method was tried in the reaction of sodium bis(trimethylsilyl)amide with $VOCl_3$. The orange crystals, which were collected by filtration with difficulty, decomposed at 0°C and turned black. If the solution was cooled to -50°C after the addition of pyridine, instead of the expected complex of $[(CH_3)Si]_2NV(O)Cl_2$ with pyridine, a greenish brown precipitate which did not contain trimethylsilyl groups formed instantaneously:

$$VOCl_3 + NaN(SiR_3)_2 \longrightarrow Cl_2V(O)N(SiR_3)_2 \xrightarrow[+4C_5H_5N]{+VOCl_3}$$
$$\longrightarrow 2R_3SiCl + Cl_3NO_2V_2(C_5H_5N)_4 \quad (R = CH_3) \quad (5\text{-}66)$$

An analogous complex is formed with quinoline.

The reaction of vanadium oxychloride with sodium bis(trimethylsilyl)amide in a molar ratio of 3 : 1 in nonpolar solvents did not give definite compounds. In ether or tetrahydrofuran the two substances reacted in the form of adducts, hypothetically according to the scheme

$$3(R_3Si)_2NNa + VOCl_3 \xrightarrow[\text{ether}]{0°} 3NaCl + OV[N(SiR_3)_2]_3 \quad (5\text{-}67)$$

TABLE 86. Properties of Compounds

Formula	m.p., °C	b.p., °C (mm)
$[(CH_3)_3Si]_2NVO(OC_3H_7\text{-iso})_2$	from —15 to —16	81 (0.5)
$[(\text{iso-}C_3H_7O)_3Si]_2NVO(OC_3H_7\text{-iso})_2$	165 decomp.	125 (0.3) subl.
$\{[(CH_3)_3Si]_2N\}_2VO(OC_3H_7\text{-iso})$	<-78	105 (0.6)
$\{[(\text{iso-}C_3H_7O)_3Si]_2N\}_2VO(OC_3H_7\text{-iso})$	—20	—
$\{[(CH_3)_3Si]_2N\}_2V\begin{smallmatrix}\nearrow NSi(CH_3)_3 \\ \searrow OSi(CH_3)_3\end{smallmatrix}$	68	130 (0.5) subl.

Analysis of the reaction product gave results which corresponded to the composition of tris[bis(trimethylsilyl)amino]vanadium oxide. However, even during sublimation the substance decomposed with the formation of a mixture of hexamethyldisiloxane and hexamethyldisilazan in a ratio of 70 : 30 and also a solid lustrous carbon-like residue. This result, and also the absence of a characteristic line at 1020 cm^{-1} (ν_{as} V=O) from the IR spectrum led the authors of [607] to the conclusion that the compound had a different structure:

$$Cl_3V{=}O + NaN(SiR_3)_2 \longrightarrow Cl_3V{=}NSiR_3 + NaOSiR_3 \xrightarrow{+2NaN(SiR_3)_2}$$

$$\longrightarrow 3NaCl + [(R_3Si)_2N]_2V({=}NSiR_3)(OSiR_3) \qquad (5\text{-}68)$$

The IR spectrum contained lines corresponding to ν_{as} V=N–Si; (1117 cm^{-1}) and ν_{as} Si–O–V (986 cm^{-1}). The NMR spectra also confirmed the structure of the substance.

The physicochemical constants of compounds containing the Si–N–V group are given in Table 86.

When $[(CH_3)_3Si]_2NVO(OC_3H_7\text{-iso})_2$ is heated to 185°C (726 mm) the compound begins to decompose with the evolution of trimethylisopropoxysilane and hexamethyldisilazan. The black residue contains 42.83% V and 7.04% N, which is close to the ratio V : N = 3 : 2. The decomposition of the compounds containing one bis(triisopropoxysilyl)amino group at the vanadium atom begins at 130°C and involves the formation of triisopropyl vanadate and hexaiso-

Containing the Si – N – V Group

n_D^{20}	d_4^t	Character of substance	Yield, %	Literature
1.4868	0.977^{20}	Lemon yellow liquid	78	604, 607
—	—	White needles	81	607
1.4952	0.961^{30}	Green-yellow viscous liquid	54	607
—	—	Red-brown liquid	68	607
—	—	Green needles	15	122, 604, 607

propoxydisiloxane in a ratio of 1 : 2. The compound with two such groups decomposes at about 100°C.

All vanadium derivatives containing bis(triorganosilyl)amino groups and isopropoxy groups dissolve readily in both polar and nonpolar solvents.

To conclude the section on organosilicon derivatives of vanadium we should mention a patent [1680], in which there is a description of a compound, which corresponds in name to $(n\text{-}C_4H_9C_5H_4)V \cdot (CO)_3Si(CH_2CH{=}CH_2)_2$ and exists in the form of a dimer, but whose true structure is as yet unknown.

NIOBIUM AND TANTALUM

13. ORGANOSILICON COMPOUNDS OF NIOBIUM AND TANTALUM

13.1. Preparation Methods

The following reaction is general for the preparation of trialkylsiloxy derivatives of niobium and tantalum [573, 575]:

$$5RR'_2SiOH + (R''O)_5M \longrightarrow (RR'_2SiO)_5M + 5R''OH \qquad (5\text{-}69)$$

(R, R', and R" are identical or different alkyl radicals; M = Nb or Ta)

The substances obtained were the first organosilicon heterocompounds in which the central hetero atom was attached to five triorganosiloxy groups. Reaction (5-69) was carried out by azeotropic distillation of the alcohol from the mixture of components. The yield of $(R_3SiO)_5M$ varied over the range of 80–95%, falling to 48–54% only when trimethylsilanol was used in the reaction.

Pentakis(trialkylsiloxy)tantalanes may also be obtained by the reaction [574, 575]:

$$5R_3SiOCOCH_3 + (R'O)_5Ta \longrightarrow (R_3SiO)_5Ta + 5R'OCOCH_3 \qquad (5\text{-}70)$$

To avoid side acetylation, a strict stoichiometric amount of the reagents is used. The process is carried out in cyclohexane, which forms azeotropes with alkyl acetates that boil at the minimal temperatures. This makes it possible to obtain the required compounds in a yield of more than 70%.

In the patent mentioned above [1680] there is a report of the preparation of derivatives of niobium–dimeric $CH_3COOC_5H_4Nb \cdot (CO)_3Si(C_5H_5)_2$, and tantalum–also dimeric $[(n\text{-}C_4H_9)_2C_5H_3]Ta(CO)_3Si(C_{18}H_{37}\text{-}n)_2$.

13.2. Properties

Table 87 gives the physicochemical constants of trialkylsiloxy derivatives of niobium and tantalum.

Pentakis(trimethylsiloxy)niobane is not very stable and in an attempt at its purification by vacuum sublimation it was converted into $[(CH_3)_3SiO]_4Nb\text{–}O\text{–}Nb[OSi(CH_3)_3]_4$, namely, octakis(trimethylsiloxy)dinioboxane.*

A peculiar modification of reaction (5-69) is the silanolysis of pentakis(trimethylsiloxy)tantalane [575]:

$$5R_3SiOH + [(CH_3)_3SiO]_5Ta \longrightarrow (R_3SiO)_5Ta + 5(CH_3)_3SiOH \qquad (5\text{-}71)$$

This reaction can also be carried out in the case of $C_2H_5 \cdot (CH_3)_2SiOH$ with an almost quantitative yield of the silanolysis product.

Careful determination of the molecular weights of $(R_3SiO)_5M$ (where M = Nb or Ta) by the ebullioscopic method indicates some association of pentakis(trimethylsiloxy)tantalane [575], while the higher trialkylsiloxy derivatives of tantalum are monomeric.†

* The name "octakis(trimethylsiloxy)diniobium oxide" [573] for this compound is incorrect.

† Pentaethoxytantalane is dimeric like its niobium analog.

TABLE 87. Trialkylsiloxy Derivatives of Niobium and Tantalum

Formula	m.p., °C	b.p., °C (mm)	Character of substance	Yield, %	Literature
$[(CH_3)_3SiO]_5Nb$	—	120 (0.1)	5-69	54	575
$[CH_3(C_2H_5)_2SiO]_5Nb$	—	140 (10^{-4}) *	5-69	90	575
$\{[(CH_3)_3SiO]_4Nb\}_2O$	—	—	—	—	573
$[(CH_3)_3SiO]_5Ta$	80	84 (0.05)	5-69	74	573
			5-69	48	575
			5-70	74	574, 575
$[C_2H_5(CH_3)_2SiO]_5Ta$	135	135 (0.1) *	5-69	82	575
			5-70	88	575
			5-71	94	575
$[n\text{-}C_3H_7(CH_3)_2SiO]_5Ta$	—	170 (0.1)	5-69	91	575
$[iso\text{-}C_3H_7(CH_3)_2SiO]_5Ta$	180	165 (0.1) *	5-69	90	575
$[CH_3(C_2H_5)_2SiO]_5Ta$	180	170 (0.1) *	5-69	86	575
			5-70	72	575
			5-71	98	575
$[(C_2H_5)_3SiO]_5Ta$	210	210 (0.1) *	5-69	96	575
			5-71	88	575

* The substance sublimed.

In the IR spectra of $(R_3SiO)_5M$ (where M = Nb or Ta) the characteristic frequencies of the Si–O bond lie in the region of 905-909 cm^{-1} for tantalum derivatives and 889 cm^{-1} for niobium compounds. Correspondingly the frequencies of the N–O bond are 575 and 619 cm^{-1} when M = Nb and 592-595 and 606 cm^{-1} when M = Ta [535].

There is as yet no information on the application of organosilicon derivatives of niobium and tantalum. Nb_2O_5 is recommended as an additive which increases the thermal stability of vulcanizates from siloxane elastomers [1433].

Chapter 6

Organosilicon Compounds of Group VI Elements

SELENIUM AND TELLURIUM

1. ORGANOSILICON DERIVATIVES OF SELENIUM AND TELLURIUM

1.1. Preparation Methods

Organosilicon derivatives of selenium and tellurium are a class of organosilicon heterocompounds which has been studied very little. The methods of synthesizing them are similar to the preparation methods of analogous sulfur compounds. To prepare compounds containing the Si–Se–C group, the reactions of halosilanes with sodium [527], magnesium [527], or mercury [666] derivatives of organoselenols are used:

$$\gt Si{-}X + M{-}Se{-}R \xrightarrow[-MX]{} \gt Si{-}Se{-}R \qquad (6\text{-}1)$$
$$(X = F,\ Cl;\ M = Na,\ Hg_{0.5})$$

Tetra(arylseleno)silanes, tetra(cyclohexylseleno)silane [527], and trifluoromethyl- and heptafluoropropylselenosilane [667] have been prepared in this way.

The organomagnesium method cannot be used for the synthesis of compounds of the type $\equiv Si-El-R$ (El = Se, Te). While phenylthiomagnesium bromide reacts with trimethylchlorosilane with the formation of $(CH_3)_3SiSC_6H_5$, in the case of $C_6H_5SeMgBr$ or its tellurium analog, bis-silyl substituted derivatives are formed. This is interpreted in the following way [846]:

$$2RElMgX + 2R'_3SiX' \xrightarrow[-2MgXX']{} [2RElSiR'_3] \longrightarrow (R'_3Si)_2El + R_2El \qquad (6\text{-}1a)$$

$$(R = C_6H_5,\ R' = CH_3;\ El = Se, Te;\ X = Br,\ X' = Cl)$$

At the same time, a compound containing the group $\equiv Si-Te-R$ is obtained in low yield (~12%) by heating a mixture of diethyl telluride with triethylsilane in evacuated ampoules at 200°C [259a]. However, with excess silane or when a compound of the type R_3MH is used in the reaction the scheme assumes a more final form:

$$R_2Te \xrightarrow[-RH]{+R_3SiH} R_3SiTeR \begin{cases} \xrightarrow[-RH]{+R_3SiH} R_3SiTeSiR_3 \\ \xrightarrow[-RH]{+R_3MH} R_3SiTeMR_3 \end{cases} \qquad (6\text{-}1b)$$

$$(R = C_2H_5;\ M = Sn)$$

Triorganoselenonium and triorganotelluronium hexafluorosilicates are formed by the schemes [676]:

$$2\,[R_3Se]\,Cl + Ag_2SiF_6 \xrightarrow[-2AgCl]{} [R_3Se]_2\,SiF_6 \qquad (6\text{-}2)$$

$$2\,[R_3Te]\,OH + H_2SiF_6 \xrightarrow[-2H_2O]{} [R_3Te]_2\,SiF_6 \qquad (6\text{-}3)$$

A general method for synthesizing compounds containing the Si–Se–Si group (silselanes),* is the reaction of halosilanes with selenides of sodium [9, 1114, 1115] and silver [674, 675, 679]:

$$2\,{>}Si{-}X + M_2Se \xrightarrow[-2MX]{} {>}Si{-}Se{-}Si{<} \qquad (6\text{-}4)$$

$$(X = Cl, I;\ M = Na, Ag)$$

Thus, heating trimethylchlorosilane with sodium selenide in boiling benzene gives hexamethyldisilselane in 80-85% yield. Under these conditions dimethyldichlorosilane forms tetramethylcyclodisilselane, which is converted on heating into hexamethylcyclotrisilselane [1114, 1115].

* Compounds of this type include silicon diselenide $SiSe_2$ [17, 55, 710a, 1050, 1250, 1252]. Together with this, silicon monoselenide SiSe is known [94, 536-541, 822, 1214-1216]. A compound containing the Si–Te–Si group is silicon ditelluride $SiTe_2$ [17, 1251]. It has also been shown that the monotelluride SiTe exists [536-539, 541, 822, 857, 1214-1216, 1252].

In contrast to triethylgermane and triethylstannane, triethylsilane does not react with selenium even with heating to 215°C for 6 hr [259b].

When methylsilane is heated with hydrogen selenide in a sealed glass ampoule at 400°C, a compound with the composition $(CH_3Si)_4Se_6$ is formed and this melts above 280°C. The NMR spectrum of this compound in deuterochloroform consists of a sharp singlet ($\tau = 8.5$ ppm). On the basis of IR, NMR, and mass spectral data the above compound was assigned a cyclic adamantane-type structure [697a]:

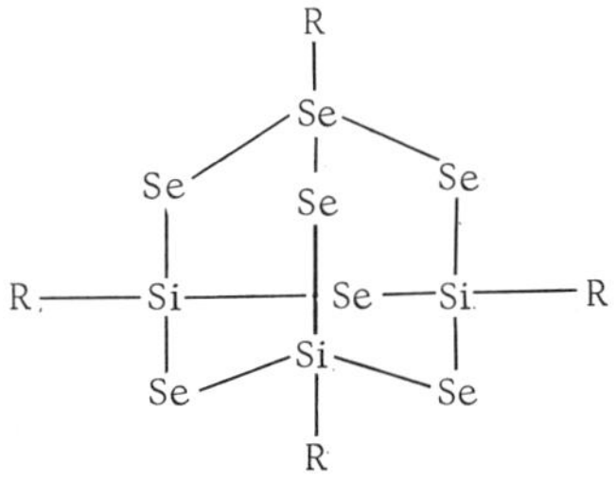

```
                R
                |
               Se
             /  |  \
          Se    Se   Se
          |     |     |
   R ---- Si ---|  Se -- Si ---- R
          |    Si     |
          Se /  |  \  Se
                R
```

A compound with an analogous structure $(C_2H_5Si)_4Se_6$ is formed by the reaction of ethyltrichlorosilane with hydrogen selenide in CCl_4 at 150°C in the presence of metallic aluminum [697a].

The reaction of trimethylchlorosilane with lithium trimethylgermanoselenolate leads to the formation of trimethyl(trimethylsilylseleno)germane [1044]:

$$(CH_3)_3\,GeSeLi + ClSi\,(CH_3)_3 \xrightarrow[-LiCl]{} (CH_3)_3\,GeSeSi\,(CH_3)_3 \qquad (6\text{-}5)$$

Bis(triorganosilyl) selenates containing the Si–O–Se group are obtained by methods which are general for the synthesis of most triorganosilyl esters of inorganic acids. They include the reaction of hexaorganodisiloxanes with anhydrides of the corresponding acids and the reaction of trialkylhalosilanes with silver salts of the acids. The reaction of selenium trioxide with hexamethyldisiloxane leads to a strong explosion. Therefore, for the preparation of bis(trimethylsilyl)selenate this reaction is carried out with cooling with liquid air. Under these conditions selenium trioxide reacts with hexamethyldisiloxane more slowly than sulfur trioxide, but it is still more vigorous than chromium trioxide [1116, 1121]:

$$(CH_3)_3\,SiOSi\,(CH_3)_3 + SeO_3 \longrightarrow [(CH_3)_3\,SiO]_2\,SeO_2 \qquad (6\text{-}6)$$

The reaction of selenium trioxide with trimethylchlorosilane forms trimethylsilyl chloroselenate quantitatively [1113, 1123]:

$$(CH_3)_3SiCl + SeO_3 \longrightarrow (CH_3)_3SiOSeO_2Cl \qquad (6\text{-}7)$$

The latter may also be obtained from trimethylchlorosilane and chloroselenic acid [1113]:

$$(CH_3)_3SiCl + HSeO_3Cl \xrightarrow[-HCl]{} (CH_3)_3SiOSeO_2Cl \qquad (6\text{-}8)$$

By the action of sodium trimethylsilanolate, trimethylsilyl chloroselenate is converted into bis(trimethylsilyl)selenate [1123]. The same compound is formed in 60-70% yield by the reaction [1116, 1121]

$$2(CH_3)_3SiCl + AgSeO_4 \xrightarrow[-2AgCl]{} [(CH_3)_3SiO]_2SeO_2 \qquad (6\text{-}9)$$

The reaction of silver selenate with dimethyldichlorosilane leads to the formation of polymeric $[(CH_3)_2SiOSeO_3]_n$ [1116].

A heterocyclic compound containing the $Si-(C)_n-Se$ group was obtained by the scheme [1122]:

$$O[Si(CH_3)_2CH_2Cl]_2 + Na_2Se \xrightarrow[-2NaCl]{C_2H_5OH} \overbrace{O\langle Si(CH_3)_2-CH_2 \rangle_2 Se}^{\text{ring}} \qquad (6\text{-}10)$$

Isoselenocyanatosilane is formed by the reaction [668]:

$$H_3SiI + AgSeCN \xrightarrow[-AgI]{} H_3SiNCSe \qquad (6\text{-}11)$$

Similarly, the reaction of phthalocyaninodichlorosilane with silver selenocyanide led to the formation of bis(isoselenocyanato)-phthalocyaninosilane [1182a].

According to the data of Voronkov trimethylsilyl isoselenocyanate is formed in good yield by heating trimethylcyanosilane with elementary selenium in a sealed tube at 200°C:

$$(CH_3)_3SiNC + Se \longrightarrow (CH_3)_3SiNCSe \qquad (6\text{-}12)$$

1.2. Physical Properties

Tetra(arylseleno)silanes are solid crystalline substances with sharp melting points [527]. Trifluoromethyl- and heptafluoropropylselenosilanes are colorless liquids, which are more stable at room temperature than their sulfur analogs [667]. Triorganoselenonium and triorganotelluronium hexafluorosilicates are readily soluble in water, difficulty soluble in ethanol, and insoluble in acetone and ether [676]. Organosilselanes are stable in the absence of oxygen. They decompose in air with the liberation of hydrogen selenide and red selenium [1114]. Bis(trimethylsilyl)selane has an odor like hydrogen selenide. Its Te analog has a weak musty odor, which strongly irritates the nasopharynx. Both compounds decompose readily in air with the liberation of free "metal" [846]. Trimethyl(trimethylsilylseleno)germane decomposes when distilled at atmospheric pressure [1044]:

$$2\,(CH_3)_3\,GeSeSi\,(CH_3)_3 \longrightarrow [(CH_3)_3\,Ge]_2\,Se + [(CH_3)_3\,Si]_2\,Se \qquad (6\text{-}13)$$

Isoselenocyanatosilane is a colorless liquid, which polymerizes slowly even in a sealed ampoule at room temperature in the dark [668]. Heating, radiation, and traces of BF_3 accelerate the polymerization.

Bis(trimethylsilyl)selenate melts and distills without decomposition and is readily soluble in inert organic solvents [1121]. Trimethylsilyl chloroselenate is less stable and decomposes gradually even at room temperature [1123]. 1,1,3,3-Tetramethyl-1,3-disila-2-oxa-5-selenacyclohexane is insoluble in water, but readily soluble in organic solvents and is volatile in ethanol vapor [1122].

The physical constants of organosilicon derivatives of selenium and tellurium are given in Table 88. The energies of the Si–Se and Si–Te bonds were calculated; they equal 55 and 46 kcal/mole, respectively [1011].

The intense polarized line at 388 cm^{-1} in the Raman spectrum belongs to the symmetrical valence vibration of the Si–Se–Si skeleton. The bands of the antisymmetrical and symmetrical valence vibrations of the Si–Se–Si group in the IR spectrum of disilselane are evidently unresolvable since they form only one band at 400 cm^{-1}. This is quite acceptable if we take into account the fact that with an increase in the mass of the central atom X the difference

TABLE 88. Organosilicon Derivatives of Selenium and Tellurium

Formula	m.p., °C	b.p., °C (mm)	Other properties	Literature
$H_3SiSeCF_3$	—125,8	—	p=15mm(—26° C); Trouton's constant 22	667
$H_3SiSeC_3F_7$-n	—98	—	p = 36mm (0° C); Trouton's constant 23	667
$(C_2H_5)_3SiTeC_2H_5$	—	52-53(1)	n_D^{20} 1.5340	259a
$(p\text{-}ClC_6H_4Se)_4Si$	182—183	—	—	527
$(C_6H_5Se)_4Si$	136.5—137.0	—	—	527
$(C_6H_{11}Se)_4Si$	93.5—94.0	—	—	527
$(p\text{-}CH_3C_6H_4Se)_4Si$	128—128.5	—	—	527
$(p\text{-}tert\text{-}C_4H_9C_6H_4Se)_4Si$	175—176	—	—	527
$[(CH_3)_3Se]_2SiF_6$	300 decomp.	—	—	676
$[(C_6H_5)_3Se]_2SiF_6$	228 decomp.	—	—	676
$[(CH_3)_3Te]_2SiF_6$	320—336 decomp.	—	—	676
$(H_3Si)_2Se$	—68.0±0.2	85.2±1	p = 20.9 mm (0° C); Trouton's constant 22.9; d_4^{20} 1.36	675, 679
$[(CH_3)_2SiSe]_2$	—	—	—	1114, 1116
$[(CH_3)_3Si]_2Se$	—	31 (2)	—	1114, 1115
		39 (4)	—	846
$[(CH_3)_3Si]_2Te$	—	40—42 (0.25)	—	846
$[(C_2H_5)_3Si]_2Te$	—	—		259a
$[(CH_3)_2SiSe]_3$	—	100 (2)	—	1114
$(CH_3Si)_4Se_6$	283—284	—	—	697a
$(C_2H_5Si)_4Se_6$	170—171	—	—	697a
$(CH_3)_3SiSeGe(CH_3)_3$	—19 — —17	79 (12)	—	1044
$(C_2H_5)_3SiTeSn(C_2H_5)_3$	—	109-112(1)	n_D^{20} 1.5680	259a
$(CH_3)_3SiOSeO_2Cl$	—7 — —5	—	—	1123
$[(CH_3)_3SiO]_2SeO_2$	29.5	62 (1)	—	1116, 1121, 1123
$[(CH_3)_2SiOSeO_3]_x$	69–79 decomp.	—	—	1116
$H_3SiNCSe$	—15.1±0.5	—	Trouton's constant 28	668
$(CH_3)_2SiCH_2SeCH_2Si(CH_3)_2$ (Si–O–Si ring)	21—23	193 (720)	n_D^{20} 1,490	1122
$(CH_3)_2SiCH_2SeCH_2Si(CH_3)_2 \cdot CH_3I$ (Si–O–Si ring)	125 decomp.	—	—	1122

between the frequencies of the antisymmetrical vibrations of the group R–X–R falls sharply, e.g. in the series $(CH_3)_2X$ from 235 cm^{-1} (X = O) to 51 cm^{-1} (X = S) and 15 cm^{-1} (X = Se) and in the series $(H_3Si)_2X$ from 500 cm^{-1} (X = O) to 28 cm^{-1} (X = S). A line at 130 cm^{-1} corresponds to the deformational vibration of the Si–Se–Si skeleton. The valence vibration ν Si–H in the IR spectrum of disilselane is represented by the frequency 2180 cm^{-1} in the spectrum of CF_3SeSiH_3 and 2200 cm^{-1} in the spectrum of $C_3F_7SeSiH_3$, while ν Si – D in the Raman spectrum of d_6-disilselane gives lines at 1590 cm^{-1} and 1568 cm^{-1} [669].

The signals of the methyl protons in the NMR spectrum of hexamethyldisilselane and hexamethyldisiltellane [and also bis(trimethylsilyl) selenate] appear at lower fields than in the case of the analogous derivatives of oxygen and sulfur, though in accordance with the order of electronegativities of these elements O > S > Se > Te, one would have expected the reverse. The same picture is observed on examining the spin-spin interaction constant which should increase with an increase in the polarity of the Si–X bond, i.e., have maximum values in the case of hexamethyldisiloxane (X = O). In actual fact, the lowest values of $J(H^1-C^{13})$ and $J(H^1-C-Si^{29})$ are observed in the case of selenium derivatives [1081, 1084, 1095]. It may be surmised that these changes in the chemical shifts and spin-

TABLE 89. NMR Spectra of $[(CH_3)_3Si]_2X$ and $[(CH_3)_3SiO]_2XO_2$ (X = S, Se, Te)*

Formula	δ_{Si-C-H}	$J(H^1-C^{13})$	$J(H^1-C-Si^{29})$	Literature
$(CH_3)_3SiOSi(CH_3)_3$	—3.5	118.0	6.86	1081
$(CH_3)_3SiSSi(CH_3)_3$	—19.5	119.5	7.10	1081
	—	—	6.9	1095
$(CH_3)_3SiSeSi(CH_3)_3$	—25.1	—	—	846
	—25.7	120.5	7.15	1081
$(CH_3)_3SiTeSi(CH_3)_3$	—36.0	—	—	846
$(CH_3)_3SiSeGe(CH_3)_3$ †	—18.8	120.0	6.9	1095
$[(CH_3)_3SiO]_2SO_2$	—23.5	121.0	—	1084
$[(CH_3)_3SiO]_2SeO_2$	—26.8	—	—	1084

* The spectra were plotted in CCl_4 solution at 60 MHz, J and δ are given in Hz; negative values of δ indicate a lower field than the internal standard $(CH_3)_4Si$.

† δ_{Ge-C-H} = 33.5; $J(H^1-C^{13}-Ge)$ = 127.6.

TABLE 90. Values of τ_{Si-H} and $J(Si^{29}-H)$ in the NMR Spectra of Organosilicon Derivatives of Selenium Containing the Si–H Bond (in cyclohexane solution)

Compound	τ_{Si-H}	$J(Si^{29}-H)$	Literature
$H_3SiSeSiH_3$	5.86	226±2	667
$H_3SiSeCF_3$	5.70	230±1	667
$H_3SiSeC_3F_7$-	5.71	—	667
$H_3SiNCSe$	5.52 *	243.2±0.4	668

* At infinite dilution.

spin interaction constants are connected with a decrease in the degree of $d\pi-p\pi$ interaction with an increase in the atomic number of the central atom X.

The characteristics of the NMR spectra of some derivatives of selenium and tellurium containing the group $(CH_3)_3Si$, are given in Table 89 and those containing the group H_3Si in Table 90. In addition, we should mention that in the NMR spectrum of bis(triethylsilyl)tellane (40 MHz, internal standard–tetramethylsilane) τ_{CH_3} = 9.0, and τ_{CH_2} = 9.19 ± 0.02 ppm, while for the spectrum of triethylsilyltriethylstannyltellane the following values were observed: $\tau_{CH_3(Si)}$ 9.03, $\tau_{CH_3(Sn)}$ 8.79, τ_{SiCH_2} 9.20, τ_{SnCH_2} 8.91 ppm [282a].

1.3. Chemical Properties

Compounds containing in the molecules the Si–Se bond or the Si–O–Se group are hydrolytically unstable [668, 679, 1114, 1116, 1121, 1123]:

$$\gt Si-Se-Si\lt + H_2O \longrightarrow \gt Si-O-Si\lt + H_2Se \quad (6\text{-}14)$$

$$[\gt Si-O-]_2 SeO_2 + H_2O \longrightarrow \gt Si-O-Si\lt + H_2SeO_4 \quad (6\text{-}15)$$

Trimethylsilyl chloroselenate, which is decomposed vigorously by water, reacts with hydrogen sulfide with the retention of the Si–O–Se group [1113]:

$$(CH_3)_3SiOSeO_2Cl + H_2S \longrightarrow (CH_3)_3SiOSeO_2SH + HCl \qquad (6\text{-}16)$$

Pyridine completely cleaves trimethylsilyl chloroselenate even at -30°C [1123]:

$$(CH_3)_3SiOSeO_2Cl + C_5H_5N \longrightarrow (CH_3)_3SiCl + C_5H_5N \cdot SeO_2 \qquad (6\text{-}17)$$

By the reaction with sodium trimethylsilanolate in methylene chloride at -20°C, trimethylsilyl chloroselenate may be converted into bis(trimethylsilyl)selenate [1123]:

$$(CH_3)_3SiOSeO_2Cl + NaOSi(CH_3)_3 \xrightarrow[-NaCl]{} [(CH_3)_3SiO]_2SeO_2 \qquad (6\text{-}18)$$

Iodine and hydrogen iodide cleave the Si–Se bond in disilsolane, forming iodosilane and selenium and hydrogen selenide, respectively [679].

Bis(trimethylsilyl)tellane reacts quantitatively with silver iodide even at room temperature according to a scheme which corresponds in character to the reverse reaction (6-4) [846]:

$$(R_3Si)_2Te + 2AgI \longrightarrow 2R_3SiI + Ag_2Te \quad (R = CH_3) \qquad (6\text{-}19)$$

On the basis of this and other reactions, the following series was proposed for the comparative reactivities of compounds of the type $(CH_3)_3Si{-}X$ in substitutions of trimethylsilyl groups:

$$(R_3Si)_2Te > R_3SiI > (R_3Si)_2Se > (R_3Si)_2S > R_3SiBr$$

Heptafluoropropylselenosilane is converted quantitatively into bromosilane by $HgBr_2$ in 48 hr [667].

Isoselenocyanatosilane is cleaved by silver chloride and bromide with the formation of chlorosilane and bromosilane [668].

Compounds containing the Si–C–Se–C group form selenonium salts with methyl chloride [1122]:

$$\equiv Si{-}\overset{|}{\underset{|}{C}}{-}Se{-}C\lessgtr + CH_3I \longrightarrow \left[\equiv Si{-}\overset{|}{\underset{|}{C}}{-}\overset{CH_3}{\overset{|}{Se}}{-}C\lessgtr\right]I \qquad (6\text{-}20)$$

ELEMENTS OF THE CHROMIUM SUBGROUPS

2. ORGANOSILICON COMPOUNDS OF CHROMIUM

Chromium hydroxide reacts exothermally with tetramethoxysilane. However, there are no detailed data in the literature on the character of the products formed by the reaction. It is possible to assess this largely by comparison of the behavior of chromium hydroxide with the behavior of aluminum and calcium hydroxides, which give the corresponding hydrosilicates [328, 481].

The first reports of the preparation of an individual organosilicon compound containing chromium appeared in 1958-59 [511, 1117, 1624]. Boiling a xylene solution of triphenylsilanol with the product from the reaction of chromium trioxide and tert-butanol gives an 85% yield of bis(triphenylsilyl)chromate $[(C_6H_5)_3SiO]_2CrO_2$ [1624]. It was found later that in this way it is also possible to prepare derivatives of tetravalent chromium [542]:

$$Cr(OC_4H_9\text{-}tert)_4 + 4(C_2H_5)_3SiOH \xrightarrow[-tert\text{-}C_4H_9OH]{} Cr[OSi(C_2H_5)_3]_4 \quad (6\text{-}21)$$

It is also possible to apply to bis(triphenylsilyl)chromate the synthesis method which has already become classical in the chemistry of organosilicon heterocompounds, in particular, through alkali metal triorganosilanolates [628]:

$$2R_3SiONa + CrO_2Cl_2 \xrightarrow[-2NaCl]{} (R_3SiO)_2CrO_2 \quad (6\text{-}22)$$
$$(R = C_6H_5)$$

The synthesis of bis(trimethylsilyl)chromate was achieved independently in Western Germany [1117] and England [511] by the same method:

$$(CH_3)_3SiOSi(CH_3)_3 + CrO_3 \longrightarrow [(CH_3)_3SiO]_2CrO_2 \quad (6\text{-}23)$$

The data on the optimal reaction conditions disagree somewhat. Abel [511] reports that the process proceeds slowly even at room temperature (in a nitrogen atmosphere), but accelerated appreciably by heating. However, at a temperature above 40°C the synthesis involved great danger due to the possibility of an explosion. This is also a possibility in vacuum distillation of bis(trimethylsilyl)chromate, even small amounts of which have a strong detonating action. Schmidt and Schmidbaur [1117] recommend that the chromium trioxide is dissolved in excess dry boiling hexamethyldisiloxane (i.e., at 100°C), that the unreacted solvent is removed by distillation, and that the bis(trimethylsilyl)chromate is isolated by distillation at 1 mm. However, in the preparation of appreciable amounts of bis(trimethylsilyl)chromate by this method, over heating during distillation leads to a strong explosion. The data on the constants of the substance likewise do not agree (see Table 91) and this is evidently connected with the difficulties of isolating it in the pure form.

It is not surprising that the action of chromium trioxide on trimethylsiloxytrimethylgermane also leads to the introduction of chromium into a heterosiloxane structure [1106]:

$$(CH_3)_3SiOGe(CH_3)_3 + CrO_3 \longrightarrow (CH_3)_3SiOCrO_2OGe(CH_3)_3 \qquad (6\text{-}24)$$

At room temperature or with slow heating this compound gradually decomposes with autooxidation to black chromium oxide. Distillation is accompanied by disproportionation:

$$2(CH_3)_3SiOCrO_2OGe(CH_3)_3 \xrightarrow{t^\circ} [(CH_3)_3SiO]_2CrO_2 + [(CH_3)_3GeO]_2CrO_2 \qquad (6\text{-}25)$$

Rapid heating of (trimethylsilyl)trimethylgermyl chromate leads, as in the case of its S-analog or bis(trimethylsilyl)chromate to a strong explosion. Hydrolysis of the compound proceeds according to the scheme

$$(CH_3)_3SiOCrO_2OGe(CH_3)_3 + 2H_2O \longrightarrow (CH_3)_3SiOH + (CH_3)_3GeOH + H_2CrO_4 \qquad (6\text{-}26)$$

Chromium trioxide dissolves in polydimethylsiloxanes at 150°C with the formation of yellow $[(CH_3)_2SiCrO_4]_n$, whose structure has not been established, in contrast to its sulfur analog [1117].

The possibility of the formation of polyorganochromasiloxanes was even pointed out in [1] and [36]. However, the routes to the preparation of these substances were described only very recently. One of them is called hydrolytic polycondensation by the authors of [140a, 229]. It is carried out by adding alkyl(aryl)trichlorosilanes to aqueous alkaline solutions of trivalent chromium nitrate with mixing in the presence of a mixed solvent, toluene – butanol. The following scheme is proposed:

$$nRSiCl_3 + Cr(NO_3)_3 + H_2O \xrightarrow{NaOH} [(RSiO_{1.5})_n CrO_{1.5}]_x + 3nNaCl + 3NaNO_3 \quad (6\text{-}27)$$

Another variant consists of dissolving in alkali the hydrolysis products of organotrichlorosilanes and then adding chromium nitrate (aqueous solution) to the reaction:

$$3[(RSiO_{1.5})_n SiR(OH)ONa] + Cr(NO_3)_3 \xrightarrow{H_2O} [(RSiO_{1.5})_n SiR(OH)O]_3Cr + 3NaNO_3 \quad (6\text{-}28)$$

In this way it was possible to prepare a series of clear green, readily soluble polychromaorganosiloxanes with Si : Cr ratios from 397 to 13 with $R = C_2H_5$ and from 493 to 9 with $R = C_6H_5$. The chromium content was less than that of the original starting mixture. Neither of the schemes presented above fully reflects the structure of the polymers obtained. In any case, the authors of [140a, 229] put forward the following formula for an elementary unit on the basis of analysis data:

$$\left\{ \cdots \left[\begin{matrix} R \\ | \\ -Si-\!\!-\!\!-O \\ | \\ O \\ | \end{matrix} \right]_n - \left[\begin{matrix} R \\ | \\ -Si-O \\ | \\ OH \end{matrix} \right]_m - \begin{matrix} \\ \\ Cr-O-\cdots \\ | \\ O \\ | \end{matrix} \right\}_x$$

The reaction of the monosodium salt of phenylsilanetriol with potassium chromium alum (20% aqueous solution) in toluene, which was proposed by Avilova and her co-workers [127a] is similar to scheme (6-28). This reaction yielded solid, green, soluble polymers with a molecular weight of 1200-1700 and a Si : Cr ratio of 5.8-7.1. These polymers did not melt before they decomposed and the following structure was suggested for them:

$$\left[-\left(\begin{matrix} C_6H_5 \\ | \\ Si-O \\ | \\ O \end{matrix} \right)_m - Cr \begin{matrix} O- \\ \\ O- \end{matrix} \right]_n$$

The chromium content of the polymer fractions increased with the molecular weight of the fraction and the reverse was true of the silicon content. Sulfuric acid does not extract the chromium from these polymers in the cold.

According to data in [140a] the thermal stability decreases with an increase in the chromium content of the polymer. Chromasiloxanes with $R = C_6H_5$ have a lower thermal stability than polyphenylsiloxanes, while polychromaethylsiloxanes are superior to polyethylsiloxanes in this respect. With the same Si : Cr ratio, chromasiloxanes with $R = C_6H_5$ have a higher thermal stability.

In the IR spectra of chromasiloxanes the bands of the vibrations of the Si – O – Cr group lie in the region of 500-600 cm^{-1}.

It has been reported that π-complexes may be formed between arylsilanes and chromium compounds, but the paper in question [86] was practically inaccessible to us.

The reaction of sodium bis(trimethylsilyl)amide with chromium trichloride in tetrahydrofuran* proceeds even in the cold. It is exothermic and is accompanied by the formation of green, crystalline tris[bis(trimethylsilyl)amino] chromine, which decomposes spontaneously in air [122, 604, 610]:

$$3\,[(CH_3)_3\,Si]_2\,NNa + CrCl_3 \longrightarrow \{[(CH_3)_3\,Si]_2\,N\}_3\,Cr + 3NaCl \qquad (6\text{-}29)$$

The attempt to prepare the analogous organosilicon derivative of hexavalent chromium was unsuccessful [610]. When CrO_2Cl_2 is treated with sodium bis(trimethylsilyl)amide, there is only reduction of the chromium to the trivalent state.

Trimethylphenylsilane reacts with chromium hexacarbonyl in an argon atmosphere with the formation of a stable complex, in which the chromium is in the form of the tricarbonyl [1142]. In this connection we should mention that chromium hexacarbonyl, like iron pentacarbonyl, is a catalyst for the reaction of triorganosilanes with olefins [396]. We cannot completely exclude the possibility that complexes of organosilicon compounds containing chromium play an active part in this case.

Phosphinomethylene compounds of silicon react with Reinecke's salt (ammonium tetrathiocyanodiammonochromate) to form reineckates [815].

* More correctly, the reaction with the adduct of $CrCl_3$ with THF (1 : 3), formed by prolonged boiling of a mixture of them in the presence of zinc dust.

3. ORGANOSILICON COMPOUNDS OF MOLYBDENUM AND TUNGSTEN

It is reported in a patent [1624] that bis(triphenylsilyl)molybdates and bis(triphenylsilyl)tungstates may be obtained in analogy with the derivatives of vanadium and chromium. However, the properties of these compounds are not described. It is reported [640] that the reaction of $MoCl_3$ with sodium triphenylsilanolate in a ratio of 1 : 3 does not form tris(triphenylsiloxy)molybdine. At the same time, this compound was isolated from the products of the reaction of MoO_2Cl_2 with sodium triphenylsilanolate though the intention was to prepare a derivative of hexavalent molybdenum. Thus, here there is a definite analogy with the reaction of sodium bis(triphenylsilyl)amide with CrO_2Cl_2.

It was not possible to isolate any definite products from the reaction of triphenylsilanol or sodium triphenylsilanolate with $MoCl_5$ [628]. In the reaction of sodium triphenylsilanolate with the complex of $MoO(OH)_2Cl_2$ with dimethylformamide or N-methyl-2-pyrrolidine (1 : 2) it was established that compounds containing the Si–O–Mo group were formed but they were not characterized.

Boiling a mixture of dimethyldivinylsilane with tungsten hexacarbonyl in a ratio of 5 : 1 in ethylcyclohexane in a nitrogen atmosphere forms the yellow crystalline adduct $(CH_3)_2Si(CH=CH_2)_2 \cdot W(CO)_4$, which decomposes only above 90°C [870, 936]. This compound is a chelate compound in character. It acquires a brown color in contact with air. The complex of analogous composition formed by the reaction of dimethyldivinylsilane with molybdenum hexacarbonyl changes at a great rate in contact with air.

The following reaction is also known [1680]:

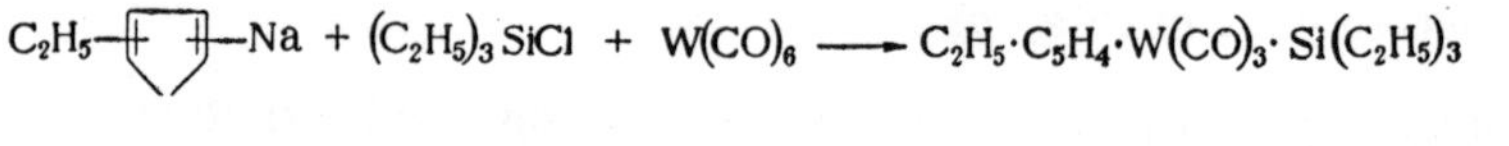

$$C_2H_5{-}C_5H_4{-}Na + (C_2H_5)_3SiCl + W(CO)_6 \longrightarrow C_2H_5 \cdot C_5H_4 \cdot W(CO)_3 \cdot Si(C_2H_5)_3 \quad (6\text{-}30)$$

4. PHYSICAL PROPERTIES AND APPLICATION

The physicochemical constants of organosilicon compounds containing chromium, molybdenum, and tungsten are given in Table 91.

TABLE 91. Organosilicon Compounds of the Chromium Subgroup

Formula	m.p., °C	b.p., °C (mm)	Other data	Literature
$[(CH_3)_3SiO]_2CrO_2$	—	75 (1)	—	1117
	~ —20	60 (0.2)	n_D^{20} 1.4937	511
$[(C_6H_5)_3SiO]_2CrO_2$	—	—	Yield 85%	1624
	138—140	—	—	628
$[(C_2H_5)_3SiO]_4Cr$	—	—	—	542
$[(CH_3)_3SiO][(CH_3)_3GeO]CrO_2$	—3+2	—	—	1106
$\{[(CH_3)_3Si]_2N\}_3Cr$	120	Subl. 13—15 (0.8)	—	122, 604, 610
$(CH_3)_3SiC_6H_5 \cdot Cr(CO)_3$	72—73	Subl. 60—65 (0.01)	Yield 20%	1142
$[(C_6H_5)_3PCH_2Si(CH_3)_3][Cr(NH_3)_2(SCN)_4]$	142—144	—	Darkens at 127°C	815
$[(C_6H_5)_3SiO]_3Mo$	—	—	—	640
$(CH_3)_2Si(CH{=}CH_2)_2 \cdot Mo(CO)_4$	—	—	—	986
$(CH_3)_2Si(CH{=}CH_2)_2 \cdot W(CO)_4$	45	—	Decomposes above 90°C	870, 936
$C_2H_5C_5H_4Si(C_2H_5)_3 \cdot W(CO)_3$	—	—	—	1680

Bis(trimethylsilyl)chromate is an orange-red heavy oily liquid, which is readily soluble in benzene and carbon tetrachloride and is similar in physical and chemical properties to tert-butyl chromate. A solution of it in hexamethyldisiloxane is recommended as a neutral oxidizing agent for organic compounds, for example, for the conversion of toluene into benzyl alcohol [1117].

It is reported that triorganosiloxy derivatives of the chromium subgroup may be used as corrosion inhibitors [1624].

Patents have appeared [1351, 1422, 1424, 1730] on the use of Werner complexes of trivalent chromium as components of waterproofing compositions based on polysiloxanes for the impregnation and surface treatment of cloths, papers, and leathers. The chromium compounds are used here as emulsifiers for the compositions and as active catalysts for the hardening of films and they also produce a substantial improvement in the waterproofing properties of coatings. Synergism is observed here. Treatment of fibrous materials with emulsions which include stearatochromium chloride and methylhydropolysiloxanes gives a better effect than treatment of the material with each of the components separately and also separate treatment with both components [1351, 1424]. This leads to the idea of chemical interaction, connected with the formation of organosilicon derivatives of chromium. Possible confirmation of this is provided by the formation of a glassy, water-soluble, black-green product by the reaction of the copolymer $\{[-(CH_3)_2SiO-]_{75}-[CH_3Si(CH_2CH_2CH_2OH)O-]_{25}\}_n$ with CrO_2Cl_2 (weight ratio 2 : 1, 5-hour reflux with anhydrous CCl_4). This product [138a, 1632] is believed to be an organosiloxane Werner complex of chromium; it forms 33% solution in methanol and a 2% methanol solution may be diluted with water until the content of the latter is 94% without decomposition of the substance. The product is recommend- for waterproofing of leather and cloth.

Chapter 7

Organosilicon Compounds of Elements of the Manganese Subgroup

1. ORGANOSILICON DERIVATIVES OF MANGANESE

There is as yet very little information on organosilicon compounds of elements of the manganese subgroup. As a prototype of compounds containing the Si–O–Mn group we have powdery $ClMnOSiCl_3$, which is formed together with the oligomers $Cl_3Si(OSiCl_2)_nCl$ (n = 1 – 4) by boiling manganese dioxide for 3 days with an acetonitrile solution of silicon tetrachloride in the presence of mercuric chloride [1555]. There is the possibility that compounds of this type are obtained by the reaction of hydrates of manganese salts with dimethyldichlorosilane [1300].

The reaction of tetraethoxysilane with manganous acetate in aqueous alcohol leads to polymeric compounds of the silicate type, which have the nominal composition $kMnO \cdot mSiO_2 \cdot n(C_2H_5)_2O \cdot pH_2O$ [32]. The content of the manganese and the organic part of the reaction product depends on the initial Si : Mn ratio. As a rule, the reaction products are white solids. According to literature data [32, 481] these are chemical compounds containing the Si–O–Mn bond and ethoxyl groups at silicon atoms, which are preserved during prolonged heating at 120°C.

The first publications on the preparation of individual organosilicon derivatives of manganese appeared only in 1964 [610, 640]. The action of sodium bis(trimethylsilyl)amide on MnI_2 in tetrahydrofuran at room temperature yielded liquid bis[bis(trimethylsilyl)-

amino]mangane, which solidifies as pink crystals:

$$MnX_2 + 2\,(R_3Si)_2NNa \longrightarrow 2NaX + [(R_3Si)_2N]_2Mn \quad (7\text{-}1)$$

There is a report [1688] of the preparation of silicon-containing manganese carbonyls of the type $R_3SiMn(CO)_5$ (R = alkyl or chlorine). In the reaction with 1,3-dienes, two carbonyl groups in them are replaced with the formation of such compounds as, for example, $(CH_3)_3SiMn(CO)_3C_4H_6$, which is obtained by the reaction with butadiene.

2. ORGANOSILICON DERIVATIVES OF RHENIUM

A boiled solution of rhenium heptoxide in hexamethyldisiloxane deposits on cooling colorless crystals of trimethylsilyl perrhenate [1120]:

$$R_3SiOSiR_3 + Re_2O_7 \longrightarrow 2R_3SiOReO_3 \quad (7\text{-}2)$$

The same compound is obtained by the reaction of trimethylchlorosilane with a suspension of silver perrhenate in hexamethyldisiloxane:

$$R_3SiCl + AgReO_4 \longrightarrow R_3SiOReO_3 + AgCl \quad (7\text{-}3)$$

Trimethylsilyl perrhenate is also formed by the reaction of rhenium heptoxide with trimethylchlorosilane, but the mixture of various rhenium oxychlorides obtained at the same time hampers the isolation of the compound and makes this route ineffective for its synthesis.

3. PROPERTIES AND APPLICATION OF ORGANOSILICON COMPOUNDS OF ELEMENTS OF THE MANGANESE SUBGROUP

The physicochemical constants of organosilicon derivatives of elements of the manganese subgroup are given in Table 92.

Bis[bis(trimethylsilyl)amino]mangane is extremely sensitive to moisture and oxygen. In the presence of traces of the latter the crystals immediately acquire a violet color and then turn black

TABLE 92. Organosilicon Derivatives of Manganese and Rhenium

Formula	m.p., °C	b.p., °C (mm)	Preparation method (scheme)	Yield, %	Literature
$\{[(CH_3)_3Si]_2N\}_2Mn$	—	100 (0.2)	7-1	73	604, 610
$(CH_3)_3SiMn(CO)_3C_4H_6$ *	—	—	—	—	1688
$Cl_2(CH_3)SiMn(CO)_3$	—	—	-	—	1688
$(n\text{-}C_{16}H_{33})_3SiMn(CO)_3C_5H_8$ †	—	—	—	—	1688
$(C_6H_5CH_2)_3SiMn(CO)_3C_4H_6$ *	—	—	—	—	1688
$(CH_3)_3SiOReO_3$	79.5—80.5	Subl. 65—80 (1)	7-2	Quant.	1120
	"	"	7-3	94	1120

* Derivative of butadiene.

† Derivative of isoprene.

and deliquesce [610]. Trimethylsilyl perrhenate dissolves without decomposition in aromatic and chlorinated hydrocarbons and ethers.. Water produces hydrolysis and the trimethylsilanol formed condenses quantitatively to hexamethyldisiloxane under the action of perrhenic acid. The reaction with alcohols yields trimethylalkoxysilanes [1120].*

Complexes of silicon-containing carbonyls of manganese with dienes are recommended as effective additives for fuels as antiknock compounds [1688]. It has been reported that various manganese compounds (salts of alkanecarboxylic, naphthenic, and resin acids, alkoxy derivatives, salt hydrates, the oxide, the hydrated oxide, and the complex of $MnCl_2$ with caprolactam) may be used as catalysts of cold vulcanization of liquid siloxane elastomers and as additives to improve the thermal stability of vulcanizates, stabilizers of fluids, and catalysts for the prepolymerization and hardening of siloxane resins [63, 110, 614, 1368, 1433, 1439, 1445, 1684, 1689]. In the latter case Noll [110, p. 147] lists manganese compounds as catalysts of low activity, which are inferior to analogous derivatives of such metals as lead or tin.

* For comparison we should point out that trimethylsilyl perchlorate $(CH_3)_3SiOClO_3$, obtained by a scheme of type (7-3) [1226, 1240], is a liquid with b.p. 35-38°C (14 mm), which explodes on rapid heating and is similar in chemical behavior to a salt rather than an acid ester.

Chapter 8

Organosilicon Compounds of Group VIII Elements

1. COMPOUNDS CONTAINING THE Si – O – Fe GROUP

1.1. Preparation Methods

1.1.1. Reaction of Halosilanes and Alkoxysilanes with Iron Compounds

The products of the reaction of $SiCl_4$ with Fe_2O_3 may be regarded as the prototype of organosilicon compounds containing the Si–O–Fe group [1555]. $ClFeOSiCl_3$ and $Cl_2FeOSiCl_3$ are obtained in acetonitrile in the presence of $HgCl_2$.

As has already been pointed out, mixing tetramethoxysilane with metal hydroxides is accompanied by an exothermic effect, which falls in the series [328]:

$$Ca(OH)_2 > Cr(OH)_3 > Cu(OH)_2 > Fe(OH)_3 > Al(OH)_3$$

The reaction products are evidently iron hydrosilicates, in analogy with calcium and aluminum hydroxides. At the same time, the reaction of tetraethoxysilane with freshly prepared iron acetate (room temperature, vacuum) yields products with an Si : Fe ratio from 1 : 4 to 4 : 1 [32, 318, 322, 335, 481], containing an organic part.

The reaction of sodium glycerosilicate (obtained by the reaction of sodium glycerate with SiO_2 in glycerol) with iron sulfate or acetate [334, 481] yields viscous ferraglycerosilicates (evidently glyceroxyferrasiloxanes). The hydrolysis of these compounds by hot water leads to artificial iron hydrosilicates such as $3.2FeO \cdot SiO_2 \cdot 7H_2O$.

1.1.2. Reaction of Silanols and Alkali Metal Silanolates with Iron Compounds

The possibility of preparing high-polymer ferrasiloxanes by the reaction of silanol compounds with highly dispersed iron or its hydroxide was mentioned in the author's certificate to which we have already referred repeatedly [1461]. However, as for most of the other elements mentioned in it, no details of the experiments have appeared in the literature since 1948 and information on the synthesis of polyorganoferrasiloxanes appeared very recently.

When polysiloxanes of the type $(RSiO_{1.5})_n$ and ferric chloride or ammonium iron alum are dissolved in alkali [152] there occur processes whose final result was represented by the authors by the following scheme:

$$RSi(OH)_2ONa + \begin{matrix} FeCl_3 \\ \text{or} \\ FeNH_4(SO_4)_2 \end{matrix} \longrightarrow \{[-RSi(OH)O-]_x[RSiO_{1.5}]_y FeO_{1.5}\}_n \quad (8\text{-}1)$$

When $R = C_6H_5$, the reaction with alum (20% solution, 16% NaOH, 78°C) yields a polymer, which is infusible but soluble in organic solvents, and has a molecular weight of 3470, $x = y \cong 2$, Si : Fe = 4 and n = 5.65. When ferric chloride is used, heating the reaction mixture for 5 hr in toluene at 80-100°C gives a powdery polymer (mol. wt. 4500, Si : Fe = 9, x and y = 4-5), which also combines solubility in most organic solvents (with the exception of benzine and decalin) with infusibility. Replacement of ferric chloride by an equimolar mixture of ferric chloride and aluminum chloride under the same conditions leads to the formation of terpolymers, namely, polyferraalumaorganosiloxanes:

$$[(C_6H_5SiO_{1.5})_x AlO_{1.5}(C_6H_5SiO_{1.5})_y FeO_{1.5}]_n$$

In this way, in particular, a polymer was prepared with mol. wt. 3770 and x = y = 6, Si : Fe = 12, Si : Al = 12, Fe : Al = 1. It retained its solubility in toluene after heating for 2 hr at 200°C, but the molecular weight of the polymer was then doubled (7430). Polymers of similar character, which melt above 300°C, are obtained by the reaction of monosodium salts of organosilanediols or triols with ferric chloride in absolute ethanol, with subsequent heating of the polymer at 150-200°C [1377]. The brown resins form clear solid films on glass.

A patent has appeared [1380] on the preparation of linear polymers of the type $[(OSiRR')_nOM]_m$, where M = Fe, Co, Ni, by the reaction of halides of these metals with disiloxanediolates;* an example is

$$NaOSi(CH_3)_2OSi(CH_3)_2ONa$$

As regards monomeric compounds of this type, the addition of a solution of ferric chloride in ether to an ether solution of sodium trimethylsilanolate at room temperature instantaneously gives a brown precipitate (NaCl contaminated with iron oxide) and after evaporation of the ether, it is possible to isolate yellow-green crystals from the brown solution, which crystals correspond in composition to $[(CH_3)_3SiO]_3Fe$ [1079, 1109]. Thus, the reaction is analogous to scheme (3-111) (p. 271).

$$3R_3SiOM + FeCl_3 \longrightarrow (R_3SiO)_3Fe + 3MCl \qquad (8-2)$$

Like the aluminum analog, this compound is dimeric.

Using four equivalents of sodium trimethylsilanolate instead of a stoichiometric ratio of the reagents does not yield tris(trimethylsiloxy)ferrane, but salt-like $Na\{Fe[OSi(CH_3)_3]_4\}$. The same type of compound is obtained in 85-93% yield by adding to a solution of tris(trimethylsiloxy)ferrane in carbon tetrachloride an equimolar amount of $(CH_3)_3SiOM$ (where M = Li, Na, and K), when an exothermal reaction occurs (cf. scheme 3-126, p. 292). Like tris (triorganosiloxy) derivatives of aluminum and gallium, tris(trimethylsiloxy)ferrane adds trimethylsiloxytetramethylstibane [1072, 1080, 1083]:

$$(R_3SiO)_3Fe + R_3SiOM \longrightarrow M[Fe(OSiR_3)_4] \qquad (8-3)$$

$$[M = Li,\ Na,\ K,\ Sb(CH_3)_4]$$

* More details of this will be given in Section 5.

1.1.3. Cleavage of Siloxanes by Iron Halides

In a series of communications (mainly patents) it is reported that cyclosiloxanes are polymerized by ferric chloride or its hexahydrate [460, 1538, 1564, 1610, 1662]. On adequate grounds the first stage is believed to be cleavage of the ring with the formation of the linear molecule $Cl[(CH_3)_2SiO]_x FeCl_2$ [460]. In the opinion of the authors, the rest of the process consists of the following reaction, whose exceptional selectivity raises some doubts:

$$Cl\left[\begin{array}{c}|\\ SiO\\ |\end{array}\right]_x FeCl_2 \xrightarrow{+H_2O} Fe(OH)Cl_2 + Cl\left[\begin{array}{c}|\\ SiO\\ |\end{array}\right]_x H \xrightarrow{-H_2O}$$

$$\longrightarrow Cl\left[\begin{array}{c}|\\ SiO\\ |\end{array}\right]_{2x-1}\!\!-\!\!\begin{array}{c}|\\ SiCl\\ |\end{array} \xrightarrow[-HCl]{+H_2O} Cl\left[\begin{array}{c}|\\ SiO\\ |\end{array}\right]_{2x} H \xrightarrow{-H_2O} Cl\left[\begin{array}{c}|\\ SiO\\ |\end{array}\right]_{4x-1}\!\!-\!\!\begin{array}{c}|\\ SiCl\\ |\end{array} \quad \text{etc.}$$

$$Fe(OH)Cl_2 + HCl \longrightarrow FeCl_3 + H_2O \qquad (8\text{–}4)$$

Stepwise polymerization of cyclosiloxanes with the participation of $FeCl_3$ is more likely:

$$[R_2SiO]_n + FeCl_3 \longrightarrow Cl[SiR_2O]_n\text{—}FeCl_2 \xrightarrow{[R_2SiO]_n} Cl[SiR_2O]_{2n}\text{—}FeCl_2 \ \text{etc.} \qquad (8\text{–}4a)$$

Hexaorganodisiloxanes are also cleaved by ferric chloride [673]. In analogy with the action of AlX_3 on $R_3SiOSiR_3$, in this case one might expect the formation of compounds of the type $R_3SiOFeCl_2$. However, no such compounds have been isolated as yet and this may be due to their inadequate stability:

$$\begin{array}{l} R_3SiOSiR_3 + FeCl_3 \rightleftarrows R_3SiOFeCl_2 + R_3SiCl \\ \qquad\qquad\qquad\qquad\qquad \llcorner\!\!\rightarrow R_3SiCl + OFeCl \end{array} \qquad (8\text{–}5)$$

Precisely this reaction may explain the gradual deactivation of the catalyst, $FeCl_3$, in the cleavage of hexaorganosiloxane by halosilanes [257, 489]. The mechanism of the main process was orginally interpreted in the following way:

$$\begin{array}{c} R'_3SiX + FeX_3 \rightleftarrows R'_3Si^+ + FeX_4^- \\ R_3SiOSiR_3 + R'_3Si^+ \rightleftarrows R'_3Si\text{—}\overset{+}{O}\begin{array}{l}\diagup SiR_3\\ \diagdown SiR_3\end{array} \rightleftarrows R_3SiOSiR'_3 + R_3Si^+ \\ R_3Si^+ + FeX_4^- \rightleftarrows R_3SiX + FeX_3 \end{array} \qquad (8\text{–}6)$$

A mechanism involving the intermediate formation of $R_3SiOFeX_2$, which then reacts with R_3SiX, seems less probable; the cleavage of a hexaalkyldisiloxane by chlorosilane is catalyzed not only by $FeCl_3$, but also by such halides as $ZnCl_2$, which are incapable of cleaving the siloxane bond under the reaction conditions. The idea of the formation of an intermediate cyclic complex is more up-to-date; essentially it extends to all cases of the cleavage of a siloxane bond by metal halides (including chlorosilanes) [24]:

$$\left[\begin{matrix} \equiv Si & & MX_{n-1} \\ & O & & X \\ \equiv Si & & FeX_2 \\ & X & & \end{matrix}\right] \longrightarrow \equiv SiX + \equiv SiOMX_{n-1} + FeX_3 \qquad (8\text{-}7)$$

Prolonged boiling of $Cl_3SiOSiCl_3$ with $FeCl_3$ or with triethylchlorosilane in the presence of $FeCl_3$ did not lead to the formation of $SiCl_4$ [257, 489]. This is explained by the $-I$ effect of the chlorine atoms, which reduces the polarity of the $Si-O$ bond and leads to loss of nucleophilic activity by the oxygen atom in $Si-O-Si$.

The character of the products from the cleavage of cyclodimethylsiloxanes by means of chlorosilanes, for example, dimethyldichlorosilane, is the same in the absence of catalyst and in the presence of $FeCl_3$ [210-212]. These are linear α,ω-dichloropolydimethylsiloxanes. However, when anhydrous $FeCl_3$ is used as the catalyst* the range of products is considerably wider:

$$[(CH_3)_2SiO]_3 + (CH_3)_2SiCl_2 \xrightarrow{FeCl_3} \begin{cases} Cl[(CH_3)_2SiO]_{3n}Si(CH_3)_2Cl; \; n = 1-3 \\ Cl[(CH_3)_2SiO]_m Si(CH_3)_2Cl; \; m = 1-6 \end{cases} \qquad (8\text{-}8)$$

Thus, while in the first case the process may be regarded as telomerization with hexamethylcyclotrisiloxane as the monomer, the catalytic reaction is more complex. In the opinion of Andrianov and Severnyi, the explanation should be sought in secondary reactions of "fragmentation" of the α,ω-dichloropolydimethylsiloxane chains by $FeCl_3$:

$$ClSiOSiO\cdots SiCl + FeCl_3 \longrightarrow \left[\begin{matrix} ClSiOSiO\cdots SiCl \\ \downarrow \\ FeCl_3 \end{matrix}\right] \longrightarrow ClSiOSiOFeCl_2 + Cl\cdots SiCl \qquad (8\text{-}9)$$

* According to the data of Sauer [1052], cleavage does not occur in the complete absence of moisture.

TABLE 93. Compounds Containing the Si–O–Fe Group

Formula	m.p., °C	Subl. temp., °C (mm)	ν_{as}SiOFe, cm^{-1}	Yield, %	Preparation method (scheme)	Literature
$\{[(CH_3)_3SiO]_3Fe\}_2$	179—181	130 (1)	947	45	8-2	1109
	—	136—160 (1)	—	40	8-2	1079
$\{[(CH_3)_3SiO]_4Fe\}Li$	—	—	893	85	8-3	1079
$\{[(CH_3)_3SiO]_4Fe\}Na$	—	—	905	53	8-2	1079
	—	—	—	90	8-3	1079, 1109
$\{[(CH_3)_3SiO]_4Fe\}K$	—	—	917	93	8-3	1079, 1109
$\{[(CH_3)_3SiO]_4Fe\}Sb(CH_3)_4$	193	190 (1) *	954	68	8-3	1080

* d_4^{25} 1.246.

Then under the action of compounds containing the Si–Cl bond there is cleavage of Si–O–Fe groups with regeneration of $FeCl_3$ and the formation of chlorosiloxanes with a shorter chain than in the original intermediate. Scheme (8-7) seems more valid here.

1.2. Properties and Application

The physicochemical constants of the few compounds containing the Si–O–Fe group are given in Table 93.

Tris(trimethylsiloxy)ferrane is dimeric, dissolves readily in nonpolar solvents, and readily undergoes heterolytic reactions, namely, hydrolysis, alcoholysis, etc. [1079]. The reaction with alkali metal silanolates is accompanied by cleavage of the 4-membered ring of the dimer.

The monomeric salt-like compounds $[(R_3SiO)_4Fe]M$ are difficultly soluble in organic solvents and do not melt up to 200°C. The position of the bands ν_{as} Si–O–Fe in their IR spectra depends on the character of the cation and indicates appreciable distortion of the completely symmetrical tetrahedral structure of the tetrakis-(trimethylsiloxy)ferrate anion, which is observed when $M = (CH_3)_4Sb$ [1256], due to the coordination interaction of the alkali metal atom with oxygen atoms of the anion. The high paramagnetism of tris-(trimethylsiloxy)ferrane leads to anomalous values of the chemical shift, broadening of the signal in the NMR spectrum, and makes it practically impossible to use this method to study the properties of the compounds. Hydrolysis of the substance (and its adducts) leads to the formation of Fe_2O_3 (brown coloration).

The hydroxyl group content of polyferraphenylsiloxane obtained by scheme (8-1) falls from 4-6% to 3-4% after heating for 5-10 hr at 200°C. However, their solubility in toluene is not decreased in this way [152]. The polymers evidently have a cyclolinear structure; condensation through the OH groups at 200°C is an intramolecular process (ring closure of sections of the chain) and does not lead to cross linking. The possibility of using these polymers in various heat resistant compositions is obvious.

There is a series of reports on the use of iron derivatives, namely, salts of carboxylic acids (acetate, isovalerate, caprylate, naphthenate, etc.), alcoholates, internal complex compounds (acetylacetonates) and complexes of $FeCl_3$ with caprolactam, as catalysts for the hardening of siloxane resins and lacquers [63, 614, 988, 1328, 1334, 1398, 1401, 1403, 1418, 1445, 1459, 1607], as components of systems producing vulcanization of organosilicon elastomers at room temperature [405, 406, 1416, 1486, 1689], as catalysts for the condensation of hydroxy- and alkoxysiloxanes [1394, 1421], and as stabilizers for organosilicon lubricants, fluids [301, 531, 532, 1684, 1696], and elastomers [1432, 1600]. For increasing the adhesion of organosilicon rubbers to solid surfaces it is recommended that the latter are treated with a mixture of the products from the hydrolysis of alkyltrialkoxysilanes with $FeCl_3$ [1458].

It is reported [63] that in their activity in shortening the drying time of varnish films based on polymethylphenylsiloxane resins, iron salts of carboxylic acids are inferior only to lead salts and surpass the derivatives of other metals. This is reflected in the following series, which shows the falling catalytic activity of the salts:

$$Pb > Fe > Na > Zn > Cu > Co > Mn > Ca$$

On the other hand, it is reported that varnish films hardened by means of iron salts retain their elasticity longer and have a higher thermal stability than those obtained by using analogous derivatives of lead [988].

2. ORGANOSILICON DERIVATIVES OF FERROCENE

2.1. Preparation Methods

2.1.1. Synthesis from Ferrocene Derivatives

If the products from the metallation of ferrocene by butyllithium in ether are treated with triphenylchlorosilane, hydrolysis of the reaction mixture is accompanied by the formation of a precipitate of 1,1-bis(triphenylsilyl)ferrocene. From the ether layer it is possible to isolate mono(triphenylsilyl)ferrocene, whose yield is higher than that of the bis derivative [551, 1614]:*

$$(C_5H_5)Fe(C_5H_5) \xrightarrow{+RM} (C_5H_4M)Fe(C_5H_5) \xrightarrow{+R_3SiX} (C_5H_4SiR_3)Fe(C_5H_5) \qquad (8\text{-}10a)$$

$$(R = C_6H_5)$$

$$(C_5H_5)Fe(C_5H_5) \xrightarrow{+2RM} (C_5H_4M)Fe(C_5H_4M) \xrightarrow{+2R_3SiX} (C_5H_4SiR_3)Fe(C_5H_4SiR_3) \qquad (8\text{-}10b)$$

By this scheme it is possible to synthesize very different triorganosilyl derivatives of ferrocene [15, 31, 235, 392, 802, 803, 1020, 1021].

However, the metallation of ferrocene by butyllithium has serious drawbacks. A large excess of butyllithium is required to obtain even a moderate yield of the metal derivatives. However, even then there is formed an inseparable mixture of mono- and dilithio derivatives with the former predominating. As a result, using lithioferrocenes for the synthesis of organosilicon compounds yields a mixture of mono- and disubstitution products (in the case of bis-silyl derivatives, heteroannular† compounds are formed). The synthesis is complicated by side reactions. Thus, by distillation of the products from the reaction of trihexylbromosilane with

* The symbol ⬠ denotes a cyclopentadienyl ring.

† Heteroannular derivatives are those with substituents in both cyclopentadienyl nuclei.

a mixture of mono- and dilithioferrocene, diferrocenyl was isolated [802]:

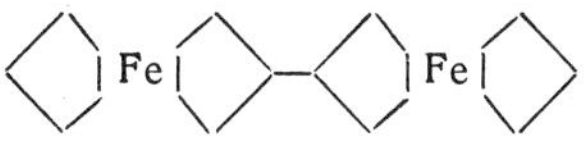

Despite these drawbacks, the lithiation of ferrocene also makes it possible to obtain organosilicon derivatives of ferrocene containing fluorine in the organic radical at the silicon atom [235], alkenyl groups [15, 31], and also reactive Si–H bonds [15, 392]. However, only one example is known of an organolithium synthesis of a compound with a methylene bridge between a ferrocene nucleus and a triorganosilyl group starting from ferrocenyl-methyllithium [395a].

It should be pointed out that in general hydrosilanes react with organolithium compounds at the Si–H bond. However, in the reaction of lithioferrocene with chlorohydrosilanes by scheme (8-10), products containing the Si–H bond are obtained in up to 70% yield. A control reaction with triethylsilane, despite a large excess of the latter, gave a total yield of silyl derivatives of 18% on the ferrocene used:

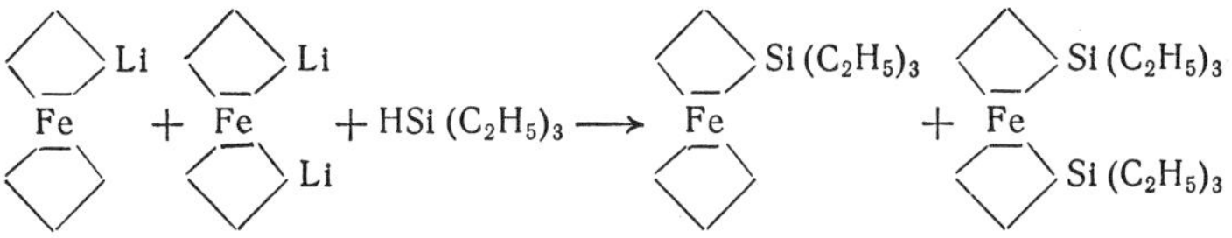

$$C_5H_4Li\text{-}Fe\text{-}C_5H_5 + LiC_5H_4\text{-}Fe\text{-}C_5H_4Li + HSi(C_2H_5)_3 \longrightarrow (C_2H_5)_3SiC_5H_4\text{-}Fe\text{-}C_5H_5 + (C_2H_5)_3SiC_5H_4\text{-}Fe\text{-}C_5H_4Si(C_2H_5)_3 \quad (8\text{-}11)$$

It is obvious that in the reaction of lithioferrocene with chlorohydrosilanes the reaction at the Si–Cl bond predominates, while the Si–H bond is affected to a slight extent.

Changing from lithium to sodium derivatives gives no advantage. On the contrary, the use of phenyl- or amylsodium forms not only a mixture of mono- and disodioferrocenes and, correspondingly, mono- and bis(triorganosilyl)ferrocenes [557, 803, 1022, 1149, 1438], but even gives a small amount of isomeric tris-silyl derivatives [557]. Therefore, an important supplement to scheme (8-10) is the preparation of lithioferrocenes through ferrocenyl derivatives of mercury [1149]. Chloromercuriferrocene is separated readily by extraction from the insoluble dimercurated compound. This makes it possible to prepare the monolithio derivative as required and then the mono(triorganosilyl)ferrocene with hardly any of the bis derivative:

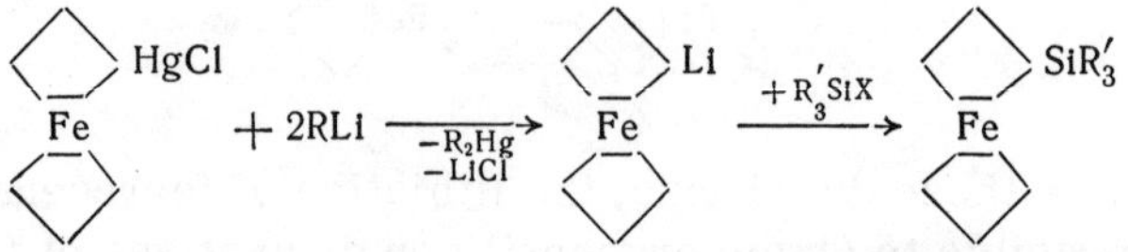

$$C_5H_4HgCl\text{–Fe–}C_5H_5 + 2RLi \xrightarrow[-LiCl]{-R_2Hg} C_5H_4Li\text{–Fe–}C_5H_5 \xrightarrow{+R'_3SiX} C_5H_4SiR'_3\text{–Fe–}C_5H_5 \quad (8\text{–}12)$$

At the same time, in the "silylation" of dilithioferrocene obtained in this way by trimethylchlorosilane an appreciable amount of trimethylsilylferrocene is obtained in addition to 1,1'-bis(trimethylsilyl)ferrocene.

Petrov and his co-workers studied the reaction of organosilicon Grignard reagents with carbonyl-containing derivatives of ferrocene [31, 271, 432].* The reaction proceeds "normally" with aldehydes and ketones:

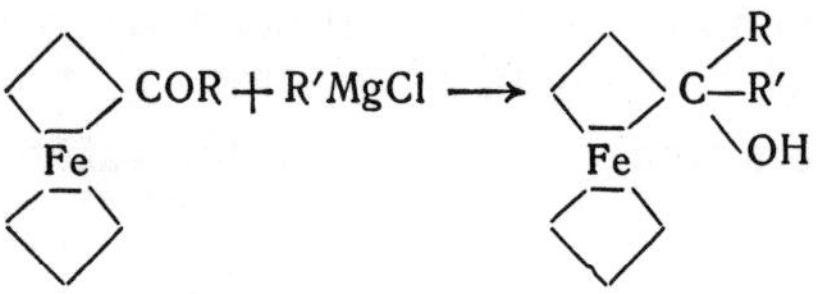

$$C_5H_4COR\text{–Fe–}C_5H_5 + R'MgCl \longrightarrow C_5H_4C(R)(R')OH\text{–Fe–}C_5H_5 \quad (8\text{–}13)$$

[$R = H, CH_3, C_6H_5$; $R' = (CH_3)_3SiCH_2—, (CH_3)_3Si(CH_2)_3—, (CH_3)_3SiOSi(CH_3)_2CH_2—$]

In the case of methyl esters of ferrocenecarboxylic or ferrocenedicarboxylic acids the reaction with γ-trimethylsilylpropylmagnesium chloride also proceeds without any complications:

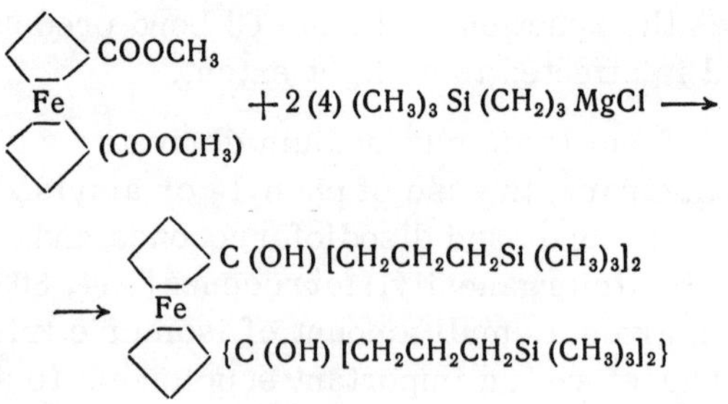

$$C_5H_4COOCH_3\text{–Fe–}C_5H_4(COOCH_3) + 2\,(4)\,(CH_3)_3Si(CH_2)_3MgCl \longrightarrow$$

$$\longrightarrow C_5H_4C(OH)[CH_2CH_2CH_2Si(CH_3)_3]_2\text{–Fe–}C_5H_4\{C(OH)[CH_2CH_2CH_2Si(CH_3)_3]_2\} \quad (8\text{–}14)$$

* A review devoted to organosilicon derivatives of ferrocene [104], which is part of the dissertation of Wu Kuan-li, is not readily accessible since it is published in Chinese.

At the same time, the methyl ester of ferrocenecarboxylic acid reacts anomalously with "siliconeopentylmagnesium chloride":

$$(C_5H_4COOCH_3)Fe(C_5H_5) + (CH_3)_3SiCH_2MgCl \longrightarrow (C_5H_4COCH_2Si(CH_3)_3)Fe(C_5H_5) \quad (8\text{-}15)$$

However, a change to bis(carbomethoxy)ferrocene does not give a diketone, but the ester of a keto acid since the reaction involves only one of the two functional groups:

$$(C_5H_4COOCH_3)Fe(C_5H_4COOCH_3) + (CH_3)_3SiCH_2MgCl \longrightarrow (C_5H_4COCH_2Si(CH_3)_3)Fe(C_5H_4COOCH_3) \quad (8\text{-}16)$$

The reasons for these anomalies have not yet been studied in detail, but it is obvious that steric factors play an important part here.

An attempt to use the hydrosilylation of vinyl derivatives for the synthesis of silicon-containing derivatives of ferrocene was unsuccessful. β-Phenylvinylferrocene is inert toward triethylsilane in the presence of Speier's catalyst [31, 271], while in the case of trichlorosilane there is strong resinification, after which it is impossible to isolate either the starting compounds or addition products.

Thus, until recently the only method of preparing organosilicon derivatives of ferrocene, starting from ferrocene and its derivatives, was organometallic synthesis. An investigation was made recently of the possibility of preparing silylferrocenes by thermal condensation of ferrocene with halohydrosilanes in the liquid phase [230].

The reaction of ferrocene with trichlorosilane in a ratio of 1 : 2 in a steel autoclave with a stirrer began at a pressure of 30-40 atm and a temperature of 285°C, but the yield of silylferrocenes after 9 hr was never higher than 8%. Raising the temperature to 320°C increased the yield to 26% (19% of monosilyl derivative and 7% of heteroannular disilylferrocene) but the optimal temperature was 340°C. Under these conditions the yield of silyl derivatives reached 42%, when the amounts of mono- and disubstituted silylferrocenes were almost the same (22 and 20%, respectively). The use of excess chlorosilane increased the yield. Thus, at 320°C with a fourfold excess of trichlorosilane the yield of silylferrocenes was increased to 38%. While there is no preferential formation of either derivative here (19 and 19%), with a sixteenfold excess of trichlorosilane only disilyl substituted ferrocene is obtained (33%):

$$C_{10}H_{10}Fe + mR_nSiX_{3-n}H \xrightarrow[-mH_2]{} (C_{10}H_{10-m})Fe(SiR_nX_{3-n})_m \qquad (8\text{-}17)$$

The reaction of ferrocene with methyldichlorosilane begins at a higher temperature (310°C). As in the case of trichlorosilane, the use of excess methyldichlorosilane raises the total yield of silylferrocenes, but in the case of a sixteenfold excess the monosilylferrocene (14%) is obtained together with the disilyl derivative (29%). In this connection a study was made of the relation of the yield of silylferrocenes to the duration of the reaction of ferrocene with hydrochlorosilanes. It was established that there is an induction period and also that the monosilyl derivative is formed first, while bis-silylferrocenes are secondary products of the reaction.

Thermal condensation of ferrocene with methylphenylchlorosilane in a ratio of 1 : 6 at 320–330°C for 6 hr led (from an identification of the products by treatment of the reaction mixture with methyllithium) to formation of 35% of monosubstituted ferrocene and 22% of heteroannular bis derivative. Since it is possible to lower the temperature of the thermal condensation of benzene with hydrochlorosilanes by 100–130°C by using a catalyst of the Lewis acid type, Babare and other investigators [15, 1510] studied the effect of such catalysts on the thermal silylation of ferrocene. In the presence of BCl_3, $AlCl_3$, and $SnCl_4$ (1.5% on ferrocene) the reaction temperature is lowered by 30–40°C. The process evidently proceeds through the formation of a complex of ferrocene with substances of the Lewis acid type. In any case, in the thermal condensation of ferrocene with trichlorosilane in the presence of BCl_3 in an equimolar ratio to ferrocene, no silyl derivatives of the latter are obtained.

Both noncatalytic and catalytic thermal condensation of ferrocene with chlorohydrosilanes are accompanied by the formation of products from decomposition of silylferrocenes at the bond of the iron to the cyclopentadienyl ring and also resinification:

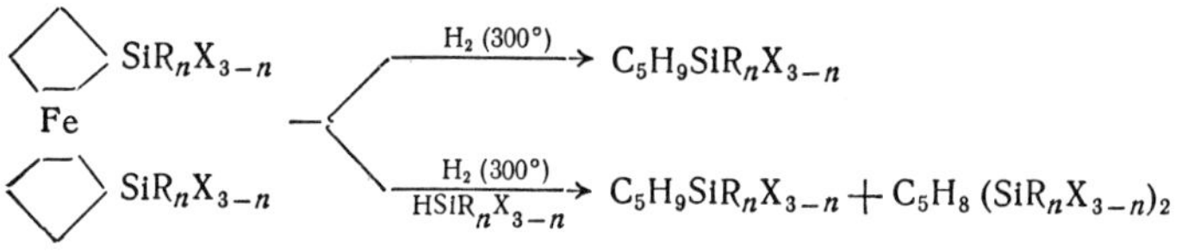

$$(C_5H_4SiR_nX_{3-n})_2Fe \begin{cases} \xrightarrow{H_2\,(300°)} C_5H_9SiR_nX_{3-n} \\ \xrightarrow[HSiR_nX_{3-n}]{H_2\,(300°)} C_5H_9SiR_nX_{3-n} + C_5H_8(SiR_nX_{3-n})_2 \end{cases} \quad (8\text{-}18)$$

The yield of mono- and bis-silyl substituted pentanes is 25-28% (up to 19% of mono and up to 9% of bis) with the exception of condensation in the presence of $AlCl_3$, when there is 63% of decomposition.

Chromatographic separation of the products from the thermal condensation of ferrocene with trichlorosilane showed that more than 50% of them contain the Si–H bond. The use of $SnCl_4$ as a catalyst raises the yield of hydrosilylferrocenes to 77%.

In the case of excess methyldichlorosilane, thermal condensation with ferrocene is accompanied by the formation of 6% of compounds containing the Si–H bond and in the presence of $SnCl_4$, 23%.

An electrophilic substitution mechanism, which is characteristic of ferrocene, has been proposed for the thermal condensation of ferrocene with halosilanes in the liquid phase. The process begins with attack on the unbound electrons of the 3d orbitals of the iron atom with subsequent rearrangement of the δ-complex into a σ-complex (explaining the induction period) and protonization of hydrogen of a cyclopentadienyl ring:

$$(C_5H_5)_2Fe + H^-Si^+R_nX_{3-n} \longrightarrow \left[(C_5H_5)_2Fe\rightarrow SiR_nX_{3-n}\right]^+ H^- \longrightarrow$$

δ-complex

$$\longrightarrow \left[(C_5H_5)Fe(C_5H_5^{+}(H)SiR_nX_{3-n})\right]^+ H^- \xrightarrow{-H_2} (C_5H_5)Fe(C_5H_4SiR_nX_{3-n}) \xrightarrow{H^-Si^+R_nX_{3-n}}$$

σ-complex

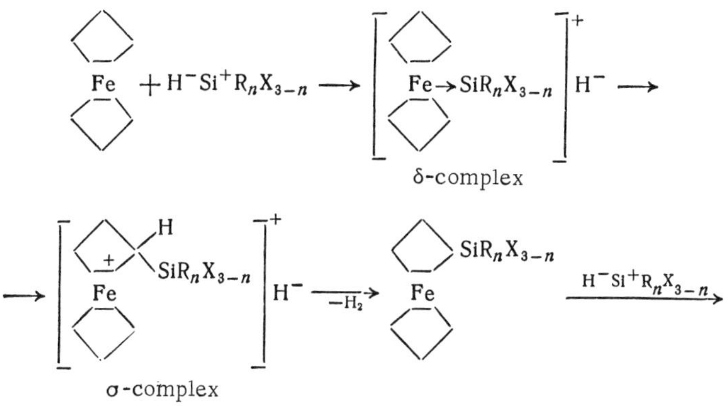

$$\longrightarrow \left[\begin{matrix} C_5H_4\text{—}SiR_nX_{3-n} \\ Fe \rightarrow SiR_nX_{3-n} \\ C_5H_5 \end{matrix} \right]^+ H^- \longrightarrow \left[\begin{matrix} C_5H_4\text{—}SiR_nX_{3-n} \\ Fe \\ \overset{+}{C_5H_5}(H)SiR_nX_{3-n} \end{matrix} \right]^+ H^- \xrightarrow{-H_2} \begin{matrix} C_5H_4\text{—}SiR_nX_{3-n} \\ Fe \\ C_5H_4\text{—}SiR_nX_{3-n} \end{matrix} \quad (8\text{–}19)$$

Thermal (and even more so, catalytic) disproportionation of halohydrosilanes which is possible under the process conditions, is the reason for the formation of compounds containing the Si–H bond:

$$2HSiCl_3 \longrightarrow SiCl_4 + H_2SiCl_2 \xrightarrow{C_{10}H_{10}Fe} C_5H_5FeC_5H_4SiHCl_2 + H_2 \quad (8\text{–}20)$$

Together with substituted ferrocenes, powdery complexes were isolated from the reaction products and their characteristic green color indicated the presence of the ferrocinium cation. The yield of the complex falls in accordance with the series

$$\underset{(25\text{—}30\%)}{HSiCl_3} > \underset{(8\text{—}12\%)}{CH_3SiHCl_2} > \underset{\text{(traces)}}{CH_3(C_6H_5)SiHCl}$$

i.e., with a decrease in the electron affinity of the selenium ion.

It has also been shown that ferrocene reacts with trichlorosilane under the influence of γ-radiation [15, 230]. In this case only dichlorosilylferrocenes are formed, but their yield is only 6-8%.

In 1963 Wilkus [124, 1718] investigated the possibility of acylation of ferrocene with acid halides of organosilicon acids. An attempt at the preliminary preparation of a complex of β-trimethylsilylpropionyl chloride with $AlCl_3$ in methylene chloride led to decomposition of the acid halide with the formation of trimethylchlorosilane and the evolution of ethylene and carbon monoxide. However, when aluminum chloride was added to a solution containing the acid halide and ferrocene at room temperature a carbonyl-containing organosilicon derivative of ferrocene was formed in 76% yield, while the amount of unchanged ferrocene was less than 5%. It is possible to carry out a similar reaction with γ-trimethylsilylbutyryl chloride, p-trimethylsilylbenzoyl chloride and also compounds containing chlorine atoms at silicon:

$$nR_mX_{3-m}SiACOCl + C_{10}H_{10}Fe \xrightarrow[-HCl]{nAlCl_3} [R_mX_{3-m}SiACO]_n C_{10}H_{10-n}Fe \quad (8\text{-}21)$$

$[R = CH_3;\ X = Cl;\ A = (CH_2)_2,\ (CH_2)_3$ or $p\text{-}C_6H_4;\ m = 1-3;\ n = 1$ or $2]$

When n = 1, monosubstituted ferrocenes are obtained, while a ratio of 2 : 1 promotes the formation of heteroannular bis derivatives, whose yield is 57-73%.

A case similar to reactions (8-10) and (8-12) was described by Wilkinson and his co-workers [1010]:

$$C_5H_5Fe(CO)_2Na + ClSi(CH_3)_3 \xrightarrow{THF} C_5H_5Fe(CO)_2Si(CH_3)_3 \quad (8\text{-}22)$$

It is believed that in the compound which is obtained in 42% yield in this way (orange yellow needles, m.p. ~70°C), the silicon and iron atoms are connected directly.

Iron trimethylsilylcyclopentadienyldicarbonyl is stable up to 200°C, soluble in organic solvents, but insoluble in water. It is oxidized in air.

2.1.2. Synthesis from Cyclopentadienylsilanes

The drawbacks of the method of preparing organosilicon derivatives of ferrocene through ferrocenylmetals by direct metallation stimulated a search for more refined routes. In 1958 Goldberg and his co-workers [803] proposed the use of cyclopentadienyl derivatives of silanes for this purpose:

$$C_5H_5SiR_3 \xrightarrow[2)\ FeCl_2]{1)\ n\text{-}C_4H_9Li} R_3SiC_5H_4FeC_5H_4SiR_3 \quad (8\text{-}23)$$

The first compound synthesized in this way, namely, 1,1'-bis(trihexylsilyl)ferrocene, was obtained in only 4% yield. However, in this case the actual demonstration of the possibility of using this route for the synthesis of heteroannular organosilicon derivatives of ferrocene was of greater value than the preparation of the particular compound in high yield. Goldberg's scheme became a preparative method after the work of Schaaf, Kan, and Lenk [863-865, 1053a, 1054-1056, 1657].

In a number of cases reaction (8–23) was found to be effective for the preparation of heteroannular silicon derivatives of ferrocene with functional groups at the silicon atom. For example, 1,1'-bis(chloromethyldimethylsilyl)ferrocene was prepared in this way. However, in the case of cyclopentadienyldimethylethoxysilane, the yield of the ferrocenyl derivative was low when butyllithium was used, while the reaction was complicated by partial replacement of ethoxyl groups by butyl radicals (even at -50°C) [1056]. Similar results were obtained with phenyllithium. When butyllithium was used in the case of $C_2H_5[Si(CH_3)_2O]_3C_2H_5$, it was found to be more effective than isopropylmagnesium chloride (see below) [1056]. However, the desired derivative was obtained in 23% yield due to the side effect of butylation. The butyllithium variant has also been used to prepare organosilicon derivatives of ferrocene containing the Si–Si bond [913, 914, 915] and also the $Si-CH_2$ group [915].

Polysiloxanes containing ferrocenyl groups $-C_5H_4-Fe-C_5H_4-$, in the main chain may be synthesized in the following way:

$$C_5H_5Li + ClSi(CH_3)_2[OSi(CH_3)_2]_nCl \longrightarrow$$

$$\longrightarrow C_5H_5Si(CH_3)_2[OSi(CH_3)_2]_nCl \xrightarrow[R_3SiCl + H_2O,\ 45-65\%]{R(CH_3)_2SiONa,\ 60-80\%}$$

$$\longrightarrow C_5H_5Si(CH_3)_2[OSi(CH_3)_2]_{n+1}-R \xrightarrow[FeCl_2]{n\text{-}C_4H_9Li} Fe\begin{cases}C_5H_4-Si(CH_3)_2[OSi(CH_3)_2]_{n+1}-R\\ C_5H_4-Si(CH_3)_2[OSi(CH_3)_2]_{n+1}-R\end{cases} \qquad (8\text{-}24)$$

$(R{=}CH_3,\ C_6H_5,\ C_6H_3Cl_2,$ while $n{=}0{-}2)$

Beginning with the second stage, the reaction may be turned in a different direction [865]:

$$2C_5H_5Si(CH_3)_2[OSi(CH_3)_2]_nCl \xrightarrow{+H_2O} \{C_5H_5Si(CH_3)_2[OSi(CH_3)_2]_n\}_2O \xrightarrow[FeCl_2]{n\text{-}C_4H_9Li}$$

$$\longrightarrow Fe\begin{cases}C_5H_4-Si(CH_3)_2[OSi(CH_3)_2]_n-\\ \qquad\qquad\qquad\qquad\qquad\quad O\\ C_5H_4-Si(CH_3)_2[OSi(CH_3)_2]_n-\end{cases} \qquad (8\text{-}25)$$

Naturally reaction (8–25) may also be used with long-chain α-cyclopentadienyl-ω-chloropolydimethylsiloxanes and also the products of their cohydrolysis with dimethyldichlorosilane [865, 1085].

Application of reaction (8-23) to bis(cyclopentadienyl)dimethylsilane [1677] leads to the preparation of polysilferrocenylenes:

$$C_5H_5\left[Si(CH_3)_2C_5H_4FeC_5H_4\right]_n-Si(CH_3)_2C_5H_5 \qquad (n=1-10)$$

In 1962 a modification of Goldberg's method was proposed [934, 1679], which consisted of treating a cyclopentadienyltriorganosilane with metallic sodium and then reacting the sodio derivative formed with $FeCl_2$ in tetrahydrofuran (yield of bis-silylferrocene ~50%).

By the action of isopropylmagnesium chloride on $C_5H_5Si(CH_3)_2\cdot OC_2H_5$ in tetrahydrofuran at 0°C it was possible to suppress the side reaction of replacement of ethoxyl groups [1056]. After addition of $FeCl_2$ to the reaction mixture, a yield of $[C_2H_5OSi(CH_3)_2C_5H_4]_2Fe$ of 62% was reached after only an hour. Heteroannular derivatives of ferrocene with $Si(OC_2H_5)_2CH_3$ groups and 1-piperidyl radicals at the silicon atom were prepared analogously. In the latter case, completion of the reaction required a long time: the yield of 1,1'-bis(dimethyl-1-piperidylsilyl)ferrocene was 22% after 1 hr, increasing to 69% after 6 hr. This may be due to steric factors. However, isopropylmagnesium chloride is not a universal agent since in the case of $C_5H_5[Si(CH_3)_2O]_3C_2H_5$ its use does not lead to the formation of a ferrocenyl derivative at all. Here, as has already been pointed out, it is necessary to resort to the use of butyllithium.

The reaction of cyclopentadienyldiorganosilanes with isopropylmagnesium bromide and subsequent addition of $FeCl_2$ to the mixture was used to demonstrate the heteroannular structure of bis-substituted ferrocenes containing the SiR_2H group [15, 230].

2.2. Physical Properties

Data on the physicochemical constants of organosilicon derivatives of ferrocene are summarized in Table 94.

Triorganosilyl derivatives of ferrocene, as a rule, have a color varying from orange to dark red and a specific mild odor and are resistant to air and light.

Table 95 gives some additional data on the properties of symmetrical heteroannular siloxane derivatives of ferrocene. Slow decomposition of all these compounds occurs only at temperature above 454°C.

TABLE 94. Organosilicon Derivatives of Ferrocene

Formula	m.p., (solidification p.), °C	b.p., °C (mm)	n_D^{20}	d_4^{20}	Preparation method	Yield, %	Literature
Monosilyl-substituted ferrocenes							
$FcSi(CH_3)_3$	24	64—65 (0.045)	1.5692^{25}	—	8-10a	19	1020
	—	—	—	—	8-10a	10	1149
	—	—	—	—	8-12	50—54	1149
$FcSi(CH_3)_2CH_2CH{=}CH_2$	—61	—	1.5570	1.0889	8-23	43	433
$FcSi(CH_3)_2C_6H_5$	83—84.5	—	—	—	8-32	60	864
	—	—	—	—	8-23	35	15
$FcSi(CH_3)_2C_6H_4CF_3$-m	50	205—232 (17)	1.5550	—	8-10a	—	235
$FcSi(CH_3)_2OC_2H_5$	—	—	—	—	8-30	37	914
$FcSi(CH_3)_2Si(CH_3)_3$	58—59.6	—	—	—	8-23	13	914
$FcSi(CH_3)_2CH_2Si(CH_3)_3$	—	—	—	—	8-10a	—	914
$FcSi(CH_3)_2(CH_2)_3Si(C_2H_5)_2CH_3$	—36	—	1.5387	1.0558	8-46	56	433
$FcSi(CH_3)_2(CH_2)_3Si(C_2H_5)_3$	—45	—	1.5402	1.0512	8-46	58	433
$FcSi(CH_3)_2(CH_2)_3Si(C_4H_9\text{-n})_2CH_3$	—39	—	1.5173	0.985	8-46	68	433
$FcSi(CH_3)_2(CH_2)_3Si(CH_3)(C_2H_5)Cl$	—	—	1.5421	1.1250	8-46	93	433
$FcSi(CH_3)_2(CH_2)_3Si(CH_3)Cl_2$	—	—	1.5389	1.2046	8-46	98	433
$FcSi(CH_3)_2OSi(CH_3)_2C_6H_5$	—	144.5—148 (0.1)	1.5569^{25}	—	Fig. 12	49	864
$FcSi(CH_3)(C_2H_5)H$	—	92—94 (0.06)	1.5015	1.1043	8-10a	—	392
$FcSi(CH_3)(C_2H_5)(C_6H_{13}\text{-n})$	—	136—138 (0.06)	1.5172	1.0380	—	63	392
$FcSi(C_2H_5)_2H$	—	112—114 (0.06)	1.5520	1.1061	8-10a	—	392
$FcSi(C_2H_5)_3$	—	102—103 (0.02)	1.5661	1.1418	8-10a	—	392
	—	155 (0.9)	1.5695	—	—	8	1438
$FcSi(CH_2CH_2CF_3)_3$	155	—	—	—	8-10a	—	235
$FcSi(C_6H_{13}\text{-n})_3$	—	180—198 (0.17—0.25)	1.5202^{25}	—	8-10a	32	803
	—	—	1.5202^{25}	—	8-10a	0.5	1149

$FcSi(C_6H_5)_3$	142—143	—	—	—	8-10a	49	551
	144—145	—	—	—	8-12	65	1149
Monosilyl-substituted organoferrocenes							
$CH_3Fc'Si(C_2H_5)_3$	—	—	—	—	—	—	557
iso-$C_3H_7Fc'Si(C_2H_5)_3$	—	140 (0.8)	1.5535	—	—	16	557
tert-$C_4H_9Fc'Si(C_2H_5)_3$	—	145 (0,3)	1,5506	—	—	—	557
Mono(silylorgano)-substituted ferrocenes							
$FeCH_2Si(CH_3)_3$	46.5-47.5	—	—	—	—	68-69	395a
$FcCH(OH)CH_2Si(CH_3)_3$	49—51	—	—	—	8-13	54	31
$FcCH(OH)CH_2Si(CH_3)_2OSi(CH_3)_3$	—26	—	—	—	8-13	8	31
$FcCOCH_2Si(CH_3)_3$	66	—	—	—	8-15	66	432
$Fc(CH_2)_3Si(CH_3)_3$	56—57	—	—	—	8-35	78—79	124, 1718
$FcCO(CH_2)_2Si(CH_3)_3$	59—60	—	—	—	8-21	76	124, 1718
$FcC[=NNHC_6H_3(NO_2)_2\text{-}2,4](CH_2)_2Si(CH_3)_3$	215 decomp.	—	—	—	—	—	124
$Fc(CH_2)_4Si(CH_3)_3$	5	—	—	—	8-35	90—91	124
$Fc(CH_2)_4Si(CH_3)_2OH$	—	—	—	—	8-34	9	124, 1718
$FcCH(OH)(CH_2)_3Si(CH_3)_3$	—28	—	—	—	8-13	75	271
	—	—	—	—	8-36	86	1718
$FcCH(N_3)(CH_2)_3Si(CH_3)_3$	—	—	—	—	8-36	—	1718
$FcC(CH_3)(OH)(CH_2)_3Si(CH_3)_3$	—41	—	—	—	8-13	60	271
$FcC(C_6H_5)(OH)(CH_2)_3Si(CH_3)_3$	98—99	—	—	—	8-13	30	271
$FcC(OH)[(CH_2)_3Si(CH_3)_3]_2$	60—62	—	—	—	8-14	90	432
$FcCO(CH_2)_3Si(CH_3)_3$	28,5—29.5	—	—	—	8-21	83	124, 1718
$FcC[=NNHC_6H_3(NO_2)_2\text{-}2,4](CH_2)_3Si(CH_3)_3$	181 decomp	—	—	—	—	—	124
$FcCOC_6H_4[Si(CH_3)_3]$-p	112—114	—	—	—	8-21	83	1718
Mono(silylorgano)-substituted organoferrocenes							
$CH_3COOFc'COCH_2Si(CH_3)_3$	106—108	—	—	—	8-16	57	432
p-$C_6H_4[COFc'CO(CH_2)_3Si(CH_3)_3]_2$	147 decomp.	—	—	—	8-47	22	124

TABLE 94. (Cont'd)

Formula	m.p., (solidification p.), °C	b.p., °C (mm)	n_D^{20}	d_4^{20}	Preparation method	Yield, %	Literature
Bis(ferrocenylorgano)disiloxanes							
$[Fc(CH_2)_3Si(CH_3)_2]_2O$	104—105	—	—	—	8-35	82	124
$[FcSi(CH_3)_2(CH_2)_3Si(CH_3)(C_2H_5)]_2O$	—	—	1.5582	1.1302	8-46	20	433
$[FcCO(CH_2)_2Si(CH_3)_2]_2O$	136—137	—	—	—	8-34	76	124
$[Fc(CH_2)_4Si(CH_3)_2]_2O$	34—35	—	—	—	8-34	65	124
	34—35	—	—	—	8-35	67	124
$[FcCO(CH_2)_3Si(CH_3)_2]_2O$	74—75	—	—	—	8-26	75	124
	74—75	—	—	—	8-34	77	124
Symmetrical bis(silyl)ferrocenes							
$Fc'[Si(CH_3)_2H]_2$	—	85—86 (0.1)	1.5637	1.1255	8-37	94	811
$Fc'[Si(CH_3)_3]_2$	16	87—88 (0.06—0.07)	1.5454[25]	—	8-10 b	27	1020, 1021
	—	153—155 (6.5)	—	—	8-10 b	23	1149, 1438
	—	96—104 (0.04—0.15)	1.5437[25]	—	8-23	50	803, 1022
$Fc'[Si(CH_3)_2CH{=}CH_2]_2$	—	104 (0.003)	1.5627	1.1019	8-43	85	811
$Fc'[Si(CH_3)_2C_2H_5]_2$	—	—	—	—	8-10 b	—	1438
$Fc'[Si(CH_3)_2CH\text{—}CH_2(\text{—O—})]_2$	—	136—138 (0.003)	1.5346	1.1192	—	70	811
$Fc'[Si(CH_3)_2C_5H_5]_2$	—	—	—	—	8-23	—	1677
$Fc'[Si(CH_3)_2C_6H_5]_2$	84	—	—	—	8-17	22	15
$Fc'[Si(CH_3)_2CH_2NH_2 \cdot HCl]_2$	decomp. 185	—	—	—	—	81	1055
$Fc'[Si(CH_3)_2CH_2N(CO)_2C_6H_4]_2$	149—150	—	—	—	—	55 (39)	1055

$Fc'[Si(CH_3)_2CH_2Cl]_2$	41—42	decomp.	—	—	8-23	39 (86)	1055
$Fc'[Si(CH_3)_2(CH_2)_3COOH]_2$	—	187—194 (0.001)	1.5208	1.1547	—	80	812
$Fc'[Si(CH_3)_2(CH_2)_3COOCH_3]_2$	—	170—175 (0.003)	1.5373	1.1923	8-38	98	812
$Fc'[Si(CH_3)_2(CH_2)_3NH_2]_2$	—	143—148 (0.001)	1.5615	1.1083	—	94	812
$Fc'[Si(CH_3)_2(CH_2)_3OH]_2$	—	—	1.5638	1.0941	—	99	812
$Fc'[Si(CH_3)_2(CH_2)_3OCH_2CH\text{—}CH_2 (\text{—O—})]_2$	—	—	1.5413	1.1467	8-38	98	812
$Fc'[Si(CH_3)_2C_6H_4CF_3\text{-}m]_2$	70—71	232—235 (17—15) decomp.	—	—	8-10b	—	235
$Fc'[Si(CH_3)_2N(CH_2)_5]_2$	—	187—195 (0.08)	1.5556^{25}	—	8-23	22	1056
	—	190—204 (0.38)	—	—	8-23	69	864
$Fc'[Si(CH_3)_2OC_2H_5]_2$	15.5—19	104—109 (0.03)	1.5262^{25}	—	8-23	62	1056
$Fc'[Si(CH_3)_2Cl]_2$	44—49	—	—	—	—	62	864
$Fc'[Si(CH_3)_2Si(CH_3)_3]_2$	58.4	—	—	—	8-23	65	914
$Fc'\{[Si(CH_3)_2]_3CH_3\}_2$	—	—	—	—	8-23	—	913, 915
$Fc'[Si(CH_3)_2CH_2Si(CH_3)_3]_2$	—	—	—	—	8-23	—	915
$Fc'[Si(CH_3)_2(CH_2)_3Si(CH_3)_2OSi(CH_3)_3]_2$	—	122 (0.003)	1.4979	1.0194	8-38	80	913
$Fc'\{Si(CH_3)_2(CH_2)_3Si(CH_3)_2[OSi(CH_3)_2]_3Cl\}_2$	—	—	1.4779	1.0485	8-38	100	812
$Fc'[Si(CH_3)_2(CH_2)_3NHSi(CH_3)_3]_2$	—	—	—	—	8-38	100	812
$Fc'[Si(CH_3)_2(CH_2)_3OSi(CH_3)_3]_2$	—	—	1.5139	1.0467	8-38	99	812
$Fc'[Si(CH_3)_2OSi(CH_3)_3]_2$	—	107—110 (0.01)	1.4940^{25}	1.0308^{25}	8-24	55	1055
$Fc'[Si(CH_3)_2OSi(CH_3)_2C_6H_5]_2$	18—19.5	200—205 (0.03)	1.5473^{25}	1.1063^{25}	8-24	59	1055
$Fe[C_5H_3(CH_3)Si(CH_3)_2OSi(CH_3)_2C_6H_5]_2$	—	205—215 (0.07)	1.5428^{25}	1.0863^{25}	8-24	52	1055
$Fc'[Si(CH_3)_2OSi(CH_3)_2C_6H_3Cl_2]_2$	—	263—265 (0.12)	1.5675^{25}	1.2591^{25}	8-24	47	1055
$Fc'[Si(CH_3)_2OSi(CH_3)_2Fc'Si(CH_3)_2C_6H_5]_2$	—	310—320 (0.008) 296—306 (0.09)	—	—	Fig. 12	41	864
$Fc'[Si(CH_3)_2OSi(C_6H_5)_3]_2$	148.5—150	—	—	—	8-24	37	1055
$Fc'\{[Si(CH_3)_2O]_2C_2H_5\}_2$	—	—	—	—	8-23	—	865
$Fc'\{Si(CH_3)_2[OSi(CH_3)_2]_2C_6H_5\}_2$	—	220—223 (0.15)	1.5162^{25}	1.0796^{25}	8-24	50	1055
$Fc'\{[Si(CH_3)_2O]_3C_2H_5\}_2$	—	164—170 (0.05)	1.4672^{25}	—	8-23	23	1056
$Fc'\{Si(CH_3)_2[OSi(CH_3)_2]_3C_6H_5\}_2$	—	245—225 (0.04)	1.4850^{25}	—	8-24	36	1055

TABLE 94. (Cont'd)

Formula	m.p., (sodification p.), °C	b.p., °C (mm)	n_D^{20}	d_4^{20}	Preparation method	Yield, %	Literature
$Fc'[Si(CH_3)(C_2H_5)H]_2$	—	138—140 (0.04)	1.4790	1.0880	8-10b	—	392
$Fc'[Si(OC_2H_5)_2CH_3]_2$	—	138—141 (0.15)	1.5076^{25}	—	8-23	57	1056
$Fc'[Si(C_2H_5)_2H]_2$	—	150—152 (0.04)	1.5330	1.0810	8-10b	—	392
	—	150—151 (0.04)	1.5326	1.0811	8-23	—	392
$Fc'[Si(C_2H_5)_3]_2$	—	148—150 (0.03)	1.5480	1.0670	8-10b	—	392
	—	—	—	—	—	12	557
$Fc'[Si(C_3H_7\text{-n})_3]_2$	—	227—233 (1)	1.5203	1.0214	8-10b	—	235
$Fc'[Si(CH_2CH_2CF_3)_3]_2$	155	—	—	—	8-10b	—	235
$Fc'[Si(C_4H_9\text{-n})_3]_2$	—	—	—	—	8-10b	—	31
$Fc'[Si(C_6H_{13}\text{-n})_3]_2$	—	—	1.5055^{25}	—	8-10b	35 (4—8)	803, 1022
$Fc'[Si(C_{12}H_{25}\text{-n})_3]_2$	—	290—295 (0.004)	1.4922	—	8-23	54	934
$Fc'[Si(C_6H_5)_3]_2$	253—254	—	—	—	8-10b	12	551
Unsymmetrical bis(silyl)ferrocenes							
$(CH_3)_3SiFc'Si(CH_3)_2OSi(CH_3)_2C_6H_5$	—	148—151 (0.09); 140—144 (0.03)	—	—	Fig. 12	52 (22)	864
$(CH_3)_3SiFc'Si(C_2H_5)_3$	—	—	—	—	8-10b	—	1438
$C_6H_5(CH_3)_2SiFc'Si(CH_3)_2OSi(CH_3)_2C_6H_5$	—	200—208 (0.03)	1.5740^{25}	—	Fig. 12	65	864
$(CH_2)_5NSi(CH_3)_2Fc'Si(CH_3)_2OSi(CH_3)_2C_6H_5$	—	204—210 (0.35)	1.5509^{25}	—	Fig. 12	49 (47)	864
Bis(silylorgano)-substituted ferrocenes							
$Fc'[(CH_2)_3Si(CH_3)_3]_2$	20—20.5	—	—	—	8-35	82	124, 1718
$Fc'[CO(CH_2)_2Si(CH_3)_3]_2$	115—116	—	—	—	8-21	55	124, 1718
$Fc'[(CH_2)_4Si(CH_3)_3]_2$	—9 — —8	—	—	—	8-35	71	124
$Fc'[CO(CH_2)_3Si(CH_3)_3]_2$	47—48	—	—	—	8-21	73	124
$Fc'\{C(OH)[(CH_2)_3Si(CH_3)_3]_2\}_2$	107—108	—	—	—	8-14	87	432

Poly(silylferrocenyloxy)organosilanes							
$[C_6H_5(CH_3)_2SiFc'Si(CH_3)_2]_2O$	—	270—280 (0,15)	—	—	Fig. 12	53	864
$[C_6H_5(CH_3)_2SiO(CH_3)_2SiFc'Si(CH_3)_2]_2O$	—	283—290 (0,025)	$1,5572^{25}$	—	Fig. 12	82	864
$[(CH_3)_3SiFc'Si(CH_3)_2O]_2Si(CH_3)_2$	—	210—218 (0.09)	—	—	Fig. 12	37	864
$[C_6H_5(CH_3)_2SiFc'Si(CH_3)_2O]_2Si(CH_3)_2$	—	283—285 (0.03)	$1,5791^{25}$	—	Fig. 12	55	864
$[C_6H_5(CH_3)_2SiFc'Si(CH_3)_2O]_3SiCH_3$	—	270—280 (0,06)	—	—	Fig. 12	37	864
$[C_6H_5(CH_3)_2SiFc'Si(CH_3)_2O]_4Si$	—	286—287 (0.3)	—	—	Fig. 12	44	864
Tris(silyl)-substituted ferrocenes							
$C_{10}H_7Fe[Si(C_2H_5)_3]_3$, various isomers	—	215 (0.8)	1.5380	—	—	2	557
	—	208 (1)	1.5477	—	—	—	557
Cyclic compounds							
$(CH_3)_2SiFc'Si(CH_3)_2$ (ring-closing bond)	—	—	—	—	—	—	913, 915
$(CH_3)_2SiFc'Si(CH_3)_2$ (bridged by CH_2)	—	—	—	—	—	—	915
$(CH_3)_2SiFc'Si(CH_3)_2$ (bridged by O)	87—88	—	—	—	8-25	22	865, 1056
	86—87	—	—	—	Fig. 12	90	1056
	87—88	—	—	—	8-28	—	914, 915
$HO(CH_3)SiFc'Si(CH_3)OH$ (bridged by O)	163,5–165	—	—	—	8-33	21—24	1056
$C_2H_5O(CH_3)SiFc'Si(CH_3)OC_2H_5$ (bridged by O)	—	128—131 (0.4)	1.5396^{25}	—	8-33	77	1056
$O(CH_3)_2SiFc'Si(CH_3)_2O$ (ring closed by $Si(CH_3)_2$)	73—75	—	—	—	8-25	6	865, 1056
$Fc'[Si(CH_3)-O-Si(CH_3)]_2Fc'$ (upper $Si(CH_3)-O-Si(CH_3)$ and lower $Si(CH_3)-O-Si(CH_3)$ chains, each Si of the upper chain linked through O to the Si below it; both chains joined at each end to Fc')	276	—	—	—	8-33	60 (11)	1056

TABLE 95. Properties of 1,1'-Bis-(siloxyanyl)ferrocenes [1657]

R⟨⟩Fe⟨⟩R

R	b.p., °C (mm)	η^{20}, cP	Flow point, °C	Weight loss after heating for 10 hr at 366°C, %
$(C_6H_5)_3SiO(CH_3)_2Si$	—*	—	—	1.0
$C_6H_5(CH_3)_2SiO(CH_3)_2Si$	200—205 (0.03)	21.7	—31	0.5
$[C_6H_5(CH_3)_2SiO(CH_3)_2Si](CH_3)$	205—215 (0.07)	30.8	—22	1.1
$C_6H_5[(CH_3)_2SiO]_2(CH_3)_2Si$	220—223 (0.15)	16.8	—40	0.4†
$C_6H_5[(CH_3)_2SiO]_3(CH_3)_2Si$	245—255 (0.04)	17.2	—57	2.6

* m.p., 149°C.
† 5.9% after 10 hr at 425°C; b.p. 463°C [1053a].

In unsubstituted ferrocene all the hydrogen atoms are equivalent. The introduction of a substitutent into a ring introduces nonequivalence of the unreplaced hydrogen atoms and this appears in the resonance spectrum. In the high-resolution NMR spectra of mono- and bis-silyl substituted ferrocenes there are [230] two signals with values of δ of 4.16-4.24 (H_α) and 3.98-4.08 (H_β). The character of the signals of bis-substituted ferrocenes confirms their heteroannular structure.

2.3. Chemical Properties

Reaction (8-25) is only one of the routes (and not the simplest) to the formation of heterocycles consisting of siloxane and ferrocenyl groups. Hydrolysis of heteroannular derivatives of ferrocene containing functional groups at the silicon atoms such as 1,1'-bis(ethoxydimethylsilyl)ferrocene gives analogous results. However, the formation of ferrocenylenesiloxane rings is only the initial stage of a diverse chain of conversions (see Fig. 12).

Hydrolysis of 1,1'-bis(ethoxydimethylsilyl)ferrocene under comparatively mild conditions leads to the formation of a ring containing one siloxane group. Under analogous conditions the hydrolysis of 1,1'-bis(5-ethoxyhexamethyltrisiloxanyl)ferrocene is accompanied by not only hydrolysis of ethoxyl groups, i.e., cleavage of Si−O−C bonds, but also partial cleavage of Si−O−Si bonds since the reaction product is 1,5-(1,1'-ferrocenylene)hexamethyltrisiloxane. At 80°C the cleavage proceeds still further and the hydrolysis gives 1,3-(1,1'-ferrocenylene)tetramethyldisiloxane. However, in neither case is rupture of the bonds between silicon and the ferrocene group observed.

A disiloxane is formed readily by hydrolysis of monosubstituted ferrocenes containing the Si–Cl bond (reaction 8-21) with 5% hydrochloric acid in methylene chloride [124]:

$$ClSi(CH_3)_2(CH_2)_3COCl + FcH \xrightarrow[CH_2Cl_2]{AlCl_3} ClSi(CH_3)_2(CH_2)_3COFc \xrightarrow{H_2O}$$
$$\longrightarrow \left[FcCO(CH_2)_3\overset{CH_3}{\underset{CH_3}{\overset{|}{\underset{|}{Si}}}}- \right]_2 O \qquad (8\text{-}26)$$

Under analogous conditions the heteroannular bis derivative gives a 57% yield of a siloxane-ferrocenylene polymer:

$$[(CH_3)_2Si(CH_2)_3COFc'CO(CH_2)_3Si(CH_3)_2O]_n$$

Hydrolysis of a ferrocene derivative containing the $SiCl_2$ group in the radical leads to a solid polysiloxane with ferrocenoylalkyl pendant groups at the silicon atoms:

$$Cl_2Si(CH_3)(CH_2)_2COFc \xrightarrow{H_2O} \left[-\overset{CH_3}{\underset{CH_2CH_2COFc}{\overset{|}{\underset{|}{Si}}}}-O- \right]_n \qquad (8\text{-}27)$$

The following copolymers were isolated from the products of cohydrolysis of methyl(β-ferrocenoylethyl)dichlorosilane with dimethyldichlorosilane:

$$\left[-\overset{CH_3}{\underset{CH_3}{\overset{|}{\underset{|}{Si}}}}-O- \right]_n - \left[-\overset{CH_3}{\underset{CH_2CH_2COFc}{\overset{|}{\underset{|}{Si}}}}-O- \right]_m$$

In one of them n = 4.5 and in another 2.3 with the latter showing the deeper orange color and higher viscosity than the polymer with n = 4.5.

According to the data of Kumada and his co-workers [913, 914, 915], ferrocenes containing organopolysilyl substituents are resistant to bases, but are very sensitive to acidic agents. Thus, 1,1'-bis(pentamethylbisilyl)ferrocene is hardly changed by boiling for many hours with a 0.01-0.1 N solution of sodium methylate in methanol. However, treatment for two hours with a dilute (10^{-3} N) methanol or ethanol solution of HCl is sufficient for quantitative cleavage:

$$\text{Fe}(C_5H_4Si(CH_3)_2Si(CH_3)_3)_2 \xrightarrow[ROH]{H^+} \text{Fe}(C_5H_5)_2 + \text{Fe}(C_5H_4)_2[Si(CH_3)_2-O-Si(CH_3)_2] + (CH_3)_3SiOR \quad (8\text{-}28)$$

Cleavage of the compound with catalytic amounts of $AgClO_4$ in dry benzene proceeds analogously [913]. The relative yield of the hydrolysis products depends substantially on the acid concentration. With a very low concentration of HCl there is preferential cleavage of Si–Si bonds. Raising the concentration promotes the rupture of bonds between silicon and ferrocene (Fig. 11).

Acid cleavage of 1,2-(1,1'-ferrocenylene)tetramethyldisilane also leads to a disiloxane and ferrocene [913, 915]:

$$\text{Fe}(C_5H_4)_2[Si(CH_3)_2-Si(CH_3)_2] \xrightarrow[H^+]{H_2O} \text{Fe}(C_5H_5)_2 + \text{Fe}(C_5H_4)_2[Si(CH_3)_2-O-Si(CH_3)_2] \quad (8\text{-}29)$$

The cleavage of 1,1'-bis(heptamethyltersilyl)ferrocene forms 1,3-(1,1 -ferrocenylene)tetramethyldisiloxane and pentamethylethoxydisilane. The cleavage of a mono(bisilyl) derivative of ferrocene proceeds similarly:

$$C_5H_5\text{Fe}C_5H_4Si(CH_3)_2Si(CH_3)_3 \xrightarrow[C_2H_5OH]{HCl\ (0{,}001n)} \text{Fe}(C_5H_5)_2 + C_5H_5\text{Fe}C_5H_4Si(CH_3)_2OC_2H_5 + (CH_3)_3SiOC_2H_5 + (CH_3)_3SiOSi(CH_3)_3 \quad (8\text{-}30)$$

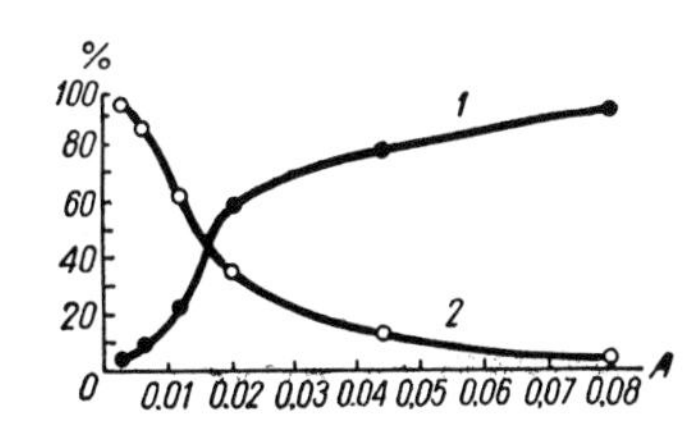

Fig. 11. Acid cleavage of 1,1'-bis(pentamethylbisilyl)ferrocene [914]. The relationship between the yields (%) of ferrocene (curve 1) and 1,3-(1,1'-ferrocenylene)tetramethyldisiloxane (curve 2) and the molar concentration of HCl in methanol (A).

Thus, if the Si–Si bond is cleaved, it is only between the α- and β-atoms of the silicon chain. The authors believe that these first examples of unusually ready acid cleavage of Si–Si bonds in hexaorganodisilanes are connected with the possibility of stabilization of the transition state through overlapping of the d-orbitals of the ferrocenyl iron atom with the p-orbitals of silicon atoms adjacent to the nucleus:

$$\underset{\text{Fe}}{\bigcirc}\text{Si(CH}_3)_2\text{Si(CH}_3)_3 \xrightarrow{H^+} \left[\underset{\text{Fe}}{\bigcirc}\overset{(\text{CH}_3)_2}{\text{Si}}\cdots\right]^+\cdots\text{Si(CH}_3)_3 \; H^+ \xrightarrow{\text{ROH}} \underset{\text{Fe}}{\bigcirc}\text{Si(CH}_3)_2\text{OR} + (\text{CH}_3)_3\text{SiOR} + \text{H}_2 \qquad (8\text{-}31)$$

Kumada and his co-workers also established that boiling mono- and bis(siliconeopentyl)dimethylsilylferrocenes in a 0.001 N ethanol solution of HCl is accompanied by desilylation. In this connection it should be pointed out that according to literature data [938] $K_1 \cdot 10^3$ (min^{-1}) for the cleavage of the bond of silicon with the ring in trimethylsilylferrocene by a solution of HCl in methanol is 3.36 and 2.45 at HCl concentrations of 0.596 and 0.477 N, respectively.

Table 96 gives data on the hydrolytic cleavage of 1-alkyl-1'-triethylsilylferrocenes.

The authors are inclined to relate the higher rate of hydrolysis of triethylsilyl derivatives of methyl- and isopropylferrocenes, which proceeds by a pseudo first order reaction, exclusively to the intraannular resonance effect. However, the result obtained with the derivative of tert-butylferrocene indicates the importance of the steric factor. This is the first case of screening of a substituent in one ferrocene ring by a substituent in the other ring.

TABLE 96. Hydrolytic Cleavage of the Silicon-Ring Bond in 1-Alkyl-1'-triethylsilylferrocenes [557]

Alkyl	$K_1 \cdot 10^{-3}$, min^{-1} (mean value)	Relative hydrolysis rate
H	2.81	1.0
CH_3	7.44	2.65
iso-C_3H_7	4.51	1.60
tert-C_4H_3	2.06	0.73

In the hydrolysis of a compound of the silazan type, formed by treatment of 1,3-(1,1'-ferrocenylene)tetramethyldisiloxane with sodium amide (see Fig. 12), the bis-substituted ferrocene is converted into a mono derivative, i.e., in this case also there is cleavage of a bond between silicon and a ring. A similar phenomenon occurs in the hydrolysis with an aqueous solution of ammonium chloride of the product from the reaction of 1,3-(1,1'-ferrocenylene)-tetramethyldisiloxane with phenyllithium [864]. Since the formation of ammonium chloride in the hydrolysis of a ferrocenylsilazan in the presence of phenyldimethylchlorosilane is quite probable, we may speak of the specific instability of the bond of silicon to a cyclopentadienyl ring toward the action of NH_4Cl. It should be noted that the course of the hydrolysis of lithium (1'-phenyldimethylsilyl-1-ferrocenyl)dimethylsilanolate indicates that the first phase is the formation of silanols, while the reactions accompanied by cleavage of bonds between silicon and the cyclopentadienyl ring are of a secondary nature [864]:

$$(C_5H_4Si(CH_3)_2R)Fe(C_5H_4Si(CH_3)_2OLi) \xrightarrow{H_2O} \left[(C_5H_4Si(CH_3)_2R)Fe(C_5H_4Si(CH_3)_2OH)\right] \rightarrow [HOSi(CH_3)_2OH] + R(CH_3)_2Si\text{-}C_5H_4FeC_5H_5$$

$$\rightarrow (C_5H_4Si(CH_3)_2R)Fe(C_5H_4\text{-}Si(CH_3)_2O[Si(CH_3)_2O]_nSi(CH_3)_2\text{-}C_5H_4)Fe(C_5H_4Si(CH_3)_2R)$$

$(R = C_6H_5;\ n = 1$ or $2)$. (8-32)

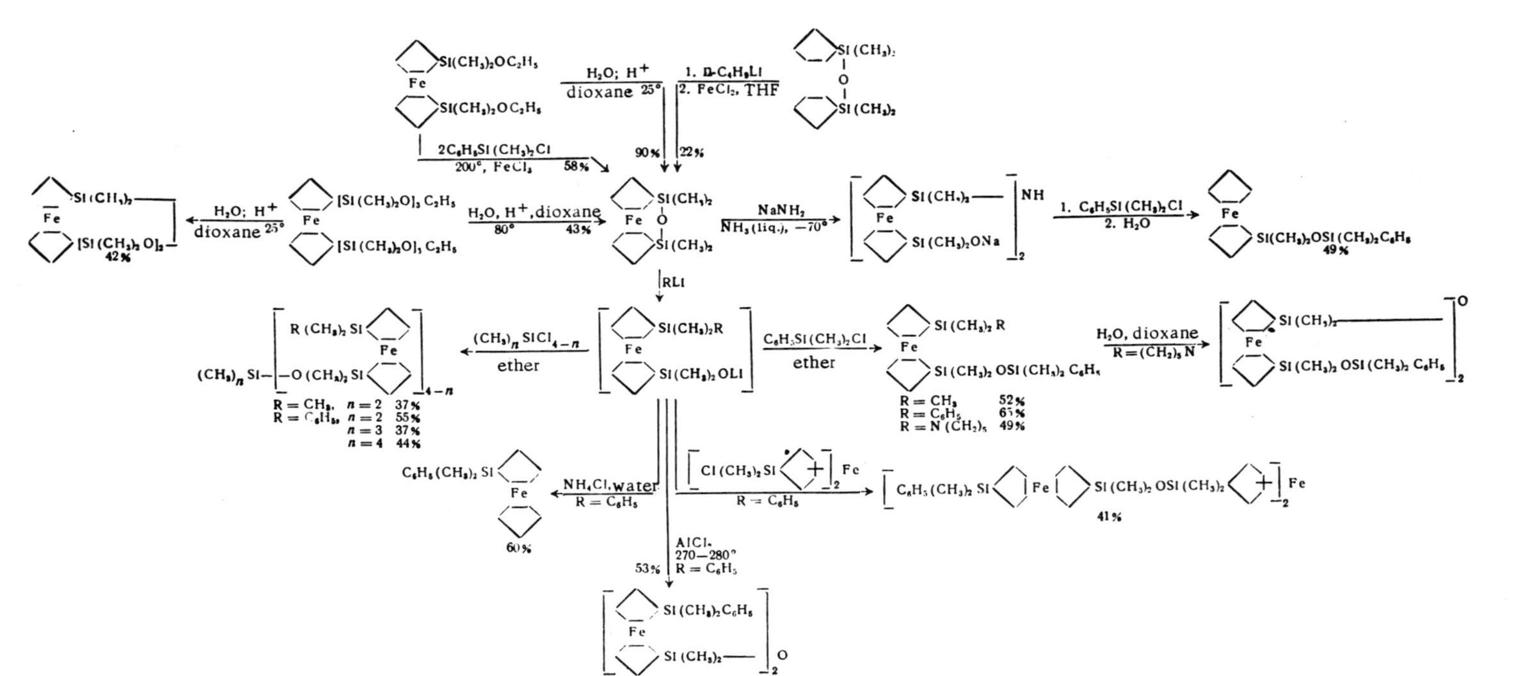

Fig. 12. Chemical conversions of ferrocenylene siloxanes.

The hydrolysis of the compound with a piperidyl radical at the silicon atom proceeds without any complications with the formation of a siloxane as a result of elimination of $(CH_2)_5N$ groups [864] (see Fig. 12). The cleavage of 1,1'-bis(piperidyldimethylsily)ferrocene by dry hydrogen chloride also proceeds smoothly. 1,1'-Bis-(chlorodimethylsilyl)ferrocene is obtained in more than 60% yield. Thus, here we again arrive at an organosilicon derivative of ferrocene containing highly reactive chlorine atoms at the silicon. By means of organometallic synthesis with 1,1'-bis(chlorodimethylsilyl)ferrocene a compound was obtained with three ferrocenylene groups in the molecule (see Fig. 12).

Scheme (8-33) illustrates hydrolytic cleavages of tetrafunctional 1,1'-bis(methyldiethoxysilyl)ferrocene [1056]:

CH_3 / Si $(OC_2H_5)_2$ / Fe / Si $(OC_2H_5)_2$ / CH_3

H_2O; H^+; 25° / dioxane → 77% → CH_3 / $SiOC_2H_5$ / Fe O / $SiOC_2H_5$ / CH_3

H_2O; H^+; 80°; ethanol 60% / dioxane 70%

H_2O; H^+ 22%: Si CH_3 OH / Fe O / Si OH CH_3 + oil

ethanol 25° 11% → CH_3 CH_3 / Si—O—Si / Fe O O Fe / Si—O—Si / CH_3 CH_3 (8-33)

Treatment of (β-trimethylsilylpropionyl)ferrocene with concentrated sulfuric acid is accompanied by strong coloration and the liberation of 1 mole of methane per mole of ketone [124]. A disiloxane was isolated from the products of hydrolysis of the reaction mixture:

$$2R_3Si(CH_2)_2COFc \xrightarrow[H_2O]{H_2SO_4} [FcCO(CH_2)_2Si(CH_3)_2]_2O + 2CH_4 \qquad (8\text{-}34)$$

The reaction is very specific and does not occur when sulfuric acid with a concentration lower than 80% is used or with such acids as phosphoric, hydrochloric, and acetic. Completely analogous behavior toward concentrated H_2SO_4 is shown by the γ-silylbutyryl derivative and the product obtained after hydrolysis is completely identical to the bis(γ-ferrocenoylpropyldimethyl)disiloxane synthesized by scheme (8-26). Replacement of the mono derivative by a bis-silylacylferrocene does not change the character of the interaction, but the final products, which are obtained in almost quantitative yield, are linear solid siloxaneferrocenylene polymers, which give soft fibers on drawing.

The reaction of (δ-trimethylsilylbutyl)ferrocene (which has a yellow color in contrast to the orange silylacyl derivatives) with sulfuric acid also proceeds according to the scheme (8-34), but to obtain the disiloxane (65% yield) it is necessary to treat the reaction mixture with 50% KOH rather than water. Moreover, in this case it was possible to isolate $HOSi(CH_3)_2(CH_2)_4Fc$ (9%), i.e., the silanol which is logically regarded as an essential intermediate in reactions of type (8-34).

Treatment of bis[ω-trimethylsilylpropyl(butyl)]ferrocenes with sulfuric acid with subsequent alkaline hydrolysis gives a ~90% yield of low-molecular viscous, fiber-forming polymers of the type $[R_2Si(CH_2)_nFc'(CH_2)_nSiR_2O]_m$, where $R = CH_3$ and $n = 3$ or 4.

Adding dimethyldichlorosilane to the products from the reaction of bis(γ-trimethylsilylpropyl)ferrocene with sulfuric acid and hydrolyzing the mixture (after the complete evolution of HCl) yielded a pale yellow liquid, which was found to be a copolymer with the composition

$$(CH_3)_2SiO : [(CH_3)_2Si(CH_2)_3Fc'(CH_2)_3Si(CH_3)_2O] = 11:1$$

which corresponded closely to the initial molar ratio of the components.

The possibility of high-temperature hydrogenation of silylferrocenes, leading to breakdown of the ferrocene structure (reaction 8-18), was mentioned above [15, 230]. At the same time, in the reaction of silylalkyl ferrocenyl ketone by Clemmensen's reduction (hydrogen chloride and zinc amalgam in hexane or benzene, refluxed for 40 hr), the yield of products from the replacement of CO groups by CH_2 groups is 70-90% and the ferrocene system is not touched [124, 1718]:

$$[R_3Si(CH_2)_n CO]_m Fc \xrightarrow[Zn/Hg]{HCl} [R_3Si(CH_2)_{n+1}]_m Fc \qquad (8\text{-}35)$$
$$(m = 1 \text{ or } 2)$$

By using this method it is possible to reduce, without touching the Si–O bonds, siloxane polymers of the type

$$[FcCO(CH_2)_n Si(CH_3)_2]_2O$$

and

$$[(CH_3)_2 Si(CH_2)_n COFc'CO(CH_2)_n Si(CH_3)_2O]_n$$

In the reaction with disiloxanes, compounds in which one of the two CO groups has been converted into CH_2 are evidently formed together with products of complete reduction. There is the possibility that with refinement, this incidental result could become a method for preparing α,ω-ferrocenyl-containing organosilicon compounds with a keto group in the chain.

The reduction of (γ-trimethylsilylbutyryl)ferrocene with lithium aluminum hydride in ether leads to an alcohol, from which it is possible to obtain an azide:

$$R_3Si(CH_2)_3 COFc \xrightarrow[ether]{LiAlH_4} R_3Si(CH_2)_3 CH(OH)Fc \xrightarrow[C_6H_6;\ Cl_3CCOOH]{HN_3}$$
$$\longrightarrow R_3Si(CH_2)_3 CH(N_3)Fc \qquad (8\text{-}36)$$

Silylferrocenes containing Si–H bonds have been obtained by using lithium aluminum hydride [15, 93, 811]:

$$Fc(SiR_2X)_n \xrightarrow[R'OR']{LiAlH_4} Fc(SiR_2H)_n \qquad (8\text{-}37)$$
$$(X = Cl^* \text{ or } OC_2H_5;\ n = 1 \text{ or } 2)$$

On the basis of these compounds it is possible to modify silicon-containing ferrocenes by means of classical addition reactions.

The addition of ferrocenylsilanes containing Si–H bonds to unsaturated compounds in the presence of Speier's catalyst requires somewhat more drastic conditions than normal to give satisfactory

* Compounds with X = Cl react readily with methyllithium to form triorganosilyl derivatives of ferrocene.

yields:

$$Fc\,(SiRR'H)_n + nCH_2{=}CHR'' \xrightarrow{H_2PtCl_6} Fc\,[SiRR'\,(CH_2)_2\,R'']_n \qquad (8\text{-}38)$$

where $R = CH_3$; $R' = CH_3$, C_2H_5; $n = 1$ or 2; $R'' =$ n-C_4H_9, $Si\,(C_2H_5)_3$ [15], CH_2COOCH_3, $CH_2OCH_2CH{-}CH_2$ (epoxide, bridged by O), $CH_2\,[Si\,(CH_3)_2\,O]_3\,Si\,(CH_3)_2\,Cl$, $CH_2OSi\,(CH_3)_3$, $CH_2NHSi(CH_3)_3$ [812], $CH_2Si\,(CH_3)_2\,OSi(CH_3)_3$ [811]

The addition of methylethylsilylferrocene to hexene-1 and triethylvinylsilane proceeds according to Farmer's rule. Reactions involving 1,1'-bis(dimethylsilyl)ferrocene evidently proceed in the same way. The addition of this compound to vinyl monomers containing functional or readily hydrolyzable groups offers new routes to the synthesis of different classes of polymers, containing ferrocenylene groups in the chain:

$R'' = CH_2COOCH_3$

$$Fc'\,[Si\,(CH_3)_2\,(CH_2)_2\,R'']_2 \xrightarrow[C_2H_5OH]{5\,N\,NaOH} Fc'\,[Si\,(CH_3)_2\,(CH_2)_3\,COOH]_2 \xrightarrow[10\,hr\,220^\circ,\,vacuum]{H_2N\,(CH_2)_6\,NH_2}$$

$$\longrightarrow [^-OOC\,(CH_2)_3\,Si\,(CH_3)_2\,Fc'Si\,(CH_3)_2(CH_2)_3\,COO^-]\,[H_3\overset{+}{N}\,(CH_2)_6\,\overset{+}{N}H_3] \xrightarrow{-H_2O}$$

$$\longrightarrow [CO\,(CH_2)_3\,Si\,(CH_3)_2\,Fc'Si\,(CH_3)_2\,(CH_2)_3\,CONH\,(CH_2)_6\,NH]_n \qquad (8\text{-}39)$$

$R'' = CH_2NHSi\,(CH_3)_3$

$$Fc'\,[Si\,(CH_3)_2\,(CH_2)_2\,R'']_2 \xrightarrow{C_2H_5OH} Fc'\,[Si\,(CH_3)_2\,(CH_2)_3\,NH_2]_2 \xrightarrow[-HCl]{ClCOC_6H_4COCl}$$

$$\longrightarrow [NH\,(CH_2)_3\,Si\,(CH_3)_2\,Fc'Si\,(CH_3)_2\,(CH_2)_3\,NHCOC_6H_4CO]_n \qquad (8\text{-}40)$$

Reaction (8-10) proceeds at the (CH_2Cl_2/H_2O) phase boundary in the presence of an HCl acceptor in the stoichiometric amount.

An attempt to prepare polyesters by direct transesterification of 1,1'-bis(γ-carbomethoxypropyldimethylsilyl)ferrocene with glycols was unsuccessful due to rupture of the silicon–ring bonds and the liberation of free ferrocene [812]. This also occurred with the use of acid (p-toluenesulfonic acid) and basic (tetrabutyl titanate) catalysts. At the same time, from 1,1'-bis(γ-hydroxypropyldimethylsilyl)ferrocene it is possible to prepare both polyesters with mol. wt. 6000 and polyurethanes:

$R'' = CH_2OSi\,(CH_3)_3$

$$(x+1)\,Fc'\,[Si\,(CH_3)_2\,(CH_2)_2\,R'']_2 \xrightarrow{2\,N\,HCl\cdot aq}$$

$$\longrightarrow (x+1)\,\mathrm{Fc'}\,[\mathrm{Si(CH_3)_2(CH_2)_3OH}]_2 \xrightarrow[\mathrm{C_5H_5N}]{x\mathrm{ClCOC_6H_4COCl}}$$

$$\longrightarrow \mathrm{HO}\left[\begin{matrix} & \mathrm{CH_3}\ \mathrm{CH_3} & & \mathrm{O} & \mathrm{O} \\ & | \quad | & & \| & \| \\ \mathrm{(CH_2)_3} & \mathrm{SiFc'Si} & \mathrm{(CH_2)_3} & \mathrm{OC} & \mathrm{C_6H_4CO} \\ & | \quad | & & & \\ & \mathrm{CH_3}\ \mathrm{CH_3} & & & \end{matrix}\right]_x \begin{matrix} & \mathrm{CH_3}\ \mathrm{CH_3} & \\ & | \quad | & \\ \mathrm{(CH_2)_3} & \mathrm{SiFc'Si} & \mathrm{(CH_2)_3OH} \\ & | \quad | & \\ & \mathrm{CH_3}\ \mathrm{CH_3} & \end{matrix} \qquad (8\text{-}41)$$

$$\mathrm{R''} = \mathrm{CH_2OH}$$

$$(x+1)\,\mathrm{Fc'}\,[\mathrm{Si(CH_3)_2(CH_2)_2R''}]_2 + x\ n\text{-}\mathrm{OCNC_6H_4NCO} \longrightarrow$$

$$\longrightarrow \mathrm{HO}\left[\begin{matrix} & \mathrm{CH_3}\ \mathrm{CH_3} & & \mathrm{O} & & \mathrm{O} \\ & | \quad | & & \| & & \| \\ \mathrm{(CH_2)_3} & \mathrm{SiFc'Si} & \mathrm{(CH_2)_3} & \mathrm{OC} & \mathrm{NHC_6H_4NH} & \mathrm{CO} \\ & | \quad | & & & & \\ & \mathrm{CH_3}\ \mathrm{CH_3} & & & & \end{matrix}\right]_x \begin{matrix} & \mathrm{CH_3}\ \mathrm{CH_3} & \\ & | \quad | & \\ \mathrm{(CH_2)_3} & \mathrm{SiFc'Si} & \mathrm{(CH_2)_3OH} \\ & | \quad | & \\ & \mathrm{CH_3}\ \mathrm{CH_3} & \end{matrix} \qquad (8\text{-}42)$$

When R" = $CH_2[Si(CH_3)_2O]_3Si(CH_3)_2Cl$, acid hydrolysis of the oligomer leads to viscous polymers with a mean molecular weight of ~8000. With p-bis(hydroxydimethylsilyl)benzene in the presence of pyridine a very peculiar heterochain polymer is obtained. The silicon atoms in it are connected successively by ferrocenylene, trimethylene, oxygen, and finally phenylene bridges. After successive heating at 200, 250, and 300°C (2 hr at each temperature in air), this polymer loses 7.7% of its weight, while the weight losses for the polyester (reaction 8-41) are 27.4% and for the polyurethane (scheme 8-42) ~54%.

The reaction of 1,1'-bis(dimethylsilyl)ferrocene with sym-divinyltetramethyldisiloxane in a ratio of 1.1 : 1 in the presence of Speier's catalyst at 120°C for 20 hr gave 66% of the polymer $[C_{22}H_{40}Si_4OFe]_7$ (mol. wt. 3320, calc. 3422). When the polymer was heated for 10 hr, its molecular weight rose to only 3460. Changing the ratio to 1.1 : 1 affected neither the degree of polymerization nor the yield of the polyaddition product [93, 811]. The degree of polymerization of 7 is also characteristic of the product from the addition of 1,1'-bis(dimethylsilyl)ferrocene to sym-diallytetramethyldisilmethane, while in the case of sym-diethynyltetramethyldisiloxane the polyadduct $[C_{22}H_{36}Si_4OFe]_{11}$ is obtained (78%). Greber and Hallensleben do not report whether α- and β-addition occurs in these processes. However, according to the data in [15, 393, 1508], the addition of mono- and 1,1'-bis(methylphenylsilyl)ferrocene to sym-divinyldimethyldiphenyldisiloxane (Speier's catalyst, 220°C) proceeds with the formation of iso structures.*

* Addition to bis[vinyl(alkyl)arylsilyl]benzenes proceeds analogously [1508].

The disiloxane $[FcSi(CH_3)(C_6H_5)CH(CH_3)Si(CH_3)(C_6H_5)]_2O$ does not change on heating to 480°C in air.

Replacement of the vinyl monomer in reaction (8-38) by acetylene leads to organosilicon derivatives of ferrocene containing vinyl groups at silicon atoms [811]:

$$Fc'(SiR_2H)_2 + 2CH{\equiv}CH \xrightarrow[THF]{H_2PtCl_6} Fc'(SiR_2CH{=}CH_2)_2 \qquad (8\text{-}43)$$

Together with the substance of this type approximately 10% of a polyaddition product is formed. Bis(vinylsilyl)ferrocene supplements hydrosilylferrocene well in the synthesis of ferrocenylene-containing organosilicon polymers:

$$nFc'(SiR_2CH{=}CH_2)_2 + (n+1)\,H|\!\!-\!\!|SiR_2H \longrightarrow$$

$$\longrightarrow H[|\!\!-\!\!|SiR_2(CH_2)_2SiR_2Fc'SiR_2(CH_2)_2]_n SiR_2|\!\!-\!\!|H \qquad (8\text{-}44)$$

$$\text{where } |\!\!-\!\!| = [Si(CH_3)_2O]_2,\ [Si(CH_3)_2CH_2]_2 \text{ or } Si(CH_3)_2C_6H_4$$

At 120°C the yield of polymer (degree of polyaddition 6-10) is 75-85% in 2-3 days. When | — | equals SiR_2Fc', a polymer is obtained with alternating dimethylene and ferrocenylene bridges between silicon atoms (molecular weight 3450), which does not change when heated at 350°C for 2 hr.

The reaction of 1,1'-bis(vinyldimethylsilyl)ferrocene with perbenzoic acid gives a 70% yield of a diepoxide, which distills without decomposition [812]. Like the substance with $R'' = CH_2OCH_2CH{-}CH_2$ (with the two last carbons bridged by O, an epoxide ring), this diepoxide is hardened by diamines or aluminum chloride with the formation of glassy insoluble polymers.

A monosilyl derivative of ferrocene containing both the Si−H bond and vinyl groups at silicon atoms is capable of autopolyaddition [15, 393]:

$$H{-}\underset{Fc}{\overset{C_2H_5}{Si}}{-}CH{=}CH_2 \xrightarrow[80\%]{Pt/C;\ 250°} \left[{-}\underset{Fc}{\overset{C_2H_5}{Si}}{-}CH_2{-}CH_2{-} \right]_n \qquad (8\text{-}45)$$

The solid soluble silcarbane polymer with mol. wt. > 6000 has a softening point of 78°C.

The hydrosilylation of allyldimethylsilylferrocene in the presence of Speier's catalyst proceeds smoothly under normal conditions in accordance with Farmer's rule [433]. When hydrochlorosilanes are used products are obtained which are capable of further conversions such as hydrolysis:

$$FcSiR_2CH_2CH{=}CH_2 + RR'SiHCl \xrightarrow{H_2PtCl_6} FcSiR_2(CH_2)_3SiRR'Cl \xrightarrow{H_2O}$$

$$\longrightarrow [FcSiR_2(CH_2)_3SiRR']_2O \qquad (R = CH_3;\ R' = C_2H_5) \qquad (8\text{-}46)$$

To complete this section we should point out that like normal ketones, organosilicon derivatives of ferrocene containing keto groups form crystalline condensation products with 2,4-dinitrophenylhydrazine and are also capable of secondary acylation [124]:

$$2R_3Si(CH_2)_2COFc' + ClOC\text{—}C_6H_4\text{—}COCl \xrightarrow[CH_2Cl_2]{4AlCl_3}$$

$$\longrightarrow R_3Si(CH_2)_2COFc'OC\text{—}C_6H_4\text{—}COFc'CO(CH_2)_2SiR_3 \qquad (8\text{-}47)$$

Together with the acylation product, the disiloxane $[FcCO(CH_2)_2 \cdot SiR_2]_2O$ is obtained unexpectedly in 11% yield and this is ascribed to the action of $HAlCl_4$.

2.4. Application

The high thermal stability, low temperature coefficient of viscosity, and good dielectric properties of silyl and siloxanyl derivatives of ferrocene make it possible to consider them as special lubricants, hydraulic fluids, and dielectrics [1054, 1679]. They may also be used as antioxidants, catalysts for carbonylation and metallation with iron, and sources of free radicals [1614]. 1,1'-Bis(trialkylsilyl)ferrocenes act as good stabilizers for organosilicon fluids [1438]. Thus, when 0.1-1% of 1,1'-bis(trimethylsilyl)ferrocene is added, a polymethylsiloxane fluid of the type $(CH_3)_3SiO[(CH_3)_2SiO]_n \cdot Si(CH_3)_3$ with a viscosity of 200 centistokes does not change when air is blown through it at 250°C for 200 hr, while a control sample gels in 113 hr.

Compounds of the type $[X(CH_3)_2Si(CH_2)_n(CO)_mC_5H_4]_2Fe$ (where $X = CH_3$; n = 2 or 3; m = 0 or 1) and siloxanes based on them [obtainable when X = Cl or by reaction (8-34)] are recommended as stabilizers for siloxane fluids and also as antioxidants and UV absorbers [1717, 1718]. Liquid polydimethylsilferrocenylenes may be used at a working temperature up to 370°C and above [1677].

These data should be regarded as preliminary.

3. ORGANOSILICON DERIVATIVES OF IRON CONTAINING NITROGEN

When Wannagat's reagent (N-sodiohexamethyldisilazan) is added to a solution of ferric chloride in ether, petroleum ether, or toluene even at 0°C, an exothermic reaction begins [122, 604, 609]. In this case, in contrast to the reaction of sodium trimethylsilanolate with ferric chloride, the chlorine is not bound in the form of NaCl, but in the form of the complex Na_3FeCl_6 and this makes it necessary to represent the course of the process in the following way:

$$2FeCl_3 + 3NaN[Si(CH_3)_3]_2 \longrightarrow Fe\{N[Si(CH_3)_3]_2\}_3 + Na_3FeCl_6 \qquad (8-48)$$

After repeated recrystallization and sublimation, tris[bis(trimethylsilyl)amino]ferrane is obtained in about 50% yield in the form of dark green, sharp needles. It melts at 135°C and sublimes at 130°C (1 mm). It is readily soluble in ether, carbontetrachloride, benzene, cyclohexane and tetrahydrofuran (dark green solution) and dissolves with difficulty in pyridine and acetonitrile with a brown coloration. Moist acetone produces decomposition with the formation of $Fe(OH)_3$. The compound is monomeric in ether, has no tendency to form complexes with ethers and amines, and is hydrolytically unstable. With liquid ammonia there is a surface reaction with the formation of a blood-red complex, which is insoluble in NH_3 and decomposes at 0°C to the starting components. It is recommended that the iron in the substance is determined analytically by heating a sample of it with a H_2SO_4/HF mixture, dissolving the residue in HCl, and precipitating the iron in the form of $Fe(OH)_3$.

Trimethylisocyanosilane reacts readily with iron pentacarbonyl at 65-75°C with the evolution of a CO molecule and the formation of $(CH_3)_3SiN{=}CFe(CO)_4$ (bright yellow crystalline substance with m.p. 47-48°C, yield 93%) [1150]. In the absence of oxygen the compound is quite stable and does not decompose up to 120-130°C, but in the presence of oxygen it immediately acquires a brownish color. It is readily soluble in ether, benzene, and chloroform. The action of iodine in pyridine produces the evolution of carbon monoxide. Hexamethyldisiloxane and iron pentacarbonyl are formed by hydrolysis.

4. COMPLEX SALTS OF IRON WITH ORGANOSILICON COMPOUNDS

According to the data of Weiss and Harvey [1249], 2-hydroxypyridine N-oxide forms a cationic complex with silicic acid, which is characterized in the form of $[(C_5H_5NO_2)_3Si]^+[FeCl_4]^-$. This pale yellow substance is resistant to the action of water, soluble in the latter and also tetrachloroethane, and undergoes changes only when heated above 220°C. It is believed that compounds of this type (see Ch. 4, Tin) are interesting in the investigation of the silicosis problem.

Concepts of the nature of the complexes of ferric chloride with organohalosilanes are analogous to those mentioned in the examination of similar complexes of aluminum halides (see Ch. 3).

5. ORGANOSILICON DERIVATIVES OF COBALT AND NICKEL

5.1. Preparation Methods

The method of preparing artificial hydrosilicates through glycerosilicates, which has already been described in Ch. 2 and in Section 1.1.1. of the present chapter, leads to the formation of glassy products with the composition $CoO \cdot SiO_2 \cdot 4H_2O$ (solution pH 8) and $NiO \cdot SiO_2 \cdot 2H_2O$ (solution pH 6) when applied to cobalt and nickel salts [384]. It is reported [481] that products of the "organosilicate" type may also be obtained by the reaction of tetraalkoxysilanes and organoalkoxysilanes with salts, oxides, and hydroxides of cobalt and nickel. However, no concrete experimental data have appeared up to now.

Heating diorganodiethoxysilanes with diacetoxycobalt at 180-200°C leads to the formation of polycobaltadiorganosiloxanes [1126, 1379]:

$$R_2Si(OC_2H_5)_2 + Co(OCOCH_3)_2 \longrightarrow CH_3COOC_2H_5 + [R_2SiOCoO]_n \quad (8\text{-}49)$$

At the same time, in the reaction of dimethyldiacetoxysilane with cobalt bromide, instead of the expected condensation with the liberation of acetyl bromide there is exchange:

$$(CH_3)_2Si(OCOCH_3)_2 + CoBr_2 \longrightarrow (CH_3)_2SiBr_2 + Co(OCOCH_3)_2 \qquad (8\text{-}50)$$

The reaction of dimethyldichlorosilane with diacetoxycobalt also yields dimethyldiacetoxysilane and cobalt chloride [1126].

Silanediols do not react with cobalt halides either at room temperature or with moderate heating, and under more drastic conditions there is only homocondensation of the diols. The use of an HCl acceptor does not promote the reaction and pyridine forms a complex with CoX_2, which does not react with silanediols.

When the silanediols are replaced by their sodium salts [1126, 1380], there is polycondensation, which is analogous in its final result to reaction [8-49]:

$$nR_2Si(ONa)_2 + nCoX_2 \xrightarrow{\text{THF}} [R_2SiOCoO]_n + 2nNaX \qquad (8\text{-}51)$$

With components of high purity, the complete absence of moisture or hyroxyl-containing solvents, and exact observation of the stoichiometric ratio, blue friable resinous polymers are obtained with mol. wt. 1200-2000 and these are soluble in benzene, tetrahydrofuran, acetone, chloroform, dimethylformamide, and dimethyl sulfoxide.

The relation of the properties of the polymer to the reaction conditions may be illustrated by the following example. Refluxing a mixture of sodium diphenylsilanediolate with $CoBr_2$ in tetrahydrofuran for 1 hr leads to the formation of a polymer with mol. wt. 1200 and a relative viscosity in a 1% dimethylformamide solution of 1.035. The polymer is soluble in most organic solvents.* An increase in the reaction time to 8 hr gives a polymer with mol. wt. 1750 and η_{rel} of 1.040. When the mixture is heated in an autoclave at 200°C for 4 hr the polymer formed is difficultly soluble or insoluble in all solvents apart from dimethylformamide (η_{rel}=1.085).

When part of the cobalt halide is replaced by aluminum chloride, in accordance with scheme (8-51) polyalumacobaltasiloxanes

* An analogous reaction with sodium methyl(phenyl)silanediolate gives a polymer which is insoluble in benzene, but readily soluble in tetrahydrofuran and dimethylformamide (η_{rel} = 1.065).

are formed (these differ little in properties from polymers which contain no aluminum). Another method of introducing a third element into a siloxane structure is reaction (8-51) when there is some deficiency of $CoBr_2$. In this case, in the first stage of the process oligomers are formed with terminal ONa groups. Then when $AlCl_3$ is introduced there is further condensation with the formation of insoluble polymers, which have elastic properties according to literature data [1126]. Their mechanical properties are low, but heating for 1 hr at 300°C in a nitrogen atmosphere produces no visible change or loss in weight. It is reported that a similar reaction may be carried out with $SnCl_2$, $SnCl_4$, and $TiCl_4$ [1380].

As has already been mentioned (see Chs. 3 and 4 and also the present chapter, Section 1.1.2.), the reaction of monosodium salts of organosilanetriols with metal halides is one method of synthesizing polyorganometallasiloxanes, including ferrasiloxanes [152] (scheme 8-1). However, an attempt to introduce atoms of other Group VIII elements (cobalt and nickel) into a siloxane chain in this way was, contrary to expectation, unsuccessful [170, 286]. In an aqueous alcohol medium the cobalt and nickel are removed from the reaction zone:

$$RSi(OH)_2ONa + H_2O \rightleftharpoons RSi(OH)_3 + NaOH \rightleftharpoons [RSiO_{1,5}] + NaOH + H_2O$$

$$\begin{array}{l} CoCl_2 + 2NaOH \longrightarrow Co(OH)_2 + 2NaCl \\ \qquad\qquad\qquad\qquad\quad \lfloor\!\!\longrightarrow CoO + H_2O \end{array} \tag{8-52}$$

At the same time, the introduction of aluminum chloride into the reaction mixture as a component for binding the alkali promotes the introduction of cobalt and nickel into the siloxane structure. However, in this case polymers containing both cobalt (nickel) and aluminum atoms in the main chain are also formed. On the example of the reaction of the monosodium salt of ethylsilanetriol with $CoCl_2$ it was shown that the amount of bound cobalt increases with an increase in the molar ratio of $CoCl_2 : AlCl_3$ from 0 to 1. With an equimolar ratio the cobalt reacts practically completely, but with a ratio $CoCl_2 : AlCl_3 > 1$ the amount of cobalt reacting falls sharply, while the aluminum reacts quantitatively (Fig. 13). An analogous picture is observed for $NiCl_2$.

An attempt to prepare polycobaltadiorganosiloxanes by the reaction of the monosodium salt of phenylsilanetriol with $CoBr_2$

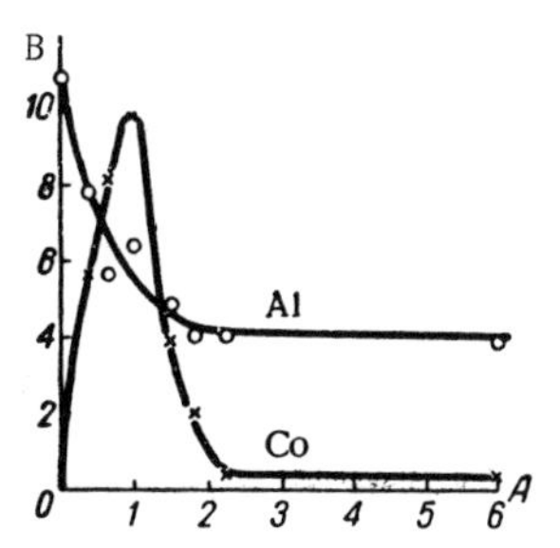

Fig. 13. Relation of the cobalt content of the polymer to the molecular ratio of cobalt chloride to aluminum chloride in the reaction medium [170]: A - Molecular ratio $CoCl_2 : AlCl_3$; B - metal content of polymer, %.

(2 : 1) with the subsequent addition of $RR'SiCl_2$ was likewise unsuccessful [1126]. In anhydrous tetrahydrofuran in the first stage a substance was formed which was close in composition to $[C_6H_5Si(OH)_2O]_2Co$. However, when this was treated with methyl(phenyl)-dichlorosilane even in the presence of pyridine (HCl acceptor) a complex of $CoCl_2$ with pyridine and methyl(phenyl)-polysiloxane (colorless resin) was formed. Nonetheless, the reaction of $C_6H_5Si(ONa)_3$ with $CoBr_2$ in a molar ratio of 2 : 3 gives a completely insoluble and infusible polymer with the composition $C_6H_5SiO_3Co_{1.5}$. With a ratio of 2 : 1 and the addition to the reaction of 2 moles (insufficient) of the halogen derivative (diphenyldichlorosilane) a soluble and fusible resin is formed and this may be represented approximately by the scheme:

$$2C_6H_5Si(ONa)_3 + CoBr_2 \longrightarrow \left[\begin{array}{c} ONa \\ | \\ C_6H_5Si{-}O{-} \\ | \\ ONa \end{array}\right]_2 Co \xrightarrow{+\,2RR'SiCl_2} \left[\begin{array}{ccc} C_6H_5 & & R \\ | & & | \\ {-}Si{-} & O & {-}Si{-}O{-} \\ | & & | \\ O & & R' \\ | & & \\ Co & & \\ | & & \\ O & & R \\ | & & | \\ {-}Si{-} & O & {-}Si{-}O{-} \\ | & & | \\ C_6H_5 & & R' \end{array}\right]_n \qquad (8\text{-}53)$$

The authors also report that cobalt atoms may be in the main chain of the polymer and that not all the bridges between the chains are cobalt bridges. In the polymers obtained in this way (mol. wt. ~4000) the Si : Co ratio varies from 4.5 to 8 instead of 4 as calculated. Nonetheless, the cross-linked linear (more accurately, cyclolinear) structure of products of this type may be regarded as demonstrated. The order of reaction plays no particular part and it is possible to carry out the reaction of $RSi(ONa)_3$ with R_2SiCl_2 (1 : 1) first and the add $CoBr_2$ (0.5).

In contrast to the data presented above on the nature of the reaction of monosodium salts of organosilanetriols with cobalt and nickel salts (reaction 8-52), there are data in the literature [397] indicating that the reaction of aqueous alcohol solutions of the preparations GKZh-10 or GKZh-11, which have the composition $[RSi(OH)_2ONa]_{1.5}$ (where R is C_2H_5 and CH_3, respectively) and an alkalinity of 15.85% and consist of a mixture of a monomer and dimer, with water-soluble salts of Co and Ni (and also metals of Groups I, II, IV, and VI) yield mixtures of siliconates of the corresponding metals and their hydroxides. If cloths are impregnated with aqueous solutions of the metals and then treated with the preparation GKZh-10, then the films formed after drying the cloth (70-80°C, 10-15 min) have an increased resistance to soap and soda treatment (see Section 5.2.).

As has already been mentioned in Chs. 1 (Section 6.), 2 (Section 5.), and 4 (Section 12.1.), linear silicon hetero-organic polymers containing metal carboxylate groups in the chain may be obtained by schemes 8-54a or b [289, 290, 1518a]. In the case of cobalt and nickel, both routes are suitable:

$$nHOOCCH_2CH_2[Si(CH_3)_2O]_mSi(CH_3)_2CH_2CH_2COOH + nM(OH)_2 \quad a)$$

$$\downarrow -2nH_2O$$

$$\{-MOOCCH_2CH_2[Si(CH_3)_2O]_mSi(CH_3)_2CH_2CH_2COO-\}_n$$

$$\uparrow -2n\ KCl$$

$$n\ KOOCCH_2CH_2[Si(CH_3)_2O]_mSi(CH_3)_2CH_2CH_2COOK + nMCl_2 \quad b)$$

$$[M = Co,\ Ni,\ \text{and also}\ Cu,\ (8\text{-}54b)\ Ca,\ Zn,\ Cd,\ Pb] \qquad (8\text{-}54)$$

A patent has been published [1639] on the preparation of cobalt and nickel salts by the reaction of their halides with organosilicon derivatives of dithiocarbamic acid $HSC(S)NH(CH_2)_nSiR_mX_{3-m}$ ($n \geq 3$; $m = 0-2$; X = OR; R = alkyl, aryl, or aralkyl).

The reaction of sodium bis(trimethylsilyl)amide with cobalt and nickel halides may be represented in the general form by the following equation [122, 604, 609, 610]:

$$2(R_3Si)_2NNa + MX_2 \longrightarrow [(R_3Si)_2N]_2M + 2NaX \qquad (8\text{-}55)$$

$$(R = CH_3;\ M = Co,\ Ni)$$

In contrast to the reaction with $FeCl_3$, which leads to the formation of the corresponding iron derivative (see p. 554), no reaction occurs with $CoCl_3$ in petroleum ether and toluene. In tetrahydrofuran the adduct $CoCl_2 \cdot 1.5C_4H_8O$ is formed and as a result of this the salt passes into solution and reacts to form black, star-shaped crystals of bis[bis(trimethylsilyl)amino]cobaltane.* In the case of a molar ratio of 1 : 1, the product of partial reaction is apparently obtained (as blue crystals):

$$(R_3Si)_2NNa + CoCl_2 \xrightarrow{THF} (R_3Si)_2NCoCl \cdot C_4H_8O \qquad (8\text{-}56)$$

Reaction (8-55) does not occur with $NiCl_2$ either in tetrahydrofuran or in toluene. It is necessary to use NiI_2 in order to obtain a low yield of liquid bis[bis(trimethylsilyl)amino]nickelane.

Patents have appeared [1614, 1677] on the preparation of silicon-containing metallocenes from metallated cyclopentadienylsilanes and cobalt and nickel dichlorides (i.e., according to a scheme analogous to 8-23). However, the properties of these compounds have not yet been described.

Dicobalt octacarbonyl $Co_2(CO)_8$ is known to be capable of homolytic cleavage of molecular hydrogen with the formation of $HCo(CO)_4$. A reaction occurs with triorganosilanes and the authors of [624, 626, 828] divide this into two stages:

$$\begin{array}{l} R_3SiH + Co_2(CO)_8 \longrightarrow R_3SiCo(CO)_4 + HCo(CO)_4 \\ R_3SiH + HCo(CO)_4 \longrightarrow R_3SiCo(CO)_4 + H_2 \\ \hline 2R_3SiH + Co_2(CO)_8 \longrightarrow 2R_3SiCo(CO)_4 + H_2 \end{array} \qquad (8\text{-}57)$$

(R = alkyl, aryl, alkoxyl, or Cl)

The compounds obtained are quite stable and change slowly in contact with air and moisture. Since dicobalt octacarbonyl is an active catalyst for the addition of hydrosilanes to unsaturated compounds it is believed that a major part is played by adducts formed as intermediates by scheme (8-57):†

$$R'CH{=}CH_2 + HCo(CO)_4 \longrightarrow$$

$$\longrightarrow R'CH_2CH_2Co(CO)_n - \begin{cases} \xrightarrow{R_3SiH} R_3SiCH_2CH_2R' + HCo(CO)_n \\ \xrightarrow{R_3SiCo(CO)_4} R_3SiCH_2CH_2R' + Co_2(CO)_{4+n} \end{cases} \qquad (8\text{-}58)$$

$(n = 3 \text{ or } 4)$

* Under certain conditions a green modification is obtained [609].

† Nickel chloride [469] and iron pentacarbonyl [69, 396, 469, 490] are also known to be catalysts of hydrosilation. In the first case the actual catalyst is colloidal nickel,

TABLE 97. Organosilicon Derivatives of Cobalt and Nickel

Formula	m.p., °C	b.p., °C (mm)	Other data	Preparation method (scheme)	Literature
$(C_2H_5)_3SiCo(CO)_4$	—	—	Liquid	8-57	626
$(C_6H_5)_3SiCo(CO)_4$	135—140	—	—	8-57	626
$C_6H_5SiCl_2Co(CO)_4$	34—35	—	—	8-57	626
$Cl_3SiCo(CO)_4$	44	—	—	8-57	626
$C_2H_3Si[Co(CO)_3]_3$	—	—	Purple-black crystals	—	870
$[Co_3(CO)_9Si]_2$	—	—	Subl. ~ 45°C (vac), violet crystals	—	870a
$\{[(CH_3)_3Si]_2N\}_2Co$	73	101 (0.6)	Yield 53%	8-55	609
$[(CH_3)_3Si]_2NCoCl \cdot nC_4H_8O$	—	—	Grey-blue crystals	8-56	609
$\{[(CH_3)_3Si]_2N\}_2Ni$	—	80 (0.2)	Red liquid	8-55	610
$\{[(CH_3)_3Si]_2N\}_2Ni \cdot nC_4H_8O$	—	—	Decomposes above 50°C (0.2 mm)	—	610

The direct addition of dicobalt octacarbonyl or its silicon derivative to an olefin is evidently excluded.

In this connection the reaction of dicobalt octacarbonyl with tetravinylsilane is of particular interest [870]. This forms tris-(tricarbonylcobalt)vinylsilane – a compound which is obtained as a result of the elimination of vinyl groups from the silicon atom and which is stable in air with cooling.

The reaction of cobalt octacarbonyl with tetraphenylsilane in ether in a nitrogen atmosphere leads to the formation of hexakis-(tricarbonylcobalt)disilane – a diamagnetic soluble substance which is stable in nitrogen or in air with cooling [870a]. Structurally this compound consists of two trigonal pyramids, connected at the apices through a Si – Si bond. The relative stability of the compound is ascribed to its partial double bonding. The mechanism of these interesting reactions has not been studied as yet.

5.2. Properties, Application, and Analysis

The physicochemical constants of organosilicon derivatives of cobalt and nickel are given in Table 97.

---

while it is believed that the process probably has an ionic mechanism. With $Fe(CO)_5$ the reaction proceeds under more drastic conditions than with $Co_2(CO)_8$, but we cannot exclude the possibility of the intermediate formation of $[R_3Si^+HFe^-(CO)_4]$, which is then capable of reacting with an olefin by a scheme like (8-58) [490].

In the infrared spectra of cobaltasiloxanes, a strong absorption band at 995-990 cm^{-1} is assigned to vibrations of the Si–O–Co group [1249].

Bis[bis(trimethylsilyl)amino] derivatives of cobalt and nickel are monomeric. They are extremely sensitive to atmospheric oxygen and moisture. Their thermal stability is insignificant and decomposition soon begins (black coloration) during vacuum distillation, even in a nitrogen atmosphere.

Polycobaltadimethylsiloxane obtained by scheme (8-49) consists of a dark blue powder; it does not melt up to 500°C, is insoluble in inorganic solvents, and contains 4.9% Si and 38.5% Co. It is resistant to the action of cold water, but is hydrolyzed on boiling in water with breakdown of the Si–O–Co groups. In the opinion of the authors, these properties indicate that the substance is not a linear polymer of the type $C_2H_5O[(CH_3)_2SiOCoO]_nOCCH_3$; but the true structure of the polymer has not yet been determined. Polycobaltdiphenylsiloxane has analogous properties and it does not change on heating to 560°C.

Some properties of cobaltasiloxanes obtained by scheme (8-15) have already been discussed. We might add that in contrast to polymers with the same composition, obtained by scheme (8-49), they decompose at 350°C with the separation of CoO. The absence of terminal functional groups (X or ONa) indicates a cyclic structure for these polymers. Heating solutions of them in benzene or tetrahydrofuran with $AlCl_3$, $ZnCl_2$, or Na silanolates is accompanied by ring opening and the formation of linear polymers of higher molecular weight.

Fractionation of the polymer obtained from $C_6H_5Si(OH)_2ONa$, $AlCl_3$ and $CoCl_2$, in a molar ratio of 1 : 2 : 2 showed that the aluminum is distributed more or less uniformly through the fractions [170]. With a mean Al content of 3.8%, the Al content of the fractions varied over the range 2.5-4.6%. At the same time, the cobalt is concentrated in the main, least soluble fraction and with a mean content in the polymer of 6.7%, the cobalt content of the fractions gradually falls from 14.3% in the first to 2.3% in the last.

The resinous polymers obtained by the reaction of monosodium salts of organosilanetriols with cobalt and nickel halides in the

TABLE 98. Polymers of the Type $\{-MOCoCH_2CH_2[Si(CH_3)_2O]_m\cdot Si(CH_3)_2CH_2CH_2COO-\}_n$ [289, 290]

M	m	Preparation method (scheme)	Yield, %	[η] Toluene	Molecular weight
Cu	74	8-54b	70(71)	–	–
Ca	29	8-54a	63	0.064	–
Zn	29	8-54a	83	0.060	–
	74	"	70	0.076	–
	106	"	75	0.070	16700*
Cd	32	"	77	–	–
Pb	29	"	72	0.068	–
Co	32	"	72(80)	0.060	8700* 8000†
	74	8-54b	76	0.150	–
	106	8-54a	88	0.065	15000* 14000†
Ni		8-54b	55	–	–

* From the viscosity in CCl_4.
† Osmometric method.

presence of $AlCl_3$ are dark violet (Co) or green (Ni) and are soluble in benzene and other solvents.

Polymers obtained by scheme (8-54a) with $m \geq 40$ dissolve in organic solvents. Reaction (8-54b) in alcohol yields polymers with mol. wt. 10,000-14,000, which are superficially rubbery. When $m = 120$ the following relation is observed between the glass transition point and the nature of M: Co > Zn > Pb, Ca > Cd; T_g of the cadmium-containing polymer is 80-90°C.

Some data on these polymers are given in Table 98.

There are as yet no definite recommendations on the use of polymetallaorganosiloxanes containing Co and Ni atoms. One would still expect that these substances would find their place among heat-resistant polymers.

Above we mentioned the improvement in the properties of water-repellent siloxane coatings with preliminary treatment of the cloth with metal salt solutions [397, 1497]. Table 99 gives comparative data on the water resistance of cloth treated with solu-

TABLE 99. Water Resistance of Cotton Cloth Waterproofed with Alkylsiliconates of Various Metals [397] (Soap-soda laundering, numerator—data after drying cloth at 60°C, denominator—at 135°C)

Salt cation	Water resistance, mm of water column			After two launderings at 100°C (30 min)
	Before laundering	After laundering for 30 min at 40°C		
		Once	Three times	
Na	170/170	130/150	40/90	0/40
Cu	200/200	210/200	170/180	160/150
Ca	195/230	185/220	80/80	20/20
Ba	200/205	170/175	90/90	30/30
Zn	195/230	185/220	80/80	20/20
Al	160/160	130/130	70/80	20/20
Pb	220/220	190/190	120/120	60/60
Co	190/190	190/190	180/180	155/160
Ni	230/225	235/220	235/220	180/180

tions of salts of various metals and then the preparation GKZh-10. They show that metal siliconates may be arranged in the following order with respect to the resistance of the film:

$$Ni\,(NiCl_2) > Co = Cu\,[Cu\,(OCOCH_3)_2] > Pb > Ba > Ca,\ Zn,\ Al > Na$$

Cobalt and nickel salts of (silylalkyl)dithiocarbamic acid are recommended for use as lubricants, fungicides, insecticides and metal extractants [1639].

Mono- and bis(trimethylsilyl)nickelocenes have been proposed as additives for certain forms of fuel [1641].

The literature contains quite a few reports on the use of salts of organic acids and cobalt (caprylate, ethylxanthate, naphthenate, carbamate, coproate, acetate, linoleate, etc.) [63, 217, 614, 988, 1303, 1326, 1328, 1330, 1334, 1401, 1403, 1416, 1439, 1449, 1652, 1705, 1706] and also nickel [1687], internal complex compounds (acetylacetonate) [1418], and some inorganic compounds of cobalt and nickel [63, 1433, 1684, 1687] as catalysts for the hardening of siloxane resins and varnishes [63, 217, 614, 988, 1303, 1426, 1334, 1403, 1418, 1439, 1449, 1709], molding materials [1330, 1705], impregnating compositions [1328, 1401], additives to improve the properties of siloxane fluids [1684], thermal stabilizers of siloxane elastomers [1433, 1652, 1687], and agents for the cold vulcanization of liquid polymers [1416]. The activity of cobalt compounds as hardeners in comparison with derivatives of other elements has already been discussed in Section 1.2. of this chapter.

It is recommended that silicon and cobalt in metallasiloxanes are determined by wet combustion with a mixture of sulfuric and nitric acids. The silicon is determined as SiO_2 and the cobalt in the filtrate after separation of the SiO_2 is determined with anthranilic acid. For determination of cobalt alone, the residue from wet combustion is heated with hydrofluoric acid and the silicon removeed in the form of SiF_4. The residue is dissolved in sulfuric acid and then the cobalt converted in the anthranilate [1126].

For nitrogen-containing organosilicon compounds of cobalt it is recommended [609] that the substance is treated with a mixture of sulfuric and hydrofluoric acids and the cobalt determined gravimetrically as $CoSO_4$ and as a check, in the form of Co_3O_4 after calcination.

6. ORGANOSILICON COMPOUNDS OF OTHER GROUP VIII ELEMENTS

In studying the mechanism of the addition of hydrosilanes to unsaturated compounds in the presence of chloroplatinic acid on the example of trichlorodeuterosilane, Speier and his co-workers [1049] established that H_2PtCl_6 is an effective catalyst for deuterium–hydrogen exchange between Cl_3SiD and, for example, methyldichlorosilane. On the other hand, data have been obtained on the homogeneous nature of addition reactions catalyzed by chloroplatinic acid. In the opinion of the authors, the true catalysts are complexes of reduced platinum with hydrosilane. This may be represented in a simplified form as follows:

$$\mathrm{Pt} + \rangle\mathrm{SiH\,(D)} \rightleftarrows \rangle\mathrm{SiPtH\,(D)} \underset{\rangle\mathrm{SiH\,(D)}}{\overset{\rangle\mathrm{SiH\,(D)}}{\rightleftarrows}} \left(\rangle\mathrm{Si}\right)_2 \mathrm{PtH_2\,(D_2)} \qquad (8\text{-}59)$$

There is then addition of the complex to the olefin with subsequent decomposition of the intermediate compound and the liberation of the primary complex. The nature of the products from the addition of trichlorodeuterosilane to allyl chloride supports these hypotheses:

$$CH_2{=}CHCH_2Cl + (Cl_3Si)_2PtD_2 \begin{cases} \xrightarrow{\beta} (Cl_3Si)_2PtDCH_2CHDCH_2Cl \xrightarrow{-Cl_3SiPtD} Cl_3SiCH_2CHDCH_2Cl \\ \xrightarrow{\alpha} (Cl_3Si)_2PtDCH(CH_2D)CH_2Cl \xrightarrow{-Cl_3SiPtD} SiCl_4 + DCH_2CH{=}CH_2 \end{cases} \quad (8\text{-}60)$$

Chalk and Harrod [625] observed the following reactions:

$$(R_3P)_2PtCl_2 + R'_3SiH \xrightarrow{N_2,\ \text{reflux}} (R_3P)_2PtClH + R'_3SiCl \quad (8\text{-}61)$$

$$(R = n\text{-}C_4H_9;\ R' = C_2H_5)$$

$$2(R_3P)_2PtCl_2 + 2R'_3SiH \xrightarrow[N_2]{145^\circ} (R_3P)_4Pt_2Cl + [R'_3SiCl + R'_3SiH] \quad (8\text{-}62)$$

$$(R = C_6H_5;\ R' = n\text{-}C_6H_{13})$$

However, bis(triphenylphosphino)iridium carbonyl chloride does not react similarly with R_3SiH. At the same time, in contact with trichlorosilane, ethyl- and phenyldichlorosilane, and triethoxysilane, the yellow color of this compound disappears at room temperature and white powdery 1 : 1 adducts are formed. They are very stable and heating to 150-200°C is required for their decomposition. The starting components are then liberated. The complex with Cl_3SiH is more stable than the adduct with $(C_2H_5O)_3SiH$. The formation of the adduct requires the simultaneous presence of the Si–H bond and a functional group in the organosilicon compound [no complex formation occurs with $(C_2H_5O)_4Si$ or $SiCl_4$]. The following scheme, which leads to the formation of an adduct with an octahedral structure, is believed to be probable:

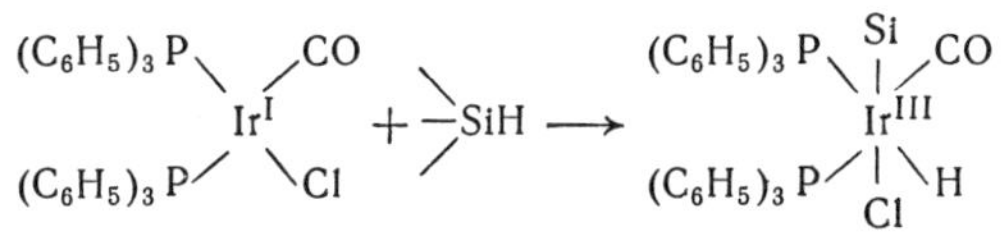

$$[(C_6H_5)_3P]_2Ir^{I}(CO)Cl + {\rangle}SiH \longrightarrow [(C_6H_5)_3P]_2Ir^{III}(Si)(CO)(H)Cl \quad (8\text{-}63)$$

Osmocene and ruthenocene may also be used in the reaction (8-21) as well as ferrocene. In particular, the mono(trimethylsilylpropionyl) derivatives of these metallocenes are obtained in yields of 60 and 50%, respectively [1718]. By carrying out a reaction of type (8-34) with mono- and bis(triorganosilylalkanoyl) osmocenes or ruthenocenes or by carrying out silylation with subsequent hydrolysis according to scheme (8-26) [1717] it is possible to obtain

TABLE 100. Rate Constants ($K \cdot 10^3$ min^{-1}) of the Cleavage of the $Si-C_{metallacene}$ Bond in $(CH_3)_3SiC_5H_4MC_5H_5$ (aqueous methanol in the presence of HCl)

HCl concentration, M	M		
	Fe	Ru	Os
0.596	3.36	156.5	62.66
0.477	2.45	109.3	48.10

polymers containing the Si–O–Si bond and osmo(rutheno)cenyl or osmo(rutheno)cenylene groups.

The rate of hydrolytic desilylation of compounds of the type $(CH_3)_3SiC_5H_4MC_5H_5$ increases in the series M = Fe < Os < Ru, as is shown by the data in Table 100 [938].

The reactivity of trimethylsilylruthenocene is 45 times as great as that of trimethylsilylferrocene.

Literature Cited

1. Monographs, Reviews, and Dissertations

1. Academy of Sciences of the USSR. Institute of Heteroorganic Compounds, Polyorganosiloxanes and Polyorganometallosiloxanes [in Russian], Pamphlet of the Exhibition of Achievements of the National Economy of the USSR (1957).
2. Alpatova, N. M. and Kessler, Yu. M., Complex Compounds of Silicon, Zh. Strukt. Khim. 5:332 (1964).
3. Andreeva, Z. A. (Borisov, S. N.), Doctoral Dissertation, Leningrad University (1962).
4. Andrianov, K. A., Organosilicon Compounds, Goskhimizdat (1956).
5. Andrianov, K. A., "Innovations in the synthesis of heterochain hetero-organic polymers," Usp. khim., 26:895 (1957).
6. Andrianov, K. A., Polymers with Inorganic Main Chains, Izd. Akad. Nauk SSSR (1962).
7. Andrianov, K. A., Heat-Resistant Organosilicon Dielectrics, Gosénergoizdat (1957).
8. Andrianov, K. A., "Heteroorganic polymeric compounds," Usp. khim., 27:1257 (1958).
9. Andrianov, K. A., Khaiduk, I., and Khananashvili, L. M., "Inorganic silicon-containing cyclic compounds and their organic derivatives," Usp. khim., 32:539 (1963).
10. Andrianov, K. A., Khaiduk, I., and Khananashvili, L. M., "The capacity of elements to form polymers with inorganic molecular chains," Usp. khim., 34:27 (1965).
11. Andrianov, K. A., Methods of Synthesis of Organometalloid Polymers, Chem. Soc. Spec. Publ. No. 15, England.
12. Andrianov, K. A., Polymers with Inorganic Primary Molecular Chains, J. Polymer. Sci., Vol. 52, No. 157, p. 257 (1961).
13. Andrianov, K. A. and Zhdanov, A.A., Synthesis of New Polymers with Inorganic Chains of Molecules, J. Polymer Sci., Vol. 30, No. 121, p. 513 (1958).
14. Asnovich, É. Z., Investigation of the Synthesis of Some Polymetalloorganosiloxanes and Their Properties in Relation to the Chemical Composition and Structure, Author's abstr of cand. dissert., All-Union Electrotechnical Institute (1964).
15. Babaré, L. V., Synthesis of Some Organosilicon Derivatives of Ferrocene, Author's abst. of cand. dissert., Institute of Petrochemical Synthesis (1965).

16. Belyakova, Z. V., Tetraacyloxysilanes and Acyloxytrichlorosilanes in Organic Synthesis, Author's abstr. of cand. dissert., Moscow University MGU, (1958).

17. Berezhnoi, A. S., Silicon and Its Binary Systems [in Russian], Izd. Akad. Nauk USSR, Kiev, (1958).

18. Berlin, A.A., "The main trends in investigations of the chemical conversions of macromolecules," Usp. khim., 29:1189 (1960).

19. Borisov, S. N., "Hetero-organic compounds of silicon," Usp. khim., 28:63 (1959).

20. Borisov, S. N., "Polysiloxane rubbers," in: Rubbers for Special Purposes ed. I. V. Garmonov, Izd. VINITI (1961), p. 67.

21. Borisov, S. N., Voronkov, M. G., and Dolgov, B. N., "Aluminum chloride in the organic chemistry of silicon," Usp. khim., 26:1388 (1957).

22. Water-Repellent Organosilicon Preparations EN, IOS Akad Nauk LatvSSR (1964).

23. Water-Repellent Organosilicon Preparations MN, IOS Akad. Nauk LatvSSR (1964).

24. Voronkov, M. G., "Heterolytic cleavage of the siloxane bond," Report on research work presented at the competition for the degree of Doctor of Chemical Sciences, Institute of Petrochemical Synthesis (1961).

25. Voronkov, M. G. and Borisov, S. N., "The nomenclature of organometallic compounds of Group IV Elements," Izv. Akad. Nauk LatvSSR, ser. khim., 1961:247.

26. Voronkov, M. G. and Shorokhov, N. V., Water-Repellent Coatings in Construction Work, Izd. Akad. Nauk LatvSSR, Riga (1963).

27. Voronkov, M. G. and Shorokhov, N. V., The Use of Sodium Alkylsiliconate Solutions for Increasing the Water-Resistance of Constructional Materials LDNTP (1956).

28. Vyazankin, N. S., Homolytic Reactions of Hetero-Organic Compounds of Group IV, Author's abstr. of doctor's dissert., Institute of Heteroorganic Compounds (1964).

28a. Vyazankin, N. S. and Kruglaya, O. A., "Covalent biheteroorganic compounds," Usp. khim., 35:1389 (1966).

29. Gar, T. K., Reaction of Trihalogermanes with Unsaturated Compounds, Author's abstr. of cand. dissert., Institute of Organic Chemistry (1964).

30. Waterproofing Organosilicon Liquids, State Comm. on Chemistry (1964).

31. Wu Kuan-li, Synthesis in the Field of Ferrocene-and Silicon-containing Organic Compounds, Author's abstr. of cand. dissert., Moscow Institute of Chemical Technology (1962).

32. Darashkevich, M. M., Preparation of New Types of Silicates Based on Organosilicon Compounds, Author's abstr. of cand dissert., Moscow Institute of Chemical Technology (1957).

33. Dzhurinskaya, N. G., Reaction of Germanium Hydrides with Unsaturated Compounds. Author's abstr. of cand. dissert., Institute of Organic Chemistry (1961).

34. Heat-Resistant Lacquer FG-9, Pamphlet of the Exhibition of Achievements of the National Economy of the USSR TsBTI Mashinostroeniya, (1960).

35. Kashireninov, O. E., Investigation of the Physicochemical Properties and Catalytic Activity of the Halides of Group IV Elements and Their Complex Compounds, Author's abstr. of cand. dissert., Rostov University, Rostov-on-Don (1962).

36. Korshak, V. V., "The main stages in the development of the chemistry of high-molecular compounds in the last 40 years," Usp. khim., 26:1295 (1957).
37. Korshak, V. V., Advances in Polymer Chemistry, Ch. 4, "New Heteroorganic high-molecular compounds," Izd. "Nauka" (1965).
38. Korshak, V.V., "Progress in the synthesis of high-molecular compounds," Usp. khim., 29:569 (1960).
39. Korshak, V. V. and Krongaus, E.S., "Progress in the synthesis of heat-resistant polymers," Usp. khim., 33:1409 (1964).
40. Kravchenko, A. L., Synthesis and Conversions of Halogen-Containing Organogermanium Compounds. Author's abstr. of cand. dissert., Institute of Organic Chemistry (1964).
41. Krasovskaya, T. A., Development of New Methods of Synthesizing Polyethylsiloxane Liquids, Author's abstr. of cand. dissert., Ministry of the Chemical Industry (1955).
42. Krasnoshchekov, V. V., Methods of Determining Silicon in Organosilicon Compounds, Author's abstr. of cand. dissert., Moscow Institute of Chemical Technology (1964).
43. Kukharskaya, É. V. and Fedoseev, A. D., "Organic derivatives of silicates with a laminar structure," Usp. khim., 32:1113 (1963).
44. Lavygin, I. A., Synthesis and Investigation of Titanoorganosiloxane Oligomers and Polymers, Author's abstr. of cand. dissert., GKKh (1965).
45. Lagzdin', É. A., The Physicochemical Properties of Sodium Aluminomethylsiliconate and Its Use for Water-Proofing Constructional Materials. Author's abstr. of cand. dissert., Riga Polytechnic Institute, Riga (1965).
46. Mal'nova, G. N., Investigation of the Catalytic Dehydrocondensation of Methyldichlorosilane with Benzene and Its Homologs and Derivatives. Author's abstr. of cand. dissert., State Insitute of Applied Chemistry (1960).
47. Motsarev, G. V., Investigation of the Halogenation of Organic Compounds of Silicon, Author's abstr. of doctor's dissert., Institute of Heteroorganic Compounds (1965).
48. Nesmeyanov, A. N. and Kabachnik, M. I., "Soviet organic chemistry in the last 40 years," Usp. khim., 27:1274 (1957).
49. Nesmeyanov, A. N. and Sokolik, R. A., Methods in Heteroorganic Chemistry – Boron, Aluminum, Gallium, Indium, and Thallium, Izd. "Nauka" (1964), p. 283.
50. Nudel'man, Z. N., Condensation of Silanols with Alkoxy Derivatives of Metals and Some Fields of Practical Applications of this Reaction, Author's abstr. of cand. dissert., Leningrad Technological Institute (1964).
51. Orlov, N. F., Synthesis of the Triorganosiloxy Derivatives of Elements of Groups III, IV, and V from Triorganosilanols and Hexaalkyldisiloxanes, Author's abstr. of cand. dissert., Leningrad University (1960).
51a. Orlov, N. F., Androsova, M. V., and Vvedenskii, N. V., Organosilicon Compounds in the Textile and Light Industries, Izd. "Legkaya Industriya," Moscow, (1966).
52. Petrashko, A. I., Investigation of the Synthesis and Conversions of Polyorganosiloxane Bonders for High-Temperature Resistant Glass-Mica Insulating Materials, Author's abstr. of cand. dissert., All-Union Electrotechnical Institute (1964).

53. Pichkhadze, Sh. V., Investigation of the Synthesis and Properties of Linear Polyorganosiloxanes. Author's abstr. of cand. dissert., Moscow Institute of Fine Chemical Technology (1962).
54. Podchekaeva, V. N. (Voronkov, M. G.), Doctoral Dissertation, Leningrad University (1953).
55. Samsonov, G. V., Silicides and Their Use in Technology, Izd. Akad Nauk UkrSSR, Kiev (1959).
56. Skorik, Yu. I. (Voronkov, M. G.), Doctoral Dissertation, Leningrad University (1957).
57. Skorik, Yu. I., Preparation and Investigation of Organic Derivatives of Kaolin and Chrysotile Asbestos, Author's abstr. of cand. dissert., Institute of the Chemistry of Silicides (1966).
58. Silicones, Izd. "Reklama," Kiev (1965).
59. Stanko, V. I., Chapovskii, Yu. A., Brattsev, V. A., and Zakharkin, L. I., "The chemistry of decaborane and its derivatives," Usp. khim., 34:1011 (1965).
60. Suvorov, A. L. and Spasskii, S. S., "Organic compounds of titanium," Usp. khim., 28:1267 (1959).
61. Tverdokhlebova, I. I., Investigation of the Properties of Solutions and the Structure of Polyorganosiloxanes and Polymetalloorganosiloxanes, Author's abstr. of cand. dissert., Institute of Heteroorganic Compounds (1963).
62. Transactions of the Conference "Chemistry and Practical Application of Organosilicon Compounds," No. 5, "Nomenclature of organosilicon compounds," (pamphlet), TsBTI Lensovnarkhoza (1958).
63. Fromberg, M. B., Heat-Resistant Insulating Coatings, Gosénergoizdat (1959).
64. Fyurst, V., Investigation of the Reaction of Organosilicon Compounds with Inorganic Compounds of Boron and Phosphorus, Author's abstr. of cand. dissert., Mosk. Khim. Technol. Inst., (1963).
65. Khaiduk, I., Investigation of the Chemistry of Inorganic Rings, Author's abstr. of cand. dissert., Moscow Institue of Chemical Technology (1963).
66. Khaiduk, I. and Andrianov, K. A., "Nomenclature of inorganic silicon heterocycles," Izv.Akad. Nauk, Otd. Tekhn. Nauk, 1963, 1537.
67. Khananashvili, L. M., Reaction of Tetraalkoxy- and Alkylalkoxysilanes and Their Derivatives with Some Inorganic Compounds, Author's abstr. of cand. dissert., Moscow Institute of Chemical Technology (1957).
68. Chamin, N. N.(Voronkov, M. G.), Doctoral Dissertation, Leningrad University (1957).
69. Chukovskaya, E. Ts., Investigation of the Telomerization of Olefins with Hydrosilanes and the Addition of Hydrosilanes to Unsaturated Compounds, Author's abstr. of cand. dissert., Institute of Organic Chemistry (1960).
70. Chumaevskii, N. A., "Vibration spectra of compounds containing elements of the carbon subgroup," Usp. khim., 32:1152 (1963).
71. Shampai, F. I., Lithium and Its Alloys, Izd. Akad. Nauk SSSR (1957), p. 117.
72. Shapatin, A. S., Some Reactions of Aluminum Alcoholates, the Synthesis and Properties of the Aluminum Derivatives of Phosphinic Acids and Polyorganophosphinate of Aluminodimethylsiloxanes, Author's abstr. of cand. dissert., Moscow Institute of Fine Chemical Technology (1965).

73. Shikhiev, I. A., Shostakovskii, M. F., and Komarov, N. V., New Oxygen-Containing Organosilicon Compounds, Baku (1960).

73a. Aitken, I. D., Sheldon, R., and Stapleton, G. B., "High temperature polymers," Brit. Plast., 34:662 (1961) (Part I); ibid., 35:39 (1962) (Part II).

74. American Chemical Society, Metal-Organic Compounds (1959); cited in [115], (15).*

75. Atlas,S. M. and Mark,H. F., Thermisch resistente Polymere, Angew. Chem. 72:249 (1960).

76. Aylett,B. J., Thesis, Cambridge (1954); cited in [85], p. 62, 68, (133), 118 (59) and in [107].

77. Barry,A. J. and Beck,H. N. N., Silicone Polymers, in: Inorganic Polymers, ed. F. G. A. Stone, W. A. G. Graham, New York-London (1962) p. 189.

78. Bažant,V. and Chvalovsky,V., Organosilicon Compounds, Vol. I. Chemistry of Organosilicon Compounds, Prague (1965).

79. Bonnot,L. and Lefebre,G., Les Polymères thermostables, Généralités, Rev. Inst. Franç. Petrole Ann. Comb. Liq., 17:1508 (1962); Bonnot L., Les polymères organomettallique, Ibid. 18:83, 284 (1963).

80. Bradley,D. C., Polymeric Metal Alkoxides, Organometalloxanes, and Organometalloxanosiloxanes in: Inorganic Polymers, New York-London (1962) p. 410.

81. Brydson,J. A., Inorganic Polymers, Plastics, 22:384 (1957).

82. Campagna,P. J., The Preparation and Properties of Certain p-Trialkylsilylphenylalkyl Alcohols and Ketones, Univ. Buffalo, Diss. Abstr., 13:312 (1953).

83. Eaborn,C., Organosilicon Compound, London, (1960).

84. Ebsworth,E. A. V., Thesis, Cambridge, (1957); cited in [85], p. 36 (114).

85. Ebsworth,E. A. V., Volatile Silicon Compounds, Oxford-London-New York-Paris, (1963) (International Series of Monographs on Inorganic Chemistry. gen. eds. H. Taube and A. G. Maddock, Vol. 4).

86. Essler,H. G., Zur π-Komplexbildung Si-substitujerter Benzolderivate und kondensierter Aromate mit Chrom, Diss., Techn. Hochschule München (1960), S. 59; cited in RZhKhim. (1961), 7V99.

87. Findeiss,W., Neue Wege zu Organogallium Verbindungen und Galliumhydriden, Diss. Univ. Marburg (1965).

88. Gaj,B. J., Diss. Abstr., 20:4521 (1960); cited in [78], 195 (DA 97).

89. George,P. D., Prober M., and Elliott J. R., Carbonfunctional Silicones, Chem. Rev., 56, 1065 (1956).

90. Gibbins,S. G., The Reactions of Silane with Triethylzinkate Anion and Dibutylborate (I) Anion, Univ. Washington, Diss. Abstr., 16:446 (1956).

91. Gilman,H. and Schwebke G. L., Organic Substituted Cyclosilanes, in: Advances in Organometallic Chemistry, Vol. 1, ed. G. A. Stone and R. West, New York-London (1964), p. 89.

92. Gilman,H. and Winkler H. J. S., Organosilylmetallic Chemistry, in: Organometallic Chemistry, ed. H. Zeiss, New York (1960); American Chemical Society Monographs No. 147, p. 270.

* Here and elsewhere the number in parentheses refers to the literature reference in the given source.

93. Hallensleben,M. L., Herstellung und Polyadditionreaktionen des 1, 1'-bis-(dimethylhydrosilyl) ferrocens, Diss., Univ. Freiburg (1965).

94. Herzberg,G., Infrared and Raman spectra of Polyatomic Molecules, New York, (1945).

95. Hockele,G., Diss., München (1960); cited in [985], p. 377 (109).

96. Hunter,D. N., Inorganic Polymers, Oxford (1963).

97. Hurd,D. T., An Introduction to the Chemistry of the Hydrides, New York-London (1952).

98. Hussek,H., Ueber Heterosiloxane mit Elementen der IV. Gruppe, Diss., Univ. Marburg (1965).

99. Ingham,R. K. and Gilman,H., Organopolymers of Silicon, Germanium, Tin, and Lead; in: Inorganic Polymers, ed. F. G. Stone and W. A. G. Graham, New York-London (1962), p. 321.

100. Ingham,R. K., Rosenberg,S. D. and Gilman,H., Organotin Compounds, Chem. Rev., 60:459 (1960); Rijkens F., Organogermanium Compounds, (1960).

101. Jones,J. I., Polymetallosiloxanes, Part I. Introduction and Synthesis of Metallosiloxanes, in: Developments in Inorganic Polymer Chemistry, ed. M. F. Lappert and G. J. Leigh, Amsterdam-London-New York (1962), p. 162.

102. Jones,J. J., Polymetallosiloxanes, Part II. Polyorganometallosiloxanes and Polyorganosiloxymetalloxanes, in: Development in Inorganic Polymer Chemistry, ed. M. F. Lappert and G. J. Leigh, Amersterdam-London-New York, (1962), p. 200.

102a. Köster,R and Grassberger M. A., Strukturen und Synthesen von Carboranen, Angew. Chem., 79:197 (1967).

103. Krüerke,U., Diss. Univ. Müchen (1953); cited in [110], p. 79 (173), 221 (110).

104. Wu Kuan-li, Organosilicon Derivatives of Ferrocene, Hua Hsueh T'ung Pao (Chem. Bull., Pekin), No. 5. p. 293 (1963); cited in C. A., 59:10116 (1963).

105. Kuivila,H. G., Reactions of Organotin Hydrides with Organic Compounds, in: Advances in Organometallic Chemistry, Vol. 1, ed. F. G. A. Stone and R. West, New York-London (1964), p. 47.

106. Lichtenwalter,G. D., Organosilylmetallic Compounds and Derivatives, Iowa State Coll., Ames., Diss Abstr., 19:2234 (1959).

107. Mac Diarmid, A. G., Silyl Compounds, Quart. Rev., 10:208 (1956).

108. McCloskey,A. L., Boron Polymers; in: Inorganic Polymers, ed. F. G. A. Stone and W. A. G. Graham, New York-London (1962), p. 159.

109. McGregor, Silicones and Their Uses, New York-Toronto-London (1954).

110. Noll W., Chemie und Technologie der Silicone, Weinheim, (1960).

111. Onyszchuk,M., Thesis, Cambridge (1956); cited in [85], p. 130 (73).

112. Rochow,E.C., Hurd, D. T., and Lewis,R. N., The Chemistry of Organometallic Compounds, New York (1957).

113. Schindler,F., Koordinationsstabilisierte Derivate des Alumosiloxans, Diss., Univ. Marburg (1964); cited in [114].

114. Schmidbaur,H., Neue Ergebnisse der Heterosiloxane-Chemie, Angew. Chem., 77, 206 (1965).

114a. Seyferth,D., The Chemistry of Halomethyl-Mercury Compounds in: The R. A. Welch Foundation Conferences on chemical research. IX. Organometallic Compounds, Houston (1965), p. 89.

115. Shiihara,I., Schwartz, W. T., and Post,H. W., The Organic Chemistry of Titanium, Chem. Rev., 61, 1 (1961).
116. Simmler,W., Diss., Univ. München (1956); cited in [110], S. 139, 222, 223 (199).
117. Stone,F. G. A., Hydrogen Compounds of the Group IV Elements, London, (1962).
118. Swisher, J. V., Silicon-substituted Styrenes: Their Synthesis, Spectra and Reactions, Univ. Missouri, Columbia, Diss. Abstr., 21:2496 (1961).
119. Thayer, J. S., Preparation and Properties of Group IVA Organometallic Azides, Univ. Wisconsin, Diss. Abstr., 25:2222 (1964).
120. Thompson, J. M. C., B-containing Polymers without N in Chain, in: Developments in Inorganic Polymer Chemistry, ed. M. F. Lappert and G. I. Leigh, Amsterdam-London-New York (1962), p. 57.
121. Wannagat, U., The Chemistry of Silicon-Nitrogen Compounds, in: Advances in Inorganic and Radiochemistry, Vol. 6, ed. H. J. Emeleus and A. G. Sharpe, New York-London (1964), p. 225.
122. Wannagat, U., Synthesen mit N-metallierten Silicium-Stickstoff-Verbindungen, Angew. Chem., 76:234 (1964).
123. Wiberg, E., Stecher, O., Andrascheck, H.-J., Kreuzbicler, L., and Staude, E., Neues aus der Chemie der Metallsilyle $M(SiR_3)_n$, Angew. Chem., 75:516 (1963).
124. Wilkus, E. V., New Compounds and Polymers Containing Silicon and Ferrocene Linked through Carbon Bridges, Diss. Abstr., 25:114 (1964).
125. Wittenberg, D. and Gilman, H., Organosilylmetallic Compounds — Their Formation and Reactions, and Comparison with Related Types, Quart. Rev., 13:116 (1959).

2. Articles and Reports of Russian Authors*

126. Aver'yanov, S. V., Poddubnyi, I. Ya., Aver'yanova, L. A., and Trenke, Yu. V., Kauchuk i Rezina, 22(8):1 (1963).
127. Avilova, T. P., Bykov, V. T., and Zolota', G. Ya., Vysokmolekul. Soedin. 7:831 (1965).
127a. Avilova, T. P., Bykov, V. T., and Kondratenko, L. A., Vysokmolekul. Soedin. 8:14 (1966).
128. Alent'ev, O. O. and Lasskaya, E. A., Khimichna promislovost', 2:21 (1962).
129. Andreev, D. N. and Kukharskaya, É. V., Izv. Akad. Nauk, Otd. Tekhn. Nauk 1958:1397.
130. Andrianov, K. A., Eighth Mendeleev Congress, Abstracts of reports and communications, section on polymer chemistry and technology, Izd. Akad. Nauk SSSR (1959), p. 24.

* References 340-344, 349, and 350 refer to work by Czech scientists published in Russian. They should have been included among the references 909-923 and their placement in this section was accidental. However, because of the consecutive numbering of the references this error in the manuscript could not be rectified during correction of the book.

131. Andrianov, K. A., Dokl. Akad. Nauk SSSR, 140:1310 (1961).
132. Andrianov, K. A., Dokl. Akad. Nauk SSSR, 151:1093 (1963).
133. Andrianov, K. A., Transactions of the Conference "Production and Application of Organosilicon Compounds," coll. 1, MDNTP (1964), p. 3.
134. Andrianov, K. A., Plastmassy, No. 1, p. 50 (1959).
135. Andrianov, K. A., Plastmassy, No. 12, p. 23 (1960).
136. Andrianov, K. A., Proceedings of the Conference "Chemistry and Practical Application of Organosilicon Compounds," No. 2, TsBTI LSNKh (1958), p. 3.
137. Andrianov, K. A., Ibid., No. 6, Izd. AN SSSR (1961), p. 213.
138. Andrianov, K. A. and Asnovich, É. Z., Vysomolekul. Soedin., 1:743 (1959).
139. Andrianov, K. A. and Asnovich, É. Z., Vysomolekul. Soedin., 2:136 (1960).
140. Andrianov, K. A. and Asnovich, É. Z., Transactions of the Conference "Production and Application of Organosilicon Compounds," Coll. 2, MDNTP (1964), p. 29.
140a. Andrianov, K. A., Asnovich, É. Z., Bebchuk, T. S., Golybenko, M. A., and Lyakhova, A. B., Proceedings of the Conference "Organosilicon Compounds," No. 4, NIITÉKhIM, Moscow (1966), p. 98.
141. Andrianov, K. A. and Astakhin, V. V., Dokl. Akad. Nauk SSSR, 127:1014 (1959).
142. Andrianov, K. A., Astakhin, V. V., and Kochkin, D. A., Izv. Akad. Nauk, Otd. Tekhn. Nauk, 1962:1852.
143. Andrianov, K. A., Astakhin, V. V., Kochkin, D. A., and Sukhanova, I. V., Zh. Obsch. Khim., 31, 3410 (1961).
144. Andrianov, K. A., Astakhin, V. V., and Sukhanova, I. V., Zh. Obsch. Khim., 31:232 (1961).
145. Andrianov, K. A., Astakhin, V. V., and Sukhanova, I. V., Zh. Obsch. Khim., 32:1637 (1962).
146. Andrianov, K. A., Astakhin, V. V., and Sukhanova, I. V., Izv. Akad. Nauk, Otd. Tekhn. Nauk, 1962:1478.
147. Andrianov, K. A., Vasil'eva, T. V., and Korotkevich, S. Kh., Zh. Obsch. Khim., 32:2311 (1962).
148. Andrianov, K. A., Vasil'eva, T. V., Nudel'man, Z. N., Khananashvili, L. M., Kocheshkova, and A. S., Cherednikova, A. G., Zh. Obsch. Khim., 32:2307 (1962).
149. Andrianov, K. A. and Volkova, L. M., Izv. Akad. Nauk, Otd. Tekhn. Nauk, 1957:303.
150. Andrianov, K. A. and Ganina, T. N., Zh. Obsch. Khim., 29:605 (1959).
151. Andrianov, K. A. and Ganina, T. N., Izv. Akad. Nauk, Otd. Tekhn. Nauk, 1956:74.
152. Andrianov, K. A., Ganina, T. N., and Sokolov, N. N., Vysokmolekul. Soedin., 4:678 (1962).
153. Andrianov, K. A., Ganina, T. N., and Khrustaleva, E. N., Izv. Akad. Nauk, Otd. Tekhn. Nauk, 1956:798.
154. Andrianov, K. A., Gashnikova, N. P., and Asnovich, É. Z., Izv. Akad. Nauk, Otd. Tekhn. Nauk, 1960:957.
155. Andrianov, K. A., Golubkov, G. E., Elinek, V. I., Kurasheva, N. A., Manucharova, I. F., Litvinova, L. F., and Artem'ev, B. K., Vysokmolekul. Soedin., 7:680 (1965).

156. Andrianov, K. A., Golubkov, G. E., Elinek, V. I., Kurasheva, N. A., Manucharova, I. F., Litvinova, L. F., and Artem'ev, B. K., Vysokmolekul. Soedin., 7:688 (1965).
157. Andrianov, K. A. and Delazari, N. V., Dokl. Akad. Nauk SSSR, 122:393 (1958).
158. Andrianov, K. A. and Delazari, N. V., Izv. Akad. Nauk, Otd. Tekhn. Nauk, 1960:1712.
159. Andrianov, K. A. and Delazari, N. V., Izv. Akad. Nauk, Otd. Tekhn. Nauk, 1961:2169.
160. Andrianov, K. A., Dzenchel'skaya, S. I., and Petrashko, Yu. K., Plastmassy, No. 7, p. 20 (1961).
161. Andrianov, K. A. and Dulova, V. G., Izv. Akad. Nauk, Otd. Tekhn. Nauk, 1958:644.
162. Andrianov, K. A. and Ermakova, M. N., Vysokmolekul. Soedin., 5:217 (1963).
163. Andrianov, K. A. and Ermakova, M. N., Zh. Obsch. Khim., 31:1310 (1961).
164. Andrianov, K. A. and Zhdanov, A. A., Vysokmolekul. Soedin., 1:894 (1959).
165. Andrianov, K. A. and Zhdanov, A. A., Vysokmolekul. Soedin., 2:1071 (1960).
166. Andrianov, K. A. and Zhdanov, A. A., Dokl. Akad. Nauk SSSR, 114:1005 (1957).
167. Andrianov, K. A. and Zhdanov, A. A., Dokl. Akad. Nauk SSSR, 138:361 (1961).
168. Andrianov, K. A. and Zhdanov, A. A., Izv. Akad. Nauk, Otd. Tekhn. Nauk, 1958:779.
169. Andrianov, K. A. and Zhdanov, A. A., Izv. Akad. Nauk, Otd. Tekhn. Nauk, 1958:1076.
170. Andrianov, K. A. and Zhdanov, A. A., Izv. Akad. Nauk, Otd. Tekhn. Nauk, 1959:1590.
171. Andrianov, K. A. and Zhdanov, A. A., Izv. Akad. Nauk, Otd. Tekhn. Nauk, 1962:615.
172. Andrianov, K. A. and Zhdanov, A. A., Izv. Akad. Nauk, Otd. Tekhn. Nauk, 1962:837.
173. Andrianov, K. A. and Zhdanov, A. A., Plastmassy, No. 7, p. 24 (1962).
174. Andrianov, K. A. and Zhdanov, A. A., Abstracts of Reports to the International Symposium on High-Molecular Compounds, Prague (1957), p. 13.
175. Andrianov, K. A. Zhdanov, A. A., and Asnovich, É. Z., Izv. Akad. Nauk, SSSR, 118:1124 (1958).
176. Andrianov, K. A., Zhdanov, A. A., and Asnovich, É. Z., Izv. Akad. Nauk, Otd. Tekhn. Nauk, 1959:1760.
177. Andrianov, K. A., Zhdanov, A. A., and Volkova, L. M., Abstracts of Reports to the Ninth Conference on High-Molecular Compounds, Izd. Akad. Nauk SSSR (1956), p. 45.
178. Andrianov, K. A., Zhdanov, A. A., and Ganina, T. N., Soobshch, VKhO, No. 3, p. 2 (1955.
179. Andrianov, K. A., Zhdanov, A. A., and Kazakova, A. A., Izv. Akad. Nauk, Otd. Tekhn. Nauk, 1959:466.
180. Andrianov, K. A., Zhdanov, A. A., and Kashutina, É. A., Dokl. Akad. Nauk SSSR, 126:1261 (1959).
181. Andrianov, K. A., Zhdanov, A. A., and Kashutina, É. A., ZhPKH, 32:463 (1959).

182. Andrianov, K. A., Zhdanov, A. A., Kurasheva, N. A., and Dulova, V. G., Dokl. Akad. Nauk SSSR, 112:1050 (1957).
183. Andrianov, K. A., Zhdanov, A. A., and Pavlov, S. A., Dokl. Akad. Nauk SSSR, 102:85 (1955).
184. Andrianov, K. A., Zubkov, I. A., Semenova, V. A., and Mikhailov, S. I., Zh. Prikl. Khim., 32:883 (1959).
185. Andrianov, K. A. and Izmailov, B. A., Zh. Obsch. Khim., 35:333 (1965).
186. Andrianov, K. A. and Kuznetsova, I. K., Izv. Akad Nauk SSSR, Ser. Khim., 1964:651.
187. Andrianov, K. A. and Kuznetsova, I. K., Izv. Akad. Nauk SSSR, Ser. Khim., 1965:945.
188. Andrianov, K. A., Kuznetsova, I. K., and Smirnov, Yu. N., Izv. Akad. Nauk SSSR, Neorg. Mat., 1:301 (1965).
189. Andrianov, K. A., Kurakov, G. A., Kopylov, V. T., and Khananashvili, L. M., Zh. Obsch. Khim., 36:105 (1966).
190. Andrianov, K. A. and Kurasheva, N. A., Dokl. Akad. Nauk SSSR, 131:825 (1960).
191. Andrianov, K. A. and Kurasheva, N. A., Dokl. Akad. Nauk SSSR, 135:316 (1960).
192. Andrianov, K. A. and Kurasheva, N. A., Izv. Akad. Nauk, Otd. Tekhn. Nauk, 1962:1011.
193. Andrianov, K. A. and Kurasheva, N. A., Abstracts of Reports to the Conference "New Organosilicon Compounds," NIITÉ Khim (1966), p. 32; Proceedings of the Conference "New Organosilicon Compounds," No. 3, NIITÉ Khim (1967), p. 11.
194. Andrianov, K. A., Kurasheva, N. A., Manucharova, I. F., and Berliner, E.M., Izv. Akad. Nauk SSSR, Neorg. Mat., 1:294 (1965).
195. Andrianov, K. A. and Lavygin, I. A., Vysokmolekul. Soedin., 7:1000 (1965).
196. Andrianov, K. A. and Lavygin, I. A., Izv. Akad. Nauk, Otd. Tekhn. Nauk, 1963:1857.
197. Andrianov, K. A., Lavygin, I. A., and Shvetsov, Yu. A., Zh. Obsch. Khim., 35:689 (1965).
198. Andrianov, K. A. and Manucharova, I. F., Izv. Akad. Nauk, Otd. Tekhn. Nauk, 1962:420.
199. Andrianov, K. A. and Petrashko, A. I., Vysokmolekul. Soedin., 1:1514 (1959).
200. Andrianov, K. A. and Petrashko, A. I. Dokl. Akad. Nauk SSSR 131:561 (1960).
201. Andrianov, K. A. and Pichkhadze, Sh. V., Vysokmolekul. Soedin., 3:577 (1966).
202. Andrianov, K. A. and Pichkhadze, Sh. V., Vysokmolekul. Soedin., 4:839 (1962).
203. Andrianov, K. A. and Pichkhadze, Sh. V., Vysokmolekul. Soedin., 4:1011 (1962).
204. Andrianov, K. A., Pichkhadze, Sh. V., and Bochkareva, I. V. Vysokmolekul Soedin., 3:1321 (1961).
205. Andrianov, K. A., Pichkhadze, Sh. V., and Bochkareva I. V., Vysokmolekul. Soedin., 4:256 (1962).
206. Andrianov, K. A., Pichkhadze, Sh. V., and Komarova, V. V., Izv. Akad. Nauk, Otd. Tekhn. Nauk, 1962:261.
207. Andrianov, K. A., Pichkhadze, Sh. V., and Komarova, V. V., Izv. Akad. Nauk, Otd. Tekhn. Nauk, 1962:724.
208. Andrianov, K. A., Pichkhadze, Sh. V., Komarova, V. V., and Vardosanidze, Ts. N., Izv. Akad. Nauk, Otd. Tekhn. Nauk, 1962:833.

209. Andrianov, K. A., Pichkhadze, Sh. V., Novikov, V. M., and Lavygin, I. A., Izv. Akad. Nauk, Otd. Tekhn. Nauk, 1962:2138.
210. Andrianov, K. A. and Severnyi, V. V., Izv. Akad. Nauk, Otd. Tekhn. Nauk, 1962:2133.
211. Andrianov, K. A. and Severnyi, V. V., Izv. Akad. Nauk, Otd. Tekhn. Nauk, 1963:82.
212. Andrianov, K. A. and Severnyi, V. V., in: Synthesis and Properties of Monomers, "Nauka" (1964), p. 160.
213. Andrianov, K. A. and Sipyagina, M. A., Izv. Akad. Nauk, Otd. Tekhn. Nauk, 1962:1392.
214. Andrianov, K. A., Sipyagina, M. A., and Fridshtein, T. I., Izv. Akad. Nauk, Otd. Tekhn. Nauk, 1963:1672.
215. Andrianov, K. A., Slonimskii, G. L., Dikareva, T. A., and Asnovich, É. Z., Vysokmolekul. Soedin., 1:244 (1959).
216. Andrianov, K. A., Tikhonov, V. S., Khananashvili, L. M., En-tsieh Han, and Shu-yü Han, Plastmassy, No. 12, p. 25 (1962).
217. Andrianov, K. A. and Fromberg, M. B., Khim. prom., No. 1, p. 12 (1958).
218. Andrianov, K. A., Fromberg, M. B., Sorokina, L. I., and Kirilenko, É. I., Izv. Akad. Nauk, Otd. Tekhn. Nauk, 1962:78.
219. Andrianov, K. A., Khananashvili, L. M., Varlamov, A. V., and Tikhonov, V. S., Plastmassy, No. 3, p. 20 (1964).
220. Andrianov, K. A., Khananashvili, L. M., Teleshova, N. A., and Tikhonov, V. S., Izv. Akad Nauk SSSR, Ser. Khim., 1965:446.
221. Andrianov, K. A., Khananashvili, L. M., Tikhonov, V. S., Shu-yü Han and En-tsieh Han, Plastmassy, No. 1, p. 21 (1963).
222. Andrianov, K. A., Shugal, Ya. L., and Asnovich, É. Z., Plastmassy, No. 2, p. 44 (1964).
223. Andrianov, K. A. and Yakushkina, S. E., Vysokmolekul. Soedin., 2:1508 (1960).
224. Andrianov, K. A. and Yakushkina, S. E., Zh. Obsch. Khim., 35:330 (1965).
225. Andrianov, K. A., in: High Temperature Resistance and Thermal Degradation of Polymers, Soc. Chem. Ind. Monograph, No. 13, London (1961), p. 89.
226. Andrianov, K. A. and Zhdanov A. A., J. Prakt. Chem (4) No. 9, 75 (1959).
227. Andrianov, K. A., Ismailov, B. A., Kononov, A. M., and Kotrelev, G. V., J. Organomet. Chem., 3, 129 (1965).
228. Asnovich, É. Z. and Andrianov, K. A., Vysokmolekul. Soedin., 4:216 (1962).
229. Asnovich, É. Z., Andrianov, K. A., Golubenko, M. A., and Lyakhova, A. B., Abstracts of Reports to the Conference "New Organosilicon Compounds," NIITÉKhim (1966), p. 33.
230. Babaré, L. V., Petrovskii, P. V., and Fedin, É. I., Zh. Strukt. Khim., 6:783 (1965).
231. Baranovskaya, N. B., Berlin, A. A., Zakharova, M. Z., and Mizikin, A. I., Proceedings of the Conference "Chemistry and Practical Application of Organosilicon Compounds," No. 2, TsBTILSNKh (1958), p. 88.
232. Baranovskaya, N. B., Berlin, A. A., Zakharova, M. Z., and Mizikin, A. I., Proceedings of the Conference "Chemistry and Practical Application of Organosilicon Compounds," No. 6, Izd. Akad. Nauk SSSR, (1961), p. 208.

233. Baranovskaya, N. B., Zakharova, M. Z., Mizikin, A. I., and Berlin, A. A., Dokl. Akad. Nauk SSSR, 122:603 (1958).
234. Baukov, Yu. I., Burlachenko, G. S., and Lutsenko, I. F., Zh. Obsch. Khim., 35:757 (1965).
235. Berdichevskaya, K. M., Chugunov, V. S., and Petrov, A. D., Dokl. Akad. Nauk SSSR, 151:1319 (1963).
236. Borisov, S. N., Proceedings of the Conference "Chemistry and Practical Application of Organosilicon Compounds," No. 6, Izd. Akad Nauk SSSR (1961), p. 161.
237. Borisov, S. N., Karlin, A. V., and Sviridova, N. G., Zh. Prikl. Khim., 35:917 (1962).
238. Borisov, S. N., Kurlova, T. V., and Sidorovich, E. A., Scientific Communications of the International Symposium on Organosilicon Chemistry, A/33-158, Prague (1965).
239. Borisov, S. N. and Sviridova, N. G., Vysokmolekul. Soedin., 3:50 (1961).
240. Bruker, A. B., Balashova, L. D., and Soborovskii, L. Z., Dokl. Akad. Nauk SSSR, 135:843 (1960).
241. Vasil'eva, T. V. and Andrianov, K. A., Abstracts of Reports to the Conference "New Organosilicon Compounds," NIITÉKhim (1966), p. 33.
242. Volkova, L. M. and Andrianov, K. A., Abstracts of reports to the Scientific and Technical Conference of MITKhT (1964), p. 54.
243. Vol'nov, Yu. N., Zh. Fiz. Khim., 29:1646 (1955).
244. Voronkov, M. G., Zh. Obsch. Khim., 29:908 (1959).
245. Voronkov, M. ., Abstracts of reports to the Scientific Session, section on chemical sciences, Iz. LGU(1954), p. 19.
246. Voronkov, M. G., Proceedings of the Conference "Chemistry and Practical Application of Organosilicon Compounds," No. 6, Izd. Akad Nauk SSSR (1961), p. 136.
247. Voronkov, M. G., Davydova, V. P., and Grishanina, N. P., Zh. Prikl. Khim., 32:1106 (1959).
248. Voronkov, M. G. and Deich, A. Ya., Zh. Strukt. Khim., 5:482 (1964).
249. Voronkov, M. G. and Deich, A. Ya., Izv. Akad. Nauk LatvSSR, Ser. Khim., 1964:145.
250. Voronkov, M. G., Dolgov, B. N., and Dmitrieva, N. A., Dokl. Akad. Nauk SSSR, 84:959 (1952).
251. Voronkov, M. G. and Zgonnik, V. N., Zh. Obsch. Khim., 27:1476 (1957).
252. Voronkov, M. G. and Lipshits, T. S., Zh. Prikl. Khim., 36:152 (1963).
253. Voronkov, M. G. and Orlov, N. F., Izv. Akad. Nauk LatvSSR, Ser. Khim., 1961:93.
254. Voronkov, M. G. and Skorik, Yu. I., Izv. Akad. Nauk SSSR, Ser. Khim., 1964:1215.
255. Voronkov, M. G. and Khudobin, Yu. I., Zh. Obsch. Khim., 26:584 (1956).
256. Voronkov, M. G. and Khudobin, Yu. I., Izv. Akad. Nauk, Otd. Tekhn Nauk, 1956:713.
257. Voronkov, M. G. and Chudesova, L. M., Zh. Obsch. Khim., 29:1534 (1959).
258. Voronkov, M. G. and Shemyatenkova, V. T., Izv. Akad. Nauk, Otd. Tekhn Nauk, 1961:178.

259. Voronkov, M. G. and Shorokhov, N. V., Stroit. Materialy, No. 7, p. 12, (1959).
259a. Vyazankin, N. S., Bochkarev, M. N., and Sanina, L. P., Zh. Obsch. Khim., 36:1154 (1966).
259b. Vyazankin, N. S., Bochkarev, M. N., and Sanina, L. P., Zh. Obsch. Khim., 36:1961 (1966).
259c. Vyazankin, N. S., Kruglaya, O. A., Razuvaev, G. A., and Semchikova, G. S., Dokl. Akad. Nauk SSSR, 166:99 (1966).
260. Vyazankin, N. S., Mitrofanova, E. D., Kruglaya, O. A., and Razuvaev, G. A., Zh. Obsch. Khim., 36:160 (1966).
261. Vyazankin, N. S., Razuvaev, G. A., and Bychkov, V. T., Zh. Obsch. Khim., 35:395 (1965).
262. Vyazankin, N. S., Razuvaev, G. A., and Gladyshev, E. N., Dokl. Akad. Nauk SSSR, 155:830 (1964).
263. Vyazankin, N. S., Razuvaev, G. A., and Gladyshev, E. N., Dokl. Akad. Nauk SSSR, 155:1108 (1964).
264. Vyazankin, N. S., Razuvaev, G. A., Korneva, S. P., Kruglaya, O. A., and Galiulina, R. F., Dokl. Akad. Nauk SSSR, 158:884 (1964).
264a. Vyazankin, N. S., Razuvaev, G. A., Kruglaya, O. A., and Semchikova, G. S., J. Organomet. Chem., 6:474 (1966).
265. Golodnikov, G. V., Dolgov, B. N., and Sedova, V. F., Zh. Obsch. Khim., 30:3352 (1960).
266. Golubkov, G. E. and Kolganova, V. A., Plastmassy, No. 1, p. 24 (1964).
267. Grinevich, K. P., Plastmassy, No. 3, p. 24 (1960).
268. Grinevich, K. P., Zubkov, I. A., and Odishariya, S. N., Plastmassy, No. 1, p. 21 (1961).
269. Gruber, V. N., Klebanskii, A. L., Degteva, T. G., Kuz'minskii, A. S., Mikhailova, T. A., and Kuz'mina, E. V., Vysokmolekul. Soedin., 7:462 (1965).
270. Gruber, V. N., Klebanskii, A. L., and Panchenko, B. I., Izv. Akad. Nauk LatvSSR, Ser. Khim., 1963:95.
271. Guan'-li, U, Sokolova, E. B., Leites, L. A., and Petrov, A. D., Izv. Akad. Nauk, Otd. Tekhn. Nauk, 1962:887.
272. Darashkevich, M. L., Abstracts of Reports to the Scientific and Technical Conference of the Moscow Institute of Chemical Technology (1957), p. 67.
273. Deich, A. Ya. and Voronkov, M. G., Izv. Akad Nauk LatvSSR, Ser. Khim., 1963:417.
274. Dolgov, B. N., Khudobin, Yu. I., and Kharitonov, N. P., Dokl. Akad. Nauk SSSR, 122:607 (1958).
275. Dzhenchel'skaya, S. I., Andrianov, K. A., and Petrashko, Yu. K., Proceedings of the Conference "Chemistry and Practical Application of Organosilicon Compounds," No. 2, TsBTI LSNKh (1958), p. 45.
276. Dolgov, B. N., Glushkova, N. E., and Kharitonov, N. P., Izv. Akad. Nauk, Otd. Tekhn. Nauk, 1960:351.
277. Dolgov, B. N. and Orlov, N. F., Dokl. Akad. Nauk SSSR, 117:617 (1957).
278. Dolgov, B. N. and Orlov, N. F., Izv. Akad. Nauk, Otd. Tekhn. Nauk, 1957:1396.
279. Dolgov, B. N., Orlov, N. F., and Voronkov, M. G., Izv. Akad. Nauk, Otd. Tekhn. Nauk, 1959:1408.

280. Dolgov, B. N. and Chugunov, V. S., Vestn. LGU, No. 16, 89 (1958).
281. Drozdov, G. V., Klebanskii, A. L., and Bartashev, V. A., Zh. Obsch. Khim., 32:2390 (1962).
282. D'yakonov, I. A., Repinskaya, I. B. and Golodnikov, G. V., Zh. Obsch. Khim., 35:199 (1965).
282a. Egorochkin, A. N., Vyazankin, N. S., Razuvaev, G. A., Kruglaya, O. A., and Bochkarev, M. N., Dokl. Akad. Nauk SSSR, 170:333 (1966).
283. Egorochkin, A. N., Khidekel', M. A., Razuvaev, G. A., Mironov, V. F., and Kravchenko, A. L., Izv. Akad. Nauk SSSR, Ser. Khim., 1964:1312.
284. Zhdanov, A. A., Eighth Mendeleev Congress, Abstracts of Reports and Communications, Section on Polymer Chem. and Tech., Izd. Akad. Nauk SSSR (1959), p. 28.
285. Zhdanov, A. A. and Andrianov, K. A., in: Heterochain High-Molecular Compounds, "Nauka" (1964), p. 45.
286. Zhdanov, A. A. and Andrianov, K. A., Proceedings of the Conference "Chemistry and Practical Application of Organosilicon Compounds," No. 2, TsBTI LSNKh (1958), p. 100; German translation: Plaste u. Kaut. 6:537 (1959).
287. Zhdanov, A. A. and Andrianov, K. A., Proceedings of the Conference "Chemistry and Practical Application of Organosilicon Compounds," No. 6, Izd. Akad. Nauk SSSR (1961), p. 220.
288. Zhdanov, A. A., Andrianov, K. A., and Bogdanova, A. A., Izv. Akad. Nauk, Otd. Tekhn. Nauk, 1961:1261.
289. A. A. Zhdanov, K. A. Andrianov, and É. A. Kashutina, Dokl. Akad. Nauk SSSR, 171:103 (1966).
290. A. A. Zhdanov, K. A. Andrianov, and É. A. Kashutina, Abstracts of Reports to the Conference "New Organosilicon Compounds," NIITÉKhIM (1966), p. 32; Proceedings of the Conference "Organosilicon Compounds," No. 3, NIITÉKhIM, Moscow (1967), p. 73.
291. Zhdanov, A. A. and Andrianov, K. A., J. Polymer. Sci., 55 (161):89 (1961).
292. Zherdev, Yu. V. Korolev, A. Ya., and Leznov, N. S., Transactions of the Conference "Production and Application of Organosilicon Compounds," coll. 2, MDNTP (1964), p. 135.
293. Zherdev, Yu. V., Korolev, A. Ya., and Leznov, N. S., Plastmassy, No. 10, p. 16 (1964).
294. Zhinkin, D. Ya., Korneeva, G. K., Korneev, N. N., and Sobolevskii, M. V., Scientific Communications of the International Symposium on Organosilicon Chemistry, V/23-311, Prague (1965).
295. Zhinkin, D. Ya., Mal'nova, G. N., and Gorislavskaya, Zh. V., Plastmassy, No. 12, p. 17 (1965).
296. Zagorovskaya, A. A. and Kreshkov, A. P., Abstracts of Reports to the Scientific and Technical Conference of Graduate Students and Assistant Lecturers of the Moscow Institute of Chemical Technology, (1960), p. 64.
297. Zagorovskaya, A. A. and Kreshkov, A. P., Abstracts of Reports to the Scientific and Technical Conference of Graduate Students and Assistant Lecturers of the Moscow Institute of Chemical Technology (1961), p. 9.

298. Zasosov, V. A. and Kocheshkov, K. A., Collection of Articles on General Chemistry, Vol. I, Izd. Akad. Nauk SSSR (1953), p. 278.

299. Zakharkin, L. I., Bregadze, V. I., and Okhlobystin, O. Yu., Izv. Akad. Nauk SSSR, Ser. Khim., 1964:1539.

300. Zakharkin, L. I. and Savina, L. I., Izv. Akad. Nauk, Otd. Tekhn. Nauk, 1962:253.

301. Zubkova, N. D., Turskii, Yu. I., Dintses, A. I. Genkin, V. I., Bystrova, T. G., and Volchinskaya, N. I., Abstracts of Reports to the Conference "New Organosilicon Compounds," NIITÉKhim (1966), p. 46; Zubkova, N. D., Dintes. A. I., Turskii, Yu.I.,Genkina, V. I., Bystrova, T. G., and Volchinskaya, N. I., Proceedings of the Conference "Organosilicon Compounds," No. 3, NIITÉKhim, Moscow (1967), p. 23.

302. Kalinina, S. P., Eighth Mendeleev Congress, Abstracts of Reports and Communications, Section of Polymer Chem. and Tech., Izd. Akad. Nauk SSSR (1959), p. 30.

303. Kalugin, N. V. and Voronkov, M. G., Zh. Prikl. Khim., 31:1390 (1958).

304. Kalugin, N. V. and Voronkov, M. G., Proceedings of the Conference "Chemistry and Practical Application of Organosilicon Compounds," No. 4, TsBTI LSNKh (1958), p. 54.

305. Karlin, A. V. and Mitrofanov, L. A., Khim. Prom., No. 3, p. 6 (1963).

306. Kartsev, G. N., Syrkin, Ya. K., Kravchenko, A. L., and Mironov, V. F., Zh. Strukt. Khim., 5:492 (1964).

307. Klebanskii, A. L., Ponomarev, A. I., Kudina, V. I., Khim. Nauka i Promy., 3:285 (1958).

308. Kozlova, N. V., Otchet VNIISK, No. 1729 (1962).

309. Kolodyazhnyi, Yu. V., Marchenko, V. N., Osipov, O. A., and Kogan, M. G., Zh. Obsch. Khim., 36:1693 (1966).

309a. Kolodyazhnyi, Yu. V., Osipov, O. A., and Kashireninov, O. E., Zh. Fiz. Khim., 39:1771 (1965).

309b. Kolodyazhnyi, Yu. V., Osipov, O. A., and Chenskaya, T. B., Zh. Obsch. Khim., 36:2189 (1966).

310. Komarov, N. V., Maroshin, Yu. V., Lebedeva, A. D., and Astaf'eva, L. N., Izv. Akad. Nauk, Otd. Tekhn. Nauk, 1963:97.

311. Komarov, N. V., Shostakovskii, M. F., and Astaf'eva, L. N., Zh. Obsch. Khim., 31:2100 (1961).

312. Komarov, N. V., Yarosh, O. G., and Astaf'eva, L. N., Scientific Communications of the International Symposium on Organosilicon Chemistry, Supplement, A/27-6, Prague (1965).

313. Konstantinov, P. A. and Shupik, R. I., Zh. Obsch. Khim., 33:1251 (1963).

314. Kopnova, N. L., Chugunov, V. S. and Klebanskii, A. L., Zh. Obsch. Khim., 34:1355 (1964).

315. Korshak, V. V., Zamyatina, V. A., Chursina, L. M., and Bekasova, N. I., Vysokmolekul. Soedin., 5:1127 (1963).

316. Korshak, V. V., Polyakova, A. M., Vdovin, V. M., Mironov, V. F., and Petrov, A. D., Dokl. Akad. Nauk SSSR, 128:960 (1959).

317. Korshak, V. V. Sultanov, A. S., Abduvaliev, A. A., Uzbeksk. Khim., Zh., No. 4, p. 39 (1959).

318. Kreshkov, A. P., Zh. Prikl. Khim, 23:545 (1950).

319. Kreshkov, A. P. and Vil'borg, S. S., Zh. Analit. Khim., 3:172 (1948).

320. Kreshkov, A. P. and Vil'borg, S. S., Tr. Mosk. Khim. Technol. Inst., 12, 40 (1947).

321. Kreshkov, A. P. and Darashkevich, M. L., Tr. Mosk. Khim. Technol. Inst., 19:3 (1954).

322. Kreshkov, A. P. and Darashkevich, M. L., Tr. Mosk. Khim. Technol. Inst., 24:327 (1957).

323. Kreshkov, A. P., Drozdov, V. A., and Tarasyants, R. R., Plastmassy, No. 7, p. 58 (1963).

324. Kreshkov, A. P., Karateev, D. A., and Fyurst, V., Zh. Obsch. Khim., 31:2139 (1961).

325. Kreshkov, A. P., Karateev, D. A., and Fyurst, V., Zh. Prikl. Khim., 34:2711 (1961).

326. Kreshkov, A. P., Karateev, D. A., and Fyurst, V., Plastmassy, No. 3, p. 63 (1962).

326a. Kreshkov, A. P. and Kuchkarev, E. A., Trudy Komis. po Analit. Khim., 13:159 (1963).

327. Kreshkov, A. P., Mikhailenko, V. A., Myshlyaeva, L. V., and Khananashvili, L. M. Zh. Prikl. Khim., 31:1746 (1958).

328. Kreshkov, A. P. and Myshlyaeva, L. V., Tr. Mosk. Khim. Technol. Inst., 13:38 (1948).

329. Kreshkov, A. P., Myshlyaeva, L. V., and Soboleva, D. A., Dokl. Akad. Nauk SSSR, 148:843 (1963).

330. Kreshkov, A. P., Myshlyaeva, and L. B., Soboleva, D. A., Zh. Obsch. Khim., 32:2190 (1962).

331. Kreshkov, A. P., Myshlyaeva, and L. V., Soboleva, D. A., Zh. Prikl. Khim., 37:2278 (1964).

332. Kreshkov, A. P., Myshlyaeva, L. V., and Khananashvili, L. M., Tr. Mosk. Khim. Technol. Inst., 24:333 (1957).

333. Kreshkov, A. P. and Chivikova, A. N., Zh. Prikl. Khim., 27:1128 (1954).

334. Kreshkov, A. P., Chivikova, A. N., and Zagorovskaya, A. A., Zh. Obsch. Khim., 32:3862 (1962).

335. Kreshkov, A. P., Chivikova, A. N., Nessonova, G. D., Matveev, V. A., and Darashkevich, M. L., Tr. Mosk. Khim. Technol. Inst., 17:16 (1952).

335a. Kruglaya, O. A., Vyazankin, N. S., Razuvaev, G. A., and Mitrofanova, E. V., Dokl. Akad. Nauk SSSR, 173:834 (1967).

336. Kudryavtsev, R. V., Kursanov, D. N., and Andrianov, K. A., Zh. Obsch. Khim., 29:1497 (1959).

337. Kukharskaya, É. V., Andreev, D. N., and Kolesova, V. A., Izv. Akad. Nauk, Otd. Tekhn. Nauk, 1958:1372.

338. Kukharskaya, É. V., Dolgov, B. N., Andreev, D. N., and Lyutyi, V. P., Proceedings of the Conference "Chemistry and Practical Application of Organosilicon Compounds," No. 1, TsBTI LSNKh (1958), p. 138.

339. Kukharskaya, É. V., Skorik, Yu. I., and Boiko, N. G., Dokl. Akad. Nauk SSSR, 148:350 (1963).

340. Kuchera, M., Vysokmolekul. Soedin., 5:938 (1963.
341. Kuchera, M., Proceedings of the Conference "Chemistry and Practical Application of Organosilicon Compounds," No. 2, TsBTI LSNKh (1958), p. 71.
342. Kuchera, M., Coll., 25, 547 (1960).
343. Kuchera, M., Elinek, M., Vysokmolekul. Soedin., 2 1860 (1960).
344. Kuchera, M. and Elinek, M., Coll., 25:536 (1960).
345. Lagzdin', É. A. and Vaivad, A. Ya., Izv. Akad. Nauk LatvSSR, Ser. Khim., 1965:21.
346. Lagzdin', É. A., Vaivad, A. Ya., and Mai, L. A., Izv. Akad. Nauk LatvSSR, Ser. Khim., 1963:641.
347. Lagzdin', É. A., Mai, L. A., and Vaivad, A. Ya., Transactions of the Conference "Production and Application of Organosilicon Compounds," Coll. 1, MDNTP (1964), p. 138.
348. Lazarev, A. N., Tenisheva, T. F., and Davydova, V. P., Dokl. Akad. Nauk SSSR, 158:648 (1964).
349. Laita, Z. and Elinek, M., Vysokmolekul. Soedin., 5:1268 (1963).
350. Laita, Z. and Elinek, M., Scientific Communications of the International Symposium on Organosilicon Chemistry, A/10-52, Prague (1965).
351. Leznov, N. S., Sabun, L. A., and Andrianov, K. A., Zh. Obsch. Khim., 29:1276 (1959).
352. Leibovich, Kh. M. and Kapkin, M. M., Preceedings of the Conference "Chemistry and Practical Application of Organosilicon Compounds," No. 4, TsBTI LSNKh (1958), p. 107.
353. Lobkov, V. D., Klebanskii, A. L., and Kogan, É. V., Izv. Akad. Nauk LatvSSR, Ser. Khim., 1965:114.
354. Lobkov, V. D., Klebanskii, A. L., and Kogan, É. V., Scientific Communications of the International Symposium on Organosilicon Chemistry, A/36-178, Prague (1965).
355. Lukevits, É. Ya. and Voronkov, M. G., Khim. Geterosikl. Soedin., 1965:31.
356. Lukevits, É. Ya. and Voronkov, M. G., Khim. Geterosikl. Soedin., 1965:36.
357. Lukevits, É. Ya. and Voronkov, M. G., Khim. Geterosikl. Soedin., 1965:179.
358. Lukevits, É. and Giller, S., Izv. Akad. Nauk LatvSSR, No. 4 (165), 95 (1961).
359. Lukevits, É. and Giller, S., Izv. Akad. Nauk LatvSSR, No. 4 (165), 99 (1961).
360. Luneva, L. N., Sladkov, A. L., and Korshak, V. V., Vysokmolekul. Soedin., 7:427 (1965).
361. Mal'nova, G. N., Proceedings of the Conference "Chemistry and Practical Application of Organosilicon Compounds," No. 6, Izd. Akad. Nauk SSSR (1961), p. 87.
362. Mal'nova, G. N. and Mikheev, E. P., Plastmassy, No. 8, p. 20 (1962).
363. Mal'nova, G. N., Mikheev, E. P., Klebanskii, A. L., and Filimonova, N. P., Dokl. Akad. Nauk SSSR, 123:693 (1958).
364. Mal'nova, G. N., Mikheev, E. P., Klebanskii, A. L., Golubtsov, S. A., and Filimonova, N. P., Proceedings of the Conference "Chemistry and Practical Application of Organosilicon Compounds," No. 1, TsBTI LSNKh (1958), p. 85.
365. Maminov, E. K., Ibid., No. 4 (1958), p. 65.
366. Maminov, E. K. and Voronkov, M. G., Zh. Prikl. Khim., 30:974 (1957).

367. Mironov, V. F., Proceedings of the Conference "Chemistry and Practical Application of Organosilicon Compounds," No. 1, TsBTI LSNKh (1958), p. 129.
368. Mironov, V. F. and Gar, T. K., Izv. Akad. Nauk SSSR, Ser. Khim., 1965:291.
369. Mironov, V. F. and Dzhurinskaya, N. S., Izv. Akad. Nauk, Otd. Tekhn. Nauk, 1963:75.
370. Mironov, V. F., Dzhurinskaya, N. G., Gar, T. K., and Petrov, A. D., Izv. Akad. Nauk, Otd. Tekhn. Nauk, 1962:460.
371. Mironov, V. F. and Kravchenko, A. L., Izv. Akad. Nauk, Otd. Tekhn. Nauk, 1963:1563.
372. Mironov, V. F. and Kravchenko, A. L., Izv. Akad. Nauk SSSR, Ser. Khim., 1964:768.
373. Mironov, V. F. and Nikishin, G. I., Proceedings of the Conference "Chemistry and Practical Application of Organosilicon Compounds," No. 3, TsBTI LSNKh (1958), p. 72.
374. Mironov, V. F., Petrov, A. D., and Maksimova, N. G., Izv. Akad. Nauk, Otd. Tekhn. Nauk., 1959:1954.
375. Mironov, V. F. and Pogonkina, N. A., Izv. Akad. Nauk, Otd. Tekhn. Nauk, 1955:182.
376. Mikhailov, B. M., Aronovich, P. M., and Tarasova, L. V., Zh. Obsch. Khim., 30:3624 (1960).
377. Mikhailov, B. M. and Blokhina, A. N., Zh. Obsch. Khim., 30:3615 (1960).
378. Mikhailov, B. M. and Shchegoleva, T. A., Izv. Akad. Nauk, Otd. Tekhn. Khim., 1959:546.
379. Mikhailov, B. M., Shchegoleva, T. A., and Blokhina, A. N., Izv. Akad. Nauk, Otd. Tekhn. Khim., 1960:1307.
380. Mikhal'chenko, V. A., Zh. Prikl. Khim., 25:803 (1952).
381. Mikheev, E. P., Klebanskii, A. L., Mal'nova, G. N., and Popkov, K. K., Plastmassy, No. 1, p. 19 (1961).
382. Mikheev, E. P. and Mal'nova, G. N., Plastmassy, No. 2, p. 29 (1961).
383. Mikheev, E. P. and Mal'nova, G. N., Plastmassy, No. 9, p. 22 (1963).
384. Mikheev, E. P., Mal'nova, G. N., Klebanskii, A. L., Golubtsov, S. A., and Filimonova, N. P., Dokl. Akad. Nauk SSSR, 117:623 (1957).
384a. Molchanov, B. V. and Borisov, M. F., Plastmassy, No. 11, p. 22 (1964).
385. Myshlyaeva, L. V. and Khananashvili, L. M., Abstracts of Reports to the Scientific and Technical Conference of the Moscow Institute of Chemical Technology, (1956), p. 33.
386. Nametkin, N. S., Vdovin, V. M., and Babich, E. D., Scientific Communications of the International Symposium on Organosilicon Chemistry, B/20-296, Prague (1965).
387. Nametkin, N. S., Vdovin, V. M., Babich, É. D., and Oppengeim, V. D., Khim. Geterotsikl. Soedin., 1965:455.
388. Nametkin, N. S., Topchiev, A. V., and Machus, F. F., Dokl. Akad. Nauk SSSR, 83:705 (1952).
389. Nametkin, N. S., Topchiev, A. V., and Machus, F. F., Dokl. Akad. Nauk SSSR, 87:233 (1952).

390. Nametkin, N. S., Topchiev, A. V., and Machus, F. F., Dokl. Akad. Nauk SSSR, 93:495 (1953).
391. Nametkin, N. S., Topchiev, A. V., Chernysheva, T. I., and Kartasheva, L. I., Izv. Akad. Nauk, Otd. Tekhn. Nauk, 1963:654.
392. Nametkin, N. S., Chernysheva, and T. I., Babaré, L. V., Zh. Obsch. Khim., 34:2258 (1964).
393. Nametkin, N. S., Chernysheva, T. I., Pritula, N. A., Kartasheva, L. I., and Babaré, L. V., Abstracts of Reports to the Conference "New Organosilicon Compounds," NIITÉKhim. (1966), p. 30; Proceedings of the Conference "Organosilicon Compounds," No. 3, NIITÉKhIM, Moscow (1967) p. 130.
394. Nesmeyanov, A. N. and Nogina, O. V., Dokl. Akad. Nauk SSSR, 117:249 (1957).
395. Nesmeyanov, A. N., Nogina, O. V., Berlin, A. M., and Kudryavtsev, Yu. P., Izv. Akad. Nauk, Otd. Tekhn. Nauk, 1960:1206.
395a. Nesmeyanov, A. N., Perevalova, É. G., and Ustynyuk, Yu. A., Dokl. Akad. Nauk SSSR, 133:1105 (1960).
396. Nesmeyanov, A. N., Freidlina, R. Kh., Chukovskaya, E. C., Petrova R. G., and Belyavsky, A. B., Tetrahedron, 17:61 (1962).
397. Nessonova, G. D., Pogosyants, E. K., Markova, G. B., and Grinevich, K. P., Plastmassy, No. 1, p. 20 (1962).
398. Nisel'son, L. A. and Voitovich, B. A., Zh. Neorgan. Khim., 7:360 (1962).
399. Nisel'son, L. A. and Petrusevich, I. V., Zh. Neorgan. Khim., 6:748 (1961).
400. Novikov, A. S., Kaluzhenina, K. F., and Nudel'man, Z. N., Kauchuk i Rezina, 18(5):16 (1959).
401. Novikov, A. S., and Nudel'man, Z. N., Kauchuk i Rezina, 19(12):3 (1960).
402. Nogina, O. V., Nesmeyanov, A. N., and Kudryavtsev, Yu. V., Eighth Mendeleev Congress, Abstracts of Reports and Communications, Section on Organ. Chem. and Tech., [in Russian], Izd. Akad. Nauk SSSR (1959), p. 363.
403. Nudel'man, Z. N., Proceedings of the Conference "Chemistry and Practical Application of Organosilicon Compounds," No. 6, Izd. Akad. Nauk SSSR (1961), p. 210.
404. Nudel'man, Z. N., Andrianov, K. A., and Kudryavitskaya, G. B., Vysokmolekul. Soedin., 4:440 (1962).
405. Nudel'man, Z. N. and Novikov, A. S., Kauchuk i Rezina, 19(5):17 (1960).
406. Nudel'man, Z. N., Novikov, A. S., and Kaluzhenina, K. F., Eighth Mendeleev Congress, Abstracts of Reports and Communications, Section of Polymer Chem. and Tech., Izd. Akad. Nauk SSSR (1959), p. 26.
407. Nudel'man, Z. N., Sviridova, A. V., and Novikov, A. S., Vysokmolekul. Soedin., 3:833 (1961).
408. Orlov, N. F., Dokl. Akad. Nauk SSSR, 114:1033 (1957).
409. Orlov, N. F., Dolgov, B. N., and Voronkov, M. G., Dokl. Akad. Nauk SSSR, 122:246 (1958).
410. Orlov, N. F. and Voronkov, M. G., Zh. Obsch. Khim., 36:347 (1966).
411. Orlov, N. F. and Voronkov, M. G., Izv. Akad. Nauk, Otd. Tekhn. Nauk, 1959:1506.
412. Orlov, N. F., Dolgov, B. N., and Voronkov, M. G., Izv. Akad. Nauk, Otd. Tekhn. Nauk, 1960:1607.

413. Orlov, N. F., Dolgov, B. N., and Voronkov, M. G., Proceedings of the Conference "Chemistry and Practical Application of Organosilicon Compounds," No. 1, TsBTI LSNKh (1958), p. 161.
414. Orlov, N. F., Dolgov, B. N., and Voronkov, M. G., Ibid., p. 172.
415. Orlov, N. F., Dolgov, B. N., and Voronkov, M. G., Ibid., No. 6, Izd. Akad. Nauk SSSR (1961), p. 123.
416. Petrashko, A. I., Transactions of the Conference "Production and Application of Organosilicon Compounds," Coll. 2, MDNTP (1964), p. 94.
417. Petrashko, A. I. and Andrianov, K. A., Vysokmolekul. Soedin., 4:221 (1962).
418. Petrashko, A. I. and Andrianov, K. A., Vysokmolekul. Soedin., 6:1505 (1964).
419. Petrashko, A. I. and Andrianov, K. A., Plastmassy, No. 5, p. 17 (1964).
420. Petrashko, A. I. and Andrianov, K. A., Plastmassy, No. 11, p. 26 (1964).
421. Petrashko, A. I., Zhdanov, A. A., and Andrianov, K. A., Izv. Akad. Nauk, Otd. Tekhn. Nauk, 1964:1276.
422. Petrov, A. A., Kormer, V. A., and Stadnichuk, M. D., Zh. Obsch. Khim., 30:2243 (1960).
423. Petrov, A. A., Kormer, V. A., and Stadnichuk, M. D., Zh. Obsch. Khim., 31:1135 (1961).
424. Petrov, A. D. and Mironov, V. F., Dokl. Akad. Nauk SSSR, 75:707 (1950).
425. Petrov, A. D. and Mironov, V. F., Izv. Akad. Nauk, Otd. Tekhn. Nauk, 1952:635.
426. Petrov, A. D., Mironov, V. F., and Dzhurinskaya, N. G., Dokl. Akad. Nauk SSSR, 128:302 (1959).
427. Petrov, A. D. and Nikishin, G. I., Izv. Akad. Nauk, Otd. Tekhn. Nauk, 1952:1128.
428. Petrov, A. D., Ponomarenko, V. A., and Voikov, V. I., Izv. Akad. Nauk, Otd. Tekhn. Nauk, 1954:504.
429. Petrov, A. D., Ponomarenko, V. A., and Snegova, A. D., Dokl. Akad. Nauk SSSR, 112:79 (1957).
430. Petrov, A. D., Sadykh-zade, S. I., and Vdovin, V. M., Dokl. Akad. Nauk SSSR, 100:711 (1955).
431. Petrov, A. D. Sadykh-zade, S. I., and Egorov, Yu. P., Izv. Akad. Nauk, Otd. Tekhn. Nauk, 1954:722.
432. Petrov, A. D., Sokolova, E. B., and Bakunchik, G. P., Dokl. Akad. Nauk SSSR, 148:598 (1963).
433. Petrov, A. D., Sokolova, E. B., Shebanova, M. P., and Golovina, N. I., Dokl. Akad. Nauk SSSR, 152:1118 (1963).
434. Petrov, A. D., Chernyshev, E. A., and Tolstikova, N. G., Dokl. Akad. Nauk SSSR, 118:957 (1958).
435. Petrov, A. D. and Chernysheva, T. I., Dokl. Akad. Nauk SSSR, 84:515 (1952).
436. Petrov, A. D. and Chernysheva, T. I., Dokl. Akad. Nauk SSSR, 89:73 (1953).
437. Petrov, A. D. and Shchukovskaya, L. L. , Zh. Obsch. Khim., 25:1128 (1955).
438. Poddubnyi, I. Ya. and Aver'yanov, S. V., Abstracts of Reports to the Twentieth International Congress on Theoretical and Applied Chemistry, V 56, "Nauka" (1965); Abbreviated English Translations of Scientific Papers Presented in Russian.

439. Poddubnyi, I. Ya., Kuzin, I. A., and Evdokimov, V. F., Dokl. Akad. Nauk SSSR, 141:1097 (1961).
440. Popkov, K. K., Plastmassy, No. 2, p. 28 (1964).
441. Razuvaev, G. A. and Vyazankin, N. S., Scientific Communications of the International Symposium on Organosilicon Chemistry, A/19-97, Prague (1965).
442. Rafikov, S. R., Andrianov, K. A., Pavlova, S. A., Tverdokhlebova, I. N., and Pichkhadze, Sh. V., Izv. Akad. Nauk, Otd. Tekhn. Nauk, 1962:1581.
443. Rodionov, A. N., Asnovich, É. Z., Shigorin, D. N., and Andrianov, K. A., in: Heterochain High-Molecular Compounds, "Nauka" (1964), p. 81.
444. Sadykh-zade, S. I., Avgushevich, I. V., and Petrov, A. D., Dokl. Akad. Nauk SSSR, 112:662 (1957).
445. Sadykh-zade, S. I., Egorov, Yu. P., and Petrov, A. D., Dokl. Akad. Nauk SSSR, 113:620 (1957).
446. Sadykh-zade, S. I., Nozdrina, L. V., and Petrov, A. D., Dokl. Akad. Nauk SSSR, 118:723 (1958).
447. Sadykh-zade, S. I. and Petrov, A. D., Zh. Obsch. Khim., 28:1542 (1958).
448. Sadykh-zade, S. I. and Petrov, A. D., Azerb. Khim. Zh., 1962:105.
449. Sanin, P. S., Zh. Obsch. Khim., 23:986 (1953).
450. Sakharovskaya, G. B., Korneev, N. N., Nazarova, D. V., and Sobolevskii, M. V., Plastmassy, No. 7, p. 21 (1964).
451. Sergeev, A. S., Andrianov, K. A., Koshelev, F. F., Khananashvili, L. M., Gridunov, I. T., Molchanova, M. V., and Demidova, L. A., Abstracts of Reports to the Scientific and Technical Conference of the Moscow Institute of Chemical Technology, (1964), p. 81.
452. Sipyagina, M. A. and Andrianov, K. A., Scientific Communications of the International Symposium on Organosilicon Chemistry, Supplement, S/2-13, Prague (1965).
453. Soboleva, D. A. and Kreshkov, A. P., Abstracts of Reports to the Scientific and Technical Conference of Graduate Students and Assistant Lecturers of the Moscow Institute of Chemical Technology, (1961), p. 10.
454. Sobolevskii, M. V., Chistyakova, L. A., Nazarova, D. V., and Kirillina, V. V., Plastmassy, No. 10, p. 17 (1962).
455. Solodovnik, V. D., Davydov, A. B., Ivanova, Z. G., Mindlin, Ya. I., and Leznov, N. S., Plastmassy, No. 3, p. 39 (1963).
456. Stavitskii, I. K. and Borisov, S. N., Vysokmolekul. Soedin., 1:1496 (1959).
457. Stavitskii, I. K., Borisov, S. N., Ponomarenko, V. A., Sviridova, N. G., and Zueva, G. Ya., Vysokmolekul. Soedin., 1:1502 (1959).
458. Stavitskii, I. K., Borisov, S. N., and Sviridova, N. G., Otchet VNIISK, No. 1077 (1959).
459. Stavitskii, I. K., Karlin, A. V., Kryukovskaya, Z. M., and Rzhendzinskaya, K. A., Otchet VNIISK No. 1382 (1960).
460. Stavitskii, I. K. and Neimark, B. E., Otchet VNIISK No. 2069 (1949).
461. Stavitskii, I. K. and Rzhendzinskaya, K. A., Otchet VNIISK No. 1022 (1959).
462. Stavitskii, I. K. and Rzhendzinskaya, K. A., Otchet VNIISK No. 1192 (1959).
462a. Stanko, V. I., Bratsev, V. A., Anorova, G. A., and Tsukerman, A. M., Zh. Obsch. Khim., 36:1865 (1966).

463. Suvorov, A. L. and Spasskii, S. S., Vysokmolekul. Soedin., 3:865 (1961).
464. Tverdokhlebova, I. I., Pavlova, S. A., and Rafikov, S. R., Izv. Akad. Nauk, Otd. Tekhn. Nauk, 1963:488.
465. Terent'ev, A. P., Luskina, B. M., and Syavtsillo, S. V., Zh. Analit. Khim., 16:635 (1961); Terent'ev, A. P., Organic Analysis [in Russian], Izd. MGU (1966), p. 56.
466. Terent'ev, E. A. and Korshun, M. O., Khim. Nauka i Prom., 4:415 (1959).
467. Tikhonov, V. S., Andrianov, K. A., Khananashvili, L. M., and Varlamov, A. V., Abstracts of Reports to the Scientific and Technical Conference of The Moscow Chemical Technology Institute, (1962), p. 52.
468. Topchiev, A. V., Nametkin, N. S., Hsiao-p'ei Ts'iu, Durgar'yan, S. G., and Kuz'mina, N. A., Izv. Akad. Nauk, Otd. Tekhn. Nauk, 1962:1497.
469. Freidlina, R. Kh., I Ts'ao, and Chukovskaya, E. Ts., Dokl. Akad. Nauk SSSR, 132:149 (1960).
470. Fromberg, M. B., Petrashko, Yu. K., Vozhova, V. D., and Andrianov, K. A., Izv. AN SSSR, Ser. Khim., 1965:660.
471. Fyurst, V. and Kreshkov, A. P., Abstracts of Reports to the Scientific and Technical Conference of Graduate Students and Assistant Lecturers of the Moscow Institute of Chemical Technology (1960), p. 63.
472. Khaidarov, Kh. F., Sultanov, A. S., and Abduvaliev, A. A., in: The Physics and Chemistry of Natural and Synthetic Polymers, No. 1, Izd. Akad. Nauk UzbSSR (1962), p. 131.
473. Khaidoarov, Kh. F., Sultanov, A. S., and Abduvaliev, A. A., in: The Physics and Chemistry of Natural and Synthetic Polymers, No. 1, Izd. Akad. Nauk UzbSSR (1962), p. 138.
474. Khananashvili, L. M., Myshlyaeva, L. V., Mikhalev, B. M., and Shkol'nyi, V. E., Zh. Prikl. Khim., 30:263 (1957).
475. Khananashvili, L. M., Chivikova, A. N., Kreshkov, A. P., and Darashkevich, M. L., Proceedings of the Conference "Chemistry and Practical Application of Organosilicon Compounds," No. 6, Izd. Akad. Nauk SSSR (1961), p. 159.
476. Chan-li Gu, Leonova, N. A., Nametkin, N. S., Topchiev, A. V., and Bazilevich, V. V., Ibid., No. 1, TsBTI LSNKh (1958), p. 249.
477. Chernyshev, E. A. and Kozhenvinkova, L. G., Dokl. Akad. Nauk SSSR, 98:419 (1964).
478. Chernyshev, E. A., Mironov, V. F., Nepomnina, V. V., and Lizgunov, S. A., Zh. Prikl. Khim., 34:458 (1961).
479. Chernyshev, E. A. and Tolstikova, N. G., Zh. Obsch. Khim., 30:4058 (1960)
480. Chivikova, A. N., Zh. Prikl. Khim., 30:454 (1957).
481. Chivikova, A. N., Kreshkov, A. P., Darashkevich, M. L., Myshlyaeva, L.V., Karateev, D. A., and Khananashvili, L. M., Proceedings of the Conference "Chemistry and Practical Application of Organosilicon Compounds," No. 1, TsBTI LSNKh (1958), p. 178.
482. Chugunov, V. S., Zh. Obsch. Khim., 27:494 (1957).
483. Chugunov, V. S., Zh. Obsch. Khim., 28:336 (1958).
484. Chugunov, V. S., Izv. Akad. Nauk, Otd. Tekhn. Nauk, 1956:1059.
485. Chugunov, V. S., Izv. Akad. Nauk, Otd. Tekhn. Nauk, 1957:1368.

486. Chugunov, V. S., Izv. Akad. Nauk, Otd. Tekhn. Nauk, 1959:1341.
487. Chugunov, V. S., Otchet IKhS Akad. Nauk SSSR (1957).
488. Chugunov, V. S., Proceedings of the Conference "Chemistry and Practical Application of Organosilicon Compounds," No. 1, TsBTI LSNKh (1958), p. 169.
489. Chudesova, L. M. and Voronkov, M. G., Proceedings of the Conference "Chemistry and Practical Application of Organosilicon Compounds," No. 1, TsBTI LSNKh (1958), p. 184.
490. Chukovskaya, E. Ts. and Freidlina, R. Kh., Scientific Communications of the International Symposium on Organosilicon Chemistry, Supplement, S/1-1, Prague (1965).
491. Shapatin, A. S., Golubtsov, S. A. Zhigach, A. F., and Siryatskaya, V. N., Abstracts of Reports to the Conference "New Organosilicon Compounds," NIITÉKhim (1966), p. 25.
492. Shapatin, A. S., Golubtsov, S. A., Solov'ev, A. A., Zhigach, A. F., and Siryatskaya, V. N., Plastmassy, No. 12, p. 19 (1965).
493. Shakhovskoi, B. G., Stadnichuk, M. D., and Petrov, A. A., Zh. Obsch. Khim., 35:1031 (1965).
494. Shelomov, I. N., Osipov, O. A., and Kashireninov, O. E., Zh. Obsch. Khim., 33:1056 (1963).
495. Shorokhov, N. V., Proceedings of the Conference "Chemistry and Practical Application of Organosilicon Compounds," No. 4, TsBTI LSNKh (1958), p. 92.
496. Shostakovskii, M. F., Komarov, N. V., Atavin, A. S., Egorov, N. V., and Yarosh, O. G., Izv. Sibirsk. Otd. Akad. Nauk SSSR, No. 7, p. 152 (1964).
497. Shostakovskii, M. F., Komarov, N. V., and Vlasova, N. N., Scientific Communications of the International Symposium on Organosilicon Chemistry, Supplement, A/28-21, Prague (1965).
498. Shostakovskii, M. F., Komarov, N. V., and Maroshin, Yu. V., Zh. Obsch. Khim., 35:335 (1965).
499. Shostakovskii, M. F., Kotrelev, V. N., Kochkin, D. A., Kuznetsova, G. I., Kalinina, S. P., and Borisenko, V. V., Zh. Prikl. Khim., 31:1434 (1958).
500. Shostakovskii, M. F., Kochkin, D. A., and Rogov, V. M., Izv. Akad. Nauk, Otd. Tekhn. Nauk, 1959:1062.
501. Shostakovskii, M. F., Sokolov, B. A., and Mantsivoda, G. P., Zh. Prikl. Khim., 33:3779 (1963).
502. Shostakovskii, M. F., Shikhiev, I. A., and Komarov, N. V., Dokl. Akad. Nauk AzSSR, 11:757 (1955).
503. Shchukovskaya, L. L., Voronkov, M. G., and Pavlova, O. V., Dokl. Akad. Nauk SSSR, 143:887 (1962).
504. Shchukovskaya, L. L., Pal'chik, R. I., and Petrov, A. D., Dokl. Akad. Nauk SSSR, 136:1354 (1961).
505. Shchukovskaya, L. L. and Petrov, A. D., Izv. Akad. Nauk., Otd. Tekhn. SSSR, 136:1354 (1961).
506. Shchukovskaya, L. L. and Petrov, A. D., Proceedings of the Conference "Chemistry and Practical Application of Organosilicon Compounds," No. 1, TsBTI LSNKh (1958), p. 134.

507. Yur'ev, Yu. K. and Belyakova, Z. V., Zh. Obsch. Khim., 29:2960 (1959).
508. Yur'ev, Yu. K. and Elyakov, G. B., Zh. Obsch. Khim., 27:176 (1957).
509. Yakovlev, B. I. and Vinogradova, N. V., Zh. Obsch. Khim., 29:695 (1959).
510. Yakubovich, A. Ya. and Motsarev, G. V., Dokl. Akad. Nauk SSSR, 99:1015 (1954).

3. Articles and Reports of Foreign Authors

511. Abel, E. W., Z. Naturf., 15b:57 (1960).
512. Abel, E. W., Armitage, D. A., Bush, R. P., and Willey, G. R., J. Chem. Soc., 1965:62.
513. Abel, E. W., Armitage, D. A., and Willey, G. R., J. Chem. Soc., 1965:57.
514. Abel, E. W. and Bush, R. P., J. Organomet. Chem., 3:245 (1965).
515. Abel, E. W. and Singh, A., J. Chem. Soc., 1959:690.
516. Abel, E. W. and Willey, G. R., J. Chem. Soc., 1964:1528.
517. Alsobrook, A. L., Collins, A. L., and Wells, R. L., Inorg. Chem., 4:253 (1965).
518. Amberger, E. and Boeters, H. D., Chem. Ber., 97, 1999 (1964).
519. Amberger, E. and Salazar, R. W., Scientific Communications of the International Symposium on Organosilicon Chemistry, Supplement, A/18-31, Prague (1965).
520. Amonoo-Neizer, E. H., Shaw, R. A., Skovlin, D. O., and Smith, B. C., J. Chem. Soc., 1965:2997.
521. Anderson, H. H., J. Org. Chem., 19:1766 (1954).
522. Anderson, R. P. and Sprung, M. M., Papers presented at the Second Conference on High Temperature Polymer and Fluid Research, Dayton, 1959; cited in [120], p. 74 (70).
523. Anderson, R. P. and Sprung, M. M., WADC Tech. Rep. 59-61, 47 (1959); cited in [101], p. 181 (91).
524. Artsdalen, Van and Gavis, J., J. Am. Chem. Soc., 74:3196 (1952).
525. Atkins, D. C., Murphy, C. M. and Saunders, C. E., Ind. Eng. Chem., 39:1395 (1947).
526. Aylett, B. J., Emeleus, H. J. and Maddock, A. G., J. Inorg. Nucl. Chem., 1:187 (1955).
527. Backer, H. F. and Hurenkamp, J. B. G., Rec. Trav. Chim., 61:802 (1942).
528. Bailey, R. E. and West, R., J. Am. Chem. Soc., 86:5369 (1964).
529. Baker, H. R., Kagarise, R. E., O'Rear, J. G. and Sniegoski, P. J., US Naval Res. Lab. Rep.; cited in [532].
530. Baker, H. R., O'Rear, J. G. and Sniegoski, P. J., J. Chem. Eng. Data, 7:560 (1962).
531. Baker, H. R. and Singleterry, C. R., Abstracts of Papers 137th Meeting Am. Chem. Soc., (1960) p. 15Q.
532. Baker, H. R. and Singleterry, C. R., J. Chem. Eng. Data, 6:146 (1961).
533. Bamford, W. R. and Fordham, S., in: High Temperature Resistance and Thermal Degradation of Polymers, Soc. Chem. Ind. Monograph No. 13, London, (1960), p. 320; [120], p. 73 (39).
534. Baney, R. H. and Krager, R. J., Inorg. Chem., 3:1657 (1964).
535. Barraclough, C. G., Bradley, D. C., Lewis, J., and Thomas, I. M., J. Chem. Soc., 1961:2601.

536. Barrow, R. F., Nature, 142:434 (1938).
537. Barrow, R. E., Proc. Phys. Soc., 51:267 (1939).
538. Barrow, R. F., Proc. Phys. Soc., 56:204 (1944).
539. Barrow, R. F., Trans. Faraday Soc., 36:1053 (1940).
540. Barrow, R. F. and Jevons, W., Proc. Phys. Soc., 52:534 (1940).
541. Barrow, R. F. and Jevons, W., Proc. Roy. Soc., A169:45 (1939).
542. Basi J. S. and Bradley, D. C., Proc. Chem. Soc., 1963:305.
543. Beck, H. N., unpublished data; cited in [77], p. 295 (56).
544. Becke-Goehring, M. and Krill, H., Chem. Ber., 94:1059 (1961).
545. Beduneau, H., Rev. Prod. Chem., 55:265 (1952).
546. Benkeser, R. A. and Brumfield, P. E., J. Am. Chem. Soc., 73:4770 (1951).
547. Benkeser, R.A. and Currie, R. B., J. Am. Chem. Soc., 70:1780 (1948).
548. Benkeser, R. A. and Currie, R. B., J. Am. Chem. Soc., 71:2493 (1949).
549. Benkeser, R. A. and Foster, D. J., J. Am. Chem. Soc., 74:4200 (1952).
550. Benkeser, R. A. and Foster, D. J., J. Am. Chem. Soc., 74:5314 (1952).
551. Benkeser, R. A., Goggin, D., and Schroll, G., J. Am. Chem. Soc., 76:4025 (1954).
552. Benkeser, R. A., Grossman, R. F., and Stanton, G. M., J. Am. Chem. Soc., 84:4727 (1962).
553. Benkeser, R. A., Hoke, D. I., and Hickner, R. A., J. Am. Chem. Soc., 80:5295 (1958).
554. Benkeser, R. A. and Krysiak, H. R., J. Am. Chem. Soc., 75:2421 (1953).
555. Benkeser, R. A. and Krysiak, H. R., J. Am. Chem. Soc., 76:599 (1954).
556. Benkeser, R. A., Landesman, H., and Foster, D. J., J. Am. Chem. Soc., 74:648 (1952).
557. Benkeser, R. A., Nagai, Y., and Hooz, J., J. Am. Chem. Soc., 86:3742 (1964).
558. Benkeser, R. A. and Riel, F. J., J. Am. Chem. Soc., 73:3472 (1951).
559. Benkeser, R. A., Robinson, R. E., and Landesman, H., J. Am. Chem. Soc., 74:5699 (1952).
560. Benkeser, R. A. and Severson, R. G., J. Am. Chem. Soc., 77:2322 (1955).
561. Birkofer, L., Ritter, A., and Goller, H., Chem. Ber., 96:3289 (1963).
562. Birkofer, L., Ritter, A., and Richter, P., Tetrahedron Letters 5:195 (1962).
562a. Blake, D., Coates, G. E., and Tate, J. M., J. Chem. Soc., 1961:618.
563. Bluestein, B. A., J. Am. Chem. Soc., 70:3068 (1948).
564. Böhm, H., Metallkunde J., 50:44 (1959).
564a. Bonamico, M., Chem. Comm., 1966:135.
565. Bonamico, M., Dessy, G., and Ercolani, C., Chem. Comm., 1966:24.
566. Bott, R. W., Eaborn, C., and Greasley, P. M., J. Chem. Soc., 1964:4804.
567. Bott, R. W., Eaborn, C., Pande, K. C., and Swaddle, T. W., J. Chem. Soc., 1962:1217.
567a. Bott, R. W., Eaborn, C., and Walton, D. R. W., J. Organomet. Chem., 2:154 (1964).
568. Bradley, D. C. and Hill, D. A. W., J. Chem. Soc., 1963:2101.
569. Bradley, D. C., Kapoor, R. N., and Smith, B. C., J. Chem. Soc., 1963:204.
570. Bradley, D. C., Kapoor, R. N., and Smith, B. C., J. Inorg. Nucl. Chem., 24:863 (1962).
571. Bradley, D. C. and Prevedorou-Demas, C., Canad. J. Chem., 41:629 (1963).

572. Bradley, D. C. and Prevedorou-Demas, C., J. Chem. Soc., 1964:1580.
573. Bradley, D. C. and Thomas, I. M., Chem. Ind., 1958:17.
574. Bradley, D. C. and Thomas, I. M., Chem. Ind., 1958:1231.
575. Bradley, D. C. and Thomas, I. M., J. Chem. Soc., 1959:3404.
576. Breed, L. W. and Elliott, R. L., Inorg. Chem., 2, 1069 (1963).
577. Breed, L. W. and Haggerty, W. J., Abstracts of Papers 139 Meeting Am. Chem. Soc. (1961), 1M-5.
578. Breed, L. W. and Haggerty, W. J., J. Org. Chem., 27:257 (1962).
579. Brook, A. G., J. Am. Chem. Soc., 74:4759 (1953).
580. Brook, A. G., J. Am. Chem. Soc., 79:4373 (1957).
581. Brook, A. G. and Gilman, H., J. Am. Chem. Soc., 76:278 (1954).
582. Brook, A. G. and Gilman, H., J. Am. Chem. Soc., 76:2333 (1954).
583. Brook, A. G. and Gilman, H., J. Am. Chem. Soc., 76:2338 (1954).
584. Brook, A. G. and Gilman, H., J. Am. Chem. Soc., 77:2322 (1955).
585. Brook, A. G., Gilman, H., and Miller, L. S., J. Am. Chem. Soc., 75:4759 (1953).
586. Brook, A. G. and Mauris, R. J., J. Am. Chem. Soc., 81:981 (1959).
587. Brook, A. G. and Schwartz, N. V., J. Am. Chem. Soc. 82:2435 (1960).
588. Brook, A. G., Tai, K. M., and Gilman, H., J. Am. Chem. Soc., 77:6219 (1955).
589. Brook, A. G., Warner, C. M., and Megriskin, M. E., J. Am. Chem. Soc., 81:981 (1959).
590. Brook, A. G. and Wolfe, S., J. Am. Chem. Soc., 79:1431 (1957).
591. Brown, H. C., Okamoto, Y., Inukai, T., J. Am. Chem. Soc., 80:4964 (1958).
592. Brown, H. C. and Subba, Rao B. C., J. Am. Chem. Soc., 78, 2582 (1956).
592a. Brown, H. C. and Subba Rao, B. C., J. Am. Chem. Soc., 78:5694 (1956).
593. Brown, J. F. and Slusarczuk, G. M. F., J. Am. Chem. Soc., 87:931 (1965).
594. Brown, M. P. and Fowles, G. W., J. Chem. Soc., 1958:2811.
595. Brynolf, S., Acta Chem. Scand., 10:883 (1956).
596. Brynolf, S., Acta Chem. Scand., 10:1143 (1956).
597. Brynolf, S., Acta Chem. Scand., 11:724 (1957).
598. Buchert, H. and Zeil, W., Angew. Chem., 73:759 (1961).
599. Buchman, O., Grosjean, M., and Nasielski, J., Bull. Soc. Chim. Belge, 71:467 (1962).
600. Burg, A. B. and Kuljian, E. S., J. Am. Chem. Soc., 72:3103 (1952).
601. Bürger, H., Angew. Chem., 75:1109 (1963).
602. Bürger, H., Monatsh., 94:574 (1963).
603. Bürger, H., Plaste u. Kaut., 10:416 (1963).
604. Bürger, H., Proceedings of the 8th International Conference on Coordination Chemistry, Vienna, 1964, Wien-New York (1964), 3B3-171.
605. Bürger, H., Forker, C., and Goubeau, J., Monatsh., 96, 597 (1965).
606. Bürger, H., Samodny, W., and Wannagat, U., J. Organomet. Chem., 3:113 (1965).
607. Bürger, H., Smrekar, O., and Wannagat, U., Monatsh., 95:292 (1964).
608. Bürger, H. and Wannagat, U., Monatsh., 94:761 (1963).
609. Bürger, H. and Wannagat, U., Monatsh., 94:1007 (1963).

610. Bürger, H. and Wannagat, U., Monatsh., 95:1099 (1964).
611. Burkhard, C. A., J. Am. Chem. Soc., 72:963 (1949).
612. Bygden, A., J. Prakt. Chem., 96:86 (1917).
613. Caglioti, V., Sartori, G., Ercolani, C., and Mele, A., Abstracts of Reports to the Twentieth International Congress on Theoretical and Applied Chemistry, D29, Izd. "Nauka" (1965).
614. Cahn, H., Ind. Fin., 32, 42 (1956).
615. Calas, R., Valade, F., and Josien, M. L., C. 249:826 (1959).
616. Caley, E. R. and Burford, M. G., Ind. Eng. Chem., Anal. Ed., 8:114 (1936).
617. Campagna, P. J. and Post, H. W., J. Org. Chem., 19:1749 (1954).
618. Campagna, P. J. and Post, H. W., J. Org. Chem., 19:1753 (1954).
619. Campagna, P. J. and Post, H. W., Rec. Trav. Chim., 74:77 (1955).
620. Canavan, A. E. and Eaborn, C., J. Chem. Soc., 1959:3751.
621. Cason, L. F. and Brooks, H. G., J. Am. Chem. Soc., 74:4582 (1952).
622. Cason, L. F. and Brooks, H. G., J. Org. Chem., 19:1278 (1954).
623. Chainani, G. and Gerrard, W., J. Chem. Soc., 1960:3168.
624. Chalk, A. J. and Harrod, J. E., Abstracts of Proceedings of the Second International Symposium on Organometallic Chemistry, Madison (1965), p. 26.
625. Chalk, A. J. and Harrod, J. F., J. Am. Chem. Soc., 87:16 (1965).
626. Chalk, A. J. and Harrod, J. F., J. Am. Chem. Soc., 87:1133 (1965).
627. Chamberlain, M. M., Techn. Rep. 1, Contract 1439 (07); Project NR 052-419, Western Reserve Univ., 1960; cited in [14].
628. Chamberlain, M. M., Kern, G., Jabs, G. A., Germanas, D., Greene, A., Brain, K. and Wayland, B., US Dept. Com., Office Techn. Serv., PB Rep 152086 (1960); C. A., 58:2508 (1963).
629. Chamberlain, M. M., Jabs, G. A., and Wayland, B. B., J. Org. Chem. 27:3321 (1962).
630. Chamberland, B. L. and Mac Diarmid, A. G., Abstracts of Papers 138th Meeting Am. Chem. Soc., 1960:20N-48.
631. Chamberland, B. L. and Mac Diarmid, A. G., J. Am. Chem. Soc., 82:4542 (1960).
632. Chamberland, B. L. and Mac Diarmid, A. G., J. Am. Chem. Soc., 83:549 (1961).
633. Champetier, G., Spassky, N., and Sigwalt, P., Rev. Chim. Acad. RPR, 7:743 (1962); Ref. Zh. Khim. 1965:2C204.
634. Chatt, J. and Williams, A. A., J. Chem. Soc., 1954:4403.
635. Chatt, J. and Williams, A. A., J. Chem. Soc., 1956:688.
636. Chem. Eng. News., 41(49):62 (1963).
637. Chih-Tang Huang and Pao-Jen Wang, Acta Chim. Sinica, 23:291 (1957).
638. Chih-Tang Huang, Pao-Jen Wang, Acta Chim. Sinica, 25:341 (1959).
639. Cohen, H. J., Abstracts of Proceedings of the Second International Symposium on Organometallic Chemistry, Madison (1965), p. 23.
640. Cohen, H. J. and Dessy, R. E., Abstracts of Papers 138th Meeting Am. Chem. Soc., 1960, 20N-49.
641. Cohen, M. S., private communication (1965).
642. Connolly, J. W. and Urry, G., J. Org. Chem., 29:619 (1964).

643. Connolly, J. W. and Urry, G., Inorg. Chem., 2:645 (1963).
644. Considine, W. J., Baum, G. A. and Jones, R. C., J. Organomet. Chem., 3:308 (1965).
645. Cordischi, D., Mele, A., and Somogyi, A., J. Chem. Soc., 1964:5281.
646. Cowley, A. H., Fairbrother, F., and Scott, N., J. Chem. Soc. 1959:717.
647. Cowley, A. H., Sisler, H. H., Ryschnewitsch, G. E., J. Am. Chem. Soc., 82:501 (1960).
648. Crain, R. D. and Koenig, P. E., Papers Presented at the Second Conference on High Temperature Polymer and Fluid Research Dayton (1959); cited in [99], p. 382 (82); [75].
649. Csakvari, B., Székely, T., and Török, F., Scientific Communications of the International Symposium on Organosilicon Chemistry, A/21-104, Prague (1965).
650. Currel, B. R., Frazer, M. J., and Gerrard, W., J. Chem. Soc., 1960:2776.
651. Danforth, J. D., J. Am. Chem. Soc., 80:2585 (1958).
652. Dannels, B. F. and Post, H. W., J. Org. Chem., 22:748 (1957).
653. Daudt, W. H. and Hyde, J. F., J. Am. Chem. Soc., 74:386 (1952).
654. Decker, Q. W. and Post, H. W., J. Org. Chem., 25:249 (1960).
655. Drake, J. E. and Jolly, W. L., Chem. Ind., 1962:1470.
656. Duck, E. W. and Thornber, M. N., Chem. Ind., 1963:1904.
657. Durkin, A. E. and Horner, A., Materials and Methods, 38:114 (1953).
658. Dyke van, C. H. and Mc Diarmid, A. G., Inorg. Chem., 3:747 (1964).
659. Eaborn, C., J. Chem. Soc., 1949, 2755.
660. Eaborn, C., Jackson, R. A., and Walsingham, R. W., Abstracts of Proceedings of the Second International Symposium on Organometallic Chemistry, Madison (1965), p. 25.
661. Eaborn, C. and Pande, K. C., J. Chem. Soc., 1960:3200.
662. Eaborn, C. and Parker, S. H., J. Chem. Soc., 1954:939.
663. Eaborn, C. and Parker, S. H., J. Chem. Soc., 1955:126.
664. Eaborn, C. and Shaw, R. A., J. Chem. Soc., 1955:1429.
665. Eaborn, C. and Webster, D. E., J. Chem. Soc., 1957:4449.
666. Ebsworth, E. A. V. and Emeleus, H. J., J. Chem. Soc.,:1958, 2150.
667. Ebsworth, E. A. V., Emeleus, H. J., and Welcman, N., J. Chem. Soc., 1962:2290.
668. Ebsworth, E. A. V. and Mays, M. J., J. Chem. Soc.,: 1963, 3893.
669. Ebsworth, E. A. V., Taylor, R., and Woodward, L. A., Trans. Faraday Soc., 55:211 (1959).
670. Eisch, J. J. and Beuhler, R. J., J. Org. Chem., 28:2876 (1963).
671. Eisch, J. J. and Husk, G. R., J. Org. Chem., 29:254 (1964).
672. Eisch, J. J. and Trainor, J. T., J. Org. Chem., 28:2870 (1963).
673. Elliott, J. R. and Boldebuck, E. M., J. Am. Chem. Soc., 74:1853 (1952).
674. Emeleus, H. J., Angew. Chem., 66:714 (1954).
675. Emeleus, H. J., Aylett, B. J., Mac Diarmid, A. C., and Maddock, A. G., IUPAC Coll., 1954:50.
676. Emeleus, H. J. and Heal, H. G., J. Chem. Soc., 1946:1126.
677. Emeleus, H. J., Maddock, A. G., and Reid, C., J. Chem. Soc., 1941:353.

678. Emeleus,H. J., Maddock,A. G., and Reid,C., Nature, 144:328 (1939).
679. Emeleus,H. J. and Onyszchuk,M., J. Chem. Soc., 1958:604.
680. Emeleus,H., Onyszchuk,M., and Kuchen, W., Z. Anorg. Chem., 283:74 (1956).
681. English,W. D. and Sommer, L. H., J. Am. Chem. Soc., 77:170 (1955).
682. Ercolani,C., Camilli, A., and De Luca,L., J. Chem. Soc., 1964:5278.
683. Ercolani,C., Camilli,A., De Luca,L., and Sartori,G., J. Chem. Soc., Part A, 1966:608.
684. Ercolani,C., Camilli, A., and Sartori,G., J. Chem. Soc., Part A, 1966:603.
685. Ercolani,C., Camilli,A., and Sartori,G., J. Chem. Soc., Part A, 1966:606.
686. Evers,E. C., Papers Presented at the Second Conference on High-Temperature Polymers and Fluid Research, Dayton (1959); cited in [75].
687. Evers, E. C., Freitag,W. O., Keith,J. N., Kriner, W. A., Mac Diarmid, A. G., and Sujishi,S., J. Am. Chem. Soc., 81:4493 (1959).
688. Evers,E. C., Freitag, W. O., Kriner,W. A., and Mac Diarmid, A. G., J. Am. Chem. Soc., 81:5106 (1959).
689. Evers,E. C., Freitag, W. O., Kriner, W. A., Mac Diarmid, A. G., and Sujishi, S., J. Inorg. Nucl. Chem., 13:239 (1960).
690. Evison,W. E. and Kipping, F. S., J. Chem. Soc., 1931:2774.
691. Fink,W., Angew. Chem., 73:467 (1961).
692. Fink,W., Angew. Chem., 73:736 (1961).
693. Fink,W., Chem. Ber., 96:1071 (1963).
694. Fink,W., Chem. Ber., 97:1433 (1964).
695. Fink,W., Helv. Chim. Acta, 45:1081 (1962).
696. Fink,W., Helv. Chim. Acta., 46:720 (1963).
697. Fink,W., Helv. Chim. Acta, 47:498 (1964).
697a. Forstner, J. A. Muetterties, E. L., Inorg. Chem., 5:552 (1966).
698. Frazer, M. J., Gerrard,W., and Strickson, J. A., J. Chem. Soc., 1960:4701.
699. Friedel, C. and Crafts, J. M., Ann. Chim. Phys. (4), 9:50 (1866).
700. Friedel,C. and Ladenburg, A., Ann., 203:251 (1880); Ann. Chim. Phys. (5), 19:401 (1880).
701. Frierel,C. and Ladenburg, A., Ber., 3:15 (1870).
702. Friedel,C. and Ladenburg, A., Compt. Rend., 68:923 (1869).
703. Frisch,K. C. and Shroff, P. D., J. Am. Chem. Soc., 75:1249 (1953).
704. Fritz,G., Z. Naturf., 10b:423 (1955).
705. Fritz,G. and Burdt,H., Z. Anorg. Chem., 314:35 (1962).
706. Fritz,G. and Kemmerling, W., Z. Anorg. Chem., 322:34 (1963).
707. Fritz,G., Kemmerling, W., Sonntag, G., Becher,H. J., Ebsworth, E. A. V., and Grobe, J., Z. Anorg. Chem., 321:10 (1963).
708. Fukukawa,S., Kogyo Kagaku Zasshi, 58:940 (1955); C. A., 50:12523 (1956).
709. Fukukawa,S. and Kohama,S., Sci. Ind., 29:70 (1955); C. A., 49:13888 (1955).
710. Fukukawa,S. and Kohama,S., Sci. Ind., 29:253 (1955); C. A., 50:1355 (1956).
710a. Gabiel, H. and Alvarez-Tostado,C., J. Am. Chem. Soc., 74:262 (1952).
711. Gai, B. J. and Gilman,H., Chem. Ind., 1960:319.

712. Gai, B. J. and Gilman, H., Chem. Ind., 1960:493.
713. George, M. V., Gai, B. J., and Gilman, H., J. Org. Chem., 24:624 (1959).
714. George, M. V. and Gilman, H., J. Am. Chem. Soc., 81:3288 (1959).
715. George, M. V., Lichtenwalter, G. D., and Gilman, H., J. Am. Chem. Soc., 81:978 (1959).
716. George, M. V., Peterson, D. J., and Gilman, H., J. Am. Chem. Soc., 82:403 (1960).
717. George, M. V., Talukdar, P. B., Gerow, C. W., and Gilman, H., J. Am. Chem. Soc., 82:4562 (1960).
718. Gerrard, W. and Strickson, J. A., Chem. Ind., 1958:860.
719. Geymayer, P., Rochow, E. G., Scientific Communications of the International Symposium on Organosilicon Chemistry, B/22-306, Prague (1965).
720. Geymayer, P., Rochow, E. G., and Wannagat, U., Angew. Chem., 76:499 (1964).
721. Ghosh, A. K., Hansing, C. E., Stutz, A. I., and Mac Diarmid, A. G., J. Chem. Soc., 1962:403.
722. Gibbs, C. F., Tucker, H., Shkapenko, G., and Park, J. C., WADC Techn. Rep. 55-453, pt. II; ASTIA Doc. No. 131036 (1957); cited in [120], p. 71 (61) and [101], p. 172 (54).
723. Gilbert, A. R. and Kantor, S. W., J. Polymer Sci., 40:35 (1959).
724. Gilman, H., Angew. Chem., 74:951 (1962).
725. Gilman, H., Bull. Soc. Chim. France, 1963:1356.
726. Gilman, H., Trans. N. Y., Acad. Sci., Ser. 2, 25:820 (1963).
727. Gilman, H. and Aoki, D., Chem. Ind., 1960:1165.
728. Gilman, H. and Aoki, D., Chem. Ind., 1961:1619.
729. Gilman, H. and Aoki, D., J. Org. Chem., 24:426 (1959).
730. Gilman, H. and Aoki, D., J. Organomet. Chem., 1:449 (1964).
731. Gilman, H. and Aoki, D., J. Organomet. Chem., 2:44 (1964).
732. Gilman, H. and Aoki, D., J. Organomet. Chem., 2:89 (1964).
733. Gilman, H. and Aoki, D., J. Organomet. Chem., 2:293 (1964).
734. Gilman, H., Aoki, D., and Wittenberg, D., J. Am. Chem. Soc., 81:1107 (1959).
735. Gilman, H. and Atwell, W. H., J. Am. Chem. Soc., 86:2687 (1964).
736. Gilman, H., Benedict, H. N., and Hartzfeld, H., J. Org. Chem., 19:419 (1954).
737. Gilman, H., Benkeser, R. A., and Dunn, G. E., J. Am. Chem. Soc., 72:1689 (1950).
738. Gilman, H., Brook, A. G., and Miller, L. S., J. Am. Chem. Soc., 75:4531 (1953).
739. Gilman, H. and Cartledge, F. K., J. Organomet. Chem., 3:255 (1965).
740. Gilman, H., Cartledge, F. K., and See-Guen Sim, J. Organomet. Chem., 1:8 (1963).
741. Gilman, H. and Diehl, J. W., J. Org. Chem., 26:2938 (1961).
742. Gilman, H. and Dunn, G. E., J. Am. Chem. Soc., 73:5077 (1951).
743. Gilman, H. and Gaj, B. J., J. Org. Chem., 26:1305 (1961).
744. Gilman, H. and Gaj, B. J., J. Org. Chem., 26:2471 (1961).

745. Gilman,H. and Gerow,C. W., J. Am. Chem. Soc., 77:4675 (1955).
746. Gilman,H. and Gerow,C. W., J. Am. Chem. Soc., 77:5509 (1955).
747. Gilman,H. and Gerow,C. W., J. Am. Chem. Soc., 78:5823 (1956).
748. Gilman,H. and Goodman,J. J., J. Am. Chem. Soc., 75:1250 (1953).
749. Gilman,H. and Gorsich,R. D., J. Am. Chem. Soc., 80:1883 (1958).
750. Gilman,H. and Gorsich,R. D., J. Am. Chem. Soc., 80:3243 (1958).
751. Gilman,H. and Gorsich,R. D., J. Org. Chem., 27:1072 (1962).
752. Gilman,H., Hawell,R., Chang, K. Y., and Cottis,S., J. Organomet. Chem., 2:434 (1964).
753. Gilman,H. and Hartzfeld,H., J. Am. Chem. Soc., 73:5878 (1951).
754. Gilman,H., Ingham,R. K., and Smith,A. G., J. Org. Chem., 18:1743 (1953).
755. Gilman,H. and Inoue,S., Chem. Ind., 1964:74.
756. Gilman,H., Klein,R. A., and Winkler, H. J. S., J. Org. Chem., 26:2474 (1961).
757. Gilman,H. and Lichtenwalter,G. D., J. Am. Chem. Soc., 80:607 (1958).
758. Gilman,H. and Lichtenwalter,G. D., J. Am. Chem. Soc., 80:608 (1958).
759. Gilman,H. and Lichtenwalter,G. D., J. Am. Chem. Soc., 80:2680 (1958).
760. Gilman,H. and Lichtenwalter,G. D., J. Am. Chem. Soc., 82:3319 (1960).
761. Gilman,H. and Lichtenwalter,G. D., J. Org. Chem., 23:1586 (1958).
762. Gilman,H. and Lichtenwalter,G. D., J. Org. Chem., 24:1588 (1959).
763. Gilman,H. and Lichtenwalter,G., J. Org. Chem., 25:1064 (1960).
764. Gilman,H., Lichtenwalter,G. D., and Wittenberg,D., J. Am. Chem. Soc., 81:5320 (1959).
765. Gilman,H. and Marrs,O. L., J. Org. Chem., 25:1194 (1960).
766. Gilman,H. and Marrs,O. L., J. Org. Chem., 27:1879 (1962).
767. Gilman,H., Marrs,O. L., Trepka,W. J., and Diehl,J. W., J. Org. Chem., 27:1261 (1962).
768. Gilman,H., McNinch,H. A., and Wittenberg,D., J. Org. Chem., 23:2044 (1958).
769. Gilman,H., Miles,D. H., Moore,L. O., and Gerow,C. W., J. Org. Chem., 24:219 (1959).
770. Gilman,H. and Nobis,J. F., J. Am. Chem. Soc., 72:2629 (1950).
771. Gilman,H. and Peterson,D. J., J. Org. Chem., 23:1895 (1958).
772. Gilman,H., Peterson,D. J., Jarwie,A. W., and Winkler,H. J. S., J. Am. Chem. Soc., 82:2076 (1960).
773. Gilman,H., Peterson,D. J., and Wittenberg,D., Chem. Ind., 1958:1479.
774. Gilman,H. and Plunkett,M. A., J. Am. Chem. Soc., 71:1117 (1949).
775. Gilman,H. and Rosenberg,S. D., J. Am. Chem. Soc., 74:531 (1952).
776. Gilman,H. and Schulze,F., J. Am. Chem. Soc., 47:2002 (1925).
777. Gilman,H. and Schwebke,G. L., J. Am. Chem. Soc., 85:1016 (1963).
778. Gilman,H. and Schwebke,G. L., J. Am. Chem. Soc., 86:2693 (1964).
779. Gilman,H. and Schwebke,G. L., J. Org. Chem., 27:4259 (1962).
780. Gilman,H. and Steudel,W., Chem. Ind., 1959:1094.
781. Gilman,H. and Tomasi,R. A., Chem. Ind., 1963:954.
782. Gilman,H. and Tomasi,R. A., J. Org. Chem., 28:1651 (1963).
783. Gilman,H. and Trepka,W. J., J. Org. Chem., 25:2201 (1960).

784. Gilman,H. and Trepka, W. J., J. Org. Chem., 27:1414 (1962).
785. Gilman,H. and Trepka, W. J., J. Org. Chem., 27:1418 (1962).
786. Gilman,H. and Trepka,W. J., J. Organomet. Chem., 1:222 (1964).
787. Gilman,H. and Trepka,W. J., J. Organomet. Chem., 3:174 (1965).
788. Gilman,H., Trepka,W. J., and Wittenberg,D., J. Am. Chem. Soc., 84:383 (1962).
789. Gilman,H. and Wittenberg, D., J. Am. Chem. Soc., 79:6339 (1957).
790. Gilman,H. and Wu T. C., J. Am. Chem. Soc., 73:4031 (1951).
791. Gilman, H. and Wu T. C., J. Am. Chem. Soc., 75:234 (1953).
792. Gilman, H. and Wu T. C., J. Am. Chem. Soc., 75:2509 (1953).
793. Gilman, H. and Wu T. C., J. Am. Chem. Soc., 75:2935 (1953).
794. Gilman, H. and Wu T. C., J. Am. Chem. Soc., 75:3762 (1953).
795. Gilman, H. and Wu T. C., J. Am. Chem. Soc., 76:2502 (1954).
796. Gilman, H. and Wu T. C., J. Org. Chem., 18:753 (1953).
797. Gilman, H. and Wu T. C., J. Org. Chem., 25:2251 (1960).
798. Gilman, H., Wu T. C., Hartzfeld, H. A., Guter, G. A., Smith A. G., Goodman J. J., and Eidt S. H., J. Am. Chem. Soc., 74:561 (1952).
799. Gilman, H. and Zuech,E. A., J. Org. Chem., 27:2897 (1962).
800. Glaser, M. A., Ind. Eng. Chem., 46:2334 (1954).
801. Gold, J. R., Sommer, L. H., and Whitmore, F. C., J. Am. Chem. Soc., 70:2874 (1948).
802. Goldberg, S. I., and Mayo,D. W., Chem. Ind., 1959:671.
803. Goldberg, S. I., Mayo, D. W., Vogel, M., Rosenberg,H., and Rausch, M., J. Org. Chem., 24:824 (1959).
804. Goodman, J. J., Iowa State Coll. J. Sci., 31:425 (1957); cited in [92], p. 274, 318, 319 (97); [77], p. 367 (148); [78], p. 101 (DA-59).
805. Goubeau,J. and Jimenez-Barbera,J., Z. Anorg. Chem., 303:217 (1960).
806. Goubeau, J., Mayer,W., Z. Anorg. Chem., 318:287 (1962).
807. Grafstein, D., Abstracts of Reports to the Twentieth International Congress on Theoretical and Applied Chemistry [Russian translation], D14, Izd. "Nauka" (1965).
808. Greber, G. and Balciunas A., Makromal. Chem., 71:62 (1964).
809. Greber, G. and Degler G., Makromal. Chem., 52:174 (1962).
810. Greber,G. and Egle,G., Makromal Chem., 62, 196 (1963).
811. Greber, G. and Hallensleben,M. L., Makromal. Chem., 83:148 (1965).
812. Greber, G. and Hallensleben, M. L., Scientific Communications of the International Symposium on Organosilicon Chemistry, A/17-91, Prague (1965).
813. Greber, G. and Metzinger,L., Makromal. Chem., 39:226 (1960).
814. Green, J., Mayer, N., Kotloby, A. P., Fein M. M., O'Brien E. L., and Cohen M. S., Polymer Letters, 2:109 (1964).
815. Grim, S. O., and Seyferth, D., Chem. Ind., 1959:849.
816. Grosse-Ruyken, H., Angew. Chem., 66:754 (1954).
817. Grosse-Ruyken, H. and Kleesaat, R., Z. Anorg. Chem., 308:122 (1960).
818. Grosse-Ruyken, H. and Kleesaat, R., Z. Chem., 1:27 (1960); cited in C. A., 55:9138 (1961).
819. Grubb, W. T. and Osthoff, R. C., J. Am. Chem. Soc., 77:1405 (1955).

820. Grüttner, G. and Cauer, M., Ber., 51:1283 (1918).
821. Grüttner, G. and Krause, E., Ber., 50:1559 (1917).
822. Guggenheimer, K. M., Proc. Phys. Soc., 58:456 (1946).
823. Guillissen, C. I. and Gancberg, A., Chem. Age, 66:976 (1952).
824. Guillissen, C. I. and Gancberg, A., Ind. Chim. Belge, 17:481 (1952).
825. Guillissen, C. I. and Gancberg, A., Ind. Chim. Belge, 20:382 (1955).
826. Gutmann, V. and Meller, A., Monatsh., 91:519 (1960).
827. Hagenmuller, P. and Pouchard, M. Bull. Soc., Chim. France, 1964:1187.
828. Harrod, J. F. and Chalk, A. J., J. Am. Chem. Soc., 87:1133 (1965).
829. Hartwell, G. E. and Brown, T. L., Inorg. Chem., 3:1656 (1964).
830. Harvey, M. C., Nebergall, W. H. and Peake, J. S., J. Am. Chem. Soc., 79:1437 (1957).
831. Hauser, C. R. and Hance, C. R., J. Am. Chem. Soc., 73:5846 (1951).
832. Hauser, C. R. and Hance, C. R., J. Am. Chem. Soc., 74:5091 (1952).
833. Hedlund, R. C., Finish, 10:45, 107 (1953).
834. Hedlund, R. C., Paint Varnish Prod., 44(11):61 (1954).
835. Hein, F. and Hecker, H., Z. Naturf., 11b:677 (1956).
836. Henglein, F. A., Chimia, 9:187 (1955).
837. Hinglein, F. A., Lang, R., and Scheinost, K., Makromal. Chem., 15:177 (1955).
838. Henglein, F. A., Lang, R., Scheinost, K., Makromal. Chem., 18:102 (1956).
839. Henglein, F. A., Lang, R., and Schmack, L., Makromal. Chem., 22:103 (1957).
840. Henry, M. C. and Noltes, J. G., J. Am. Chem. Soc., 82:558 (1960).
841. Henry, M. C. and Noltes, J. G., J. Am. Chem. Soc., 82:561 (1960).
841a. Herbstman, S., J. Org. Chem., 29:986 (1964).
842. Heying, T. L., Ager, J. W., Clark, S. L., Alexander, R. P., Papetti, S., Reid, J. A., and Trotz, S. I., Inorg. Chem., 2:1097 (1963).
843. Höfler, F., Bürger, H., and Wannagat, U., Scientific Communications of the International Symposium on Organosilicon Chemistry, B/30-335, Prague (1965).
844. Hohmann, E., Z. Anorg. Chem., 257:113 (1948).
845. Homeyer, H. N., Preston, J. H., Casapulla, S., and Beekman, E. M., Ind. Eng. Chem., 46:2349 (1954).
846. Hooton, K. A. and Allread, A. L., Inorg. Chem., 4:671 (1965).
847. Hornbaker, E. D. and Conrad, F., J. Org. Chem., 24:1858 (1959).
848. Hunter, M. J., Hyde, J. F., Warrick, E. L. and Fletcher, H. J., J. Am. Chem. Soc., 68:667 (1946).
849. Hurd, D. T., Osthoff, R. C., and Corrin, M. L., J. Am. Chem. Soc., 76:249 (1954).
850. Hyde, J. F., J. Am. Chem. Soc., 75:2166 (1953).
851. Hyde, J. F., Johannson, O. K., Daudt, W. H., Fleming, R. F., Laudenslager, H. B., and Roche, M., J. Am. Chem. Soc., 75:5615 (1953).
852. Ind. Eng. Chem., No. 9, I, 37A (1959).
853. Jarwie, A. W. P. and Gilman, H., J. Org. Chem., 26:1999 (1961).

854. Jarwie, A. W. P., Winkler, H. J. S., Peterson, D. J., and Gilman, H., J. Am. Chem. Soc., 83:1921 (1961).

855. Jenkner, H., Z. Naturf., 14b:133 (1959).

856. Jenne, H. and Niedenzu, K., Inorg. Chem., 3:68 (1964).

857. Jevons, W., Proc. Phys. Soc., 56:211 (1944).

858. Johnson, W. K. and Pollart, K. A., J. Org. Chem., 26:4092 (1961).

859. Jolly, W. L., J. Am. Chem. Soc., 85:3083 (1963).

860. Joyner, R. D. and Kenney, M. E., J. Am. Chem. Soc., 82:5790 (1960).

861. Joyner, R. D., Linck, R. G., Esposito, J. N., and Kenney, M. E., J. Inorg. Nucl. Chem., 24:299 (1962).

862. Joyner, R. D., Linck, R. L., and Kenney, M. E., Abstracts of Papers 139th Meeting Am. Chem. Soc. (1961), 16M-46.

863. Kan, P. T., Lenk, C. T., and Schaaf, R. L., Abstracts of Papers 138th Meeting Am. Chem. Soc. (1960), 30 O-42.

864. Kan, P. T., Lenk, C. T., and Schaaf, R. L., J. Org. Chem., 26:4038 (1961).

865. Kan, P. T., Schaaf, R. L., and Lenk, C. T., Abstracts of Papers 136th Meeting Am. Chem. Soc. (1959), 113P.

866. Kantor, S. W., Grubb, W. T., and Osthoff, R. C., J. Am. Chem. Soc., 76:5190 (1954).

867. Kary, R. M. and Frisch, K. C., J. Am. Chem. Soc., 79:2140 (1957).

868. Kather, W. S. and Torkelson, A., Ind. Eng. Chem. 46P:381 (1954).

869. Kautsky, H. and Bartocha, B., Z. Naturf., 10b:422 (1955).

870. Kettle, S. F. A. and Khan, J. A., Proc. Roy. Soc., 1962:354L; Proc. Chem. Soc., 1962, 82.

870a. Kettle, S. F. A. and Khan, I. D., J. Organomet. Chem., 5:588 (1966).

871. Khotinsky, E. and Seregenkoff, B., Ber., 41:2946 (1908).

872. Kipping, F. S., J. Chem. Soc., 93:457 (1908).

873. Kipping, F. S., J. Chem. Soc., 101:2108 (1912).

874. Kipping, F. S., J. Chem. Soc., 101:2125 (1912).

875. Kipping, F. S., J. Chem. Soc., 119:647 (1921).

876. Kipping, F. S., J. Chem. Soc., 119:830 (1921).

877. Kipping, F. S., J. Chem. Soc., 123:2590 (1923).

878. Kipping, F. S., J. Chem. Soc., 125:2291 (1924).

879. Kipping, F. S., J. Chem. Soc., 1927:2719.

880. Kipping, F. S. and Blackburn, J. C., J. Chem. Soc., 1932:2205.

881. Kipping, F. S. and Hackford, J. E., J. Chem. Soc., 99:138 (1911).

882. Kipping, F. S. and Lloyd, L. L., J. Chem. Soc., 79:449 (1901).

883. Klejnot, O., Abstracts of Papers XVII Congress IUPAC, Münich 1959, Vol. 1, A220; cited in [985], p. 377 (114).

884. Klemm, W. and Struck, M., Z. Anorg. Chem., 278:117 (1955).

885. Kloppe, A., Silikat Techn., 13:446 (1962).

886. Knoth, W. H. and Lindsey, R. V., J. Org. Chem., 23:1392 (1959).

887. Koenig, P. E. and Crain, P. D., WADC Techn. Rep. 58-44, Part II (1959).

888. Koenig, P. E. and Hutchinson, J. H., WADC Techn. Rep. 58-44 Part I, (1958); ASTIA Doc. No. 151197; cited in [101], p. 182 (92).

889. Kohama, S., J. Chem. Soc. Japan, Ind. Chem. Sec., 63:1439 A79 (1960); C. A., 57:11226 (1962).

890. Kohama, S., J. Chem. Soc. Japan, Pure Chem. Sec., 80:284 (1959); C. A., 55:4399 (1961).
891. Kohama, S., J. Chem. Soc. Japan, Pure Chem. Sec., 81:1602, A107 (1960); Ref. Zh. Khim., 1962, 5Zh334.
892. Kohama, S., J. Chem. Soc. Japan, Pure Chem. Sec., 81:1760 (1960); C. A., 56:3505 (1962).
893. Kohama, S., J. Chem. Soc. Japan, Pure Chem. Sec., 81:1874, A126 (1960); Ref. Zh. Khim., 1962, 15Zh290.
894. Kohama, S., J. Chem. Soc. Japan, Pure Chem. Sec., 81:1889, A127 (1960); Ref. Zh. Khim., 1962, 17Zh311.
895. Kohama, S., J. Chem. Soc. Japan, Pure Chem. Sec., 83:188 (1962); C. A., 58, 11392 (1963).
896. Kohama, S. and Fukugawa, S., J. Chem. Soc. Japan, Ind. Chem. Sec., 63:1019, A55 (1960); Ref. Zh. Khim., 1961, 19Zh192.
897. Kraus, C. A. and Eatough, H., J. Am. Chem. Soc., 55:5008 (1933).
898. Kraus, C. A. and Nelson, W. K., J. Am. Chem. Soc., 56:195 (1934).
899. Kraus, C. A. and Rosen, R., J. Am. Chem. Soc., 47:2739 (1925).
900. Krieble, R. H. and Elliott, J. R., J. Am. Chem. Soc., 68:2291 (1946).
901. Kriegsmann, H. and Licht, K., Z. Electrochem., 62:1163 (1958).
902. Kriner, W. A., Mac Diarmid, A. G., and Evers, E. C., J. Am. Chem. Soc., 80:1546 (1958).
903. Krüger, C. R. and Rochow, E. G., Angew. Chem., 74:491 (1962).
904. Krüger, C. R. and Rochow, E. G., Angew. Chem., 75:793 (1963).
905. Krüger, C. and Rochow, E. G., Inorg. Chem., 2:1295 (1963).
906. Krüger, C. R. and Rochow, E. G., J. Organomet. Chem., 1:476 (1964).
907. Krüger, C., Rochow, E. G., and Wannagat, U., Chem. Ber., 96:2132 (1963).
908. Krüger, C., Rochow, E. G., and Wannagat, U., Chem. Ber., 96:2138 (1963).
909. Kucera, M., Jelinek, M., Lanikova, J., and Vesely, K., Polymer Sci., 53:311 (1961).
910. Kucera, M. and Lanikova, J., J. Polymer Sci., 54:375 (1961).
911. Kucera, M. and Lanikova, J., J. Polymer Sci., 59:79 (1962).
912. Kucera, M. and Lanikova, J., Jelinek, M., J. Polymer Sci., 53:301 (1961).
913. Kumada, M., In Report at the International Symposium on Organosilicon Chemistry, Prague (1965).
913a. Kumada, M., Imaki, N., and Yamamoto, K., J. Organomet. Chem., 6:490 (1966).
913b. Kumada, M. and Ishikawa, M., J. Organomet. Chem., 1:411 (1964).
913c. Kumada, M., Ishikawa, M., and Maeda, S., J. Organomet. Chem., 2:146 (1966).
914. Kumada, M., Mimura, K., Ishikawa, M., and Shiina, K., Tetrahedron Letters, 1965:83.
915. Kumada, M., Mimura, K., Ishikawa, M., and Tsunemi, H., Abstracts of Proceedings of the Second International Symposium on Organometallic Chemistry, Madison (1965), p. 13.
916. Kummer, D. and Rochow, E. G., Z. Anorg. Chem., 321:21 (1963).
917. Ladenburg, A., Ann., 164:300 (1872).
918. Ladenburg, A., Ann., 173:143 (1874).

919. Ladenburg,A., Ber., 4:726 (1871).
920. Ladenburg,A., Ber., 4:901 (1871).
921. Ladenburg,A., Ber., 6:379 (1873).
922. Ladenburg,A., Ber., 6:1029 (1873).
923. Lanning,F. C., Trans. Kansas Acad. Sci., 57:374 (1954); C. A., 49:9900 (1955).
924. Lappert,M. F. and Srivastava,G., Proc. Chem. Soc., 1964:120.
925. Lengyel, B., Székely, T., Jenei, S., and Garzo, G., Z. Anorg. Chem., 323:65 (1963).
926. Lienhard,K. and Rochow,E. G., Angew. Chem., 75:638 (1963).
927. Lienhard,K. and Rochow, E. G., Z. Anorg. Chem., 331:307 (1964).
928. Lienhard,K. and Rochow, E. G., Z. Anorg. Chem., 331:316 (1964).
929. Lissner, A. and Schäfer, H. G., Angew. Chem., 67:89 (1950).
930. Lissner, A. and Schäfer, H. G., Chem. Techn., 2:181 (1950).
931. Lorberth,J. and Kula, M. R., Chem. Ber., 98:520 (1965).
932. Mac Diarmid, A. G., Moscony, J. J., Russ C. R., and Yoshioka T., Scientific Communications of the International Symposium on Organosilicon Chemistry, A/20-100 Prague (1965).
933. Maier, L., Helv. Chim. Acta, 46:2667 (1963).
934. Mailey, E. A., Dickey, C. R., Goodale, G. M., and Matthews, V. E., J. Org. Chem., 27:616 (1962).
935. Manasevit, H. M., US Dept. Com. Office Techn. Serv. PB Rep. 143572, (1959); cited in [121], p. 232 (95).
936. Manuel, T. A. and Stone, F. G. A., Chem. Ind., 1960:231.
937. Marr, G. and Webster, D. E., J. Organomet. Chem., 2:93 (1964).
938. Marr, G. and Webster, D. E., J. Organomet. Chem., 2:99 (1964).
939. Martin, G., Ber., 45:403 (1912).
940. Martin, G., Ber., 46:2442 (1913).
941. Martin, G., Ber., 46:3294 (1913).
942. Masdupuy, E., Compt. Rend., 244:2390 (1957).
943. Massey, A. G., Abstracts of Reports to the Twentieth International Congress on Theoretical and Applied Chemistry [Russian translation], D12, Izd. "Nauka" (1965).
944. Massey, A. G. and Urch, D. S., Proc. Chem. Soc., 1964:284.
945. McCloskey A. L. et al., WADC Techn. Rep. 59-761 (1960); cited in [77], p. 306 (270).
946. McCusker, P. A. and Ostdick, T., J. Am. Chem. Soc., 80:1103 (1958).
947. McCusker, P. A. and Ostdick, T., J. Am. Chem. Soc., 81:5550 (1959).
948. McDiarmid, A. G., Abedini, M., Spanier, E. J., Urenovitch, and van Dyke, C. H., Abstracts of Papers 142nd Meeting Am. Chem. Soc., 1962, 15N.
949. Mac Diarmid, A. G., Sternbach, B., and Ward, L. G. L., Abstracts of Papers 139th Meeting Am. Chem. Soc. (1961), 7M-17.
950. Meads, J. A. and Kipping, F. S., J. Chem. Soc., 105:679 (1914).
951. Meads, J. A. and Kipping, F. S., J. Chem. Soc., 107:459 (1915).
952. Meen, R. H. and Gilman, H., J. Org. Chem., 20:73 (1955).
953. Meen, R. H. and Gilman, H., J. Org. Chem., 22:564 (1957).
954. Meen, R. H. and Gilman, H., J. Org. Chem., 22:684 (1957).

955. Mehrotra,R. C. and Pant,B. C., Indian J. Appl. Chem., 26:109 (1963); C. A., 60:13265 (1964).
956. Meller, A., Monatsh., 94:183 (1963).
957. Melzer, W., Ber., 41:3390 (1908).
958. Merker, R. L. and Scott,M. J., J. Am. Chem. Soc., 81:975 (1959).
959. Merker, R. L. and Scott,M. J., J. Polymer Sci., A2:15 (1964).
960. Merker, R. L., Scott,M. J., and Haberland,G. G., J. Polymer Sci., A2:31 (1964).
961. Milligan,J. G. and Kraus,C. A., J. Am. Chem. Soc., 72:5297 (1950).
962. Minné, R. and Rochow,E. G., J. Am. Chem. Soc., 82:5625 (1960).
963. Minné, R. and Rochow,E. G., J. Am. Chem. Soc., 82:5628 (1960).
964. Moissan,H., Bull. Soc. Chim. France, 29 [3]:443 (1903).
965. Moissan,H., Compt. Rend., 134:1083 (1902).
966. Moissan,H., Compt. Rend., 135:1284 (1902).
967. Moody,L. S., J. Am. Chem. Soc., 72:5754 (1950).
968. Müller,R., Dathe,C., and Köhne,R., Z. Chem., 3:427 (1963).
969. Müller,R., Meier, G., and Rotzsche, H., Z. Anorg. Chem., 314:291 (1962).
970. Murphy,C. M. and Ravner, H., Polym. Lett., 2:715 (1964).
971. Murphy,C. M., Saunders,C. E., and Smith, D. C., Ind. Eng. Chem., 42:2462 (1950).
972. Nagy, J. and Borbély-Kuszmann,A., Scientific Communications of the International Symposium on Organosilicon Chemistry, A/40-201, Prague (1965).
973. Nakaido,Y., Kogyo Kagaku Zasshi, 67:236 (1964); C. A., 61:9519 (1964).
974. Nakaido,Y. and Takiguchi,T., J. Org. Chem., 26:4144 (1961).
975. Nakazima,I., Kogyo Kagaku Zasshi, 65:1693 (1963); C. A., 58:7010 (1963).
976. Neville,R. G., J. Am. Chem. Soc., 25:1063 (1960).
977. Newlands,M. J., Proc. Chem. Soc., 1960:123.
978. Niederprüm,H., Angew. Chem., 75:165 (1963).
979. Nitzsche,S., Makromal. Chem., 34:231 (1959).
980. Nitzsche,S. and Wick,M., Angew. Chem., 69:96 (1957).
981. Nitzsche,S. and Wick,M., Kunstst., 47:431 (1957).
982. Noller,D. C. and Post,H. W., J. Org. Chem., 17, 1393 (1952).
983. Noltes,J. G. and van der Kerk,G. J. W., Rec. Trav. Chim., 81:39 (1962).
984. Noltes,J. G. and van der Kerk,G. J. M., Rec. Trav. Chim., 81:565 (1962).
985. Nöth,H., Angew. Chem., 73:371 (1961).
986. Nöth,H., Z. Naturf., 16b:618 (1961).
987. Nöth,H. and Höllerer, G., Angew. Chem., 74:718 (1962).
988. Nowak,P. and Rickling,E., Kunstst., 44:191 (1954).
989. Nowotny, H. and Scheil,E., Metallforsch., 2:76 (1947).
990. O'Brien,J. F., WADC Techn. Rep. 57-502, ASTIA Doc. No. 142100 (1957); cited in [101], p. 182 (94).
991. Okawara,R., Proc. Chem. Soc., 1961:383.
992. Okawara,R. and Sugita,K., J. Am. Chem. Soc., 83:4480 (1961).
993. Okawara,R., White,D. G., Fujitani,K., and Sato, H., J. Am. Chem. Soc., 83:1342 (1961).

994. Onyszchuk,M., Canad. J. Chem., 39:808 (1961).
995. Otani,S., Ide,T., Kogure,T., and Okada,Y., Kogyo Kagaku Zasshi, 65:1898 (1962); C. A., 58:8472 (1963).
996. Otani S., Nakaido,Y., Kojima,A., and Arai,C., Kogyo Kagaku Zasshi, 67:239. (1964); C. A., 61:12935 (1964).
997. Owen,J. E., Joyner,R. D., and Kenney,M. E., Abstracts of Papers 139th Meeting Am. Chem. Soc. (1961), 17M-47.
998. Owen,J. E. and Kenney,M. E., Inorg. Chem., 1:331 (1962).
999. Owen,J. E. and Kenney, M. E., Inorg. Chem., 1:334 (1962).
1000. Pape,C., Ann., 222:354 (1884).
1001. Papetti,S. and Heying,T. L., Inorg. Chem., 2:1105 (1963).
1002. Papetti,S. and Heying,T. L., Inorg. Chem., 3:1448 (1964).
1003. Papetti,S. and Post,H. W., J. Org. Chem., 22:526 (1957).
1004. Papetti,S., Schaeffer, B., Troscianiec,H., and Heying,T., Inorg. Chem., 3:1444 (1964).
1005. Patnode,W. and Schmidt,F. C., J. Am. Chem. Soc., 67:2272 (1945).
1006. Petree,H. E., Abstracts of Papers 136th Meeting Am. Chem. Soc., 51N (1959).
1007. Piekoš,R., Roczniki. Chem., 37:301 (1963).
1008. Piekoš,R. and Radecki,A., Z. Anorg. Chem., 309:258 (1961).
1009. Pink,H. S. and Kipping,F. S., J. Chem. Soc., 123:2830 (1923).
1010. Piper, T. S., Lemal,D., and Wilkinson,G., Naturwiss., 43:129 (1956).
1011. Portner,C., Helv. Chem. Acta, 32:1438 (1949).
1012. Prober,M., J. Am. Chem. Soc., 77:3224 (1955).
1013. Prober,M., J. Am. Chem. Soc., 77:5180 (1955).
1014. Pump, J. and Rochow, E. G., Chem. Ber., 97:627 (1964).
1015. Pump,J., Rochow,E. G., and Wannagat,U., Angew. Chem., 75:374 (1963).
1016. Pump,J. and Wannagat,U., Angew. Chem., 74:117 (1962).
1017. Pump,J. and Wannagat,U., Ann. Chem., 652:21 (1962).
1018. Radecki,A., Piekos, R., Roczinki. Chem., 35:869 (1961).
1019. Radecki,A. and Szyrmulewicz,R., Wiadomosci Chemi., 14:23 (1960).
1020. Rausch,M., Vogel,M., and Rosenberg,H., J. Org. Chem., 22:900 (1957).
1021. Rausch,M., Vogel,M., and Rosenberg,H., WADC Techn. Rep. 57-62, pt. I, 1958; cited in [934].
1022. Rausch,M., Vogel,M., Rosenberg,H., Mayo,D., and Shaw, P., WADC Techn. Rep. 57-62 pt. II, 1958; cited in [934].
1023. Reuther,H., J. Prakt. Chem. [4], 5:310 (1958).
1024. Reuther,H., Munkelt,S., Technik, 12:704 (1957).
1025. Reynolds,H. H., Bigelow,L. A., and Kraus,C. A., J. Am. Chem. Soc., 51:3067 (1929).
1026. Ring,M. A., Freeman,L. P., and Fox,A. P., Inorg. Chem., 3:1200 (1964).
1027. Ring,M. A. and Ritter,D. M., J. Am. Chem. Soc., 83:802 (1961).
1028. Roberts,J. D., McElhill,E. A., and Armstrong,R., J. Am. Chem. Soc., 71:2923 (1949).
1029. Robinson,D. E. and Schreihans,F. A., Mater. 7 Symp. Natl. Soc., Aerospace Mater. Process Engrs, Los Angeles (1964), p. 29; C. A., 65:5422 (1965).

1030. Robinson, R. and Kipping, F. S., J. Chem. Soc., 101:2142 (1912).
1031. Robinson, R. and Kipping, F. S., J. Chem. Soc., 101:2156 (1912).
1032. Robinson, R. and Kipping, F. S., J. Chem. Soc., 105:40 (1914).
1033. Rochow, E. G., Bull. Soc. Chim. France, 1963:1360.
1034. Rochow, E. G., Chim. Ind. (Paris), 85:897 (1961); C. A., 55:20491 (1961).
1035. Roedel, G. F., J. Am. Chem. Soc., 71:269 (1949).
1036. Rosenberg, S. D. and Rochow, E. G., J. Am. Chem. Soc., 77:2907 (1955).
1037. Rosnati, L., Gazz. Chim. Ital., 78:516 (1948).
1038. Royen, P. and Rocktäschel, C., Angew. Chem., 76:302 (1964).
1039. Rühlman, K., Chem. Ber., 94:2311 (1961).
1040. Ruidisch, I., Mitteilungsblatt Chem. Ges. DDR, 10:53 (1963).
1041. Ruidisch, I. and Schmidbaur, H., Angew. Chem., 75:1108 (1963).
1042. Ruidisch, I. and Schmidt, M., Angew. Chem., 75:575 (1963).
1043. Ruidisch, I. and Schmidt, M., Chem. Ber., 96:1424 (1963).
1044. Ruidisch, I. and Schmidt, M., J. Organomet. Chem., 1:160 (1963).
1045. Russ, C. R. and Mac Diarmid, A. G., Angew. Chem., 76:500 (1964).
1046. Rust, J. B., Segal, C. L., and Takimoto, H. H., US Dept. Com., Office Techn. Serv. PB Rep. 171522 (1959); cited in C. A., 58:14113 (1963); [101], p. 211 (93).
1047. Rust, J. B., Takimoto, H. H., and Denault, G. C., J. Org. Chem., 25:2040 (1960).
1048. Rust, J. B., Takimoto, H. H., and Segal, C. L., Abstracts of Papers 136th Meeting Am. Chem. Soc. (1959), 3N-7.
1049. Ryan, J. W. and Speier, J. L., J. Am. Chem. Soc., 86:895 (1964).
1050. Sabatier, P., Compt. Rend., 113:132 (1891).
1051. Sakata, Y. and Hashimoto, T., J. Pharm. Soc., Japan., 79:872 (1959); C. A., 54, 357 (1960).
1052. Sauer, R. O., Communication in Discussion on the International Symposium on Organosilicon Chemistry, Prague, Sept., 1965.
1053. Sauer, R. O., J. Am. Chem. Soc., 66:1707 (1944).
1053a. Schaaf, R. L. and Kan, P. T., WADC Tech. Rep. 59-427; cited in [73a], p. 655(17).
1054. Schaaf, R. L., Kan, P. T., Lenk, C., T., and Deck, E. P., Abstracts of Papers 135th Meeting Am. Chem. Soc. (1959), 85-O-136.
1055. Schaaf, R. L., Kan, P. T., Lenk, C. T., and Deck, E. P., J. Org. Chem., 25:1986 (1960).
1056. Schaaf, R. L., Kan, P. T., and Lenk, C. T., J. Org. Chem., 26:1790 (1961).
1057. Schatz, M., Sb. Vysoke Skoly Chem. Techn. Praze, Oddil. Fak. Anorg. Org. Tehn., 4:437 (1960); C. A., 60:14710 (1964).
1058. Scherer, O., Abstracts of Reports to the Twentieth International Congress on Theoretical and Applied Chemistry [Russian translation], D95, Izd. "Nauka" (1965).
1059. Scherer, O. and Schmidt, M., Angew. Chem., 75:139 (1963).
1060. Scherer, O. and Schmidt, M., Angew. Chem., 75:642 (1963).
1061. Scherer, O. and Schmidt, M., Angew. Chem., 75:1115 (1963).
1062. Scherer, O. J., and Schmidt, M., Angew. Chem., 76:144 (1964).
1063. Scherer, O. J. and Schmidt, M., Angew. Chem., 76:787 (1964).

1064. Scherer, O. J. and Schmidt, M., J. Organomet. Chem., 1:490 (1964).
1065. Scherer, O. J. and Schmidt, M., J. Organomet. Chem., 3:156 (1965).
1066. Scherer, O. J. and Schmidt, M., Naturwiss., 50:302 (1963).
1067. Scherer, O. J. and Schmidt, M., Scientific Communications of the International Symposium on Organosilicon Chemistry, B/24-315, Prague (1965).
1068. Scherer, O. and Schmidt, M., Z. Naturf., 18b:415 (1963).
1069. Scherer, O. J., Schmidt, J., Wokulat, J., and Schmidt, M., Z. Naturf., 20b:183 (1965).
1070. Schlenk, W., Renning, J., and Racky, G., Ber., 44:1178 (1921).
1071. Schmidbaur, H., Abstracts of Papers XIX Congress IUPAC, London, (1963) AB-4-2.
1072. Schmidbaur, H., Angew. Chem., 75:137 (1963).
1073. Schmidbaur, H., Angew. Chem., 76:752 (1964).
1074. Schmidbaur, H., Angew. Chem., 77:169 (1965).
1075. Schmidbaur, H., Chem. Ber., 96:2696 (1963).
1076. Schmidbaur, H., Chem. Ber., 97:270 (1964).
1077. Schmidbaur, H., Chem. Ber., 97:469 (1964).
1078. Schmidbaur, H., Chem. Ber., 97:830 (1964).
1079. Schmidbaur, H., Chem. Ber., 97:836 (1964).
1080. Schmidbaur, H., Chem. Ber., 97:842 (1964).
1081. Schmidbaur, H., J. Am. Chem. Soc., 85:2336 (1963).
1082. Schmidbaur, H., J. Organomet. Chem., 1:28 (1963).
1083. Schmidbaur, H., Mitteilungsblatt Chem. Ges. DDR, 10:55 (1963).
1084. Schmidbaur, H., Z. Anorg. Chem., 326:272 (1964).
1085. Schmidbaur, H., Arnold, H. S., and Beinhofer, E., Chem. Ber., 97:449 (1964).
1086. Schmidbaur, H. and Findeiss, W., Angew. Chem., 76:753 (1964).
1087. Schmidbaur, H. and Findeiss, W., unpublished work; cited in [114].
1088. Schmidbaur, H., Findeiss, W., and Gast, E., Angew. Chem., 77:170 (1965).
1089. Schmidbaur, H. and Hussek, H., Angew. Chem., 75:575 (1963).
1090. Schmidbaur, H. and Hussek, H., J. Organomet. Chem., 1:235 (1963).
1091. Schmidbaur, H. and Hussek, H., J. Organomet. Chem., 1:244 (1963).
1092. Schmidbaur, H. and Hussek, H., J. Organomet. Chem., 1:257 (1963).
1093. Schmidbaur, H., Hussek, H., and Schindler, F., Chem. Ber., 97:255 (1964).
1094. Schmidbaur, H., Perez-Garcia, J. A., and Arnold, H. S., Z. Anorg. Chem., 328:105 (1964).
1095. Schmidbaur, H. and Ruidisch, I., Inorg. Chem., 3:599 (1964).
1096. Schmidbaur, H. and Ruidisch, I., Proceedings of the 8th International Conference on Coordination Chemistry, Vienna, 1964; Vienna-New York (1964), 3B2-168.
1097. Schmidbaur, H. and Schindler, F., Abstracts of Proceedings of the Second Symposium on Organometallic Chemistry, Madison (1965), p. 11.
1098. Schmidbaur, H. and Schindler, F., Angew. Chem., 75:1115 (1963).
1099. Schmidbaur, H. and Schindler, F., Chem. Ber., 97:952 (1964).
1100. Schmidbaur, H. and Schmidt, M., Angew. Chem., 73:655 (1961).
1101. Schmidbaur, H. and Schmidt, M., Angew. Chem., 74:327 (1962).
1102. Schmidbaur, H. and Schmidt, M., Angew. Chem., 74:328 (1962).

1103. Schmidbaur,H. and Schmidt,M., Angew. Chem., 74:589 (1962).
1104. Schmidbaur,H. and Schmidt,M., Chem. Ber., 94:1138 (1961).
1105. Schmidbaur,H. and Schmidt,M., Chem. Ber., 94:1349 (1961).
1106. Schmidbaur,H. and Schmidt,M., Chem. Ber., 94:2137 (1961).
1107. Schmidbaur,H. and Schmidt,M., J. Am. Chem. Soc., 83:2963 (1961).
1108. Schmidbaur,H. and Schmidt,M., J. Am. Chem. Soc., 84:1069 (1962).
1109. Schmidbaur,H. and Schmidt,M., J. Am. Chem. Soc., 84:3600 (1962).
1110. Schmidbaur,H. and Waldmann,S., Angew. Chem., 76:753 (1964).
1111. Schmidbaur,H. and Waldmann,S., Chem. Ber., 97:3381 (1964).
1112. Schmidt,M., In report M/2 on the International Symposium on Organosilicon Chemistry, Prague (1965).
1113. Schmidt,M., Bornmann,P., and Wilhelm, I., Angew. Chem., 75:1024 (1963).
1114. Schmidt,M. and Ruf,H., Angew. Chem., 73:64 (1961).
1115. Schmidt,M. and Ruf,H., Z. Anorg. Chem., 321:270 (1963).
1116. Schmidt,M., Ruidisch,I., and Schmidbaur,H., Chem. Ber., 94:2451 (1961).
1117. Schmidt,M. and Schmidbaur,H., Angew. Chem., 70:704 (1958).
1118. Schmidt,M. and Schmidbaur,H., Angew. Chem., 71:220 (1959).
1119. Schmidt,M. and Schmidbaur,H., Angew. Chem., 71:553 (1959).
1120. Schmidt,M. and Schmidbaur,H., Chem. Ber., 92:2667 (1959).
1121. Schmidt,M., Schmidbaur,H. and Ruidisch,I., Angew. Chem., 73:408 (1961).
1122. Schmidt,M. and Wieber,M., Chem. Ber., 94:1426 (1961).
1123. Schmidt,M. and Wilhelm,I., Chem. Ber. 97:876 (1964).
1124. Schmitz-Du Mont,O.. and Bungard,G., Chem. Ber., 92:2399 (1959).
1125. Schmitz-Du Mont,O., Merten,D., and Eiding,D., Z. Anorg. Chem., 319:362 (1963).
1126. Schneider,C. and Berg,G., Makromal. Chem., 54:171 (1962).
1127. Schumb,W. C. and Saffer,C. M., J. Am. Chem. Soc., 61:363 (1939).
1128. Schwartz, N. N., O'Brien,E., Karlan,S., and Fein,M. M., Inorg. Chem., 4:661 (1965).
1129. Schwartz,N. N. and Brook,A. G., J. Am. Chem. Soc., 82:2439 (1960).
1130. Schwarz,R. and Konrad,E., Ber., 55:3242 (1922).
1131. Schwarz,R. and Sexauer,W., Ber., 59:333 (1926).
1132. Segal,C. L., Takimoto,H. H.,and Rust,J. B., Abstracts of Papers 137th Meeting Am. Chem. Soc., (1960), 3S; [102], p. 211 (19); [101], p. 182 (95).
1133. Selin,T. G., West,R., Abstracts of Papers 135th Meeting Am. Chem. Soc, (1959), 24M-64.
1134. Selin,T. G. and West,R., Tetrahedron, 5:97 (1959).
1135. Senear,A. E., Wirth,J., and Neville,R. G., J. Org. Chem., 25:807 (1960).
1136. Seyferth,D., J. Am. Chem. Soc., 79:5881 (1957).
1137. Seyferth,D., J. Am. Chem. Soc., 80:1336 (1958).
1138. Seyferth,D., J. Am. Chem. Soc., 81:1844 (1959).
1139. Seyferth,D., J. Am. Chem. Soc., 83:1610 (1961).
1140. Seyferth,D., J. Inorg. Nucl. Chem., 7:152 (1958).
1141. Seyferth,D., J. Org. Chem., 22:1252 (1957).

1142. Seyferth, D. and Alleston, D. L., Inorg. Chem., 2:417 (1963).
1143. Seyferth, D. and Alleston, D. L., Inorg. Chem., 2:418 (1963).
1144. Seyferth, D. and Burlitsch, J. M., J. Am. Chem. Soc., 85:2667 (1963).
1145. Seyferth, D. and Cohen, H. M., J. Inorg. Nucl., Chem., 20:73 (1961).
1146. Seyferth, D., Dertouzos, H., and Todd, L. J., J. Organomet. Chem., 4:18 (1965).
1147. Seyferth, D. and Freyer, W., J. Org. Chem., 26, 2604 (1961).
1148. Seyferth, D., Freyer, W. R., and Takamizawa, M., Inorg. Chem., 1:710 (1962).
1149. Seyferth, D., Hofmann, H. P., Burton, R., and Helling, J. F., Inorg. Chem., 1:227 (1962).
1150. Seyferth, D. and Kahlen, N., J. Am. Chem. Soc., 82:1080 (1960).
1151. Seyferth, D. and Kögler, H. P., J. Inorg. Nucl. Chem., 15:99 (1960).
1152. Seyferth, D., Raab, G., and Grim, S. O., J. Org. Chem., 26:3034 (1961).
1153. Seyferth, D. and Rochow, E. G., J. Am. Chem. Soc., 77:906 (1955).
1154. Seyferth, D. and Rochow, E. G., J. Org. Chem., 20:250 (1955).
1155. Seyferth, D. and Rochow, E. G., J. Polymer Sci., 18:543 (1955).
1156. Seyferth, D. and Takamizawa, M., Inorg. Chem., 2:731 (1963).
1157. Seyferth, D. and Wada, T., Inorg. Chem., 1:78 (1962).
1158. Seyferth, D., Wada T., and Raab G., Tetrahedron Lett., 1960:20.
1159. Seyferth, D., Weiner, M. A., Grim, S. O., and Kahlen, N., Abstracts of Papers 135th Meeting Am. Chem. Soc. (1959), 20-O-31.
1160. Seyferth, D., Weiner, M. A., Vaughan, L. G., Raab, G., Welch, D. E., Cohen H. M., and Alleston, D. L., Bull. Soc. Chim. France, 1963:1364.
1161. Seyferth, D., Yamazaki, H., and Sata, Y., Inorg. Chem., 2:734 (1963).
1162. Schackelford, J. M., De Schmertzing, H., Heuther, C. H., and Podall, H., J. Org. Chem., 28:1700 (1963).
1163. Sheng-Lieh Liu, and Chin-Hsia Ho, J. Chin. Chem. Soc. (Taiwan), 6:141 (1960); C. A., 55:4673 (1961).
1164. Sheng-Lieh Liu, J. Chin. Chem. Soc. (Taiwan), Ser. II, 8:226 (1961); C. A., 58, 8234 (1963).
1165. Sidlow, R., Chem. Prod., 16:215 (1953).
1166. Sidlow, R., Ind. Finish., 5:850 (1953).
1167. Sidlow, R., J. Oil. Coll. Chem. Assoc., 39:415 (1956).
1168. Simmler, W. and Wiberg, E., Angew. Chem., 67:709 (1955).
1169. Sisler, H. H., Pfahler, B., and Wilson, W. J., J. Am. Chem. Soc., 70:3825 (1948).
1170. Sommer, L. H. and Ansul, G. R., J. Am. Chem. Soc., 77:2482 (1955).
1171. Sommer, L. H. and Frye, C. L., J. Am. Chem. Soc., 82:3796 (1960).
1172. Sommer, L. H., Gold, J. R., Goldberg, G. M., and Marans, N. S., J. Am. Chem. Soc., 71:1509 (1949).
1173. Sommer, L. H., Goldberg, G. M., Gold, J., and Whitmore, F. C., J. Am. Chem. Soc., 69:980 (1947).
1174. Sommer, L. H., Green, L. Q., and Whitmore, F. C., J. Am. Chem. Soc., 71:3253 (1949).
1175. Sommer, L. H., Mitch, F. A., and Goldberg, G. M., J. Am. Chem. Soc., 71:2746 (1949).

1176. Sommer,L. H., Murch,R. M., and Mitch,F. A., J. Am. Chem. Soc., 76:1619 (1954).
1177. Sommer,L. H., Pietrusza,E. W., and Whitmore,F. C., J. Am. Chem. Soc., 68:2282 (1946).
1178. Sommer,L. H., Pioch,R. P., Marans,N. S., Goldberg,G. M., Rockett,J., and Kerlin,J., J. Am. Chem. Soc., 75:2932 (1953).
1179. Sommer,L. H., Strien van,R. E., and Whitmore,F. C., J. Am. Chem. Soc. 71:3056 (1949).
1180. Sommer,L. H., and Whitmore,F. C., J. Am. Chem. Soc., 68:485 (1946).
1181. Spanier,E. J. and Mac Diarmid,A. G., Inorg. Chem., 2:215 (1963).
1182. Spialter,L., Harris,C. W., Abstracts of Papers 126th Meeting Am. Chem. Soc. (1954), p. 86; cited in [99], p. 331 (424).
1182a. Starshak, A. J., Joyner, R. D., and Kenney, M. E., Inorg. Chem., 5:330 (1966).
1183. Steel, 138, 92 (1956).
1184. Steele,A. R. and Kipping,F. S., J. Chem. Soc., 1928:1431.
1185. Steele,A. R. and Kipping,F. S., J. Chem. Soc., 1929:2549.
1186. Sternbach,B. and Mac Diarmid,A. G., J. Am. Chem. Soc., 83:3384 (1961).
1187. Steudel,W. and Gilman,H., J. Am. Chem. Soc., 82:6129 (1960).
1188. Stock,A. and Somieski,C., Ber., 49:147 (1916).
1189. Stock,A. and Somieski,C., Ber., 52:695 (1919).
1190. Stock,A. and Somieski,C., Ber., 54:524 (1921).
1191. Stokes,H. N., Am. Chem. J., 14:438, 545 (1892).
1192. Stolberg,U., Angew. Chem., 74:696 (1962).
1193. Stolberg,U., Angew. Chem., 75:206 (1963).
1194. Sujishi,S. and Ando,W., US Dept. Com., Office Techn. Serv., PB. Rep. 143572 (1959); cited in [85], p. 130 (26a).
1195. Sujishi,S. and Manasevit,H. M., Abstracts of Papers 135th Meeting Am. Chem. Soc. (1959), 48M-120.
1196. Sujishi,S. and Manasevit,H. M., US Dept. Com., Office Techn. Serv. PB Rep. 143572 (1959); cited in [85], p. 113 (44a).
1197. Sujishi,S. and Witz,S., J. Am. Chem. Soc., 76:4631 (1954).
1198. Sujishi,S. and Witz,S., J. Am. Chem. Soc., 79:2447 (1957).
1199. Takiguchi,T., J. Org. Chem., 23:1216 (1958).
1200. Takiguchi,T. and Sakurai,M., Kogyo Kagaku Zasshi, 63:1476 (1960); C. A., 57:7297 (1962).
1201. Takiguchi,T., Sakurai,M., Kishi,T., Ichimura,J., and Iizuka,Y., J. Org. Chem., 25:310 (1960).
1202. Takimoto,H. H. and Rust,J. B., Abstracts of Papers 137th Meeting Am. Chem. Soc. (1960), 86-O.
1203. Takimoto,H. H. and Rust,J. B., J. Org. Chem., 26:2467 (1961).
1204. Tamborski,C., Ford,F. E., Lehn,W. L., Moore,G. J., and Soloski,E., J. Org. Chem., 27:619 (1962).
1205. Tamborski,C., Ford,F. E., and Soloski,E. J., J. Org. Chem., 28:181 (1963).
1206. Tatlock,W. S. and Rochow,E. G., J. Org. Chem., 17:1555 (1952).
1207. Taurke,F., Ber., 38:1661 (1905).
1208. Thayer,J. S. and West,R., Inorg. Chem., 4:114 (1965).

1209. Thies, C. and Kinsinger, J. B., Inorg. Chem., 3:551 (1965).
1210. Timms, P. L., Simpson, C. C., and Phillips, C. S. G., J. Chem. Soc., 1964:1467.
1211. Trepka, W. J., J. Polymer Sci., B1:683 (1963).
1212. Tse-Cheng Wu, Iowa State Coll. J. Sci., 27:282 (1953); Ref. Zh. Khim., 1953, 8498.
1213. Tyler, L. J., Sommer, L. H., and Whitmore, F. C., J. Am. Chem. Soc., 70:2876 (1948).
1214. Vago, E. E. and Barrow, R. F., J. Chim. Phys., 45:9 (1948).
1215. Vago, E. E. and Barrow, R. F., Nature, 157:77 (1947).
1216. Vago, E. E. and Barrow, R. F., Proc. Phys. Soc., 58:538 (1946).
1217. Vale, R. L., J. Chem. Soc., 1960:2252.
1218. Varma, R. and Cox, A. R., Angew. Chem., 76:649 (1964).
1219. Viehe, H. G., Chem. Ber., 92:3064 (1959).
1220. Waack, R. and Doran, M. A., Chem. Ind., 1965:563.
1221. WADC Techn. Rep. 58-160, ASTIA Document No. 155675; cited in [102], p. 212 (21).
1222. Wannagat, U., Angew. Chem., 72:586 (1960).
1223. Wannagat, U., Angew. Chem., 75:173 (1963).
1224. Wannagat, U., Behmel, K., Wolf, H., and Bürger, H., Z. Anorg. Chem., 333:62 (1964).
1225. Wannagat, U., Bogusch, E., and Geymayer, P., Monatsh., 95:801 (1964).
1226. Wannagat, U., Brandmair, F., Liehr, W., and Niederprüm H., Z. Anorg. Chem., 302:185 (1959).
1227. Wannagat, U. and Brandstätter, O., Angew. Chem., 75:345 (1963).
1228. Wannagat, U. and Bürger, H., Angew. Chem., 75:95 (1963).
1229. Wannagat, U. and Bürger, H., Angew. Chem., 76:497 (1964).
1230. Wannagat, U. and Bürger, H., Z. Anorg. Chem., 326:309 (1964).
1231. Wannagat, U., Bürger, H., Geymayer, P., and Torper, G., Monatsh, 95:39 (1964).
1232. Wannagat, U., Bürger, H., Peach, M. E., Hensen, K., and Lebert, K. H., Z. Anorg. Chem., 336:129 (1965).
1233. Wannagat, U., Geymayer, P., and Schreiner, G., Angew. Chem., 76:99 (1964).
1234. Wannagat, U. and Krüger, C., Z. Anorg. Chem., 326:288 (1964).
1235. Wannagat, U., Krüger, C., and Niederprüm, H., Z. Anorg. Chem., 321:198 (1963).
1236. Wannagat, U. and Kuchertz, H., Angew. Chem., 74:117 (1962).
1237. Wannagat, U. and Kuchertz, H., Angew. Chem., 75:95 (1963).
1238. Wannagat, U. and Kuchertz, H., unpublished data, 1962; cited in [121], p. 242 (165).
1239. Wannagat, U., Kuchertz, H., Krüger, C., and Pump, J., Z. Anorg. Chem., 333:54 (1964).
1240. Wannagat, U. and Liehr, W., Angew. Chem., 69:783 (1957).
1241. Wannagat, U. and Niederprüm, H., Angew. Chem., 71:574 (1959).
1242. Wannagat, U. and Niederprüm, H., Chem. Ber., 94:1540 (1961).

1243. Wannagat, U. and Niederprüm, H., Z. Anorg. Chem., 308:337 (1961).
1244. Wannagat, U. and Niederprüm, H., Z. Anorg. Chem., 310:32 (1960).
1245. Wannagat, U., Pump, J., and Bürger, H., Monatsh., 94:1013 (1963).
1246. Wannagat, U. and Schreiner, G., Monatsh., 95:46 (1964).
1247. Wannagat, U., Veigl, W., and Bürger, H., Monatsh., 96:593 (1965).
1248. Wartik, T. and Pearson, R. K., J. Inorg. Nucl. Chem., 5:250 (1958).
1249. Weiss, A. and Harvey, D. R., Angew. Chem., 76:818 (1964).
1250. Weiss, A. and Weiss, A., Angew. Chem., 66:714 (1954).
1251. Weiss, A. and Weiss, A., Z. Anorg. Chem., 273:124 (1953).
1252. Weiss, A. and Weiss, A., Z. Naturf., 8b:104 (1953).
1253. Weyenberg, D. R. and Toporcer, L. H., J. Am. Chem. Soc., 84:2843 (1962).
1254. Weyenberg, D. R. and Toporcer, L. H., J. Org. Chem., 30:943 (1965).
1255. Wheatley, P. J., J. Chem. Soc., 1963:2562.
1256. Wheatley, P. J., J. Chem. Soc., 1963:3200.
1257. Whitmore, F. C. and Sommer, L. H., J. Am. Chem. Soc., 68:481 (1946).
1258. Whitmore, F. C., Sommer, L. H., Gold, J., and Strien van, R. E., J. Am. Chem. Soc., 69:1551 (1947).
1259. Wiberg, E. and Höckele, G., unpublished work; cited in [985], p. 377 (109).
1260. Wiberg, E. and Krüerke, U., Angew. Chem., 66:339 (1954).
1261. Wiberg, E. and Krüerke, U., Z. Naturf., 8b:608 (1953).
1262. Wiberg, E. and Krüerke, U., Z. Naturf., 8b:609 (1953).
1263. Wiberg, E. and Krüerke, U., Z. Naturf., 8b:610 (1953).
1264. Wiberg, N. and Gieren, A., Angew. Chem., 74:942 (1962).
1265. Wiberg, N. and Schmid, K. H., Angew. Chem., 76:380 (1964).
1266. Wiberg, N. and Schmid, K. H., Angew. Chem., 76:381 (1964).
1267. Wick, M., Kunstst., 50:433 (1960).
1268. Wieber, M. and Schmidt, M., Angew. Chem., 74:903 (1962).
1269. Wieber, M. and Schmidt, M., Chem. Ber., 96:1016 (1963).
1270. Willis, W. O., Soil Sci. Soc. Am., Proc., 19:263 (1955); C. A., 49:16296 (1955).
1271. Wilson, G. and Smith, A., J. Org. Chem., 26:557 (1961).
1272. Winkler, H. J. S. and Gilman, H., J. Org. Chem., 27:254 (1962).
1273. Winkler, H. J. S., Jarvie, A. W. T., Peterson, D. J., and Gilman, H., J. Am. Chem. Soc., 83:4089 (1961).
1274. Wittenberg, D., Aoki, D., and Gilman, H., J. Am. Chem. Soc., 80:5933 (1958).
1275. Wittenberg, D., George, M. V., and Gilman, H., J. Am. Chem. Soc., 81:4812 (1959).
1276. Wittenberg, D., George, M. V., Wu T. C., Miles, D. H., and Gilman, H., J. Am. Chem. Soc., 80:4532 (1959).
1277. Wittenberg, D. and Gilman, H., Chem. Ind., 1958:390.
1278. Wittenberg, D. and Gilman, H., J. Am. Chem. Soc., 80:2677 (1958).
1279. Wittenberg, D. and Gilman, H., J. Am. Chem. Soc., 80:4529 (1958).
1280. Wittenberg, D. and Gilman, H., J. Org. Chem., 23:1063 (1958).
1281. Wittenberg, D., McNinch, H. A., and Gilman, H., J. Am. Chem. Soc., 80:5418 (1958).

1282. Wittenberg,D., Talukdar,P. B., and Gilman,H., J. Am. Chem. Soc., 82:3608 (1960).
1283. Wittenberg,D., Wu T. C., and Gilman,H., J. Org. Chem., 23:1898 (1958).
1284. Wittenberg,D., Wu T. C., and Gilman,H., J. Org. Chem., 24:1349 (1959).
1285. Woods,W. G. and Iverson,M. L., Abstracts of Papers 136th Meeting Am. Chem. Soc. (1959), 2N-6; cited in [101], p. 172 (56); [102], p. 211 (16).
1286. Woods,W. G., Iverson,M. L., Papers Presented at the Second Conference on High Temperature Polymer and Fluid Research, Dayton, (1959); cited in [101], p. 173 (57).
1287. Wu,T. C. and Gilman,H., J. Org. Chem., 23:913 (1958).
1288. Wu,T. C., Wittenberg,D., and Gilman,H., J. Org. Chem., 25:596 (1960).
1289. Zeitler,V. A. and Brown,C. A., J. Am. Chem. Soc., 79:4616 (1957).
1290. Zeitler, V. A. and Brown,C. A., J. Am. Chem. Soc., 79:4618 (1957).
1291. Zeitler, V. A. and Brown,C. A., J. Phys. Chem., 61:1174 (1957).

4. Author's Certificates and Patents*

Australian Patents
1292. 160036 (1951); Ref. Zh. Khim., 1955, 47599 (see 1392 and 1589).

Austrian Patents
1293. 218149 (1961); Zbl., 1962, 18341.
1294. 218150 (1960); Zbl., 1962, 18341 (see 1492 and 1672).

British Patents
1295. 450256 (1936); C. A., 30, 8532 (1936) (see 1390 and 1699).
1296. 549081 (1942); C. A., 38, 552 (1944) (see 1402 and 1520); Paint Techn., 11, 277 (1946).
1297. 563995 (1944); C. A., 40, 2137 (1946) (see 1521).
1298. 583875 (1947); C. A., 41, 3115 (1947).
1299. 583878 (1947); C. A., 41, 3115 (1947).
1300. 592456 (1947); C. A., 42, 1762 (1948).
1301. 594481 (1947); C. A., 42, 2809 (1948).
1302. 594506 (1947); C. A., 42, 2809 (1948).
1303. 597834 (1948); C. A., 42, 4793 (1948).
1304. 609324 (1948); C. A., 43, 2221 (1949).
1305. 627136 (1949); C. A., 44, 4284 (1950).
1306. 631506 (1949); C. A., 44, 4490 (1950).
1307. 643298 (1950); C. A., 45, 7819 (1951) (see 1528).
1308. 648840 (1951); C. A., 45, 5455 (1951).
1309. 651699 (1951); C. A., 45, 7818 (1951).
1310. 667333 (1952); C. A., 46, 9342 (1952).
1311. 668234 (1952); C. A., 47, 2766 (1953).
1312. 668903 (1952); C. A., 48, 3063 (1954).
1313. 669790 (1952); C. A., 46, 10681 (1952).
1314. 670923 (1952); C. A., 46, 8894 (1952).

*The cross references indicate equivalent patents of other natuions.

1315. 675233 (1952); C. A., 47, 2539 (1953).
1316. 682541 (1952); C. A., 47, 4647 (1953).
1317. 685183 (1952); C. A., 48, 2761 (1954) (see 1396 and 1574).
1318. 694526 (1953); C. A., 48, 10765 (1954).
1319. 698300 (1953); Ref. Zh. Khim., 1955, 44645.
1320. 705639 (1954); Zbl., 1954, 7302 (see 1573).
1321. 706781 (1954); Ref. Zh. Khim., 1956, 8243.
1322. 713233 (1954); Zbl., 1955, 9922 (see 1405 and 1596).
1323. 716035 (1954); Ref. Zh. Khim., 1955, 56734 (see 1552).
1324. 716293 (1954); Ref. Zh. Khim., 1956, 11219.
1325. 718905 (1954); C. A., 49, 5878 (1955) (see 1397 and 1553).
1326. 722593 (1954); Ref. Zh. Khim., 1956, 63230.
1327. 723989 (1955); Ref. Zh. Khim., 1956, 56266; Paint. Oil. a. Colour J., 127, 1121 (1955).
1328. 726224 (1955); Zbl., 1956, 14563 (see 1401).
1329. 728751 (1955); C. A., 49, 12878 (1955) (see 1292 and 1589).
1330. 729605 (1955); Zbl., 1957, 14203 (see 1705).
1331. 736184 (1955); Zbl., 1956, 7410.
1332. 759013 (1956); C. A., 51, 6213 (1957).
1333. 766810 (1957); C. A., 51, 10950 (1957).
1334. 767226 (1957); C. A., 51, 10950 (1957) (see 1403).
1335. 771167 (1957); C. A., 52, 1204 (1958) (see 1411).
1336. 783302 (1957); C. A., 52, 2451 (1958) (see 1607).
1337. 786434 (1958); C. A., 52, 7757 (1958).
1338. 788599 (1958); C. A., 52, 9651 (1958).
1339. 788653 (1958); C. A., 52, 9646, (1958).
1340. 791370 (1958); C. A., 52, 15525 (1958).
1341. 791991 (1958); C. A., 52, 16745 (1958).
1342. 792339 (1958); C. A., 52, 19237 (1958) (see 1623).
1343. 792470 (1958); C. A., 52, 19237 (1958).
1344. 796141 (1958); C. A., 53, 2640 (1959).
1345. 797235 (1958); C. A., 53, 17568 (1959).
1346. 797598 (1958); C. A., 52, 21163 (1958).
1347. 798669 (1958); C. A., 53, 3756 (1959).
1348. 799372 (1958); C. A., 53, 2512 (1959).
1348a. 822862 (1959); C. A., 54, 10366 (1960) (see 1632).
1349. 826620 (1960); C. A., 55, 2570 (1961).
1350. 841825 (1960); C. A., 55, 4028 (1961) (see 1711).
1351. 844985 (1960); C. A., 55, 5981 (1961) (see 1424).
1352. 845651 (1960); C. A., 55, 5982 (1961).
1353. 847082 (1960); C. A., 55, 17079 (1961).
1354. 847414 (1960); C. A., 55, 5981 (1961).
1355. 851497 (1960); C. A., 55, 13315 (1961).
1356. 867066 (1961); C. A., 55, 22891 (1961).
1357. 870646 (1961); C. A., 55, 27967 (1961).
1358. 871784 (1961); C. A., 55, 26523 (1961).

1359. 875728 (1961); C. A., 56, 4950 (1962).
1360. 876708 (1961); C. A., 56, 6001 (1962).
1361. 881179 (1961); C. A., 56, 8745 (1962) (see 1434).
1362. 883266 (1961); C. A., 56, 15685 (1962).
1363. 886140 (1962); C. A., 57, 2254 (1962).
1364. 888938 (1962); C. A., 57, 7310 (1962).
1365. 889125 (1962); C. A., 56, 14472 (1962).
1366. 890007 (1962); C. A., 57, 1014 (1962).
1367. 891087 (1962); C. A., 59, 11560 (1963) (see 1671).
1368. 894758 (1962); C. A., 57, 7444 (1962).
1369. 914332 (1963); C. A., 58, 7012 (1963).
1370. 916508 (1963); C. A., 58, 12737 (1963).
1371. 917226 (1963); C. A., 58, 11560 (1963).
1372. 930470 (1963); C. A., 59, 10307 (1963).
1373. 936408 (1963); C. A., 59, 15466 (1963).
1374. 940545 (1963); C. A., 60, 7010 (1964).
1375. 945810 (1964); C. A., 61, 5816 (1964).
1376. 947847 (1964); C. A., 60, 12214 (1964).
1377. 978666 (1964); C. A., 62, 6650 (1965).

Belgian Patents
1378. 500963 (1951); C. A., 48, 9080 (1954).
1379. 583614 (1959); C. A., 54, 14777 (1960).
1380. 589008 (1960); C. A., 55, 3113 (1961).
1381. 616597 (1962); C. A., 58, 8100 (1963).
1382. 618862 (1962); C. A., 58, 5867, 9304 (1963).
1383. 620132 (1963); C. A., 58, 14250 (1963).
1384. 623602 (1963); C. A., 60, 12216 (1964).
1385. 624744 (1963); C. A., 59, 828 (1963).
1386. 630249 (1963); C. A., 60, 15911 (1965).
1387. 636777 (1963); C. A., 62, 2900 (1965).

Hungarian Patents
1388. 147040 (1960); C. A., 58, 3557 (1963).
1389. 150947 (1963); C. A., 60, 8214 (1964).

German Patents
General
1390. 665149 (1938); C. A., 33, 1881 (1939) (see 1295 and 1699).

(East German Patents)
1391. 9957 (1955); C. A., 53, 1804 (1959).
1392. 15855 (1958); C. A., 54, 3797 (1960).
1393. 17062 (1959); C. A., 54, 18977 (1960).

West German
Patents

1394. 831098 (1952); Zbl., 1952, 6450.
1395. 833126 (1952); Zbl., 1952, 6942.
1396. 873085 (1953); Zbl., 1954, 417 (see 1317 and 1574).
1397. 885312 (1953); C. A., 51, 1642 (1957) (see 1325 and 1553).
1398. 888614 (1953); Zbl., 1954, 4004 (see 1459 and 1558).
1399. 891330 (1953); Zbl., 1954, 4503 (see 1524).
1400. 924533 (1955); Zbl., 1955, 11722 (see 1575).
1401. 924678 (1955); Zbl., 1956, 833 (see 1328).
1402. 926811 (1955); Zbl., 1955, 8982 (see 1296 and 1520).
1403. 935455 (1955); Zbl., 1956, 5421 (see 1334).
1404. 937557 (1956); C. A., 53, 222 (1959).
1405. 942886 (1956); Zbl., 1956, 14553 (see 1322 and 1596).
1406. 951889 (1956); C. A., 53, 3767 (1959).
1407. 951890 (1956); C. A., 53, 2676 (1959).
1408. 954645 (1956); Zbl., 1957, 8952.
1409. 958702 (1957); C. A., 53, 13657 (1959) (see 1616).
1410. 959320 (1957); C. A., 53, 16586 (1959).
1411. 962077 (1957); C. A., 53, 13057 (1959) (see 1335).
1412. 967031 (1957); C. A., 53, 18406 (1959).
1413. 1008434 (1957); C. A., 53, 14579 (1959).
1414. 1009631 (1957); C. A., 53, 21666 (1959).
1415. 1016870 (1957); C. A., 54, 11561 (1960).
1416. 1019462 (1957); C. A., 54, 10378 (1960).
1417. 1030660 (1958); C. A., 54, 10862 (1960).
1418. 1035360 (1958); C. A., 54, 23402 (1960).
1419. 1045092 (1958); C. A., 54, 23424 (1960).
1420. 1048226 (1958); cited in [855].
1421. 1058254 (1959); C. A., 55, 6011 (1961).
1422. 1073740 (1960); C. A., 55, 13860 (1961).
1423. 1077355 (1960); C. A., 55, 10920 (1961).
1424. 1080061 (1960); C. A., 55, 18132 (1961) (see 1351).
1425. 1080089 (1960); C. A., 55, 13317 (1961).
1426. 1083820 (1960); C. A., 55, 22906 (1961).
1427. 1087810 (1960); C. A., 55, 15998 (1961).
1428. 1097133 (1961); C. A., 55, 22904 (1961).
1429. 1097134 (1961); C. A., 55, 22904 (1961) (see 1294 and 1672).
1430. 1099743 (1961); C. A., 55, 27875 (1961).
1431. 1104705 (1961); C. A., 55, 24103 (1961).
1432. 1108429 (1961); C. A., 57, 1014 (1962).
1433. 1110410 (1961); C. A., 55, 26497 (1961).
1434. 1111183 (1961); C. A., 56, 8746 (1962) (see 1361).
1435. 1111400 (1961); C. A., 56, 11804 (1962).
1436. 1113813 (1961); C. A., 56, 10354 (1962).

1437. 1115250 (1961); C. A., 56, 8743 (1962).
1438. 1116396 (1961); C. A., 57, 6131 (1962).
1439. 1117248 (1961); C. A., 56, 6119 (1962).
1440. 1117250 (1961); C. A., 56, 11774 (1962).
1440a. 1117869 (1961); C. A., 56: 13095 (1962).
1441. 1117872 (1961); C. A., 57, 16832 (1962) (see 1673).
1442. 1118454 (1961); C. A., 56, 11772 (1962).
1443. 1118455 (1961); C. A., 57, 7310 (1962).
1444. 1120689 (1961); C. A., 56, 10360 (1962).
1445. 1123465 (1962); C. A., 57, 1071 (1962).
1446. 1127586 (1962); C. A., 57, 2388 (1962).
1447. 1131009 (1962); C. A., 57, 8741 (1962).
1448. 1138933 (1962); C. A., 58, 2556 (1963).
1449. 1150168 (1963); C. A., 59, 7755 (1963).
1450. 1153167 (1963); C. A., 59, 13001 (1963).
1451. 1158972 (1963); C. A., 60, 9311 (1964).
1452. 1158973 (1963); C. A., 60, 9312 (1964).
1453. 1163021 (1964); C. A., 60, 12218 (1964).
1454. 1176137 (1964); C. A., 61, 12035 (1964).

Dutch Patents

1455. 99386 (1961); C. A., 57, 9983 (1962).

Canadian Patents

1456. 497728 (1953); Ref. Zh. Khim., 1955, 22533 (see 1525).
1457. 504282 (1954); Ref. Zh. Khim., 1955, 53696.
1458. 517150 (1955); Ref. Zh. Khim., 1956, 73155.
1459. 519188 (1955); Ref. Zh. Khim., 1957, 24897 (see 1398 and 1558).
1460. 523955 (1956); Ref. Zh. Khim., 1957, 42851.

Author's Certificates of the USSR

1461. 71115 (1947); Summary of USSR Inventions, No. 5, 1948, Gostoptekhizdat (1950), p. 65.
1462. 102041 (1955); Byull. Izobr., No. 12, 13 (1955).
1463. 107334 (1957); Byull. Izobr., No. 7, 40 (1957).
1464. 110206 (1958); Byull. Izobr., No. 1, 77 (1958).
1465. 110915 (1958); Byull. Izobr., No. 2, 16 (1958).
1466. 110972 (1958); Byull. Izobr., No. 2, 16 (1958).
1467. 110975 (1958); Byull. Izobr., No. 2, 17 (1958).
1468. 110979 (1958); Byull. Izobr. No. 2, 17 (1958).
1469. 110992 (1958); C. A., 52, 19228 (1958).
1470. 111485 (1958); Byull. Izobr., No. 3, 76 (1958).
1471. 112939 (1958); Byull. Izobr., No. 5, 36 (1958).
1472. 115167 (1958); Byull. Izobr., No. 9, 13 (1958).
1473. 115461 (1958).
1474. 116609 (1958); Byull. Izobr., No. 12, 30 (1958).
1475. 116680 (1958); Byull. Izobr., No. 12, 30 (1958).
1476. 117658 (1959); Byull. Izobr., No. 2, 28 (1959).

1477. 121963 (1959); Byull. Izobr., No. 16, 57 (1959).
1478. 122280 (1959); Byull. Izobr., No. 17, 40 (1959).
1479. 125251 (1960); Byull. Izobr., No. 1, 14 (1960).
1480. 125681 (1960); Byull. Izobr., No. 2, 35 (1960).
1481. 126115 (1960); Byull. Izobr., No. 4, 15 (1960).
1482. 126496 (1960); Byull. Izobr., No. 5, 17 (1960).
1483. 126881 (1960); Byull. Izobr., No. 6, 19 (1960).
1484. 127023 (1960); Byull. Izobr., No. 6, 44 (1960).
1485. 128020 (1960); Byull. Izobr., No. 9, 19 (1960).
1486. 128461 (1960); Byull. Izobr., No. 10, 19 (1960).
1487. 130181 (1960); Byull. Izobr., No. 14, 39 (1960).
1488. 132634 (1960); Byull. Izobr., No. 20, 17 (1960).
1489. 132636 (1960); Byull. Izobr., No. 20, 17 (1960).
1490. 134668 (1961); Byull. Izobr., No. 1, 11 (1961).
1491. 135635 (1961); Byull. Izobr., No. 3, 48 (1961).
1492. 139431 (1961); Byull. Izobr., No. 13, 42 (1961).
1493. 139434 (1961); Byull. Izobr., No. 13, 43 (1961).
1494. 139438 (1961); Byull. Izobr., No. 13, 43 (1961).
1495. 141154 (1961); Byull. Izobr., No. 18, 17 (1961).
1496. 141156 (1961); Byull. Izobr., No. 18, 17 (1961).
1497. 145349 (1962); Byull. Izobr., No. 5, 46 (1962).
1498. 148048 (1964); Byull. Izobr., No. 12, 26 (1962).
1499. 148403 (1962); Byull. Izobr., No. 13, 15 (1962).
1500. 150959 (1962); Byull. Izobr., No. 20, 41 (1962).
1501. 151686 (1962); Byull. Izobr., No. 22, 26 (1962).
1502. 152573 (1963); Byull. Izobr., No. 18, 106 (1962).
1503. 154387 (1963); Byull. Izobr., No. 9, 51 (1963).
1504. 154397 (1963); Byull. Izobr., No. 9, 52 (1963).
1504a. 156742 (1963); Byull. Izobr., No. 16, 20 (1963).
1505. 162530 (1964); Byull. Izobr., No. 10, 21 (1964).
1506. 162531 (1964); Byull. Izobr., No. 10, 21 (1964).
1507. 165539 (1964); Byull. Izobr., No. 19, 34 (1964).
1508. 165718 (1964); Byull. Izobr., No. 20, 14 (1964).
1509. 165897 (1964); Byull. Izobr., No. 18, 60 (1964).
1510. 166336 (1964); Byull. Izobr., No. 22, 20 (1964).
1511. 166338 (1964); Byull. Izobr., No. 22, 20 (1960).
1512. 166834 (1964); Byull. Izobr., No. 23, 60 (1964).
1513. 168444 (1965); Byull. Izobr., No. 4, 62 (1965).
1514. 168689 (1965); Byull. Izobr., No. 5, 22 (1965).
1515. 170688 (1965); Byull. Izobr., No. 9, 72 (1965).
1516. 170968 (1965); Byull. Izobr., No. 10, 28 (1965).
1517. 173762 (1965); Byull. Izobr., No. 16, 33 (1965).
1518. 175479 (1965); Byull. Izobr., No. 20, 15 (1965).
1518a. 176066 (1965); No. 21, 48 (1965).

US Patents

1519. 2346155 (1944); C. A., 38, 5395 (1944).

1520. 2371068 (1945); C. A., 39, 4889 (1945) (see 1296 and 1402).
1521. 2375998 (1945); C. A., 40, 245 (1946); reissue 23060 (1948); C. A., 43, 2469 (1949) (see 1297).
1522. 2410737 (1946); C. A., 41, 623 (1947).
1523. 2415389 (1947); C. A., 41, 3116 (1947).
1524. 2431878 (1948); C. A., 42, 1761 (1948) (see 1399).
1525. 2434953 (1948); C. A., 42, 2985 (1948) (see 1456).
1526. 2436220 (1948); C. A., 42, 3205 (1948).
1527. 2438055 (1948); C. A., 42, 4601 (1948).
1528. 2440101 (1948); C. A., 42, 6376 (1948) (see 1307).
1529. 2441422 (1948); C. A., 42, 7903 (1948).
1530. 2442613 (1948); C. A., 42, 6168 (1948).
1531. 2443353 (1948); C. A., 42, 7573 (1948).
1532. 2444858 (1948); C. A., 42, 7317 (1948).
1533. 2448756 (1948); C. A., 43, 432 (1949).
1534. 2452895 (1948); C. A., 43, 3656 (1949).
1535. 2453092 (1948); C. A., 43, 3657 (1949).
1536. 2455880 (1948); C. A., 43, 2032 (1949).
1537. 2459387 (1948); cited in [110], p. 137 (131).
1538. 2464231 (1948); C. A., 43, 8210 (1949).
1539. 2472799 (1949); C. A., 43, 7500 (1949).
1540. 2489138 (1949); C. A., 44, 4490 (1950).
1541. 2489139 (1949); C. A., 44, 4490 (1950).
1542. 2490357 (1949); C. A., 44, 4490 (1950).
1543. 2491833 (1949); C. A., 44, 2547 (1950).
1544. 2504388 (1950); C. A., 44, 6196 (1950).
1545. 2507200 (1950); C. A., 44, 7069 (1950).
1546. 2507518 (1950); C. A., 45, 3410 (1951).
1547. 2507551 (1950); C. A., 45, 790 (1951).
1548. 2512058 (1950); C. A., 44, 8698 (1950).
1549. 2517945 (1950); C. A., 44, 10375 (1950).
1550. 2541851 (1951); C. A., 45, 4485 (1951).
1551. 2546036 (1951); C. A., 45, 5966 (1951).
1552. 2558560 (1951); C. A., 45, 8802 (1951) (see 1323).
1553. 2558561 (1951); C. A., 45, 8802 (1951) (see 1325 and 1397).
1554. 2567110 (1951); C. A., 45, 10676 (1951).
1555. 2571884 (1951); C. A., 46, 6434 (1952).
1556. 2574265 (1951); C. A., 46, 2342 (1952).
1557. 2587636 (1952); C. A., 47, 347 (1953).
1558. 2588393 (1952); Zbl., 1954, 4004 (see 1398 and 1459).
1559. 2598402 (1952); C. A., 46, 9342 (1952).
1560. 2601237 (1952); C. A., 46, 10681 (1952).
1561. 2609201 (1952); C. A., 47, 347 (1953).
1562. 2610199 (1952); C. A., 47, 9346 (1953).
1563. 2611775 (1952); C. A., 47, 8092 (1953).
1564. 2615033 (1952); C. A., 47, 9344 (1953).
1565. 2626957 (1953); C. A., 47, 6132 (1953).

1566. 2634252 (1953); C. A., 47, 3415 (1953).
1567. 2635084 (1953); C. A., 47, 8413 (1953).
1568. 2640067 (1953); C. A., 48, 5206 (1954).
1569. 2641589 (1953); C. A., 47, 9054 (1953).
1570. 2642411 (1953); C. A., 47, 10894 (1953).
1571. 2642415 (1953); C. A., 47, 9058 (1953).
1572. 2642453 (1953); C. A., 48, 4581 (1954).
1573. 2644805 (1953); C. A., 48, 310 (1954) (see 1320).
1574. 2645654 (1953); C. A., 48, 7050 (1954) (see 1317 and 1396).
1575. 2652385 (1953); C. A., 48, 1060 (1954) (see 1400).
1576. 2654710 (1953); C. A., 48, 995 (1954).
1577. 2658882 (1953); C. A., 48, 3060 (1954).
1578. 2672455 (1954); C. A., 48, 7924 (1954).
1579. 2673843 (1954); C. A., 48, 8580 (1954).
1580. 2680723 (1954); C. A., 49, 16463 (1955).
1581. 2684957 (1954); C. A., 48, 12452 (1954).
1582. 2687388 (1954); C. A., 49, 666 (1955).
1583. 2689843 (1954); C. A., 49, 643 (1955).
1584. 2703294 (1955); Ref. Zh. Khim., 1956, 37765.
1585. 2712533 (1955); C. A., 49, 12812 (1955).
1586. 2713063 (1955); C. A., 50, 5742 (1956) (see 1708).
1587. 2713564 (1955); C. A., 51, 12532 (1957) (see 1707).
1588. 2714585 (1955); C. A., 49, 14344 (1955).
1589. 2716656 (1955); C. A., 50, 1372 (1956) (see 1292 and 1329).
1590. 2721855 (1955); C. A., 50, 6090 (1956).
1591. 2721857 (1955); C. A., 50, 3794 (1956).
1592. 2724704 (1955); C. A., 50, 6072 (1956).
1593. 2727875 (1955); C. A., 50, 6065 (1956).
1594. 2732318 (1956); C. A., 50, 17517 (1956).
1595. 2736721 (1956); C. A., 50, 7502 (1956).
1596. 2742368 (1956); C. A., 50, 11713 (1956) (see 1322 and 1405).
1597. 2743192 (1956); C. A., 50, 11686 (1956).
1598. 27511314 (1956); C. A., 50, 17517 (1956).
1599. 2755261 (1956); C. A., 50, 15122 (1956).
1600. 2759904 (1956); C. A., 51, 4750 (1957).
1601. 2763675 (1956); C. A., 51, 5848 (1957).
1602. 2774674 (1956); C. A., 52, 730 (1958).
1603. 2774690 (1956); C. A., 51, 4730 (1957).
1604. 2778793 (1957); C. A., 51, 6214 (1957).
1605. 2784139 (1957); C. A., 51, 9077 (1957).
1606. 2791511 (1957); C. A., 51, 10868 (1957).
1607. 2803614 (1957); C. A., 51, 18698 (1957) (see 1336).
1608. 2807554 (1957); C. A., 52, 2432 (1958).
1609. 2816610 (1957); C. A., 52, 5803 (1958).
1610. 2818089 (1957); C. A., 52, 5018 (1958).
1611. 2818906 (1958); C. A., 52, 5030 (1958).
1612. 2830968 (1958); C. A., 52, 13313 (1958).

1613. 2831009 (1958); C. A., 52, 14653 (1958).
1614. 2831880 (1958); C. A., 52, 14694 (1948).
1615. 2837550 (1958); C. A., 52, 15958 (1958).
1616. 2842521 (1958); C. A., 52, 19238 (1958) (see 1409).
1617. 2843555 (1958); C. A., 53, 765 (1959).
1618. 2844435 (1958); C. A., 52, 19238 (1958).
1619. 2850473 (1958); C. A., 53, 3767 (1959).
1620. 2850514 (1958); C. A., 53, 4166 (1959).
1621. 2851439 (1958); C. A., 53, 10803 (1959).
1622. 2855378 (1958); C. A., 53, 1823 (1959).
1623. 2855380 (1958); C. A., 53, 3767 (1959) (see 1342).
1624. 2863391 (1958); C. A., 53, 9147 (1959).
1625. 2863893 (1958); C. A., 53, 9148 (1959).
1626. 2868750 (1959); C. A., 53, 7617 (1959).
1627. 2868751 (1959); C. A., 53, 13625 (1959).
1628. 2879919 (1959); cited in [115], (341).
1629. 2886460 (1959); C. A., 53, 16492 (1959).
1630. 2887371 (1959); C. A., 53, 19883 (1959).
1631. 2888475 (1959); C. A., 53, 15637 (1959).
1632. 2894967 (1959); C. A., 54, 946 (1960) (see 1348a).
1633. 2902505 (1959); C. A., 54, 5466 (1960).
1634. 2905562 (1959); C. A., 54, 1825 (1960).
1635. 2915543 (1959); C. A., 54, 6649 (1960).
1636. 2920060 (1960); C. A., 54, 16386 (1960).
1637. 2928799 (1960); C. A., 54, 19024 (1960).
1638. 2934550 (1960); C. A., 54, 14777 (1960).
1639. 2938046 (1960); C. A., 55, 1066 (1961).
1640. 2951864 (1960); C. A., 55, 5345 (1961).
1641. 2956045 (1960); C. A., 55, 5552 (1961).
1642. 2957900 (1960); C. A., 55, 5424 (1961).
1643. 2962446 (1960); C. A., 55, 20266 (1961).
1644. 2964550 (1960); C. A., 55, 6439 (1961).
1645. 2967877 (1961); C. A., 55, 12357 (1961).
1646. 2977336 (1961); C. A., 55, 17079 (1961).
1647. 2983697 (1961); C. A., 55, 18160, 25324 (1961).
1648. 2994684 (1961); C. A., 55, 27968 (1961).
1649. 2994711 (1961); C. A., 56, 4795 (1962).
1650. 2998407 (1961); C. A., 56, 6170 (1962).
1651. 2998440 (1961); C. A., 56, 591 (1962).
1652. 2999076 (1961); C. A., 56, 2544 (1962).
1653. 3002986 (1961); C. A., 56, 4974 (1962).
1654. 3002988 (1961); C. A., 56, 2472 (1962).
1655. 3004053 (1961); C. A., 56, 4795 (1962) (see 1735).
1656. 3007986 (1961); C. A., 56, 4974 (1962).
1657. 3010982 (1961); C. A., 57, 2253 (1962).
1658. 3013992 (1961); C. A., 57, 1091 (1962).

1659. 3015637 (1962); C. A., 56, 11787 (1962).
1660. 3018299 (1962); C. A., 57, 13803 (1962).
1661. 3019993 (1961); C. A., 56, 13098 (1962).
1662. 3021297 (1962); C. A., 56, 13104 (1962).
1663. 3024262 (1962); C. A., 57, 2254 (1962).
1664. 3030394 (1962); C. A., 57, 3589 (1962).
1665. 3032532 (1962); C. A., 57, 11236 (1962).
1666. 3033807 (1962); C. A., 57, 3604 (1962).
1667. 3033885 (1962); C. A., 57, 13803 (1962).
1668. 3035071 (1962); C. A., 57, 6153 (1962).
1669. 3036985 (1962); C. A., 57, 6149 (1962).
1670. 3041363 (1962); C. A., 57, 15362 (1962).
1671. 3043858 (1962); C. A., 58, 1489 (1963) (see 1367).
1672. 3050490 (1962); Zbl., 1963, 22182 (see 1294 and 1429).
1673. 3050491 (1962); C. A., 58, 14266 (1963) (see 1441).
1674. 3057469 (1962); C. A., 58, 2548 (1963).
1675. 3057822 (1962); C. A., 59, 10313 (1963).
1676. 3060150 (1962); C. A., 58, 5869 (1963).
1677. 3060215 (1962); C. A., 58, 6865 (1963).
1678. 3061587 (1962); C. A., 58, 1550 (1963).
1679. 3062854 (1962); C. A., 58, 9145 (1963).
1680. 3069445 (1962); C. A., 58, 10237 (1963).
1681. 3070560 (1962); C. A., 58, 5868 (1963).
1682. 3070566 (1962); C. A., 58, 7012 (1963).
1683. 3070567 (1962); C. A., 58, 7012 (1963).
1684. 3082181 (1963); C. A., 58, 14275 (1963).
1685. 3086983 (1963); C. A., 60, 553 (1964).
1686. 3094436 (1963); C. A., 59, 7753 (1963).
1687. 3098836 (1963); C. A., 59, 10331 (1963).
1688. 3099667 (1963); C. A., 60, 549 (1964).
1689. 3109826 (1963); C. A., 60, 749 (1964).
1690. 3110689 (1963); C. A., 60, 4319 (1964).
1691. 3112331 (1963); C. A., 60. 5554 (1964).
1692. 3125637 (1964); C. A., 60, 13271 (1964).
1693. 3137719 (1964); C. A., 61, 5692 (1964).
1694. 3146248 (1964); C. A., 61, 13120 (1964).
1695. 3152999 (1964); C. A., 61, 16190 (1964).
1696. 3154515 (1964); C. A., 62, 1831 (1965).
1697. 3154520 (1964); C. A., 62, 1763 (1965).
1698. 3159661 (1964); C. A., 62, 5297 (1965).

French Patents
1699. 784329 (1935); Zbl., 1936, I, 1094 (see 1295 and 1390).
1700. 979605 (1951); C. A., 47, 7824 (1953).
1701. 979606 (1951); C. A., 47, 7824 (1953).
1702. 1042019 (1953); C. A., 52, 9652 (1958).

1703. 1054423 (1954); Ref. Zh. Khim., 1956, 7974.
1704. 1061035 (1954); Zbl., 1955, 2567.
1705. 1073050 (1954); Zbl., 1956, 9006 (see 1330).
1706. 1096609 (1955); C. A., 53, 1823 (1959).
1707. 1104454 (1955); C. A., 51, 14281 (1957) (see 1587).
1708. 1114148 (1956); Zbl., 1957, 3395 (see 1586).
1709. 1193721 (1959); C. A., 55, 3114 (1961).
1710. 1213488 (1960); C. A., 55, 14938 (1961).
1711. 1266528 (1961); C. A., 56, 11771 (1962) (see 1350).
1712. 1308055 (1962); C. A., 58, 7044 (1963).
1713. 1329088 (1963); C. A., 59, 8956 (1963); supplement 81770 (1963); C. A., 60, 9459 (1964).
1714. 1349887 (1964); C. A., 61, 10796 (1964).
1715. 1352325 (1964); C. A., 62, 2890 (1965).
1716. 1359396 (1964); C. A., 62, 4197 (1965).
1717. 1396271 (1965); C. A., 63, 3077 (1965).
1718. 1396273 (1965); C. A., 63, 7045 (1965).
1719. M1069 (1962); C. A., 59, 12911 (1963).

Czechoslovakian Patents
1720. 92206 (1959); C. A., 55, 6030 (1961).
1721. 105217 (1962); C. A., 59, 15406 (1963).

Swiss Patents
1722. 287225 (1953); Ref. Zh. Khim., 1955, 22520.

Japanese Patents
1723. 1441 (1953); C. A., 47, 6179 (1953).
1724. 3343 (1953); cited in [114].
1725. 3493 (1953); C. A., 48, 7932 (1954).
1726. 4791 (1953); C. A., 48, 11109 (1954).
1727. 6639 (1953); C. A., 48, 12426 (1954).
1728. 7568 (1955); C. A., 51, 18396 (1957).
1729. 889 (1959); C. A., 53, 12739 (1959).
1730. 8996 (1959); C. A., 54, 7174 (1960).
1731. 12945 (1960); C. A., 55, 10319 (1961).
1732. 12946 (1960); C. A., 55, 10319 (1961).
1733. 12948 (1960); C. A., 55, 10319 (1961).
1734. 13245 (1960); C. A., 55, 11301 (1961).
1735. 14617 (1961); C. A., 56, 10190 (1962) (see 1655).

Index